Solid Edge 2019

Basics and Beyond

Online Instructor

Contents

Introduction

Welcome to the *Solid Edge 2019 Basics and Beyond* book. This book is written to assist students, designers, and engineering professionals. It covers the important features and functionalities of Solid Edge using relevant examples and exercises.

This book is written for new users, who can use it as a self-study resource to learn Solid Edge. In addition, experienced users can also use it as a reference. The focus of this book is part modeling, assembly modeling, drawings, sheet metal design, and surface design.

Topics covered in this Book

- Chapter 1, "Getting Started with Solid Edge 2019", gives an introduction to Solid Edge. The user interface and terminology are discussed in this chapter.
- Chapter 2, "Sketch Techniques", explores the sketching commands in Solid Edge. You will learn to create parametric sketches.
- Chapter 3, "Extrude and Revolve features", teaches you to create basic 3D geometry using the Extrude and Revolve commands.
- Chapter 4, "Placed Features", covers the features which can be created without using sketches.
- Chapter 5, "Patterned Geometry", explores the commands to create patterned and mirrored geometry.
- Chapter 6, "Sweep Features", covers the commands to create swept and helical features.
- Chapter 7, "Loft Features", covers the Loft command and its core features.
- Chapter 8, "Additional Features and Multibody Parts", covers additional commands to create complex geometry. In addition, the multibody parts are also covered.
- Chapter 9, "Modifying Parts", explores the commands and techniques to modify the part geometry.
- Chapter 10, "Assemblies", explains you to create assemblies using the bottom-up and top-down design approaches.
- Chapter 11, "Drawings", covers how to create 2D drawings from 3D parts and assemblies.
- Chapter 12, "Sheet Metal Design", covers how to create sheet metal parts and flat patterns.
- Chapter 13, "Surface Design", covers how to create complex shapes and designs using surface modeling tools.

Chapter 1: Getting Started with Solid Edge 2019

Introduction to Solid Edge 2019

Solid Edge 2019 is a parametric and feature-based system that allows you to create 3D parts, assemblies, and 2D drawings. The design process in Solid Edge is shown below.

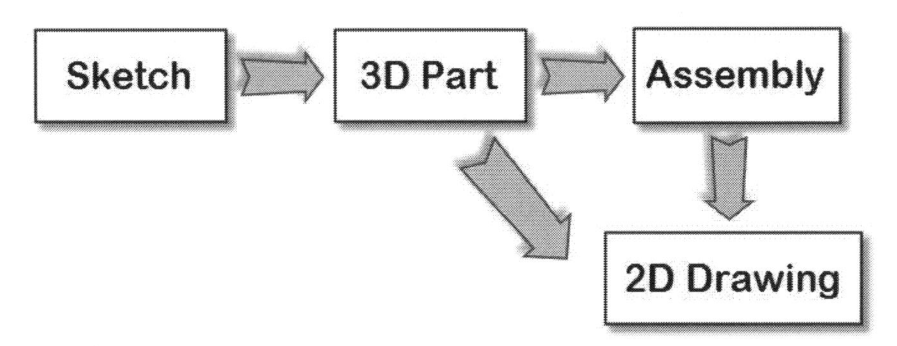

In Solid Edge, everything is controlled by parameters, dimensions, or relationships. For example, if you want to change the position of the hole shown in figure, you need to change the dimension or relation that controls its position.

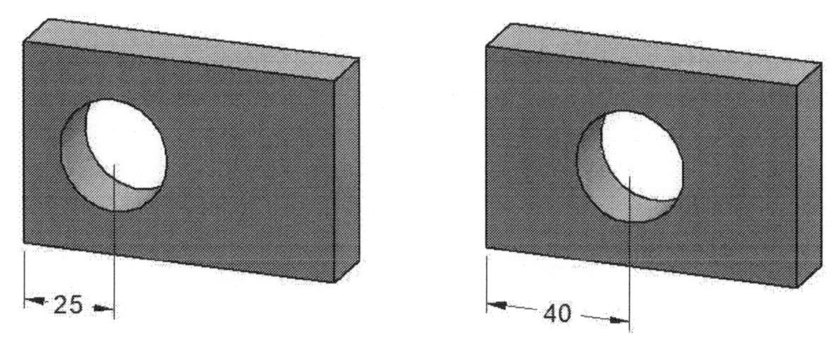

The parameters and relationships that you set up allow you to have control over the design intent. The design intent describes the way your 3D model will behave when you apply dimensions and relationships to it. For example, if you want to position the hole at the center of the block, one way is to add dimensions between the hole and the adjacent edges. However, when you change the size of the block, the hole will not be at the center.

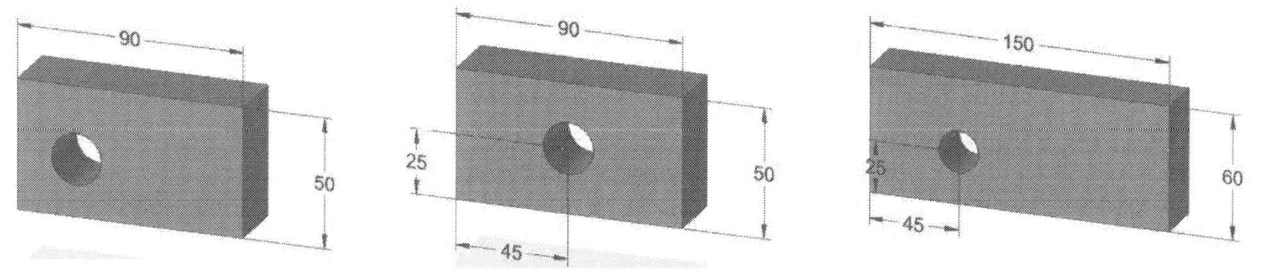

You can make the hole to be at the center, even if the size of the block changes. You need to apply the **Horizontal/Vertical** relationships between the hole and midpoints of the adjacent edges. Now, even if you change the size of the block, the hole will always remain at the center.

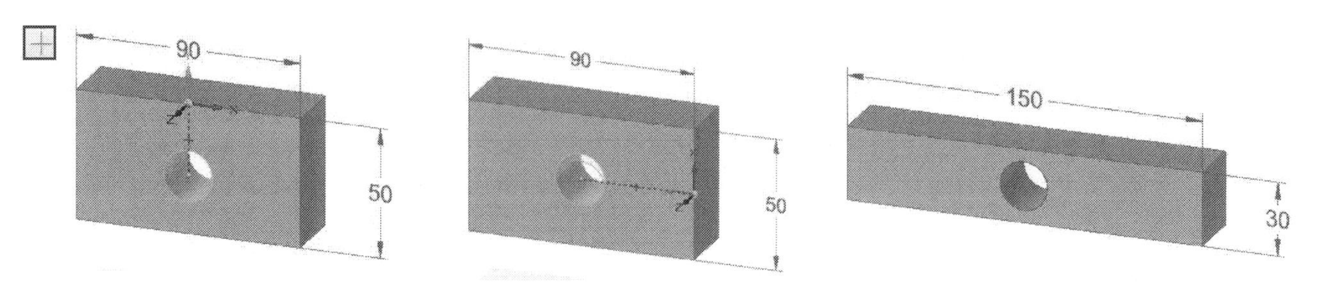

The other big advantage of Solid Edge is the associativity between parts, assemblies and drawings. When you make changes to the design of a part, the changes will take place in any assembly that it's a part of. In addition, the 2D drawing will update automatically.

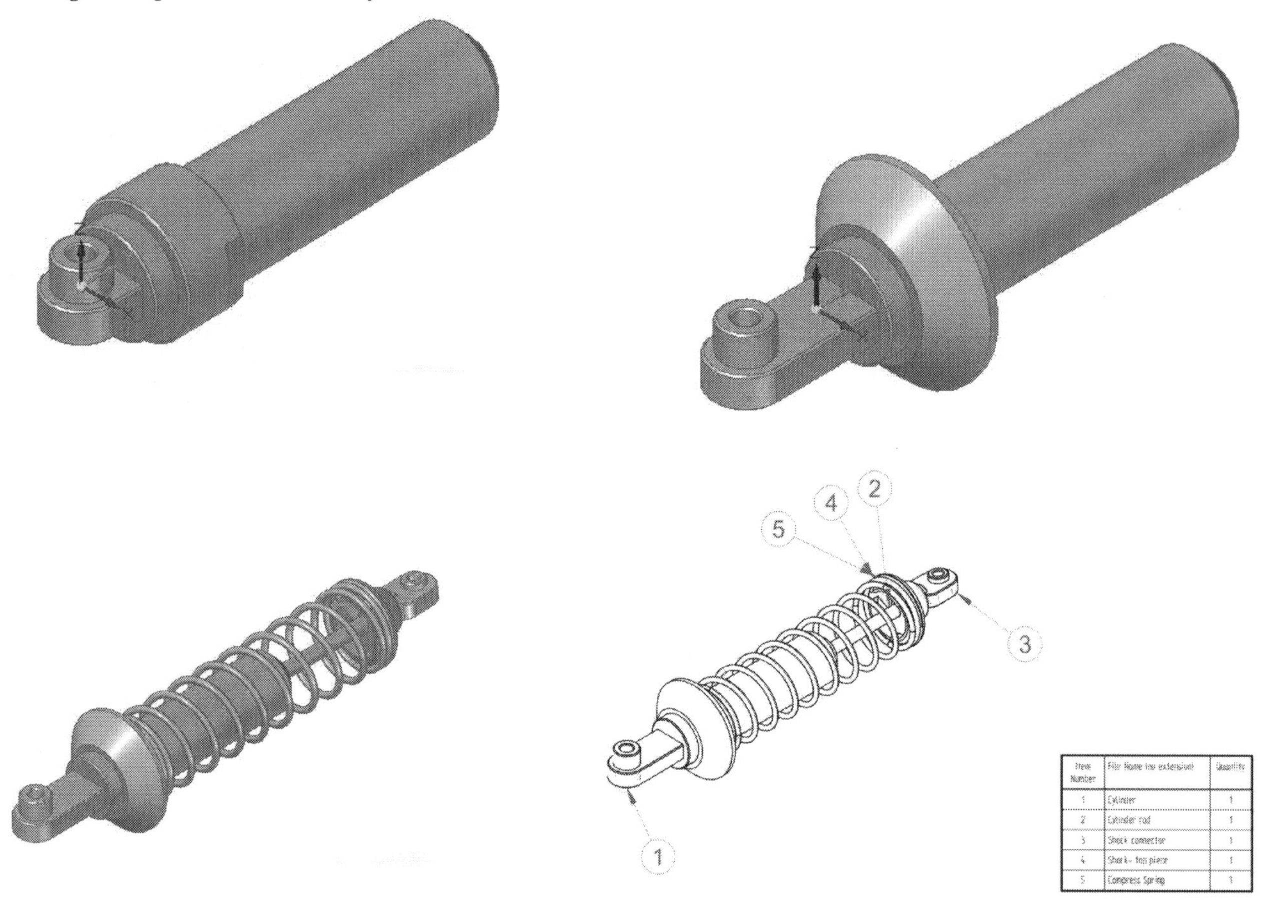

Item Number	File Name (no extension)	Quantity
1	Cylinder	1
2	Cylinder rod	1
3	Shock connector	1
4	Shock- top piece	1
5	Compress Spring	1

Installing Solid Edge 2019

To install **Solid Edge 2019**, click the **autostart** icon in the Solid Edge 2019 disc; the **Solid Edge** window appears. Click the **Solid Edge** link on the **Solid Edge** window; the **Solid Edge Installation Wizard** starts. On the **Solid Edge 2019** window, type-in the **User name** and **Organisation**, and then select the **Modeling standard**. You can select a modeling standard, which your company or client uses. In this book, we use the **ISO Metric** modeling standard to create all parts, assemblies, and drawings. Click **Install** after selecting the modeling standard. Close the **Solid Edge** window after the installation is complete.

Starting Solid Edge 2019

To start **Solid Edge 2019**, click the **Solid Edge 2019** icon on your computer screen; the **Solid Edge** message box pops up showing, "Your copy of Solid Edge must be licensed for first-time use". Click **OK**. Select your license option and specify the license code or file. Click **OK** after specifying the license; the theme selection window appears. A theme is a predefined user-interface layout, which suits your working style. This window displays four user-interface themes: **Some Assistance**, **Maximum Assistance**, **Maximum Workspace**, and **Balanced (Solid Edge Default)**. Users who are familiar with other CAD packages can use the **Some Assistance** theme. Users who are new to CAD can use the **Maximum Assistance** theme. The **Maximum Workspace** theme is for users who have already used Solid Edge. The **Balanced (Solid Edge Default**) theme is the predefined workspace, which is similar to the previous versions of Solid Edge.

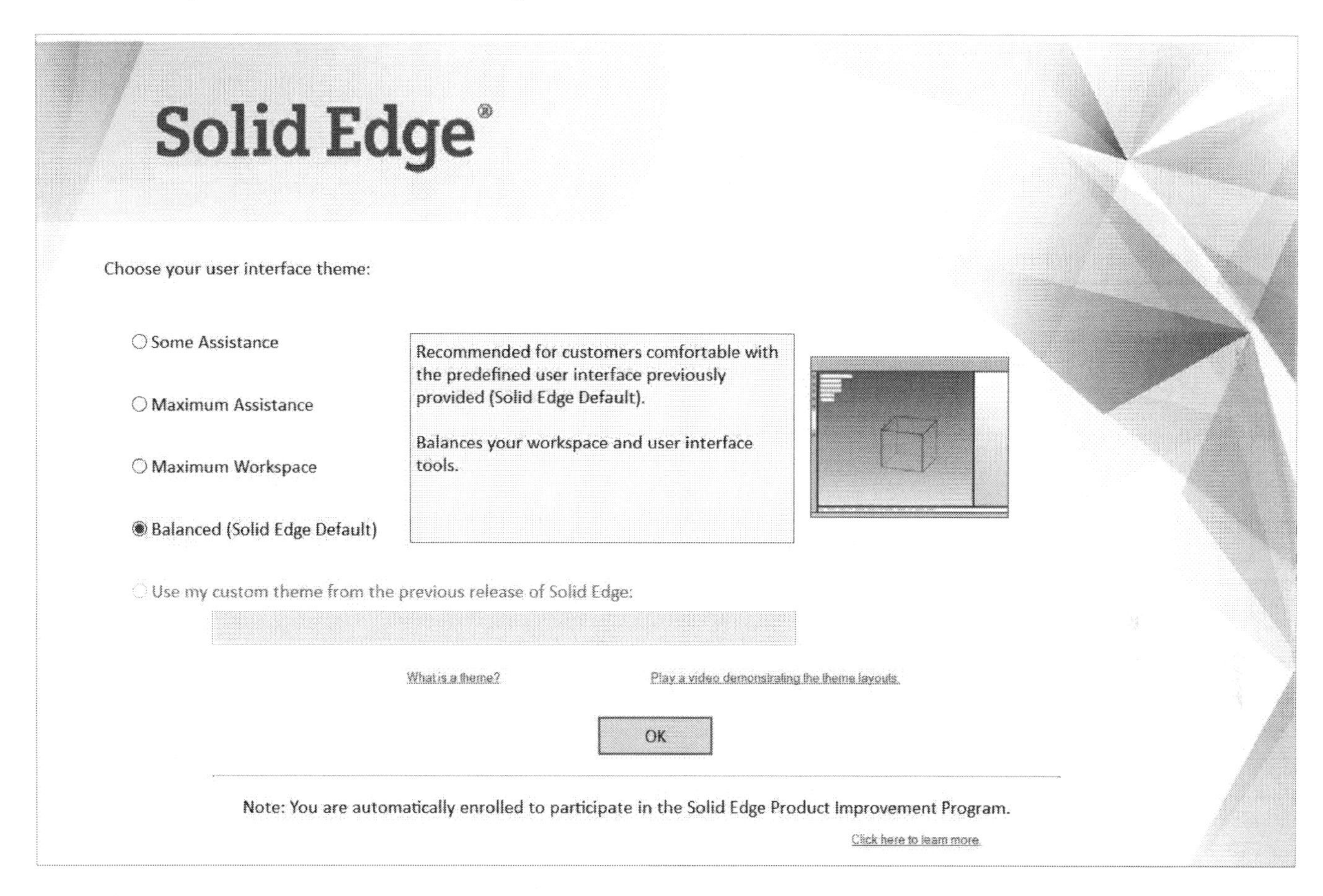

Select the **Balanced (Solid Edge Default)** theme and click **OK**. The **Solid Edge 2019** application window appears. On this window, click the **Application Menu** located at the top left corner; the application menu appears. You can use this menu to start a new document, open an existing one, learn Solid Edge test drive, print drawings and change other settings.

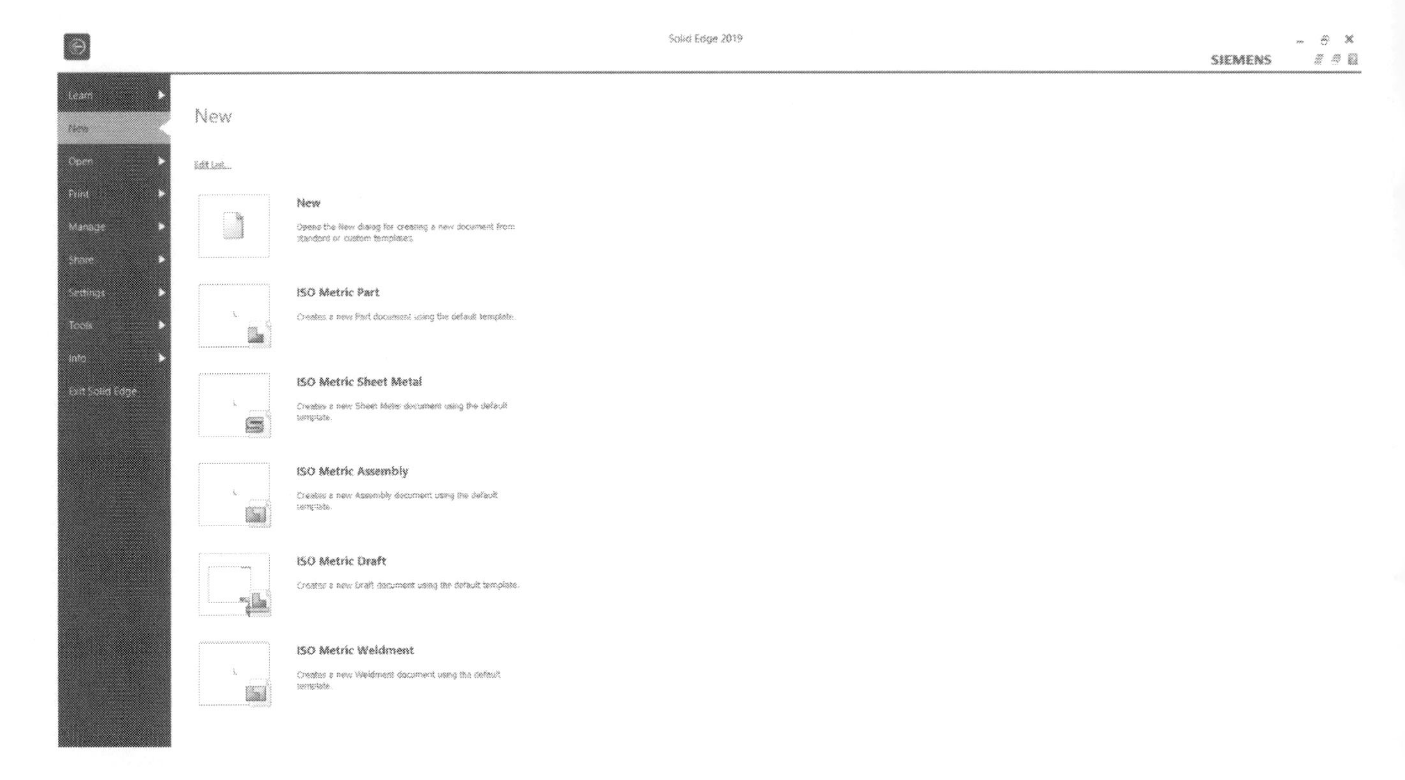

On the Application Menu, click the **New** option, and then click **ISO Metric Part** under the **Create** section to start a new part document.

*You can change the templates displayed in the **New** page by clicking **Edit List**. On the **Template List Creation** dialog, select a modelling standard from the **Standard Templates** section. You can change the order of the templates by selecting them from the **Templates** section and clicking the **Move Up** and **Move Down** arrows. Likewise, you can change the **Name** and **Description** of the template and click **Apply**. Click **OK** on the **Template List Creation** dialog to apply the changes.*

File Types

Various file types that can be created in Solid Edge are given below.

- **Part (.par)**
- **Assembly (.asm)**
- **Draft (.dft)**
- **Sheet Metal (.psm)**
- **Weldment (.pwd)**

User Interface

The following image shows the **Solid Edge 2019** application window.

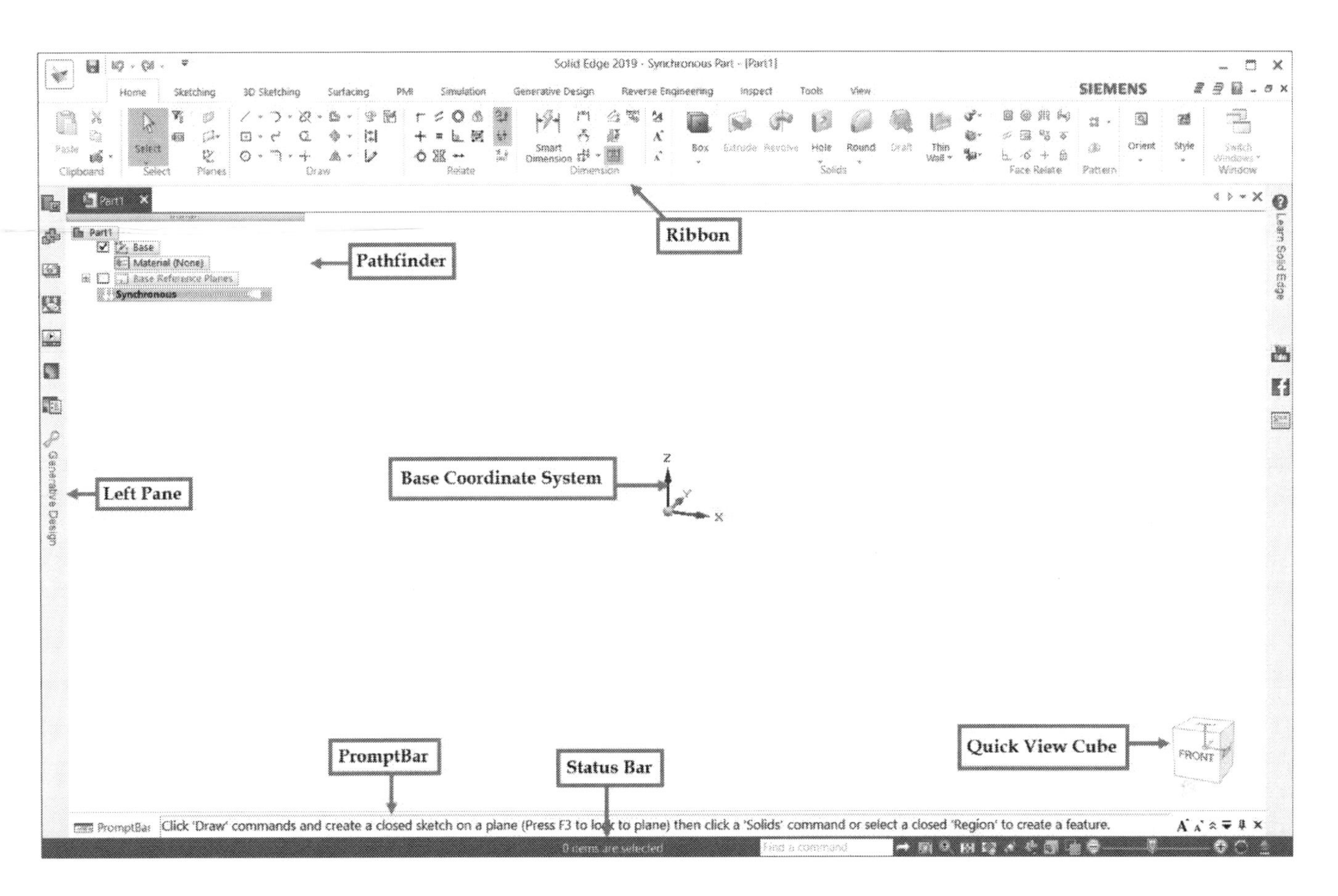

Environments in Solid Edge

There are main five environments available in Solid Edge: **Part (Synchronous** and **Ordered)**, **Assembly**, **Draft**, **Weldment**, and **Sheet Metal (Synchronous** and **Ordered)**. In addition, there some additional environments to create exploded views, renderings, structures, piping, wire harnesses, and so on.

Part environment (Synchronous and Ordered)

This environment has all the commands to create a 3D part model. It is available in two modes: **Synchronous** and **Ordered**. The **Synchronous** mode allows you to create and edit models directly. The **Ordered** mode allows you to create History-based models. In this mode, every feature or sketch that you create is stored in the Pathfinder. You can always go back and edit the feature or sketch. It has a ribbon located at the top of the screen. The ribbon is arranged in a hierarchy of tabs, panels, and commands. Panels such as **Draw**, **Relate**, and **Dimension** consists of commands, which are grouped based on their usage. Panels in turn are grouped into various tabs. For example, the panels such as **Draw**, **Relate**, and **Dimension** are located in the **Home** tab.

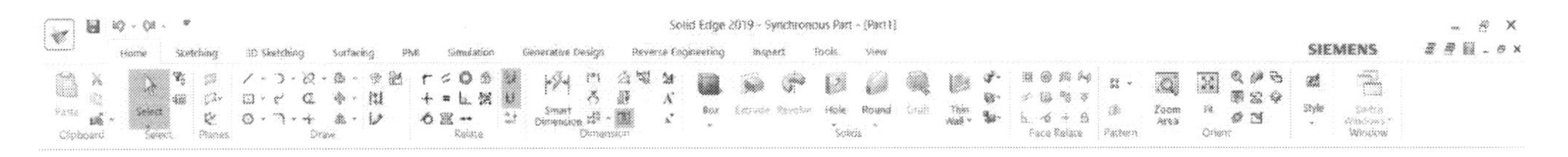

Assembly environment

This environment is used to create assemblies. The **Home** tab of the Ribbon has various commands, which will allow you to assemble and modify the components.

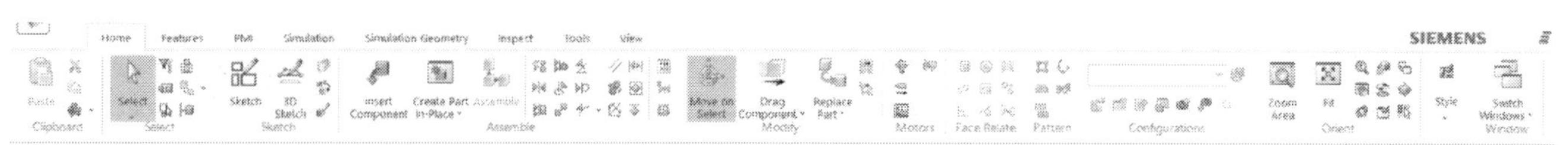

The **Features** tab has commands, which will help you to create cutouts, holes and other features at the assembly level.

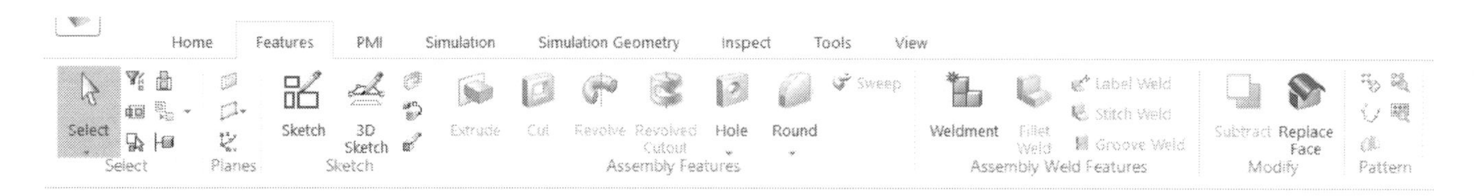

The **Inspect** tab helps you to inspect the assembly geometry.

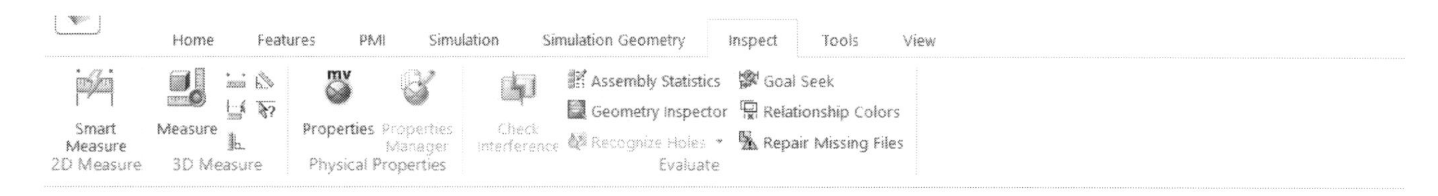

The **Tools** tab has some advanced commands, which will help you to switch to other environments.

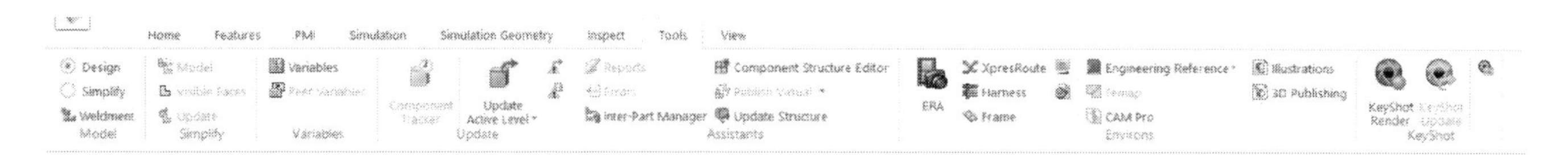

Draft environment

This environment has all the commands to generate 2D drawings of parts and assemblies.

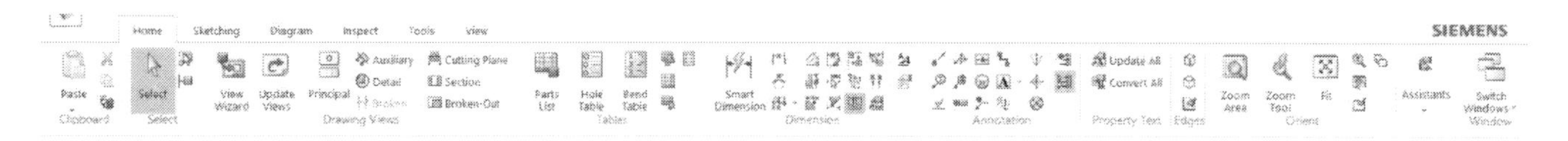

Sheet Metal environment

This environment has commands to create sheet metal parts.

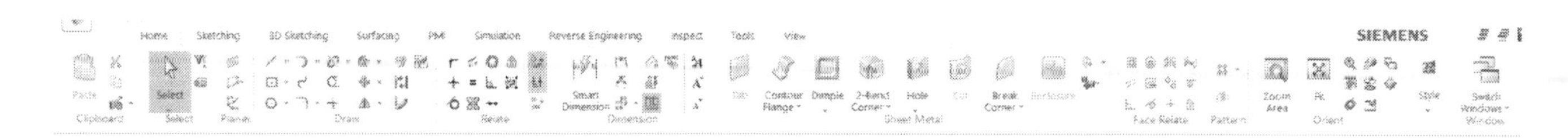

The other components of the user interface are discussed next.

Application Menu

The **Application Menu** appears when you click on the icon located at the top left corner of the window. The **Application Menu** consists of a list of self-explanatory menus. Click on the **Open** menu to see a list of recently opened documents under **Recent Files** menu.

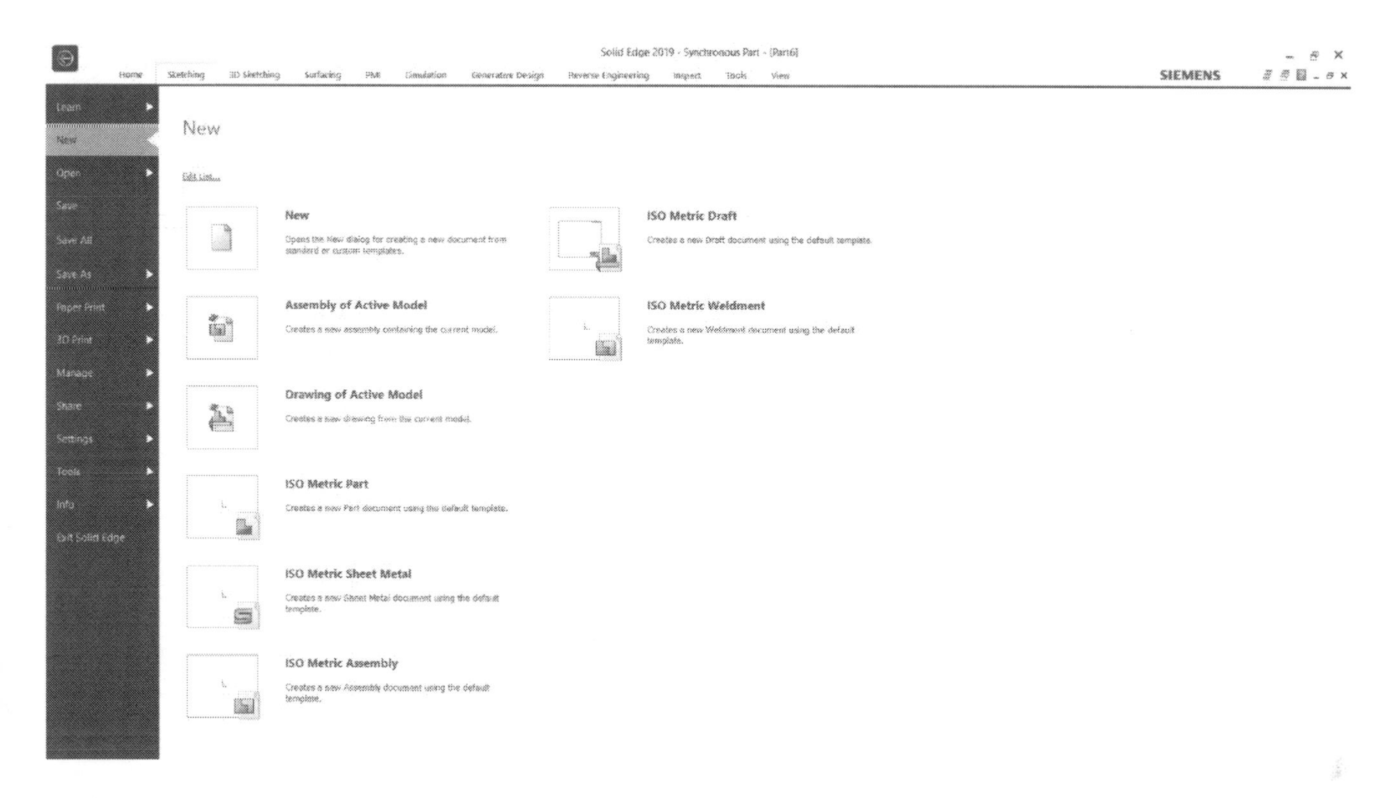

Quick Access Toolbar

This is located at the top left corner of the window. It consists of commonly used commands such as **New**, **Save**, **Open**, **Save As**, and so on. You can add more commands to the **Quick Access Toolbar** by clicking on the down-arrow next to it, and then selecting commands from the pop-up menu.

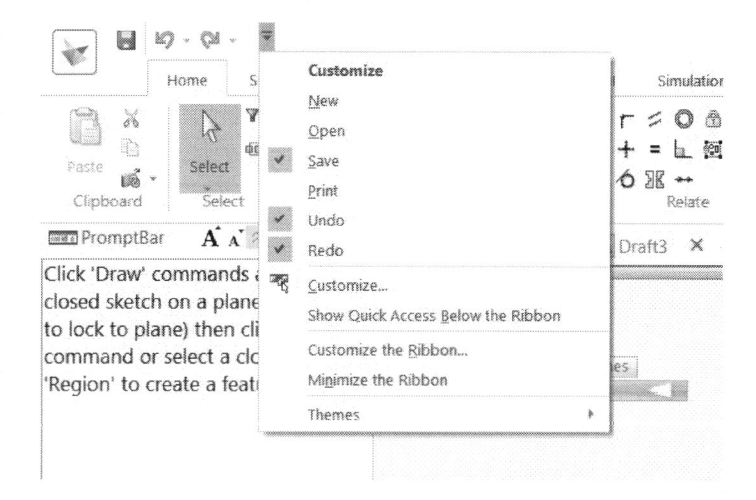

Graphics Window

Graphics window is the blank space located below the ribbon. You can draw sketches and create 3D geometry in the Graphics window. The left corner of the graphics window has a **Pathfinder**. Using the **Pathfinder**, you can access the features of the 3D model.

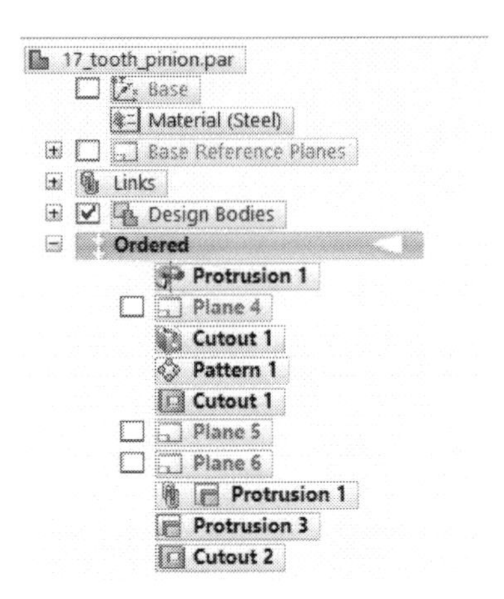

Prompt Bar

Prompt Bar is located below the Graphics Window. It is useful when you activate a command. It displays various prompts while working with any command. These prompts are series of steps needed to create a feature successfully.

Status Bar

Status Bar is located at the bottom of the Solid Edge window. It contains many icons, which help you to visualize the 3D model. You can use the **Record** and **Upload to Youtube** icons to create and upload videos. To add more icons to the Status Bar, click the right mouse button on it and select options from the pop-up menu.

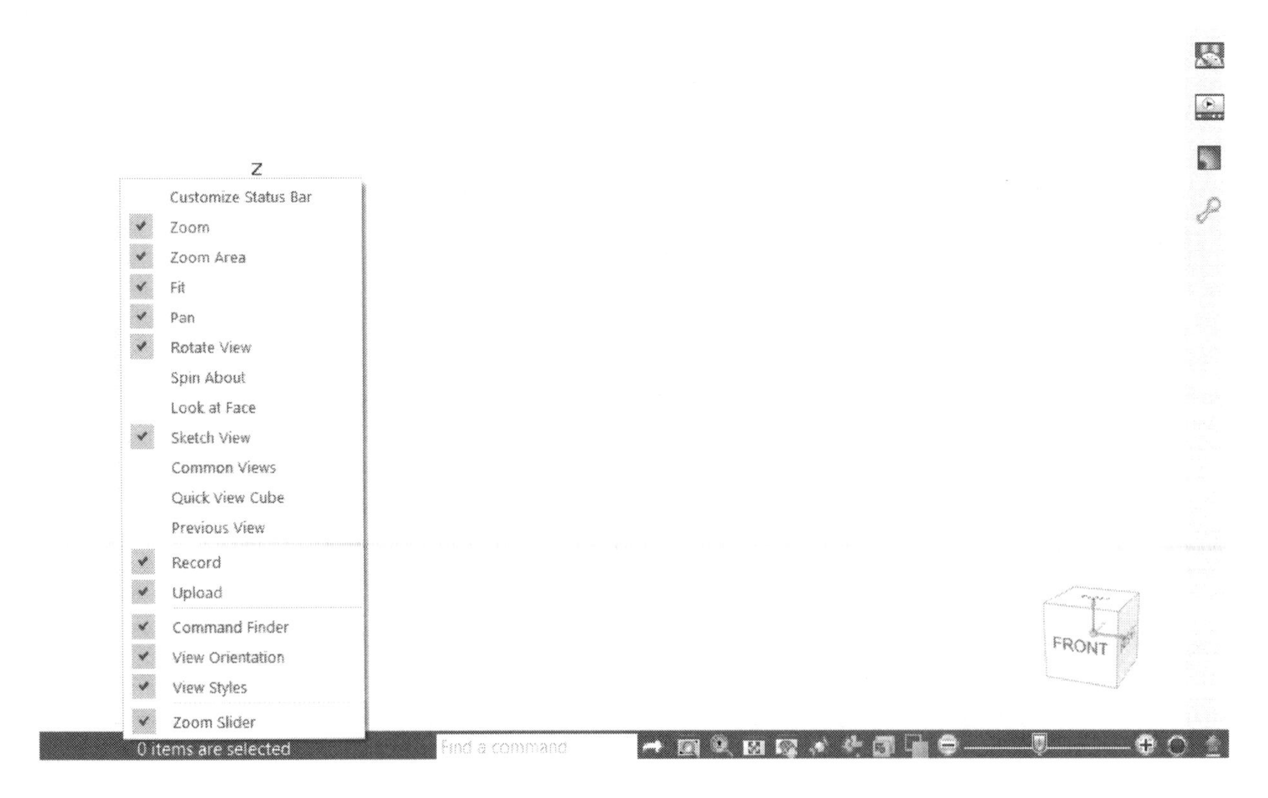

The **Command Finder** bar is used to search for any command available in Solid Edge 2019. You can type any keyword in the **Command Finder** bar and find a list of commands related to it.

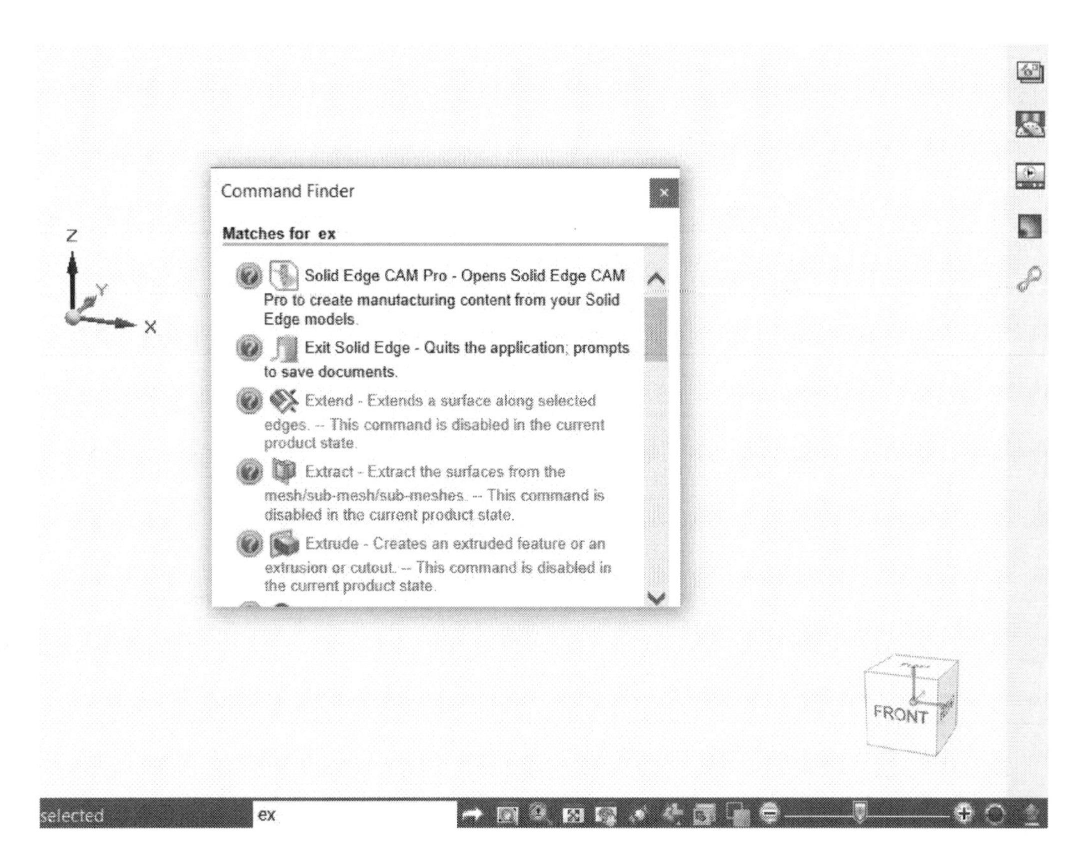

Quick View Cube

It is located at the bottom right corner of the graphics window. It is used to set the view orientation of the model.

Command bar

When you activate any command in Solid Edge, a contextual productivity tool called the command bar pops up on the screen. It displays the options and steps to complete the execution of the command.

Changing the display of the Ribbon

You can add or remove more commands to the ribbon by clicking the right mouse button on it and selecting **Customize the Ribbon**. On the **Customize** dialog, click on the options in the right-side box, and then click **Add** or **Remove**. After making the required changes, close the dialog and click **Yes** to save the changes.

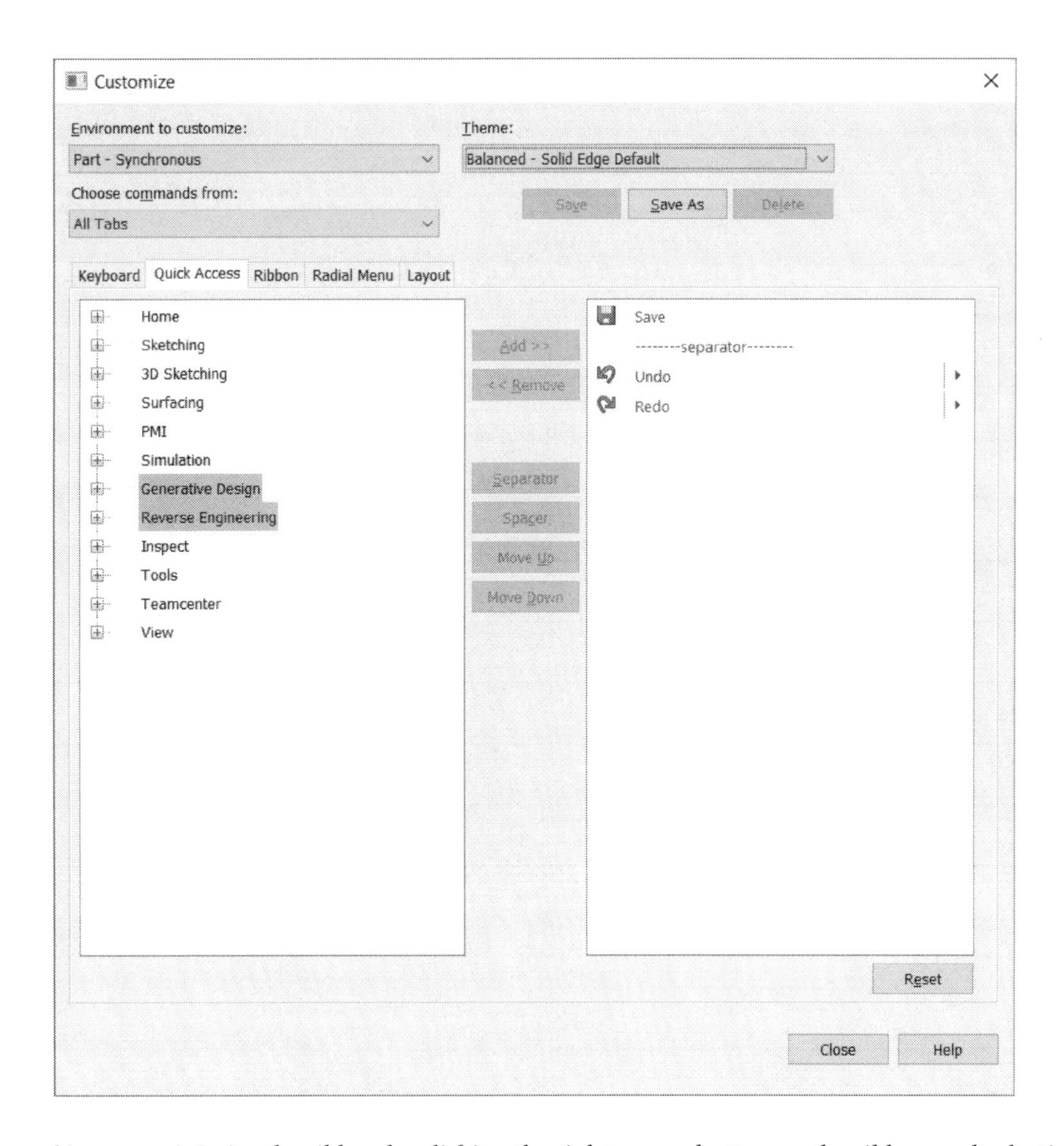

You can minimize the ribbon by clicking the right mouse button on the ribbon and selecting **Minimize the Ribbon**.

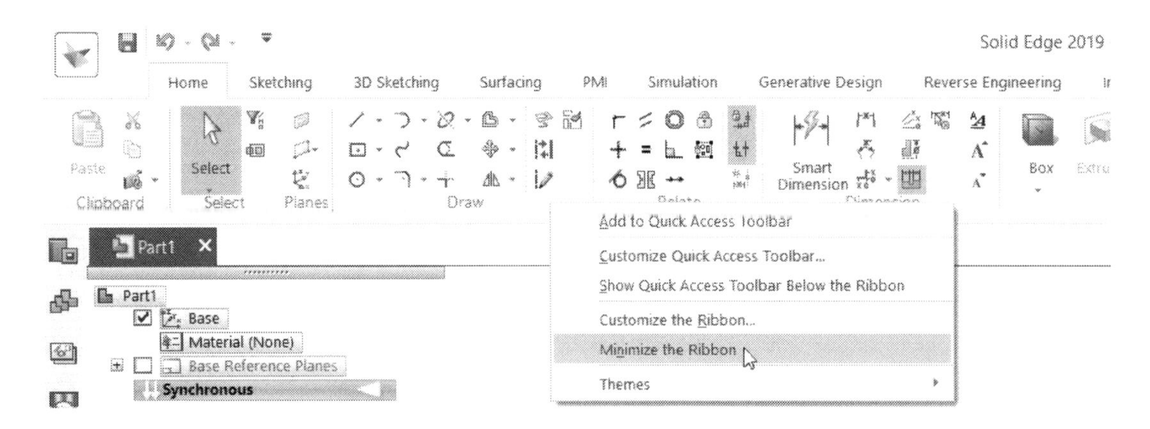

Dialogs

Dialogs are part of Solid Edge user interface. Using a dialog, you can easily specify many settings and options. Examples of dialogs are shown below.

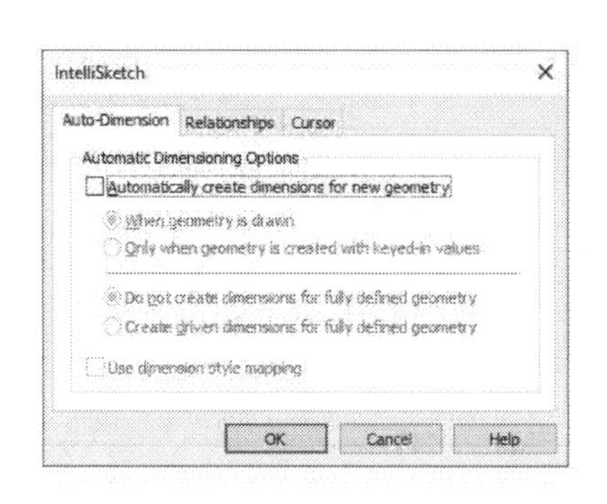

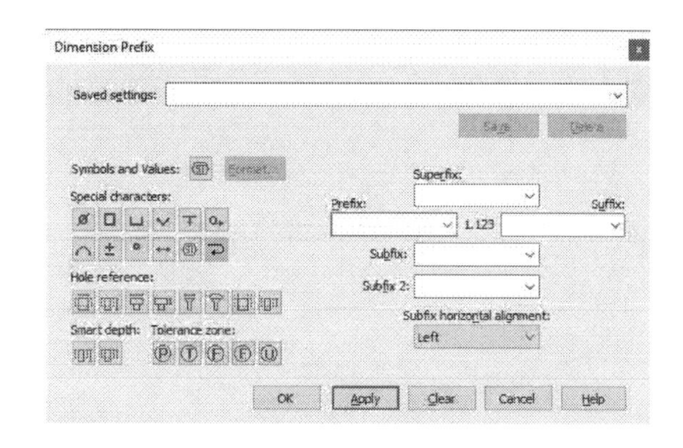

Radial Menus

Radial Menus provide you with another way of activating commands. You can display Radial Menus by clicking the right mouse button and dragging the pointer. A Radial Menu has various commands arranged in a radial manner. You can add or remove commands to the Radial Menu by using the **Customize** dialog.

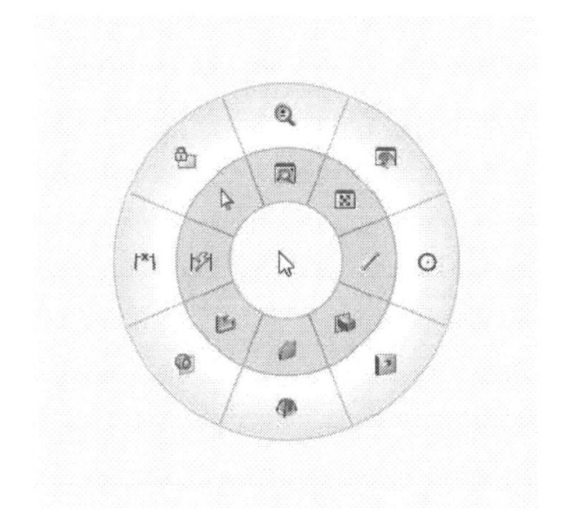

Shortcut Menus

Shortcut Menus are displayed when you right-click in the graphics window. Solid Edge provides various shortcut menus in order to help you access some options very easily and quickly. The options in shortcut menus vary based on the environment.

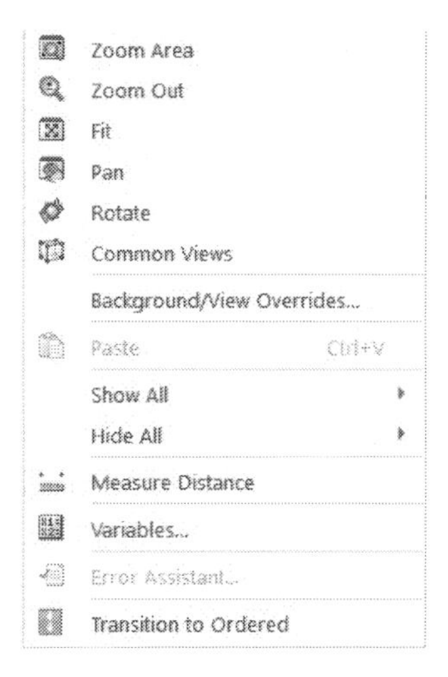

Starting a new document

You can start a new document directly from the Application Menu or by using the **New** dialog. On the initial screen, Application menu button located at the top left corner. On the Application Menu, click the **New** option and click on the required template to start a part, assembly, drawing, weldment or sheet metal document.

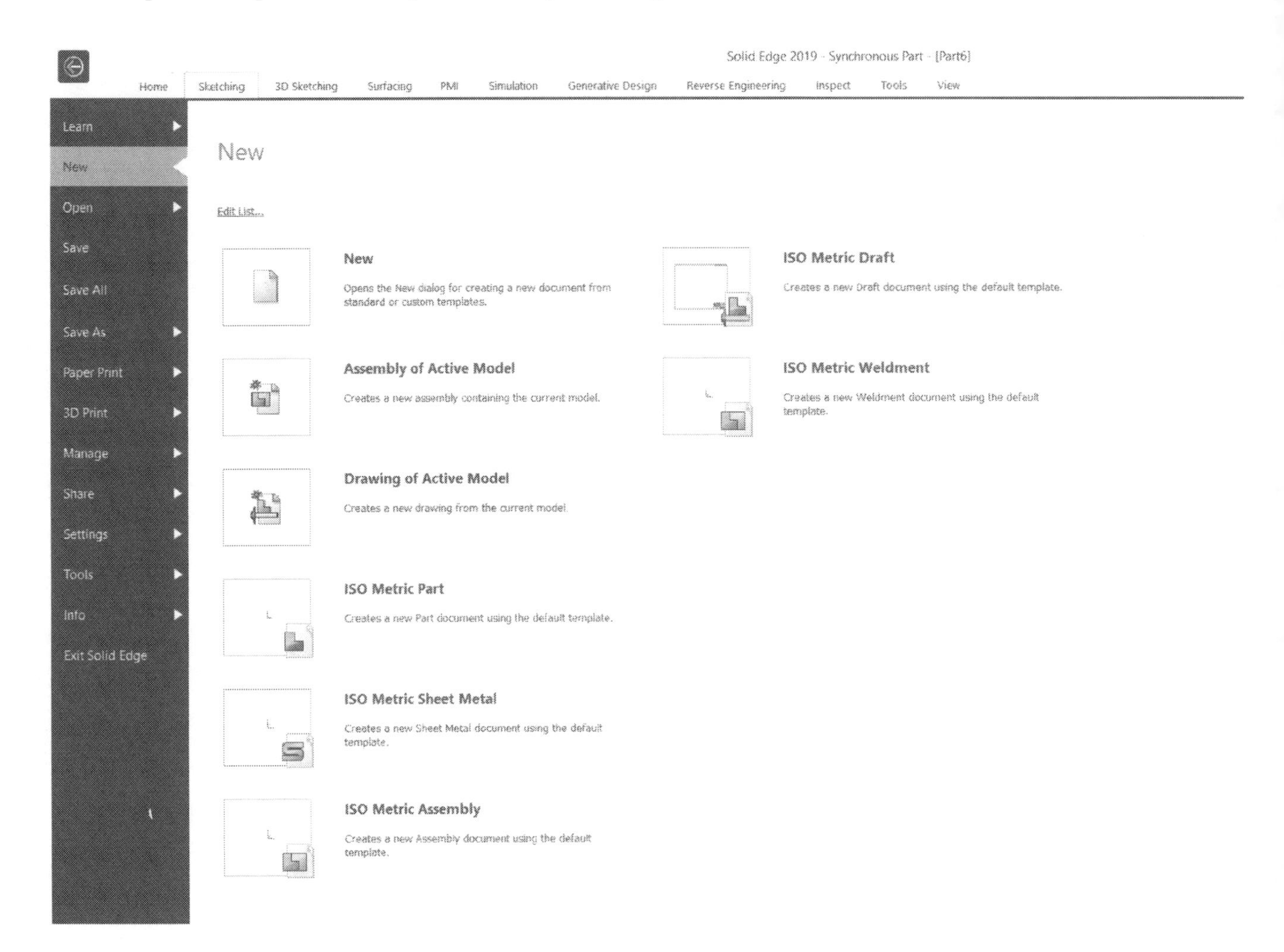

The New dialog

To start a new document using the **New** dialog, click the **New** button on anyone of the following:

- **Quick Access Toolbar**
- **Application Menu**

The **New** dialog appears when you click the **New** button. In this dialog, select the standard from the **Standard Templates** section. The templates related to the selected standard will appear. Select the .asm, .dft, .par, or .psm to start an assembly, drafting, part, or sheetmetal file, respectively.

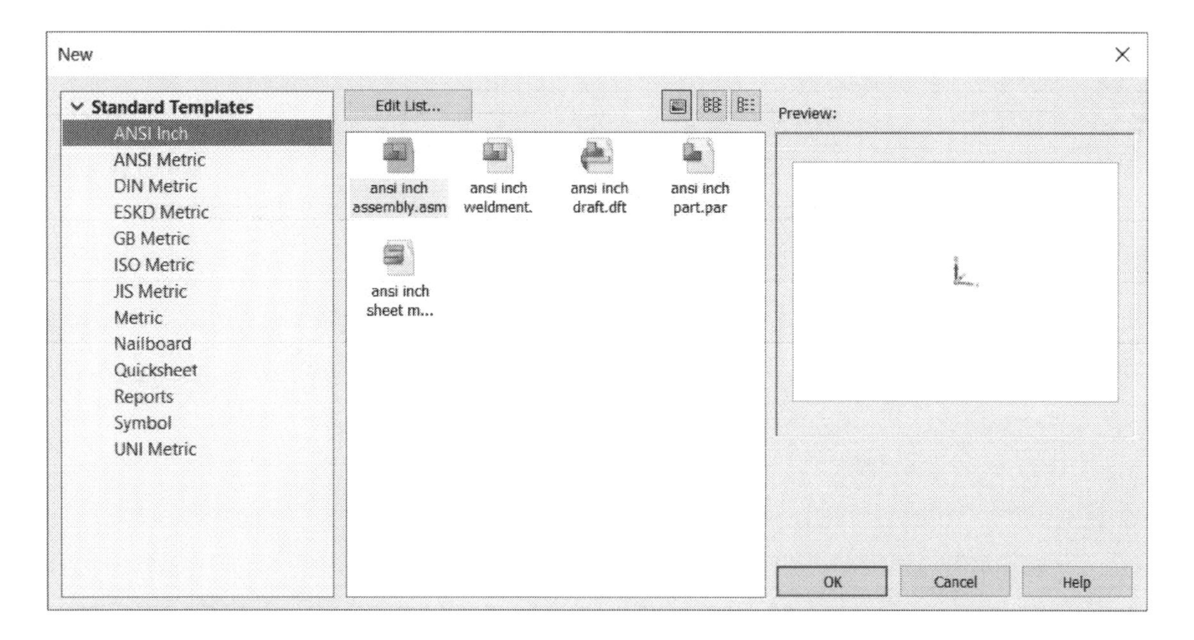

Solid Edge Options

You can customize Solid Edge as per your requirement. On the **Application Menu**, click **Settings** > **Options** to open the **Solid Edge Options** dialog. On this dialog, you can set options on each of the pages. The options on this dialog vary depending upon the environment that you are in.

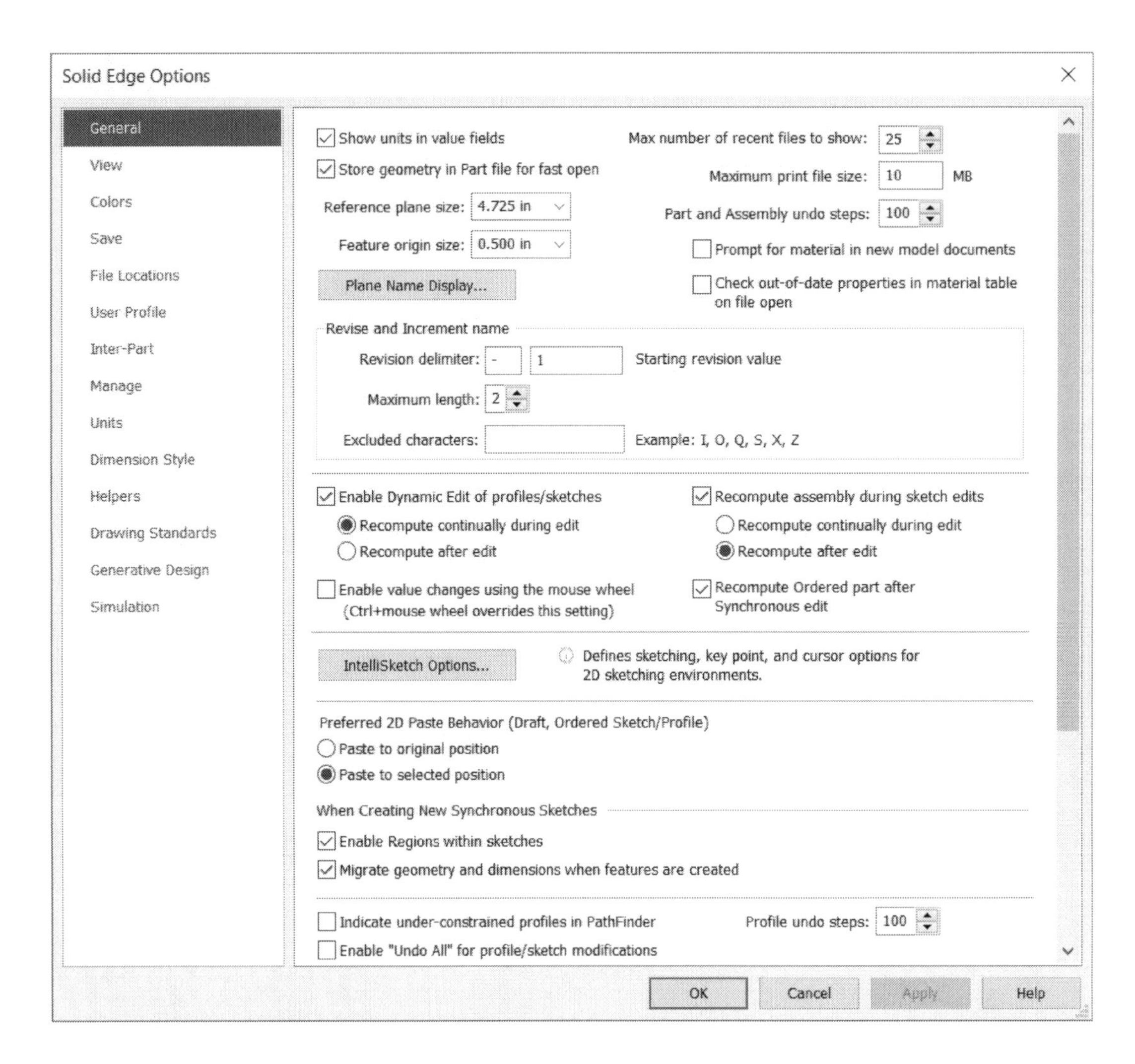

View Overrides dialog

The **View Overrides** dialog helps you to change the background color, rendering, and light settings. On the ribbon, click **View > Style > View Overrides** to open this dialog. On this dialog, click the **Background** tab and set the **Type** to **Solid**. Next, select **White** from the drop-down; the background color is changed to white.

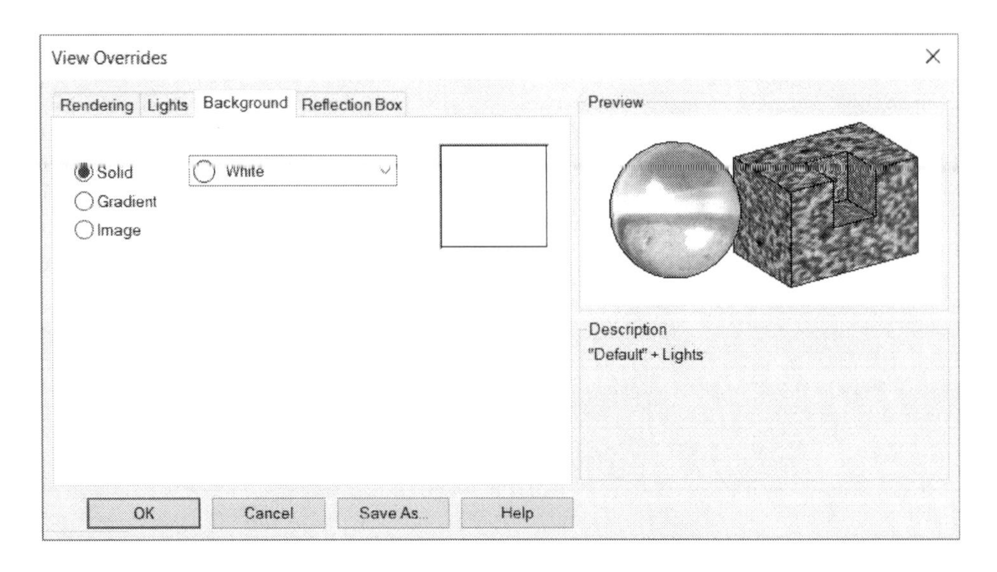

Solid Edge Help

Solid Edge offers you with the help system that goes beyond basic command definition. You can access Solid Edge help by using any of the following methods:

- Press the F1 key.
- Click on the **Solid Edge Help** option on the right-side of the window.

Questions

1. Explain how to customize the Ribbon.
2. What is design intent?
3. Give one example of where you would establish a relationship between a part's features.
4. Explain the term 'associativity' in Solid Edge.
5. List any two procedures to access Solid Edge Help.
6. How can you change the background color of the graphics window?
7. How can you activate the Radial Menu?
8. How is Solid Edge a parametric modeling application?

Chapter 2: Sketch Techniques

This chapter covers the methods and commands to create sketches in the part environment. The commands and methods are discussed in context to part environment. In Solid Edge, the part environment is divided into two separate modes: Synchronous and Ordered.

In Solid Edge, you create a rough sketch, and then apply dimensions and constrains that define its shape and size. The dimensions define the length, size, and angle of a sketch element, whereas constrains define the relations between sketch elements.

The topics covered in this chapter are:

- Create sketches in the Part environment (Synchronous and Ordered mode)
- Use relationships and dimensions to control the shape and size of a sketch
- Learn sketching commands
- Learn commands and options that help you to create sketches easily

Create Sketches in the Synchronous mode

Synchronous is the default mode activated in the Part environment. The process to create sketches in this mode is very simple. You need to select a sketch command, and then define a plane on which you want to create the sketch. The sketch commands are available in the **Sketching** or **Home** tab of the ribbon. To create a sketch, check the **Base Reference Planes** option under the **PathFinder** to display the **Base Reference Planes**. Next, select any of the sketch command (For example, the **Line** command) from **Sketching > Draw** panel and place the pointer on anyone of the Base Reference Planes. You will notice that a lock symbol appears on the plane. Click on the lock symbol (or) press F3 on your keyboard to lock the plane. You can now start drawing sketches on the locked plane. After creating the sketch, press the Esc key and click on the lock icon at the top right corner. The plane will be unlocked.

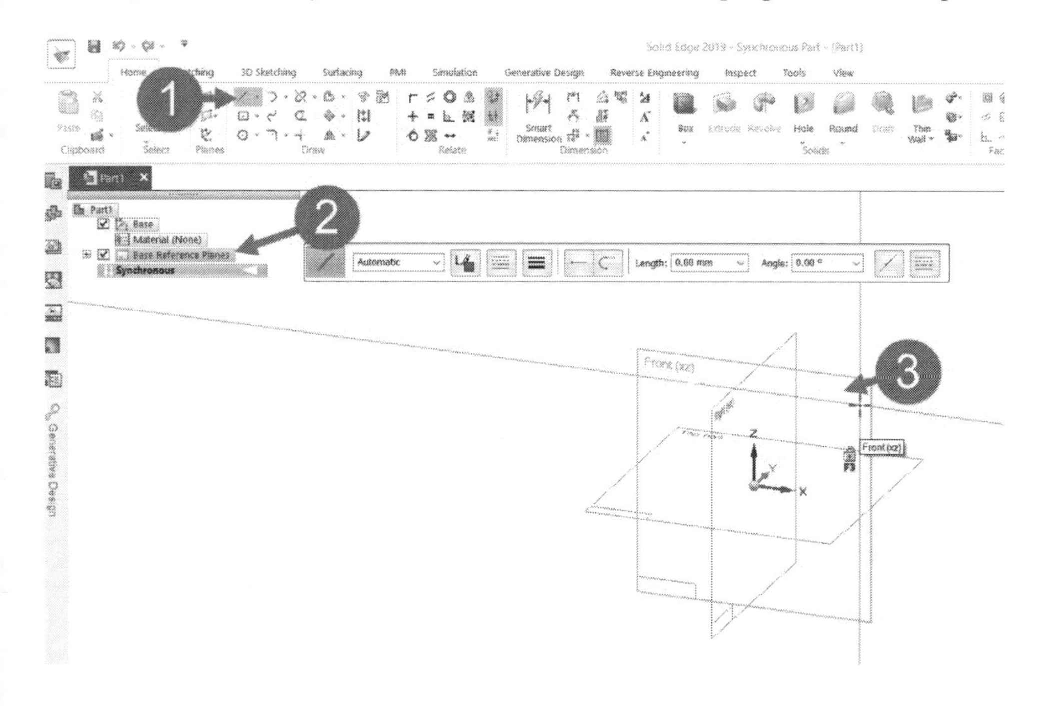

Create Sketches in the Ordered mode

Ordered mode was previously called as traditional environment. You can activate this mode by right-clicking and selecting **Transition to Ordered** or by selecting **Tools > Model > Ordered** on the ribbon.

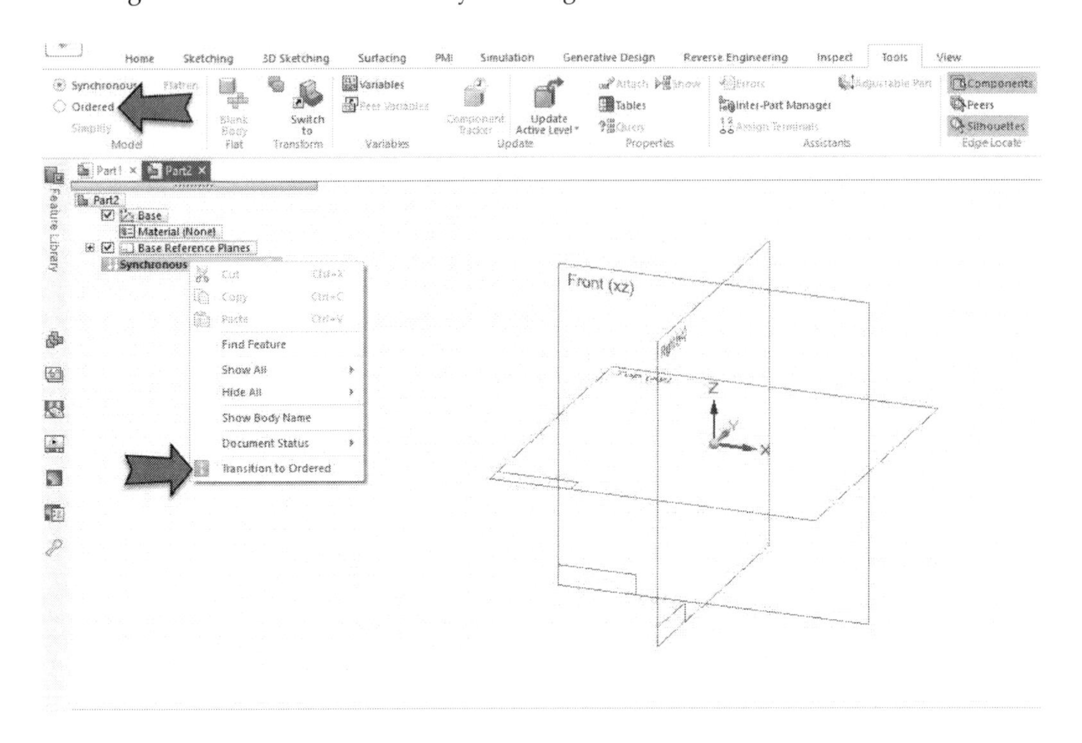

To create sketches, this mode offers a separate environment called the Sketching environment. To open this environment, select **Home > Sketch > Sketch** on the ribbon, and then click on a **Base Reference Plane** from the graphics window. You will notice that the **Line** command is activated, by default. You can start sketching lines or select any other sketching command. After completing the sketch, select **Home > Close > Close Sketch** on the ribbon. Next, enter the name of the sketch, and then click the **Finish** button on the **Sketch** command bar.

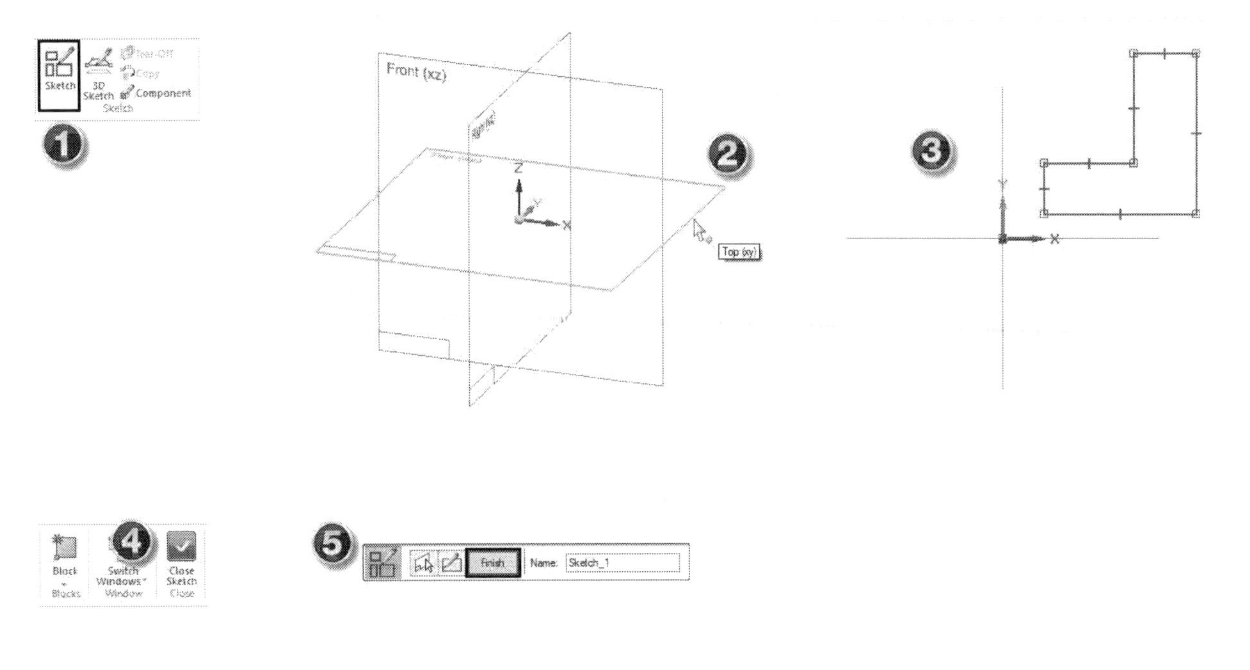

Draw Commands

Solid Edge provides you with a set of commands to create sketches. These commands are located on the **Draw** panel of the **Home** ribbon.

The Line command

This is the most commonly used command while creating a sketch. To activate this command, you need to click **Home > Draw > Line** on the ribbon. As you move the pointer in the graphics window, you will notice that it is changed to a set of crosshairs. This indicates that the command is active. To create a line, click in the graphics window and move the pointer. You will notice that the length and angle dimensions are attached to the line. Type-in the length value and press **Tab** on your keyboard. Type-in the angle value and press **Enter** to create the line. This creates a line with the precise length and orientation. However, you can simply select points to create lines, and then apply dimensions. After creating the lines, you can press **Esc** to deactivate the **Line** command. You can also click **Home > Select > Select** on the ribbon to deactivate a command.

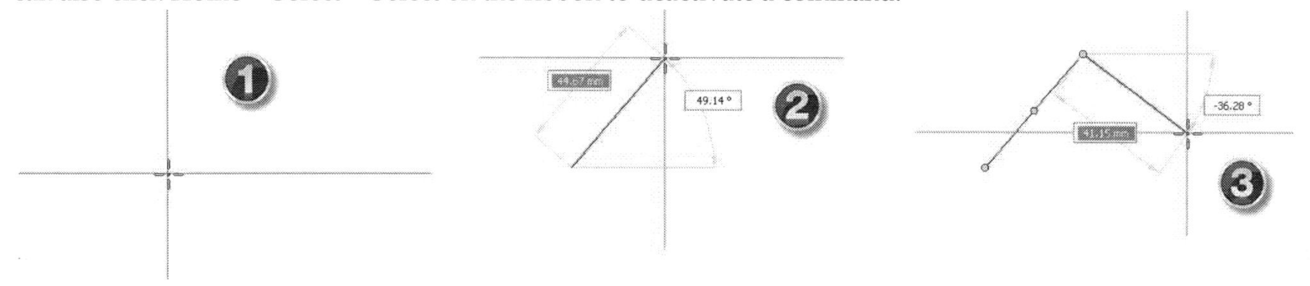

The **Line** command can also be used to draw arcs continuous with lines. Click the **Arc** icon from the command bar to draw this type of arc. The figure below shows the procedure to draw arcs connected to lines.

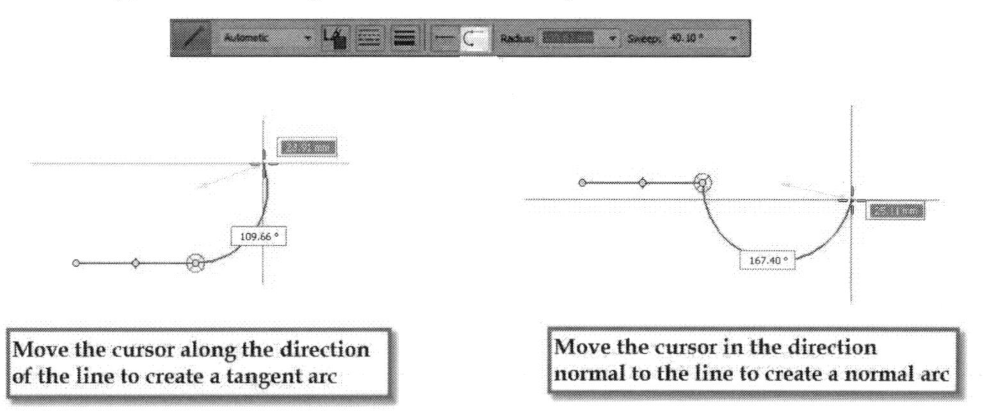

To delete a line, select it and press the **Delete** key. To select more than one line, press and hold the **Ctrl** key and then click on the line segments; the lines will be highlighted. You can also select multiple lines by dragging a box from left to right. Press and hold the left mouse button and drag a box from left to right; the lines inside or crossing the box boundary will be selected. Dragging a box from right to left will only select the lines that are inside the box.

Using Grid and Snap settings

If you are new to Solid Edge, the grid and snap settings will help you to create sketches easily. A grid is similar to a graph paper on your computer screen, whereas the snap mode forces the pointer to select the grid points. You can locate sketch points easily and accurately using the grid and snap settings. To use these settings, you need to activate the **Show Grid** and **Snap to Grid** icons on the **Draw** panel of the **Sketching** tab. Next, activate a drawing command and start drawing the sketch. You will notice that you can select the grid points easily. This makes it easy to create sketches.

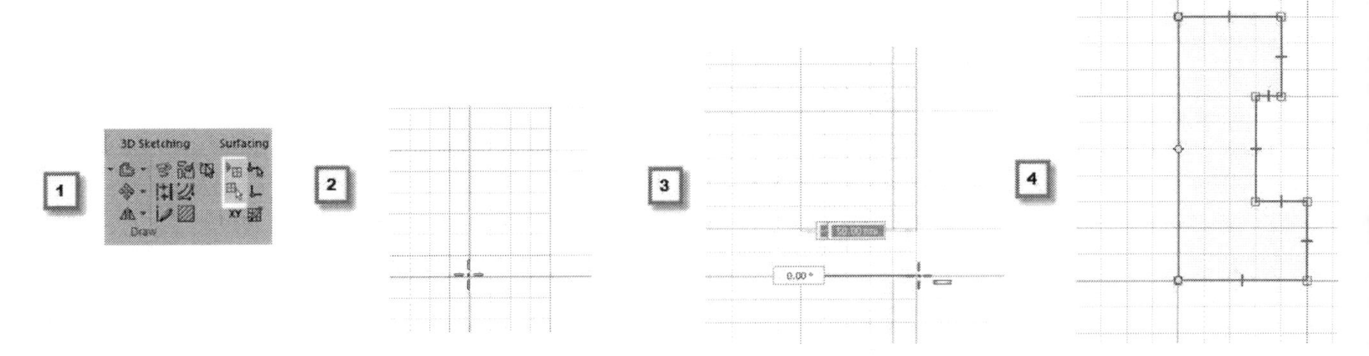

You can change the spacing between the grid points using the **Grid Options** dialog. Select **Sketching > Draw > Grid Options** on the ribbon to open this dialog. Next, modify the **Major line spacing value** to change the distance between the dark grid lines. Change the **Minor spaces per major** value to change the number of lighter grid lines between two major lines (dark lines). Examine the other options in this dialog. Most of them are self-explanatory.

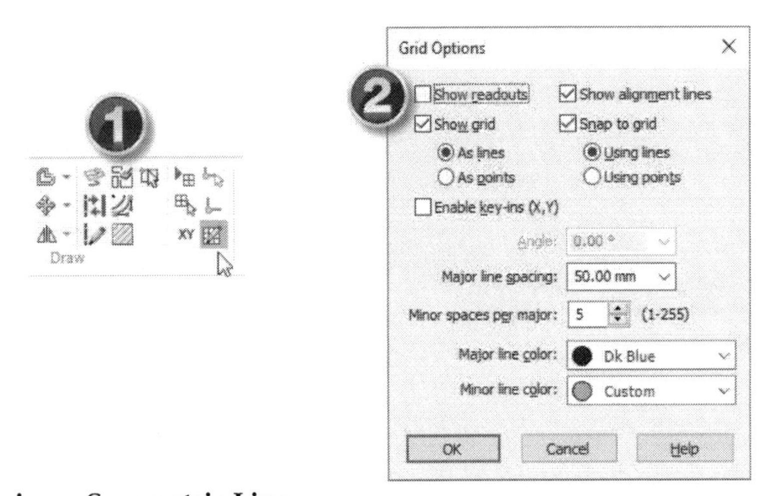

Drawing a Symmetric Line

You can create a symmetric line using the **Line** command. Activate this command, press **S**, and click to define the midpoint of the line. Move the pointer and click to define the end point; a symmetric line is created about the specified midpoint.

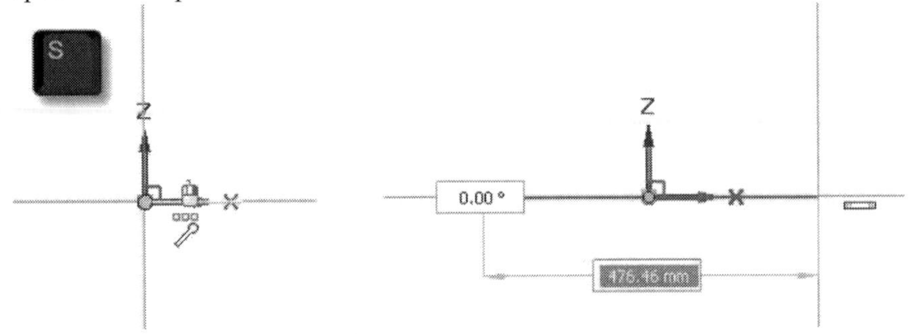

The Tangent Arc command

This command creates an arc tangent to another entity. The working of this command is same as the **Arc** icon on the **Line** command bar. You have to select the endpoint of a line and create a tangent or normal arc.

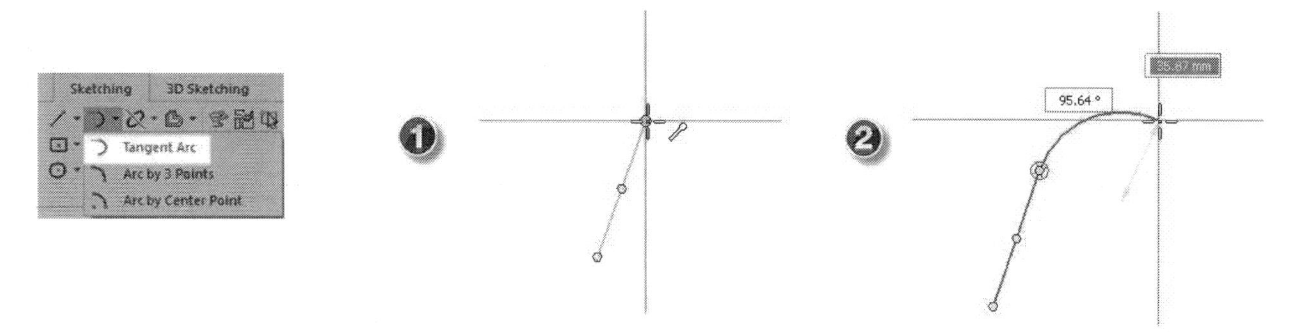

The Arc by 3 Points command

This command creates an arc by defining its start, end, and radius. Activate this command (click **Home > Draw > Tangent Arc > Arc by 3 Points** on the ribbon) and click to define the start point of the arc. Click again to define the endpoint. After defining the start and end of the arc, you have to the define size and position of the arc. Move the pointer and click to define the radius and position of the arc (or) type-in the radius value in the dimension box attached to the pointer.

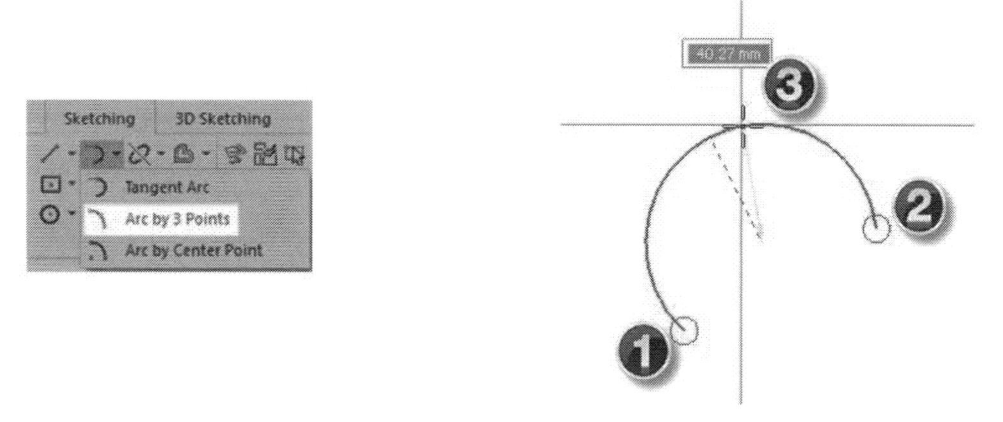

The Arc by Center Point command

This command creates an arc by defining its center, start and end points. Activate this command (click **Home > Draw > Tangent Arc > Arc by Center Point** on the ribbon) and click to define the center point. Next, move the pointer and you will notice that a line appears between center and the mouse pointer. This line is the radius of the arc. Now, click to define the start point of the arc and move the pointer. You will notice that an arc is drawn from the start point. Now, type-in the radius value and press Tab. Type-in the arc angle and press Enter (or) simply move the pointer and click.

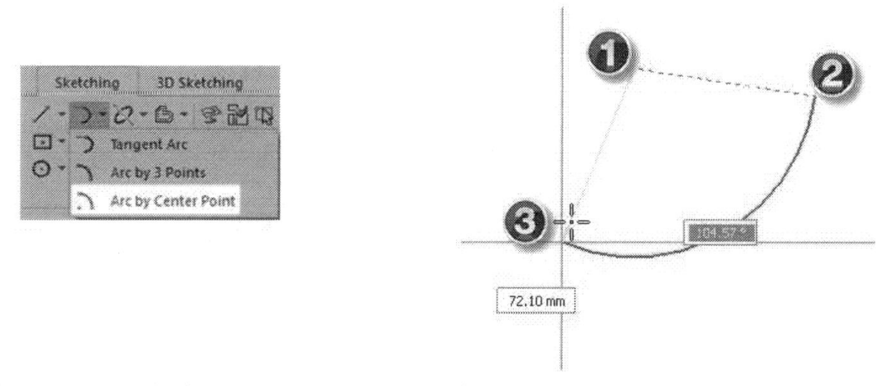

The Rectangle by Center command

This command creates a rectangle by defining its center and one corner point.

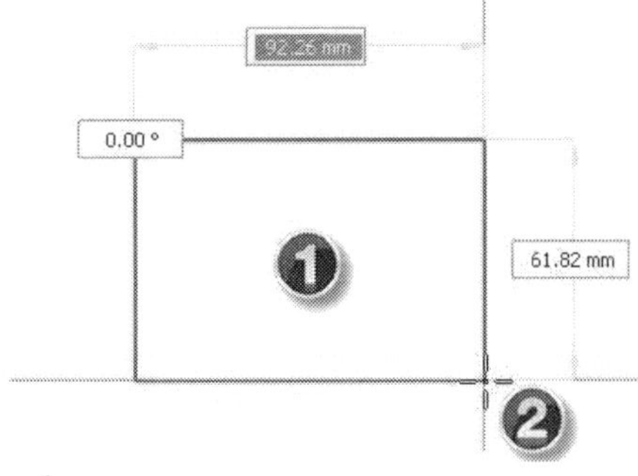

The Rectangle by 2 Points command

This command creates a rectangle by defining its diagonal corners.

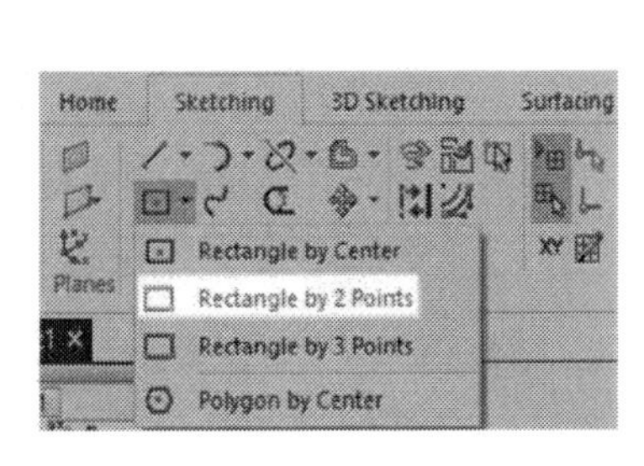

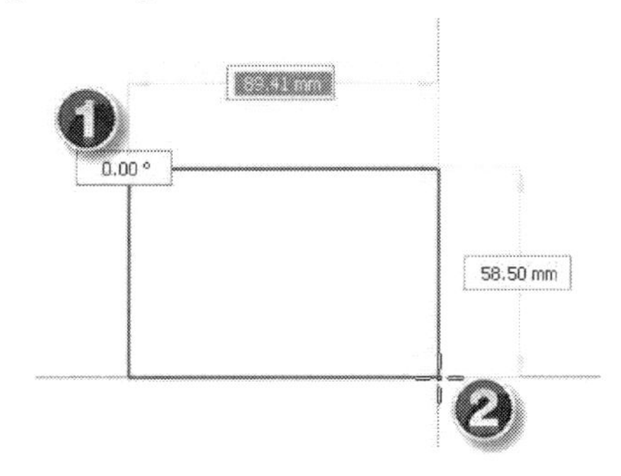

The Rectangle by 3 Points command

This command creates an inclined rectangle. The first two points define the length and inclination angle of the rectangle. The third point defines its width.

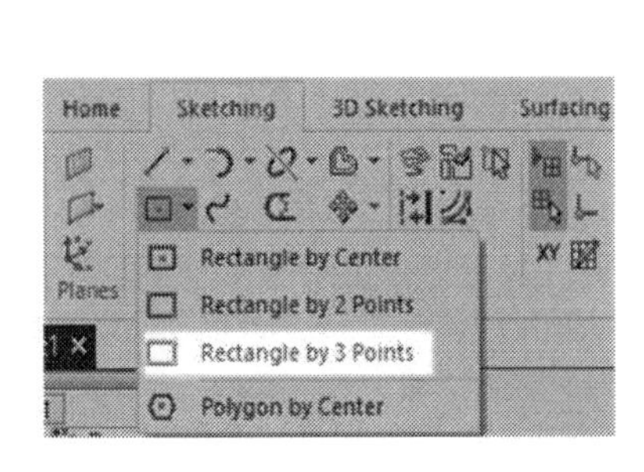

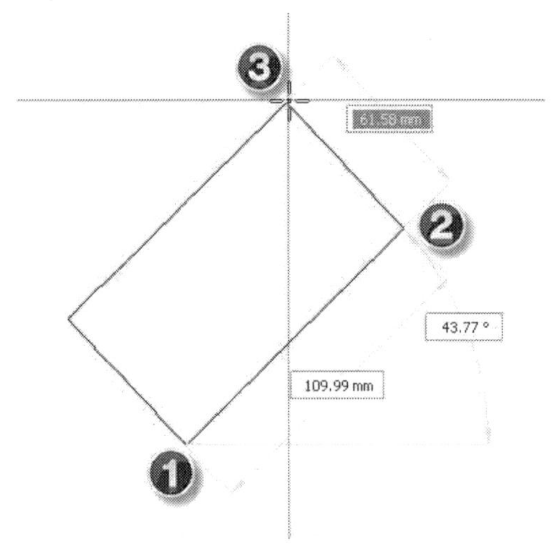

The Polygon by Center command

This command provides a simple way to create a polygon with any number of sides. As soon as you activate this command, a command bar pops up. Now, click in the graphics window to define the center of the polygon. As you move the pointer away from the center, you will see a preview of the polygon. To change the number of sides of the polygon, just click in the **Sides** field on the command bar and type a new number. Next, press the ENTER key to update the preview. You will notice that there are two icons available on the command bar: **By Vertex** and **By Midpoint**. If you select the **By Vertex** icon, a vertex of the polygon will be attached to the pointer. If you select the **By Midpoint** icon, the pointer will be on one of the flat sides of the polygon. Next, click in the graphics window to define the size and angle of the polygon. You can also define the size and angle of the polygon by entering values in the **Distance** and **Angle** fields attached to the pointer. After creating a polygon, you will notice that a dashed circle is created touching its vertices. You can change the polygon size by changing the size of this circle.

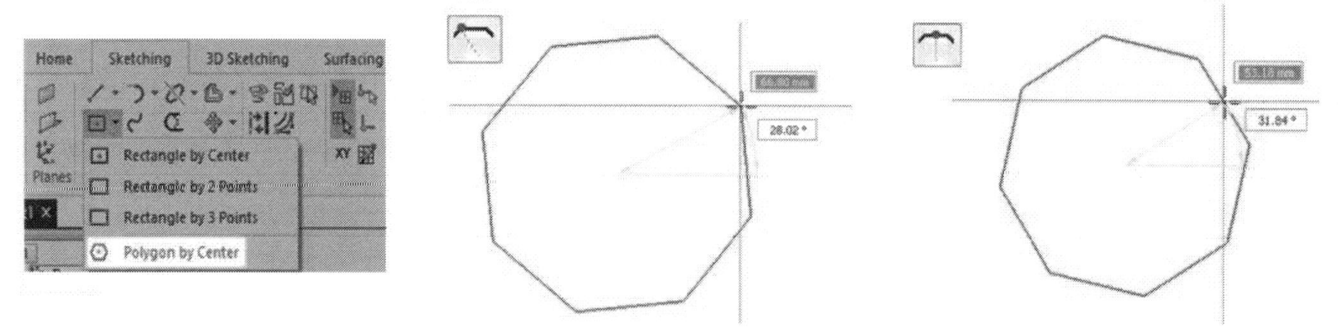

The Circle by Center Point command

This command is the common way to draw a circle. Activate this command (click **Home > Draw > Circle by Center Point** on the ribbon) and click to locate the center of the circle. Next, move the pointer, and then click again to define the diameter of the circle. You can also enter the diameter or radius value of the circle on the command bar.

The Circle by 3 Points command

This command creates a circle by using three points. Activate this command and the select three points from the graphics window. You can also select existing points from the sketch geometry. The first two points define the location of the circle and the third point defines its diameter.

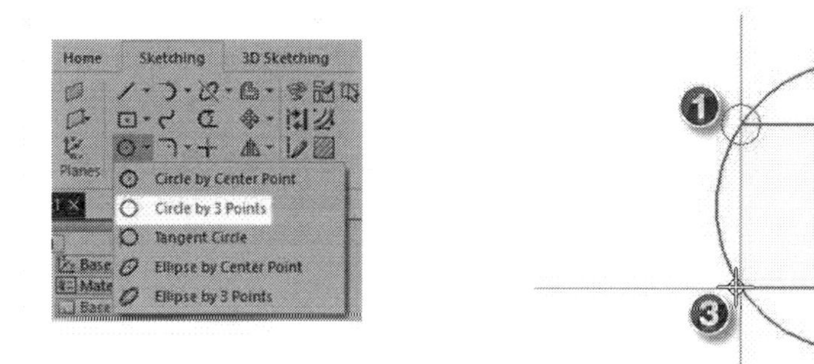

The Tangent Circle command

This command creates a circle by using two tangent points. Activate this command and select two lines, arcs, or circle; a circle will be drawn tangent to them.

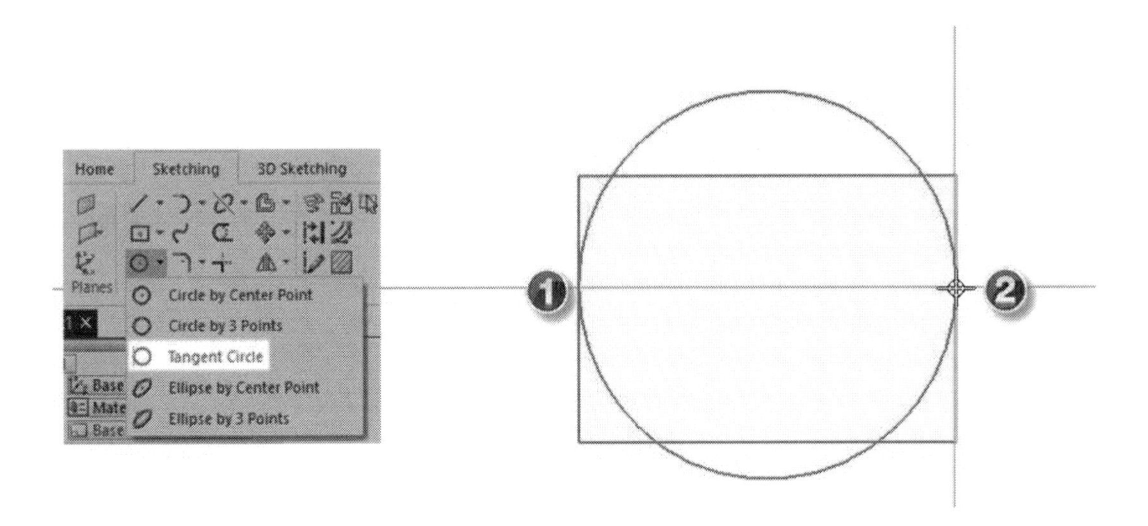

The Ellipse by Center Point command

This command creates an ellipse using a center point, and major and minor axes. Activate this command and click to define the center of the ellipse. As you move the pointer away from the center, you will notice that an axis is displayed. This can be either the major or the minor axis of the ellipse. When you click to place it, a preview of the ellipse appears and you can define the other axis. Note that you can also enter the radius and angle values of the axis. After defining the first axis, click to define the other axis (or) enter the axis radius; the ellipse will be drawn.

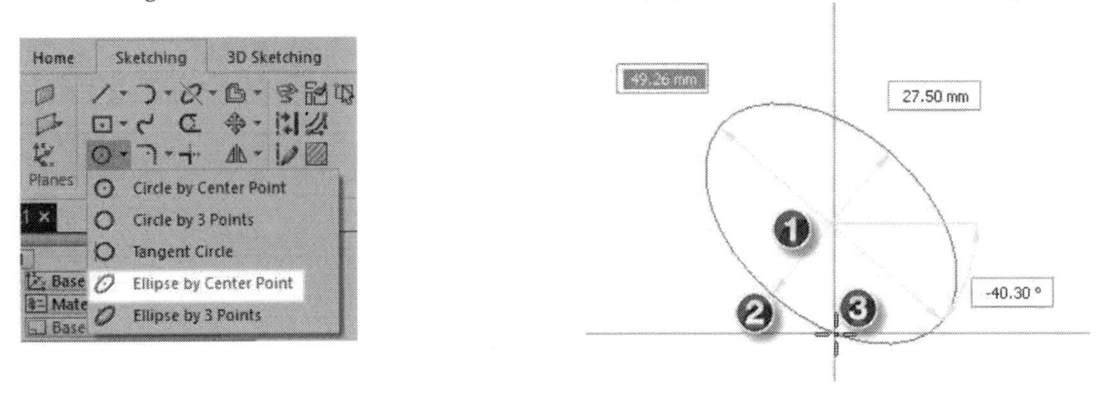

The Ellipse by 3 Points command

This command creates an ellipse by using three points. The first two points define the location and angle of the first axis of the ellipse. The third point defines the second axis of the ellipse.

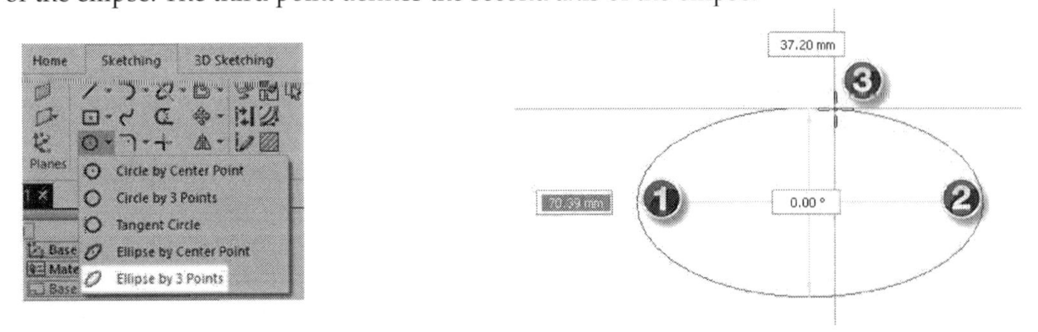

The Curve command

This command creates a smooth B-spline curve along the selected points. B-Splines are non-uniform curves, which are used to create irregular shapes. You can select points or press the left mouse button and drag to create a curve.

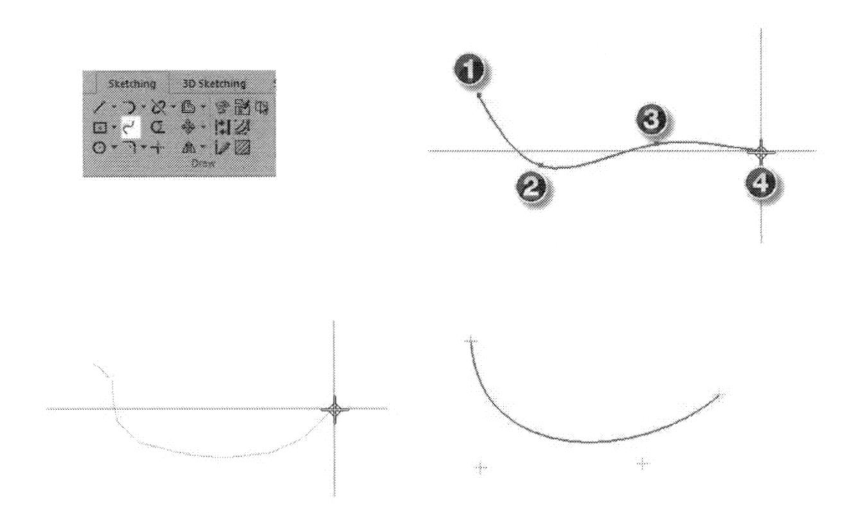

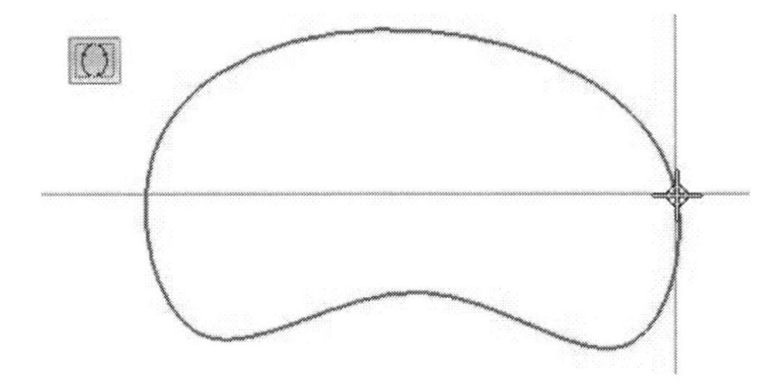

You also use the **Close Curve** option to create a closed curve.

Press **Esc** to deactivate this command, and then select the curve; you will notice that control vertices are displayed on the curve. Click and drag the control vertices to edit the curve shape. You can use the **Add/Remove points** option on the command bar to add more points or remove points from the curve.

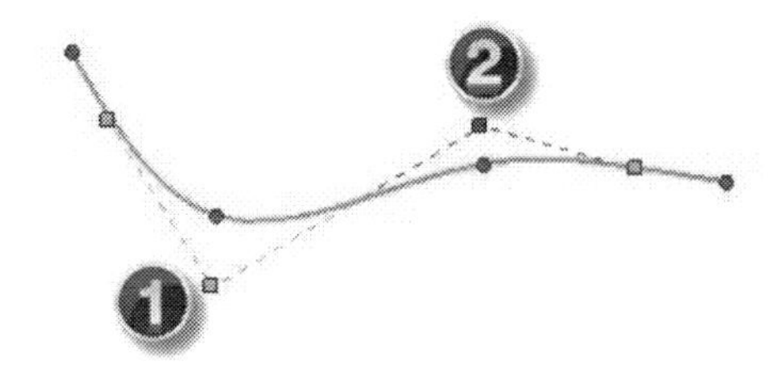

The Smart Dimension command

It is generally considered a good practice to ensure that every sketch you create is fully-constrained before moving on to create features. The term, 'fully-constrained' means that the sketch has a definite shape and size. You can fully-constrain a sketch by using dimensions and relations. You can add dimensions to a sketch by using the **Smart Dimension** command. You can use this command to add all types of dimensions such as length, angle, and diameter and so on. This command creates a dimension based on the geometry you select. For instance, to dimension a circle, activate the **Smart Dimension** command, and then click on the circle. Next, move the pointer and click again to position the dimension; you will notice that a box pops up. You can type-in a value in this box, and then press Enter to update the dimension.

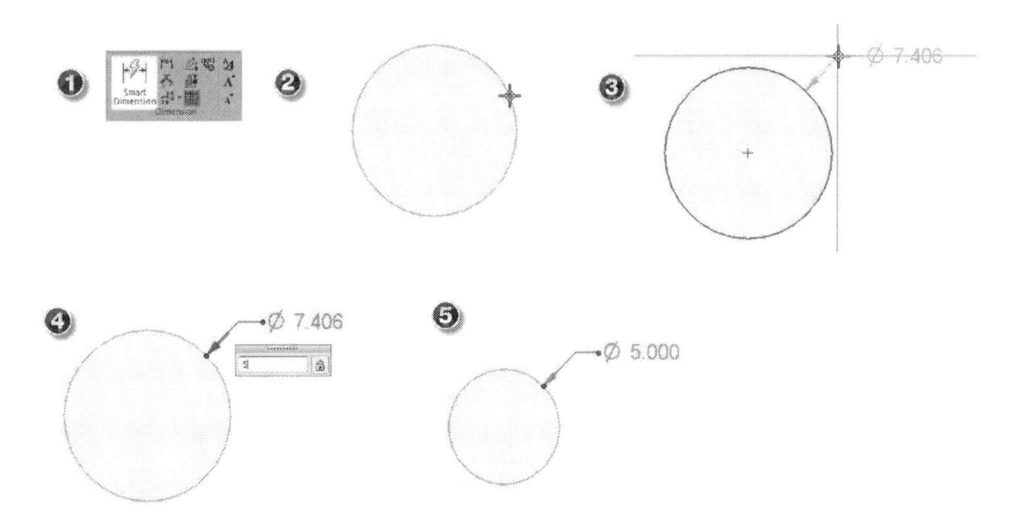

If you click a line, this command automatically creates a linear dimension. Click once more to position the dimension, and then type-in a value and press Enter; the dimension will be updated.

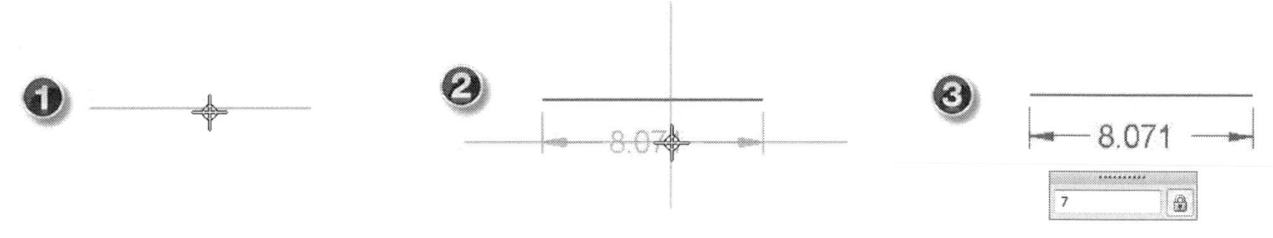

You can use the **Angle** option on the **Smart Dimension** command bar to add an angle dimension.

The Distance Between command

This command creates a linear dimension between two points. Activate this command and select the **Horizontal/Vertical** option on the command bar. Select the endpoints of a line and move the pointer to establish a vertical or horizontal dimension.

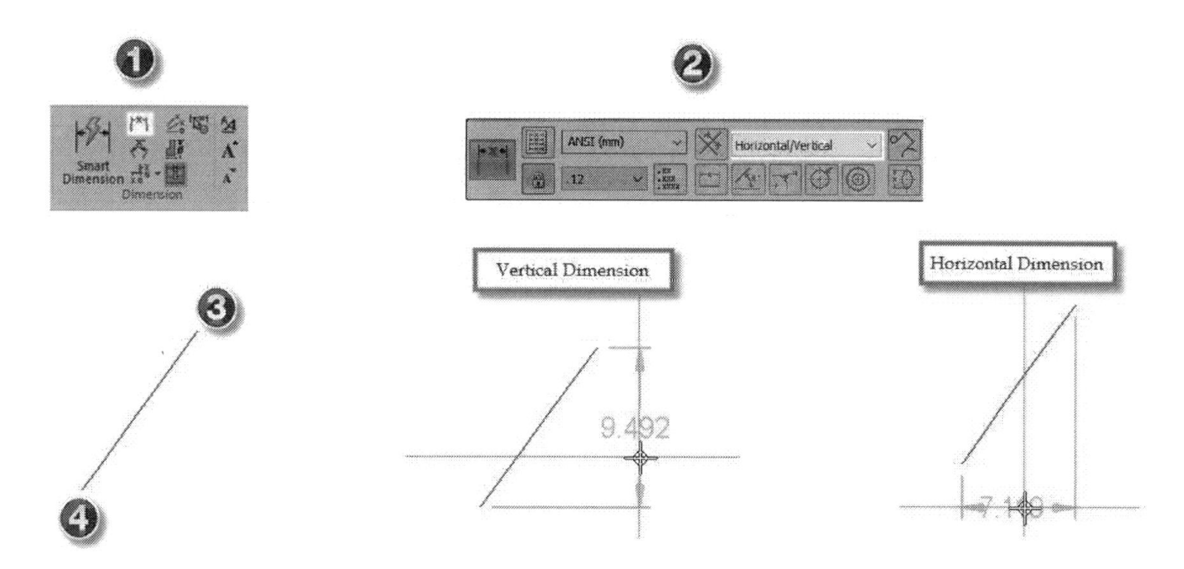

If you want the true length of the line, select the **By 2 Points** option on the command bar. Next, select the end points of a line, and the move the pointer and position the dimension. Type-in a value in the box and press Enter to update the dimension.

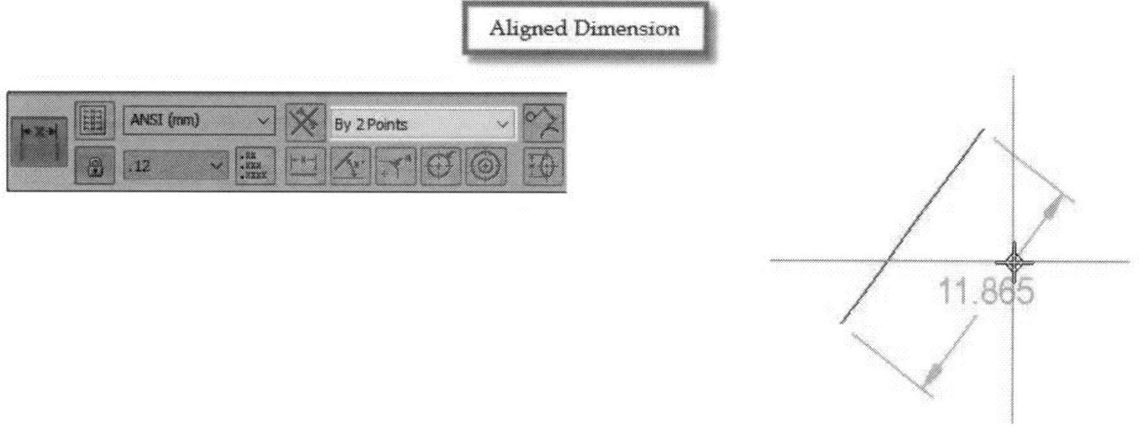

The Angle Between command

This command creates an angle dimension between two selected elements. Activate this command and select the elements positioned at an angle with each other. Next, move the pointer and position the dimension. Type-in a value and press Enter to update angle.

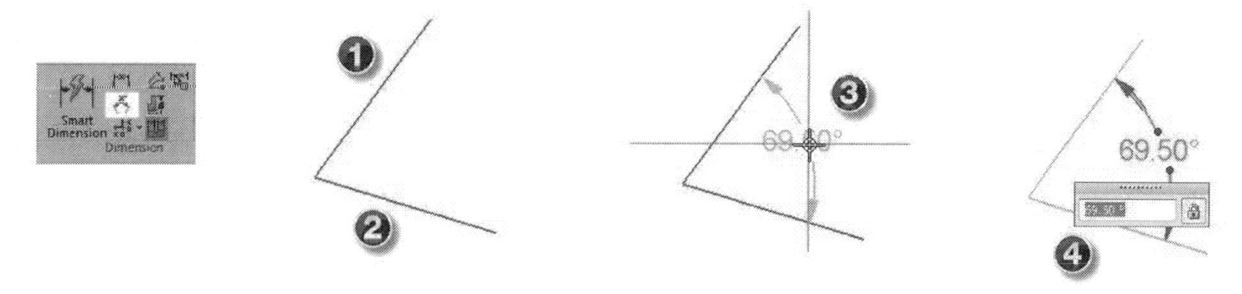

Driving Vs Driven dimensions

When creating sketches for a part, Solid Edge will not allow you to over-constrain the geometry. The term 'over-constrain' means adding more dimensions than required. The following figure shows a fully constrained sketch. If you add another dimension to this sketch (e.g. diagonal dimension), it appears in blue color. This type of dimension is a driven dimension. You cannot double-click and edit this dimension because it is redundant.

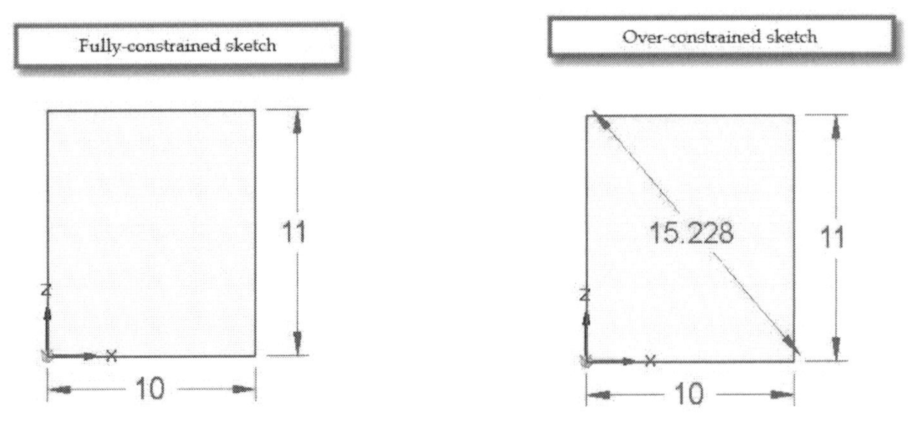

Existing driving dimensions (dimensions in red color) already define the sketch geometry. Driving dimensions are so named because they drive the geometry of the sketch. Double-clicking one of the driving dimensions and changing the value will change the geometry of the sketch. For example, if you change the value of the width, the driven dimension along the diagonal updates, automatically. Also, note that the dimensions, which are initially created, will be driving dimensions, whereas the dimensions created after fully defining the sketch are driven dimensions.

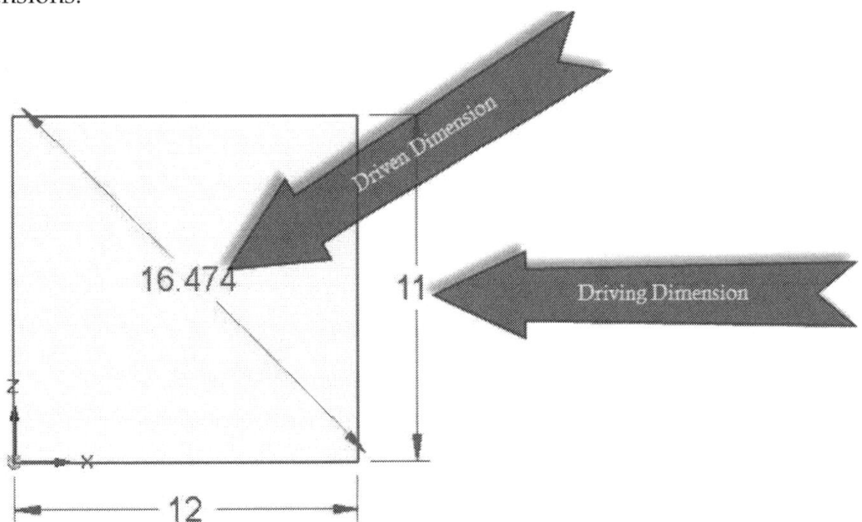

IntelliSketch Auto-Dimensions

Solid Edge provides you with an option to create dimensions, automatically. You can do it using the **IntelliSketch** dialog. Click **Sketching > IntelliSketch > IntelliSketch Options** to activate this dialog. On this dialog, select the **Auto-Dimension** tab, and then select the **Automatically create dimensions for new geometry** option.

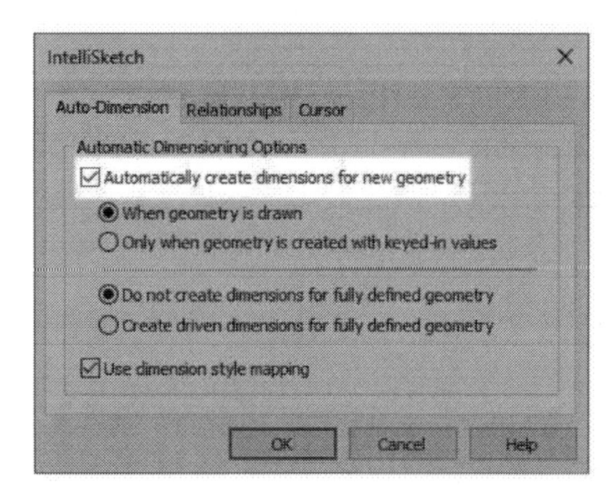

In addition, there are other options on the **IntelliSketch** dialog to define the conditions to create automatic dimensions. These options are self-explanatory. Click **OK** after defining the settings in this dialog.

Geometric Relations

Geometric Relations are used to control the shape of a sketch by establishing relationships between the sketch elements. The geometric relations are available on the **Relate** panel of the **Home** tab and are explained next.

Connect

This relation connects a point to another point or element. Activate this button, and then select two points; the selected points will be connected.

Parallel

This relation makes two lines parallel to each other. Activate this button, and then select two lines from the sketch; the first line is made parallel to the second line.

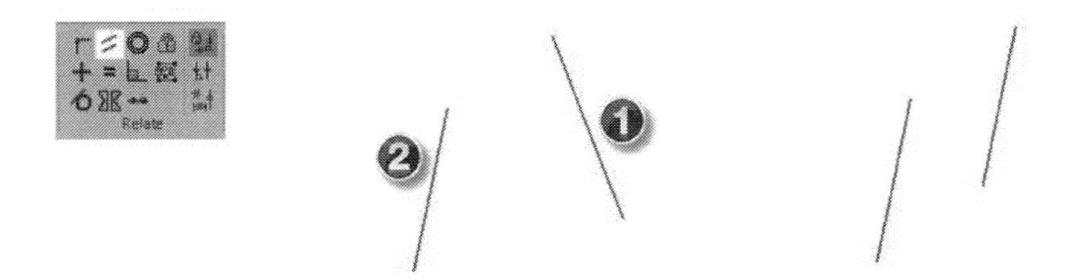

Concentric

This relation makes the center points of arcs, circles or ellipses coincident. Activate this button and select a circle or arc from the sketch. Select another circle or arc. The first circle/arc will be concentric with the second circle/arc.

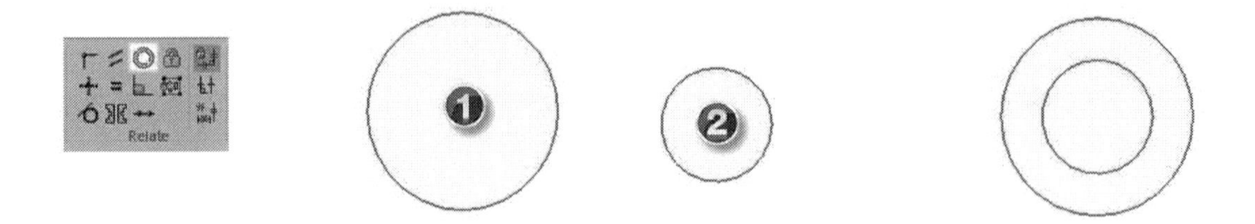

Lock

This relation locks a sketch element or dimension so that it cannot be moved or modified. Activate this button and select an element or dimension; it will be locked at its current position. In addition, you cannot change the shape and size of the element.

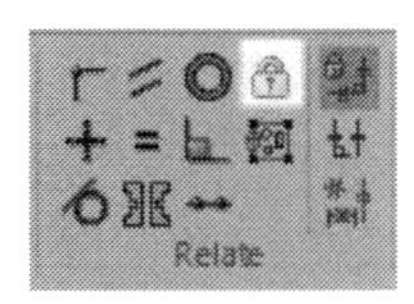

Horizontal/Vertical

This relation makes a line horizontal or vertical. The lines positioned at an angle below 45-degrees will be made horizontal. The lines positioned at an angle above 45-degrees will be made vertical. You can also make two points horizontally or vertically aligned with each other.

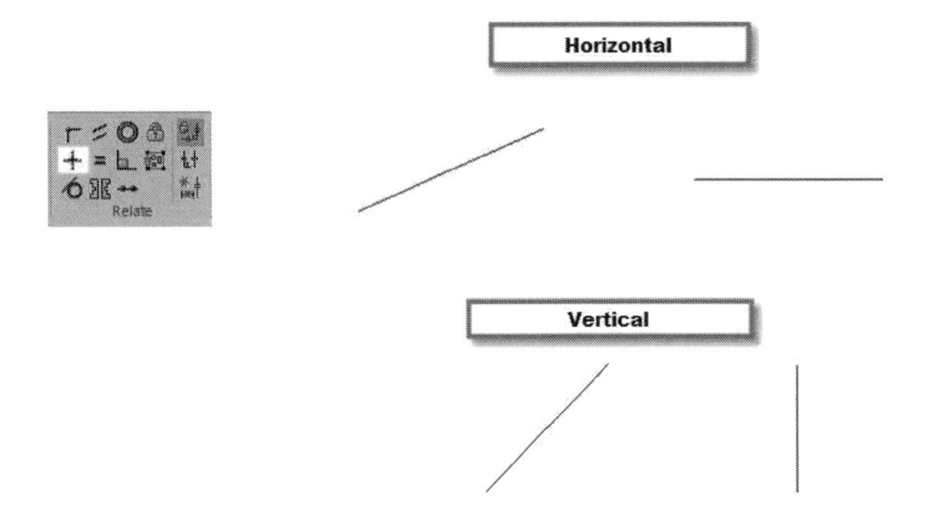

Equal

This relation makes the selected objects equal in magnitude. For example, if you select two circles, the diameter of the selected circles will become equal. If you select two lines, the length of the two lines will be equal.

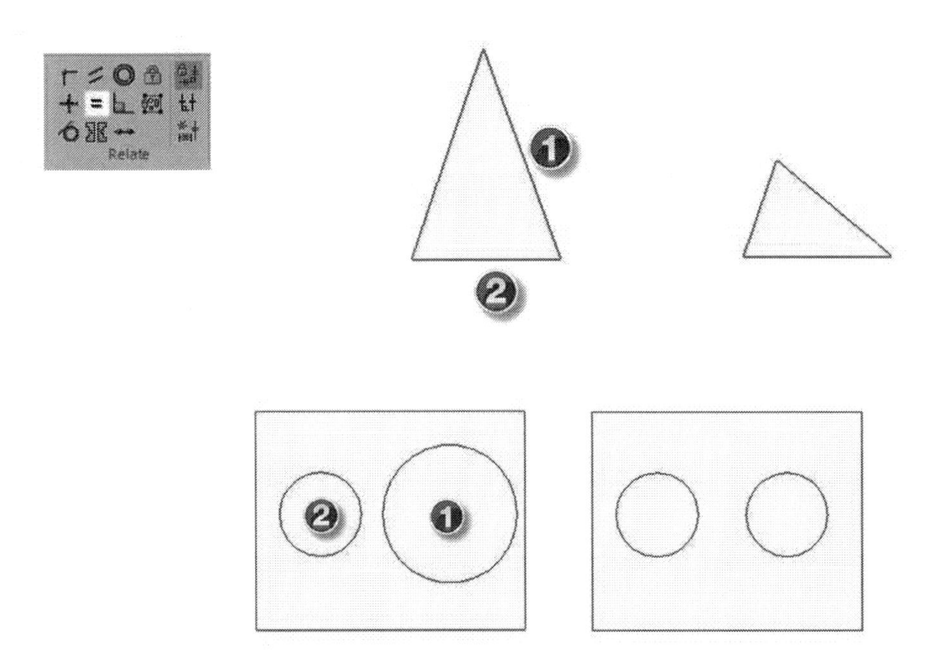

Perpendicular

This relation makes two lines perpendicular to each other. Activate this button and select two lines from the sketch. The first line will be made perpendicular to the second line.

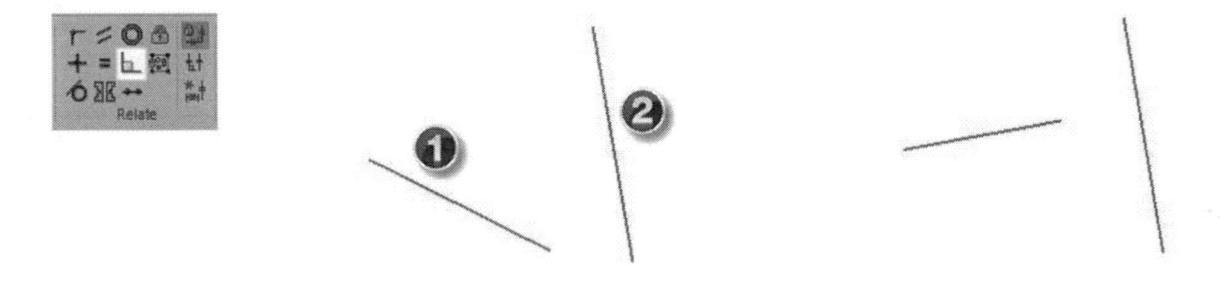

Rigid Set

This relation makes the selected objects act as a single unit. Activate this button and select two or more objects from the sketch. Click the green check on the command bar. The selected objects will be made into a rigid set.

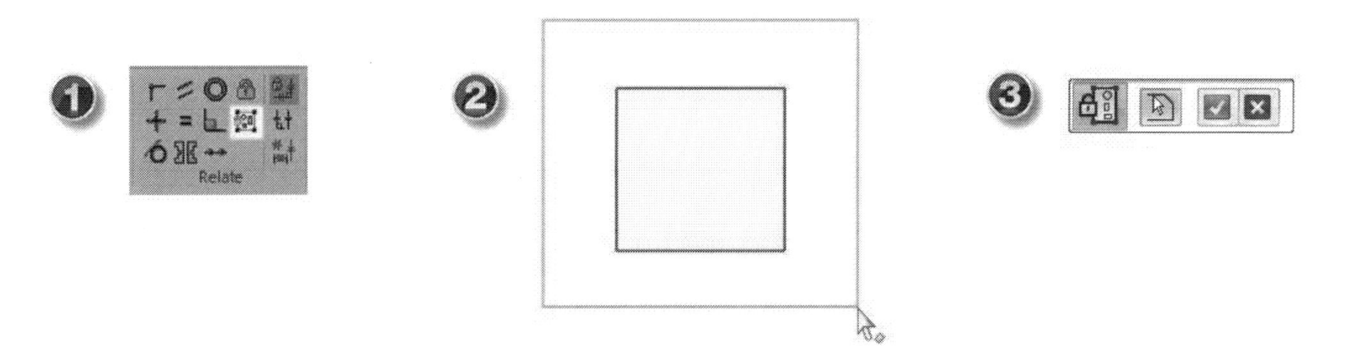

Now, click and drag anyone of the object from the rigid set. You will notice that entire set will be dragged.

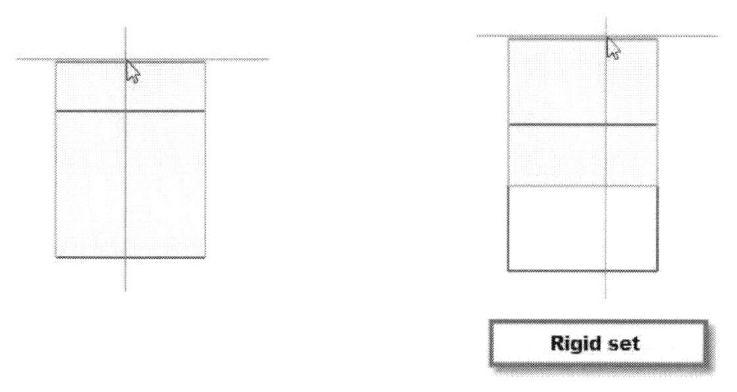

Tangent

This relation makes an arc, circle, or line tangent to another arc or circle. Activate this button and select a circle, arc, or line. Select another circle, arc, or line. The first object will be tangent to the second object.

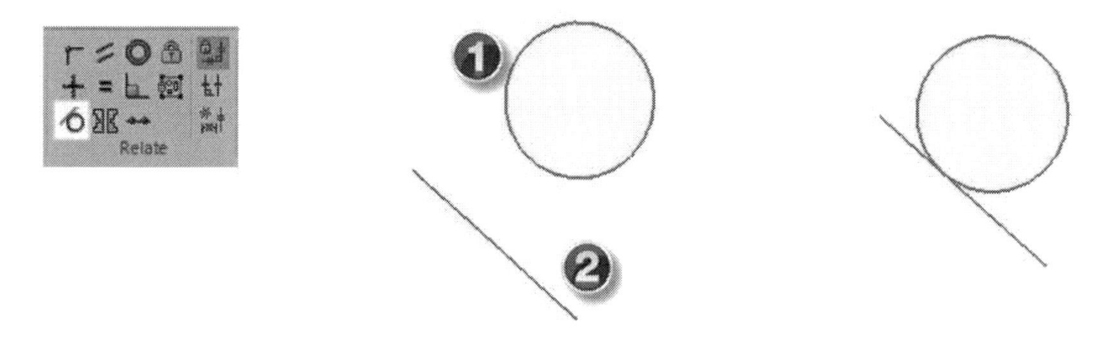

You can also make a curve continuous with another curve or arc using the **Tangent** relation.

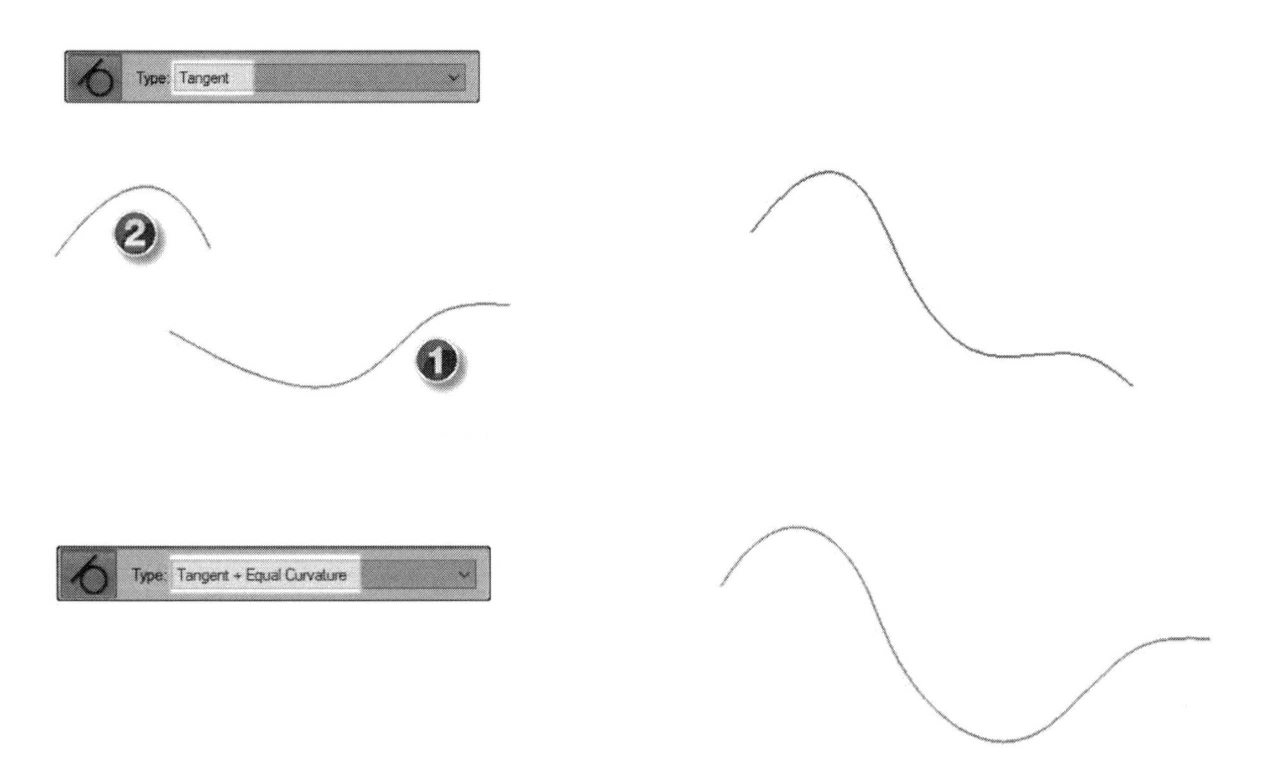

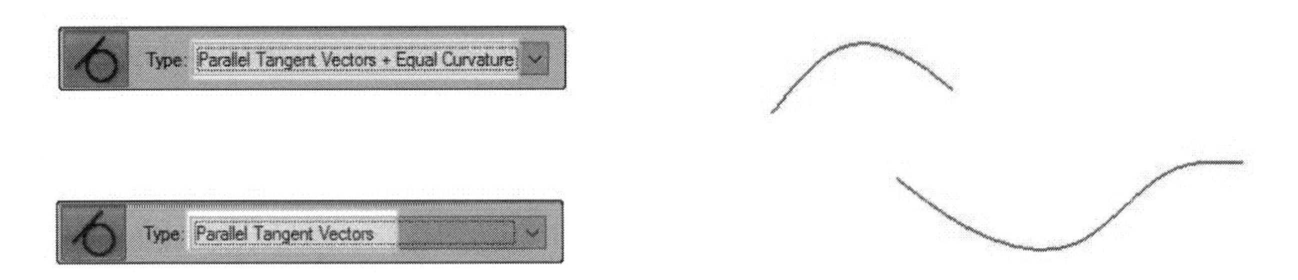

Symmetric

This relation makes two objects symmetric about a line. The objects will have same size, position and orientation about a line. For example, if you select two circles about a line, they will become equal in size, aligned horizontally, and positioned at an equal distance from the line. Activate the **Symmetric** button and select the symmetry line. Select two objects from the sketch. They will be made symmetric about the selected line.

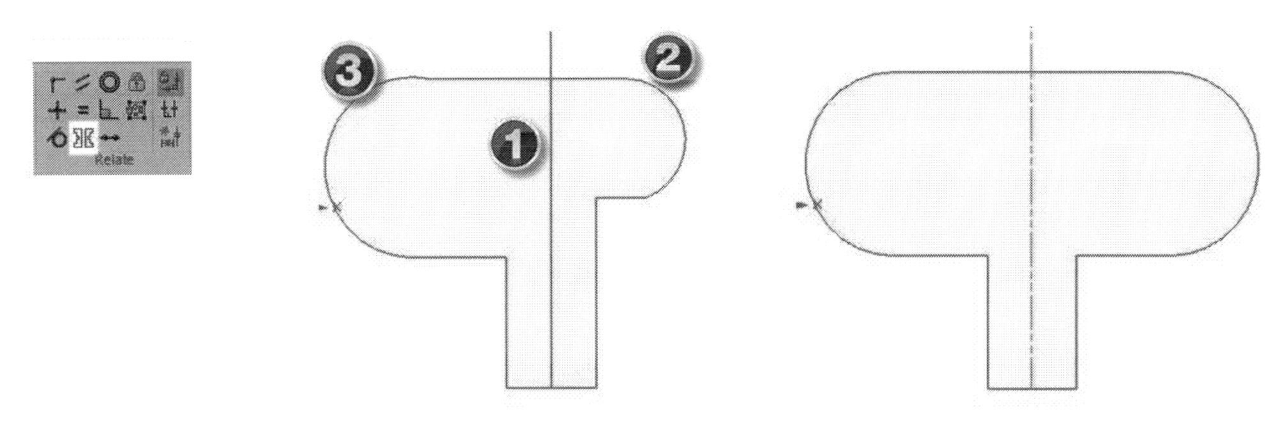

Collinear

This relation forces a line to be collinear to another line. The lines are not required to touch each other.

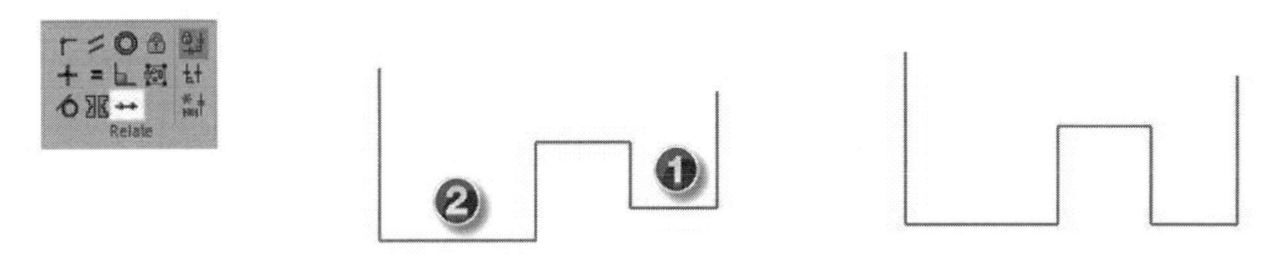

Maintain Relationships

Relations can also be applied automatically by activating the **Maintain Relationships** command. Activate or deactivate **Maintain Relationships** by picking the **Maintain Relationships** button on the **Relate** panel. With this command on, relations are applied automatically when the sketch elements are created. You can define which relations to apply automatically by using the **IntelliSketch Options** dialog. Click **Sketching > IntelliSketch > IntelliSketch Options**, and then select the **Relationships** tab on the **IntelliSketch Options** dialog. In this tab, select the relations to be created while sketching elements, and then click **OK**.

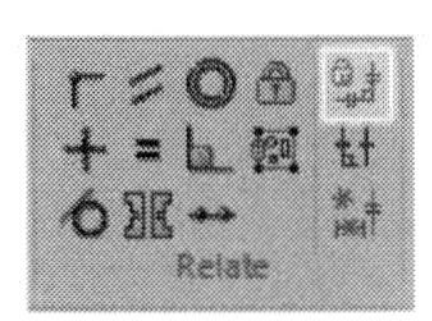

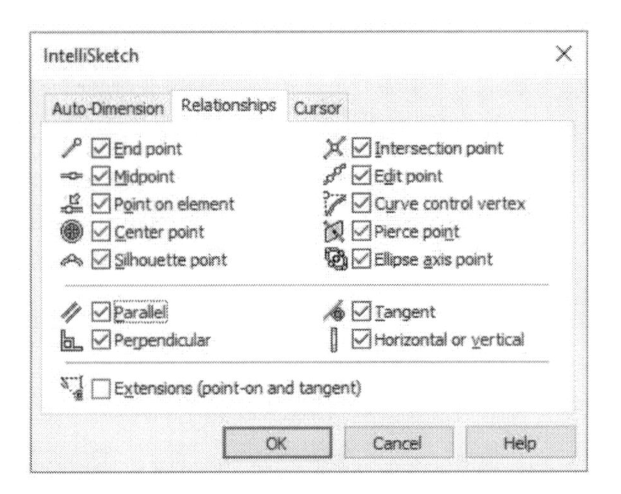

Relationship Handles

As relations are created, they can be viewed using the **Relationship Handles** button located on the **Relate** panel. When dealing with complicated sketches involving numerous relations, you can deactivate this button to turn off all relationship handles.

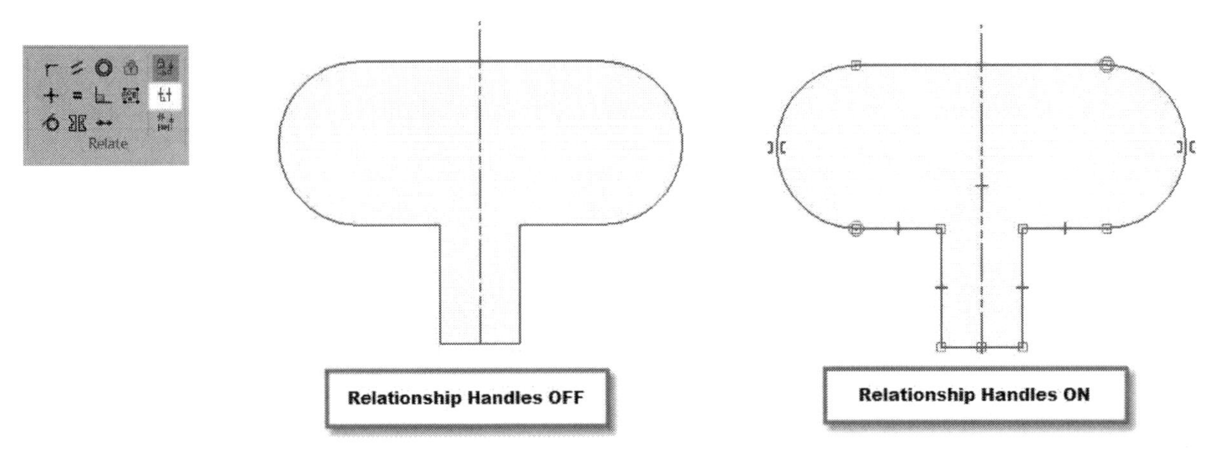

Relationship Assistant

In addition to the **Smart Dimension** command and other geometric relations, Solid Edge provides you with the **Relationship Assistant** command. This command automatically applies relations and dimensions to fully-constrain a sketch. To activate this command, click **Home > Relate > Relationship Assistant** on the ribbon. A command bar pops up. Select the **Options** icon on the command bar to open the **Relationship Assistant** dialog. On this dialog, click the **Geometry** tab, and then select the relations to be applied. Similarly, click the **Dimension** tab and select the dimensions to be applied. You can also select the **Dimension Scheme** such as **Stack**, **Chain**, and **Coordinate**. Click **OK** on the **Relationship Assistant** dialog and select the objects to apply relations and dimensions. Next, click the green check on the command bar and then select the horizontal and vertical dimension origins. The relations and dimensions will be created, automatically.

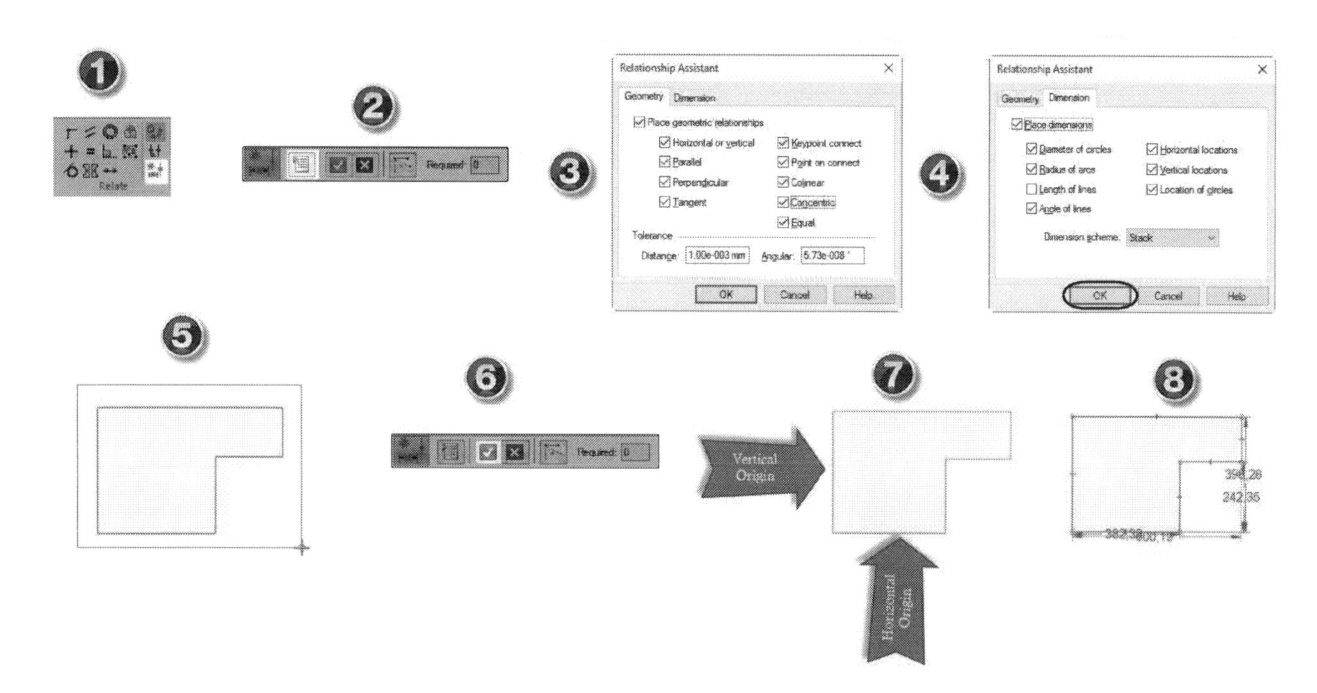

The Construction command

This command converts a sketch element into a construction element. Construction elements are reference elements, which support you to create a sketch of desired shape and size. Activate the **Construction** command from the **Draw** panel and click on a sketch element. The selected element will be converted to a construction element. You can also convert back the construction element to a sketch element by activating the **Construction** command and clicking on the element.

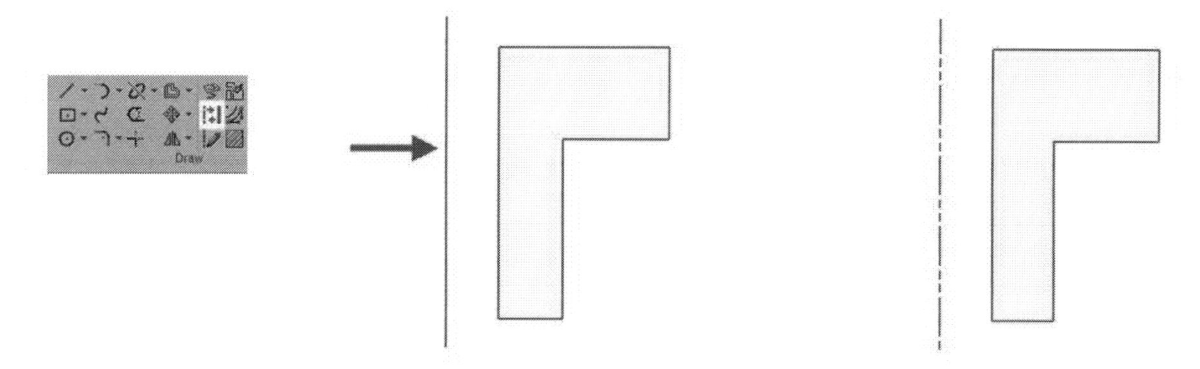

The Symmetric Diameter command

This command is very useful while creating a sketch for a revolved feature. It creates a dimension by measuring the distance between two lines or points, and then multiplying it by two. Activate this command from the **Dimension** panel, and then select the dimension origin. You need to ensure that the dimension origin is locked at its location. Now, select the line up to which the dimension is to be created. Click to position the dimension, and then change the value.

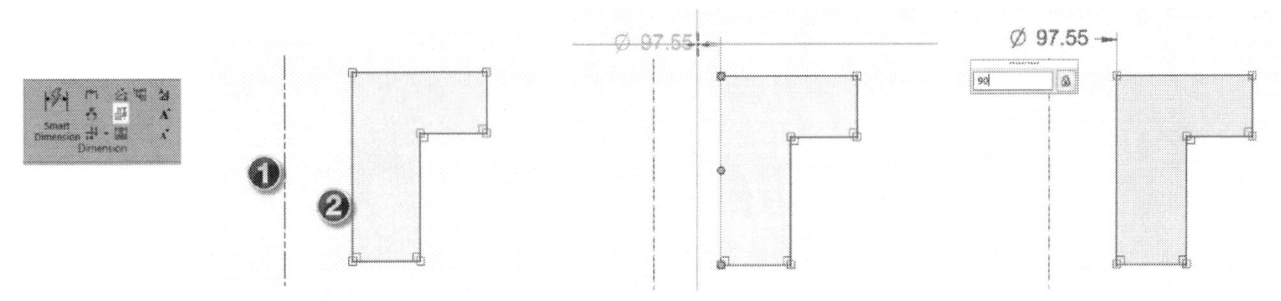

The Fillet command

This command rounds a sharp corner created by intersection of two lines, arcs, circles, and rectangle or polygon vertices. Activate this command from the **Draw** panel and select the elements' ends to be filleted. Type-in a radius value in the **Radius** box of the command bar and press Enter. The elements to be filleted are not required to form an intersection.

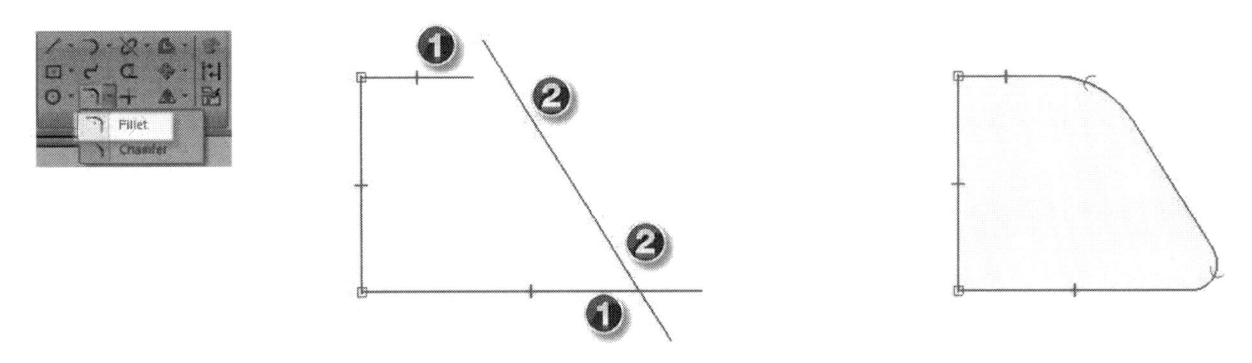

You can also drag the pointer across the elements to fillet the corner.

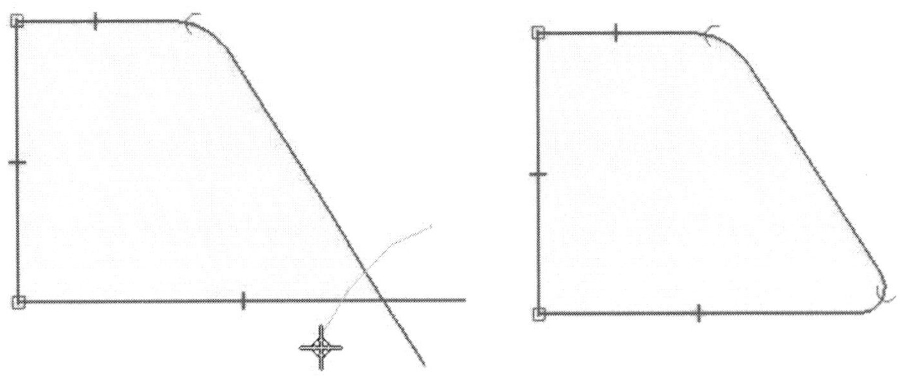

By default, the elements are automatically trimmed or extended to meet the end of the new fillet radius. You can use the **No Trim** option on the command bar, if you do not want to trim or extend the elements as necessary.

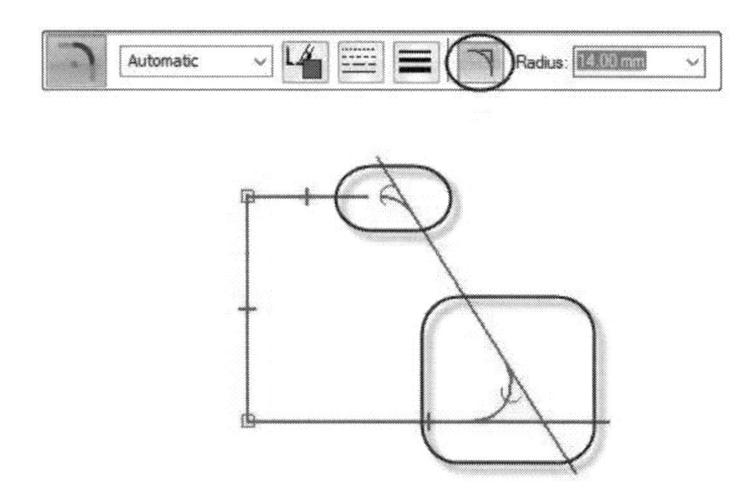

The Chamfer command

This command replaces a sharp corner with an angled line. Activate this command from the **Fillet** drop-down on the **Draw** panel and select the elements' ends to be chamfered. Type-in the chamfer angle in the **Angle** box on the command bar and press Enter. Next, move the pointer and click to create the chamfer. Instead, you can also use the **Setback A** and **Setback B** boxes on the command bar to define the chamfer size.

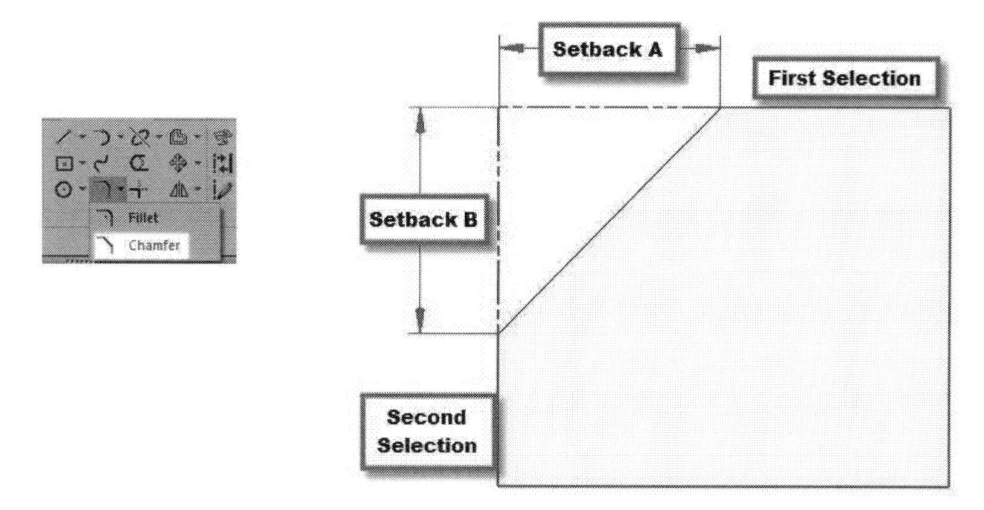

The Split command

This command splits an element into two elements. Activate this command from the **Draw** panel and click the element to split. Next, select a split point on the element. In case of a circle or ellipse, you must select two split points.

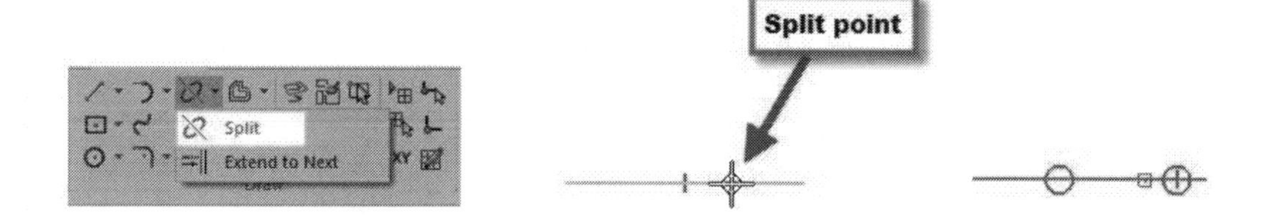

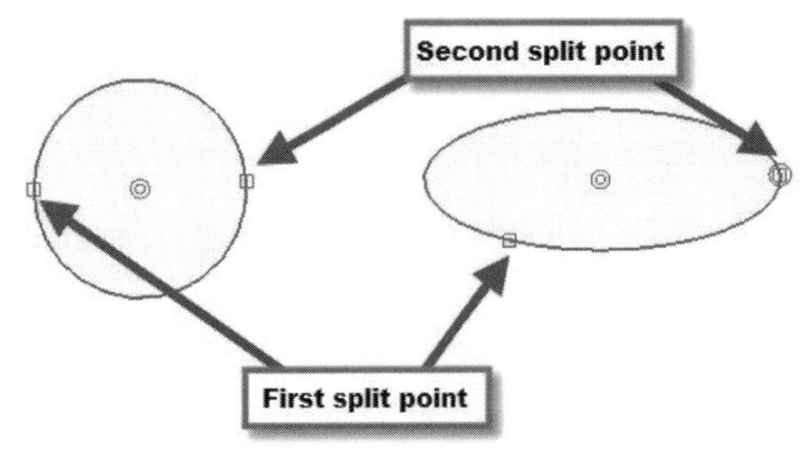

The **Split** command in also available in the 3D Sketching environment.

The Extend to Next command

This command extends elements such as lines, arcs, and curves until they intersect another element called a boundary edge. Activate this command from the **Split** drop-down on the **Draw** panel and click on the element to extend. It will extend up to the next element.

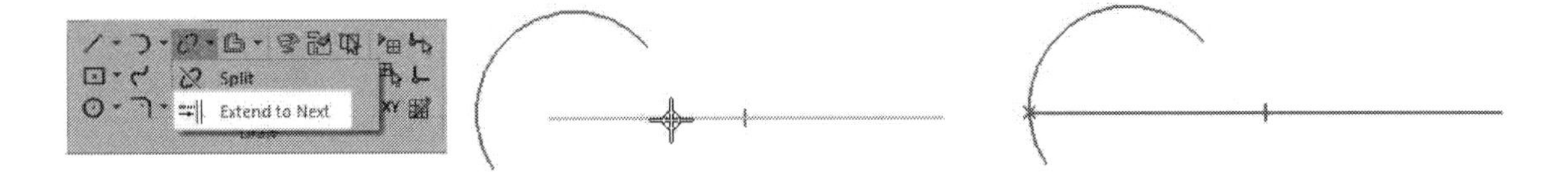

The Trim command

This command trims the end of an element back to the intersection of another element. Activate this command from the **Draw** panel and select the element or elements to trim. You can also drag the pointer across the elements to trim.

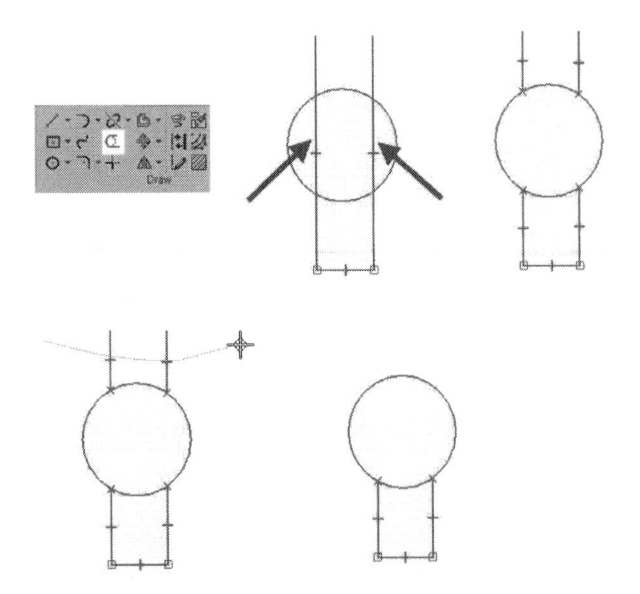

The Trim Corner command

This command trims and extends elements to form a corner. Activate this command from the **Draw** panel and select two intersecting elements. The elements will be trimmed and extended to form a closed corner.

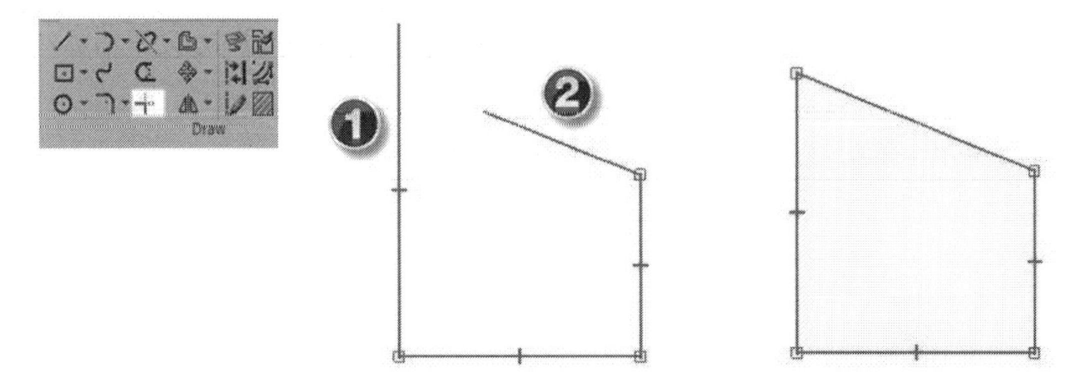

The Offset command

This command creates a parallel copy of a selected element or chain of elements. Activate this command from the **Draw** panel and select an element or chain of elements to offset. You can use the **Single** or **Chain** option from the **Select** drop-down on the command bar to select a single element or chain of elements. After selecting the element, type-in a value in the **Distance** field on the command bar and click the green check. Click to define the side of the offset. The parallel copy of the elements will be created and you can click again to create another parallel copy. Click the right mouse button after creating parallel copies. Press **Esc** to deactivate this command.

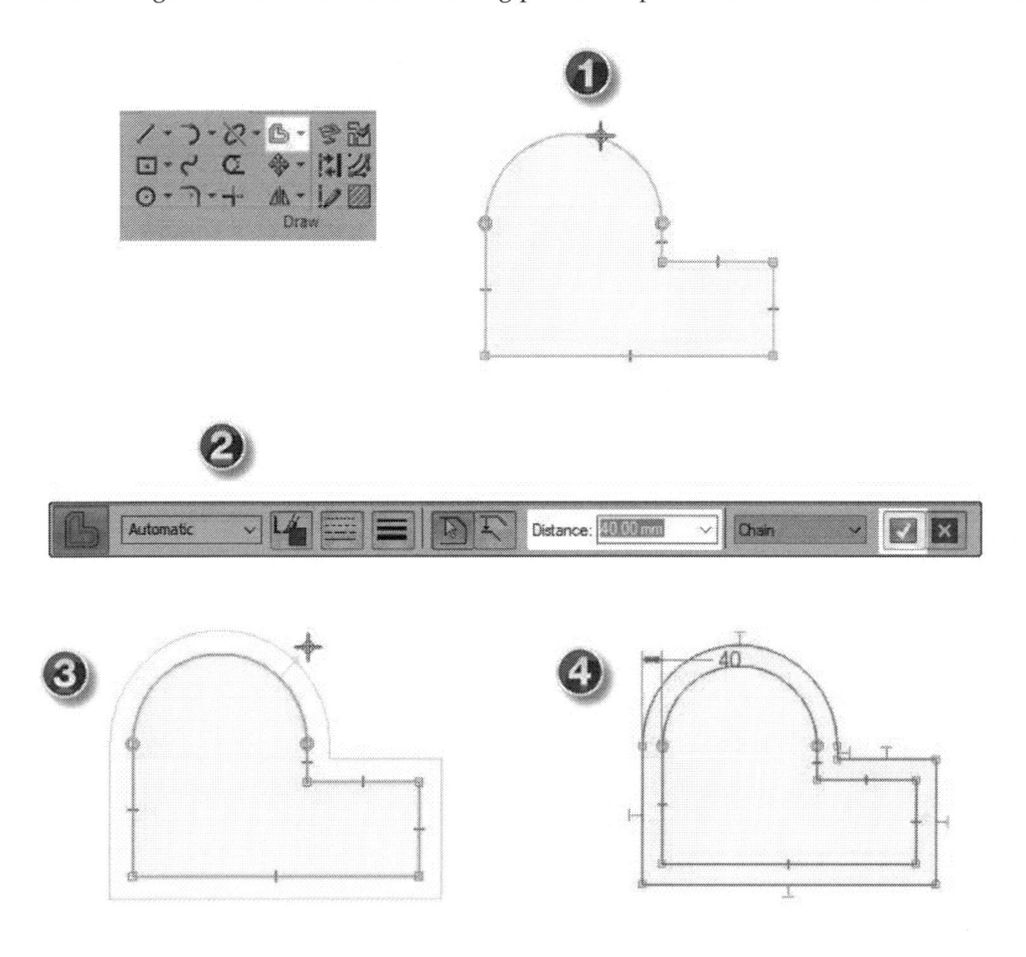

The Symmetric Offset command

This command creates a parallel copy on both sides of a selected element or chain of elements. It is helpful while creating a sketch slot. Activate this command from the **Offset** drop-down on the **Draw** panel. The **Symmetric Offset Options** dialog pops ups on the screen.

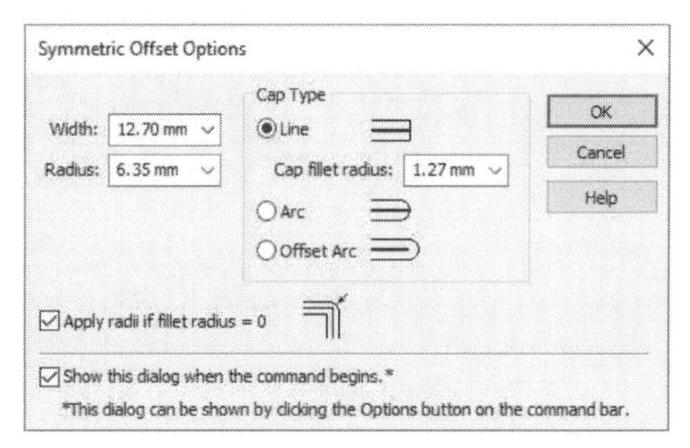

The options on this dialog are illustrated below. Set the required options on the dialog and click **OK**. Next, select an open sketch (line or arc) and click the green check to create the symmetric offset.

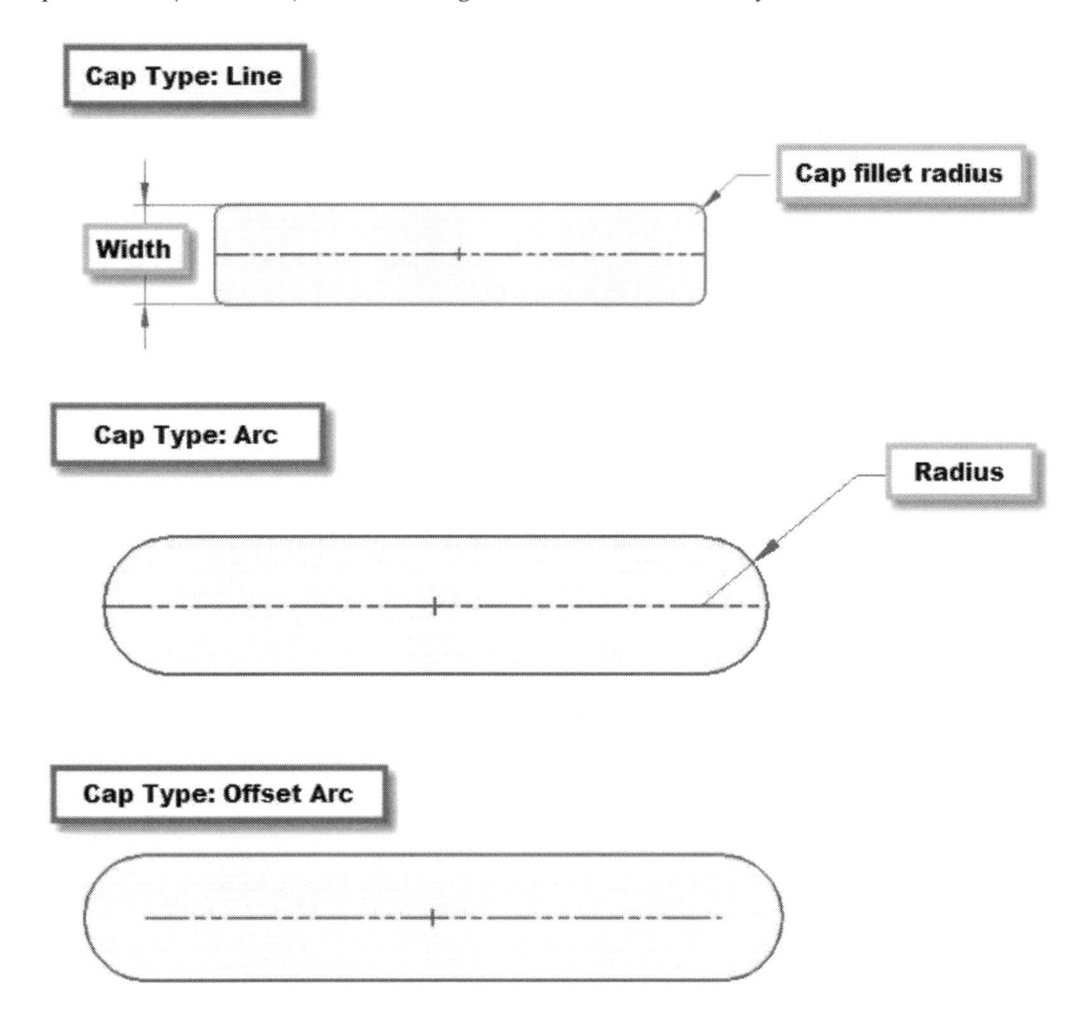

The Move command

This command relocates one or more elements from one position in the sketch to any other position you specify. Activate this command from the **Draw** panel, and then click on the elements to move. Next, you must select a base point and click at a new location.

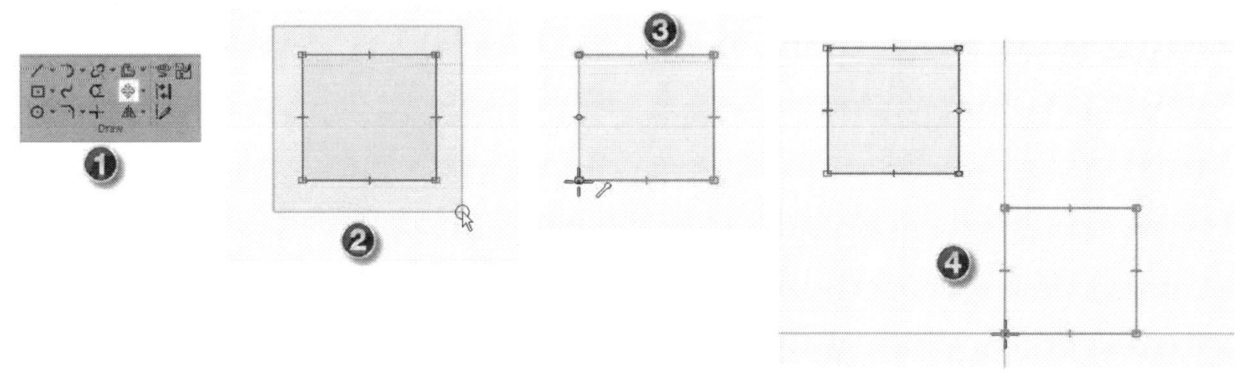

The **Copy** option on the **Move** command bar can be used to copy and move the selected elements.

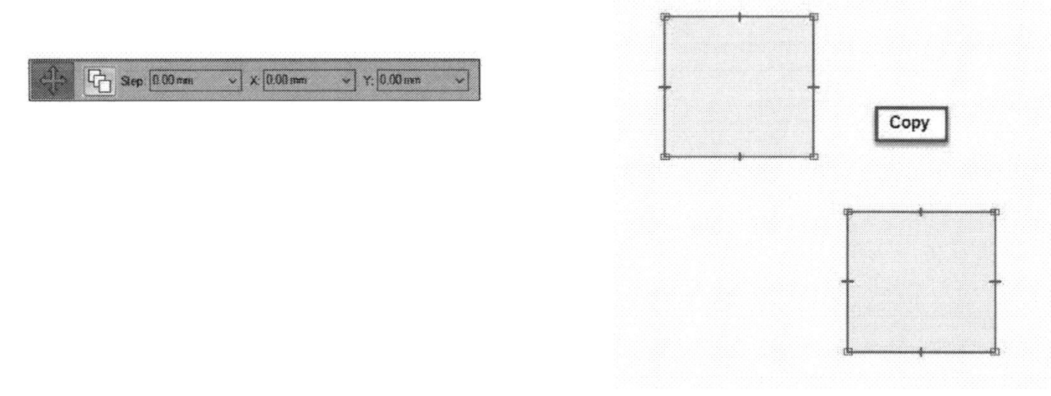

The Rotate command

This command rotates the selected elements to any position. Activate this command from the **Move** drop-down of the **Draw** panel, and then select the elements to rotate. Next, you must define a base point and a point from which the object will be rotated. Move the pointer and click to define the rotation angle. You can use the **Copy** option on the command bar to copy and rotate the selected elements.

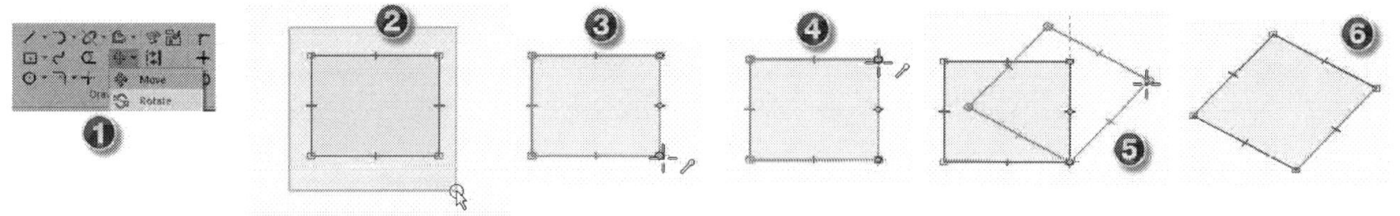

The Mirror command

This command creates a mirror image of the selected elements. You have the option to retain or delete the original elements. Activate this command from the **Draw** panel, and then select the elements to mirror. Next, you have to create two points defining the mirror-line. To retain the original elements, you must ensure that the **Copy** option is active on the command bar.

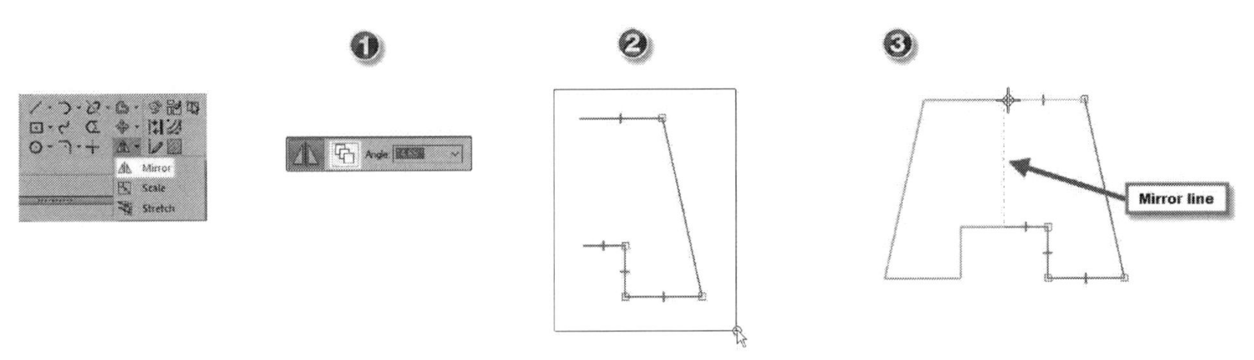

The Scale command

This command increases or decreases the size of elements in a sketch. Activate this command from the **Mirror** drop-down of **Draw** panel, and then select the elements to scale. After selecting the elements, you must select a base point. You can then scale the size of the selected elements by moving the pointer or entering a scale value in the **Scale** field on the command bar.

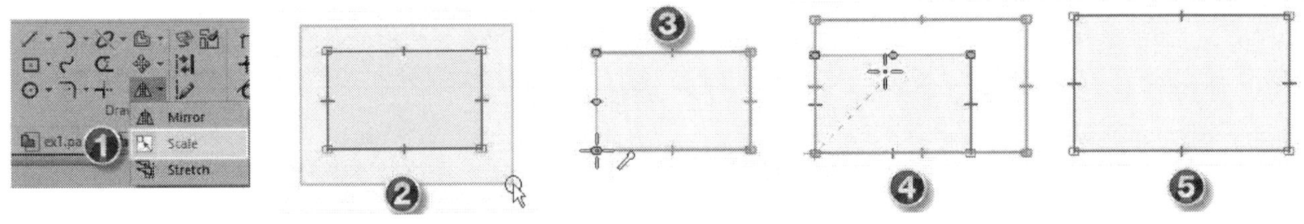

The Stretch command

This command moves a portion of the sketch while still preserving other parts of it. Activate this command from the **Mirror** drop-down on the **Draw** panel, and then drag a box to select the elements to be stretched. Select a base point and move the pointer to stretch the selected elements.

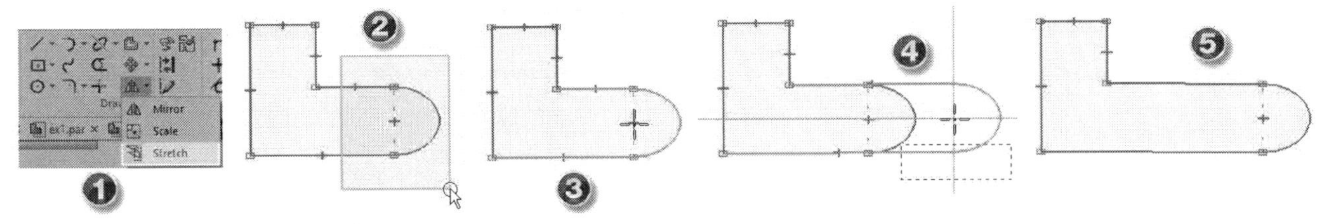

The Draw command

This command helps you to create sketches by using a pen, mouse, or finger. It converts a freehand sketch into an accurate sketch. Activate this command (On the ribbon, click **Home** tab > **Draw** panel > **Line** drop-down > **Draw**), and then click on a plane to activate the sketch mode; the selected plane orients parallel to the screen. On the command bar, click the **Create Geometry** icon. Next, draw a freehand sketch on your touch enabled screen using a pen, or finger. If you do not have a touch screen, you can use the mouse to draw a freehand sketch. If you use a mouse, then press and hold the left mouse button, and then drag it on the screen to create a sketch. Release the mouse after drawing the sketch; the freehand sketch is converted into an accurate drawing using lines, arcs, circles, ellipses, or splines. Next, you can fully define the sketch by adding dimensions to it.

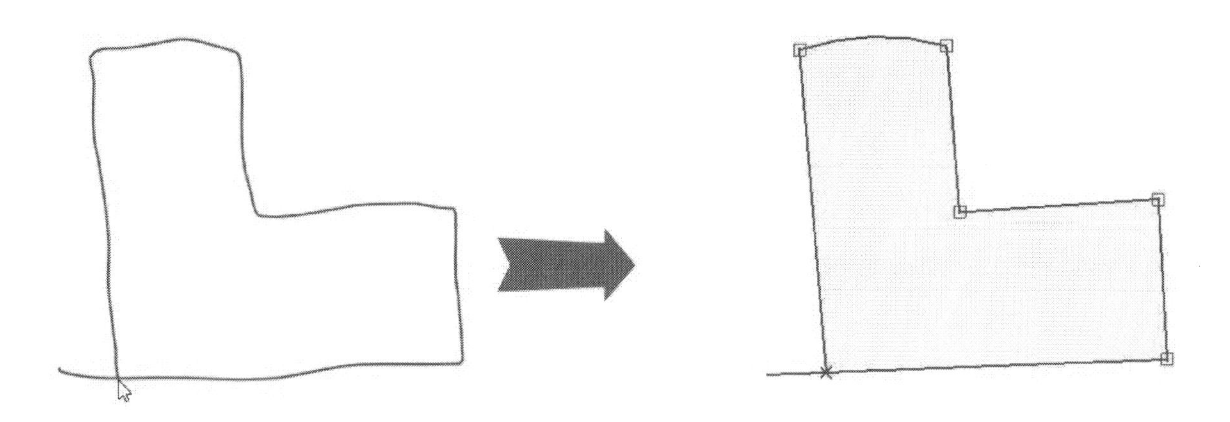

On the command bar, click the **Trim Geometry** icon. Next, press and hold the left mouse button, and then drag the pointer across the unwanted sketch elements; the elements will be deleted. Press Esc to deactivate the command.

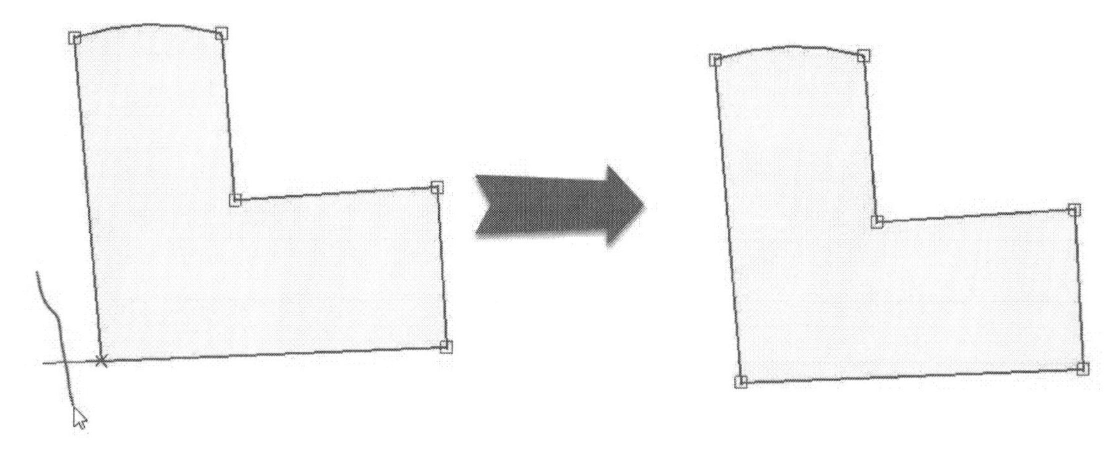

3D Sketching

3D Sketching in Solid Edge helps you to design things like piping, tubing, weldments, and so on. You can create a 3D sketch by using the commands available in the 3D Sketching tab. To start a 3D sketch in the Synchronous mode, click **3D Sketching > New Sketch > New 3D Sketch** on the ribbon. The 3D Sketches entry is listed in the Pathfinder. Now, you can create a 3D sketch using the commands available on the **3D Draw** panel. Most of the commands are similar to the 2D **Draw** commands.

Creating a 3D Line

The **3D Line** command is similar to the **Line** command except that it creates a chain of lines from selected points in the 3D space. You can create 3D lines without selecting any plane. Activate this command by clicking the **3D Line** button on the **3D Draw** panel. You will notice that the pointer is turned into a 3D crosshair. In addition, a triad appears in the graphics window. Now, you can create 3D lines by clicking in the graphics window. For example, select the origin point of the Base Coordinate system to define the start point of the line. Move the pointer vertically upward and you will notice that a parallel symbol appears on the line. In addition, the Z-axis of the triad is highlighted in orange. This means that the line is drawn parallel to the Z-axis. Click to define the endpoint of the line.

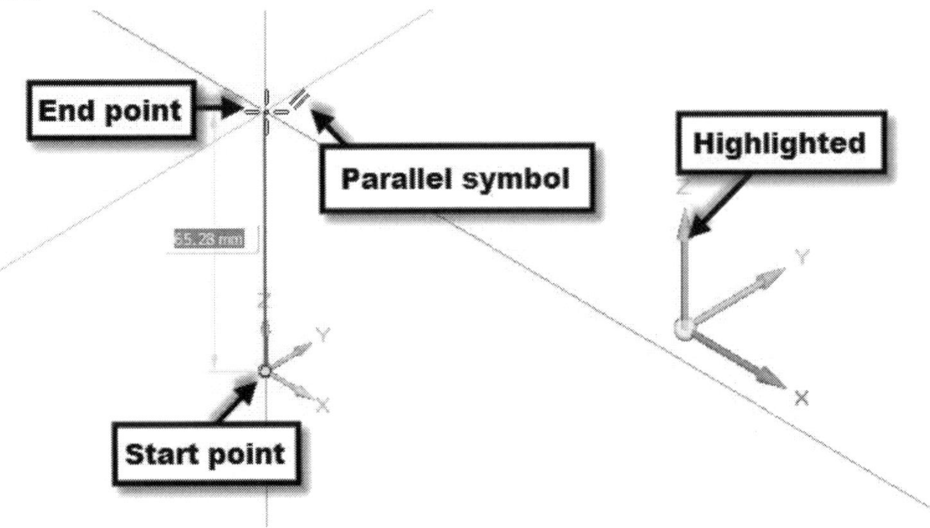

Now, move the pointer along the red line of the crosshairs. You will notice that the line is drawn parallel to the X-axis. Click to define the endpoint of the line.

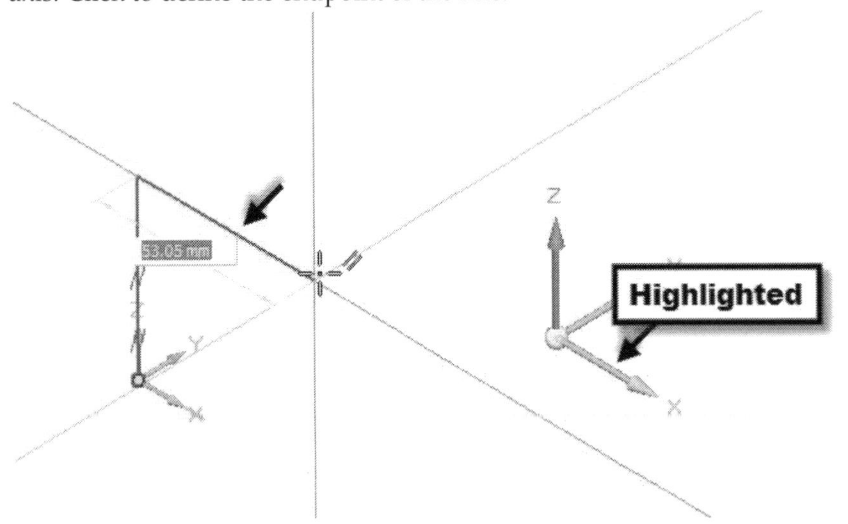

Move the pointer along the green line of the crosshairs. This draws a line parallel to the Y-axis. Click to define the endpoint of the line.

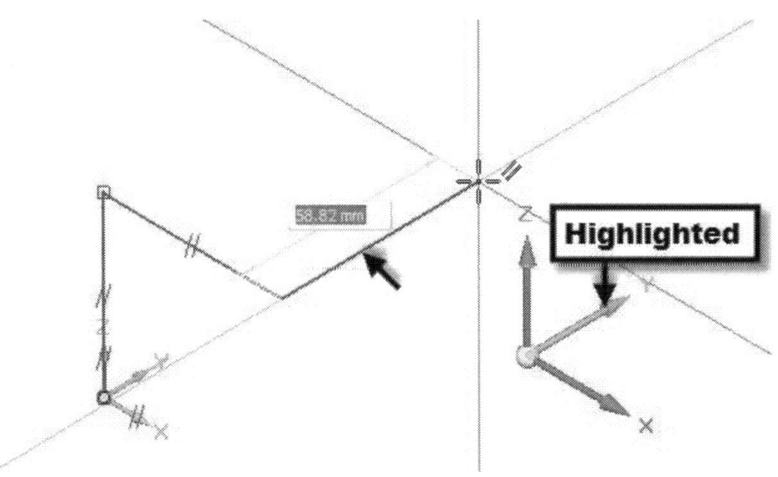

Likewise, create some more lines in the Z, X, and Y-axes. Press Esc to deactivate the **3D Line** command. When you rotate the view, you can see the 3D Line created.

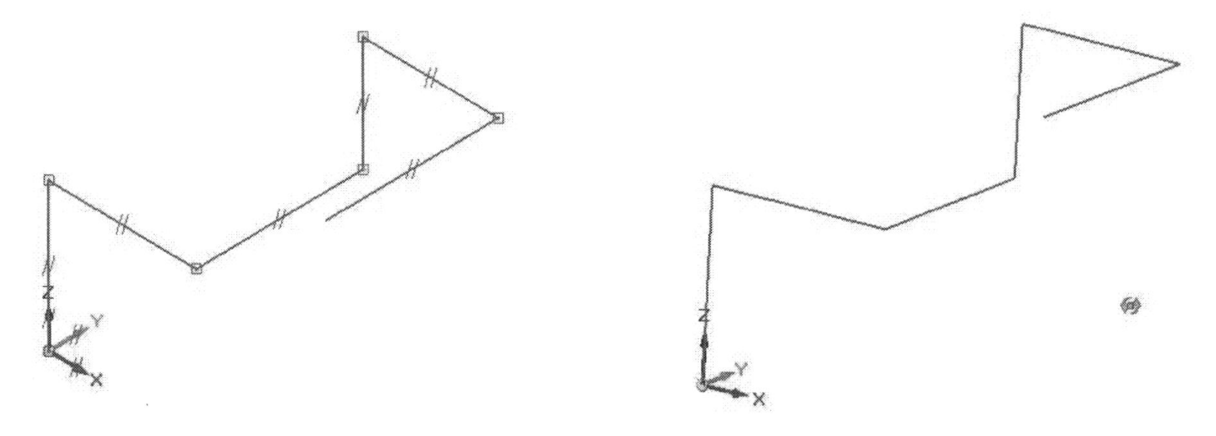

Adding Relationships and Dimensions

Adding Relationships in 3D Sketching is similar to that in 2D sketching except that there are two additional commands: **On Plane** and **Coaxial**. The **On Plane** command moves the selected sketch element onto a plane. Click the **On Plane** button on the **3D Relate** panel. Select the 3D sketch element and the plane. The selected sketch element will be move to the selected plane.

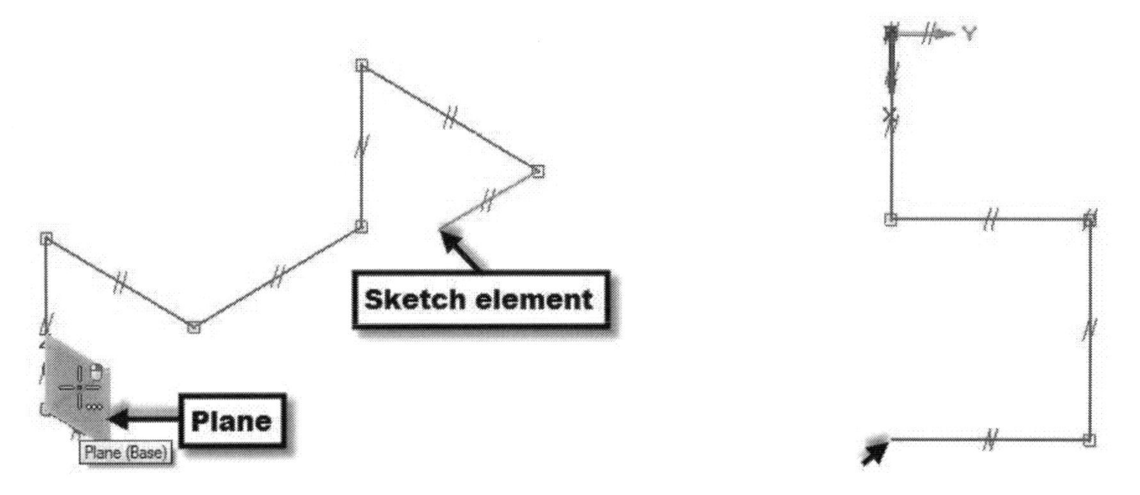

The **Coaxial** command makes the center of circle coaxial with a line. Click the **Coaxial** button on the **3D Relate** panel and select an arc or circle. After selecting the first element, select the second element.

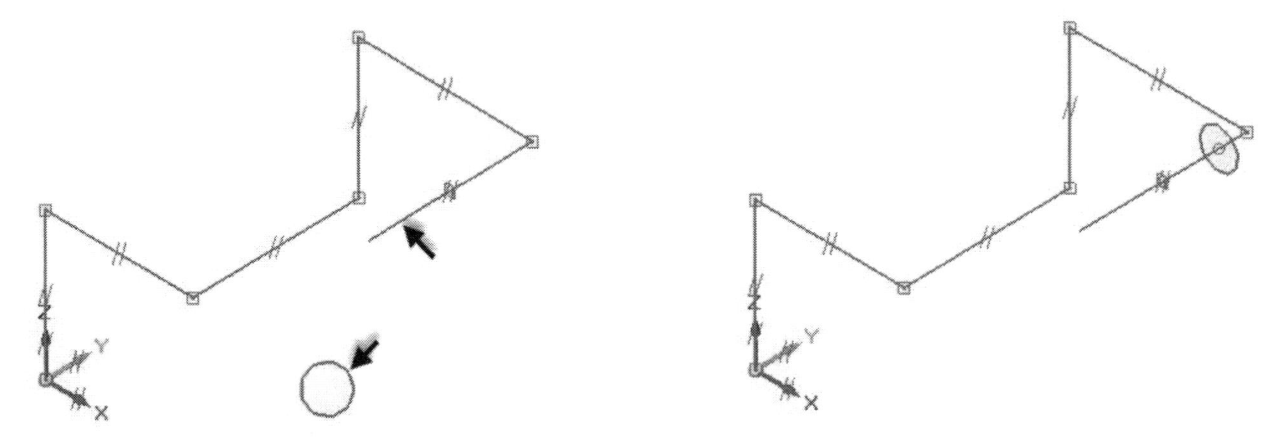

Use the **Smart Dimension** command to apply dimensions to the 3D sketch. Activate this command and select the sketch element. However, if you select the endpoints of two elements, the dimension will be displayed aligned to them. Press **N** on your keyboard to change the orientation of the dimension.

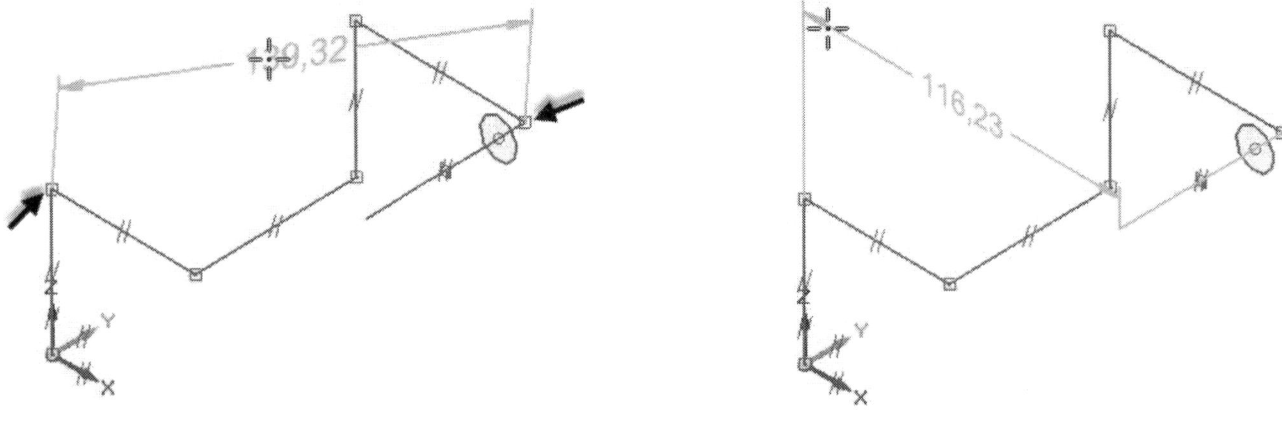

Drawing a 3D Sketch element by Locking a plane

To create 3D sketch by locking a plane, you can use the default planes or create new planes. For this example, you will create a new plane and use it to draw a 3D sketch. On the ribbon, click **3D Sketching > Planes > Coincident Plane**, and then select the XZ plane from the Base coordinate system. You will notice the steering wheel on the new plane. Click on the torus of the steering wheel and move the pointer. Type-in 135 and press Enter. The plane will be rotated by 135 degrees.

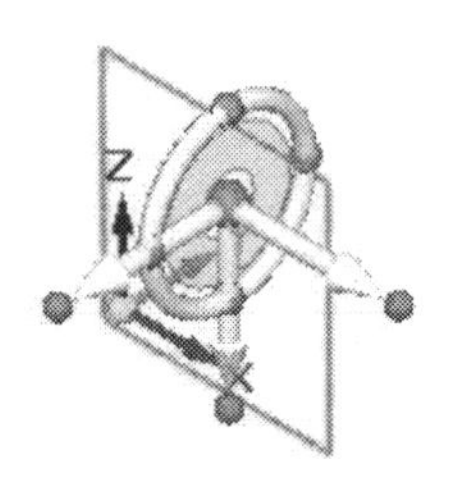

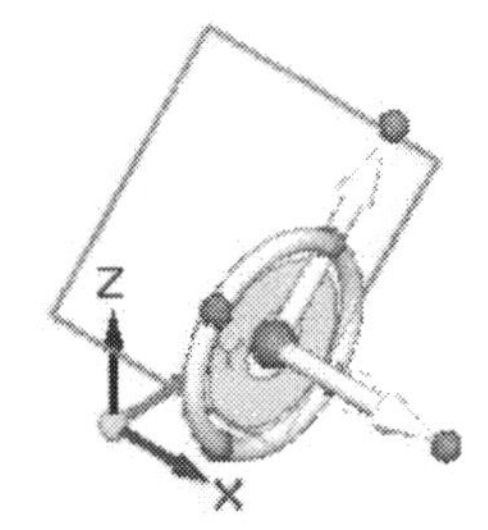

Click the **3D Line** button on the **3D Draw** panel. On the command bar, click the **Lock Sketch Plane** icon and select the plane. Select the origin point of the Base coordinate system to define the start point. Move the pointer upward, and then type-in 50 and -90 in the Length and Angle boxes, respectively and press Enter. Move the pointer along the horizontal crosshair and enter 50 and 0 in the length box and angle boxes, respectively. Likewise, create another vertically inclined line of 50 mm length. Press F3 to unlock the plane.

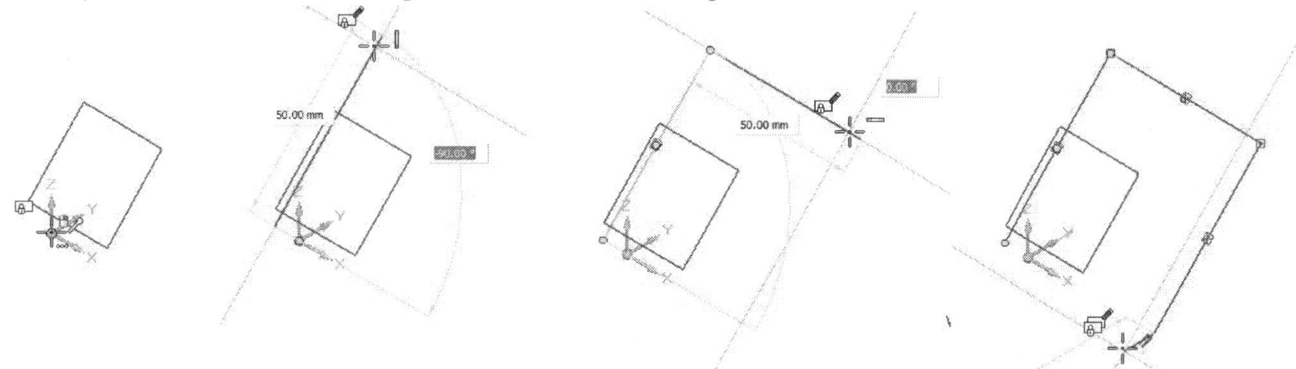

Now, move the pointer along the green line of the crosshairs, type 100 in the length box, and press Enter . Complete the sketch by creating other lines, as shown below. Press Esc to deactivate the command.

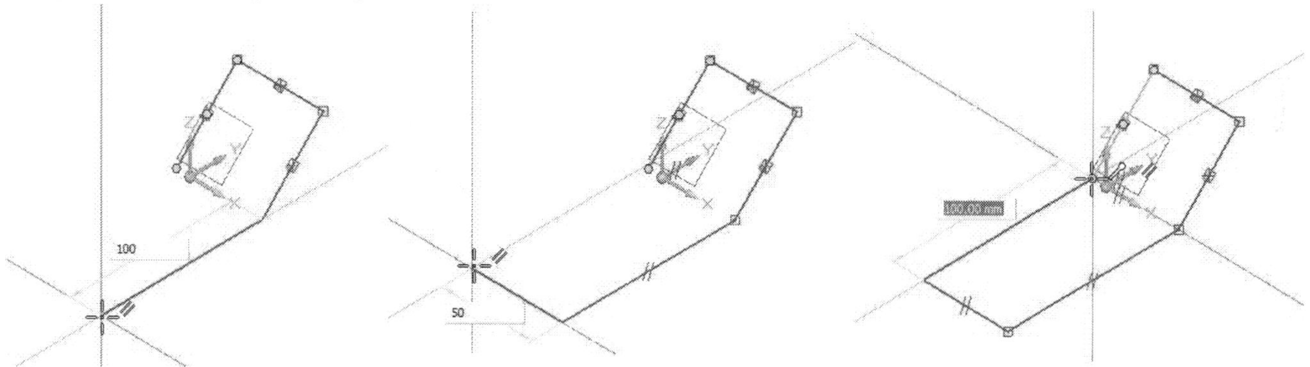

To add fillets to the sketch, click **3D Fillet** button on the **3D Draw** panel. Type-in a value in the radius box and press Enter. Select the corners of the sketch to fillet them.

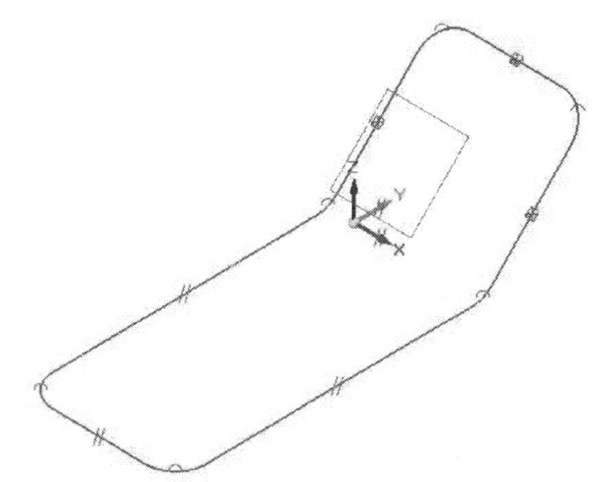

In the Ordered Environment, you have to open the 3D Sketching environment to create a 3D sketch. On the ribbon, click **Home > Sketch > 3D Sketch**. Use the drawing commands and create the 3D sketching. Click the **Close 3D Sketch** button after completing the sketch.

The Routing Path command

The **Routing Path** command creates a 3D sketch path between the two selected points. Activate this command (on the ribbon, click **3D Sketching** tab > **Draw** panel > **Routing Path**) and then select two points; Solid Edge displays a 3D sketch path between the selected points. On the command bar, click the **Next** icon to display the next path between the selected points. Click **Accept** if you are satisfied with result.

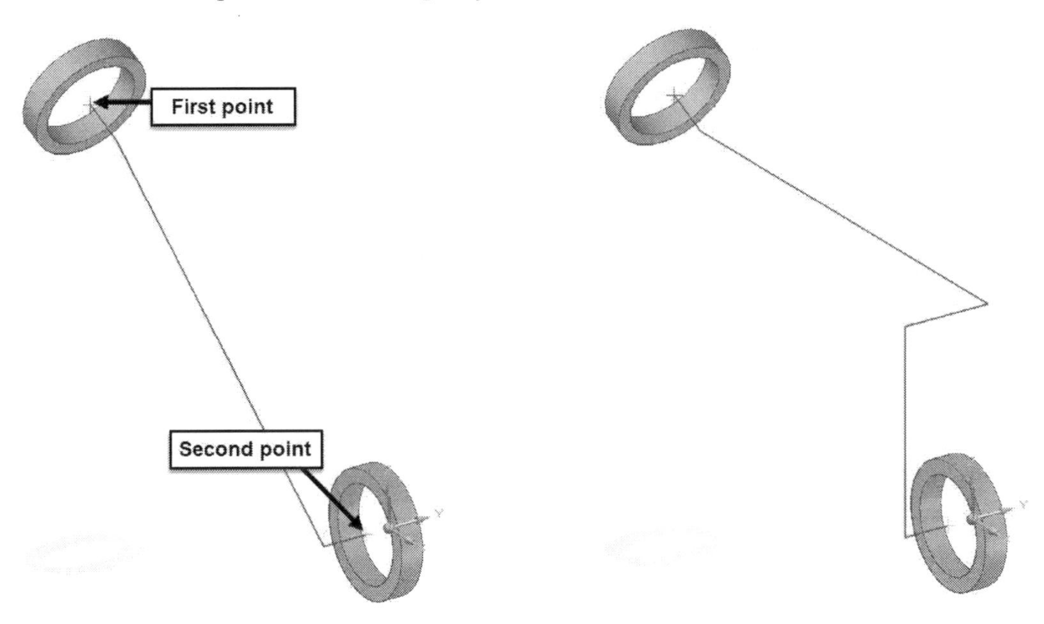

Examples

Example 1 (Millimetres)

In this example, you will draw the sketch shown below.

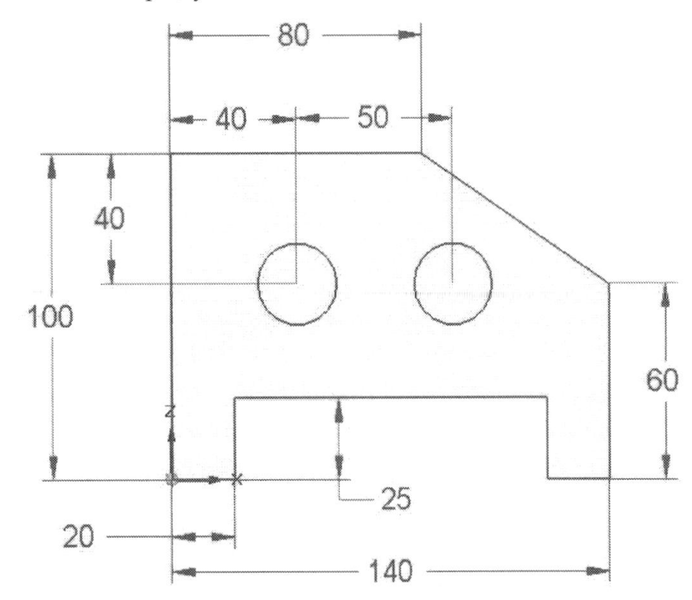

1. Start **Solid Edge 2019** by clicking the **Solid Edge 2019** icon on your desktop.
2. Click **New** button on the **Quick Access Toolbar**; the **New** dialog is opened.

3. Select **ISO Metric > iso metric part.par**. Next, click **OK**; a new part file is opened.

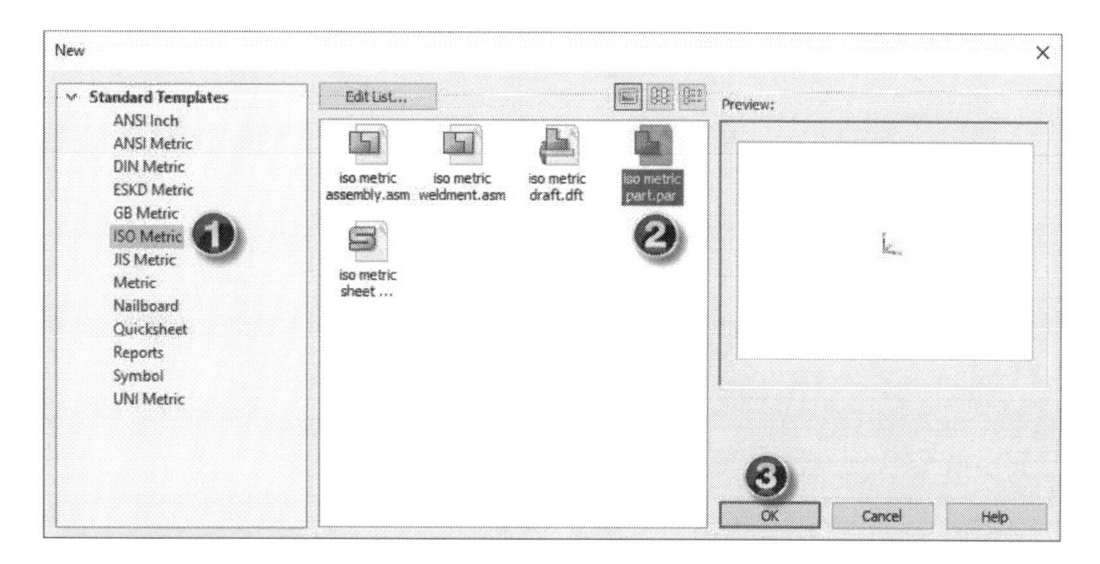

4. To start a new sketch, click **Home > Draw > Line** on the ribbon. The pointer turns into a crosshair.
5. Place the mouse pointer on the coordinate system; the XZ plane is highlighted.
6. Click the lock icon on the XZ plane (or) press F3 to lock the plane.
7. Click the **Sketch View** icon located at the bottom of the window; this orients the sketch plane normal to the screen.

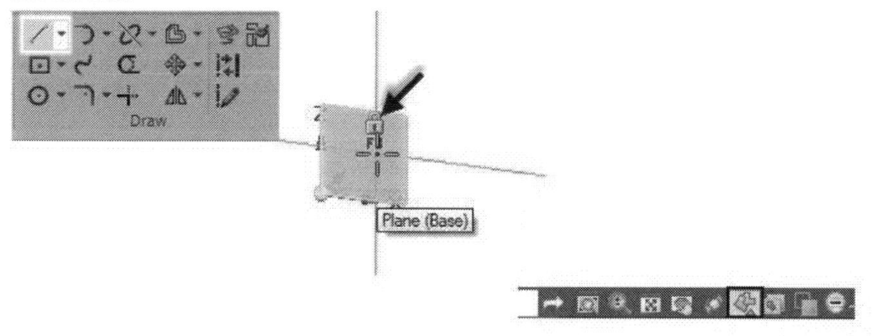

8. Click on the origin point to define the first point of the line.
9. Move the pointer along the horizontal line of the crosshair and toward right.
10. Click to define the endpoint of the line.
11. Move the pointer along the vertical line of the crosshair and upwards. Click to define the second line.

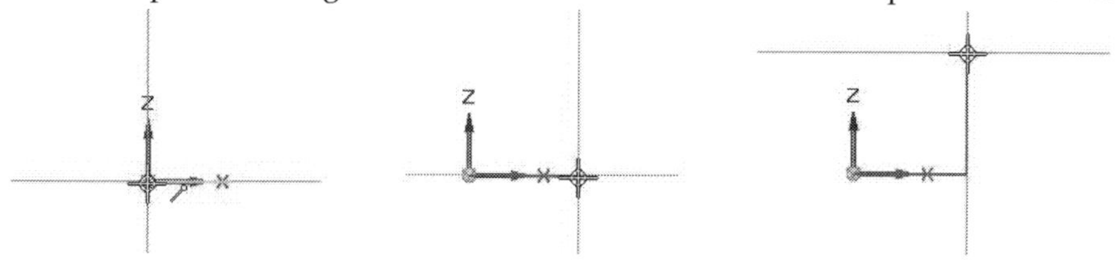

12. Create a closed loop by selecting points in the sequence, as shown below.

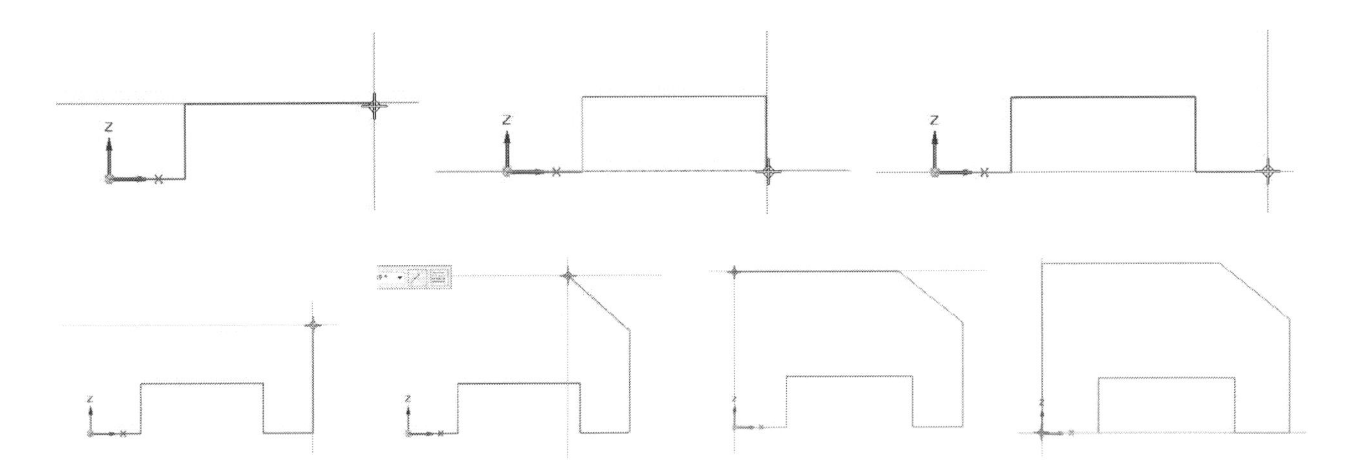

13. Click **Home > Relate > Collinear** on the ribbon and click on the two horizontal lines at the bottom; they become collinear.

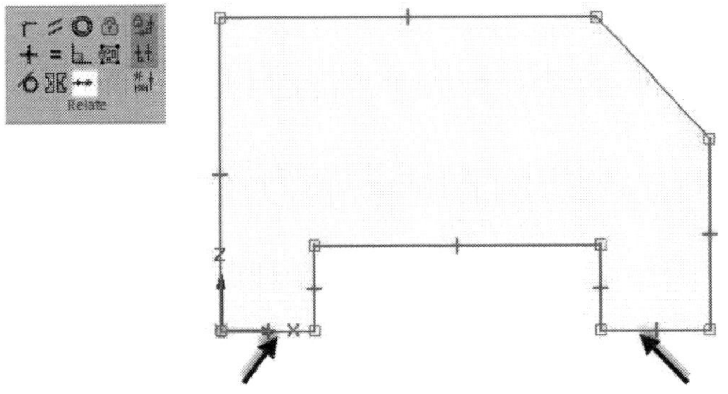

14. Click **Home > Relate > Equal** on the ribbon and click on the two horizontal lines at the bottom; they become equal in length.

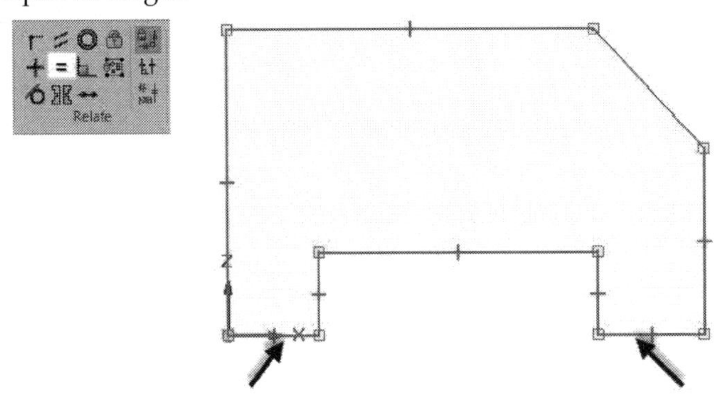

15. Select the small vertical lines to make their lengths equal.

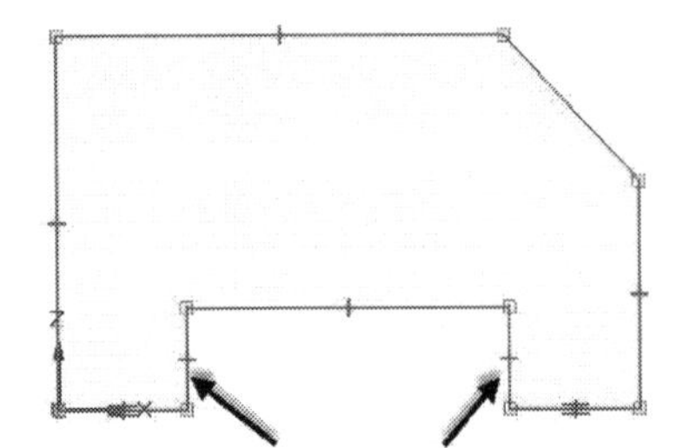

16. Click **Home > Dimension > Smart Dimension** on the ribbon and click on the lower left horizontal line. Move the mouse pointer downward and click to locate the dimension.
17. Type-in **20** in the dimension box and press Enter.

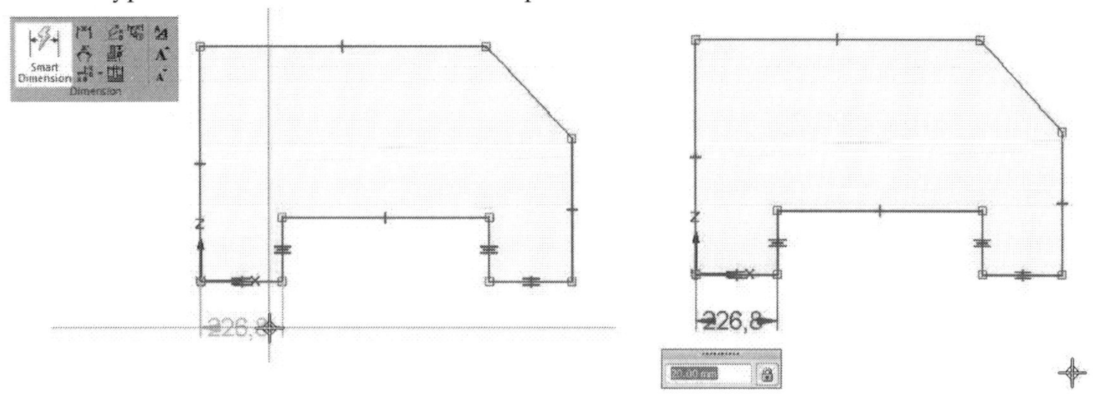

18. Click on the small vertical line located at the left side. Move the mouse pointer towards right and click to position the dimension.
19. Type-in **25** in the dimension box and press Enter.

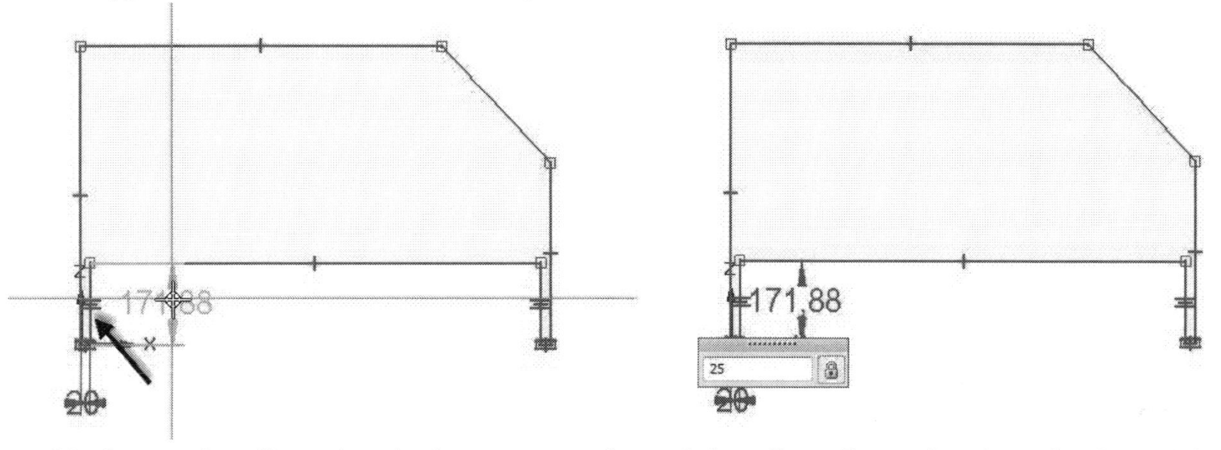

20. Create other dimensions in the sequence, shown below. Press Esc to deactivate the **Smart Dimension** command.

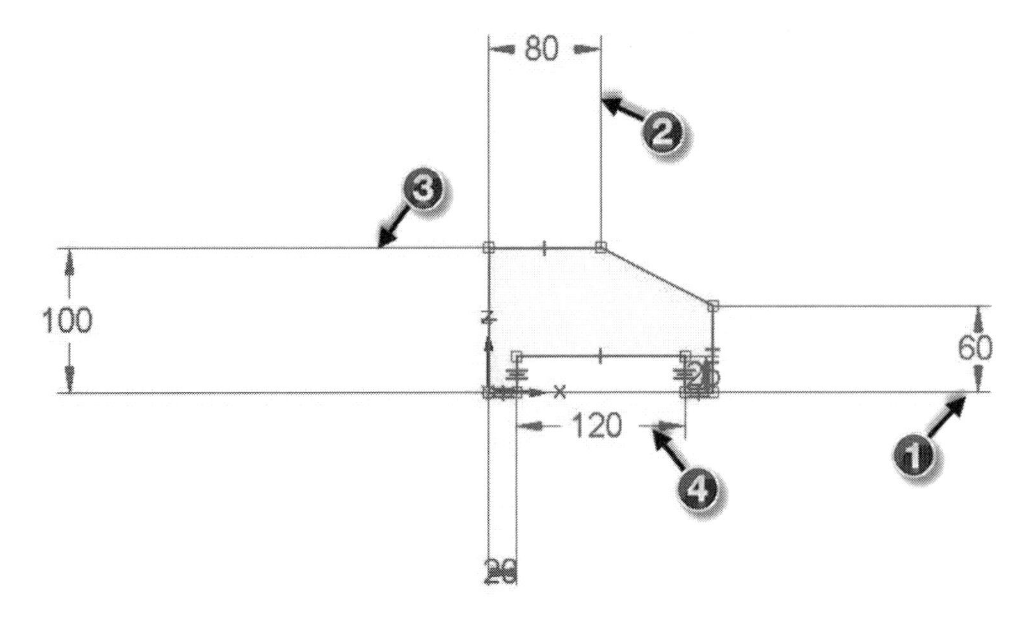

21. Click on the portion between the dimension value and the arrow. Press the left mouse button and drag the dimension near to the sketch.

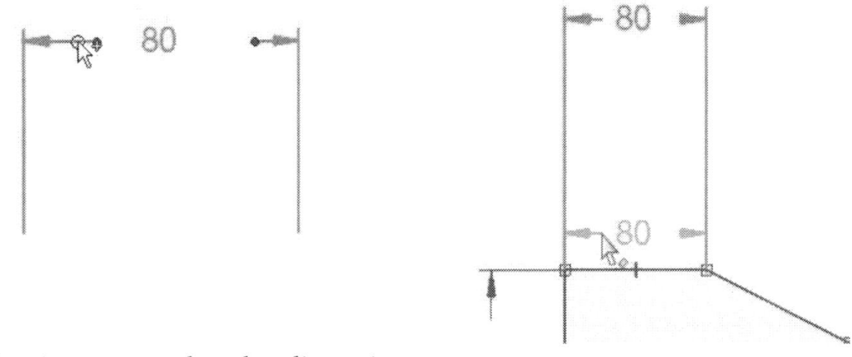

22. Likewise, arrange the other dimensions.

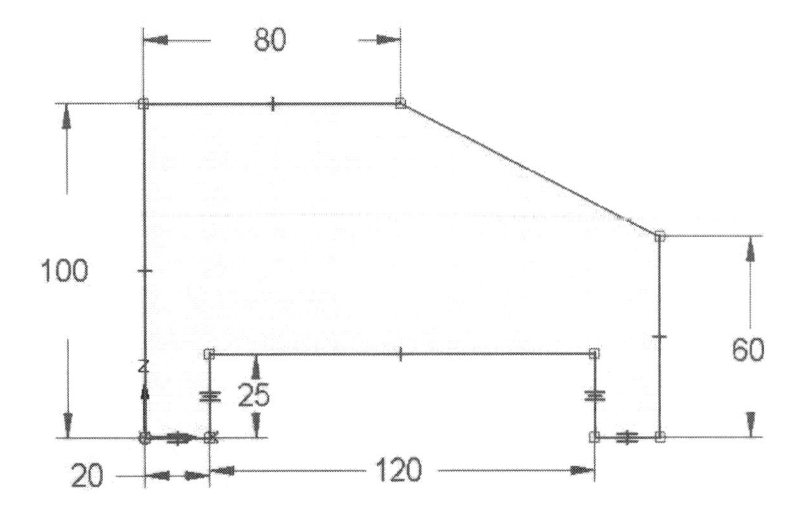

23. On the ribbon, click **Home > Draw > Circle by Center Point**. Click inside the sketch region to define the center point of the circle. Move the mouse pointer and click to define the diameter. Likewise, create another circle.

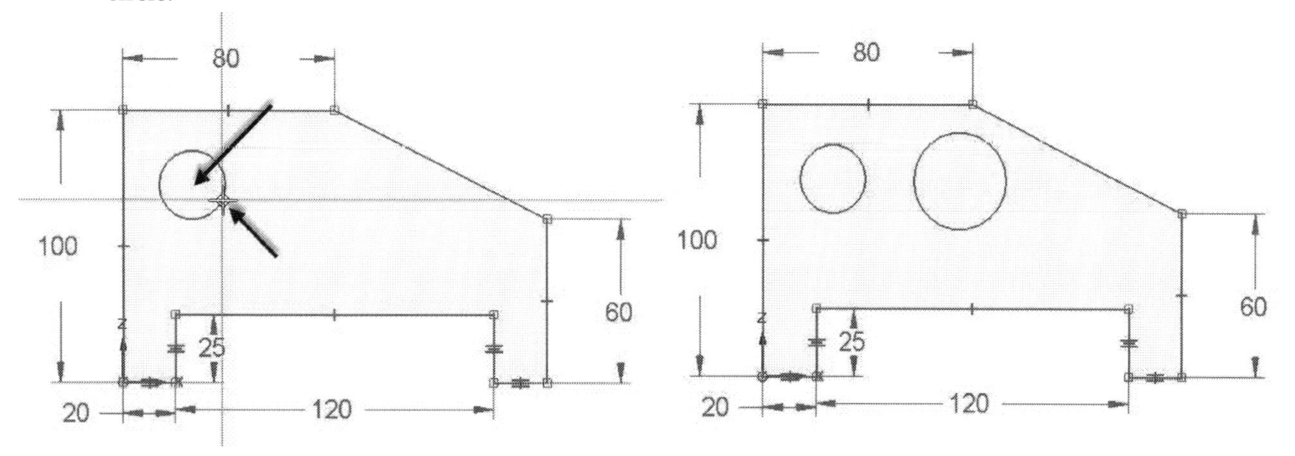

24. On the ribbon, click **Home > Relate > Horizontal/Vertical**. Click on the center points of the two circles to make them horizontally aligned.

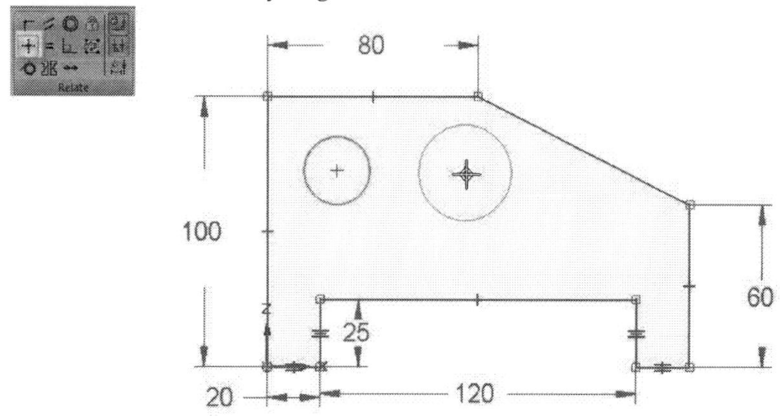

25. On the ribbon, click **Home > Relate > Equal**, and then click on the two circles. The diameters of the circles will become equal.

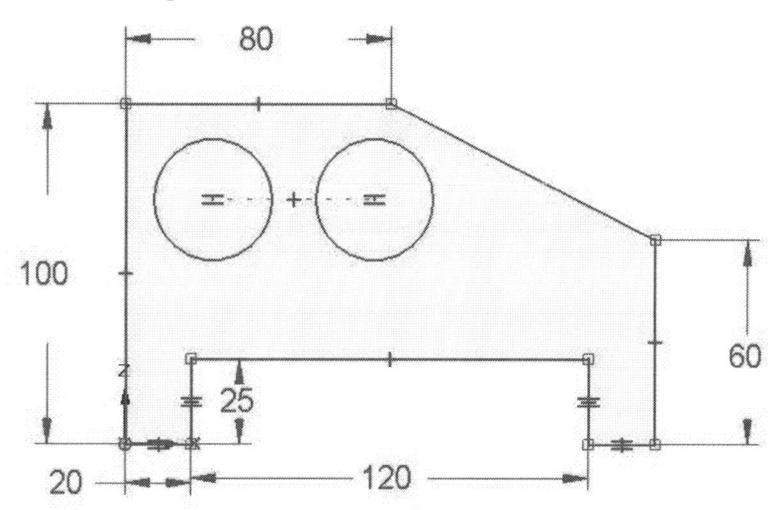

26. Activate the **Smart Dimension** command and click on anyone of the circles. Move the mouse pointer and click to position the dimension. Type 25 in the dimension box and press Enter.

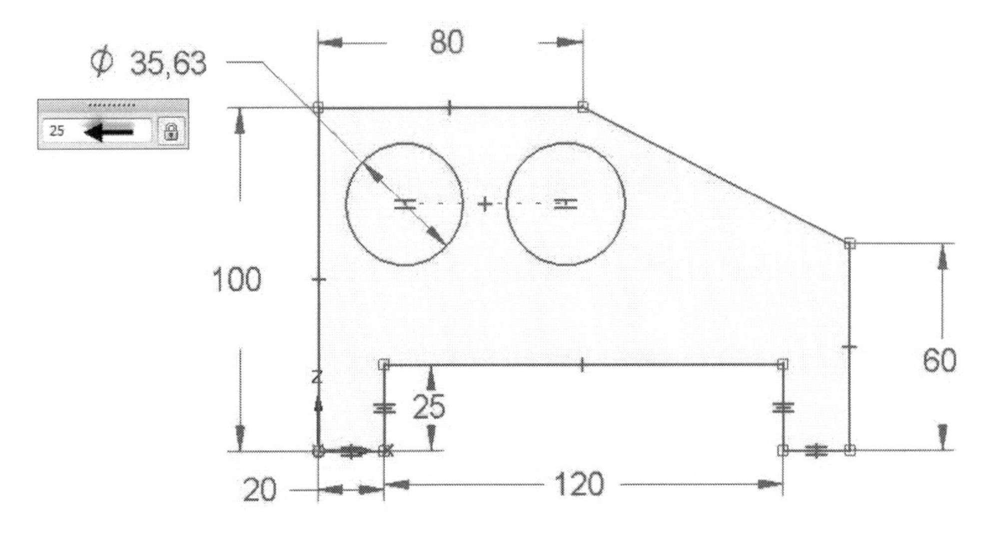

27. Create other dimensions between the circles and the adjacent lines, as shown below.

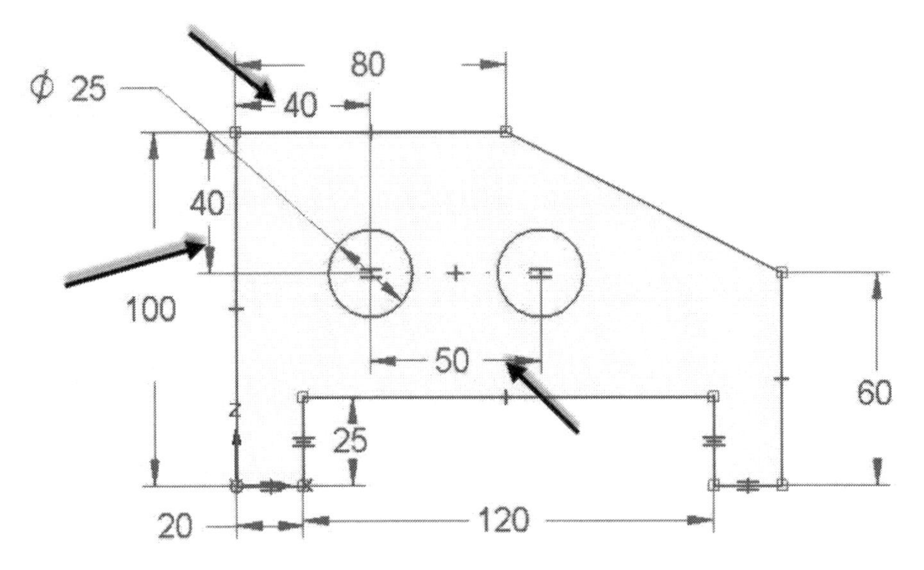

28. Click the Lock icon on the right-side of the graphics window to unlock the sketch plane.

29. Click the **Save** icon on the **Quick Access Toolbar**. Define the location and file name and click **Save** to save the part file.

30. Click **Close Window** on the top right corner to close the part file.

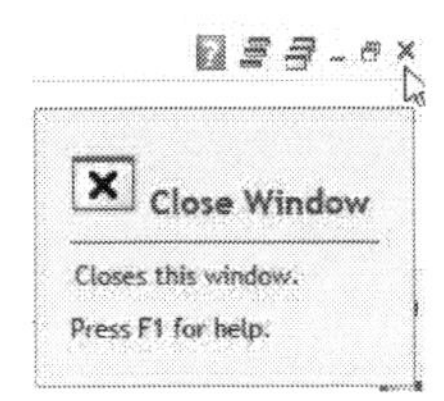

Example 2 (Inches)

In this example, you will draw the sketch shown below.

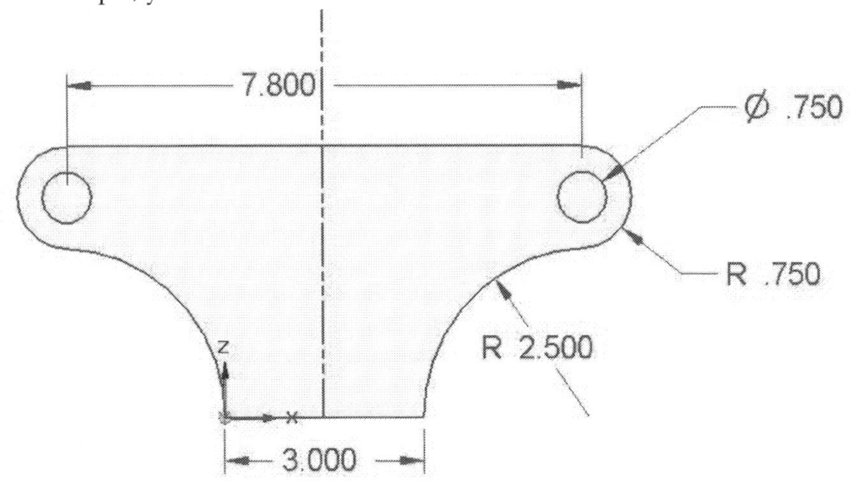

1. Start **Solid Edge 2019** by clicking the **Solid Edge 2019** icon on your desktop.
2. On the Application Menu, click the **New** > **New** icon; the **New** dialog is opened.
3. On the **New** dialog, click **Standard Templates > ANSI Inch** and select the **ansi inch part.par** template. Click **OK** to start a new part file.

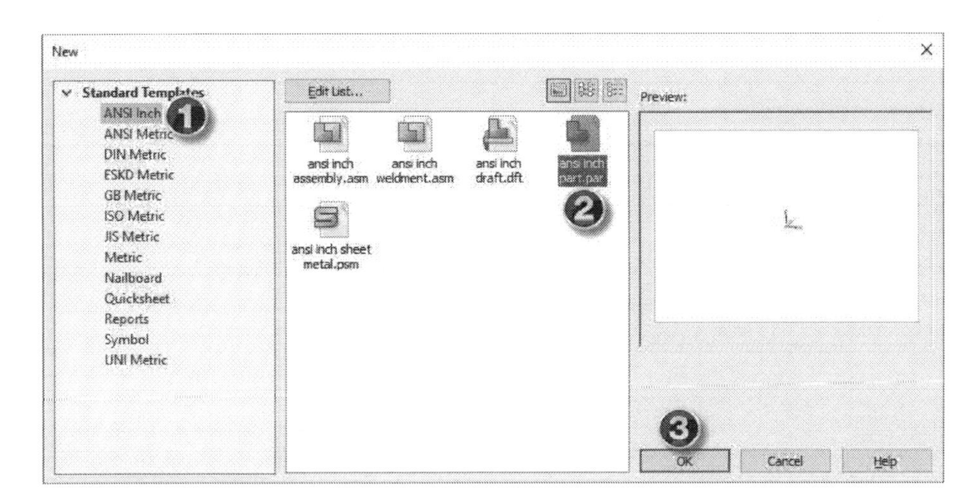

4. To start a new sketch, click **Home > Draw > Line** on the ribbon.
5. Place the mouse pointer on the coordinate system; the XZ plane is highlighted and a mouse icon appears.
6. Click the right mouse button to display the **QuickPick** box.
7. On the **QuickPick** box, select the XY plane.

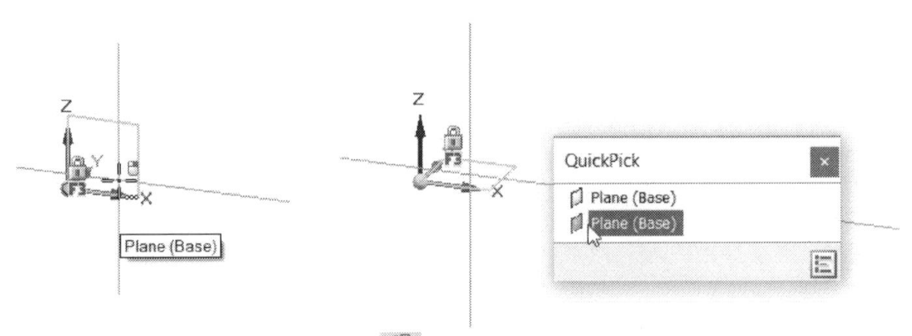

8. Click the **Sketch View** icon located at the bottom of the window. This orients the sketch plane normal to the screen.
9. Click on the origin point to define the first point of the line. Move the mouse pointer horizontally and click to draw a line.

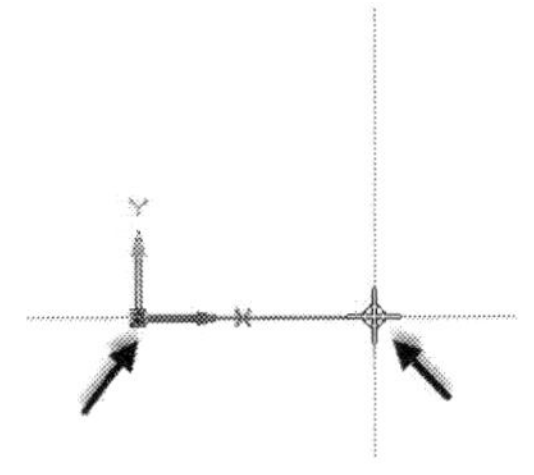

10. On the **Line** command bar, click the **Arc** icon.
11. Take the mouse pointer to the end point of the line.
12. Move it upwards and right. Next, click to create the arc.

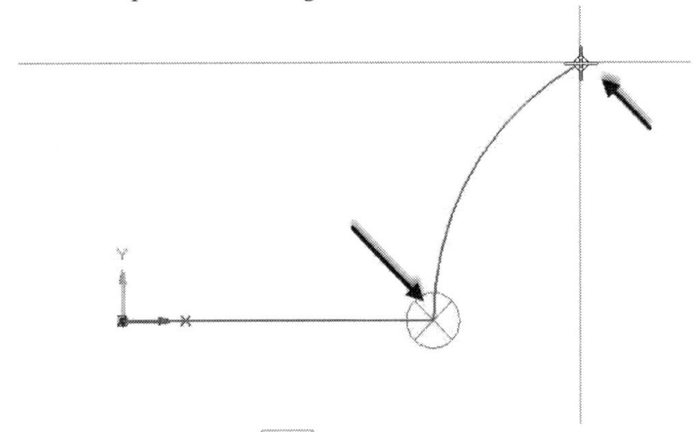

13. Again, click the **Arc** icon the command bar.
14. Move the pointer to the end point of the arc, and then move it upwards right.
15. Move the pointer toward left and click when a vertical dotted line appears, as shown below.

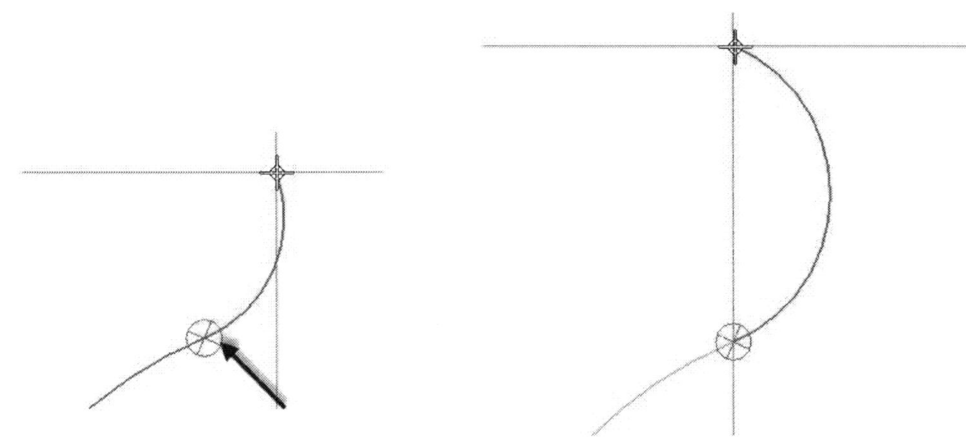

16. Move the mouse pointer horizontally toward left and click to create a horizontal line. Note that the length of the new line should be greater than that of the lower horizontal line.
17. Click the **Arc** icon on the command bar and move it downward left.
18. Move the pointer toward right and click when a vertical dotted line appears, as shown below.

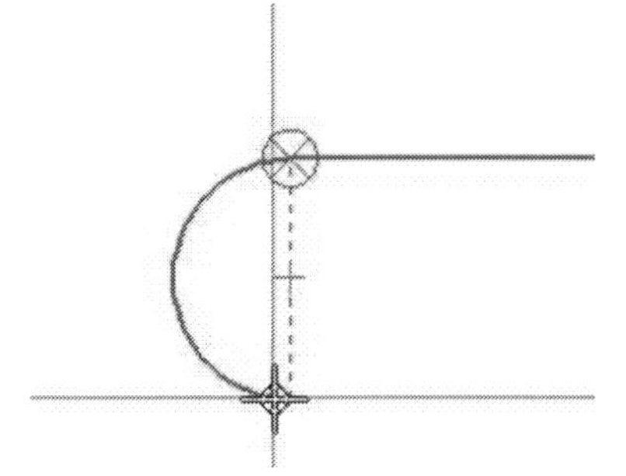

19. Click the **Arc** icon on the command bar. Next, move the pointer to the end point of the previous arc.
20. Move the mouse pointer downwards right and click on the origin to close the sketch.

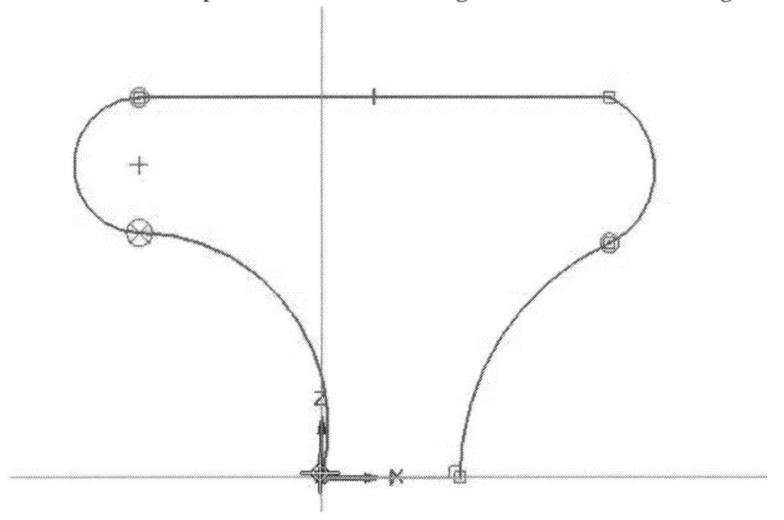

21. Click the right mouse button to end the chain.
22. Click on the midpoint of the lower horizontal line. Move the mouse pointer vertically up and click to create a vertical line.

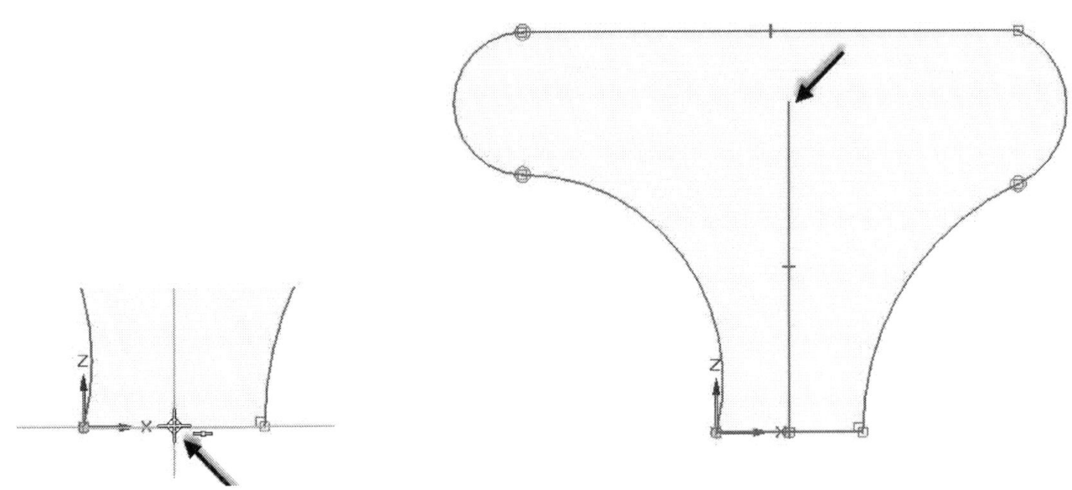

23. On the ribbon, click **Home > Draw > Construction**. Click on the vertical line located at the center. The line is converted into a construction element.

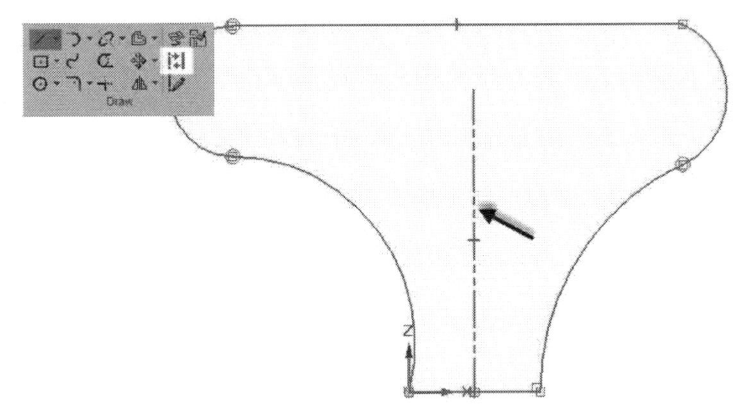

24. Activate the **Circle by Center Point** command and draw a circle on the right side of the construction line.
25. On the ribbon, click **Home > Relate > Concentric.** Click on the circle and the small arc on the right side. The circle and arc are made concentric.

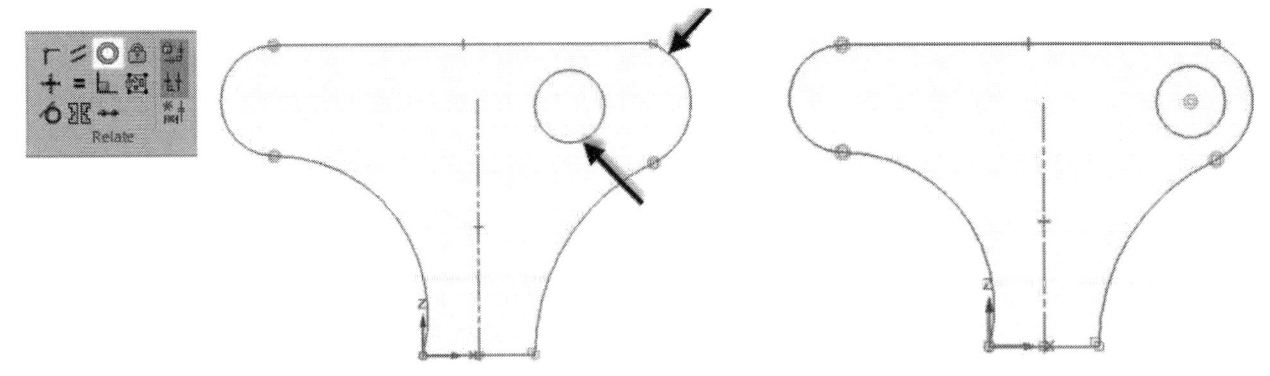

26. Likewise, create another circle concentric to the small arc located on the left side of the construction line.
27. On the ribbon, click **Home > Relate > Symmetric**. Click on the construction line located at the center. The line will act as a symmetry line.
28. Click on the large arcs on both sides of the symmetry line. The arcs are made symmetric about the construction line.
29. Likewise, make the small arcs and circles symmetric about the construction line.

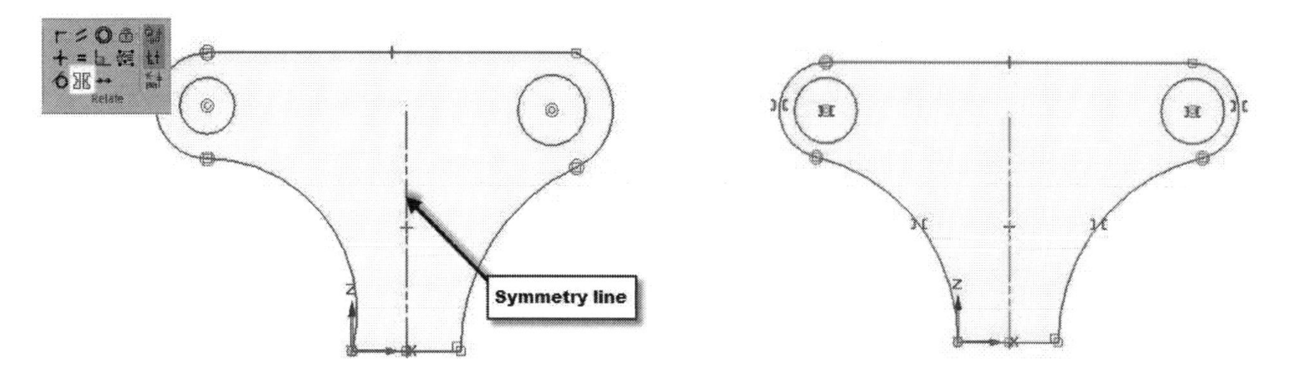

30. Activate the **Smart Dimension** command and apply dimensions to the sketch in the sequence, as shown below.

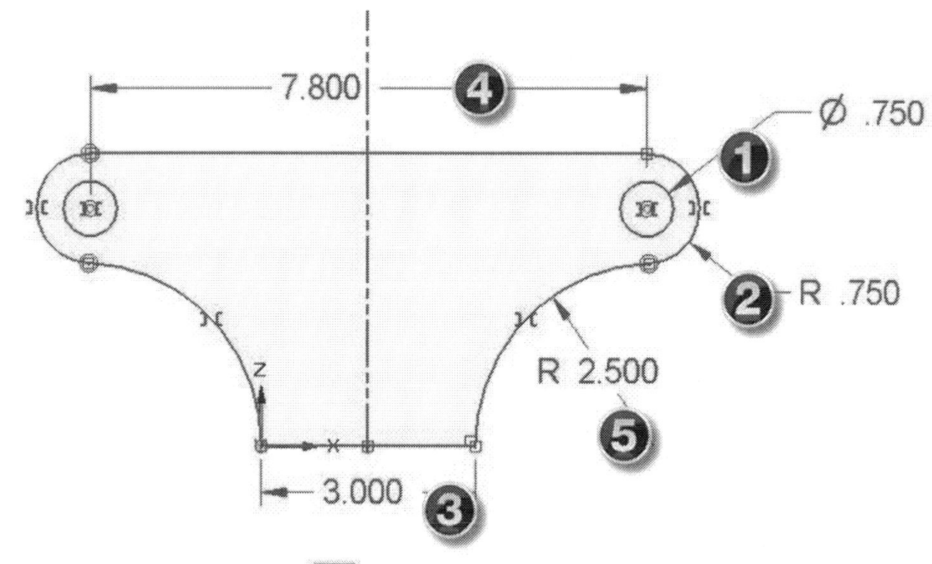

31. On the status bar, click the **Fit** icon to fit the drawing in the graphics window.
32. To save the file, click **Application Menu > Save**. Define the location and file name, and then click **Save**; the part file is saved.

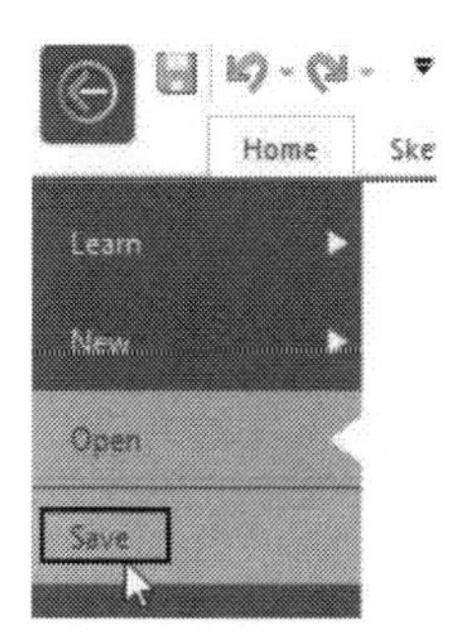

33. To close the file, click **Close** on the top right corner of the graphics window.

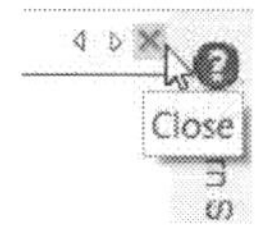

Example 3 (Millimetres)

In this example, you will draw the sketch shown below.

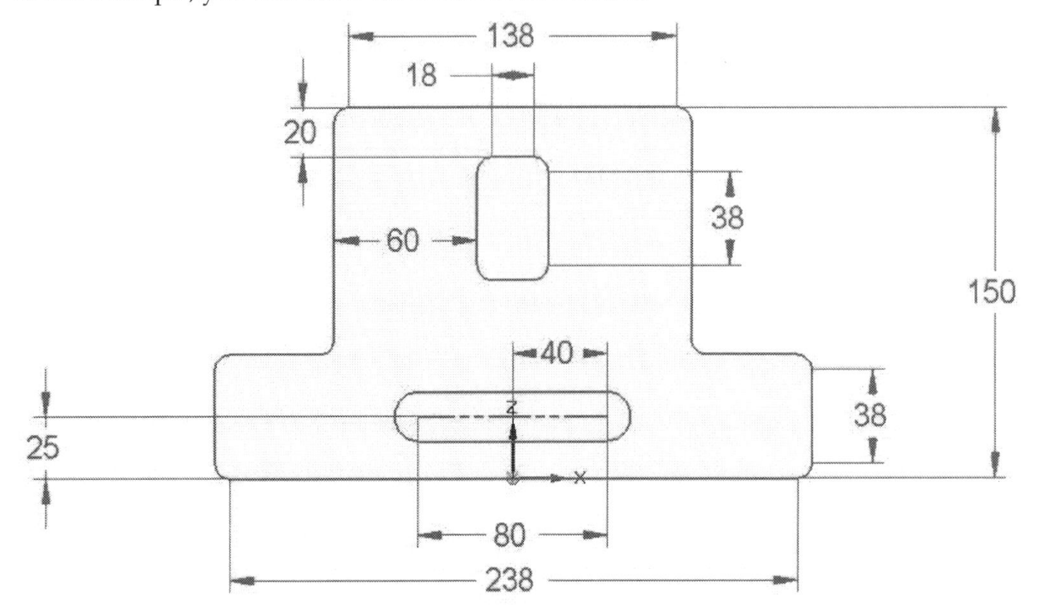

1. Start **Solid Edge 2019**, if not already opened.
2. To start a new part file, click **Application Menu > New > ISO Metric Part**.

Tip: You can change the templates displayed in the ***New*** *page by clicking* ***Edit List****. On the* ***Template List Creation*** *dialog, select a modelling standard from the* ***Standard Templates*** *section. You can change the order of the templates by selecting them from the* ***Templates*** *section and clicking the* ***Move Up*** *and* ***Move Down*** *arrows. Likewise, you can change the* ***Name*** *and* ***Description*** *of the template and click* ***Apply****. Click* ***OK*** *on the* ***Template List Creation*** *dialog to apply the changes.*

3. To start sketching, activate the **Line** command and click on the XZ plane. Press F3 to lock the sketch plane.
4. Click the **Sketch View** icon located at the bottom of the window. This orients the sketch plane normal to the screen.
5. Place the mouse pointer on the origin and move it toward left; a dotted line appears.

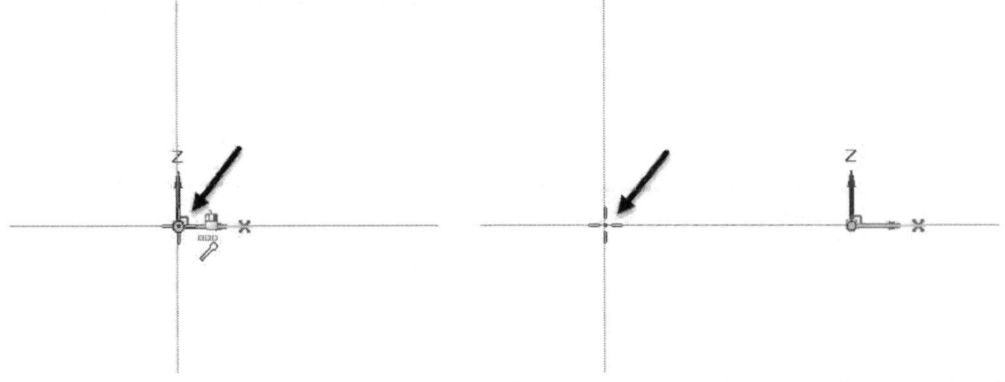

6. Click to define the first point. Move the mouse pointer horizontally toward right and click to define the second point.
7. Create a closed loop by clicking points in the sequence shown below.

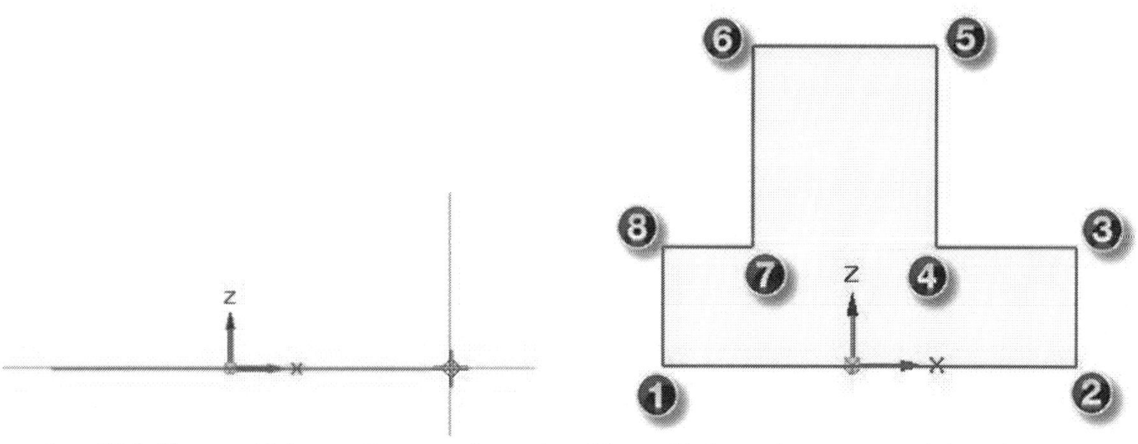

8. Click **Home > Relate > Symmetric** on the ribbon. Click on the Z-axis to define the symmetric axis.
9. Click on the small vertical lines to make them symmetric.
10. Click on the other vertical lines to make them symmetric.

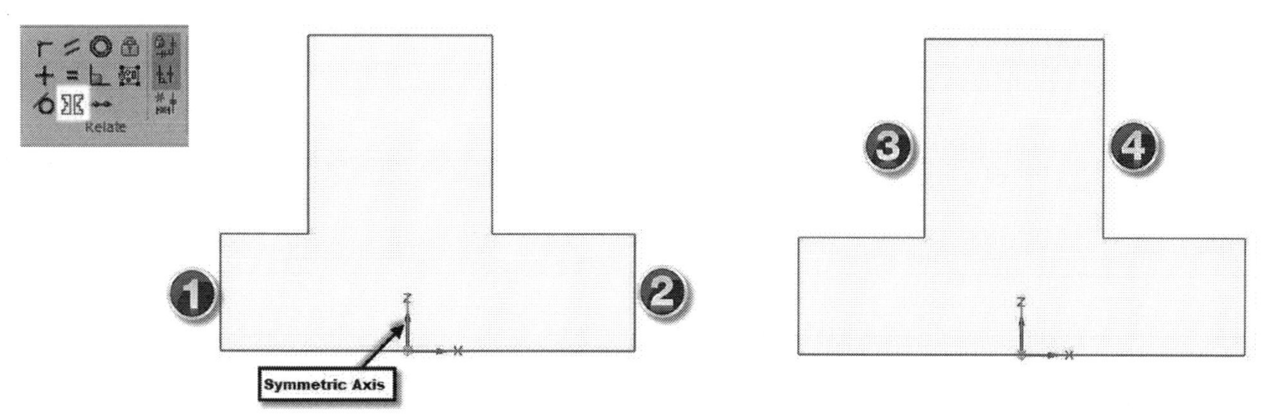

11. On the ribbon, click **Home > Relate > Collinear**. Click on the lower horizontal line and the X-axis to make them collinear.
12. Activate the **Smart Dimension** command and apply dimensions in the sequence shown below.

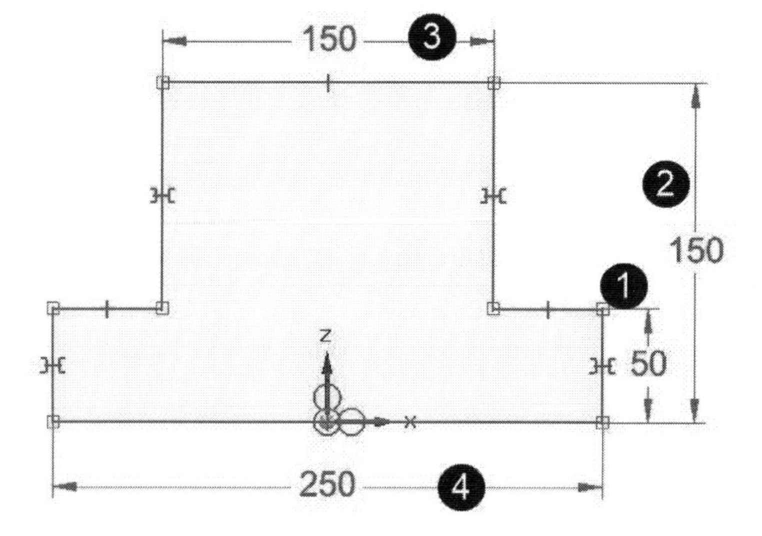

13. Click **Home > Draw > Rectangle by Center** on the ribbon. Click in the sketch region to define the center of the rectangle.
14. Move the mouse pointer toward top right and click to define the corner of the rectangle.

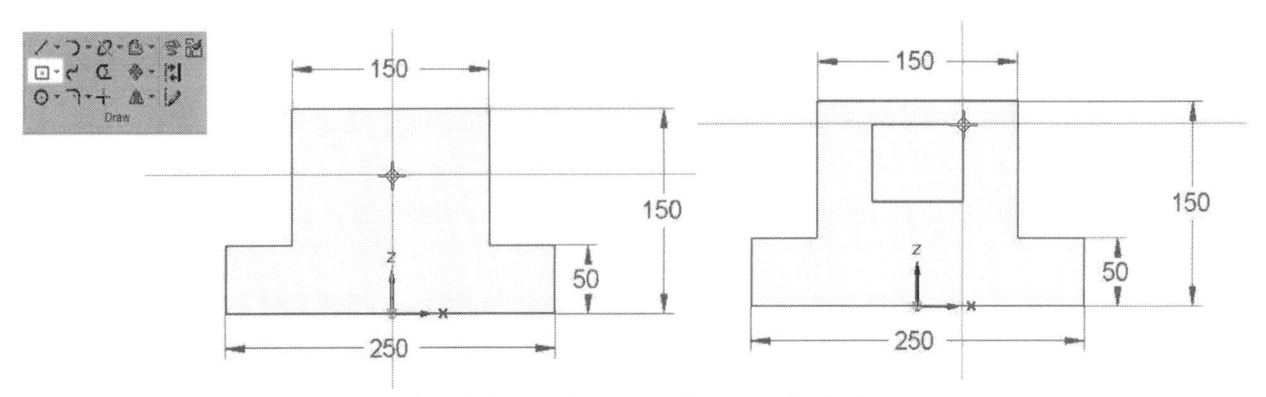

15. Activate the **Line** command and draw a horizontal line inside the loop.

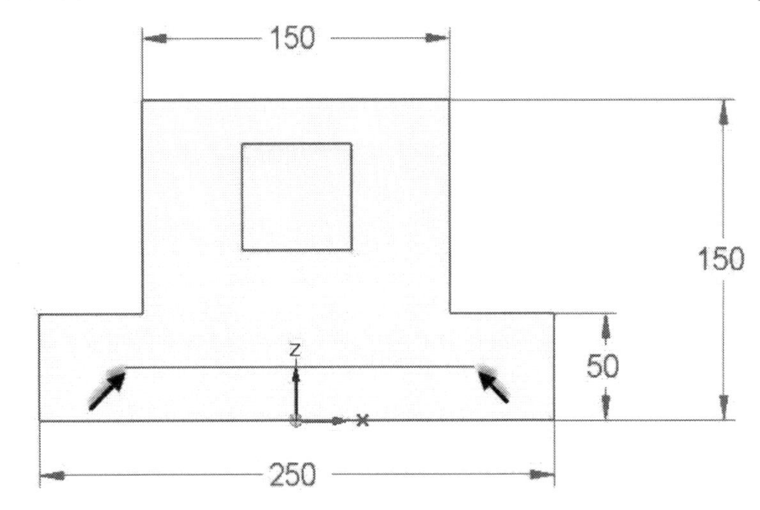

16. Activate the **Smart Dimension** command and apply dimensions in the sequence shown below.

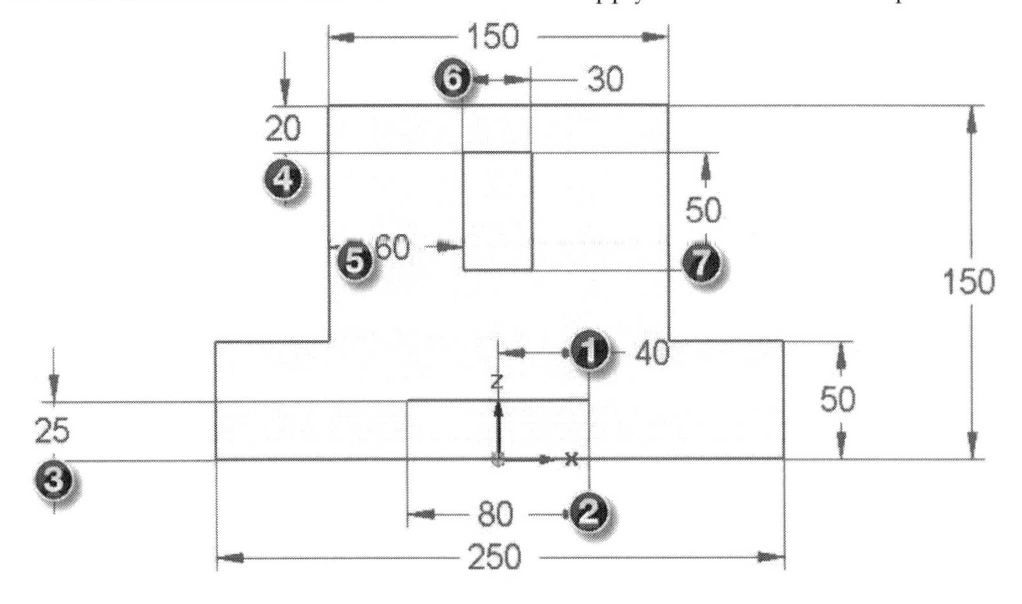

17. Click **Home > Draw > Offset > Symmetric Offset** on the ribbon; the **Symmetric Offset Options** dialog pops up.

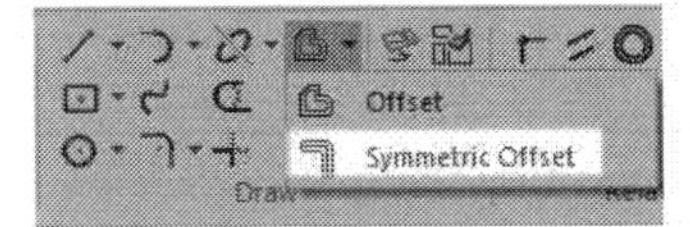

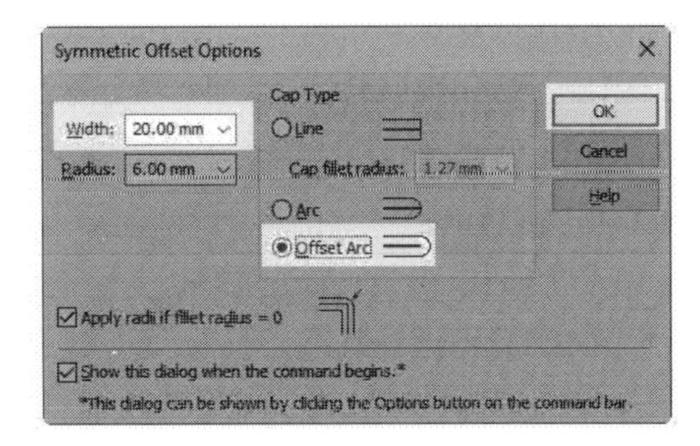

18. On this dialog, type-in 20 in the **Width** box and select the **Offset Arc** option. Click **OK** to close the dialog.
19. Select the horizontal line and click the green check on the command bar.

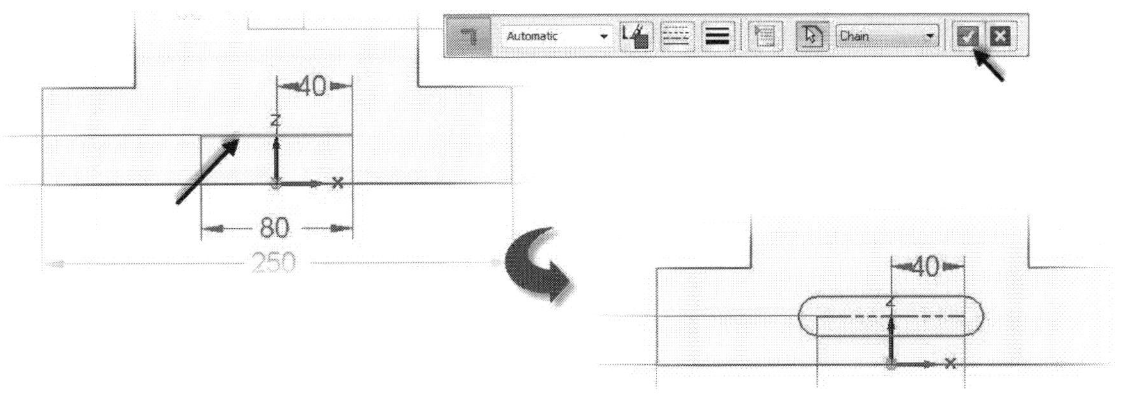

20. Click **Home > Draw > Fillet** on the ribbon. Type-in 6 in the **Radius** box on the command bar and press Enter.
21. Create fillets by clicking on corners of the sketch.

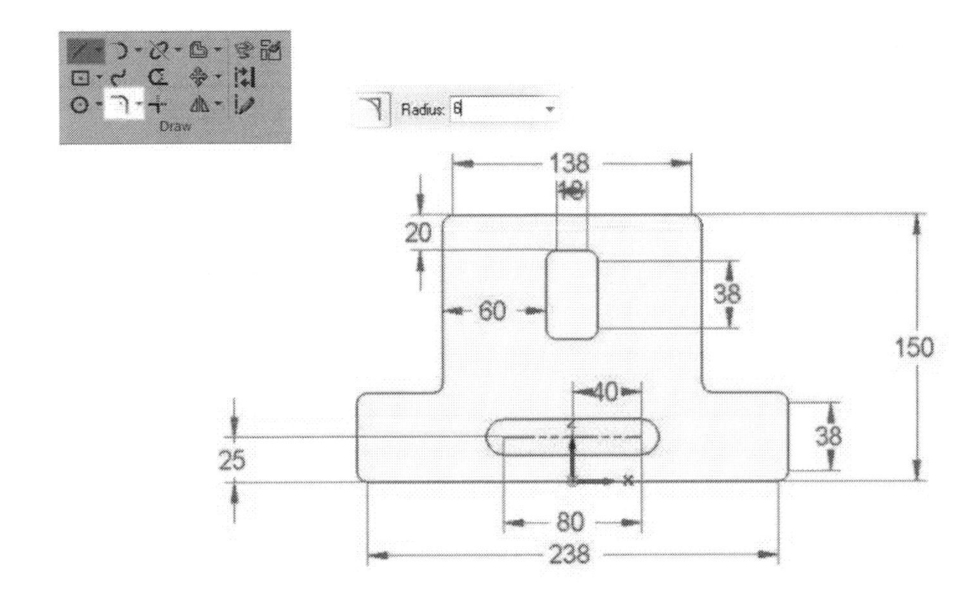

22. Save and close the file.

Questions

1. What is the procedure to create sketches in Synchronous mode?
2. List any two sketch *Relationships* in Solid Edge.
3. Which command orients the sketch normal to the screen?
4. What is the procedure to create sketches in Ordered mode?
5. Which command allows you to apply dimensions to a sketch automatically?
6. Describe the two methods to create ellipses.
7. How do you define the shape and size of a sketch?
8. How do you create a tangent arc using the **Line** command?
9. Which command is used to apply multiple types of dimensions to a sketch?
10. List any two commands to create circles?

Exercises

Exercise 1

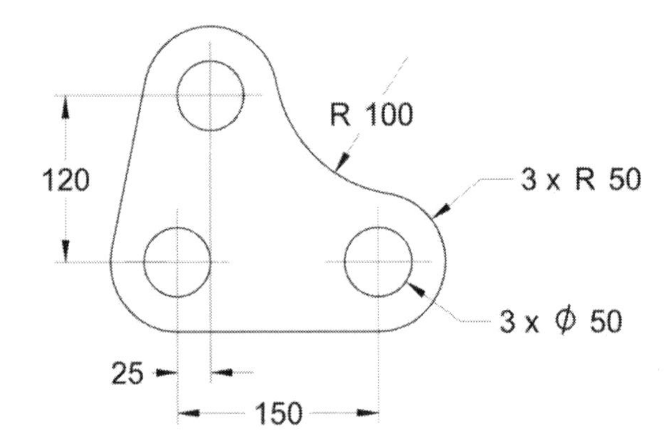

Exercise 2

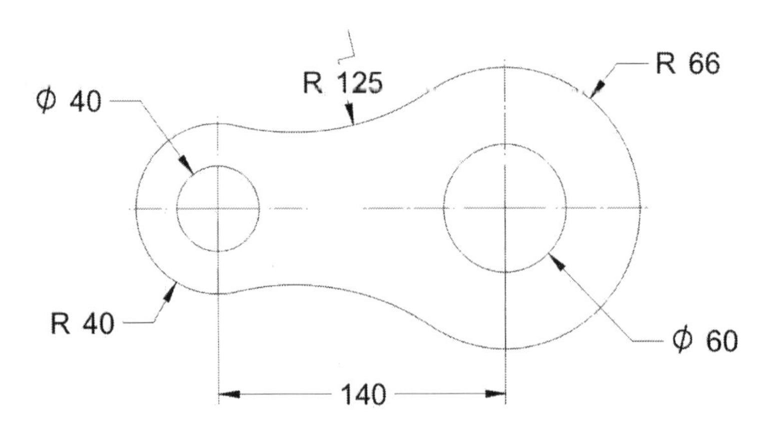

Exercise 3

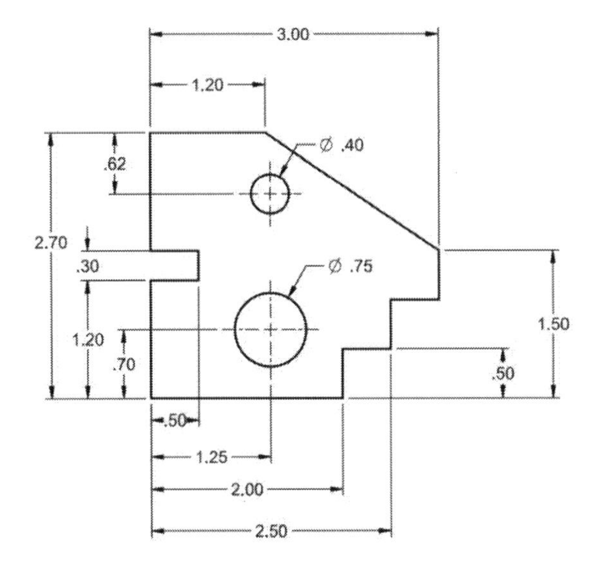

Chapter 3: Extrude and Revolve Features

This chapter covers the methods and commands to create extruded and revolved features.

The topics covered in this chapter are:

- Constructing *Extrude* and *Revolve* features in the Part environment (Synchronous and Ordered mode)
- Creating Reference Planes
- Additional Options in the *Extrude* and *Revolve* commands

Extrude Features (Synchronous)

Extrude is the process of taking a two-dimensional profile and converting it into 3D by giving it some thickness. A simple example of this would be taking a circle and converting it into a cylinder. Once you have created a sketch profile or profiles you want to *Extrude*, click inside the sketch to display a two-sided arrow. Click the arrow and move the pointer. You will notice that a thickness is added to the sketch profile. Use the **Symmetric** option on the command bar, if you want to add thickness to both sides of the sketch. Next, type-in a value in the box that appears on the extrusion, and then press Enter to create the *Extrude* feature.

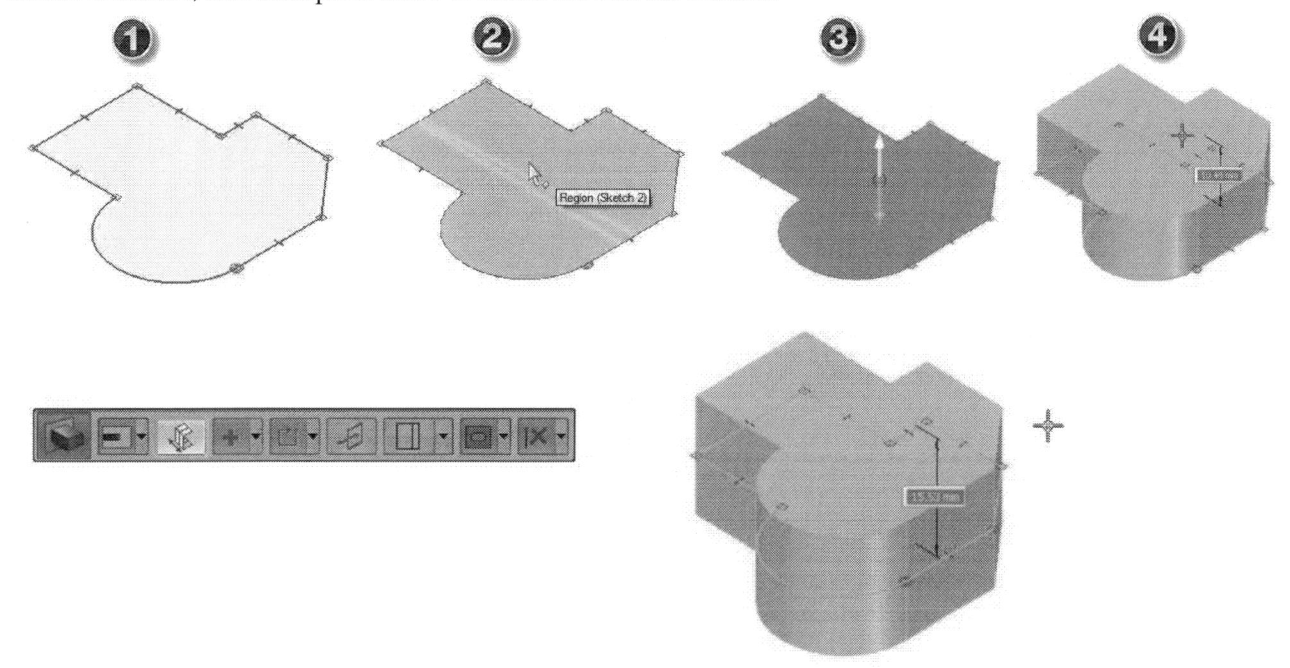

Extrude Features (Ordered)

The process of creating *Extrude* features in the **Ordered** mode is similar to **Synchronous**, but it includes few additional steps. Activate the **Extrude** command from the **Solids** panel on the **Home** tab of the ribbon. The **Extrude** command bar pops up on the screen. Select the **Select from Sketch** and **Chain** options on the command bar. Click on the sketch profile and click the green check on the command bar. Move the pointer upward or downward to add thickness to the sketch profile. Next, type-in a value in the **Distance** box on the command bar and press Enter to create the *Extrude* feature.

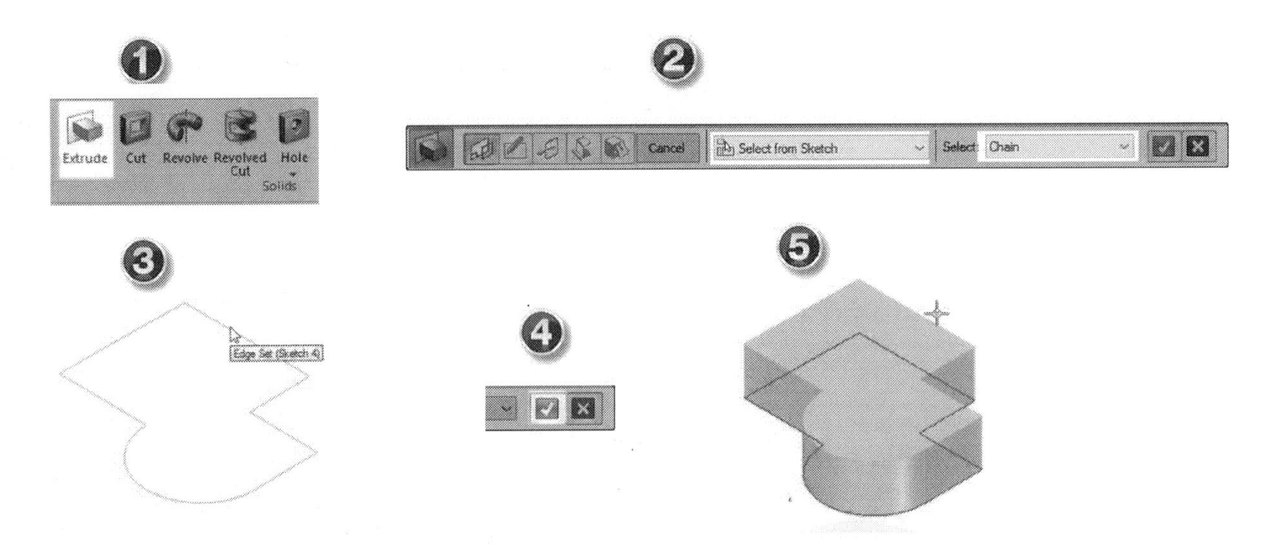

Use the **Symmetric Extent** option on the command bar to add equal thickness on both sides of the sketch. Use the **Non-Symmetric Extent** option to add separate thickness on either sides of the sketch profile. You need to click the **Extent Step** icon to view the **Non -Symmetric Extent** option.

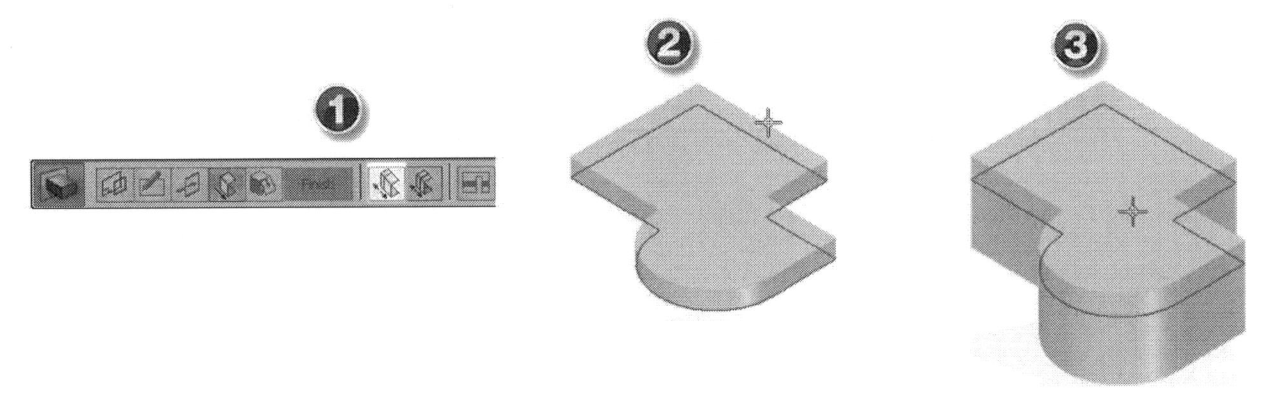

Tip: You can create a part model with a combination of the Synchronous and Ordered features. Right-click in the graphics window and select the ***Transition to Synchronous*** *option; the Synchronous mode is activated and the Ordered feature appears transparent. However, you can completely hide the Ordered feature. To do this, right click in the graphics window and select* ***Hide All > Ordered Body****. To show the hidden Ordered body, right click in the graphics window and select* ***Show All > Ordered Body****.*

Revolve Features (Synchronous)

Revolve is the process of taking a two-dimensional profile and revolving it about a centerline to create a 3D geometry (shapes that are axially symmetric). While creating a sketch for the *Revolve* feature, it is important to think about the cross-sectional shape that will define the 3D geometry once it's revolved about an axis. For instance, the following geometry has a hole in the center. This could be created with a separate *Cut* or *Hole* feature. But in order to make that hole part of the *Revolve* feature, you need to sketch the axis of revolution so that it leaves a space between the profile and the axis.

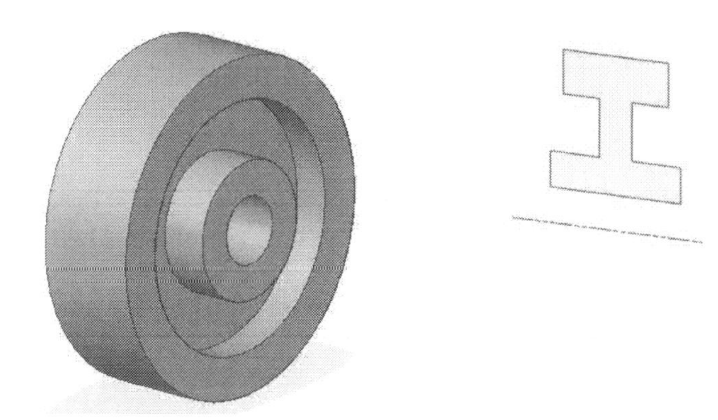

After completing the sketch, click inside the sketch region. The *Extrude Handle* appears on the sketch region. Activate the **Revolve** command from the command bar (click the down arrow next to the **Extrude** icon and select **Revolve**). You will notice that the *Extrude Handle* is changed to *Revolve Handle* (a two-sided arrow with a torus and spear in the middle). Click the spear on the *Revolve Handle*, and drag and place it on the axis of revolution. Click the torus on the *Revolve Handle* and move the pointer to revolve the sketch. Type-in an angle value and press Enter to create the *Revolve* feature. Select **Finite > 360** on the command bar to revolve the sketch up to 360 degrees.

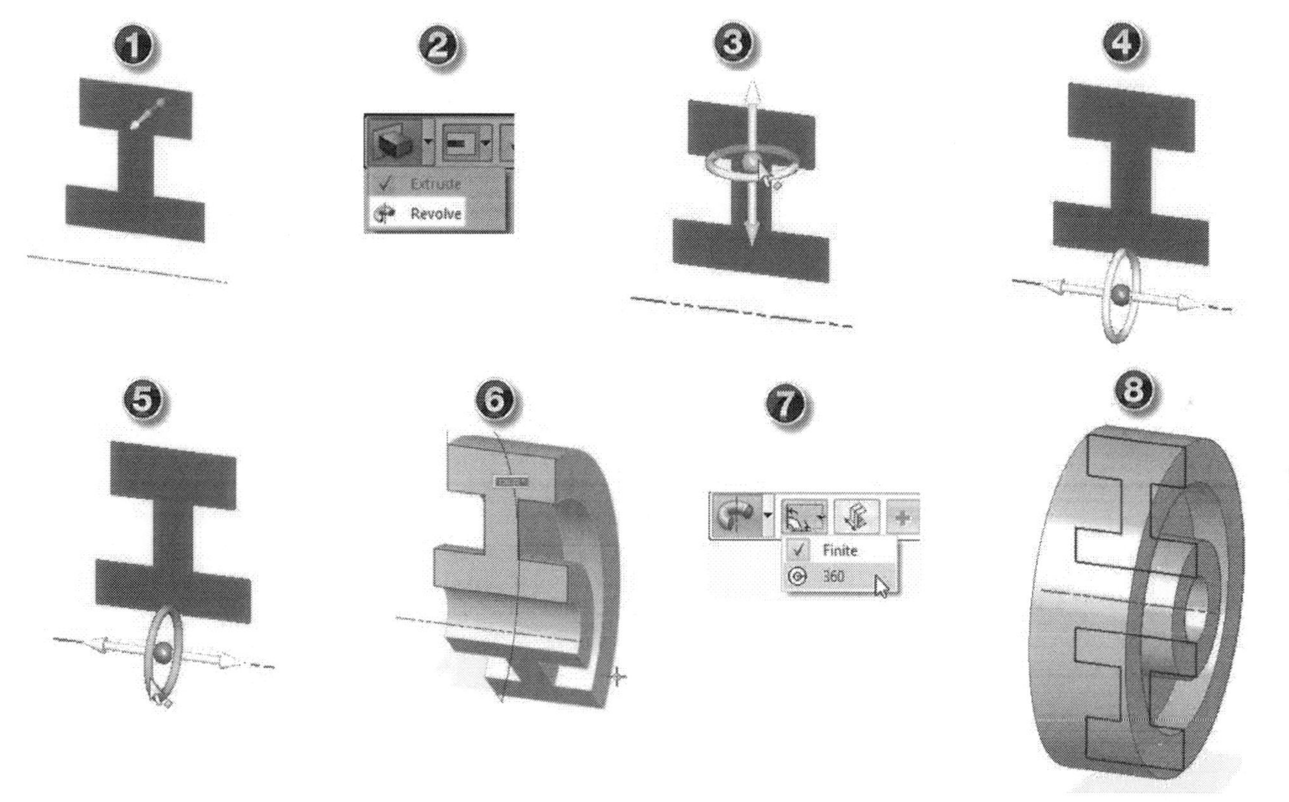

Revolve Features (Ordered)

The process of creating the *Revolve* feature in the **Ordered** mode is little bit different from **Synchronous**. First, you must activate the **Revolve** command (click **Home > Solids > Revolve** on the ribbon), and then select the sketching plane. Next, draw the cross-section and axis of revolution. Click **Home > Draw > Axis of Revolution** on the ribbon and select a line to define the axis of revolution. Close the sketch and type-in a value in the **Angle** field on the command bar (or) click the **Revolve 360** icon on the command bar to revolve up to 360 degrees. Next, click **Finish** to create the *Revolve* feature. Click **Cancel** to deactivate this command.

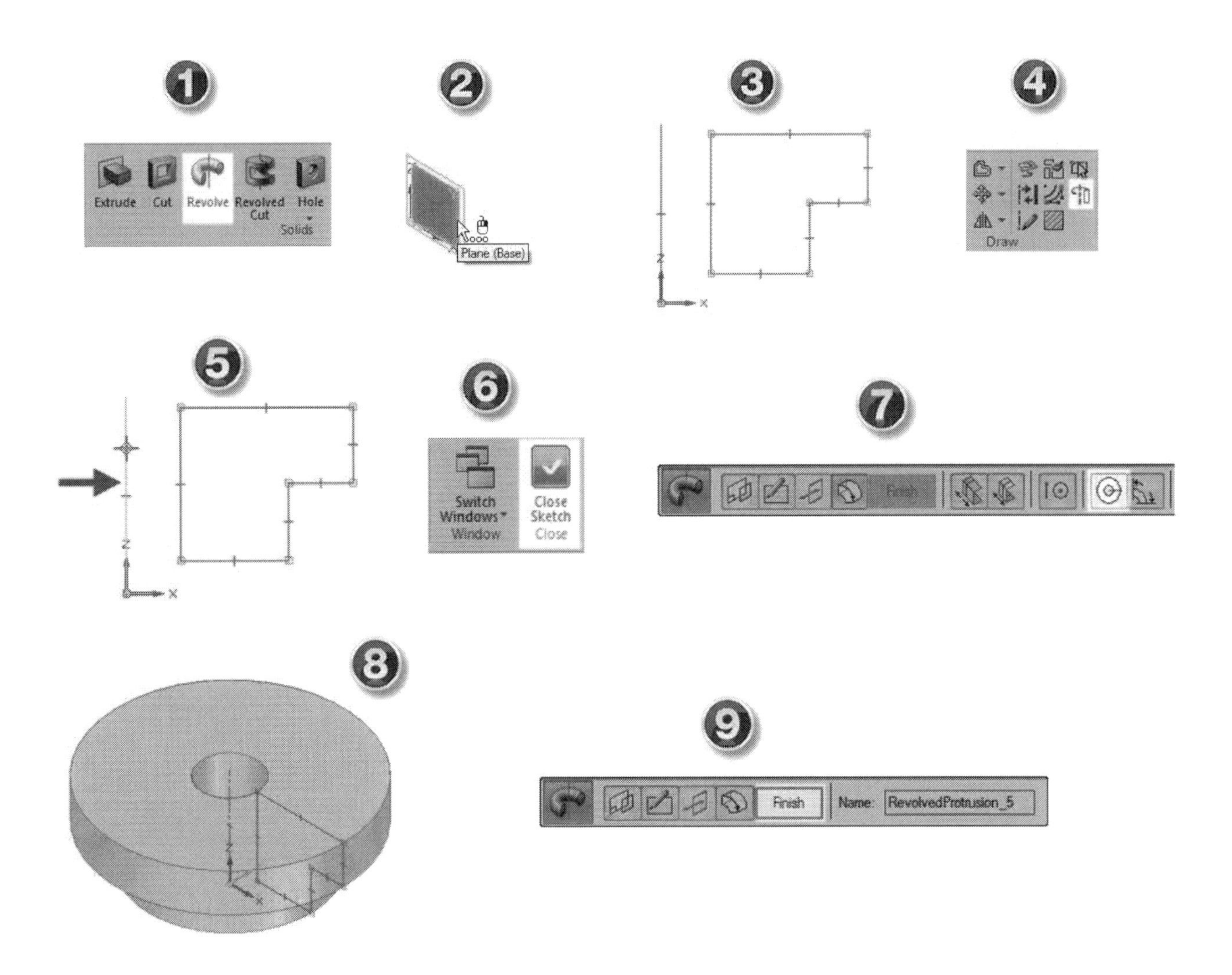

Primitive Shapes (Synchronous only)

Solid Edge provides you with commands to create primitive shapes such as boxes, cylinders, and shapes. These commands are available in the Synchronous mode only.

Box

This command creates a box by using a rectangular sketch. You can create rectangles by using three options: **by Center**, **by 2 points**, and **by 3 points**. These options are discussed earlier in Chapter 2 in the **Rectangles** section.

Activate this command (on the ribbon, click **Home > Solids > Box drop-down** > **Box**) and set the **Selection Type** on the command bar. For example, set the **Selection type** to **by Center** and select a plane. Click to define the center point of the rectangle. Move the pointer and click to define the corner point (or) type-in values in the length, width, and angle dimension boxes by pressing the Tab key. Move the pointer and click (or) type-in the extrusion depth and press Enter.

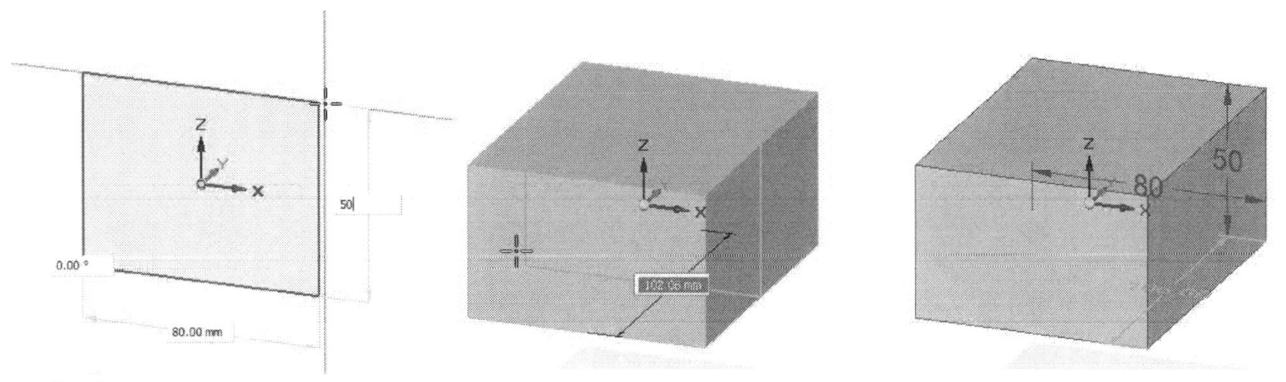

Cylinder

Creating a cylinder is similar to that of a box. Activate the **Cylinder** command (on the ribbon, click **Home > Solids > Primitives** drop-down **> Cylinder**) and select a plane. Click to define the center point of the cylinder. Next, define the extrusion depth by entering a value (or) by moving the pointer and clicking.

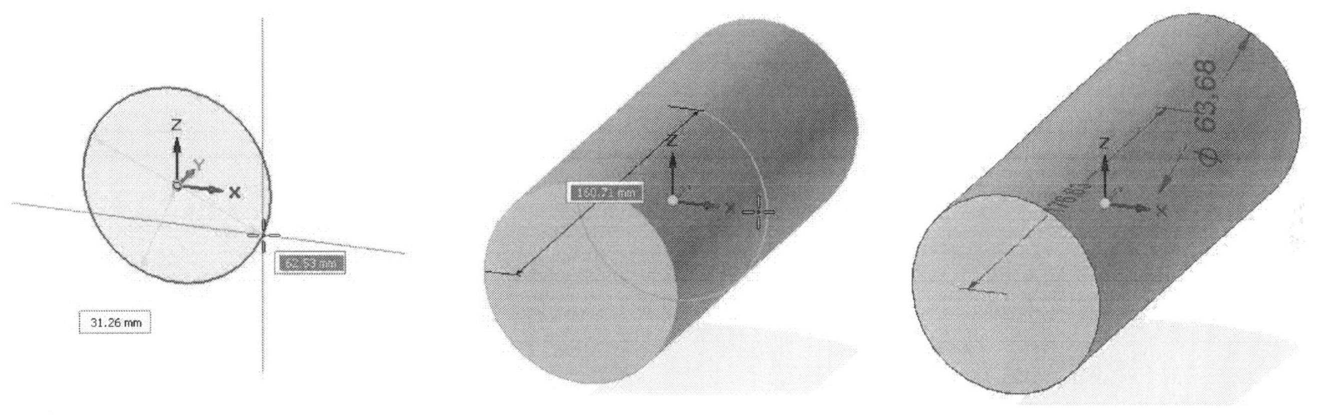

Sphere

This command creates a sphere by defining the center and radius. Activate the **Sphere** command (on the ribbon, click **Home > Solids > Primitives** drop-down **> Sphere**) and click to define the center point of the sphere. You can also select a plane and define a point on it. Next, define the radius by entering a value (or) moving the pointer and clicking.

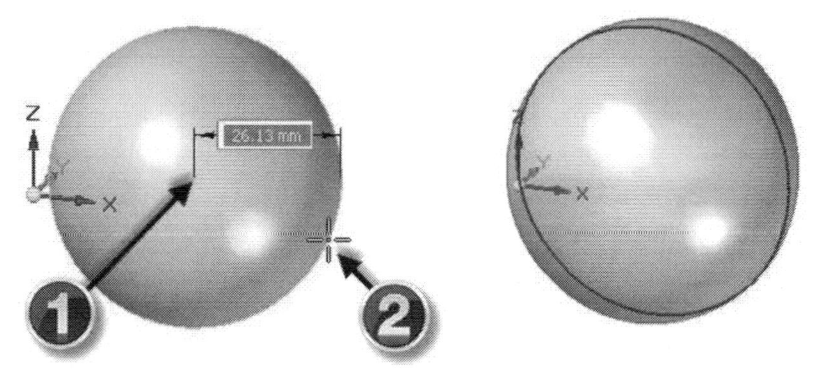

Creating Planes (Synchronous)

Each time you start a new part file, Solid Edge automatically creates default reference planes (Base Reference Planes) along with the default coordinate system. Planes and coordinate system make up a specific type of features in Solid Edge, known as Reference features. These features act as supports to your 3D geometry. In addition to the default reference features, you can create your own additional planes and coordinate systems too. Until now, you have known to create sketches on any of the default reference planes. If you want to create sketches and geometry at locations other than default reference planes, you can create new reference planes manually. You can do so by using the commands available in the **Planes** panel of the **Home** tab.

Coincident Plane

This command creates a reference plane, which is coincident with a selected face or plane. Activate this command (click **Home > Planes > Coincident Plane** on the ribbon) and click on a face or plane. A plane coincident with the selected face will be placed. In addition, the *Steering Wheel* tool appears on the plane.

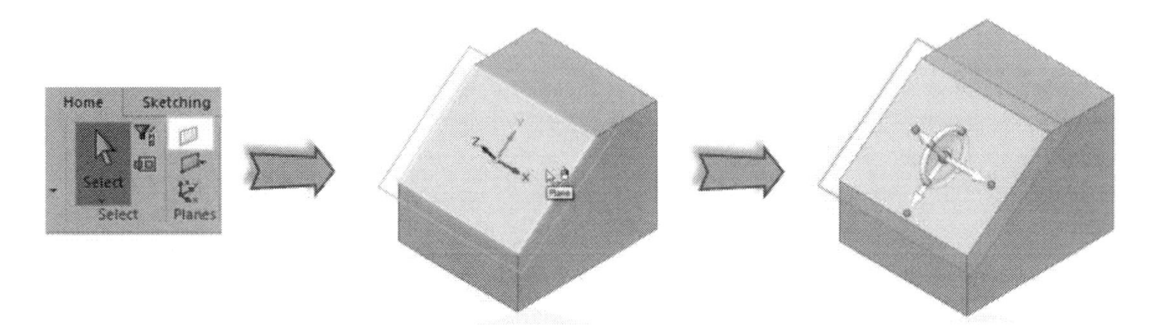

Click on any of the arrows of the *Steering Wheel* tool and drag the pointer to change the location of the plane.

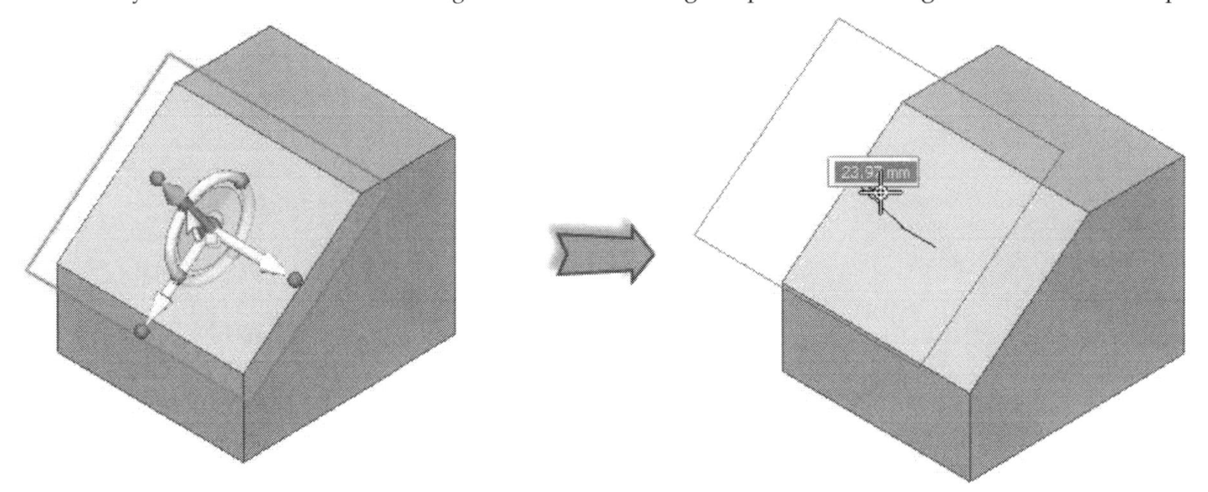

Click the torus of the *Steering Wheel* tool and drag the pointer to rotate the plane.

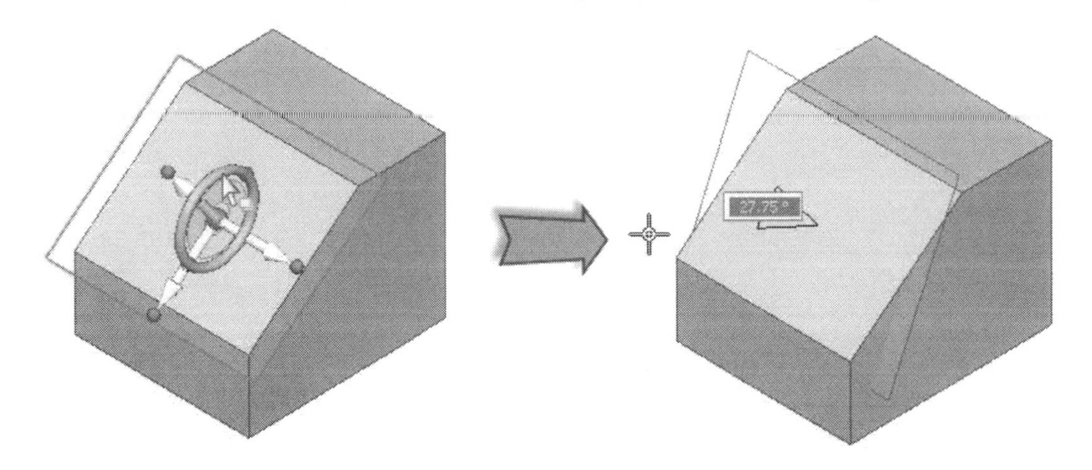

Normal to Curve

This command creates a reference plane, which will be normal (perpendicular) to a line, curve, or edge. Activate this command (click **Home > Planes > More Planes > Normal to Curve** on the ribbon), and then select an edge, line, curve, arc, or circle. Drag the pointer and click on a point to define the location of the plane (or) type-in a value in the **Position** box (or) type-in a distance value in the **Distance** box and press Enter.

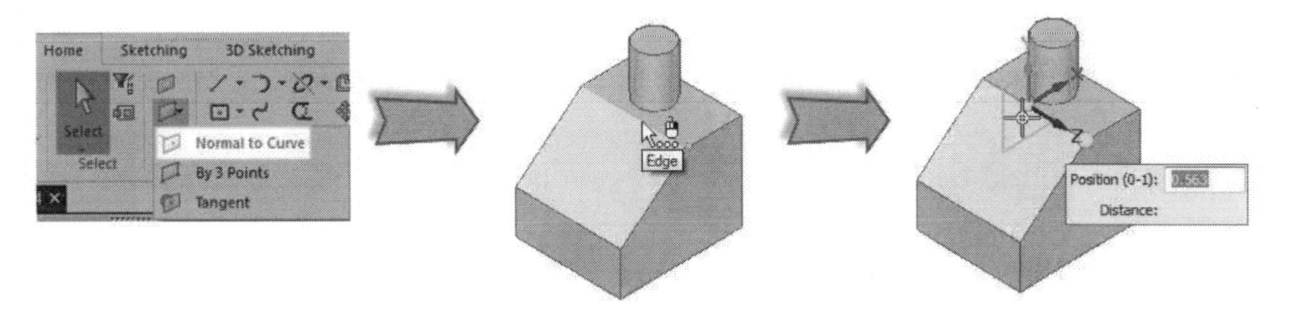

By 3 Points

This command creates a reference plane passing through three points. Activate this command (click **Home > Planes > More Planes > By 3 Points** on the ribbon), and then select three points from the model geometry. A plane will be placed passing through these points.

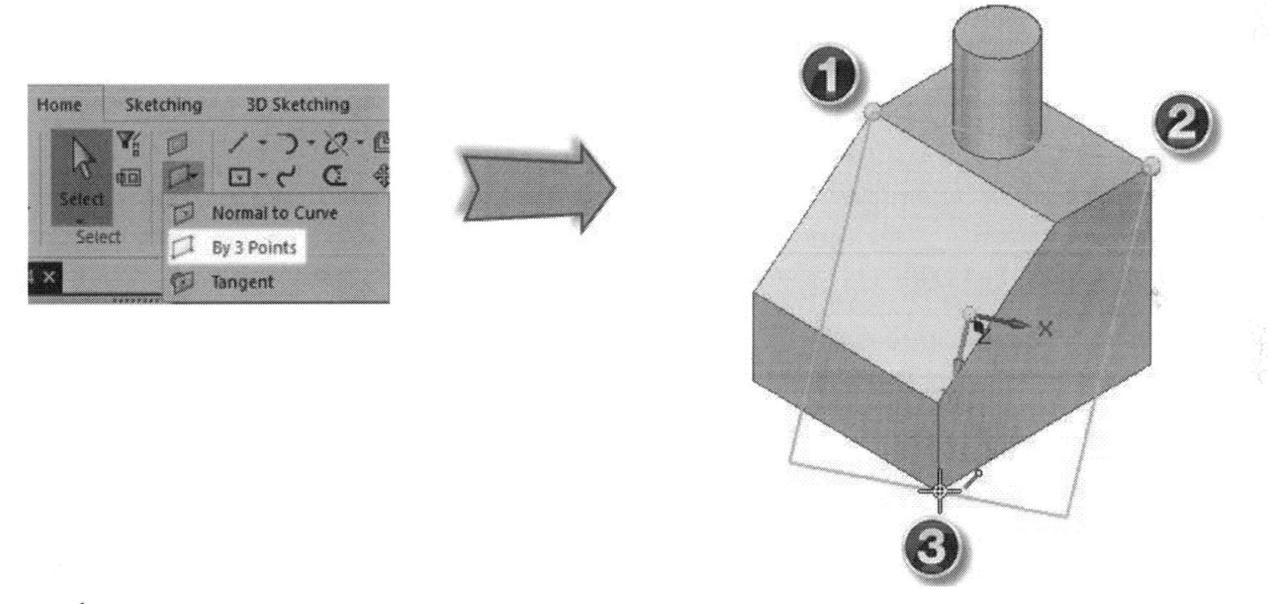

Tangent

This command creates a plane tangent to a curved face. Activate this command (click **Home > Planes > More Planes > Tangent** on the ribbon) and select a curved face. A plane tangent to the selected face appears. Drag the pointer and click to define the position of the tangent plane (or) type-in an angle value and press Enter.

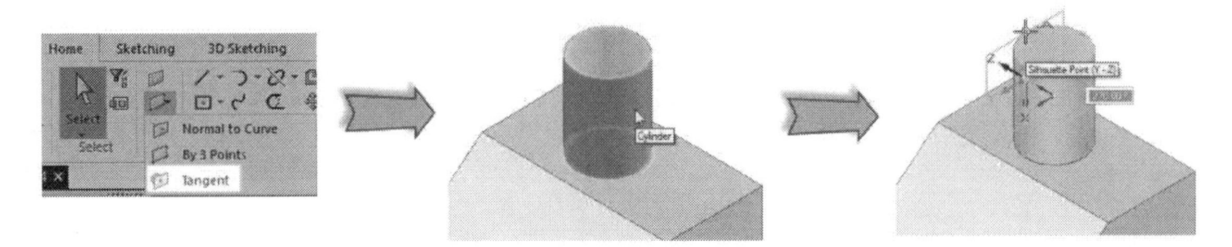

Coordinate System

This command creates a new coordinate system in addition to the default one. Activate this command (click **Home > Planes > Coordinate System** on the ribbon) and position the pointer on a point or face or line. Use the orientation keys, if you want to change the orientation of the coordinate system. For example, press F to flip the coordinate system by reversing the direction of the z-axis. Press T to flip the coordinate system by reversing the x-axis. Press G to return to the default orientation. After re-orienting the coordinate system, click to define its location.

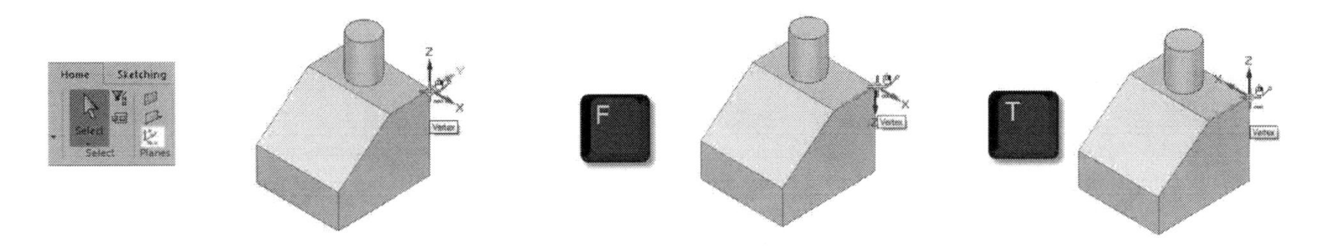

Creating Planes (Ordered)

The Ordered mode offers some additional methods to create reference planes.

Parallel

This command creates a reference plane, which will be parallel to a face or another plane. Activate this command (click **Home > Planes > More Planes > Parallel** on the ribbon) and select a flat face or plane. Drag the pointer and click to define the location of the plane (or) type-in a value in the **Distance** box on the command bar and press Enter.

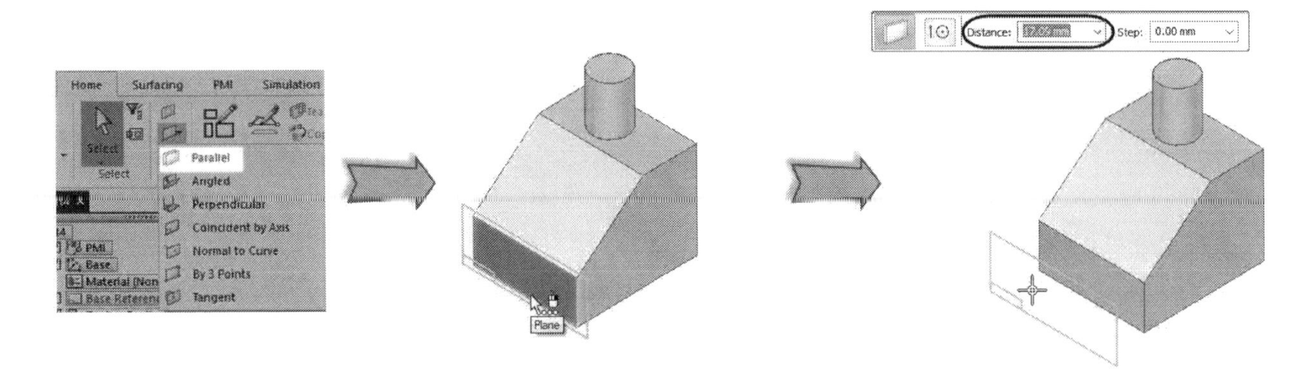

Angled

This command creates a plane, which will be positioned at an angle to a face or plane. Activate this command (click **Home > Planes > More Planes > Angled** on the ribbon) and select a flat face or plane. Next, select another face, which acts as a base reference. Select a point to define the orientation of the plane, and then type-in a value in the **Angle** box on the command bar.

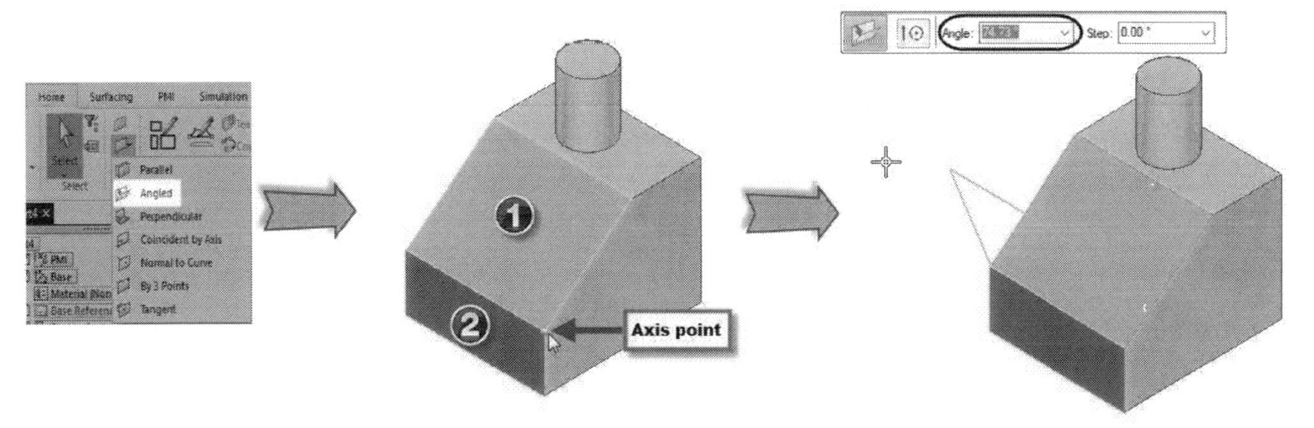

Perpendicular

This command creates a plane, which will be perpendicular to a face or plane. Activate this command (click **Home > Planes > More Planes > Perpendicular** on the ribbon) and select a flat face or plane. Next, select another face which acts as a base reference. Select a point to define the origin of the plane, and then click to specify the side of the plane.

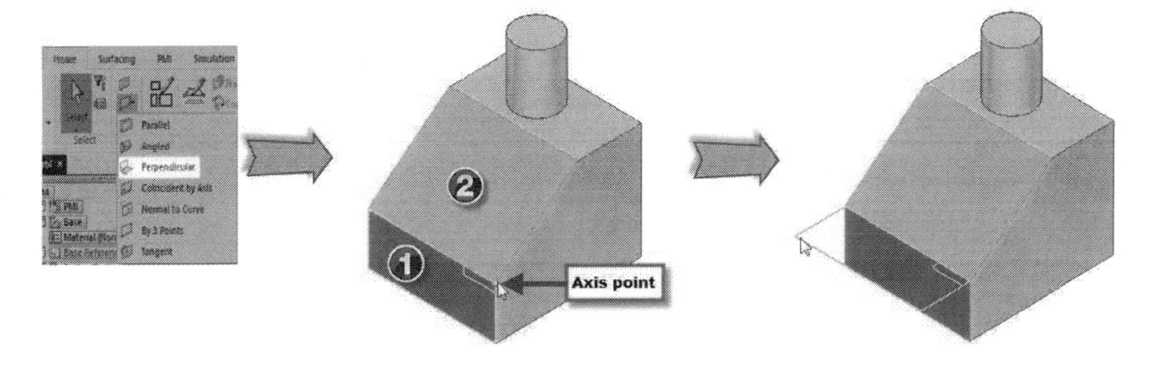

Coincident by Axis

This command creates a plane, which is coincident with a selected face or plane. Activate this command (click **Home > Planes > More Planes > Coincident by Axis** on the ribbon) and select a flat face or plane. Next, select a part edge to define the x-axis of the plane, and then select a point to define the x-axis origin.

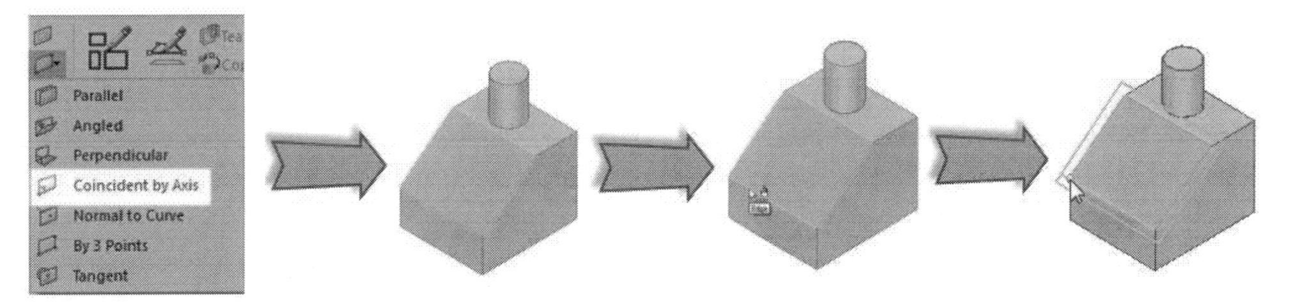

Additional options of the Extrude command

The **Extrude** command has some additional options to create a 3D geometry, complex features, and so on. These options are inactive by default and are activated after you have created the first feature of the part.

Selection Type options

The **Selection Type** drop-down menu on the command bar has three options: **Single**, **Chain**, and **Face**. Note that the Selection Type drop-down menu is available only when you activate the **Extrude** command from the ribbon. The **Single** option selects the individual elements of a sketch, whereas the **Chain** option selects the complete loop. The **Face** option selects the region enclosed by the sketch. Note that the **Face** option is available only in the **Synchronous** Mode.

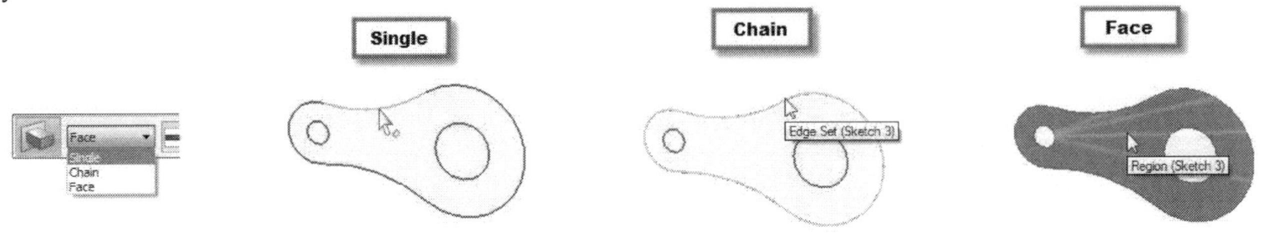

Include Internal Loops

This option is useful while working with a sketch having internal loops. If you select this option, the internal loops of the sketch will be detected while creating the *Extrude* feature.

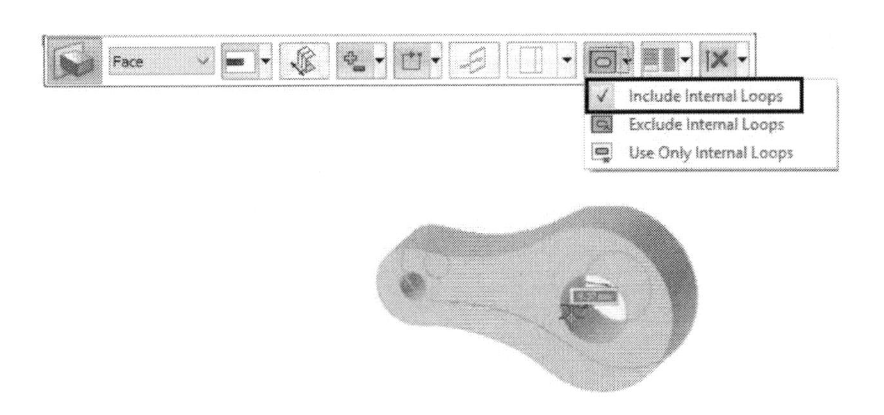

Exclude Internal Loops

This option ignores the internal loops of a sketch.

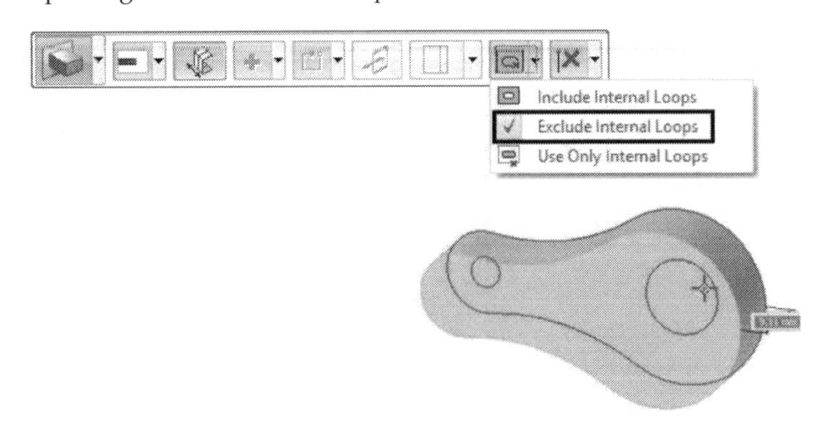

Use Only Internal Loops

This option considers only the internal loops of the sketch.

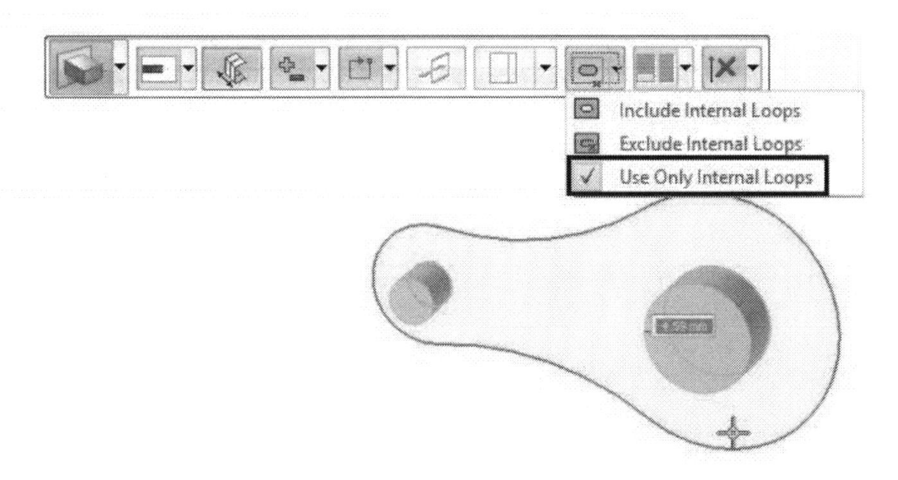

Add

This option adds material to the geometry.

Cut

This option removes material from the part geometry.

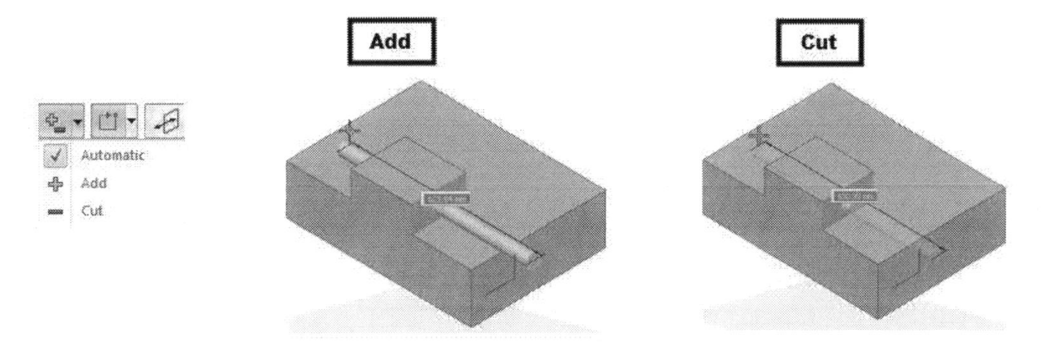

Automatic

This option adds or removes material based on the extrusion direction.

Open

This option extrudes an open sketch without using the adjacent edges.

Closed

Converts an open sketch into a closed one by using the adjacent edges, and then extrudes the sketch.

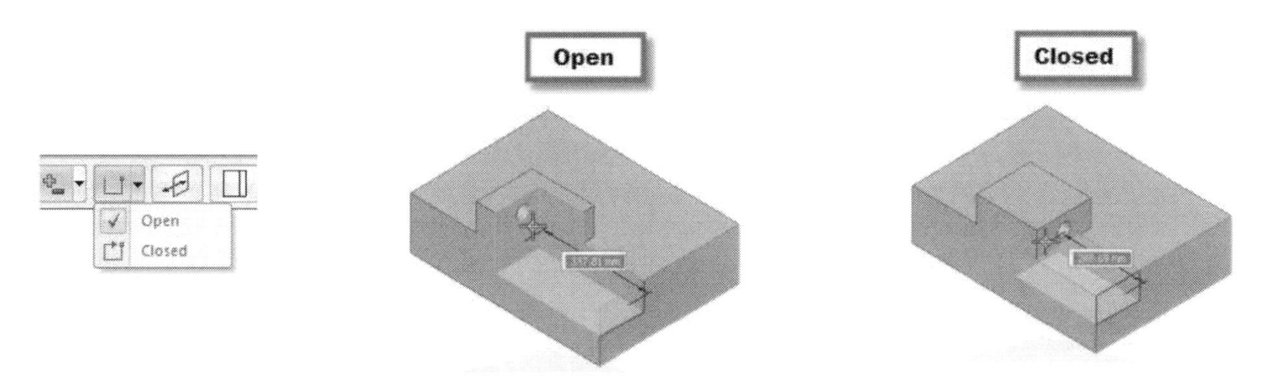

Side Step

This option defines the side of the sketch to extrude. On the command bar, set the **Selection Type** to **Chain** and select an open sketch. Right click to accept the sketch. Use the arrow that appears on the selected sketch to define the side of the sketch to be extruded. Move the pointer to extrude the sketch. You can click the **Side Step** button to change the side to be extruded.

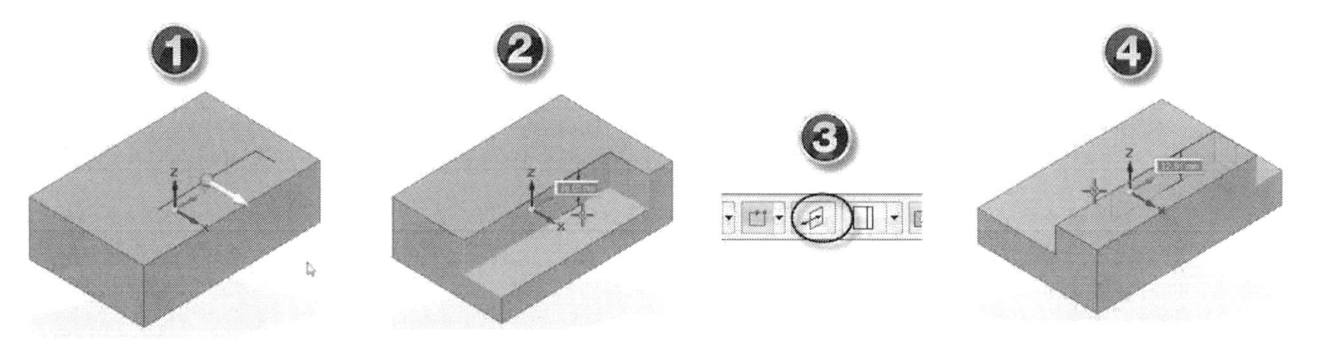

Extent Type options

The **Extent Type** drop-down menu on the command bar has four options: **Finite**, **Through All**, **Through Next**, and **From-To**. The **Finite** option extrudes the sketch up to the distance that you specify. The **Through All** option extrudes the sketch throughout the 3D geometry.

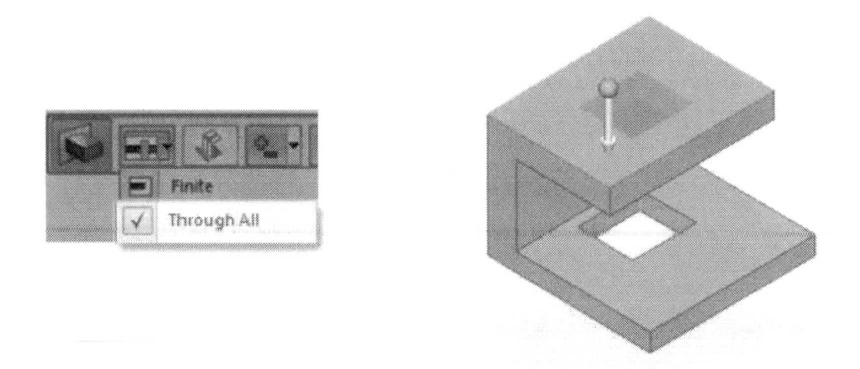

The **Through Next** option extrudes the sketch through the face next to the sketch plane.

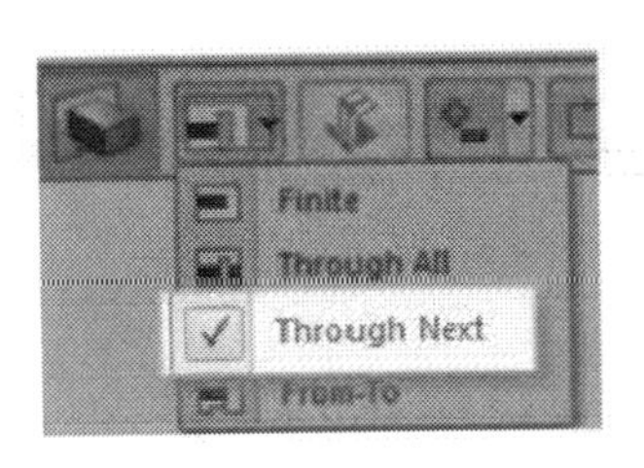

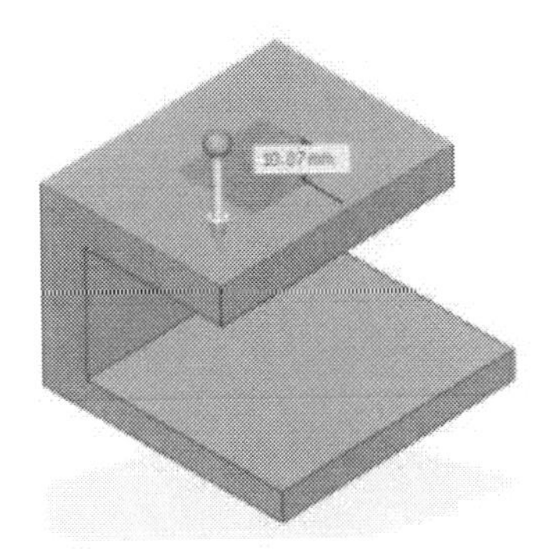

The **From-To** option extrudes the sketch from the sketch plane up to a selected face. Click inside the sketch region, and then click on the arrow handle. On the **Extrude** command bar, select **From-To** from the **Extent Type** drop-down. The 'From' surface is selected automatically and you need to select the 'To' surface. Select the 'To' Surface from the model geometry; the sketch will be extruded up to the selected surface.

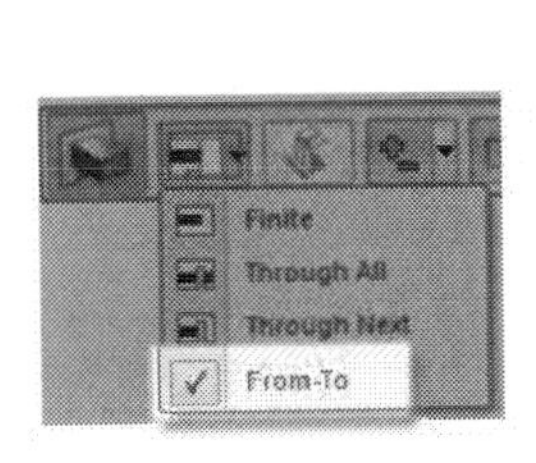

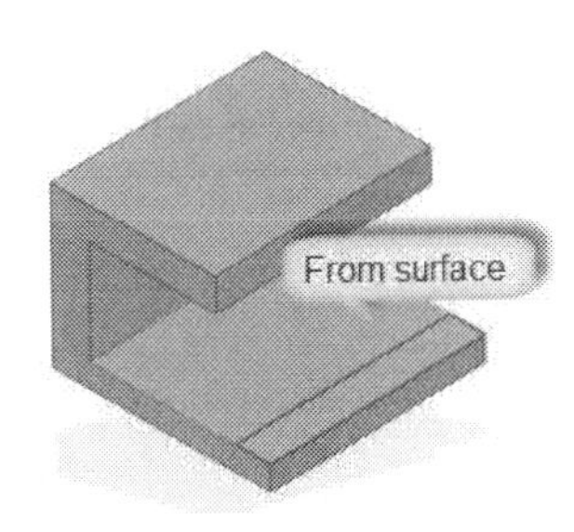

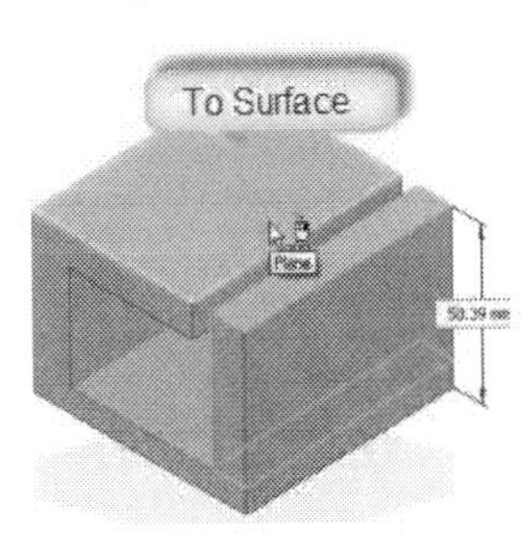

Treatments options

The **Treatments** drop-down menu on the command bar has three options: **No Treatment**, **Draft**, and **Crown**. The **No Treatment** option creates the *Extrude* feature without any treatment. The **Draft** option applies draft to the *Extrude* feature. Activate this option and type-in a draft angle value in the **Draft Parameters** dialog. Click the **Flip** button next to the **Angle 1** box to flip the draft angle. Click **OK** on the **Draft Parameters** dialog and define the *Extrude* distance. If you activate the **Symmetric** option on the command bar, you can define draft in the second direction as well.

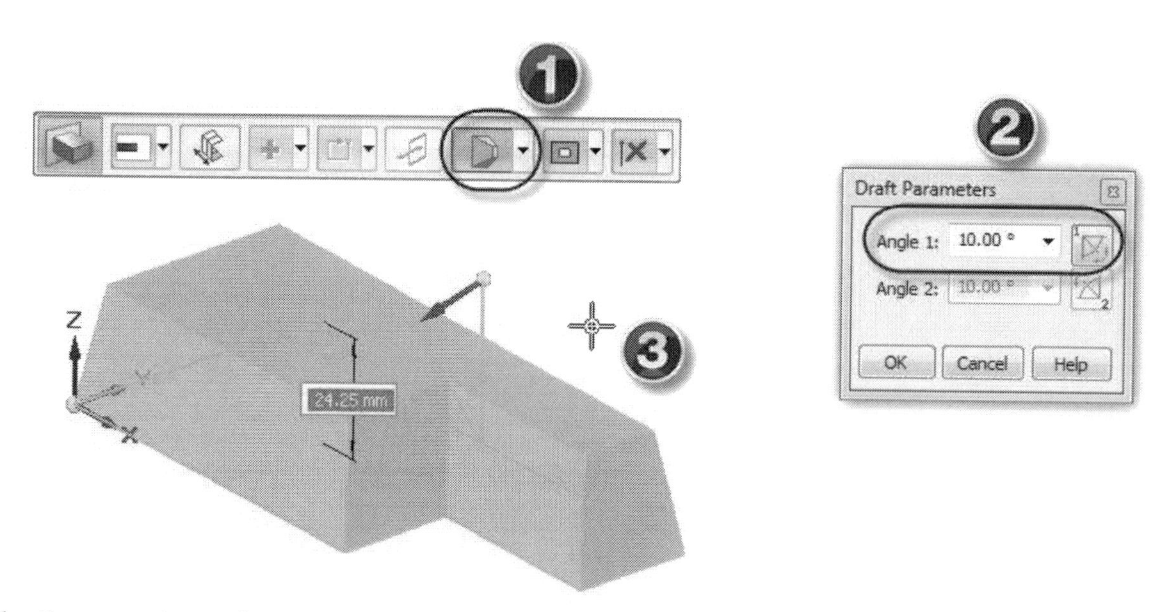

The **Crown** option in the **Treatments** drop-down menu adds a crown to the *Extrude* feature. As you activate this option, the **Crown Parameters** dialog appears. Type-in a value in the **Radius** box and click **OK**. Drag the pointer and click to define the thickness of the *Extrude* feature.

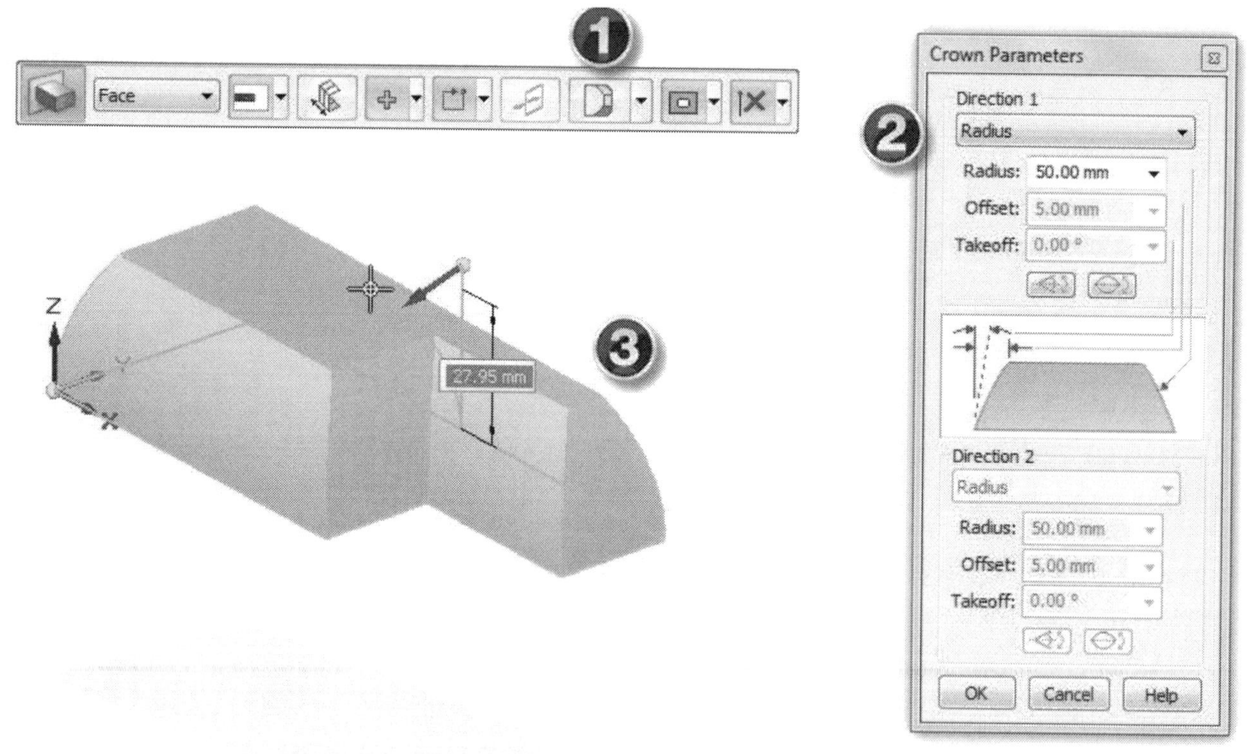

The **Direction 1** drop-down has five options: **No Crown**, **Radius**, **Radius and take-off**, **Offset**, **Offset and take-off**. The **Radius** option defines the crown by using the crown radius that you specify in the **Radius** box. The **Radius and take-off** option creates the crown by using the radius and take-off angle values. The take-off value is the starting angle crown measured from the sketch plane. The **Offset** option creates a crown by using the offset value (difference between the start and end sections of the crown). The **Offset and take-off** option uses both the offset and take-off angle values.

Applying Material to the Model

Solid Edge allows you to apply a material to the model easily. To apply material to the geometry, double-click the **Material** option under the **PathFinder** tree. On the **Solid Edge Material Table** dialog, expand the **Material** tree and select a material. The **Material Properties** of the selected material appear. Click the **Apply to Model** button. The selected material will be applied to the model. If you want to remove the material, right-click on the **Material** option under **PathFinder** and select **Remove Material**.

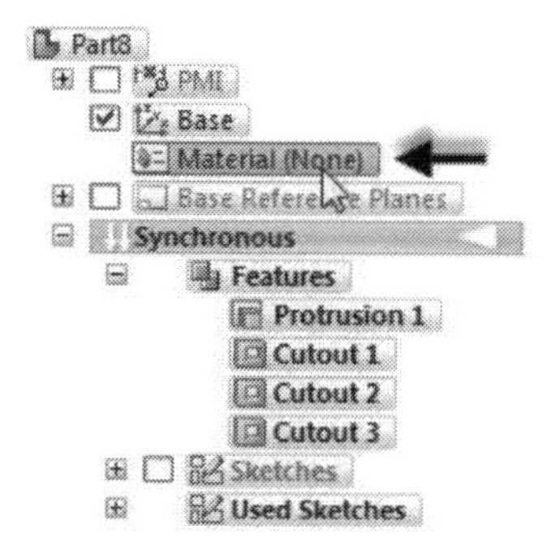

Examples

Example 1 (Millimetres)

In this example, you will create the part shown below.

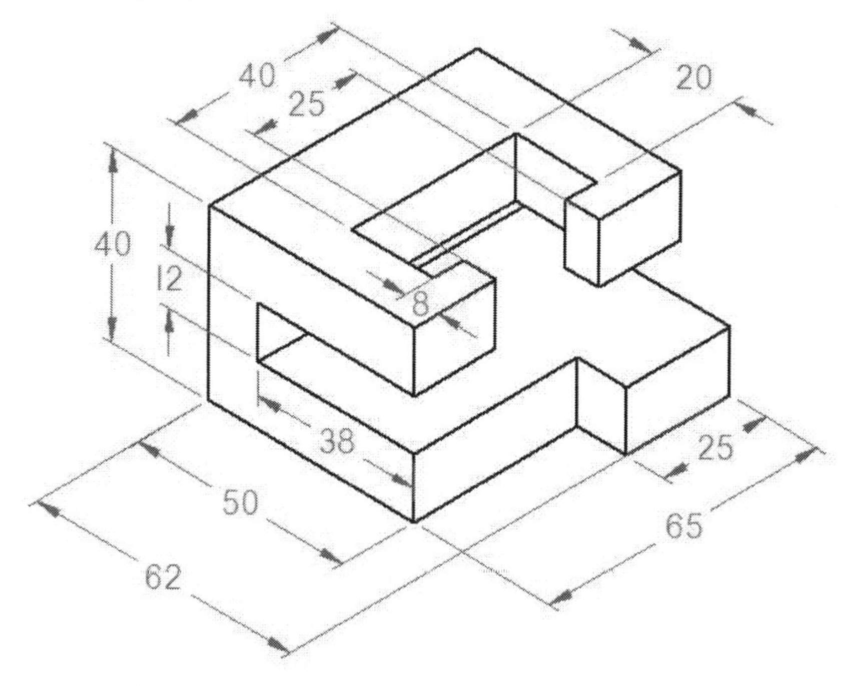

1. Start **Solid Edge 2019**.
2. On the Application Menu, click **New > ISO Metric Part**; a new part file is opened.
3. On the ribbon, click **Home > Draw > Rectangle by Center > Rectangle by 2 Points**.

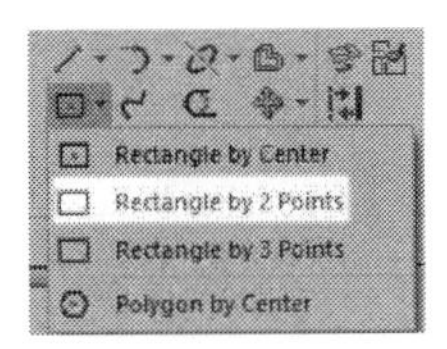

4. Place the mouse pointer on the coordinate system; the XZ plane gets highlighted.
5. Click the lock icon on the XZ plane to lock the plane.
6. Click the origin point to define the first corner of the rectangle.
7. Move the mouse pointer toward top right corner and click to define the second corner.
8. Use the **Smart Dimension** command and apply dimensions to the rectangle.

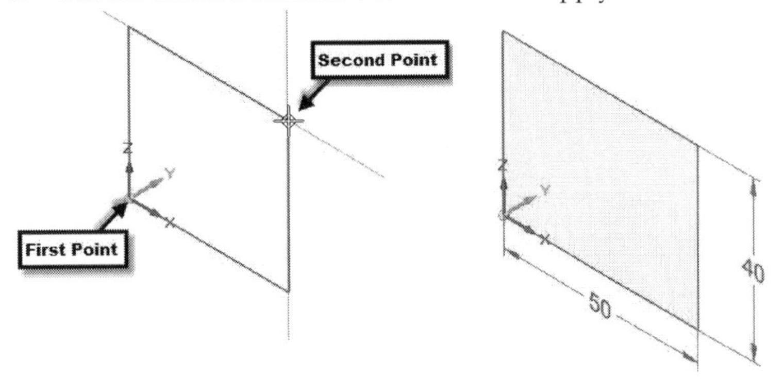

9. Click the lock icon on the screen to unlock the sketch plane. Press Esc to deactivate the **Smart Dimension** command.
10. Click inside the rectangle to display a two-sided arrow. In addition, a command bar pops up.
11. On the command bar, activate the **Symmetric** icon. Click the arrow on the sketch.

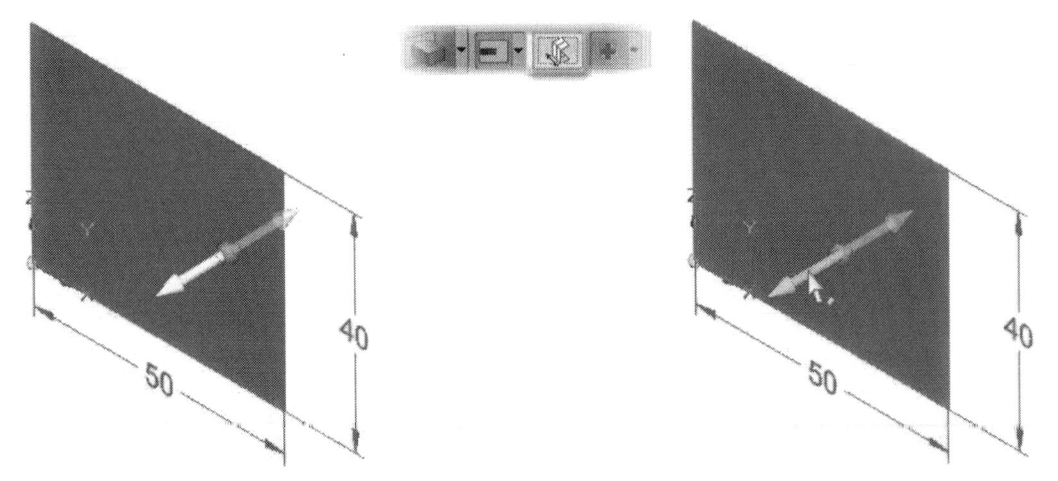

12. Type-in **65** in the box that appears on the model and press Enter.

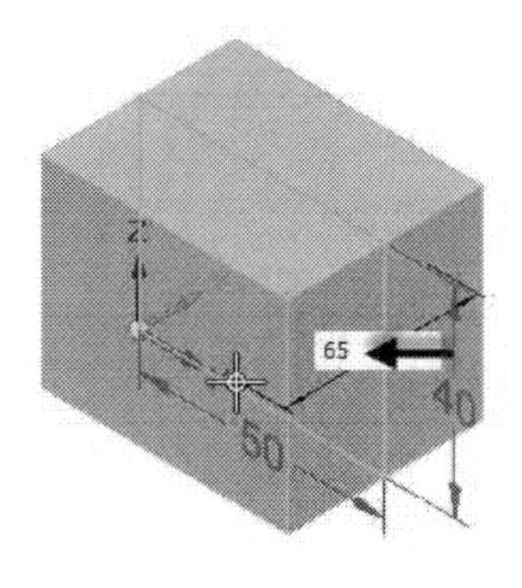

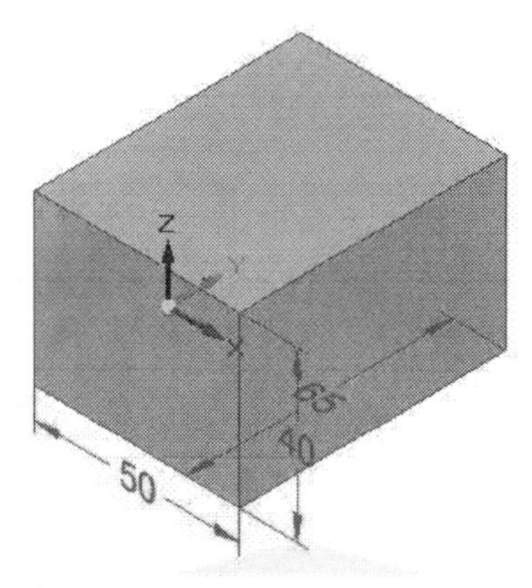

13. Activate the **Line** command and place the mouse pointer on the front face of the part geometry.
14. Click the lock icon to lock the sketching plane.
15. Click **Sketch View** on the status bar.
16. Draw the sketch and apply dimensions to it. Unlock the sketch plane.

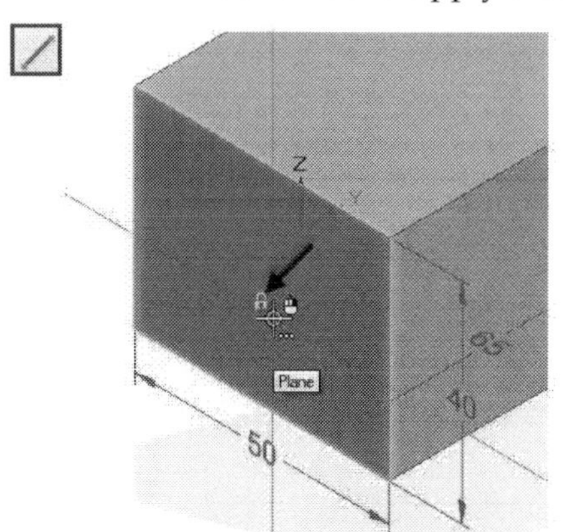

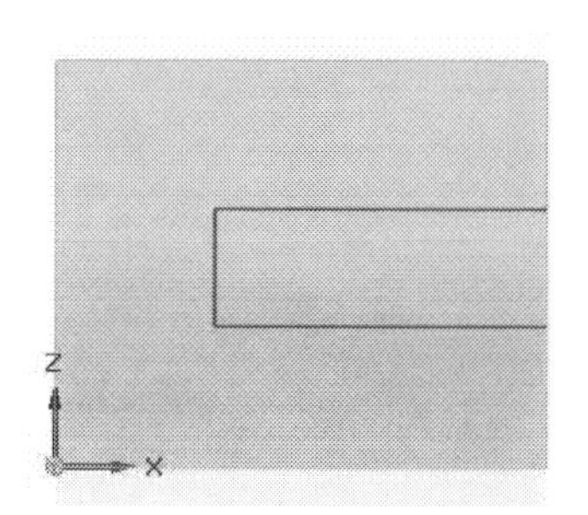

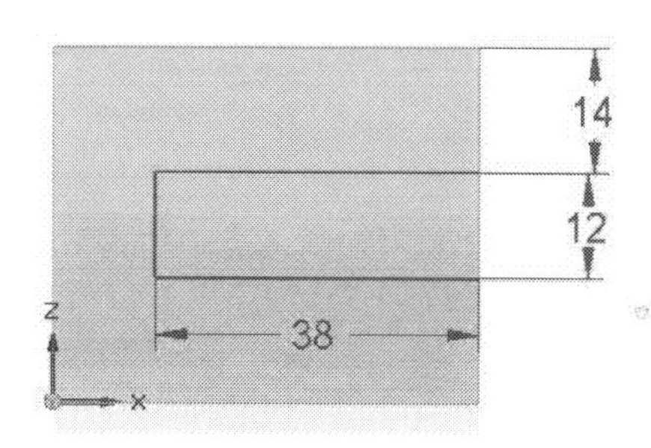

17. On the status bar, click **View Orientation > Dimetric View**; the model orientation is changed to dimetric.
18. On the ribbon, click **Home > Select > Select** and click inside the region enclosed by the sketch.
19. On the command bar, click **Extent Type > Through All**.
20. Click on the arrow pointing toward the part geometry.
21. Move the mouse pointer toward the part geometry and click to create the extruded cut.

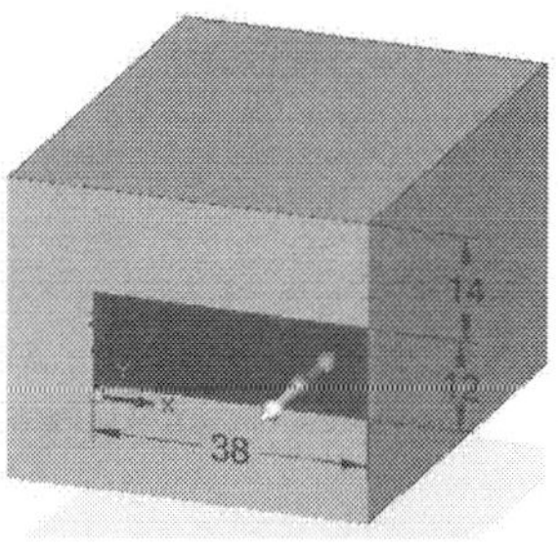

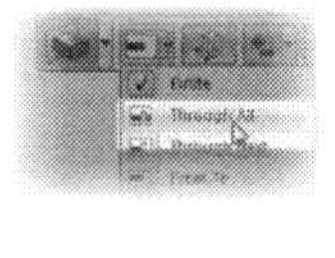

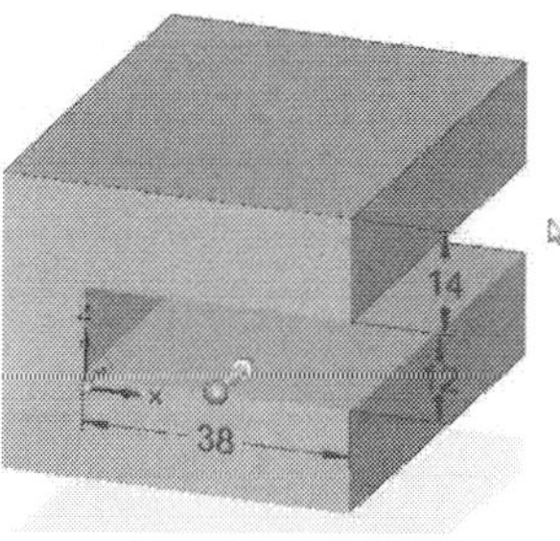

22. Click **Home > Draw > Line** on the ribbon. Place the pointer on the top face of the part geometry and click the lock icon.
23. Draw the sketch on the top face.
24. Click **Home > Relate > Symmetric** on the ribbon. Click on the X-axis to define the symmetric axis.
25. Click on the horizontal lines in the sequence shown in figure.
26. Use the **Smart Dimension** command to apply dimension to the sketch.

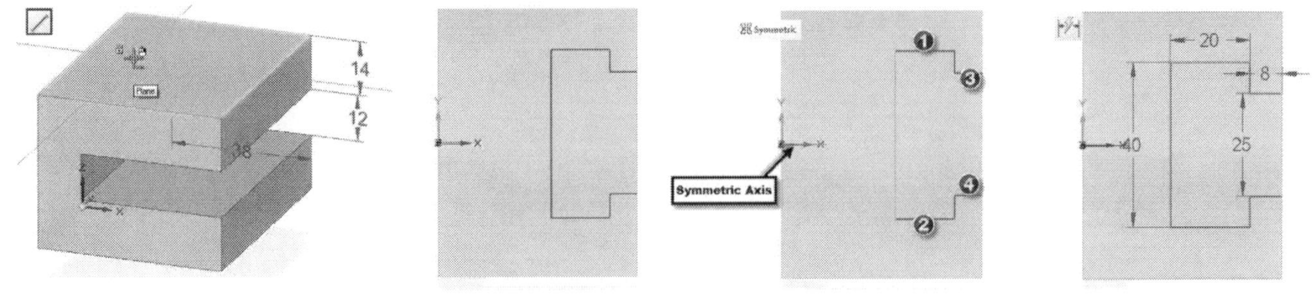

27. Unlock the sketch plane and change the view orientation to Dimetric.
28. On the ribbon, click **Home > Select > Select** and click inside the region enclosed by the sketch.
29. On the command bar, click **Extent Type > Through Next**.
30. Click on the arrow and move the mouse pointer downward.
31. Click to create the extruded cut.

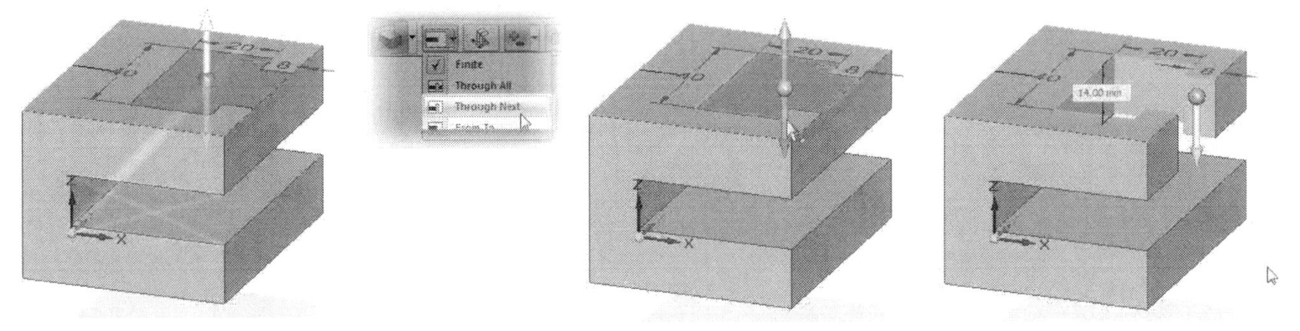

32. Activate the **Line** command and place the pointer on the horizontal face, as shown in figure.
33. Lock the face and draw the sketch. Apply dimensions to the sketch. Next, unlock the sketch plane.

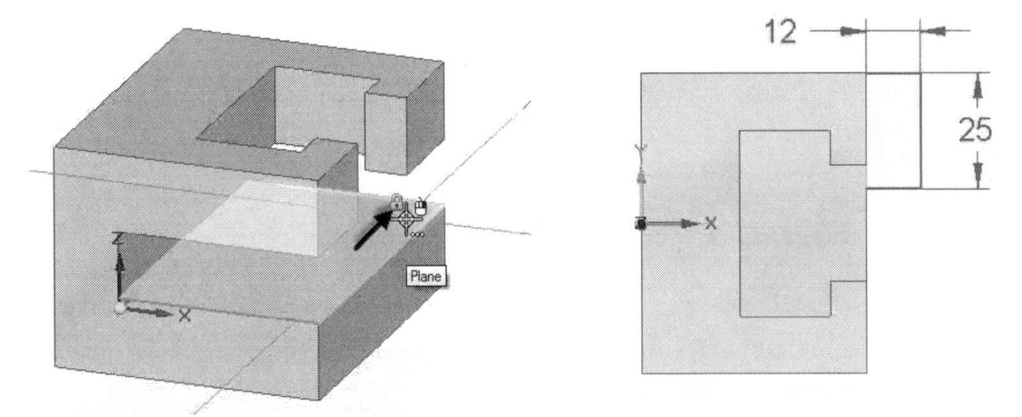

34. On the status bar, click **View Orientation > ISO View**; the view orientation changes to isometric.
35. Click **Home > Solids > Extrude** on the ribbon and click on the sketch region. Right-click to accept.
36. On the command bar, click **Extent Type > From-To**.
37. Place the mouse pointer on the side face of the geometry and right-click when the mouse icon appears. The **QuickPick** box appears.
38. From the **QuickPick** box, select the bottom face of the geometry to define the face up to which the sketch is extruded.

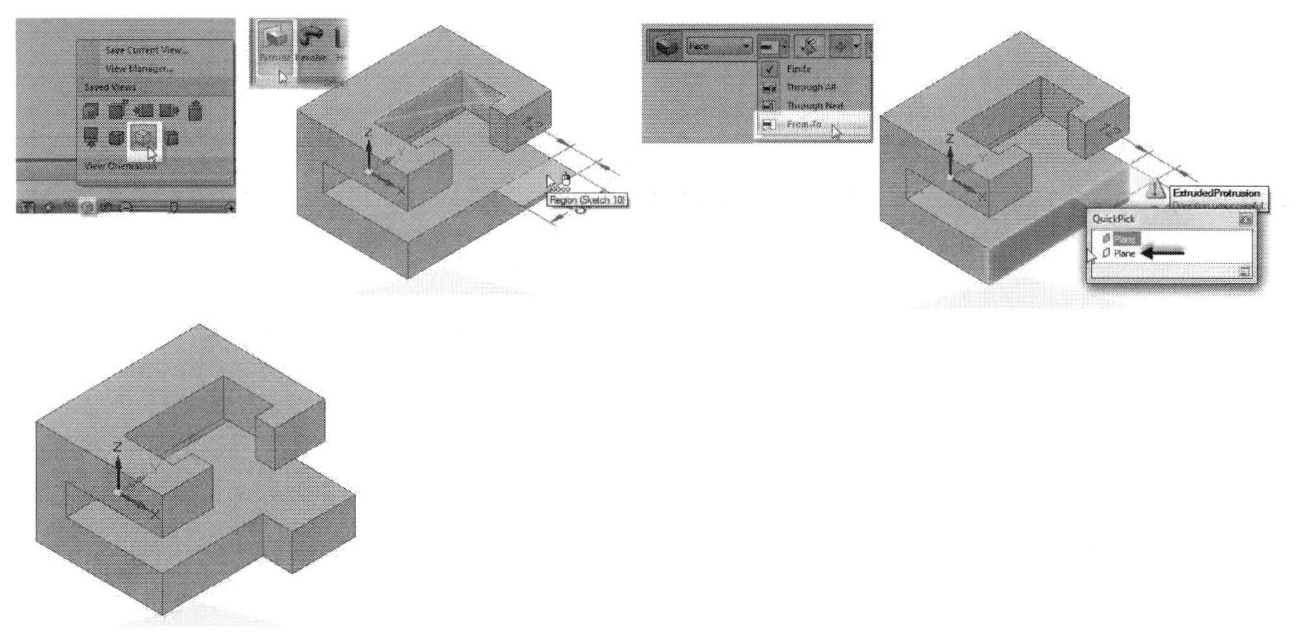

39. Save and close the file.

Example 2 (Inches)

In this example, you will create the part shown below.

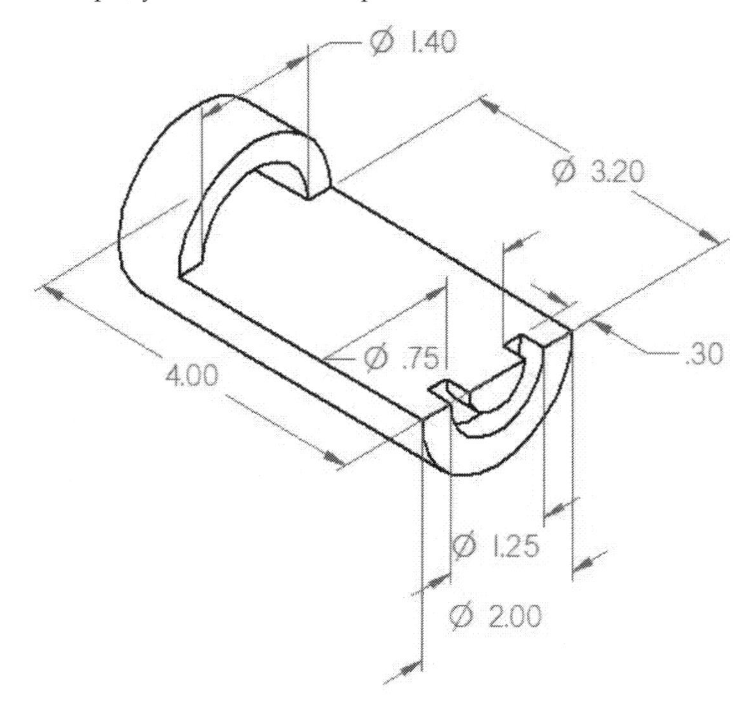

1. Start **Solid Edge 2019**.
2. On the **Quick Access Toolbar**, click **New**; the **New** dialog appears.
3. On the **New** dialog, click **Standard Templates > ANSI Inch** and select the **ansi inch part.par** template. Click **OK** to start a new part file.
4. Draw a sketch on the XZ plane, as shown below.

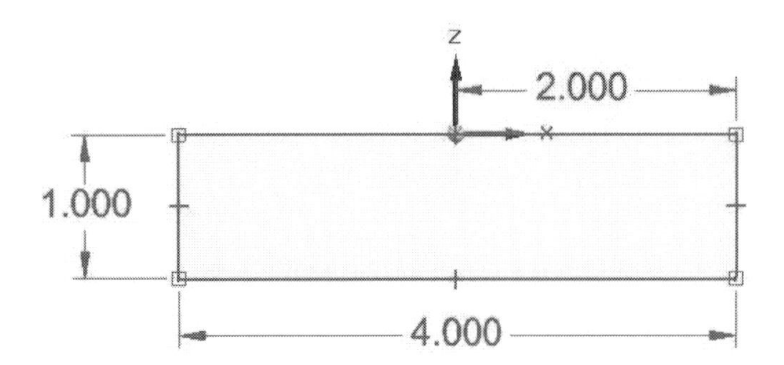

5. Unlock the sketch plane and click the **Home** icon below the Quick View Cube.

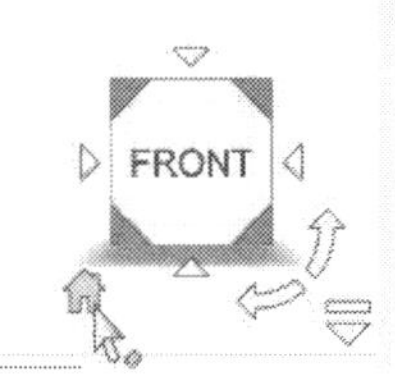

6. Activate the **Select** command and click inside the region enclosed by the sketch.
7. On the command bar, click **Extrude > Revolve**; the *Extrude* handle is replaced by the *Revolve* handle.
8. Click the spear of the *Revolve* handle, and then drag and align it with the top horizontal line.
9. Click the torus of the *Revolve* handle. Activate the **Symmetric** icon on the command bar.

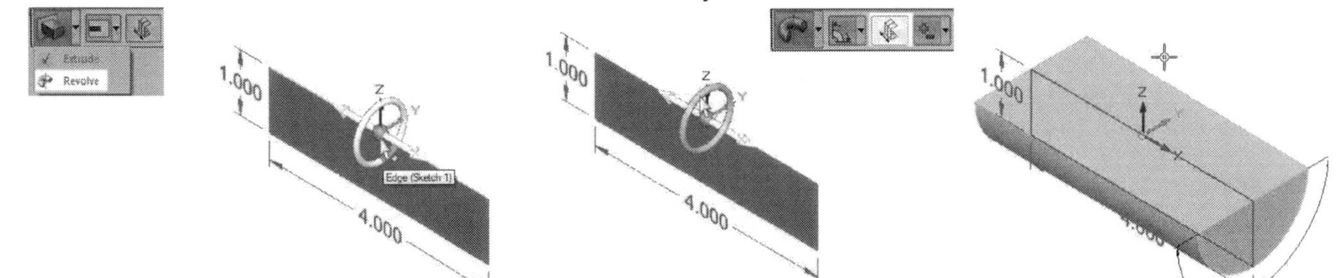

10. Type-in **180** in the box displayed on the model. Press **Enter** and click to create the *Revolve* feature.

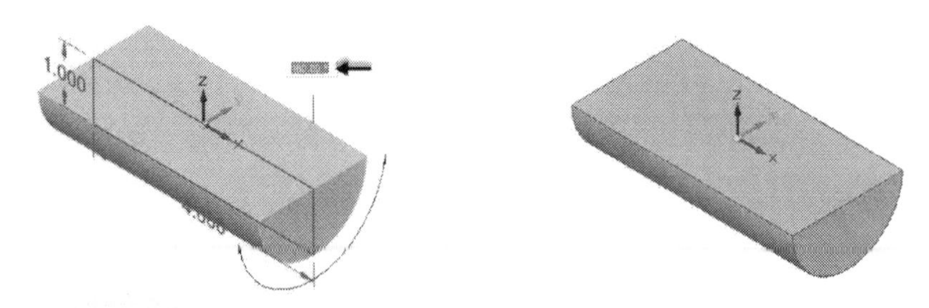

11. Activate the **Line** command and lock the top face of the part geometry.
12. Draw the sketch on top face and apply dimensions.

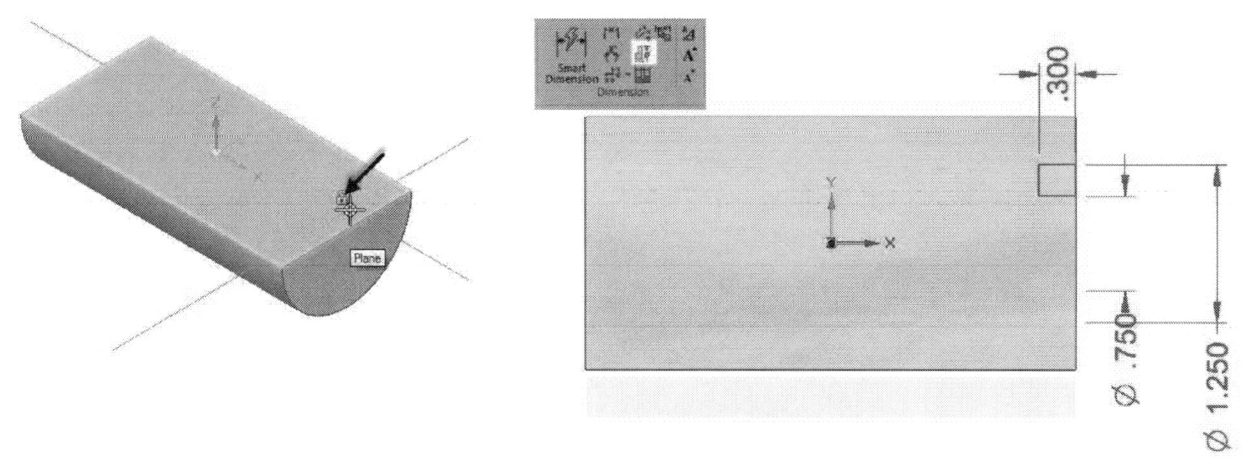

13. Unlock the sketch plane and change the model orientation to ISO.
14. Click **Home > Solids > Revolve** on the ribbon. Click inside the region enclosed by the sketch, and right-click to accept the selection.
15. Click on the X-axis to define the axis of the revolution.
16. On the command bar, deactivate the **Symmetric** icon.
17. Move the pointer and type-in **180** in the box displayed on the model.
18. Move the pointer downward and click to create the revolved cut.

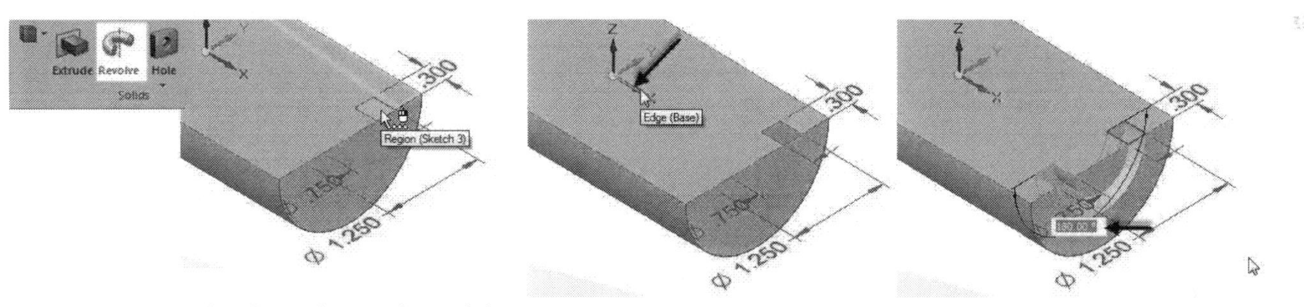

19. Draw a sketch on the top face of the part geometry.
20. Revolve the sketch to create the third feature.

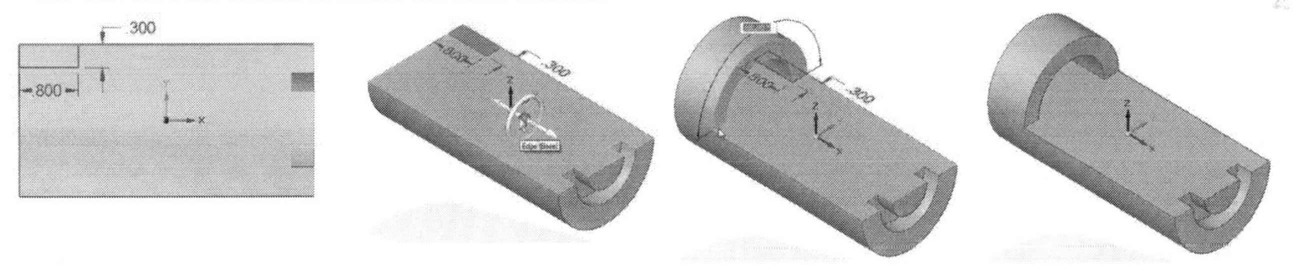

21. Save and close the file.

Questions

1. List the two methods to create *Extrude* features in Synchronous mode.
2. List the two methods to create *Revolve* features in Synchronous mode.
3. How do you create parallel planes in Synchronous mode?
4. List the three options to extrude sketches containing internal loops.

5. What are the treatment options available on the **Extrude** command bar?
6. List the four extent types available on the **Extrude** command bar.

Exercises

Exercise 1 (Millimetres)

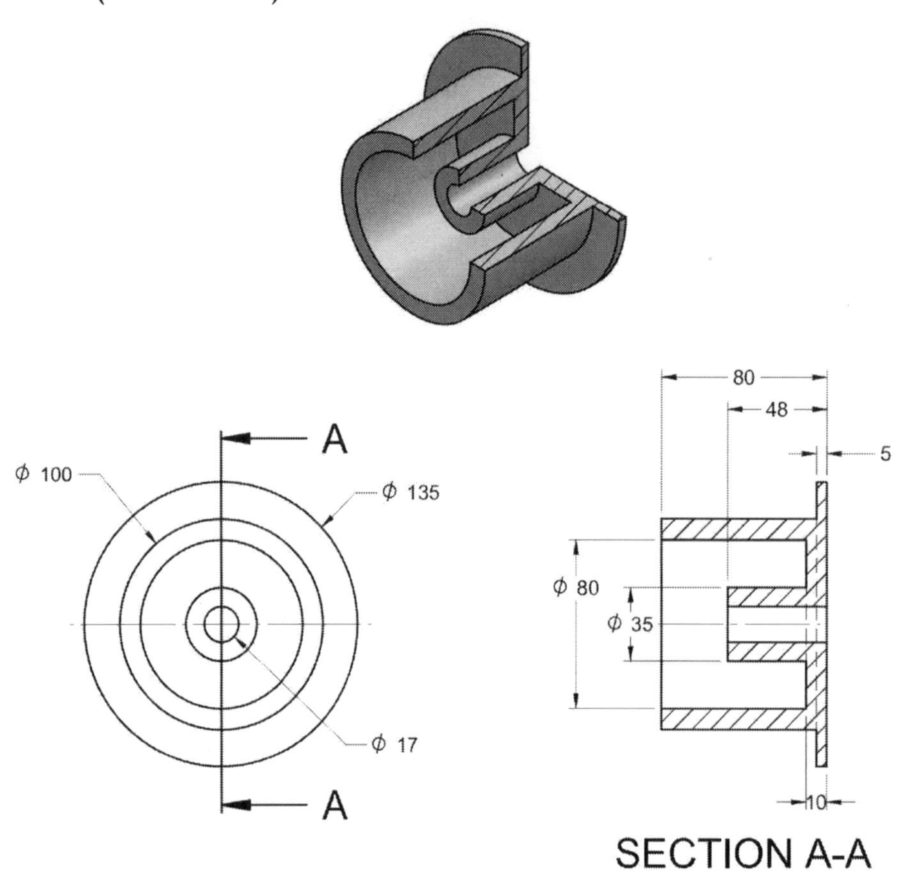

Exercise 2 (Inches)

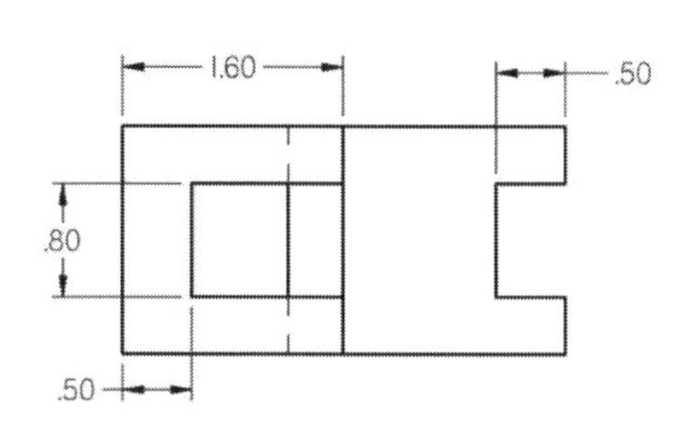

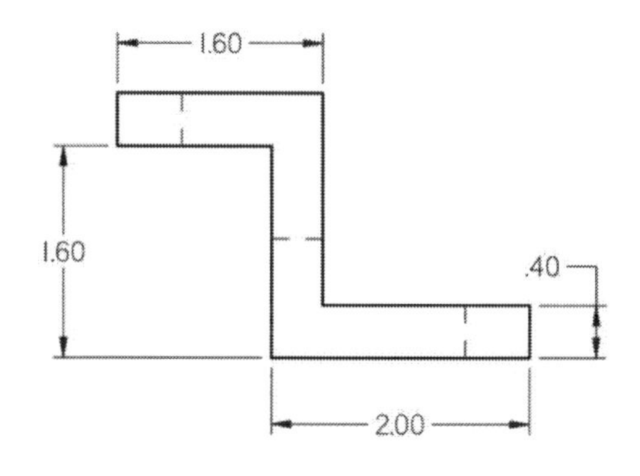

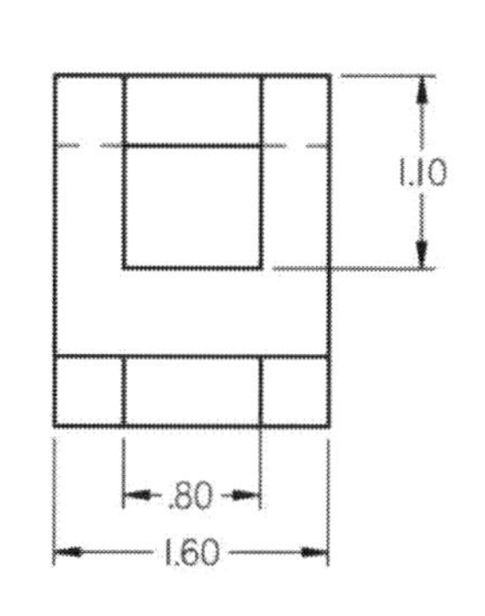

Exercise 3 (Millimetres)

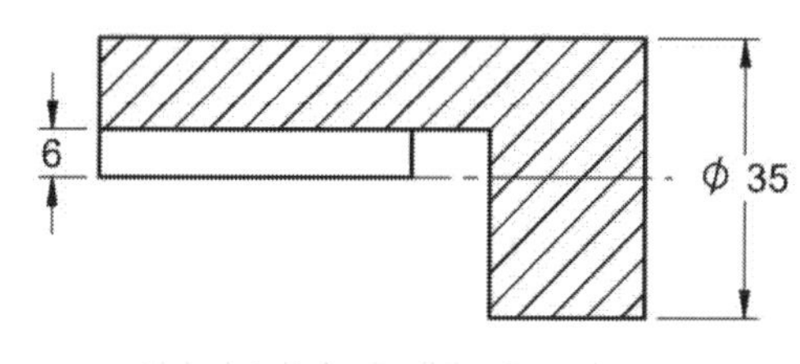

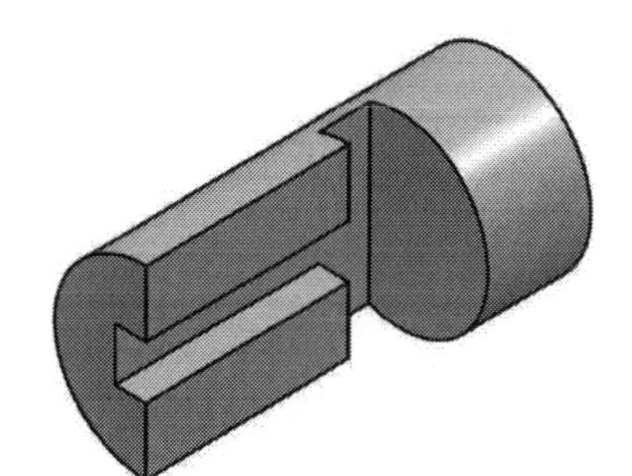

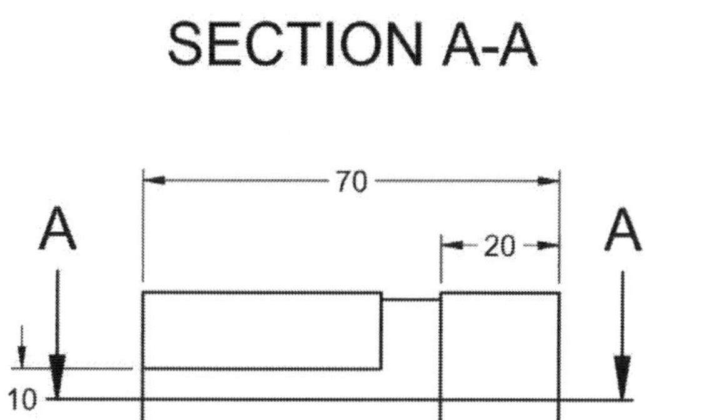

Chapter 4: Placed Features

So far, all of the features that were covered in previous chapters were based on two-dimensional sketches. However, there are certain features in Solid Edge that do not require a sketch at all. Features that do not require a sketch are called placed features. You can simply place them on your models. You must have some existing geometry to create placed features. Unlike a sketch-based feature, you cannot use a placed feature as a first feature of a model. For example, in order to create a *Fillet* feature, you must have an already existing edge. Now, you will learn how to add placed features to your design.

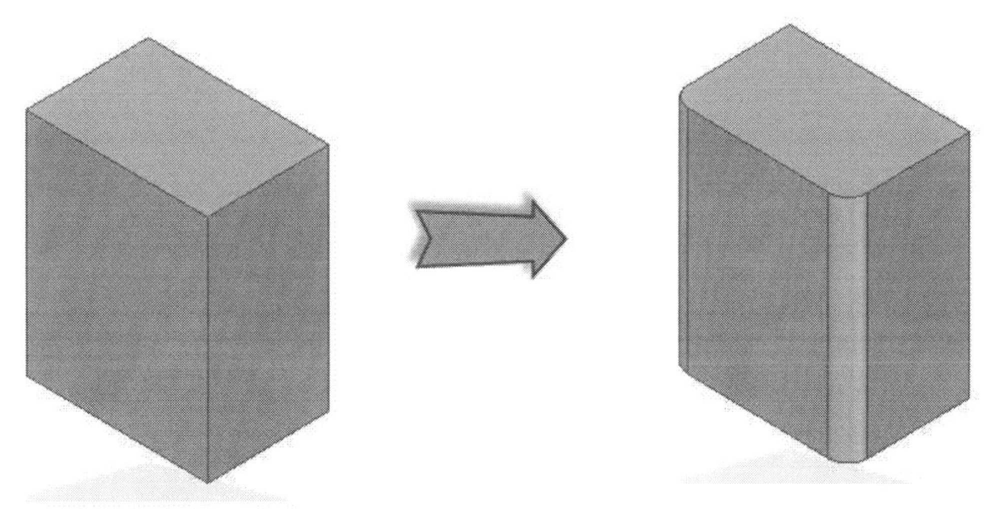

The topics covered in this chapter are:

- *Holes*
- *Threads*
- *Slots*
- *Rounds* and *Blends*
- *Chamfers*
- *Drafts*
- *Shells*

Hole

You know it is possible to use the *Extrude* command to create cuts and remove material. But, if you want to drill holes that are of standard sizes, the **Hole** command is a better way to do this. The reason for this is it has many hole types already predefined for you. All you have to do is choose the correct hole type and size. The other benefit is when you are going to create a 2D drawing, Solid Edge can automatically place the correct hole annotation. Activate this command (Click **Home > Solids > Hole** on the ribbon) and you will notice that a command bar pops up. Click the **Hole Options** icon on the command bar to open the **Hole Options** dialog. The options on this dialog help you to create different types of holes.

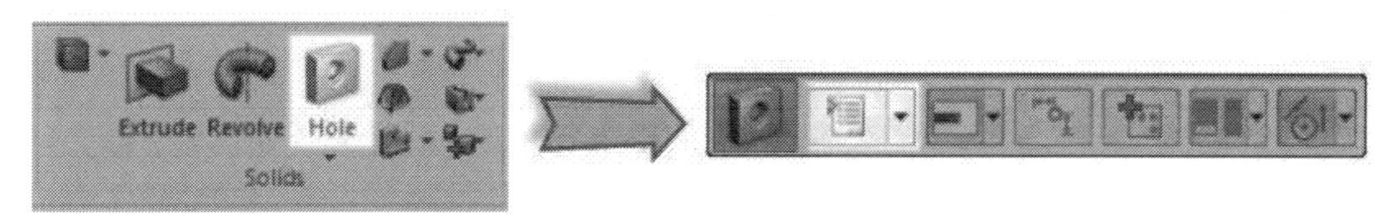

Create a Simple Hole feature

To create a simple hole feature, select the **Simple** button from the top left corner and set the **Standard** of the hole. Set the **Size** of the hole and **Hole Extents** type. If you have selected **Finite Extent**, type-in a value in the **Hole depth** box. If you want a V-bottom hole, check the **V-bottom angle** option and type-in a value in the angle box. If you want to add a chamfer to the hole, check the **Start Chamfer** option. Type-in the chamfer offset and angle values. Click **OK** to close the dialog.

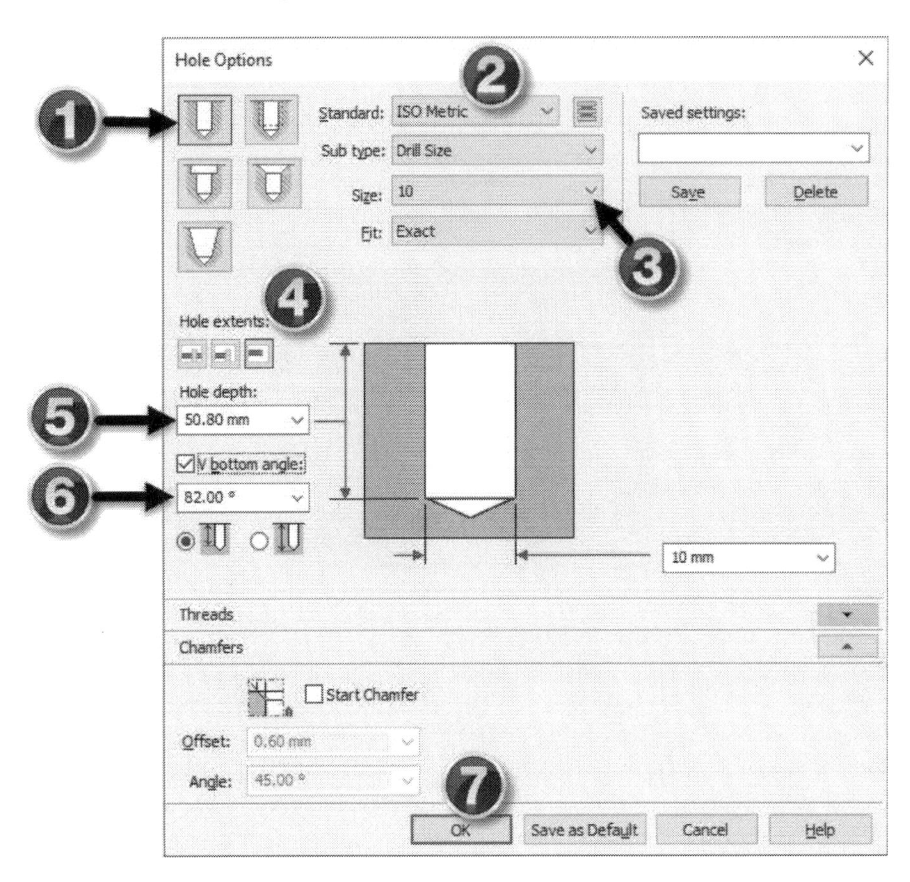

Set the **Keypoints** option on the command bar to **All** and start selecting points from the model. You can select an endpoint of a line, edge or curve (or) center point of an arc or circular edge by placing the pointer on the edge. After placing two holes continuously on a same face, you will notice that the face will be locked and all the future holes will be placed on the locked plane. Click the lock icon on the screen, if you want to unlock the face. After placing the holes, you can use the **Smart Dimension** command to position the hole.

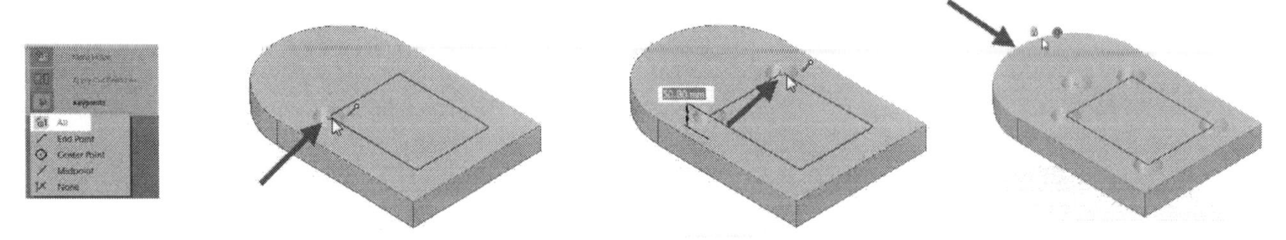

Note: *The hole size in the example should be adjusted based on the model size.*

You can also specify the location of the hole by placing the pointer with the hole preview on the adjacent edge, and then press E. Likewise, place the pointer on the edge perpendicular to previously selected, and then press Enter. Type-in values in the dimensions attached to the hole and press Enter. Press Tab to switch between the dimensions.

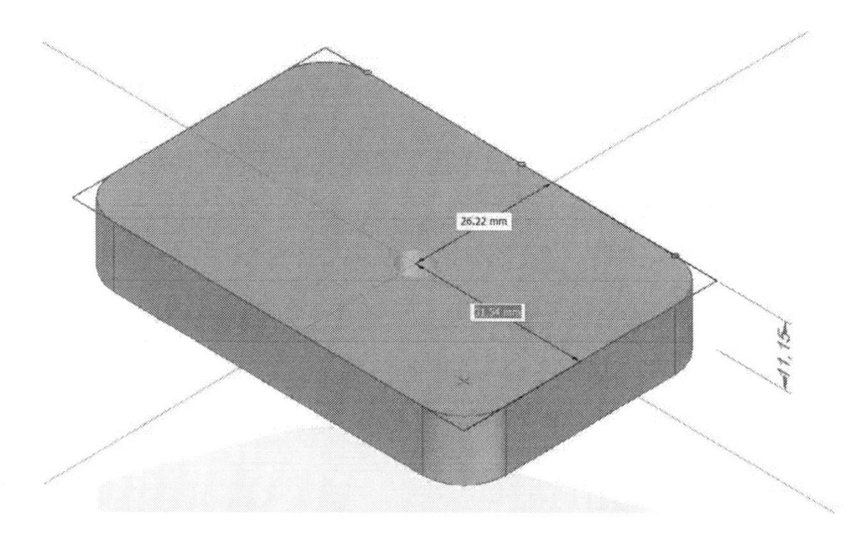

If you want to place a hole on a cylindrical or curve surface, there is an easy technique to do this. Position the pointer on the cylindrical face and press F3 to lock the face. A plane tangent to the face will appear. Drag the pointer and type-in an angle value (or) select a key point to define the plane orientation. Now, place holes on the locked plane.

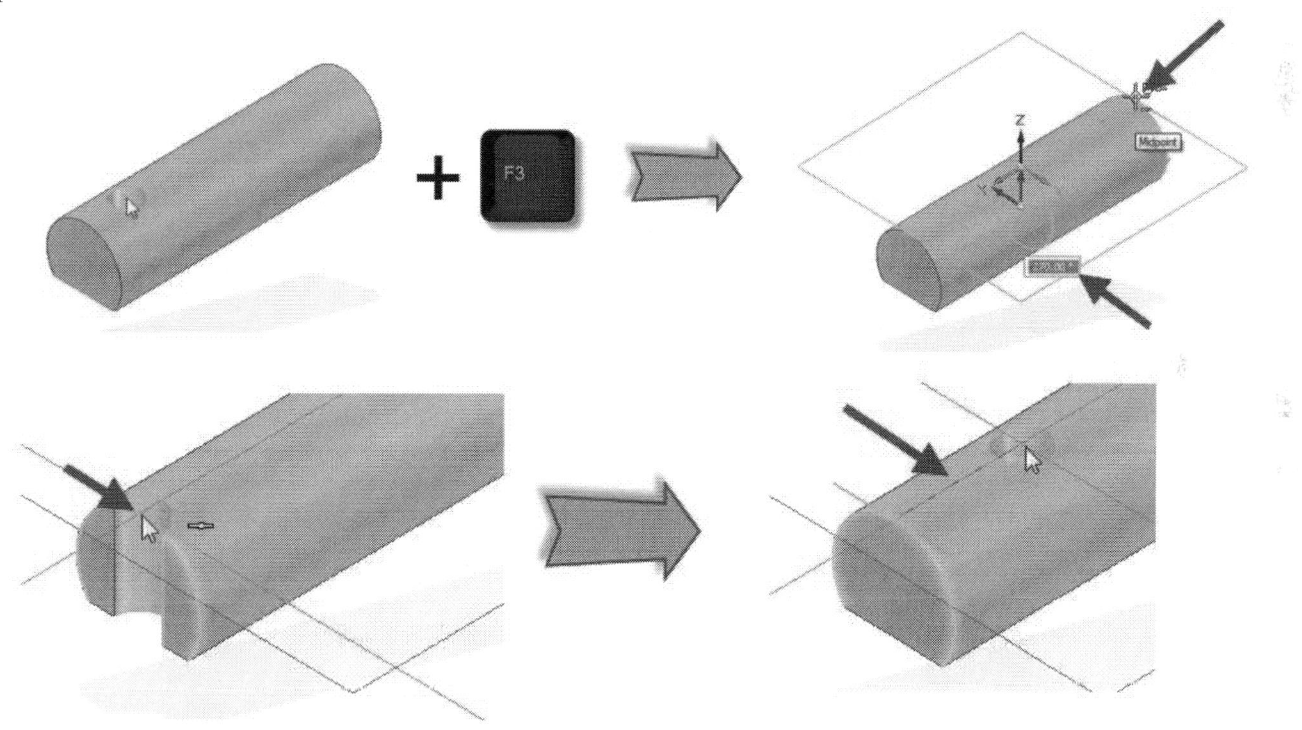

Create a Threaded Hole feature

To create a threaded hole feature, activate the **Hole** command and open the **Hole Options** dialog. On the **Hole Options** dialog, select the **Threaded** button and define the hole **Standard**. Define the other parameters such as sub type, size, and hole extents. Under the **Threads** section, select the **Tap drill diameter**, **Internal diameter**, or **Nominal diameter** option. For example, if you select the **Nominal diameter** option, the hole size on the geometry will be equal to the nominal diameter of the thread. Next, define the **Thread extent**. You can define it up to the hole extent or by entering a value. Specify the thread pitch value and click **OK** on the dialog and place the hole.

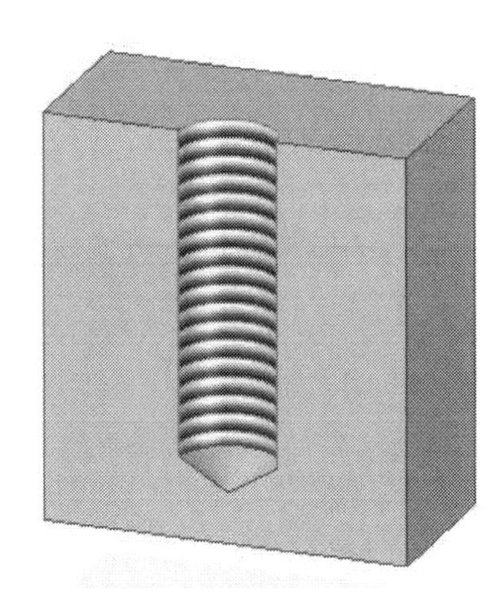

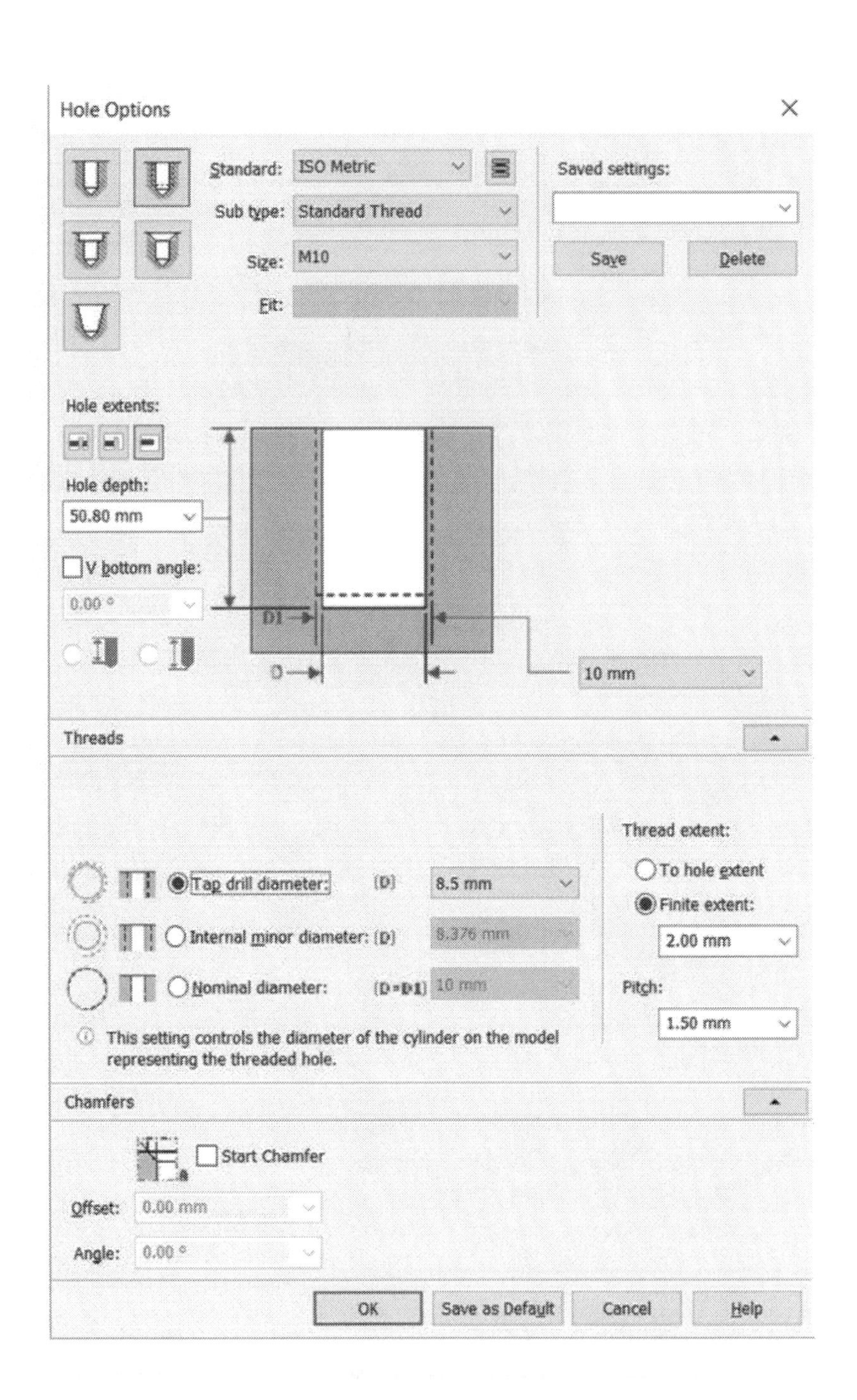

Create a Tapered Hole feature

Tapering is the process of decreasing the hole diameter toward one end. A tapered hole has a smaller diameter at the bottom. To create a tapered hole, select the **Tapered** button on the **Hole Options** dialog. Next, select the option to define the bottom diameter or top diameter. Type-in a value in the **Hole Diameter** box, and then define the taper ratio. The taper ratio is the rate of decrease in the diameter for a specific length. You can define the taper by using the **Decimal (R/L**), **Ratio (R:L),** or simply enter the taper angle in the **Angle** box. After defining the taper, specify the hole depth and end condition in the **Hole Extents** section. Click **OK** and place the hole feature.

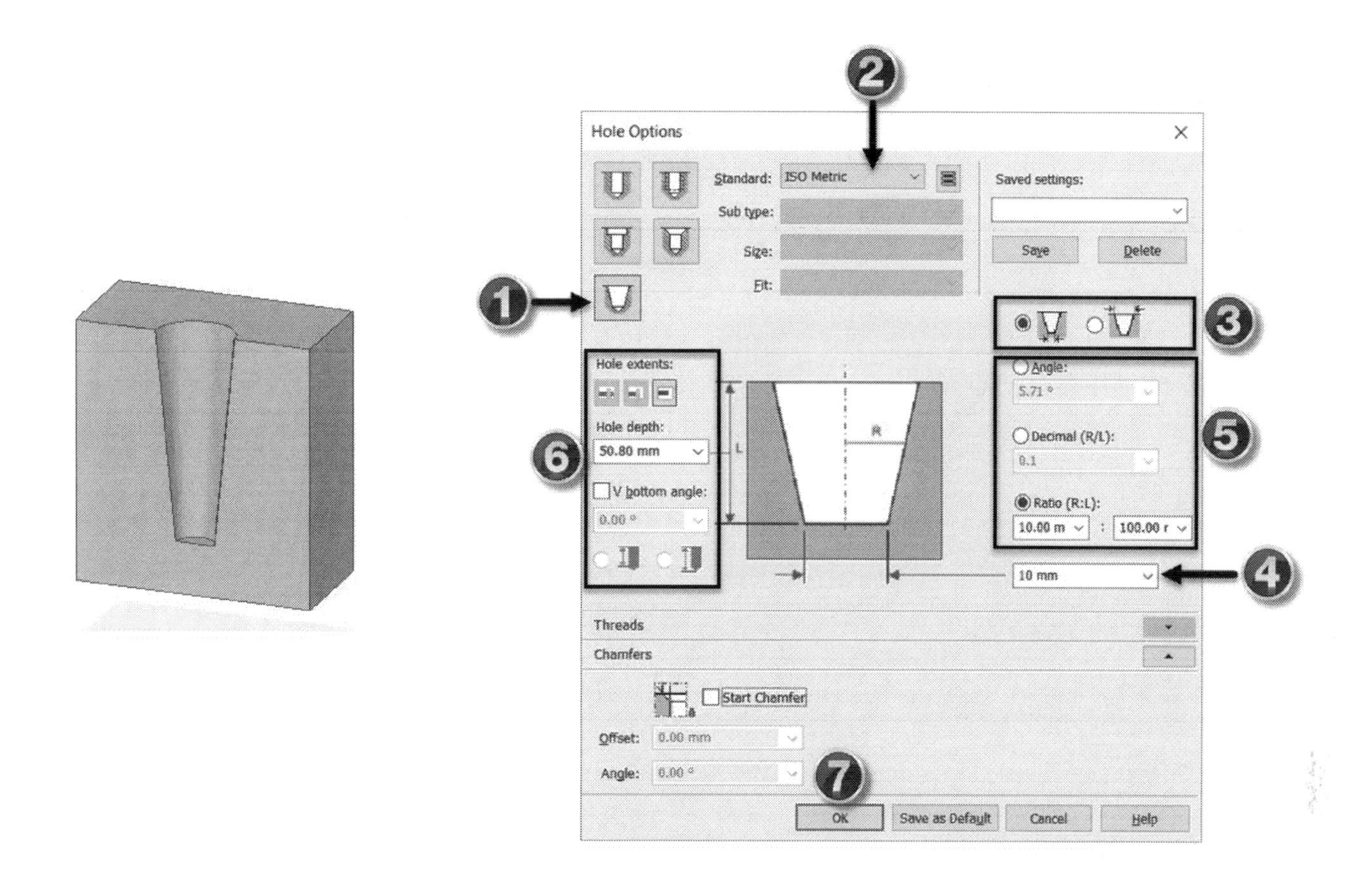

Create a Counterbore Hole feature

A counterbore hole is a large diameter hole added at the opening of another hole. This counterbore hole is used to accommodate a fastener below the level of workpiece surface. The three types of counterbore holes that can be created in Solid Edge are shown in figure.

To create a counterbore hole, select the **Counterbore** button on the **Hole Options** dialog. Next, define the counterbore sub type, hole size, fit, counterbore diameter, and counterbore depth. Check the **Neck Chamfer** option under the **Chamfer** section, if you want a V-bottomed counterbore hole. Check the **Thread** option under the **Thread** section and define the thread parameters, if you want to add a thread to the hole. Click **OK** and place the hole.

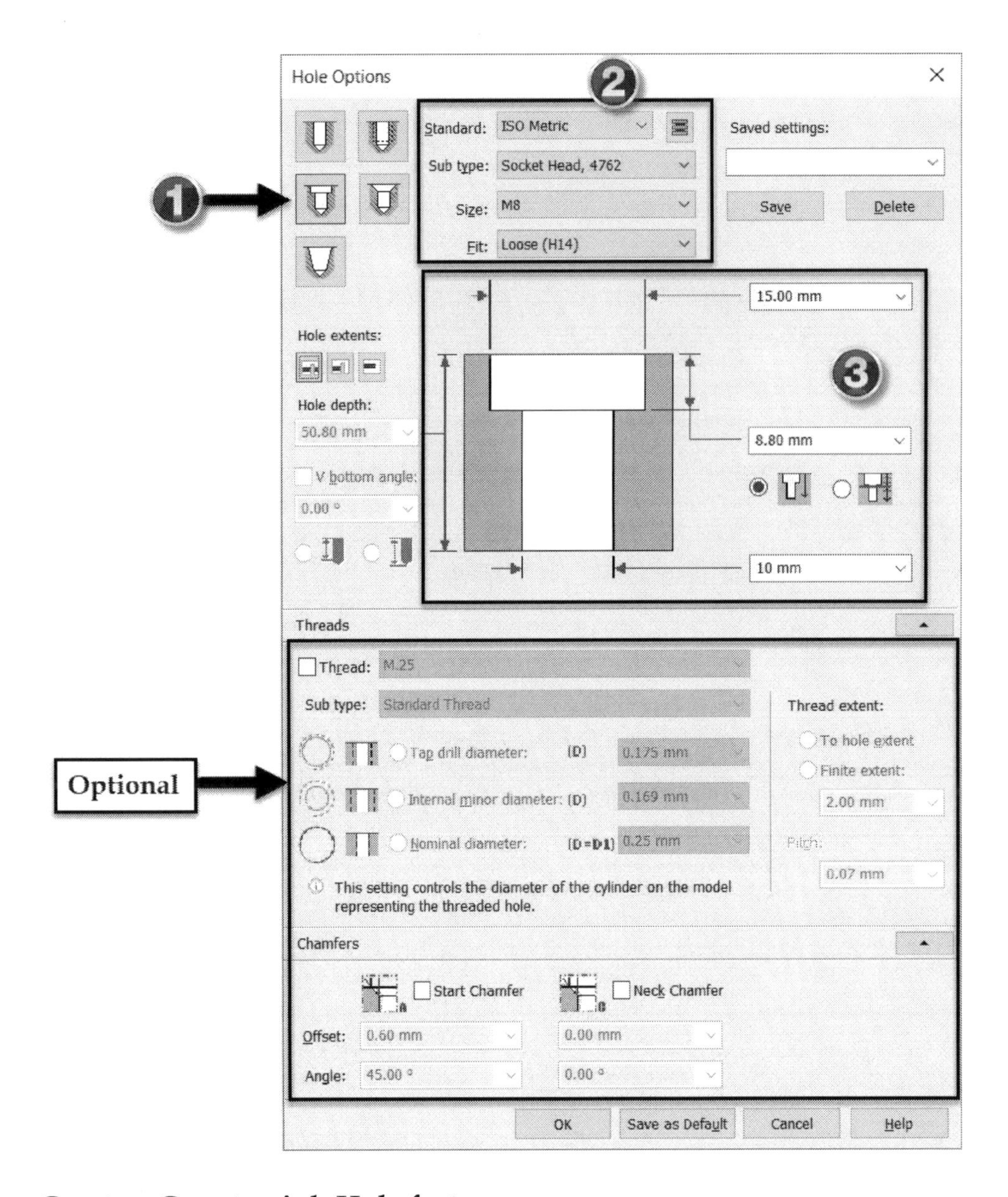

Create a Countersink Hole feature

A countersink hole has an enlarged V-shaped opening to accommodate a fastener below the level of work piece surface. To create a countersink hole, select the **Countersink** button on the **Hole Options** dialog. Type-in values in the **Diameter**, **Countersink diameter**, and **Countersink angle** boxes. You can also check the **Head clearance** option, if you want to provide head clearance. Set the hole depth and end condition in the **Hole Extents** section. Click **OK** and place the hole.

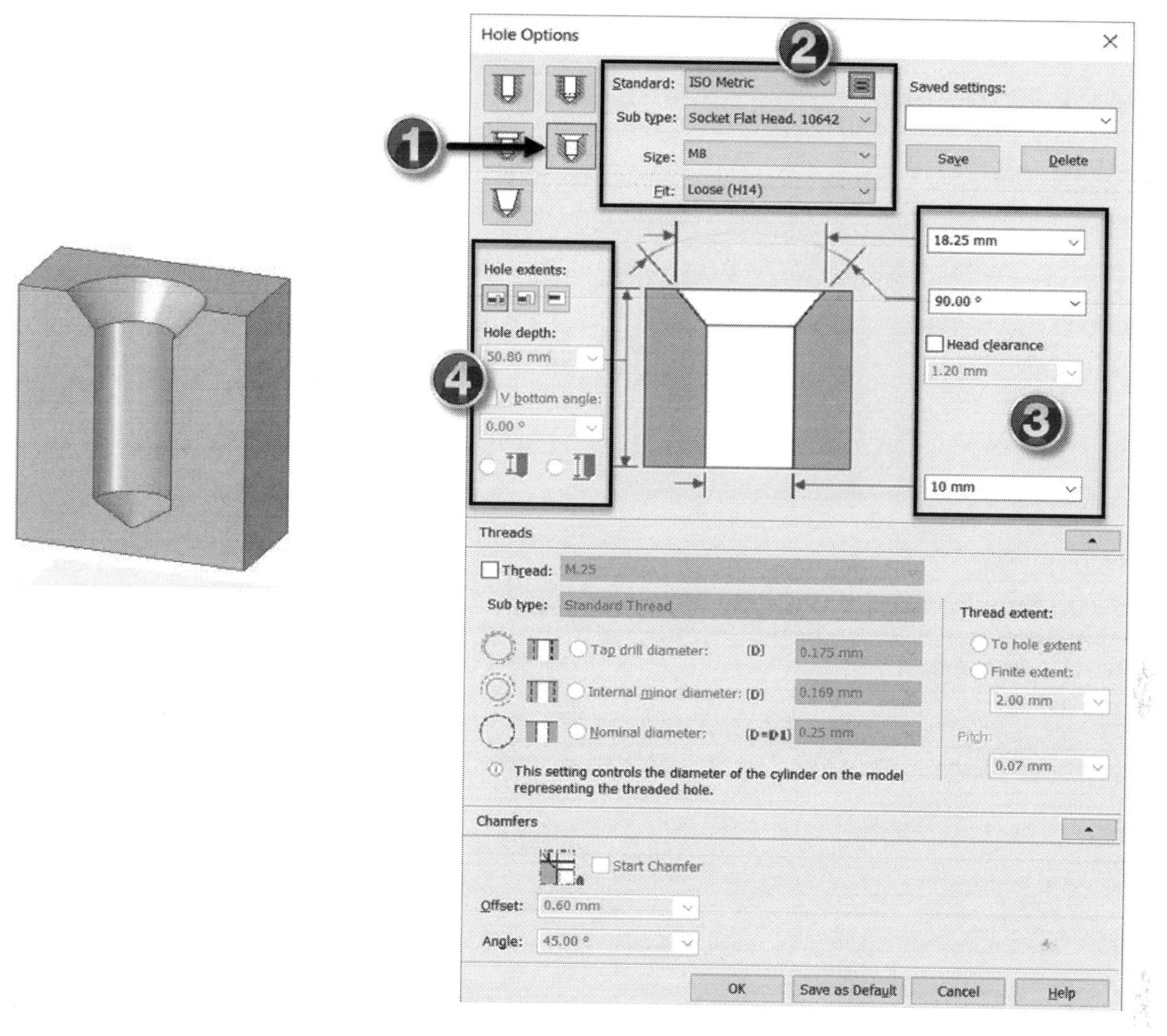

Modify Holes

After placing holes, you may be required to modify them or add more holes to the set. To modify a hole, you must select it and click on the hole diameter. A box appears with the hole parameters. Change the hole parameters by entering new values in the box. You can use the command bar options to change the hole type. Click and drag the arrows displayed on the holes to change the location of the hole.

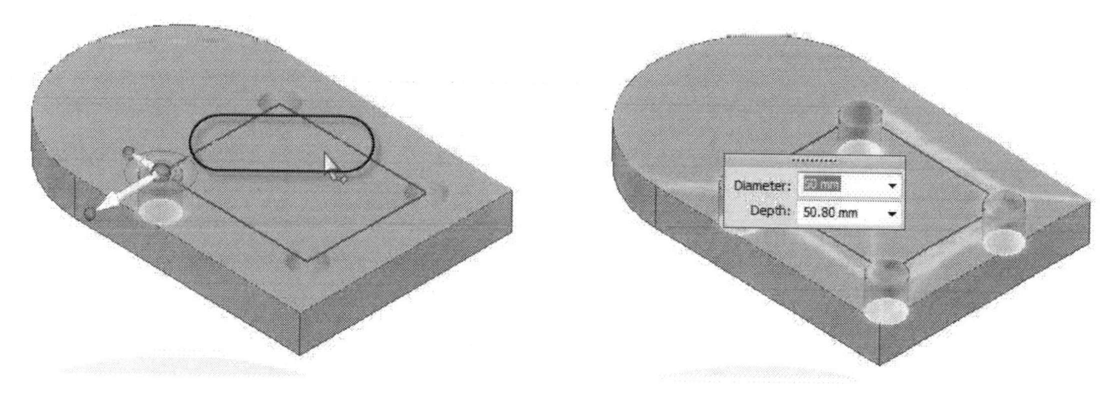

You will notice that, all the holes that are placed at a time are grouped under one set in the Pathfinder. If you modify one hole in the set, all the other holes will also be modified. Use the **More holes** option on the command bar to add more holes to the hole set. If you want to remove a hole from the set, click the right mouse button on it in the Pathfinder and select **Separate**. The hole will be separated.

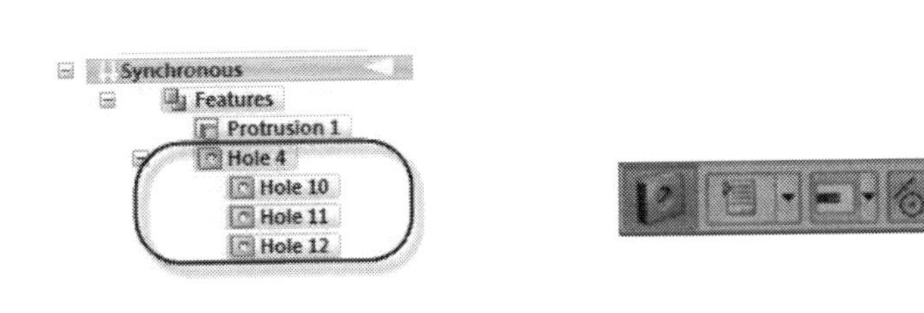

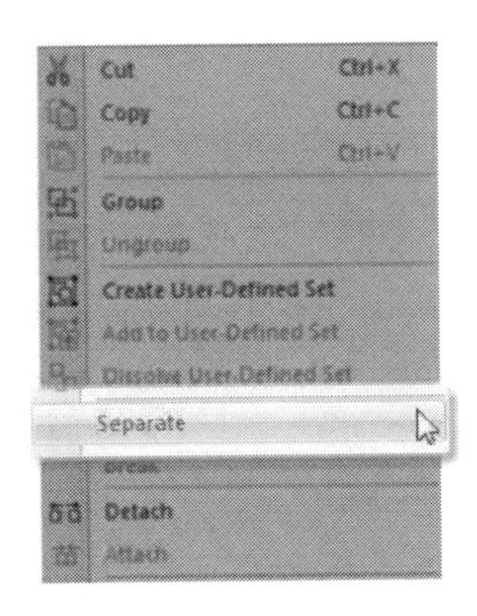

Recognize Holes

This command converts the cylindrical features created by using the cutting operation into the **Hole** features. This command is also helpful to convert the cylindrical cut features in the imported geometry into holes. Activate this command (click **Home > Solids > Hole > Recognize Holes** on the ribbon). The **Hole Recognition** dialog pops up and displays all the cylindrical cut features that are recognized as holes. As you place the pointer over the holes in the dialog, they will be highlighted in the model. Click the **Hole Options** buttons on the dialog to open the **Hole Options** dialog of individual holes. Change the hole type and diameter (if required) in this dialog and click **OK**. Uncheck the **Recognize** options, if you do not want to recognize the holes. Click **OK** on the **Hole Recognition** dialog to convert the cut features into holes.

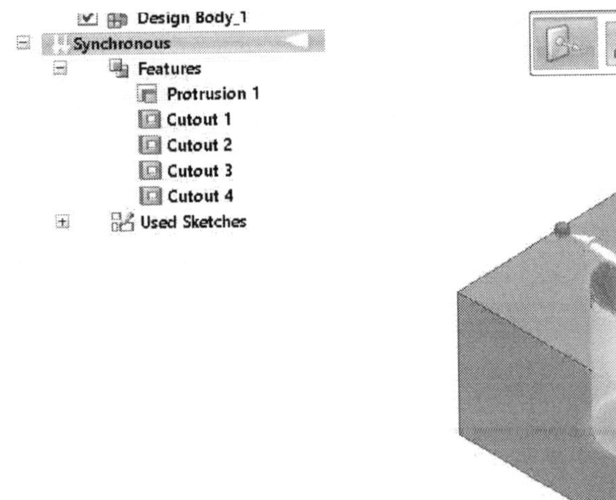

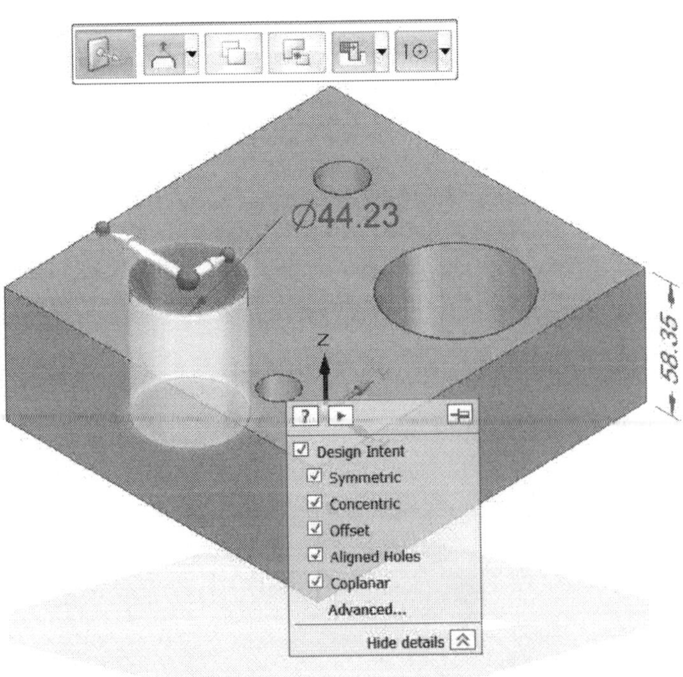

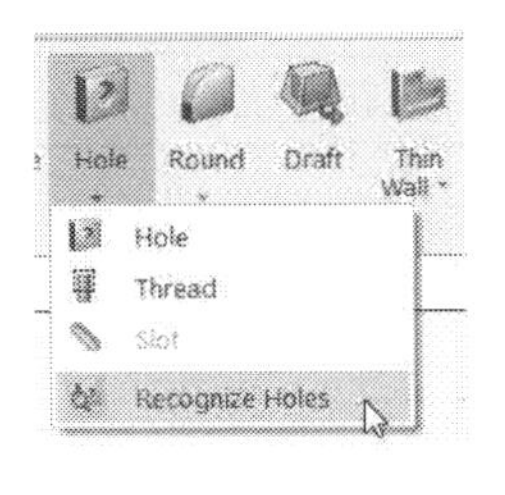

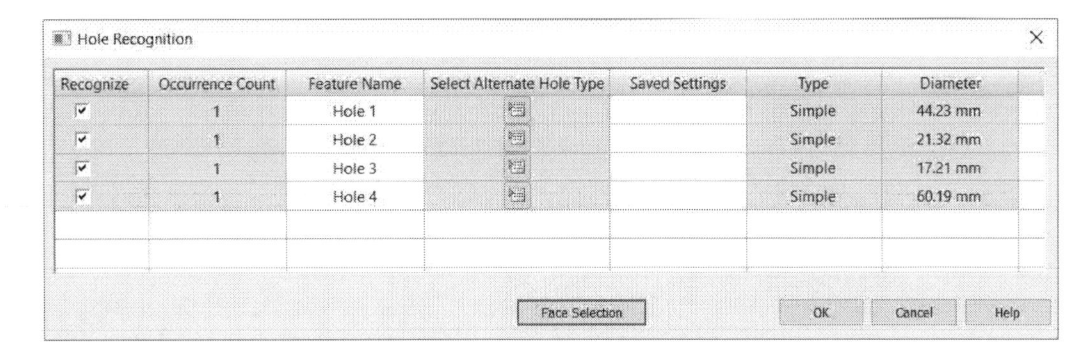

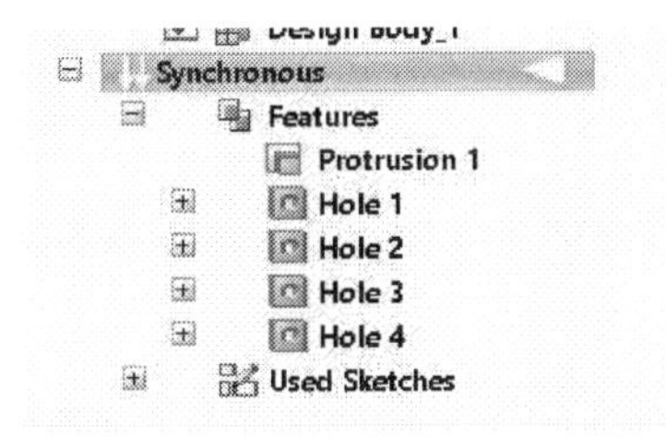

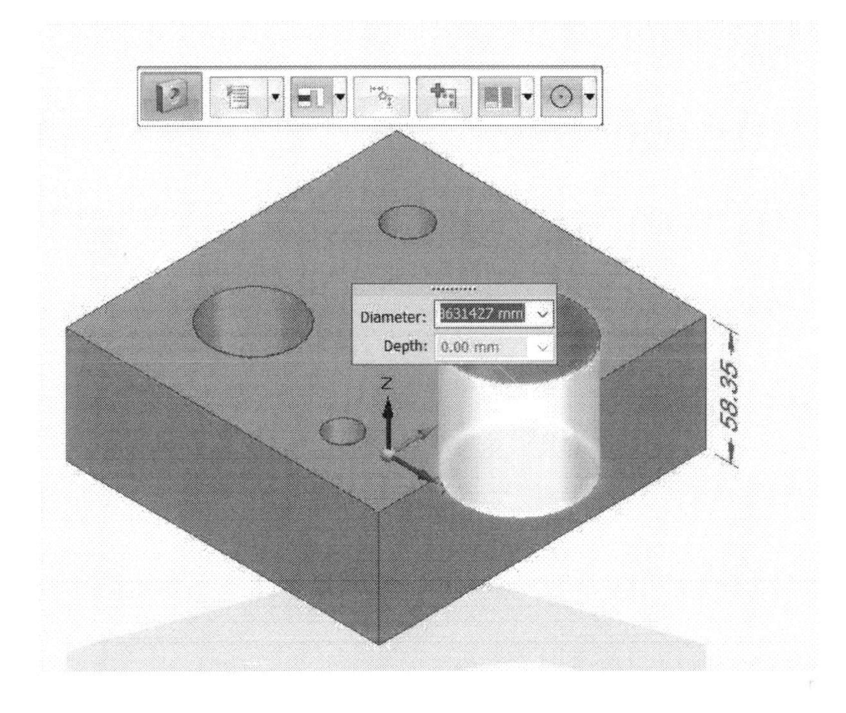

Thread

This command adds a reference thread feature to a cylindrical face. The thread features are added to a 3D geometry so that when you create a 2D drawing, Solid Edge can automatically place the correct thread annotation. Activate this command (click **Home > Solids > Hole > Thread** on the ribbon) and click the **Options** icon on the command bar. The **Thread Options** dialog pops up. Set the thread parameters such as type, standard, size, thread diameter, and so on, and then click the **OK** button. Set the **Extent Type** on the command bar and select a cylindrical face. The **Change Diameter** message appears. Click **OK** to change the diameter of the cylindrical face to suit the selected thread size. Type-in the thread length and press Enter, if you have set the **Extent Type** to **Finite Value**.

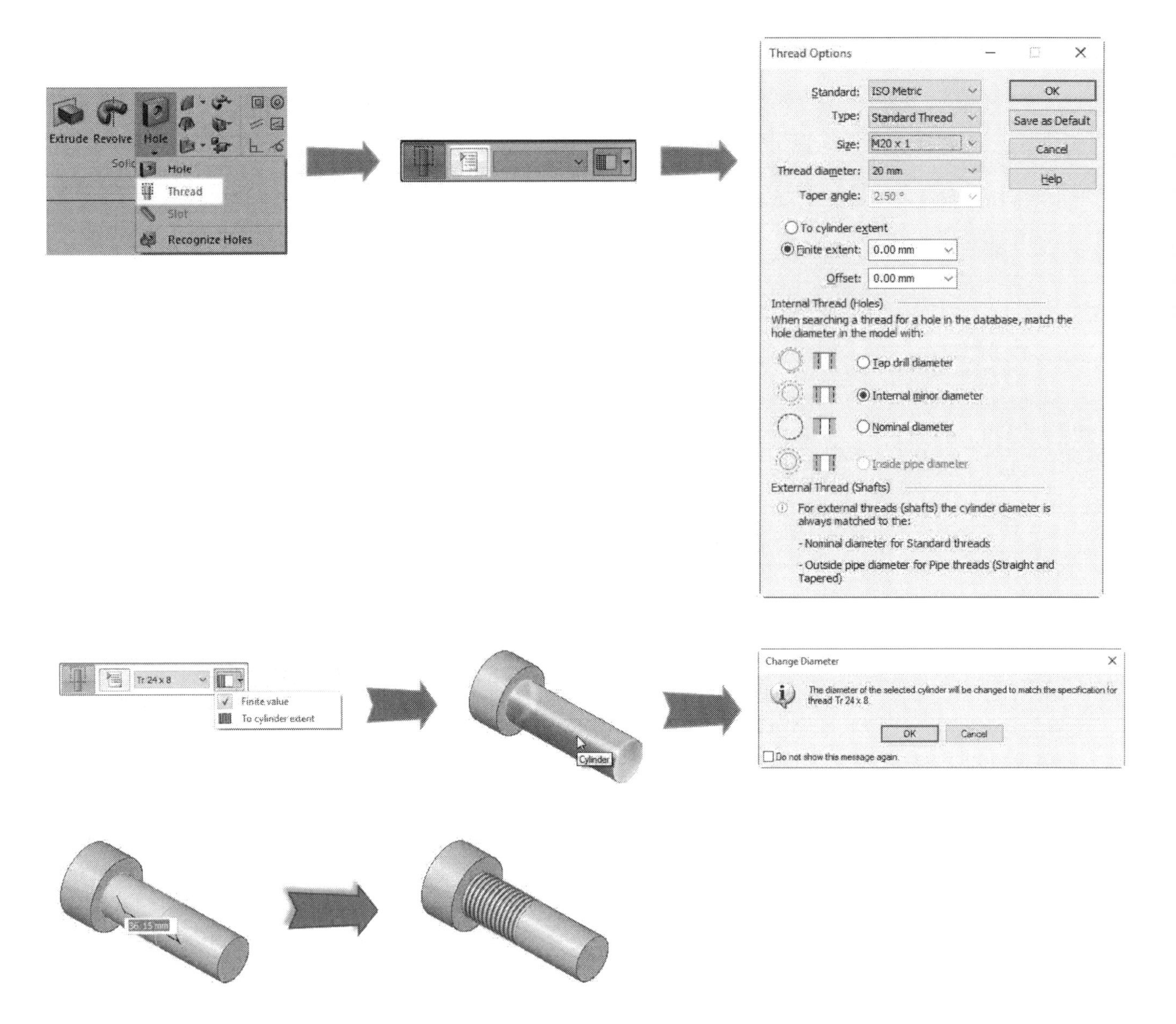

Round

This command breaks the sharp edges of a model and rounds them. It does not need a sketch to create a round. All you need to have is model edges. Activate this command (click **Home > Solids > Round** on the ribbon) and select edges. As you start selecting edges, you will see a preview of the geometry. You can select the edges, which are located at the back of the model without rotating it. By mistake, if you have selected a wrong edge you can deselect it by holding the CTRL key and selecting the edge again. You can change the radius by typing a value in the box displayed on selected edge. As you change the radius, all the selected edges will be updated. This is because they are all part of one instance. If you want the edges to have different radii, you must create rounds in separate instances. Select the required number of edges and right-click to finish this feature. The *Round* feature will be listed in the Pathfinder.

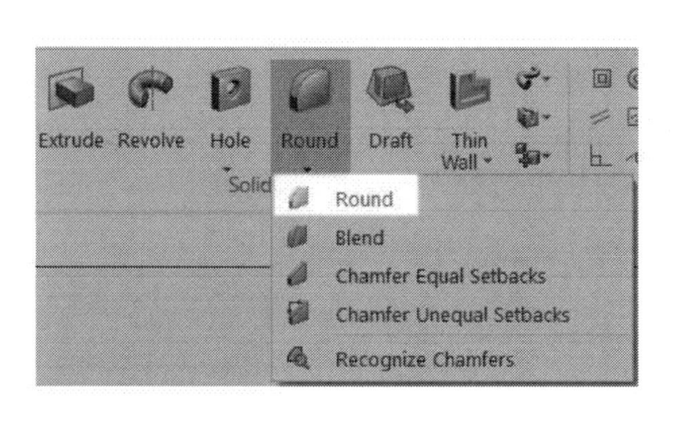

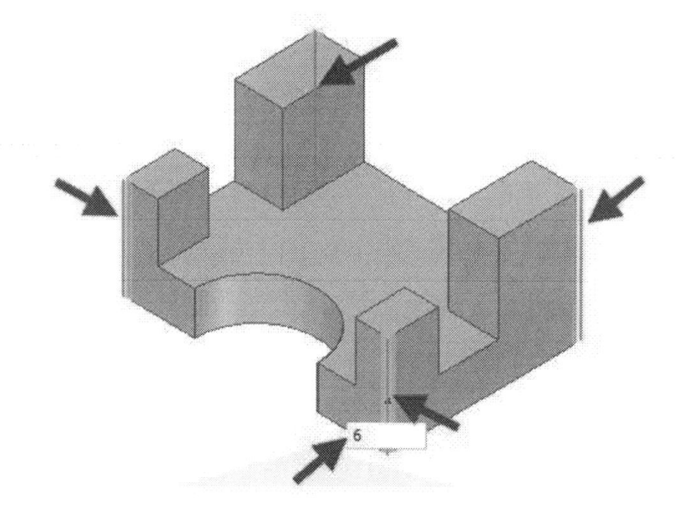

After creating the *Round* feature, the command will be still active so that you can create more *Round* features. Now, if you select the **Loop** option on the command bar, the pointer will be able to select a loop of edges on a face. Select a loop and change the radius. As you press Enter, all of the edges will be rounded.

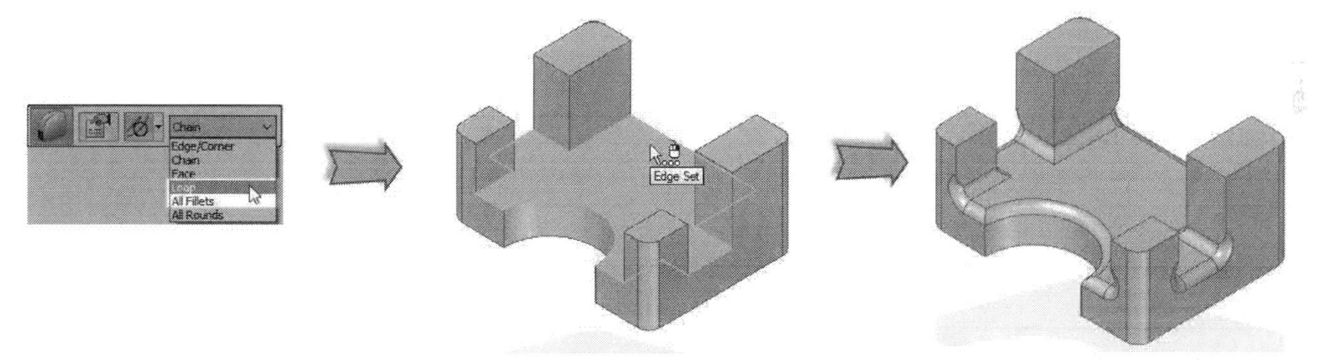

If you select the **All Fillets** option on the command bar, all fillets (concave corners) will be created on the model.

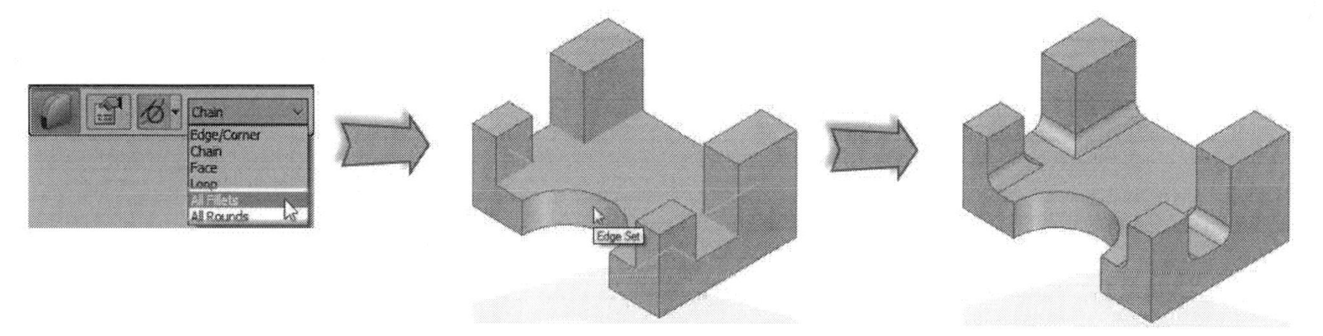

If you select the **All Rounds** option on the command bar, all rounds (convex corners) will be created on the model.

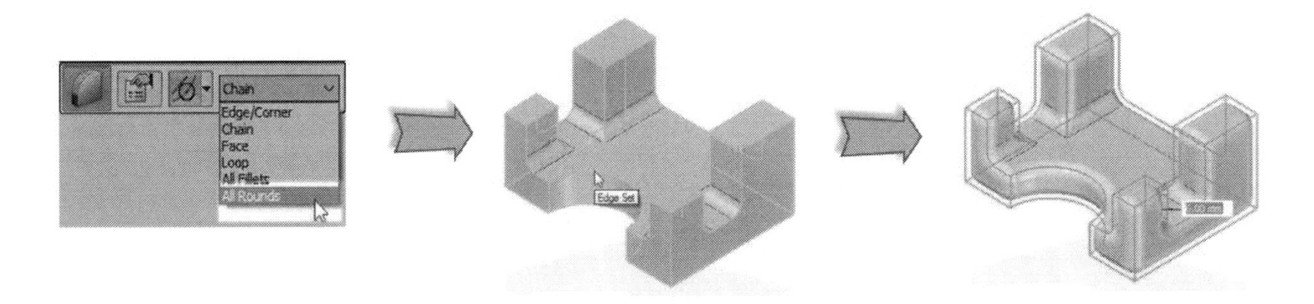

If you ever needed to change the radius of a *Round* feature, select it from the Pathfinder or from model, and then click the diameter value appearing on the feature. Next, type-in a new value in the box that pops up on the *Round* feature and press Enter. To remove a *Round* feature, right-click on it, and then select **Delete**.

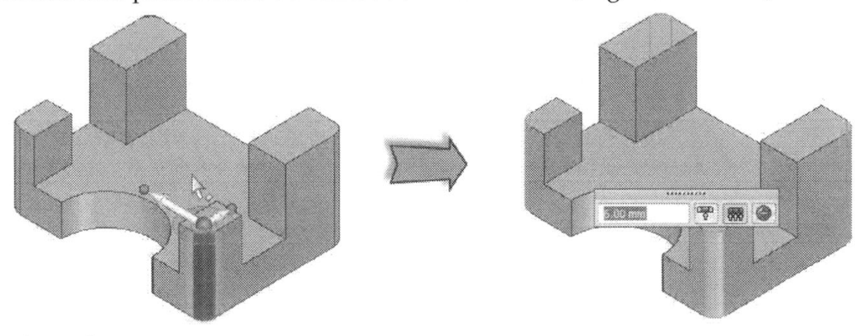

Blend

This command creates a variable radius blend, blend between two faces, and surface blend. These three types of blends are explained next.

Variable Radius Blend

The process to create a variable radius blend is illustrated below.

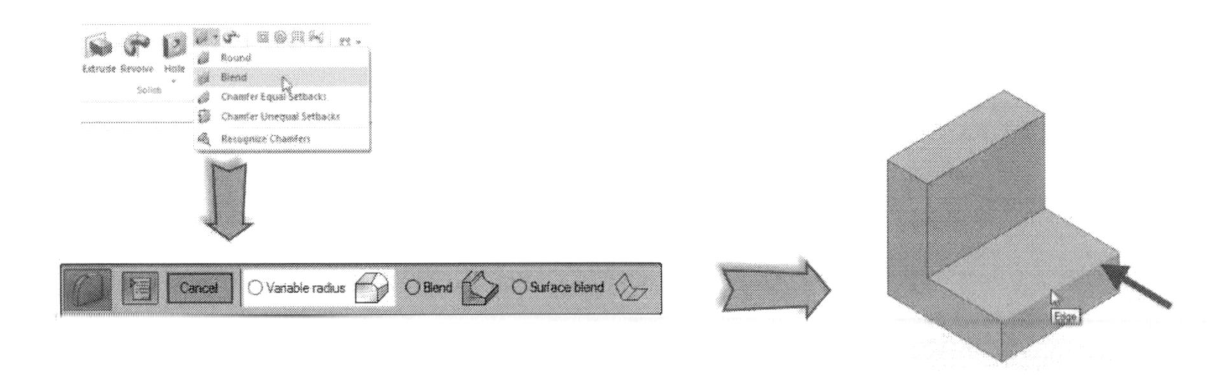

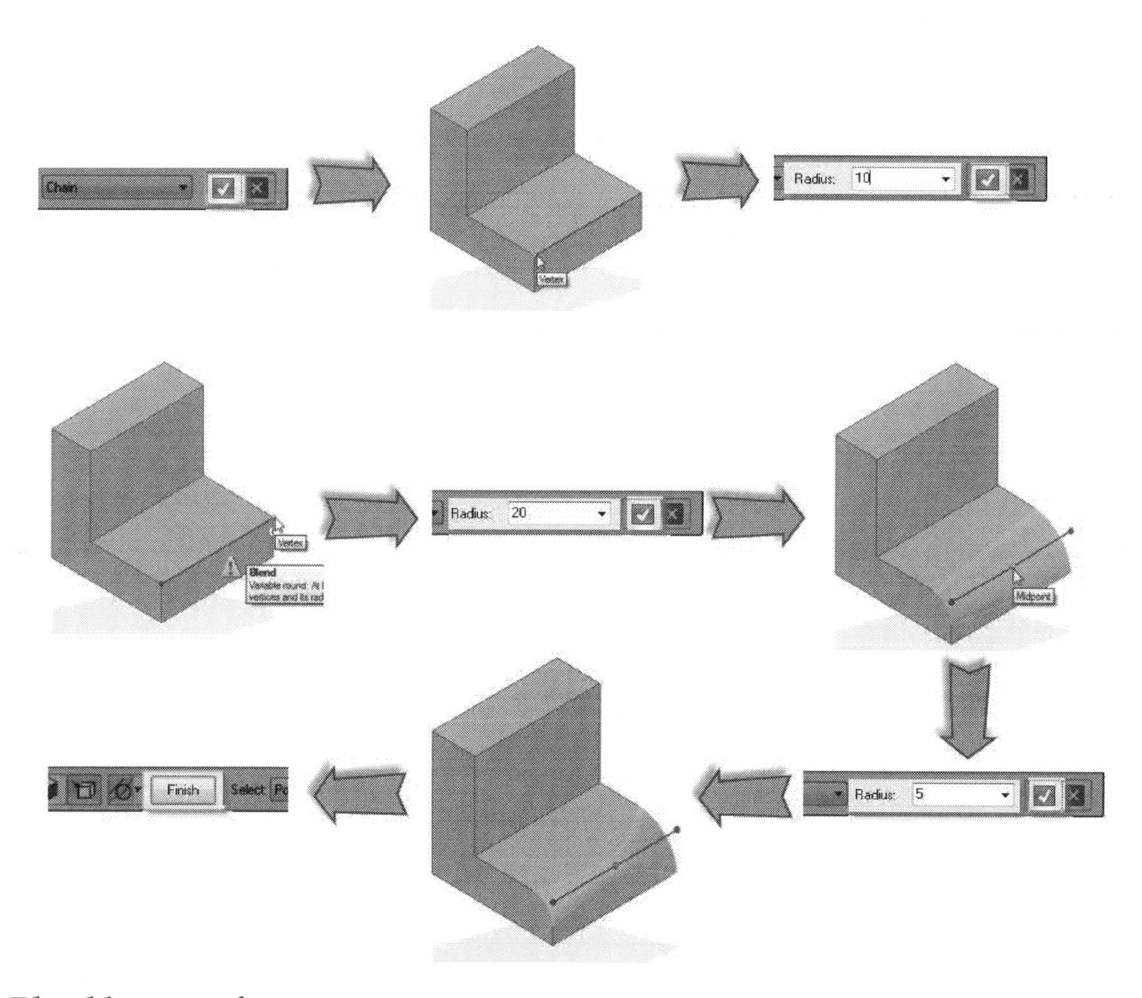

Blend between faces

The process to create a blend between two faces is illustrated in the figure.

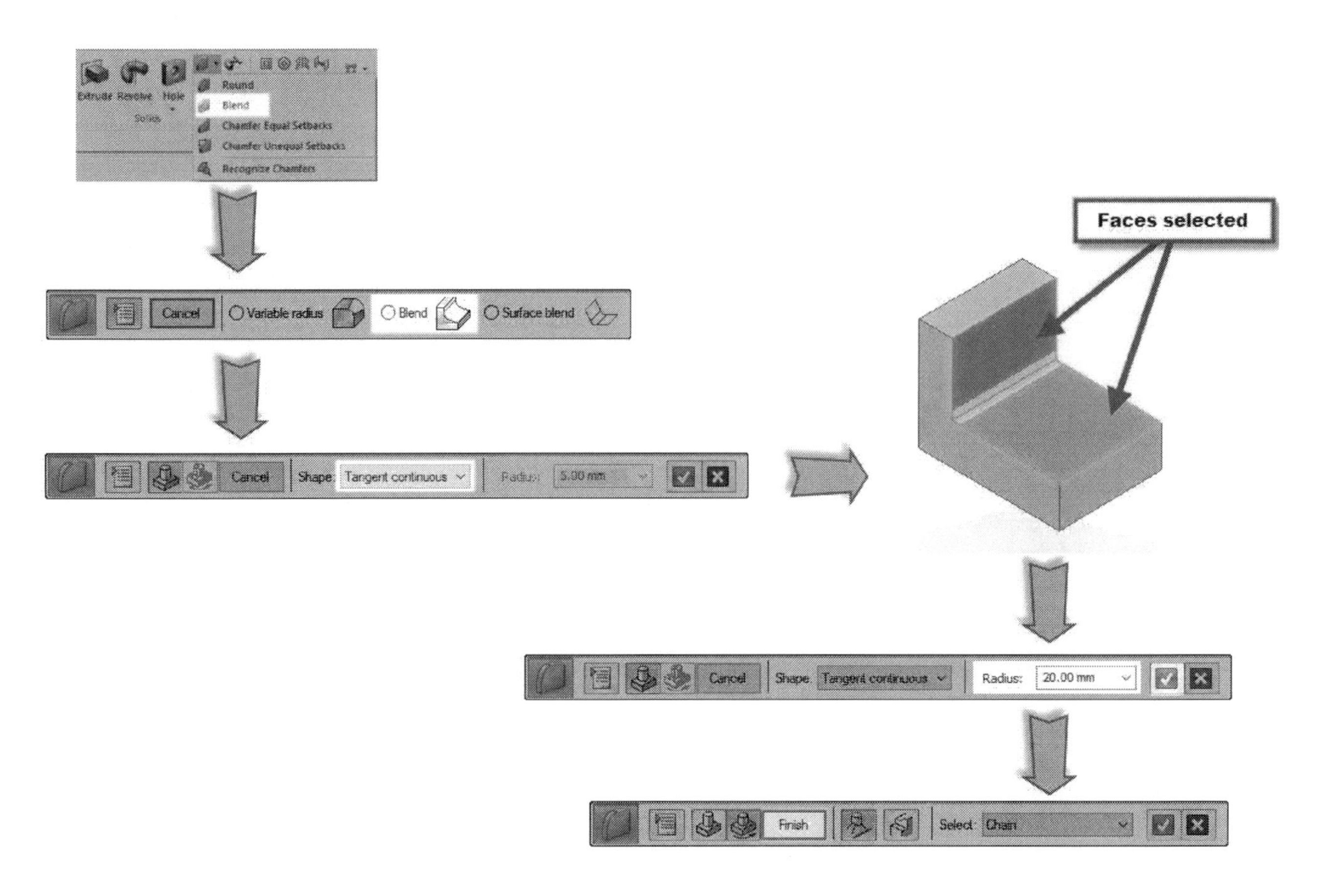

If you want the blend to be tangent to an edge, select the **Tangent Hold Line** option on the command bar and follow the steps given next. The blend will be created tangent to the selected hold line.

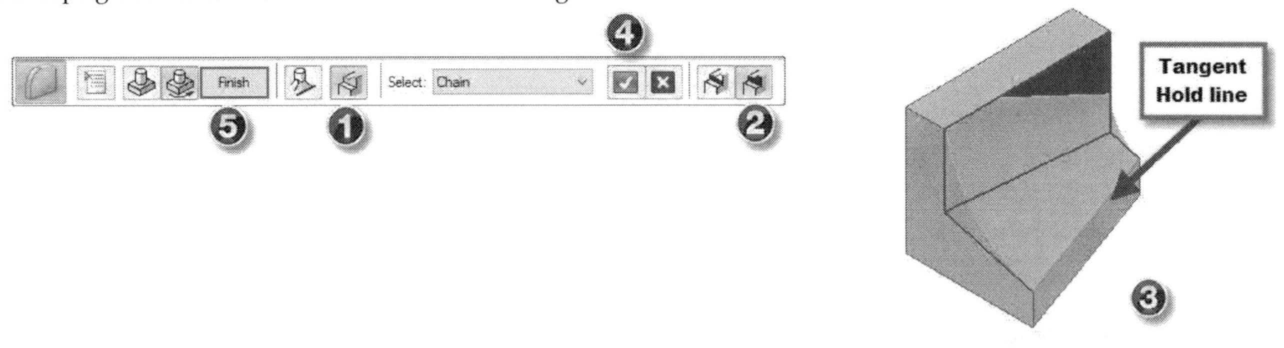

Chamfer Equal Setbacks

The **Chamfer** and **Round** commands are commonly used to break sharp edges. The difference is that the **Chamfer Equal Setbacks** command adds a 45-degree bevel face to the model, whereas the Round command adds a curved face. A chamfer is also a placed feature. Activate this command (click **Home > Round > Chamfer Equal Setbacks** on the ribbon) and select an edge to chamfer. Type-in the distance value in the box attached to the chamfer and press Enter to create the chamfer.

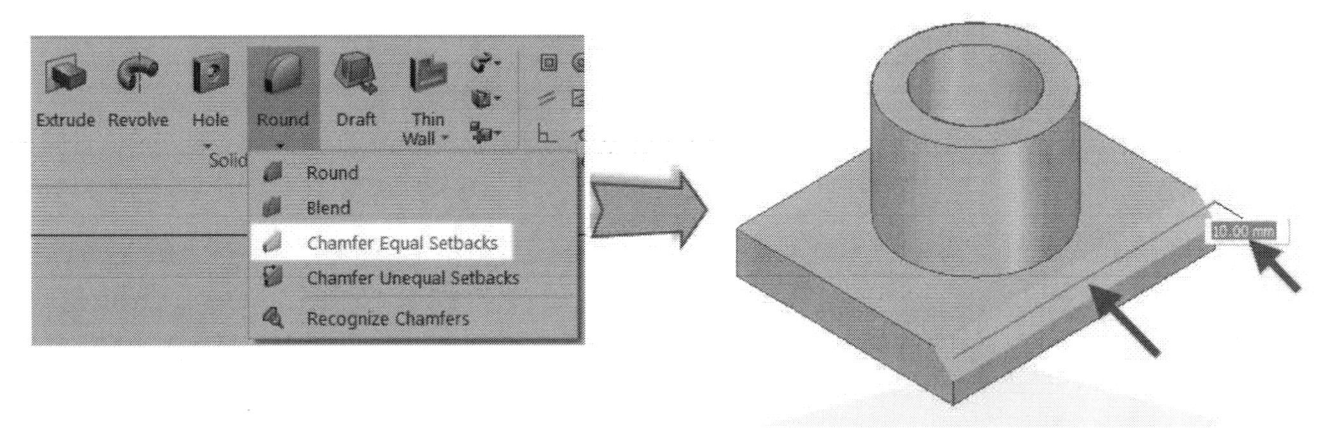

In Solid Edge 2019, the **Chamfer Equal Setbacks** command gives a good result for non-planar edges.

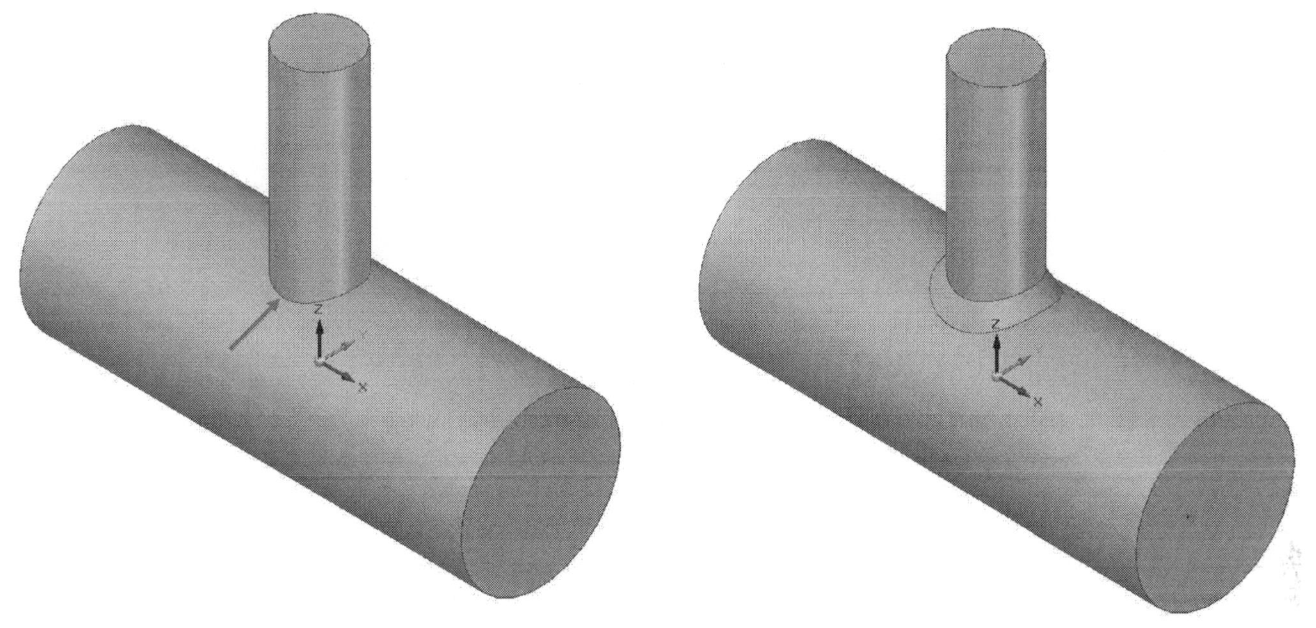

Chamfer Unequal Setbacks

This command will be useful, if you want a chamfer to have different setbacks on both sides of the edge. As you activate this command, you need to select both a face and an edge. First, you need to select a face, which acts as the reference. Click the green check on the command bar, and then type-in the **Setback** and **Angle** value. Solid Edge measures the setback distance and angle with reference to the selected face. Select the edge to be chamfered, click the green check, and then click **Finish**.

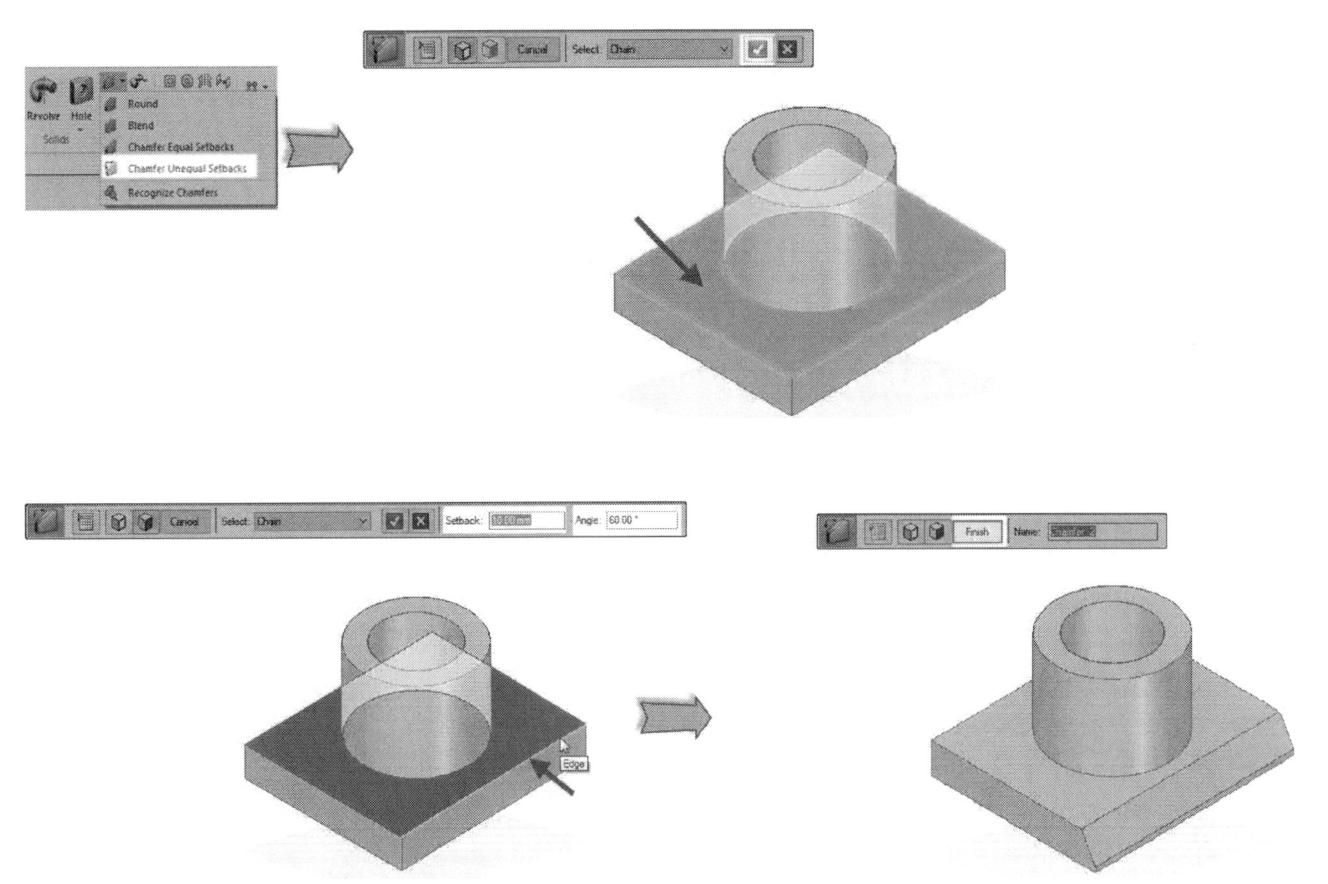

Tip: *You can create rounds and chamfers by selecting the edges of multiple bodies of a part.*

Draft

When creating cast or plastic parts, you are often required to add draft on them so that they can be molded. A draft is an angle or taper applied to the faces of parts to make it easier to remove them from a mold. When creating *Extrude* features, you can predefine the draft angle. However, most of the time, it is easier to apply the draft after the features are created. Activate the **Draft** command from the **Solids** panel. Select a face that will act as a reference plane for the draft. The draft angle will be measured with reference to this face. After selecting the reference plane, select the faces to draft. There are four options (**Chain**, **Face**, **Loop**, and **All Normal Faces**) on the command bar, which will help you to select the faces to draft. As you select the faces to draft, a two-sided arrow will appear along with a box. Use this two-sided arrow to define the direction of pull, and then type-in a value (angle) in the box.

Press Enter to create the *Draft* feature.

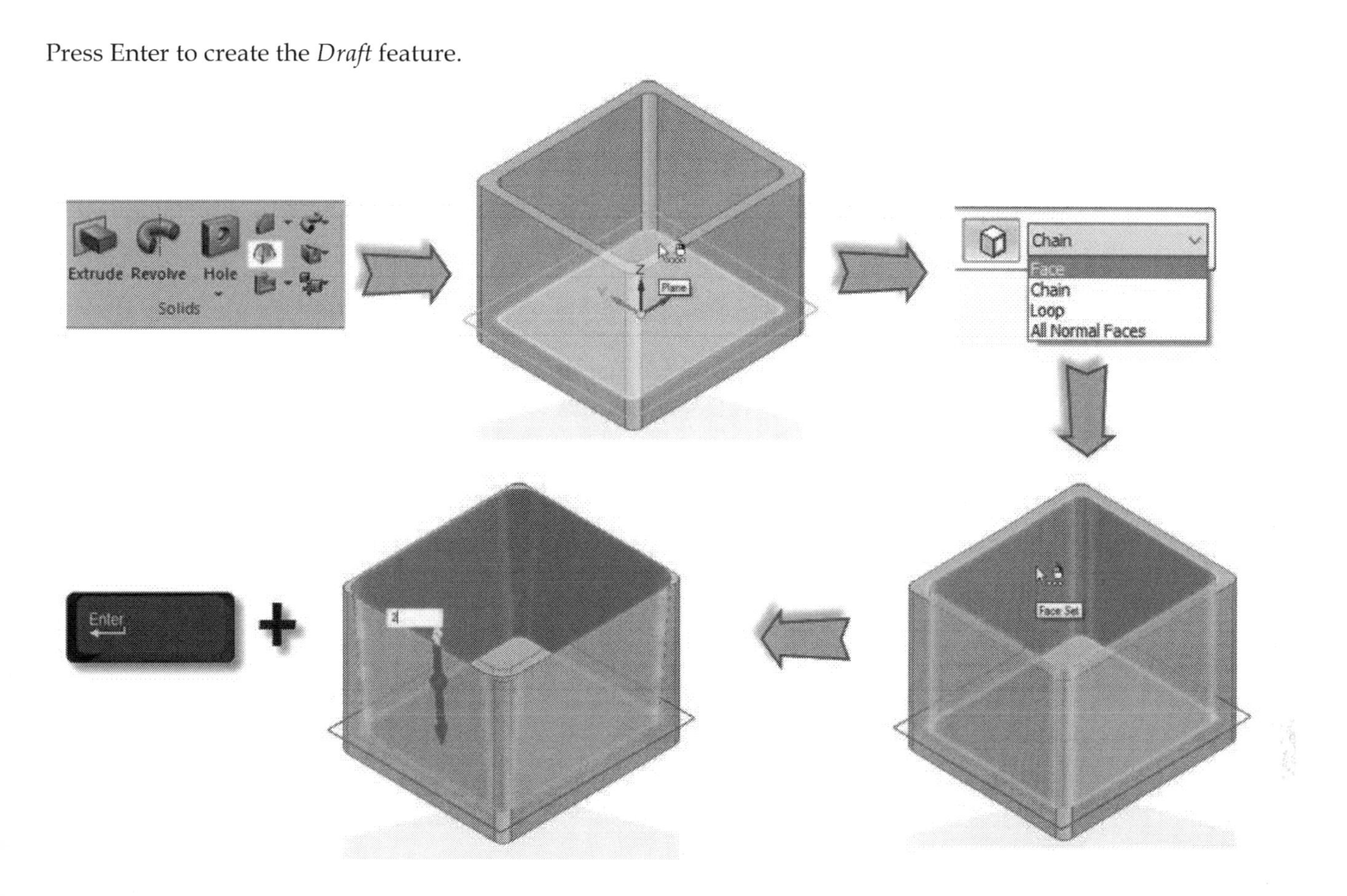

Thin Wall

The **Thin Wall** is another useful command that can be applied directly to a solid model. It allows you to take a solid geometry and make it hollow. This can be a powerful and timesaving technique, when designing parts that call for thin walls such as bottles, tanks, and containers. This command is easy to use. You should have a solid part, and then activate this command from the **Solids** panel. Now, select the faces to remove, and then type-in the wall thickness in the box that appears on the model. Click the arrow on the model to specify whether the thickness is added inside or outside the model. Right-click to finish the feature.

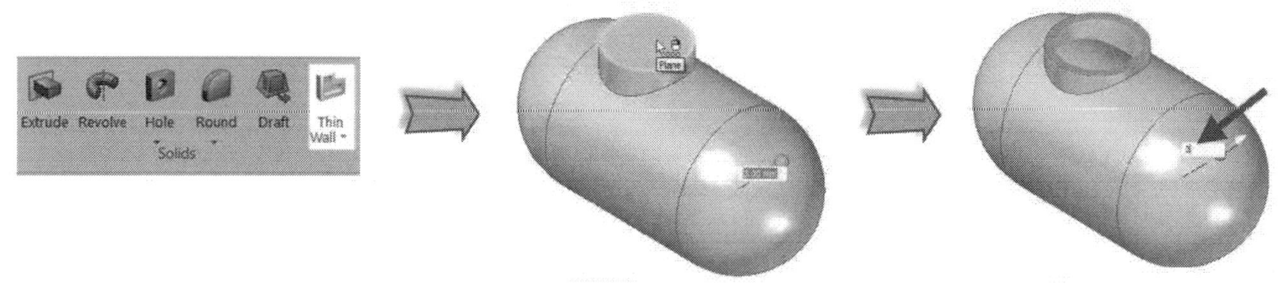

Examples

Example 1 (Millimetres)

In this example, you will create the part shown below.

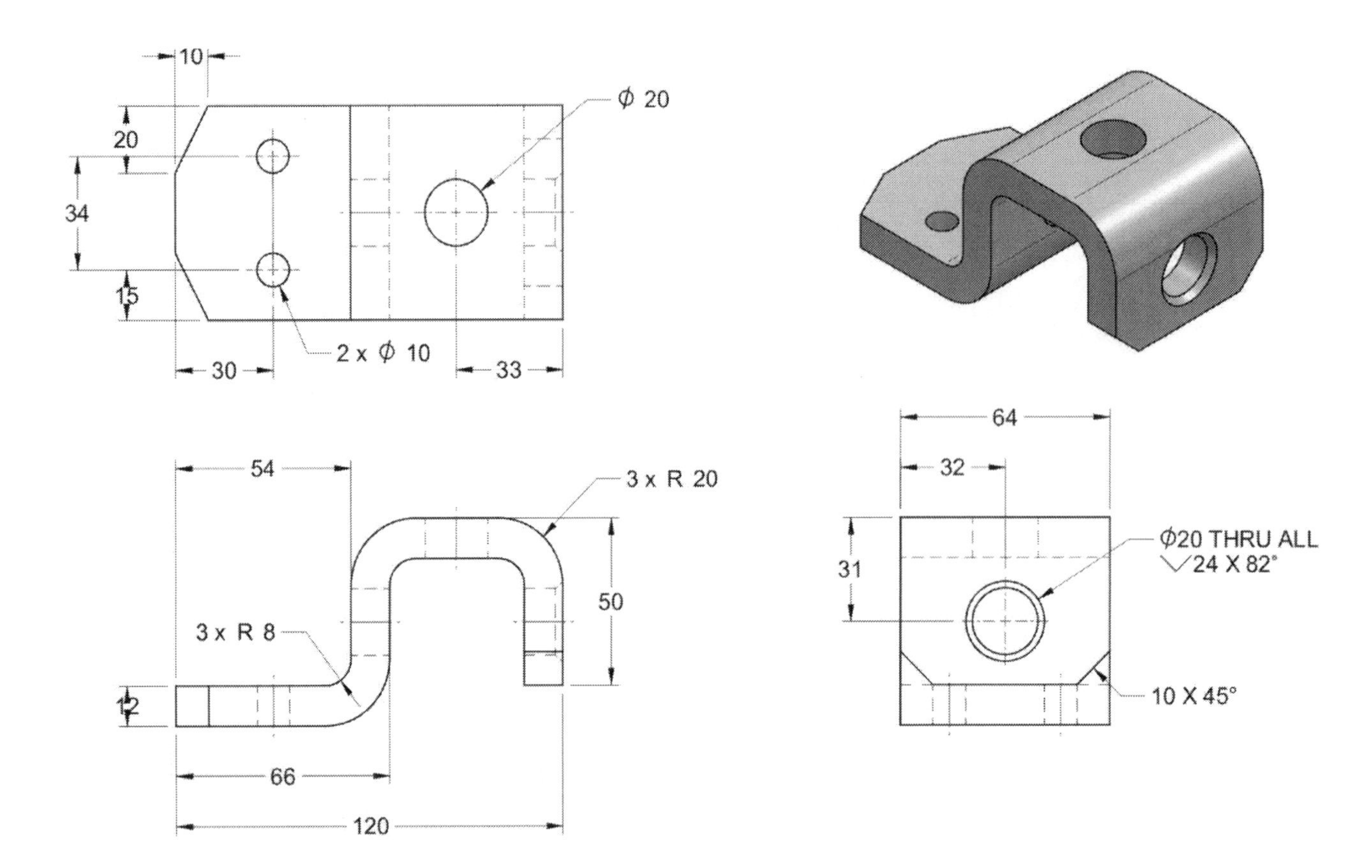

1. Start **Solid Edge 2019**.
2. On the Application Menu, click **New > ISO Metric Part**; a new part file is opened.
3. On the ribbon, click **Home > Draw > Line**.
4. Lock the XZ plane and draw the sketch, as shown below.
5. Create the *Extrude* feature of 64 mm thickness.

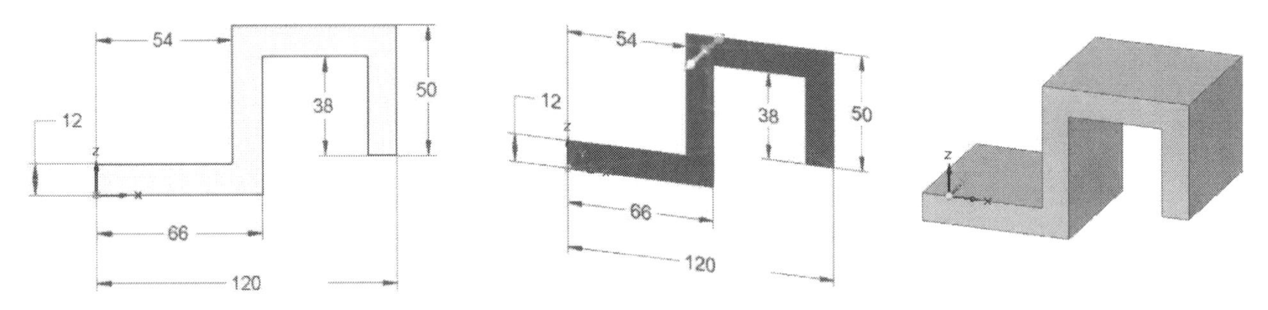

6. On the ribbon, click **Home > Solids > Hole**.
7. Place the mouse pointer on the right side face and click the lock icon.
8. On the command bar, click the **Hole Options** icon; the **Hole Options** dialog pops up.
9. On this dialog, select the **Countersink** button and set the **Standard** to **mm**.
10. Type-in 20, 24, and 82 in the **Hole Diameter**, **Countersink Diameter**, **Countersink Angle** boxes.
11. Set the **Hole Extents** type to **Through All**. Click **OK** to close the dialog.

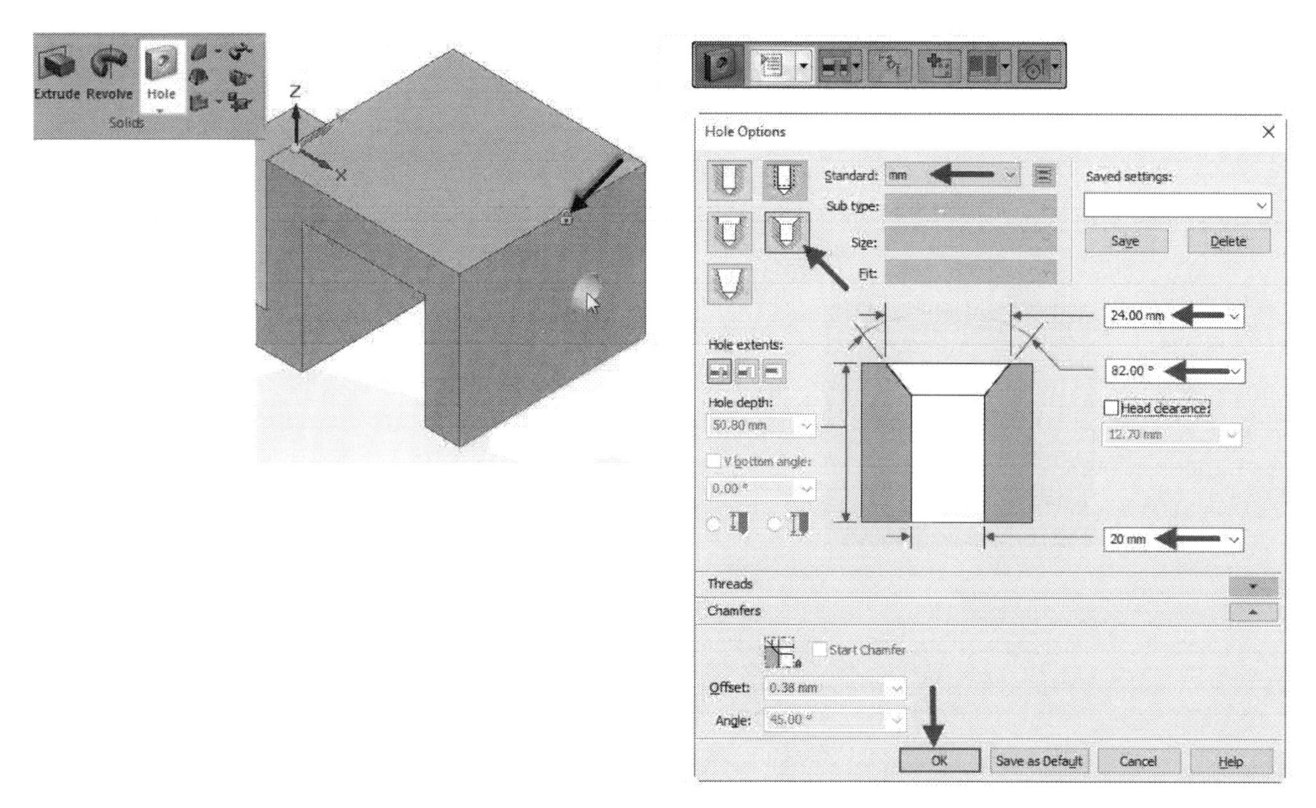

12. Place the mouse pointer on the top edge of the locked face and press E on the keyboard; a dimension appears between the edge and the hole.
13. Place the mouse pointer on the side edge of the locked face and press E on the keyboard; a dimension appears between the edge and the hole.
14. Set the dimension between the hole and top edge to 31 and press Tab on the keyboard.
15. Set the dimension between the hole and side edge to 32 and press Enter on the keyboard.

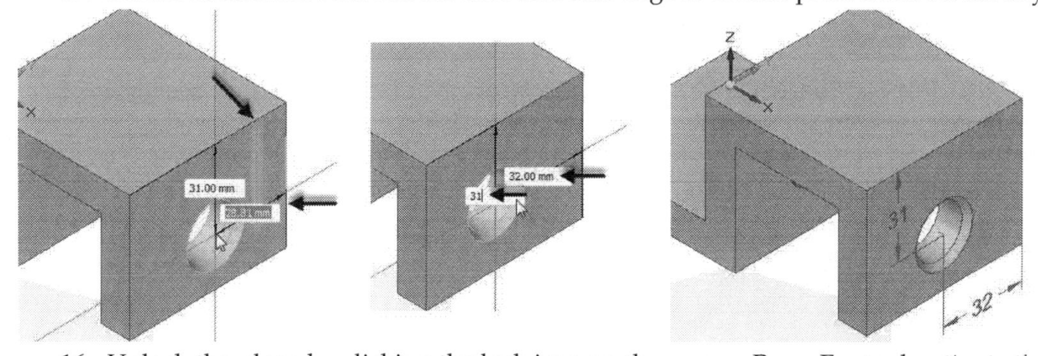

16. Unlock the plane by clicking the lock icon on the screen. Press Esc to deactivate the **Hole** command.
17. Activate the **Hole** command and place the mouse pointer on the top face of the part geometry. Lock the face.
18. Click the **Hole Options** icon on the command bar; the **Hole Options** dialog pops up.
19. On this dialog, select the **Simple** button and set the **Standard** to **mm**.
20. Enter 20 in the hole diameter box.
21. Set the **Hole extents** type to **Through All**. Click **OK** to close the dialog.
22. Place the mouse pointer on the midpoint of front edge.
23. Move the pointer on the locked face and notice a dotted line from the midpoint of the front edge.

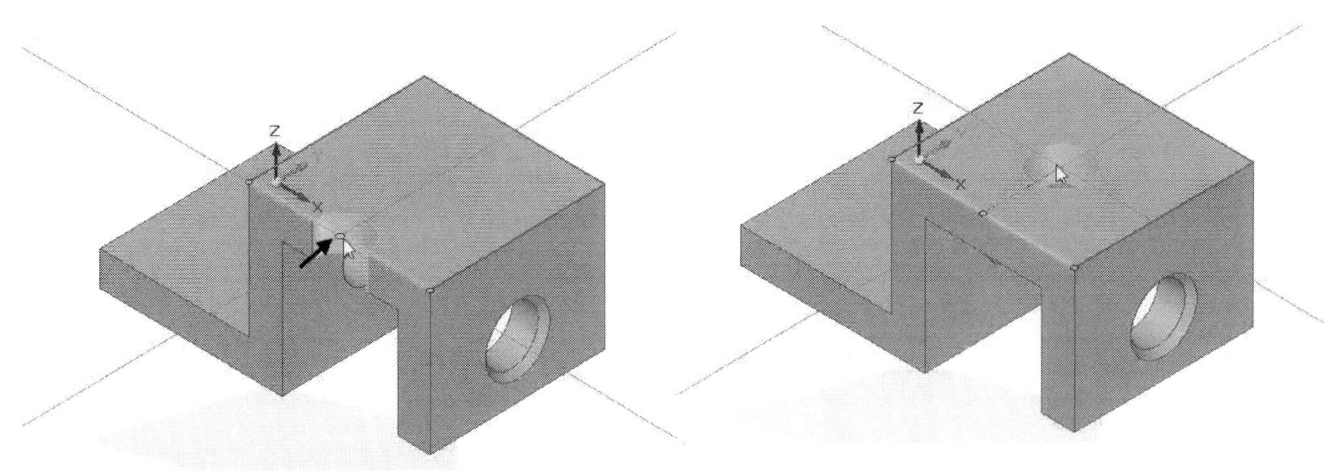

24. Likewise, place the pointer on the midpoint of the side edge and move the pointer.
25. Click when both the dotted lines intersect with each other.

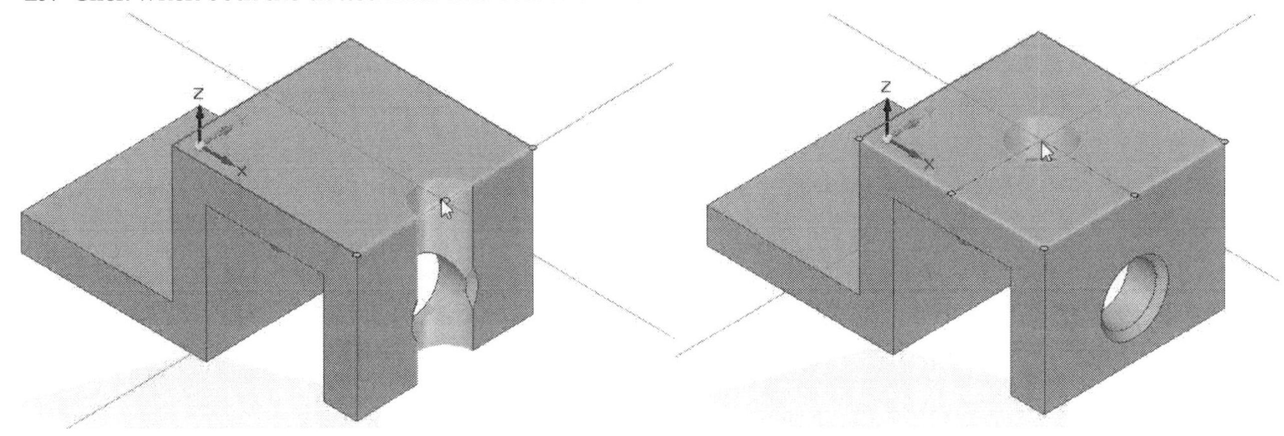

26. Unlock the face by clicking the lock icon. Press Esc to deactivate the **Hole** command.
27. On the **Quick View Cube**, click on the top left corner; the view orientation of the model changes.

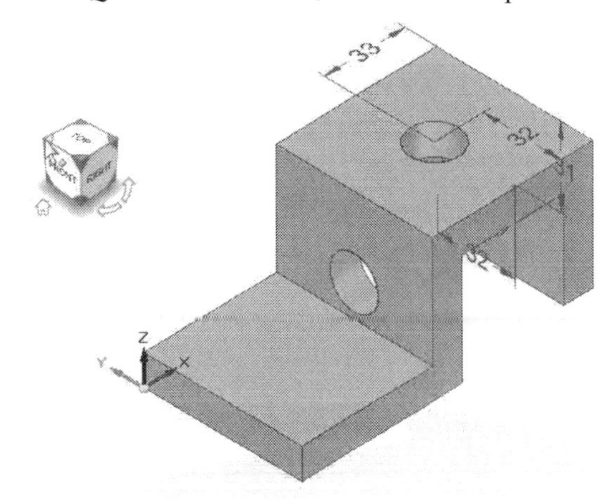

28. Activate the **Hole** command.
29. On the command bar, click the **Hole Options** icon to open the **Hole Options** dialog.
30. Set the **Standard** to **mm**
31. Select the **Simple** button and set the **Size** to 10. Click **OK** on the dialog.
32. Place the mouse pointer on the lower top face, and then press F3.

33. Place the mouse pointer on the front edge of the locked face and press E on the keyboard; a dimension appears between the edge and the hole.
34. Place the mouse pointer on the side edge of the locked face and press E on the keyboard; a dimension appears between the edge and the hole.
35. Set the dimension between the hole and front edge to 30 and press Tab on the keyboard.
36. Set the dimension between the hole and side edge to 15 and press Enter on the keyboard; a hole is created and another hole is attached to the mouse pointer.

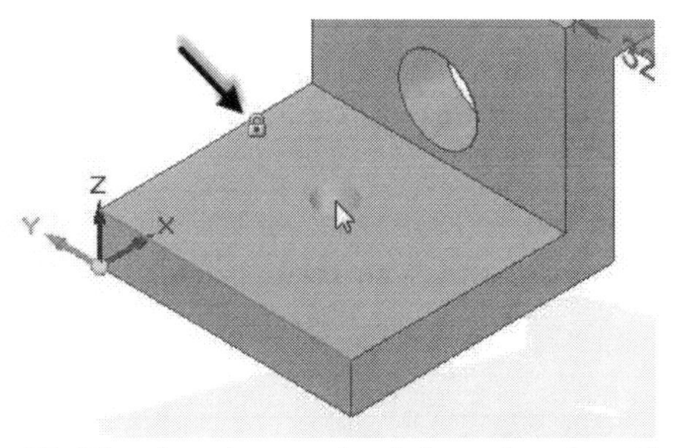

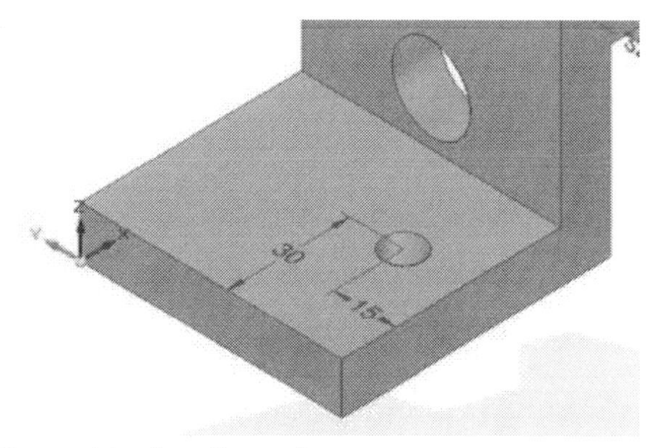

37. Likewise, place another hole on the other side. The positioning dimensions are same.

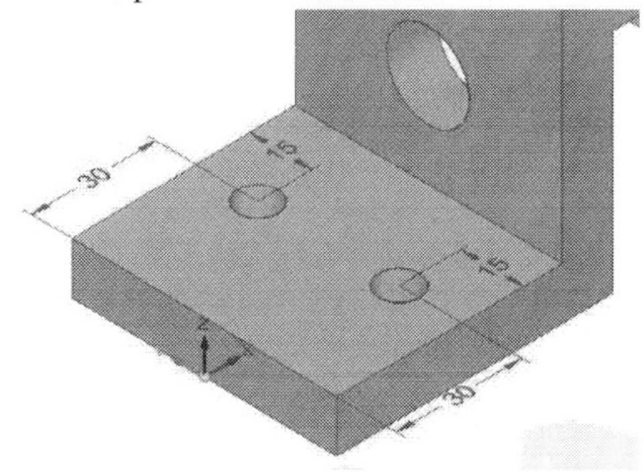

38. Click **Home > Solids > Round > Chamfer Unequal Setbacks** on the ribbon.
39. On the command bar, click the **Chamfer Options** icon; the **Chamfer Options** dialog pops up.
40. On this dialog, select **2 Setbacks** and click **OK**.
41. Click on the front face of the model and click the green check on the command bar.

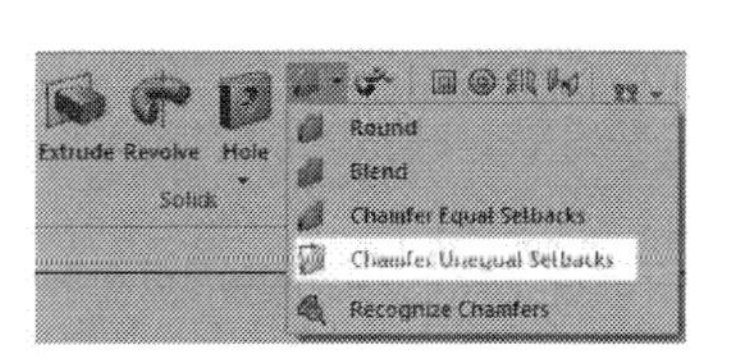

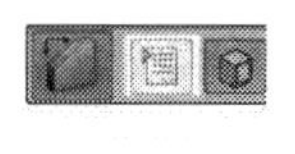

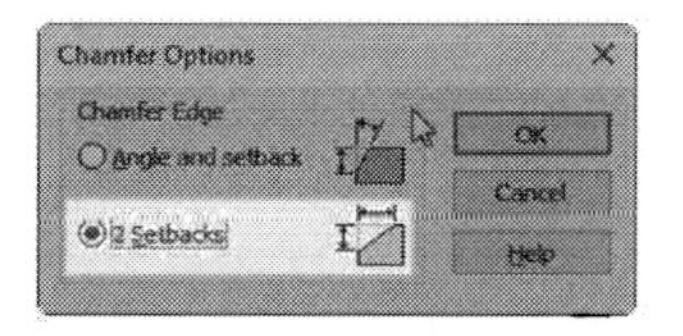

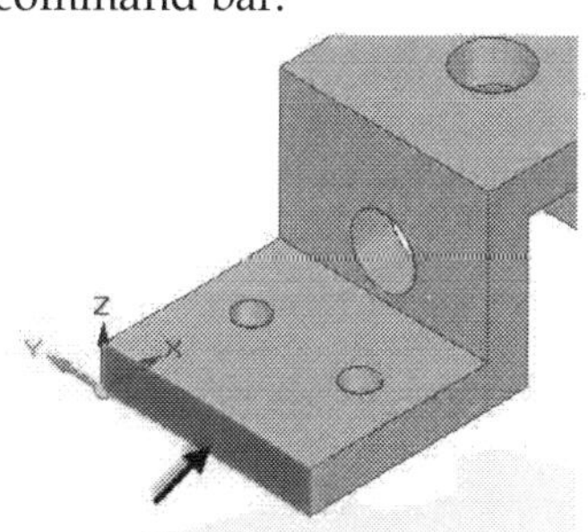

42. Set the **Setback 1** and **Setback 2** to **20** and **10**, respectively.
43. Click on the side edges of the selected face, as shown in figure.
44. Click the green check, and then **Finish** on the command bar.

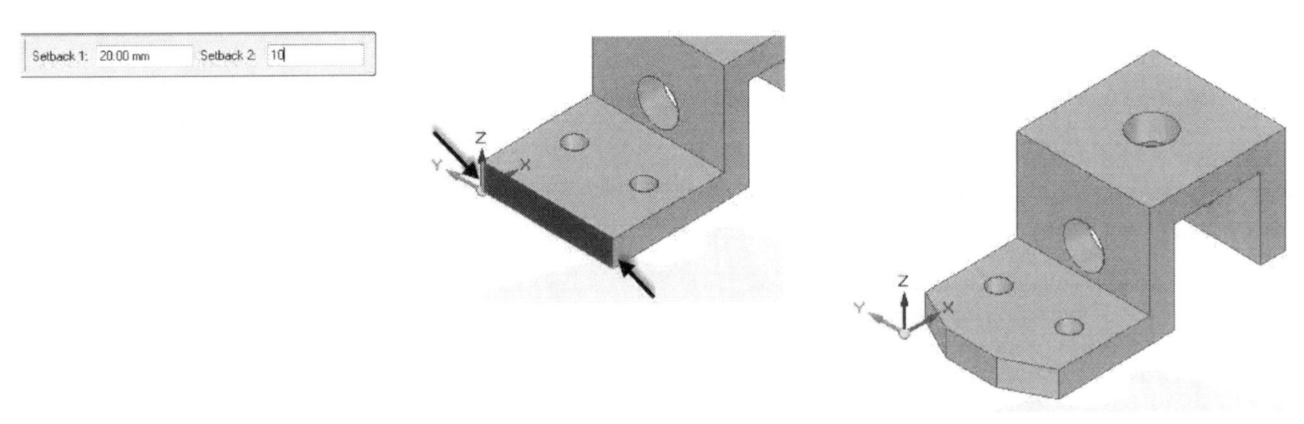

45. Press Esc to deactivate the command.
46. Click **Home > Solids > Round** on the ribbon. Set the **Selection type** to **Edge/Corner**.
47. Click on the horizontal edges of the geometry, as shown below.
48. Type-in **8** in the box that appears on the geometry, and then press Enter.

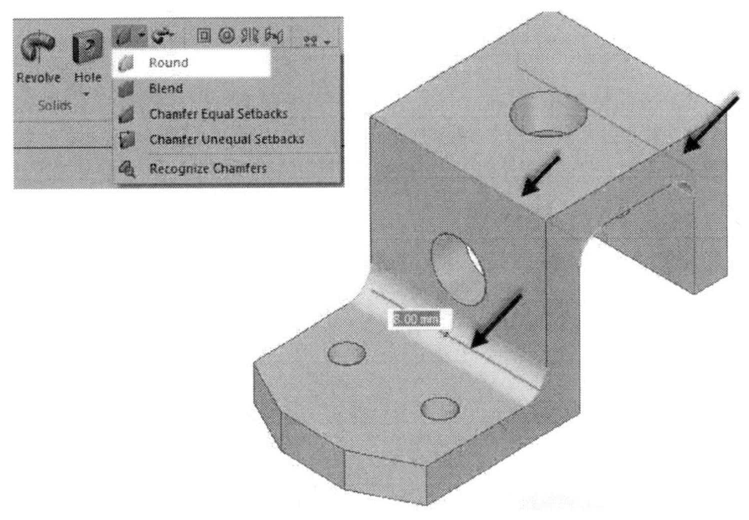

49. Click on the outer edges of the model, as shown below.
50. Type-in **20** in the box that appears on the geometry, and then press Enter.

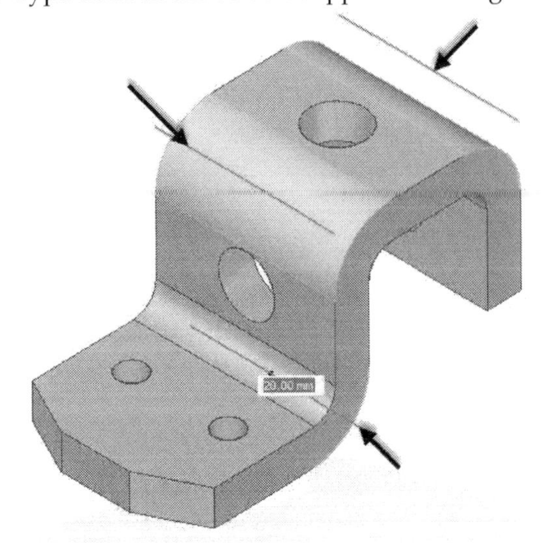

51. Change the orientation of the model view to Isometric by clicking the **Home** icon below the **Quick View Cube**.
52. Click **Home > Solids > Round > Chamfer Equal Setbacks** on the ribbon. Set the **Selection type** to **Edge/Corner**.
53. Click on the lower corners of the part geometry.
54. Type-in **10** in the box that appears on the part geometry. Press Enter to chamfer the edges.
55. In the Pathfinder, uncheck the PMI option to hide all dimensions of the model geometry.

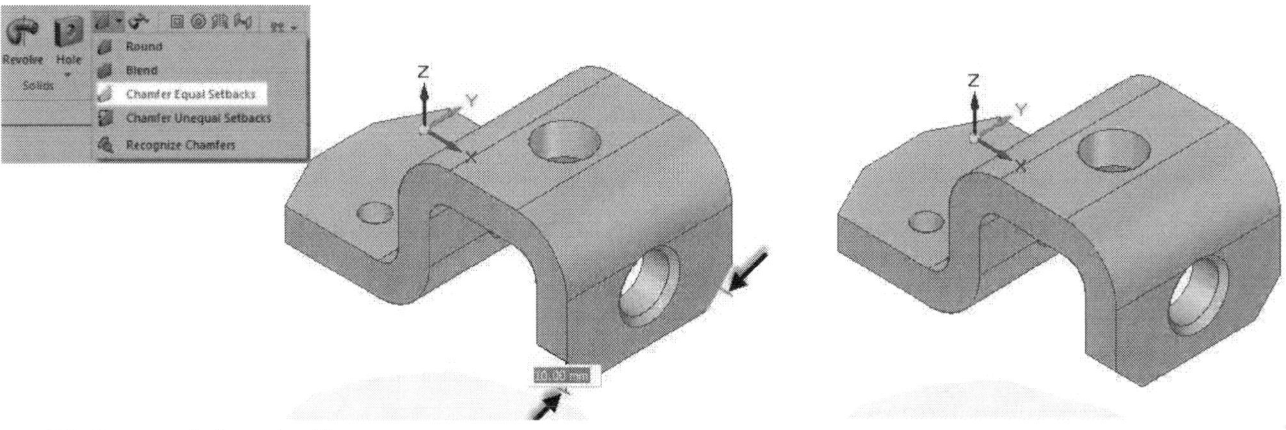

56. Save and close the file.

Questions

1. What are placed features?
2. How do you create a hole on a cylindrical face?
3. Which command allows you to create chamfer with unequal setbacks?
4. Which command allows you to create a variable radius blend?
5. When you create a thread on a cylindrical face, will the diameter of the cylinder remains the same or not?

Exercises

Exercise 1 (Millimetres)

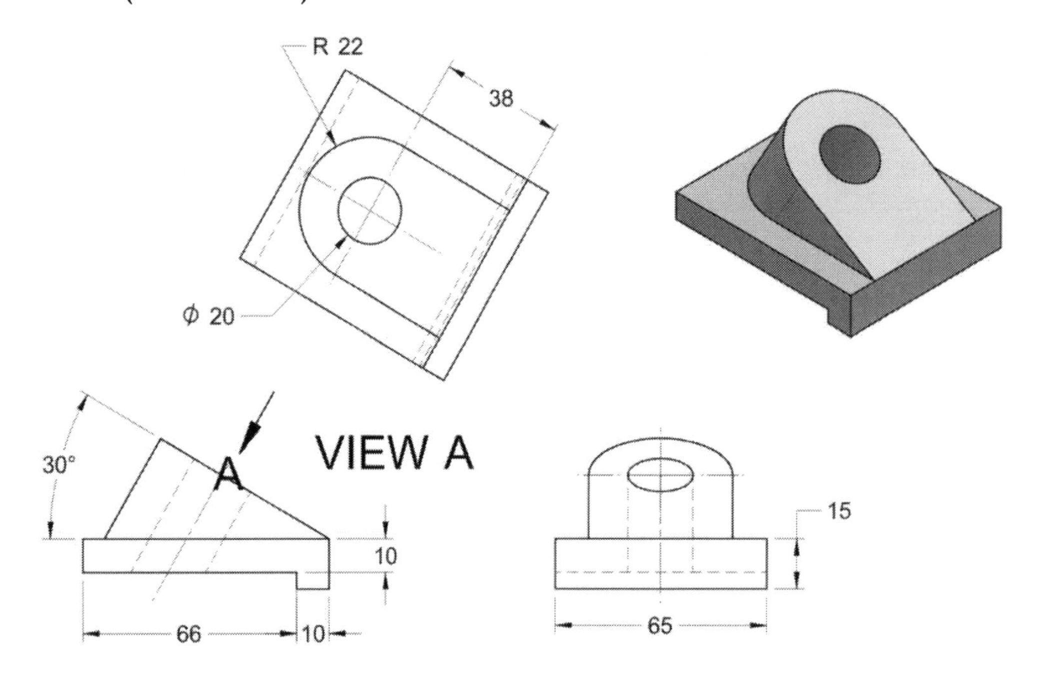

Exercise 2 (Inches)

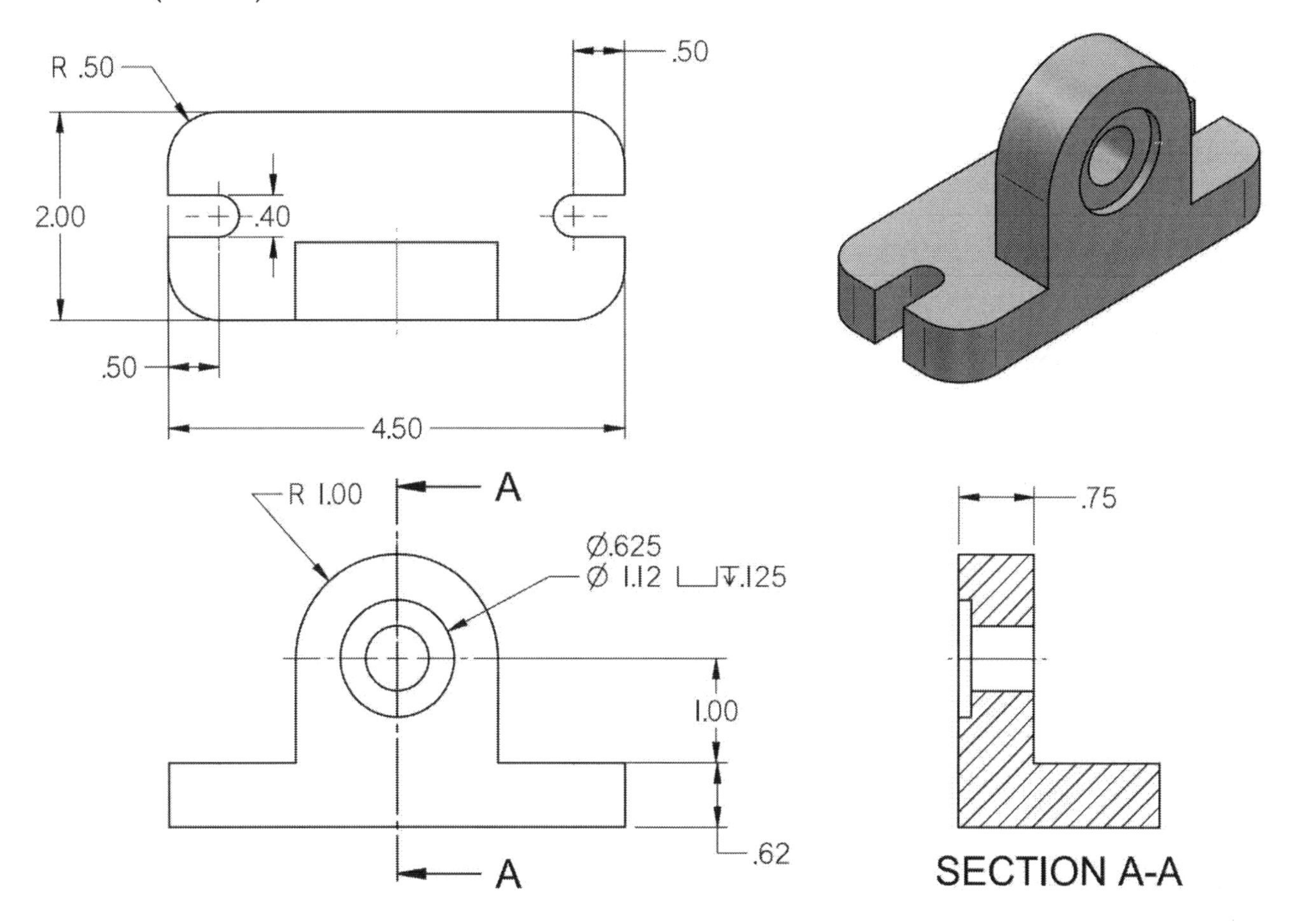

Chapter 5: Patterned Geometry

When designing a part geometry, oftentimes there are elements of symmetry in each part or there are at least a few features that are repeated multiple times. In these situations, Solid Edge offers you some commands that save your time. For example, you can use mirror features to design symmetric parts, which makes designing the part quicker. This is because you only have to design a portion of the part and use the mirror feature to create the remaining geometry.

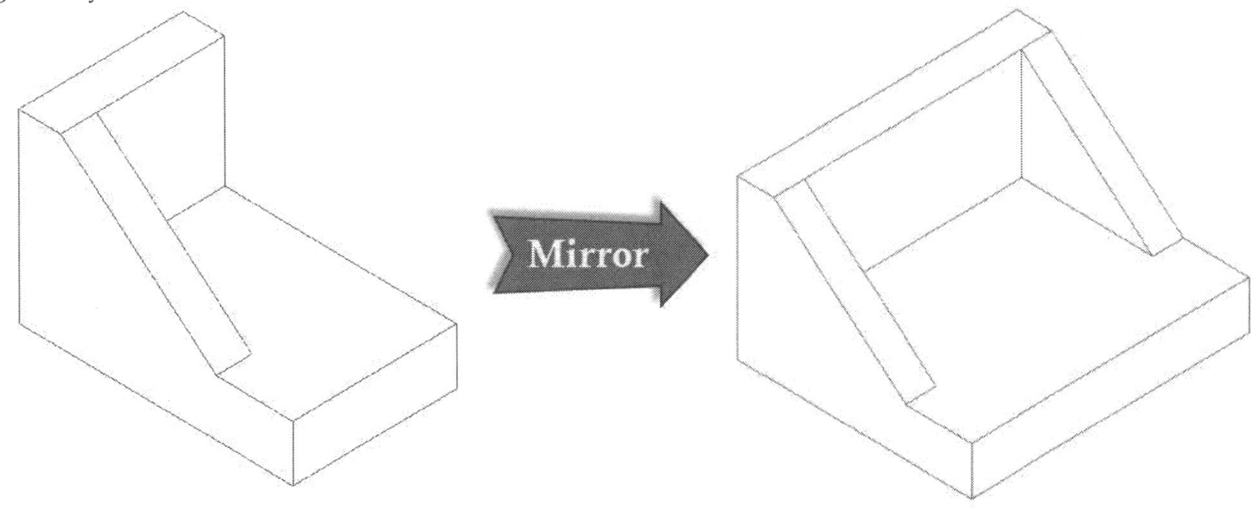

In addition, there are some pattern commands to replicate a feature throughout a part quickly. They save you from creating additional features individually and help you modify the design easily. If the design changes, you only need to change the first feature and the rest of the pattern features will update, automatically. In this chapter, you will learn to create the mirrored and pattern geometries using the commands available in Solid Edge.

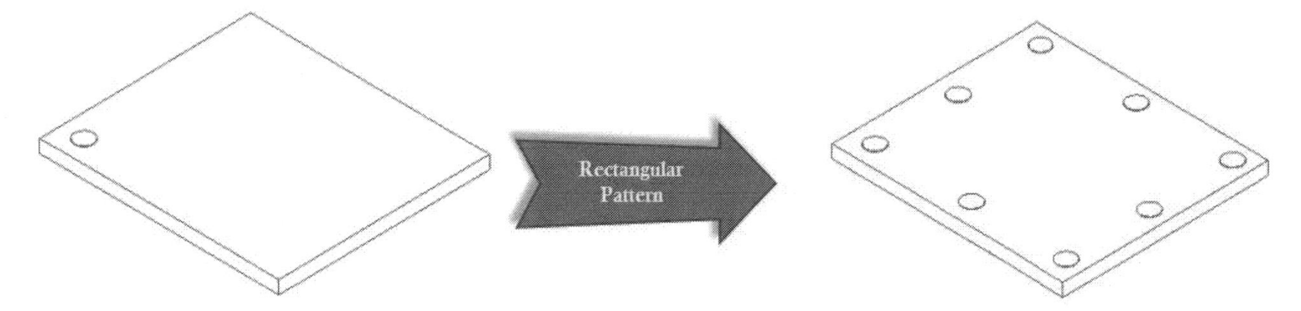

The topics covered in this chapter are:

- *Mirror* features
- *Rectangular Patterns*
- *Circular Patterns*
- *Along Curve Patterns*
- *Fill Patterns*
- *Recognize Hole Patterns*

Mirror

If you are designing a part that is symmetric, you can save time by using the **Mirror** command. Using this command, you can replicate individual features of the entire body. To mirror features (3D geometry), you need to have a face or plane to use as a reference. You can use a model face, default plane, or create a new plane, if it does not exist where it is needed.

Click on the features to be mirrored in the Pathfinder, and then activate the **Mirror** command (click **Home > Pattern > Mirror** on the ribbon). Now, select the reference plane about which the features are to be mirrored.

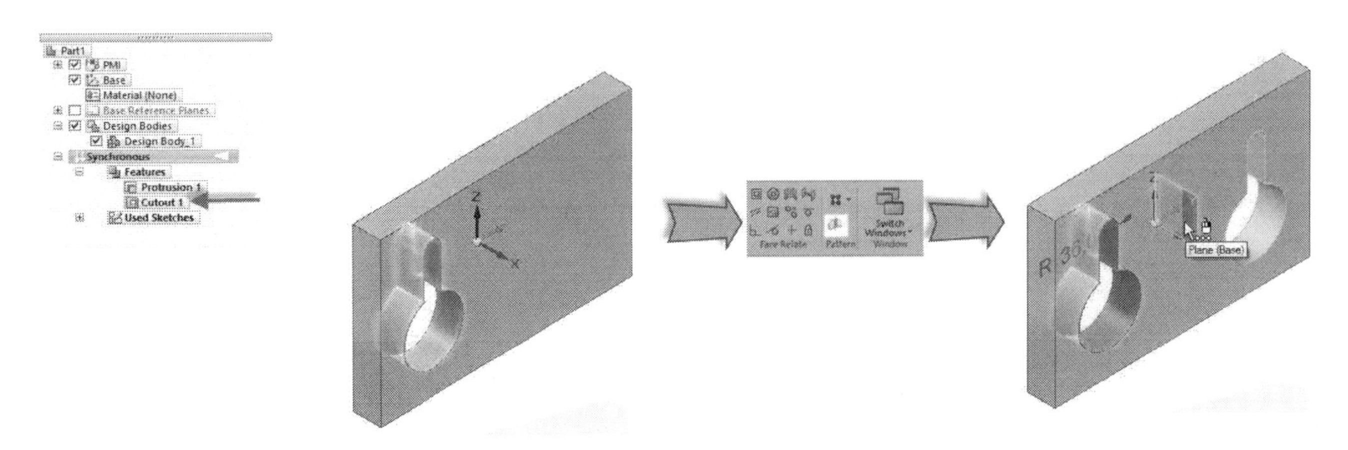

Now, if you make changes to the original feature, the mirror feature will be updated automatically.

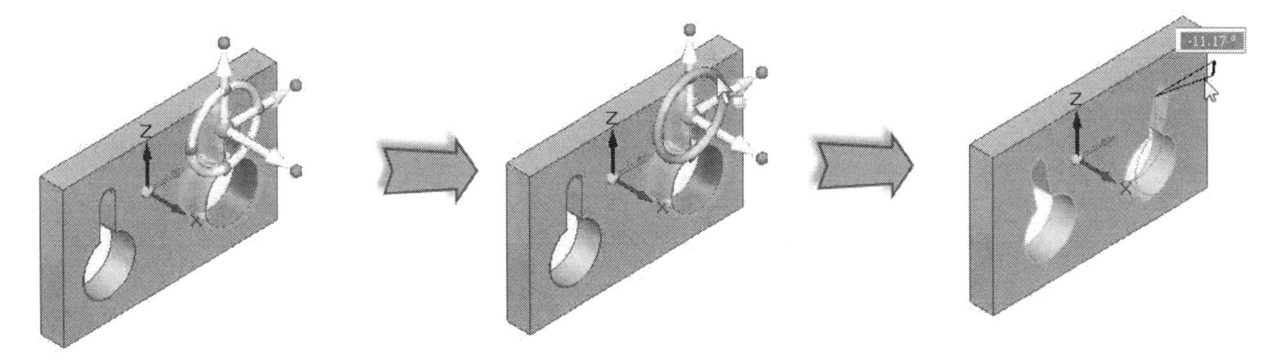

If you select the **Detach Faces** option on the command bar, the faces of the feature will be mirrored, but will be detached from rest of the model.

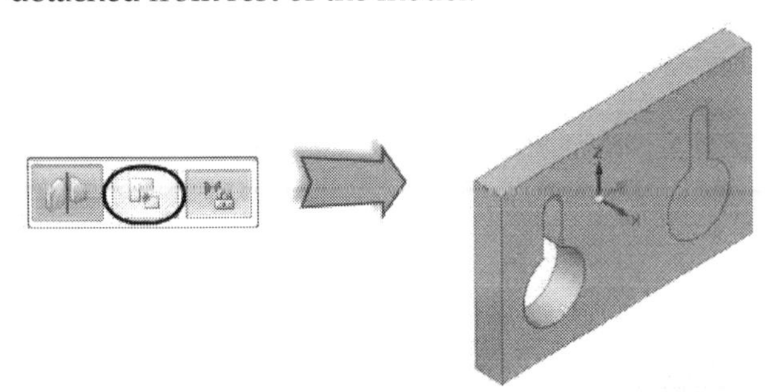

Rectangular Pattern

This command creates a rectangular pattern of a feature. To create a rectangular pattern, you must first select the feature to pattern, and then activate the **Rectangular** command (click **Home > Pattern > Rectangular** on the command bar). Next, define the second corner of the rectangular pattern by moving the pointer diagonally and clicking on the face of the model. You will notice that a pattern preview appears on the model. Now, select the **Fit** option on the command bar and set the parameters of the pattern (Total Spacing along X-axis and Y-axis, X Count, and Y Count). If you want to suppress some instances, click the **Suppress Instance** option on the command bar and select the green dots from the pattern preview. Next, click the green check on the command bar.

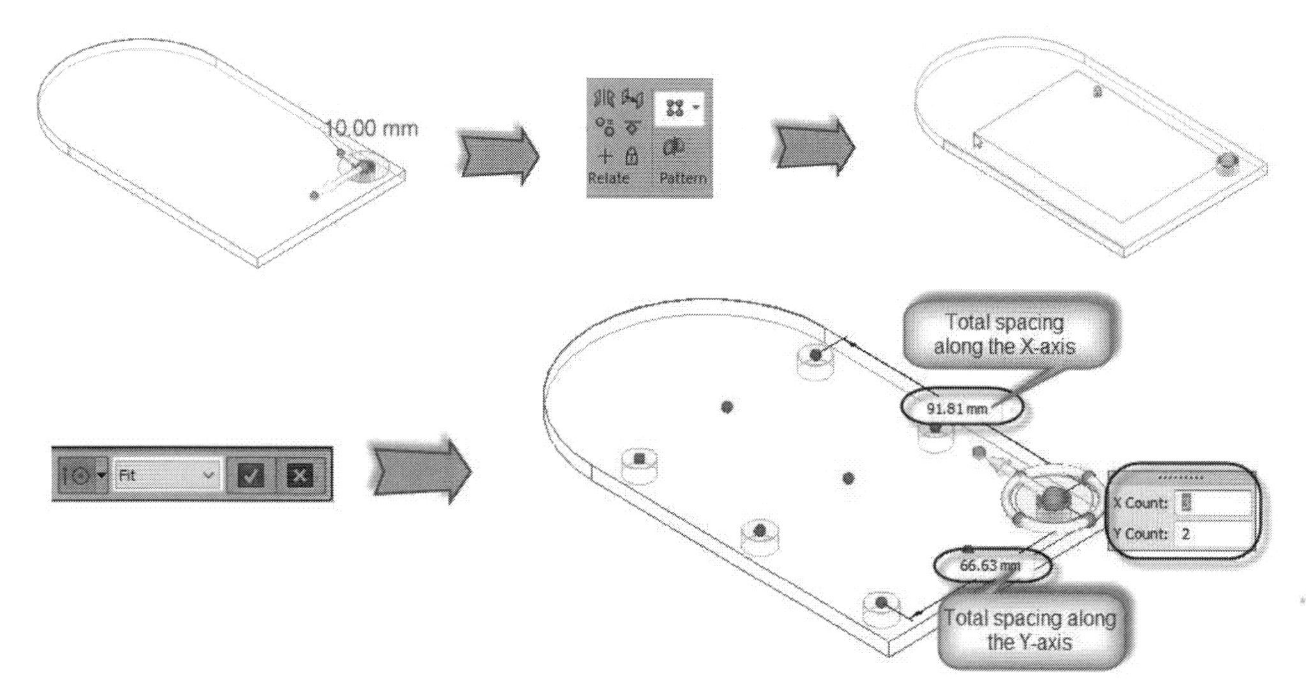

Select the **Fixed** option on the command bar, if you want to enter the spacing between individual instances of the pattern. Click the green check on the command bar to finish the rectangular pattern.

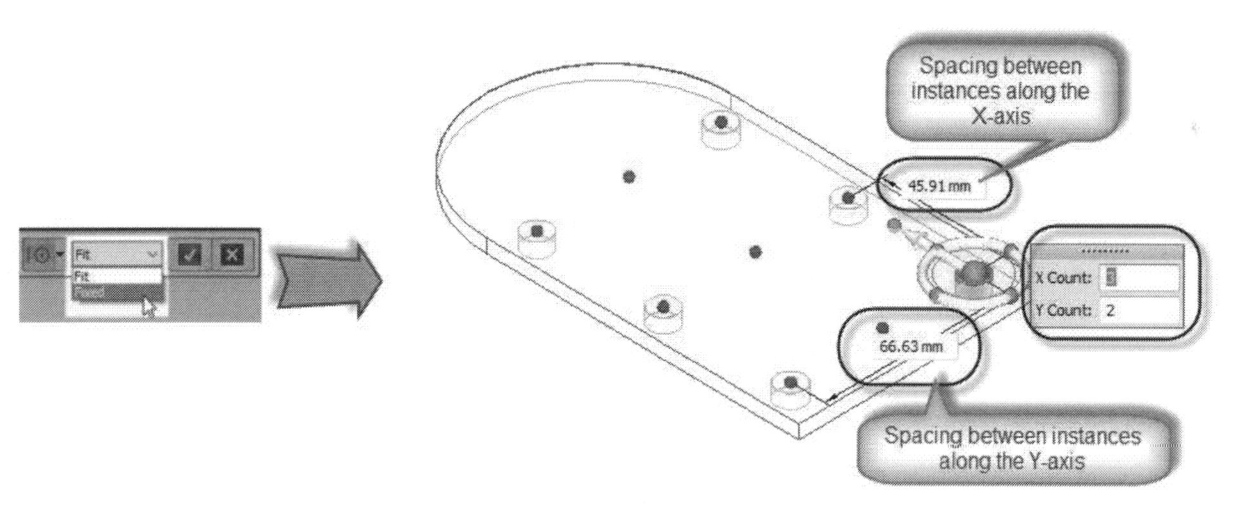

If you want to modify the rectangular pattern, just select it from the model or Pathfinder. A pattern annotation (for example, Pattern 3X2) appears on it. Select the annotation, and then modify the pattern parameters. You can use the **Add to Pattern** option on the command bar to add more features to the pattern.

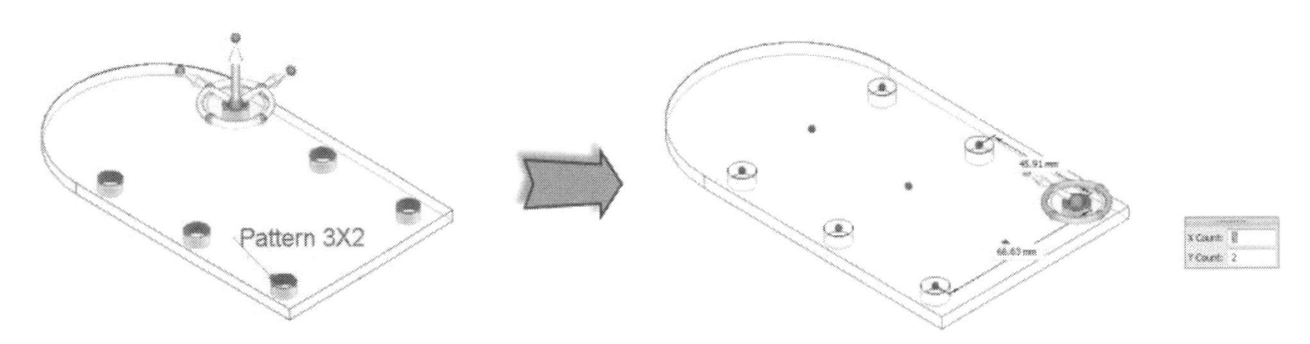

Circular Pattern

This command creates a pattern of selected features in a circular fashion. Select the feature to pattern and activate the **Circular** command (click **Home > Pattern > Rectangular > Circular** on the ribbon). Next, define the axis of the circular pattern by selecting a key point.

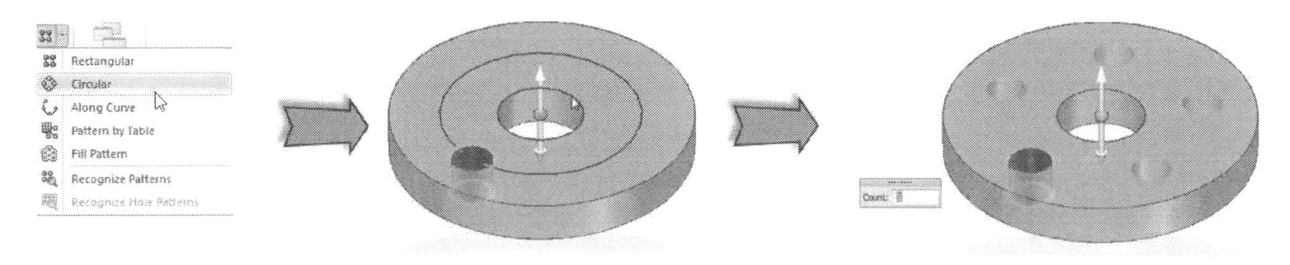

Use the **Circle/Arc Pattern** option on the command bar, if you want to create an arc pattern. Type-in values in the **Count** and **Angle** boxes, and click the green check to create the circular pattern.

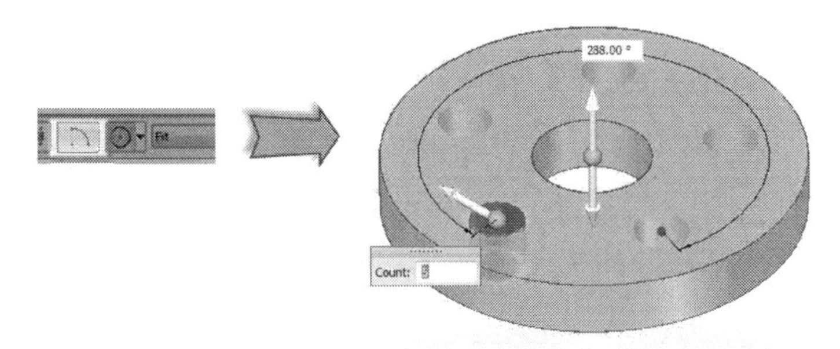

Along Curve Pattern

This command creates a pattern along a selected curve or edge. Activate this command (click **Home > Pattern > Rectangular > Along Curve** on the ribbon). Next, set the **Selection Type** on the command bar and click on a curve or edge. Click the green check on the command bar to accept the selection.

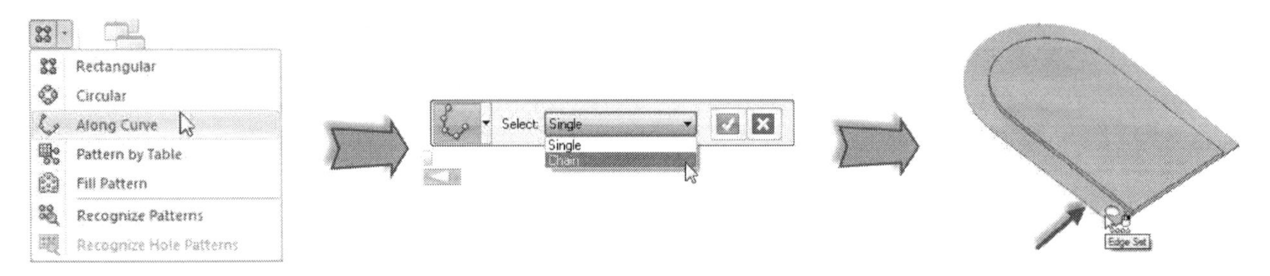

Select a point on the selected curve/edge to define the anchor point of the pattern. Click to define the side of the pattern. On the command bar, click the **Advanced** icon to display a box.

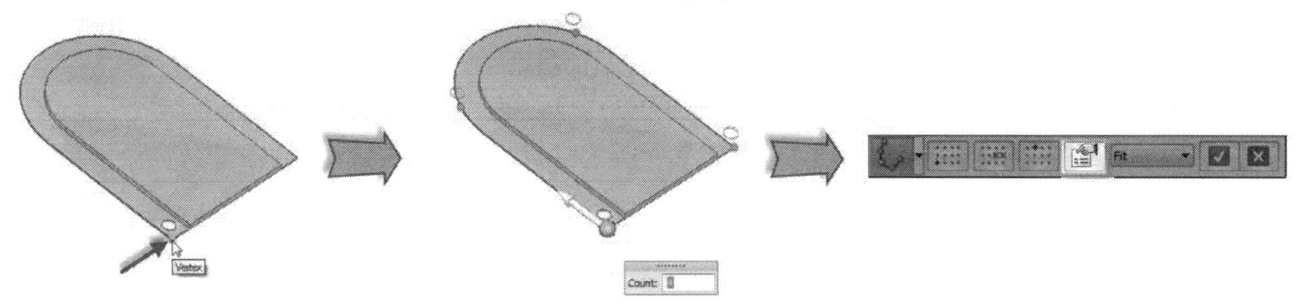

On this box, set the **Transformation Type** to **Follow Curve** and **Rotation Type** to **Curve Position**. Click the green check on the box and type-in a value in the **Count** box. On the command bar, set the **Fill Style** to **Fit** and click the green check to create the pattern along the curve.

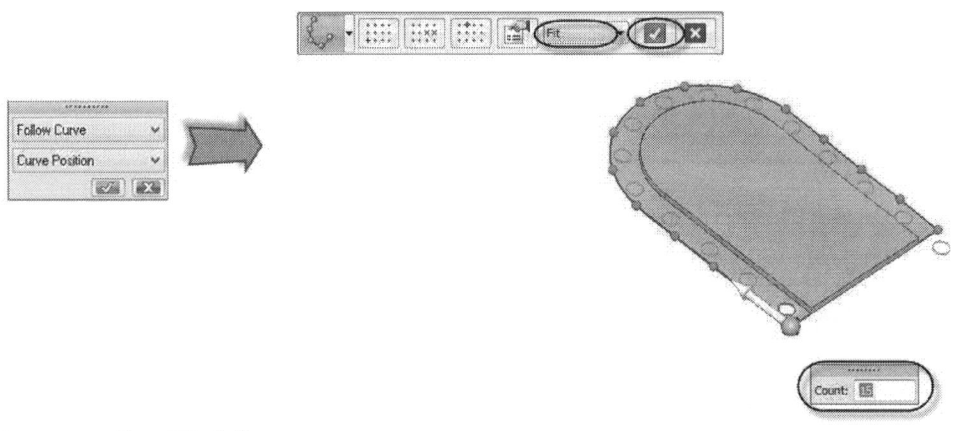

Pattern by Table

The **Pattern by Table** command creates a pattern of a feature by using a spreadsheet to define the location of the pattern instances. This command uses the first two columns of a spreadsheet to define the X and Y coordinates of the instance. The value in the third column is used to define the orientation angle of the pattern instance. Select the feature to pattern from the model and activate the **Pattern by Table** command (On the ribbon, click **Home > Pattern > Pattern drop-down > Pattern by Table**). On the command bar, click the **Pattern by Table Options** icon to open the **Pattern by Table Options** dialog. On this dialog, specify the units by using the **Use units from current document** or **Use custom units** options. If you select the **Use custom units** option, then you can define **Units** and **Round-off** values from the table available on the dialog. Click **OK** after defining the units.

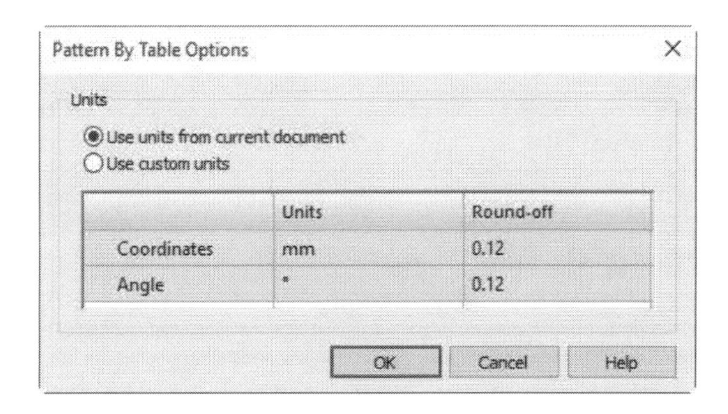

Select the coordinate system from the graphics window; the **Instance Table** dialog pops up on the screen. On this dialog, click the **Browse** button, select the spreadsheet, and click **Open**. You can use the **Edit** and **Update** buttons to make changes to the spreadsheet. On the **Instance Table** dialog, select the **Excel file** or **Selected keypoint** option from the **Parent Reference Point** section. Select a key point from the model to define the parent reference point, if you have set the **Parent Reference Point** to **Selected keypoint**. Click **Close** on the dialog. Next, click the **Accept** button on the command bar to create the pattern.

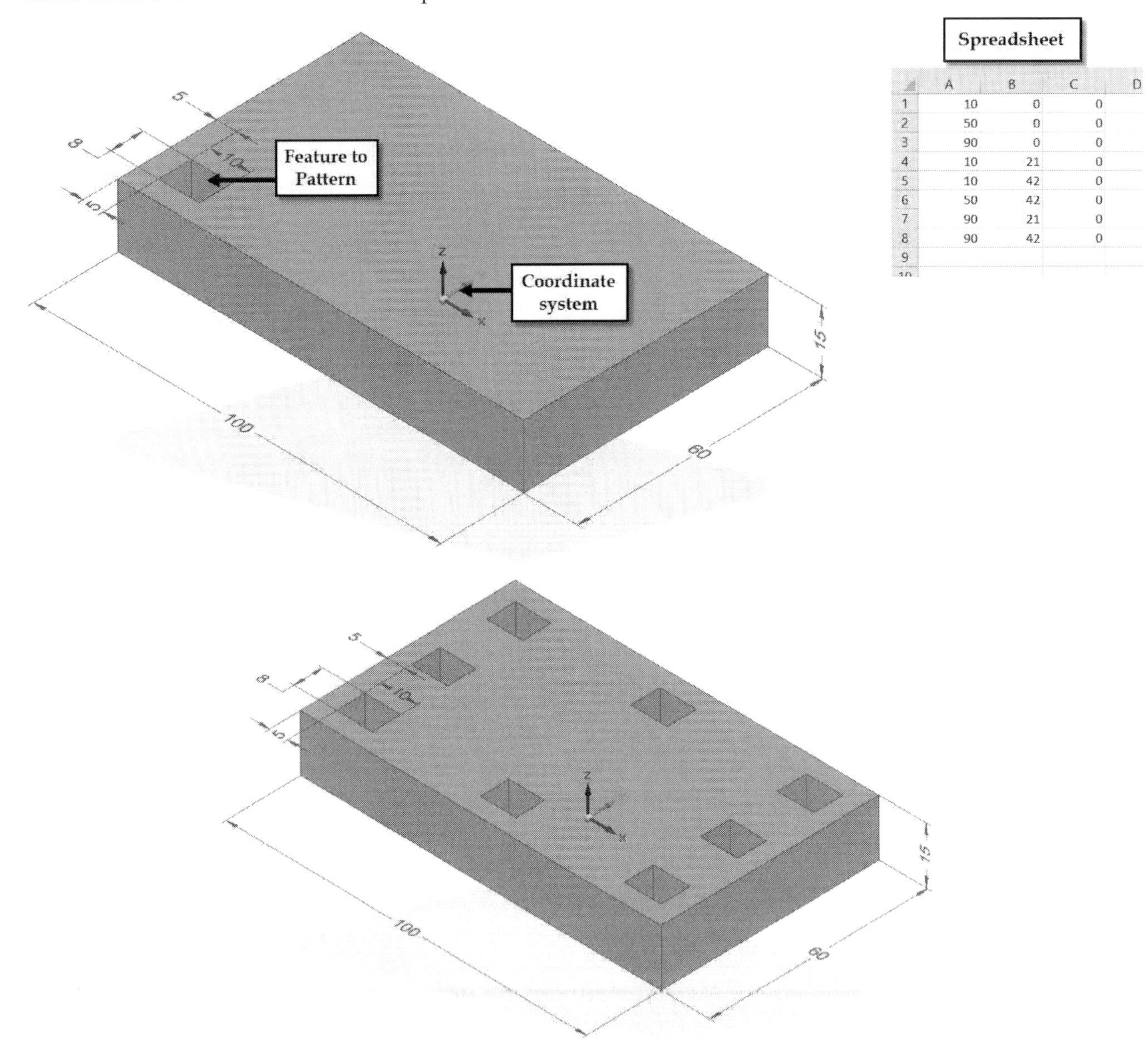

	A	B	C	D
1	10	0	0	
2	50	0	0	
3	90	0	0	
4	10	21	0	
5	10	42	0	
6	50	42	0	
7	90	21	0	
8	90	42	0	
9				

Fill Pattern

This command creates a pattern of a feature by filling it on a defined region. You can create three different types of fill patterns: **Rectangular**, **Staggered**, and **Radial**.

Rectangular Fill Pattern

Select the feature from the geometry and activate the **Fill Pattern** command (click **Home > Pattern > Rectangular > Fill Pattern** on the command bar). Select the face or region on which to create the fill pattern. Set the **Fill Style** to

Rectangular and click the green check on the command bar. Type-in the spacing values between the pattern instances.

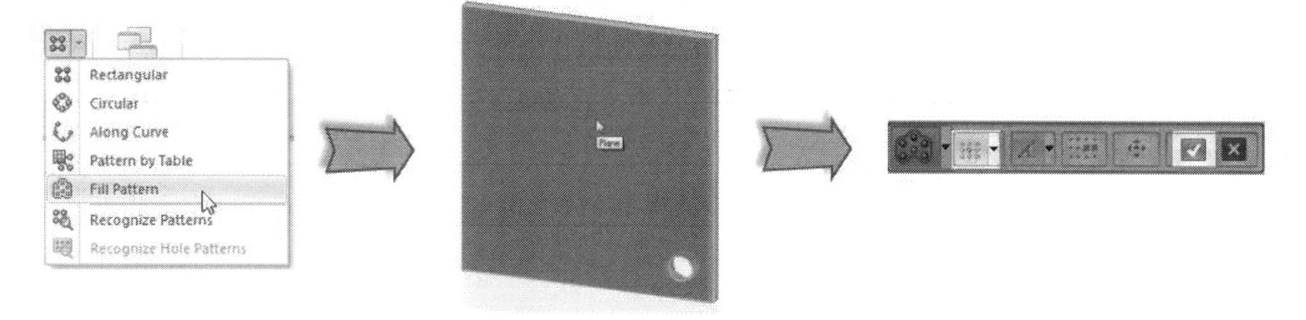

On the command bar, click the **Suppress Instance** icon to suppress the unwanted instances.

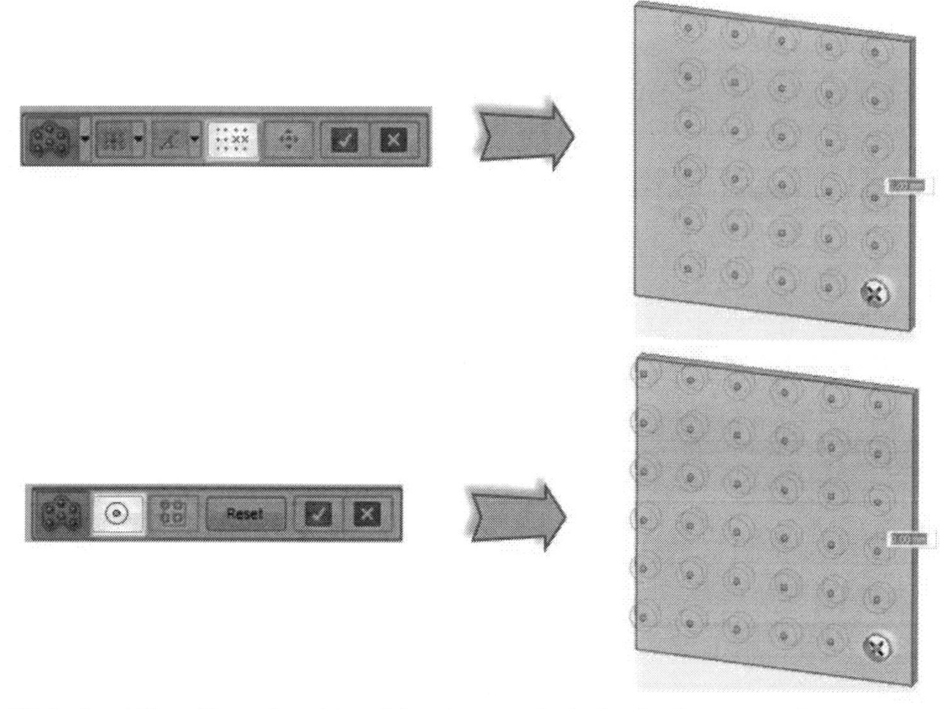

Click the **Allow Boundary Touching** icon to include the instances that are outside the region and touching the boundary. Click the green dots on the instances to suppress them. On the command bar, click the green check after suppressing the instances.

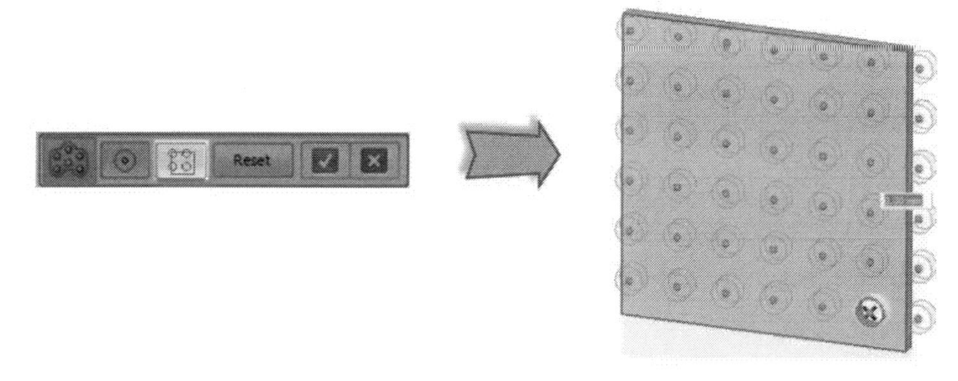

Click and drag the pattern boundaries to increase or decrease the fill pattern region. After defining the required settings, right-click to create the rectangular fill pattern.

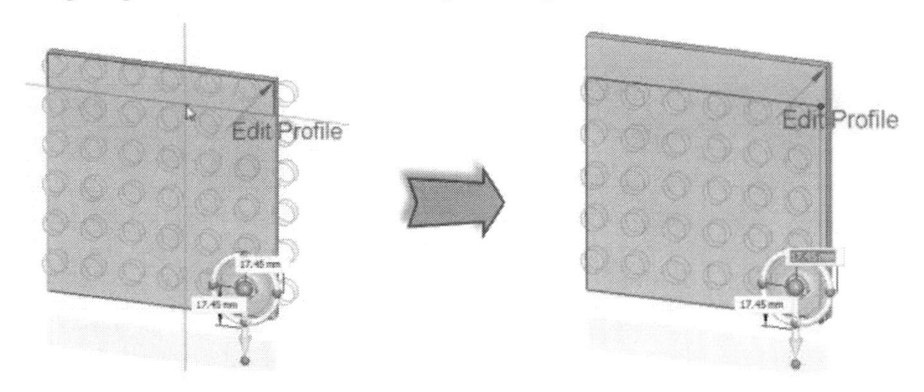

Staggered Fill Pattern

In this type of pattern, the features are arranged in a perforation fashion. To create a staggered fill pattern, set the **Fill Style** to **Stagger** on the command bar.

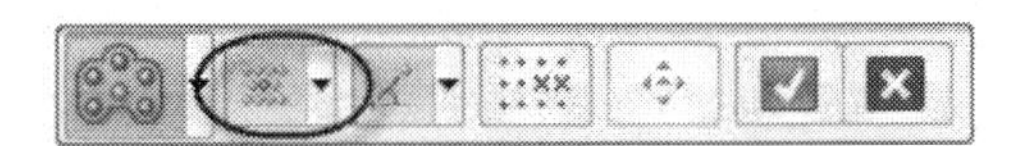

Click on the face fill and define the spacing between the instances. This can be done by using the **Fill spacing** methods. There are three methods to define the spacing between the instances: **Polar**, **Linear Offset**, and **Complex Linear Offset**. The **Polar** method creates a pattern by using (a) rotation angle between two rows and (b) distance between the instances. The **Linear Offset** method creates a pattern by using (c) spacing between two rows and (d) stagger offset. The **Complex Linear Offset** method creates a pattern by using (e) spacing between two instances in a row, (f) spacing between rows, and (g) stagger offset. Select a **Fill spacing method** and click the green check.

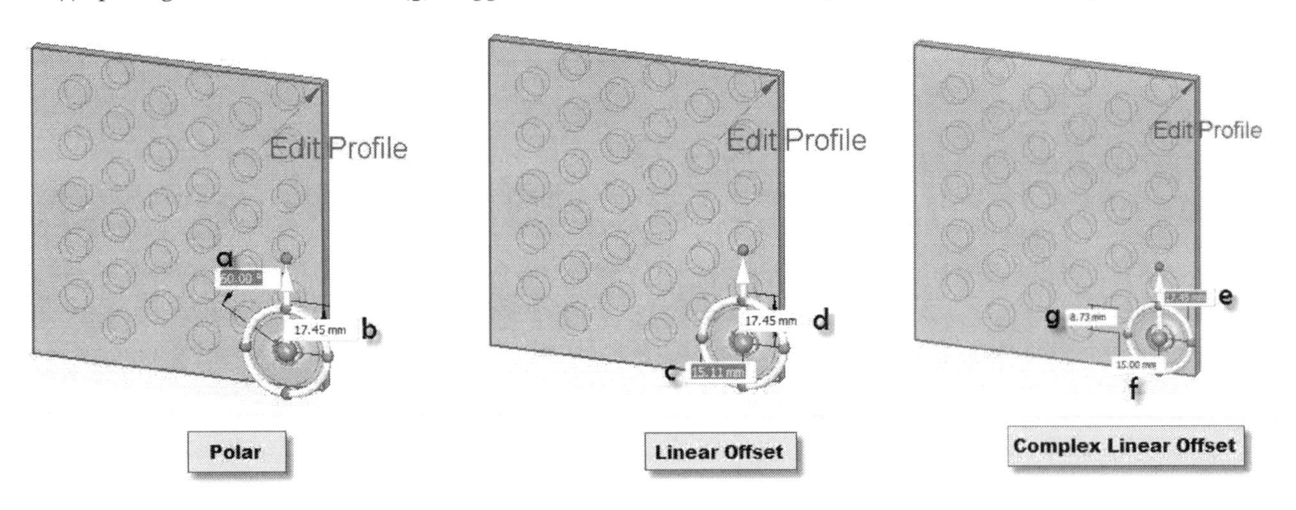

Specify the spacing parameters and click the green check to create the staggered pattern.

Radial Fill Pattern

In this type of pattern, the features are filled in a radial fashion inside the selected region. To create a radial fill pattern, set the **Fill Style** to **Radial** on the command bar.

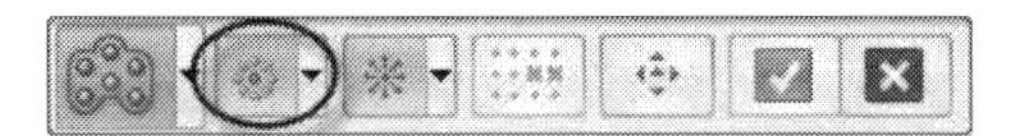

Next, define the spacing between the instances. This can be done by using the Fill spacing methods. There are two methods to define the spacing between the instances: **Target Spacing** and **Occurrence Count**. The **Target Spacing** method creates a pattern by using (a) spacing between the rings and (b) spacing between the instances. The **Occurrence Count** method creates a pattern by using (a) number of instance per ring and (b) spacing between the rings.

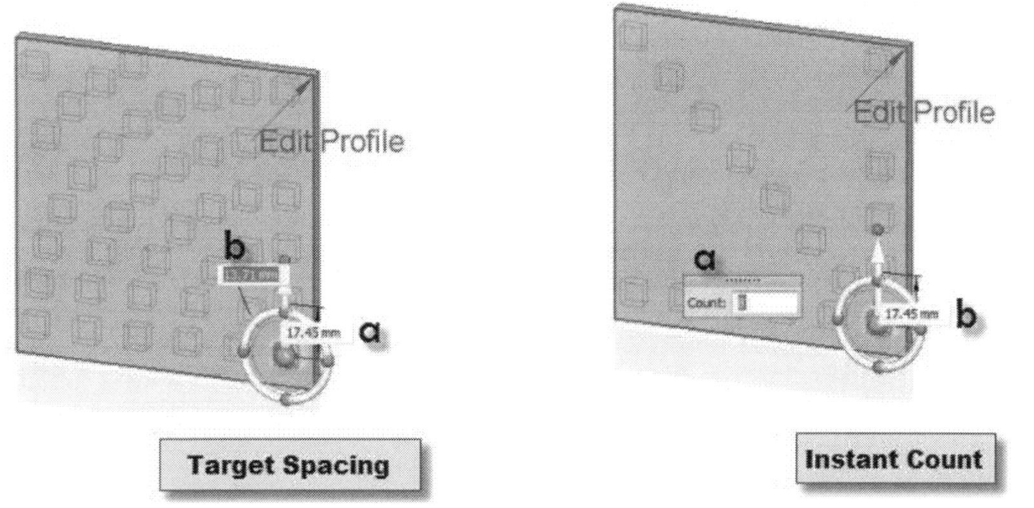

Use the **Center Orient** option to orient the feature towards the center of the radial fill pattern. Specify the spacing parameters and click the green check to create the fill pattern.

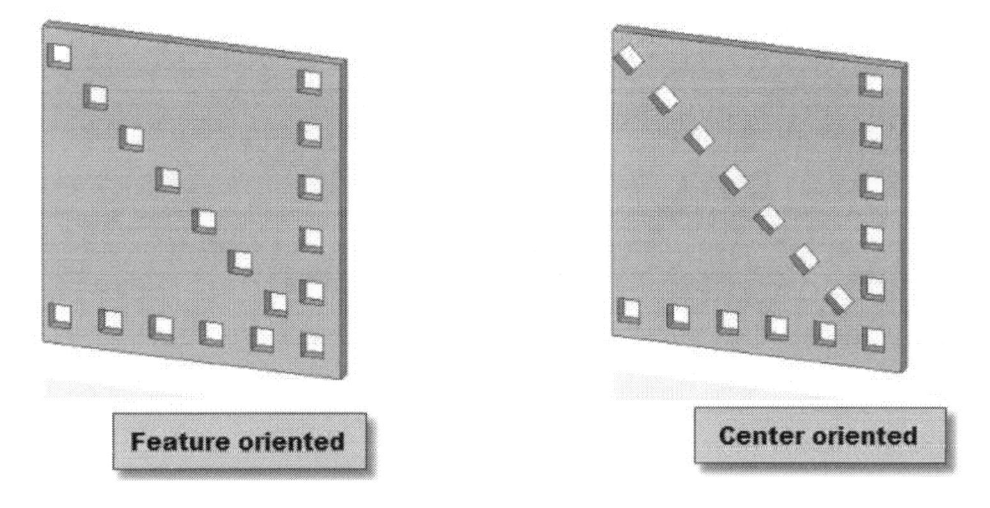

Recognize Hole Patterns

This command converts the holes arranged in a circular or rectangular fashion into patterns. It will be easier for you to modify patterns than individual features. Activate this command (click **Home > Pattern > Recognize Hole Patterns** on the ribbon) and select the holes that are arranged in a circular or rectangular fashion. Right click to accept the selection. Click the **Define Master Occurrence** button on the **Hole Pattern Recognition** dialog. Next, select a hole from the selected hole set to define the master occurrence of the pattern. Click **OK** to convert the group of holes into a pattern.

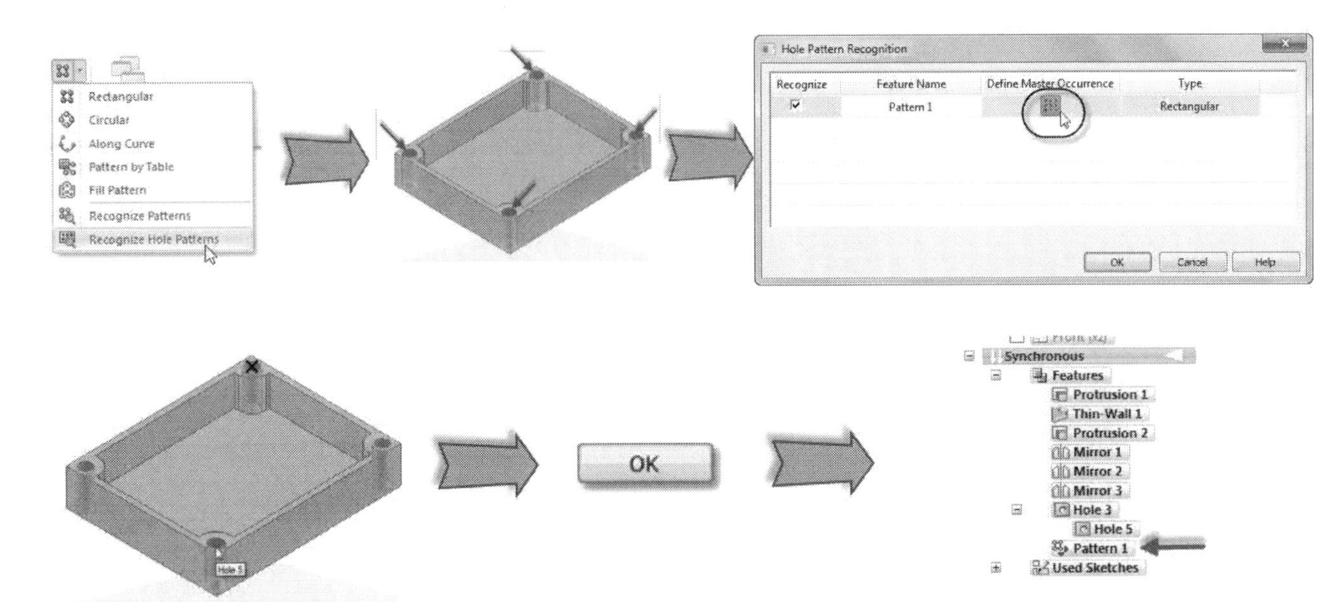

Recognize Patterns

This command is similar to the **Recognize Hole Pattern** command except that is converts any type of feature arranged in rectangular or circular fashion into a pattern feature. Activate the **Recognize Patterns** command (On the ribbon, click **Home > Pattern > Pattern drop-down > Recognize Patterns**) and select all the faces of anyone of the features arranged in the rectangular or circular fashion. Right click to accept the selection. On the **Pattern Recognition** dialog, click the **Define Master Occurrence** icon and select a feature from the pattern. The selected feature will act as the master occurrence. Click **OK** to recognize the arrangement as a pattern feature.

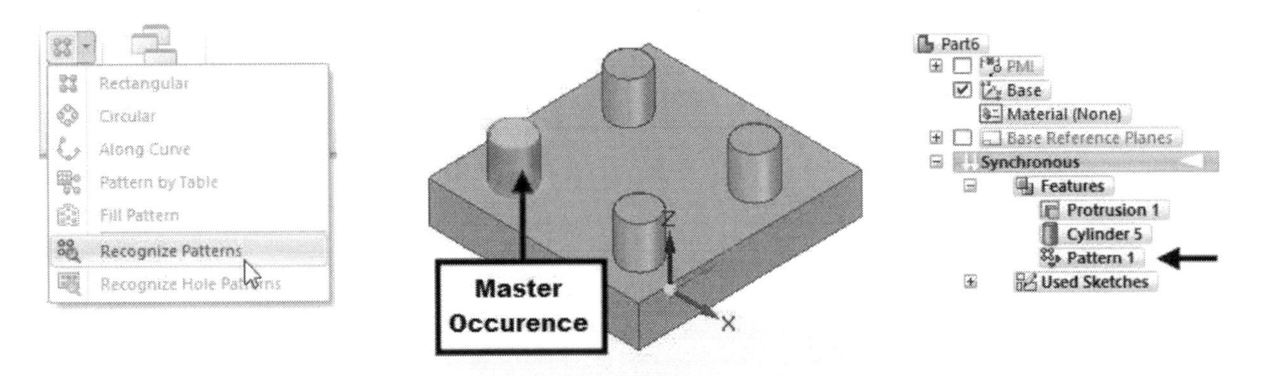

Examples

Example 1 (Millimetres)

In this example, you will create the part shown below.

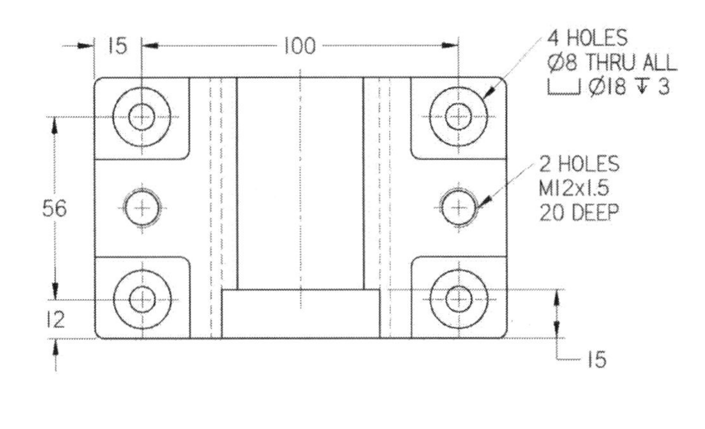

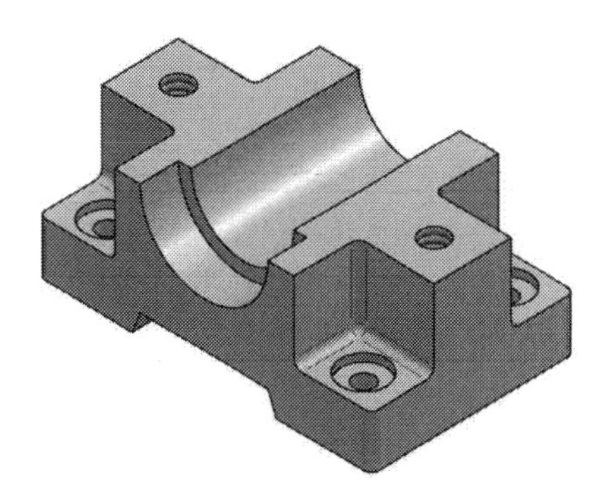

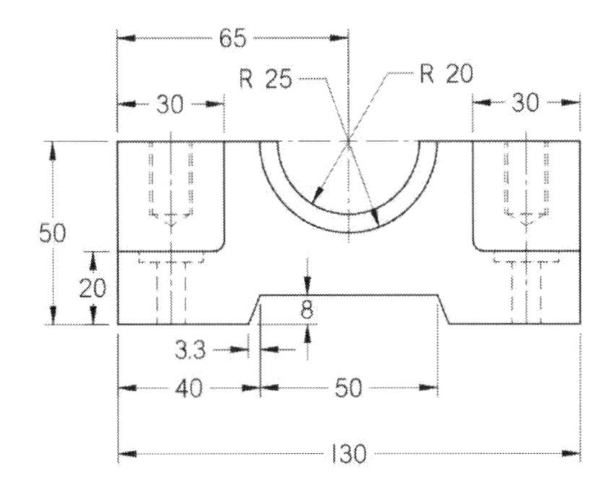

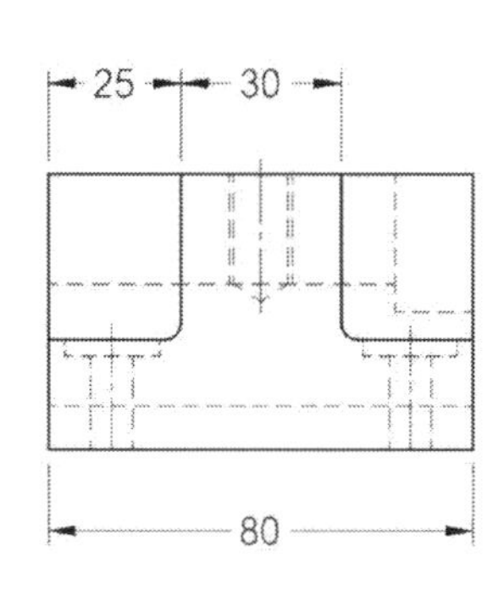

1. Start **Solid Edge 2019**.
2. On the Application Menu, click **New > ISO Metric Part**; a new part file is opened.
3. To start a new sketch, click **Home > Draw > Rectangle by Center** on the ribbon.
4. Lock the XZ plane and draw the sketch, as shown below.

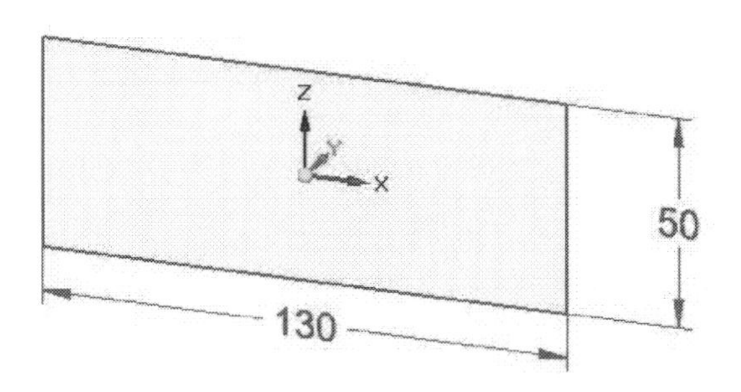

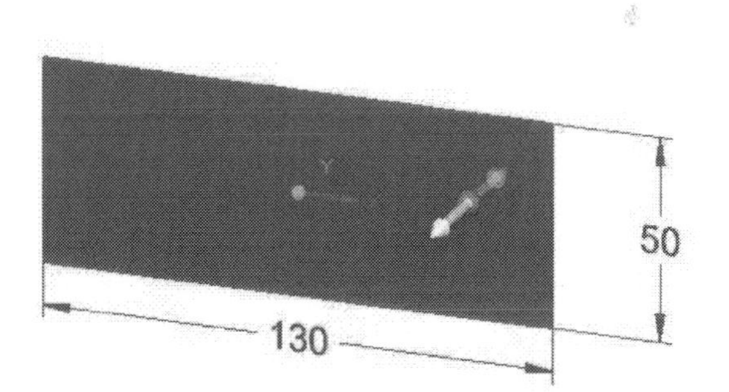

5. Create the *Extrude* feature of 80 mm thickness.

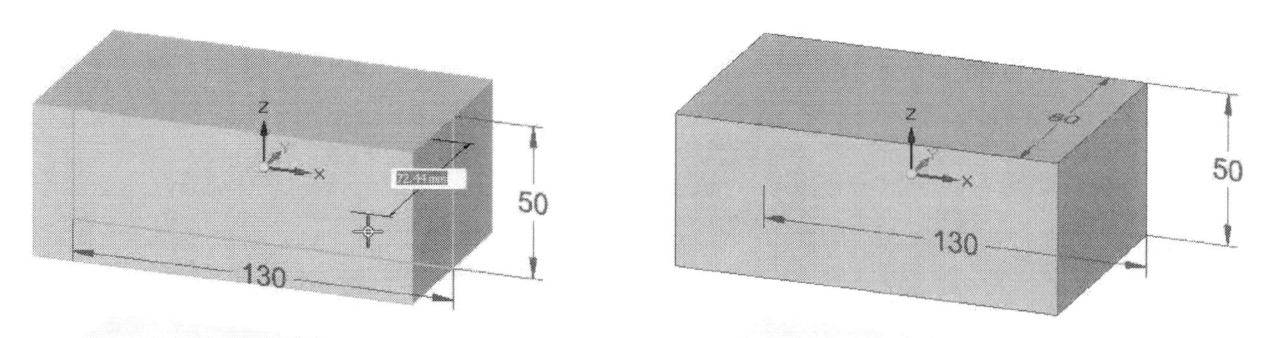

6. Click **Home > Draw > Rectangle by Center > Rectangle by 2 Points** on the ribbon.
7. Lock the top face of the part geometry and draw the sketch.
8. Create the *Cutout* feature of **30 mm** depth.

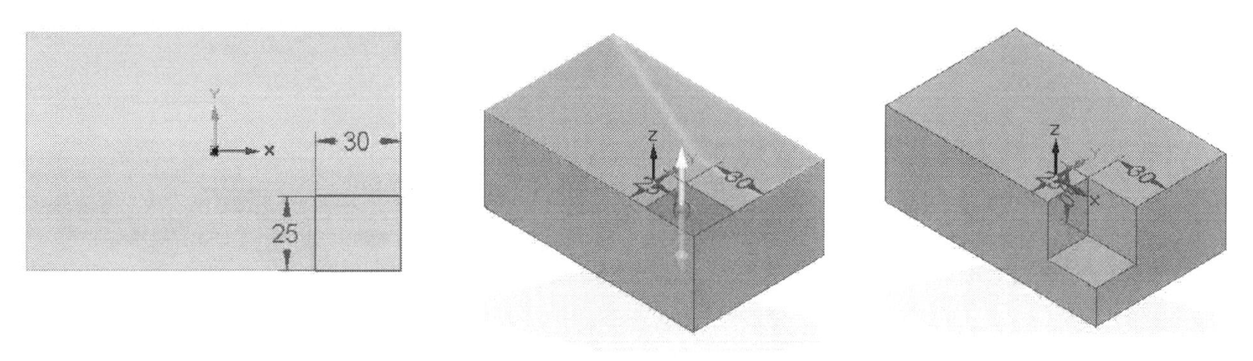

9. Click on the *Cutout* in the Pathfinder.
10. Click **Home > Pattern > Mirror** on the ribbon.
11. Check the **Base Reference Planes** option in the Pathfinder and click on the **Right (yz)** plane. The selected geometry is mirrored.

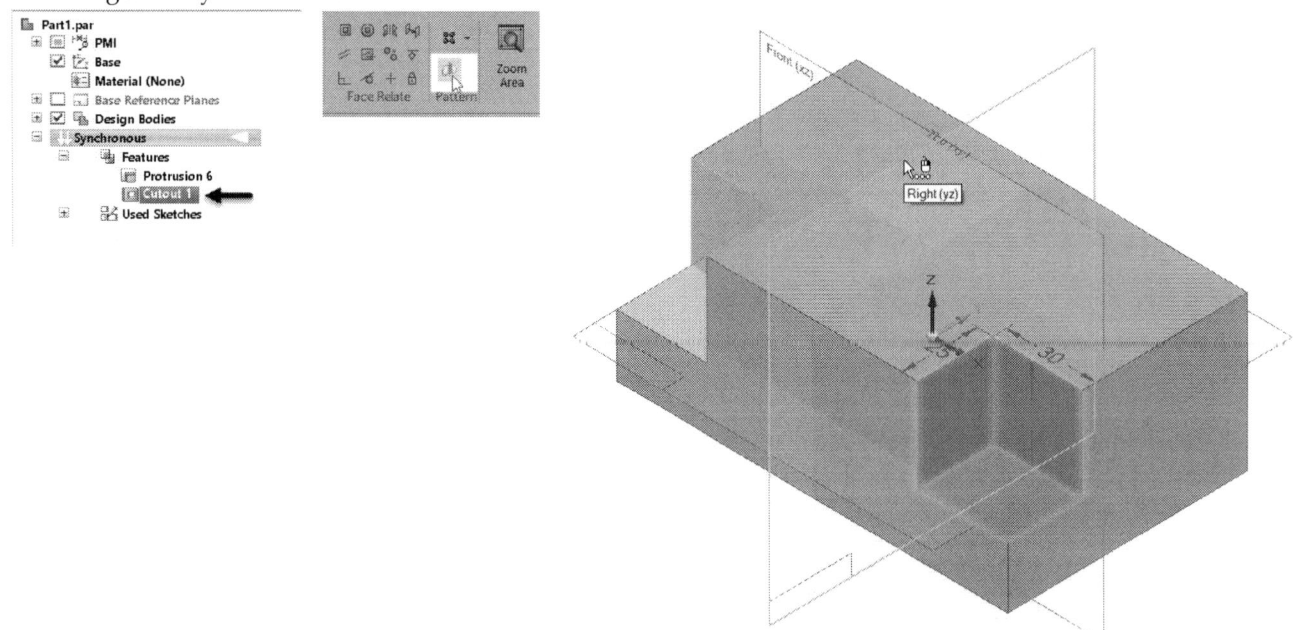

12. Press Shift on the keyboard and click on the *Cutout*, and *Mirror* in the Pathfinder.

13. Activate the **Mirror** command.
14. Click on the **Front (xz)** plane to mirror the selected geometry.

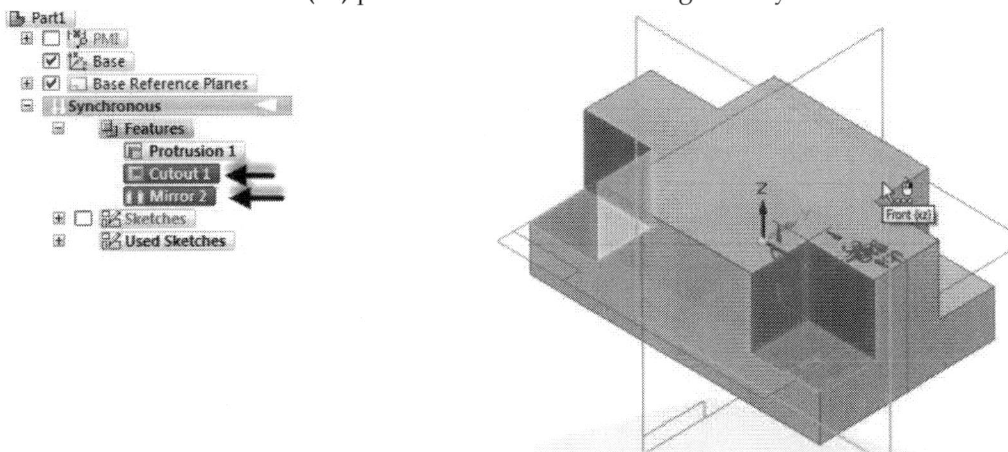

15. Activate the **Hole** command and place a counterbore hole on the *Cutout* feature. Press Esc to deactivate the **Hole** command. Also, unlock the locked face. (refer to the Create a Counterbore Hole feature topic in Chapter 4: Placed Features)

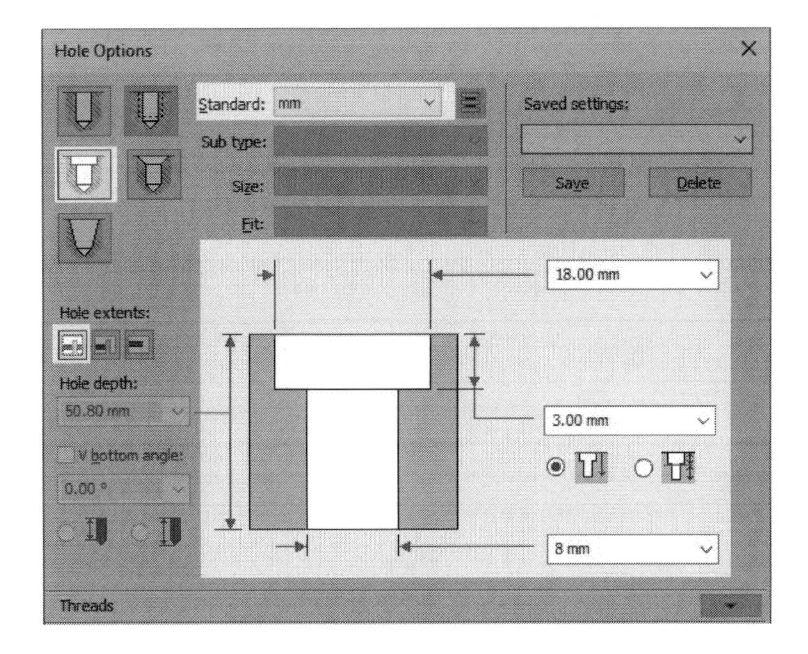

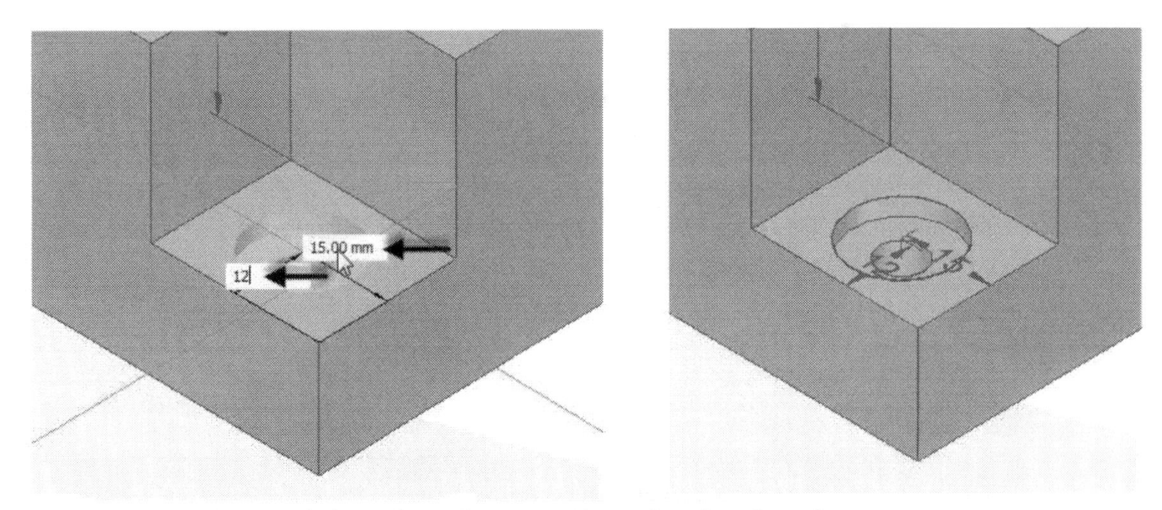

16. In the Pathfinder, click on the *Hole* option, if not already selected.
17. Click **Home > Pattern > Rectangular** on the ribbon.
18. Rotate the model such that it is displayed, as shown in figure.
19. Click on the opposite corner of the part geometry to define the rectangular pattern.
20. On the command bar, set the **Fill Style** option to **Fit**.
21. Type-in **100** in the spacing box along the X-axis, and then press Tab.
22. Type-in **56** in the spacing box along the Y-axis.
23. Type **2** in the X and Y boxes, respectively.

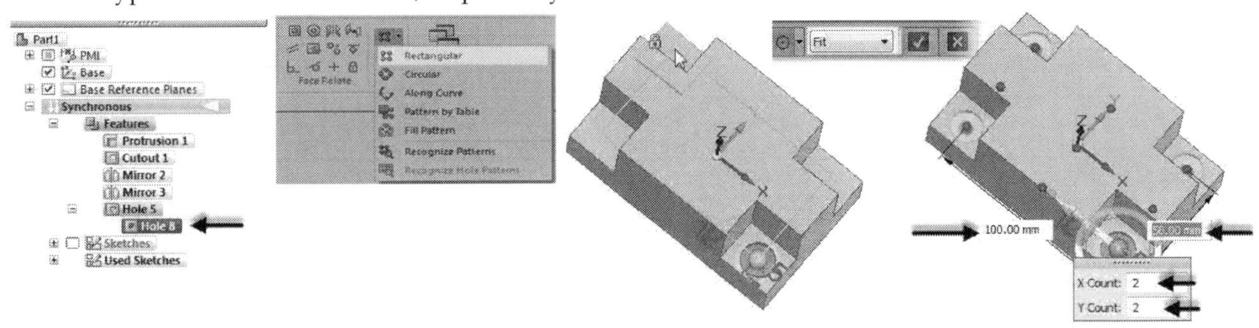

24. Click the green check on the command bar to create the rectangular pattern.
25. Activate the **Hole** command and lock the front face of the part geometry.
26. Activate the **Hole Options** dialog and set the parameters of the counterbore hole, as shown in figure.
27. Click on the midpoint of the top of the model to place the hole.

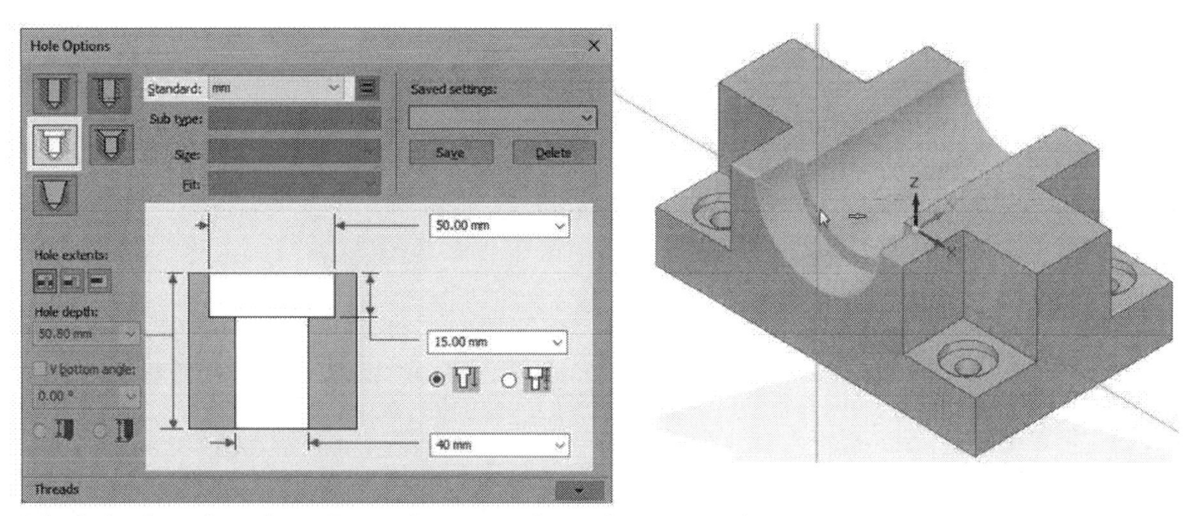

28. Unlock the front face. Press Esc to deactivate the **Hole** command.
29. Activate the **Hole** command and lock the top face of the part.
30. Activate the **Hole Options** dialog and set the parameters of the threaded hole, as shown in figure. Click **OK** to close the dialog.
31. Create the threaded hole and deactivate the **Hole** command.

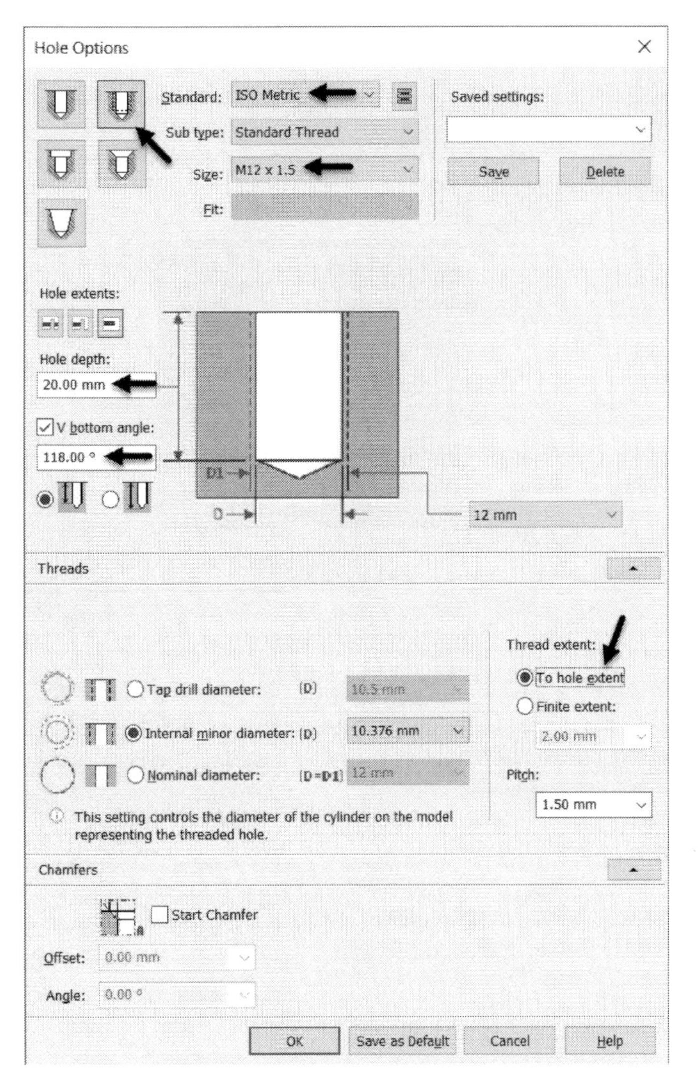

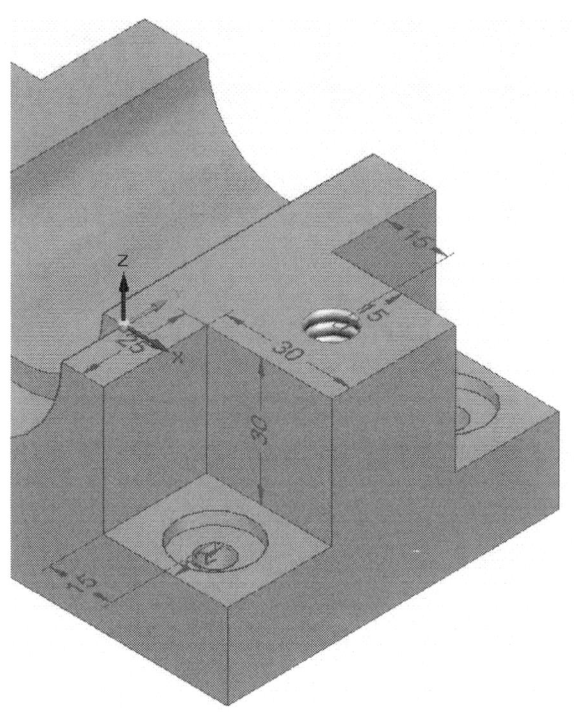

32. Mirror the threaded hole about the YZ plane.

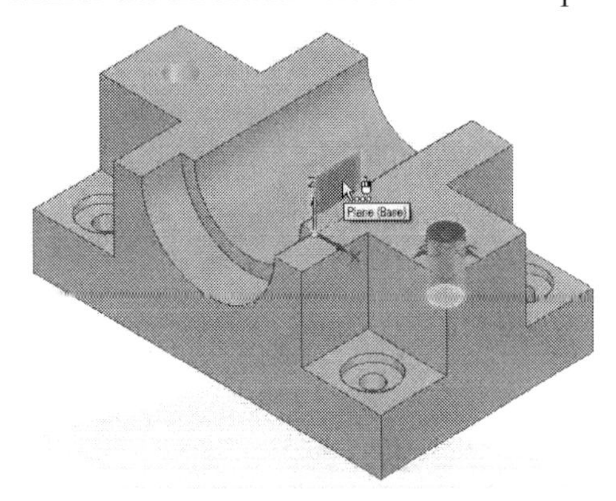

33. Draw a sketch on the front face of the part geometry and create a *Cutout* throughout the model.

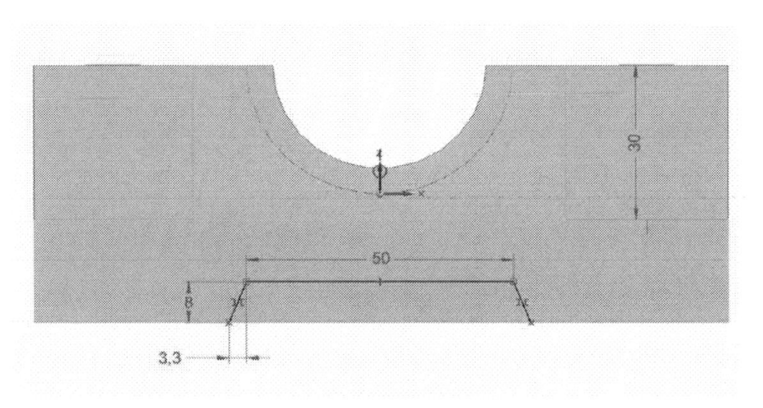

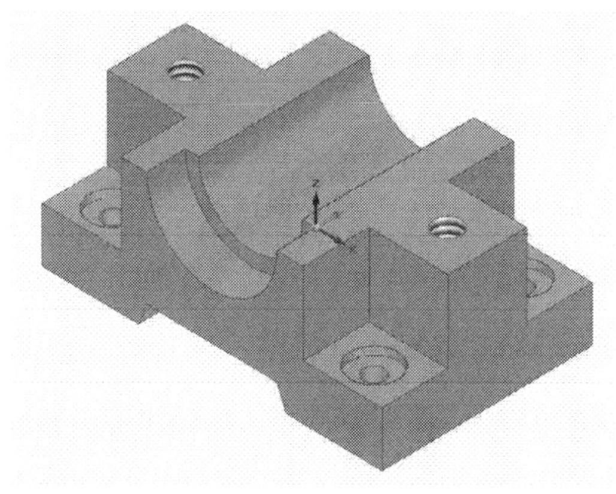

34. Round the sharp edges of the geometry. The round radius is 2 mm.

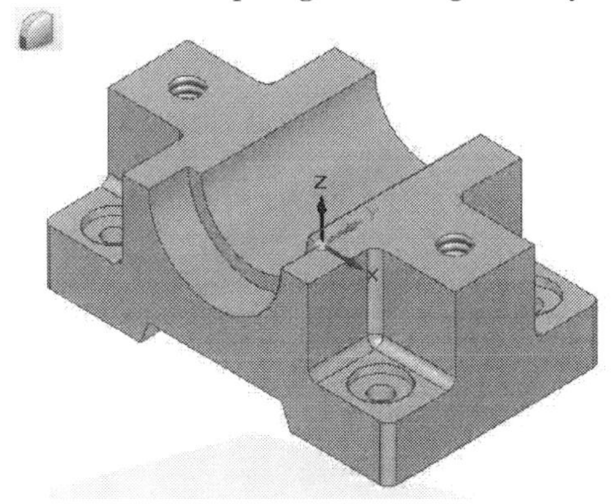

35. Save and close the part file.

Questions

1. Describe the procedure to create a mirror feature.
2. List any two commands to create patterns.
3. Why it is important to convert a set of holes into a pattern?
4. How do you add more features to an existing pattern?
5. List the options that define the orientation of the feature in a fill pattern.

Exercises

Exercise 1 (Millimetres)

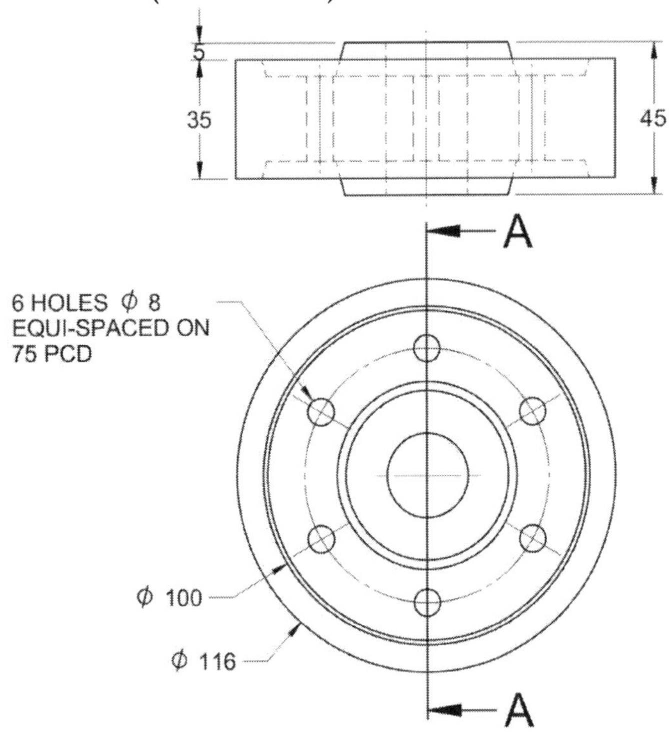

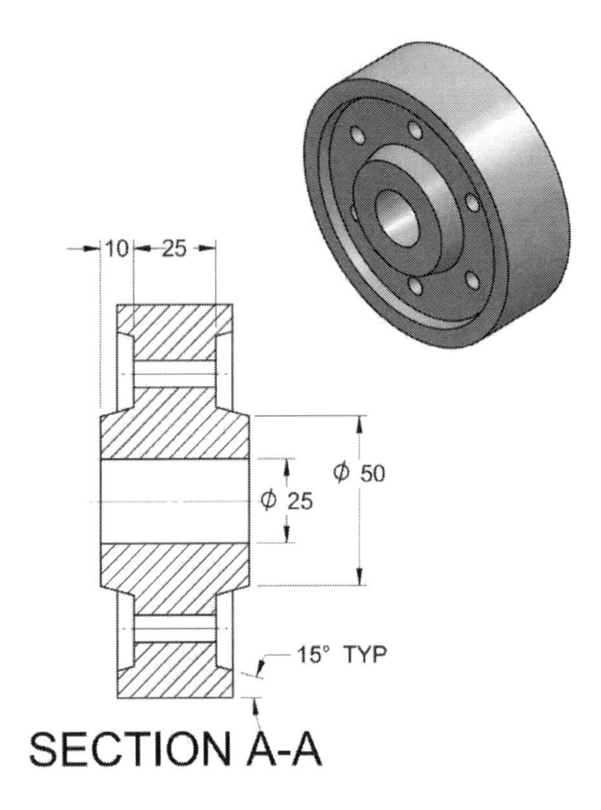

Exercise 2 (Inches)

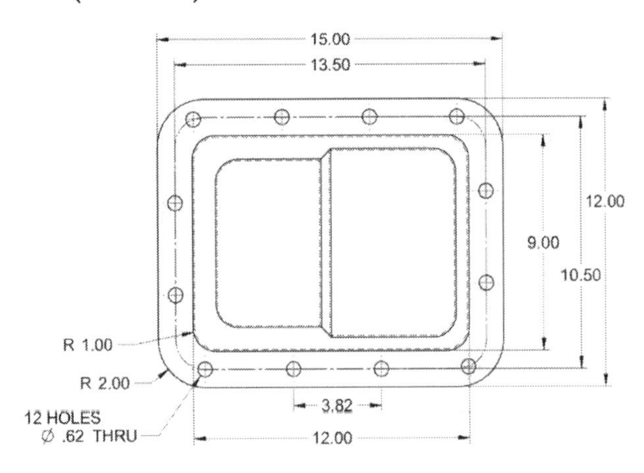

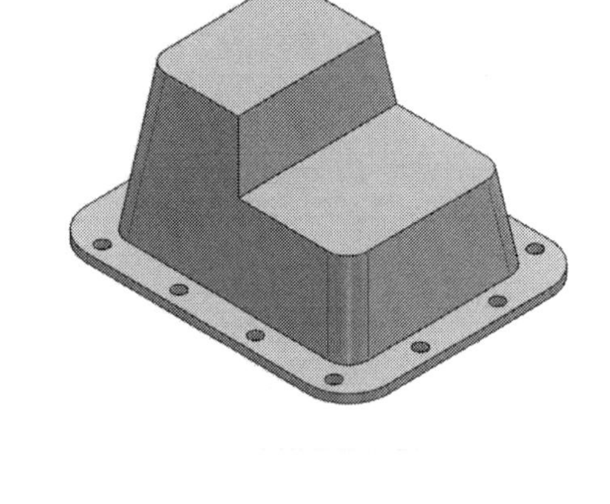

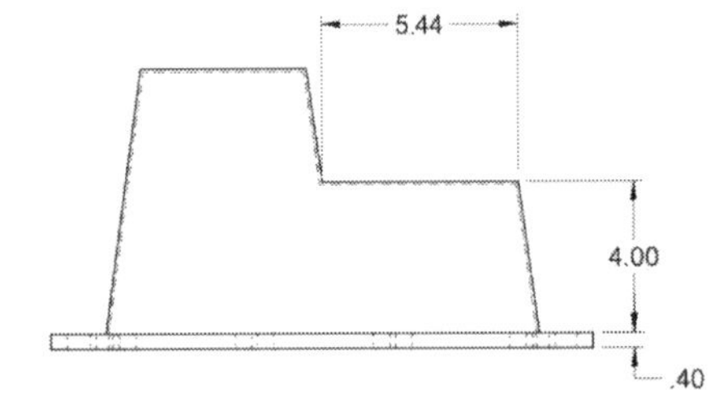

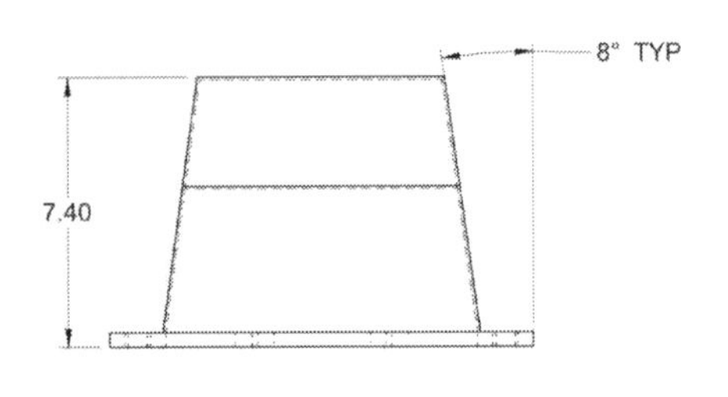

SHEET THICKNESS = 0.079 in

Chapter 6: Sweep Features

The **Sweep** command is one of the basic commands available in Solid Edge that allow you to generate solid geometry. It can be used to create simple geometry as well as complex shapes. A sweep is composed of two items: a cross-section and a path. The cross-section controls the shape of sweep while the path controls its direction. For example, take a look at the angled cylinder shown in figure. This is created using a simple sweep with the circle as the profile and an angled line as the path.

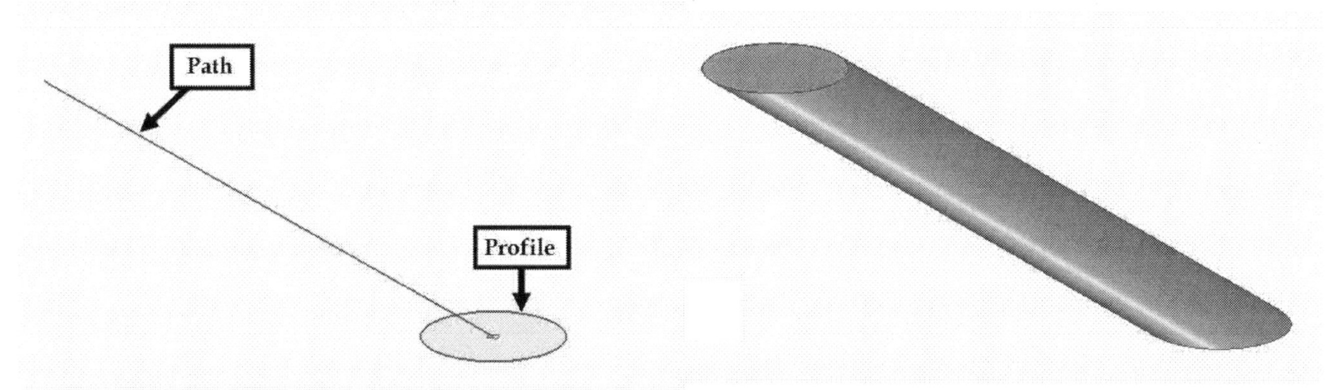

By making the path a bit more complex, you can see that a sweep allows you to create shapes you would not be able to create using commands such as Extrude or Revolve.

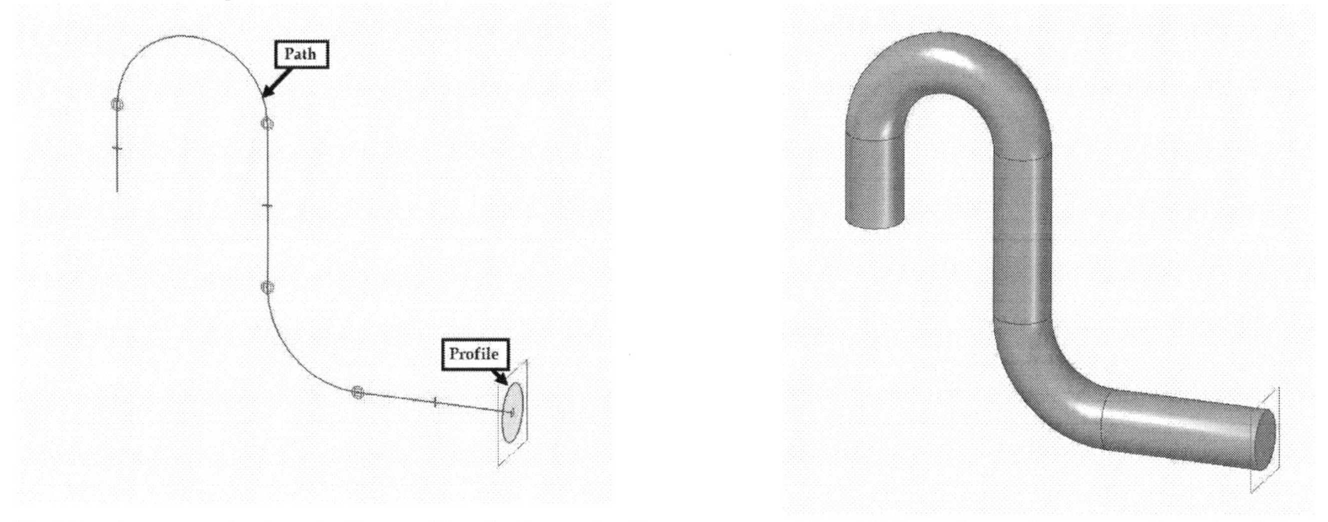

To take the sweep feature to the next level of complexity, you can add multiple paths and cross-sections. By doing so, the shape of the geometry is controlled by multiple cross-sections and paths. For example, the elliptical cross-section in figure varies in size along the path because an additional path controls it.

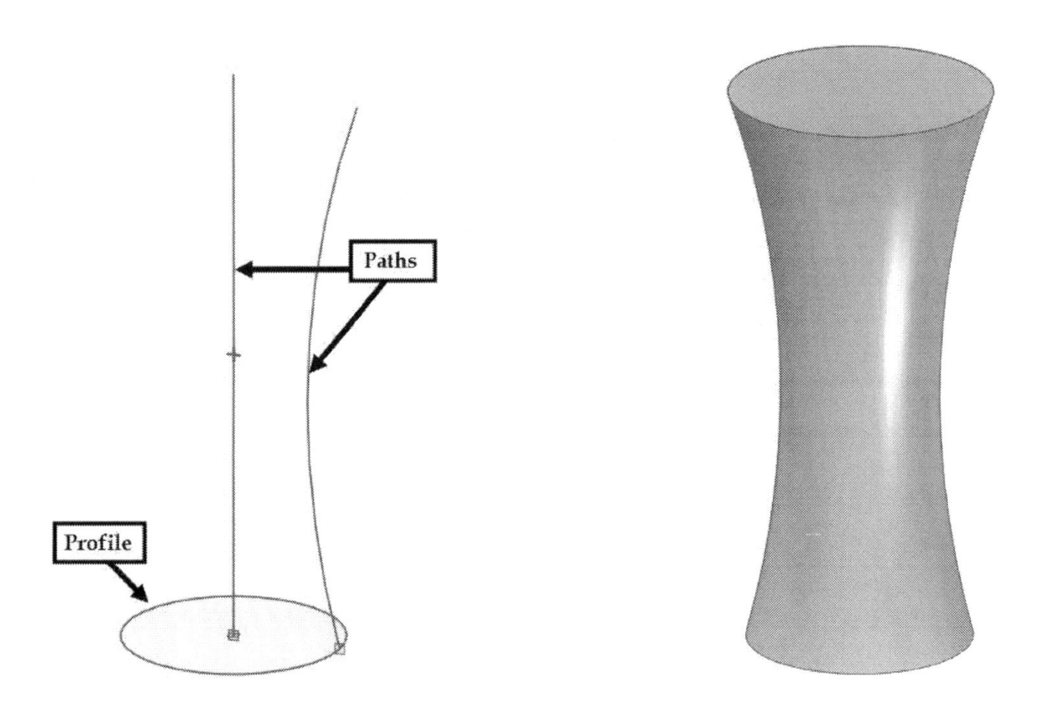

The topics covered in this chapter are:

- *Single Path and cross-section sweeps*
- *Multiple path and cross-section sweeps*
- *Scaling and twisting the cross-section along the path*
- *Swept Cutouts*
- *Helical sweeps and cutouts*

Single path and cross-section sweeps

This type of sweep requires two elements: a path and cross-section. The cross-section defines the shape of the sweep along the path. A path is used to control the direction of the cross-section. A path can be a sketch or an edge. To create a sweep, you must first create a path and a cross-section. Create a path by drawing a sketch. It can be an open or closed sketch. Next, click **Home > Planes > More Planes > Normal to Curve** on the ribbon, and then create a plane normal to the path. Sketch the cross-section on the plane normal to the path.

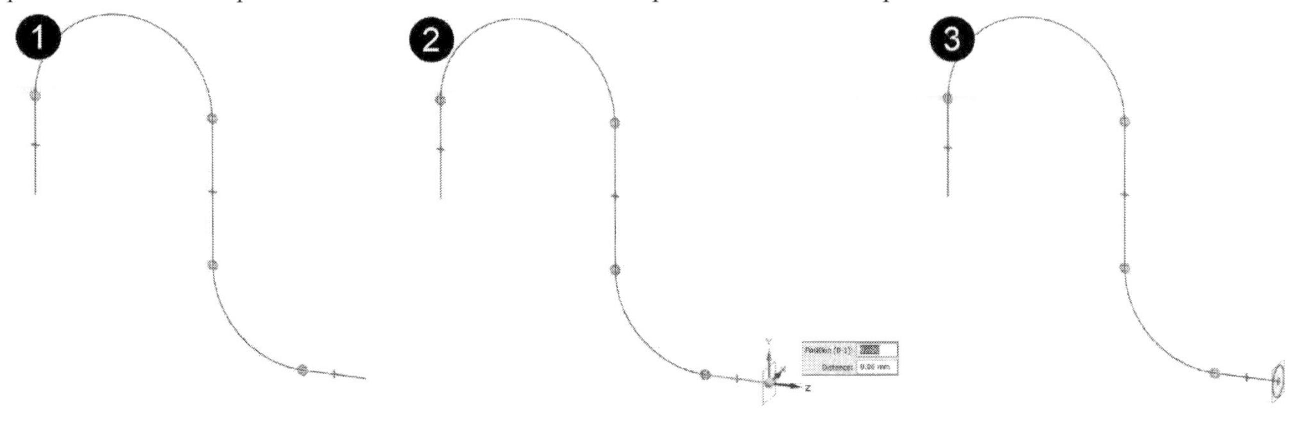

Activate the **Sweep** command (click **Home > Solids > Sweep** on the ribbon). As you activate this command, a dialog appears showing different options to create the sweep. Select the **Single path and cross-section** option on the dialog and click **OK**.

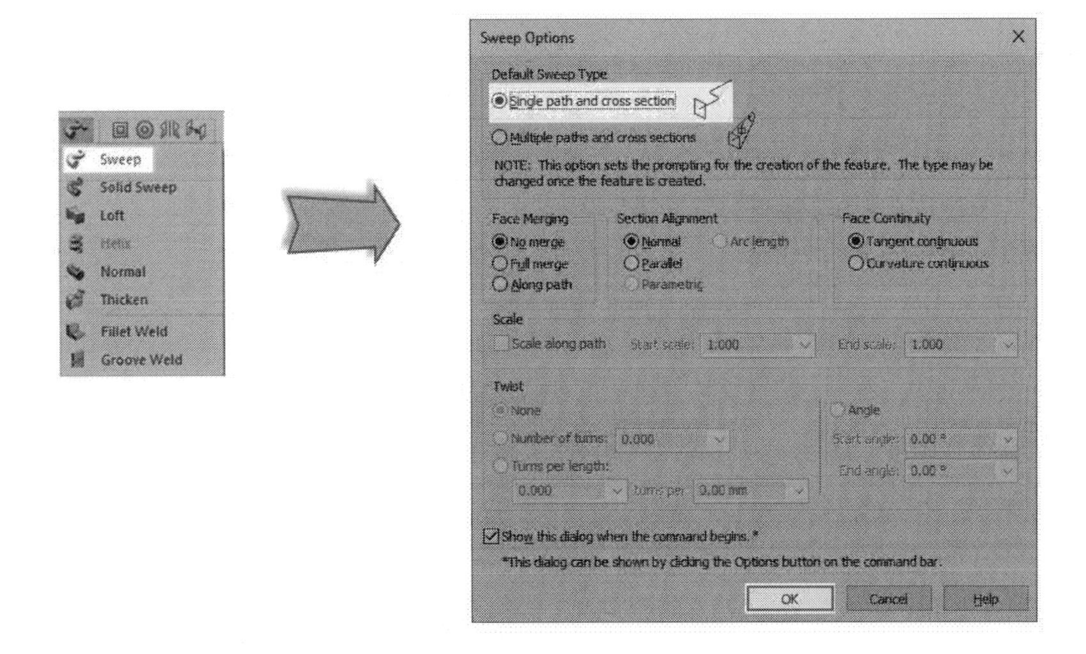

Select the path and click the green check on the command bar.

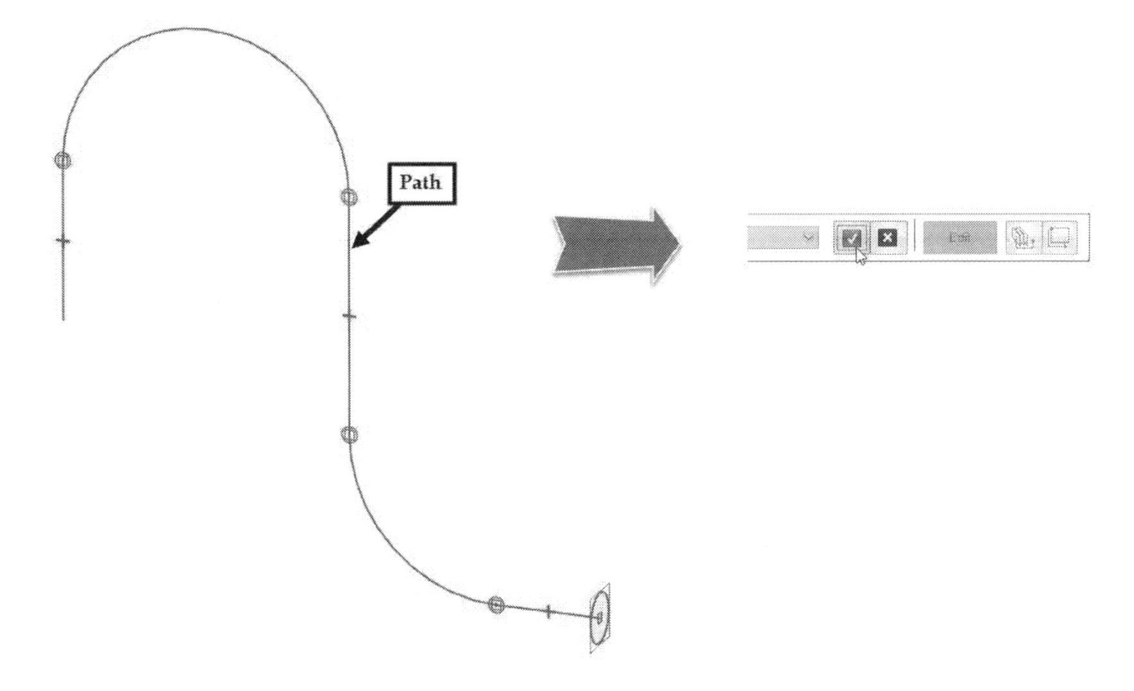

Select the cross-section and click **Finish** on the command bar. Click **Cancel** to deactivate the command.

Sweep Features

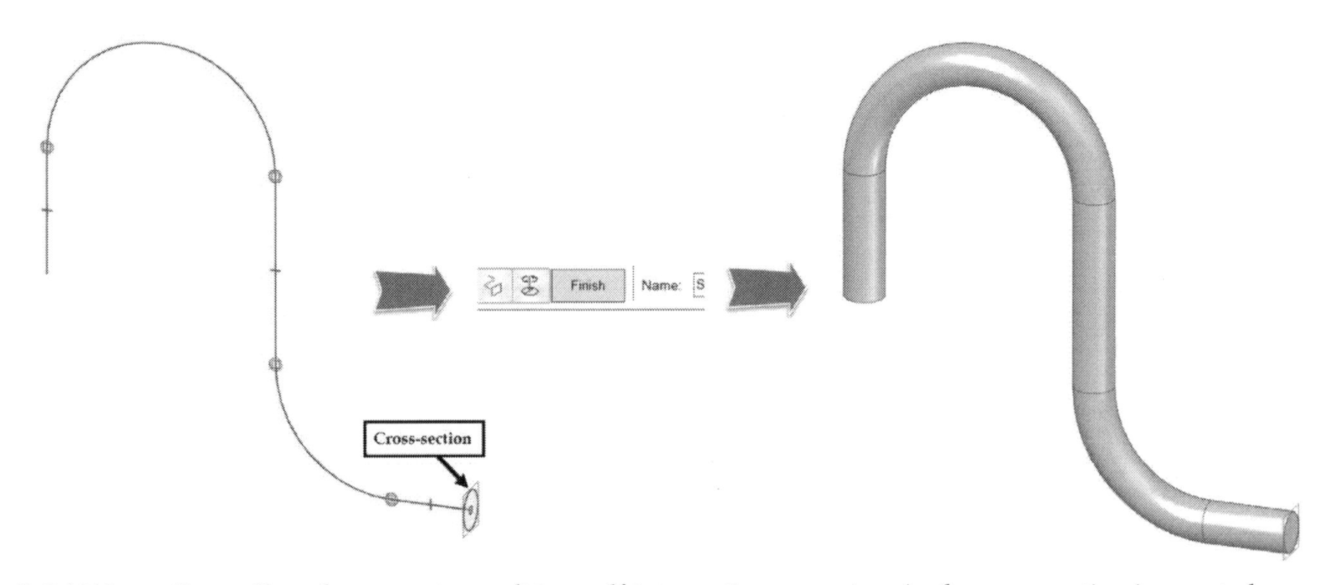

Solid Edge will not allow the sweep to result in a self-intersecting geometry. As the cross-section is swept along a path, it cannot comeback and cross itself. For example, if the cross-section of the sweep is larger than the curves on the path, the resulting geometry will intersect and the sweep will fail.

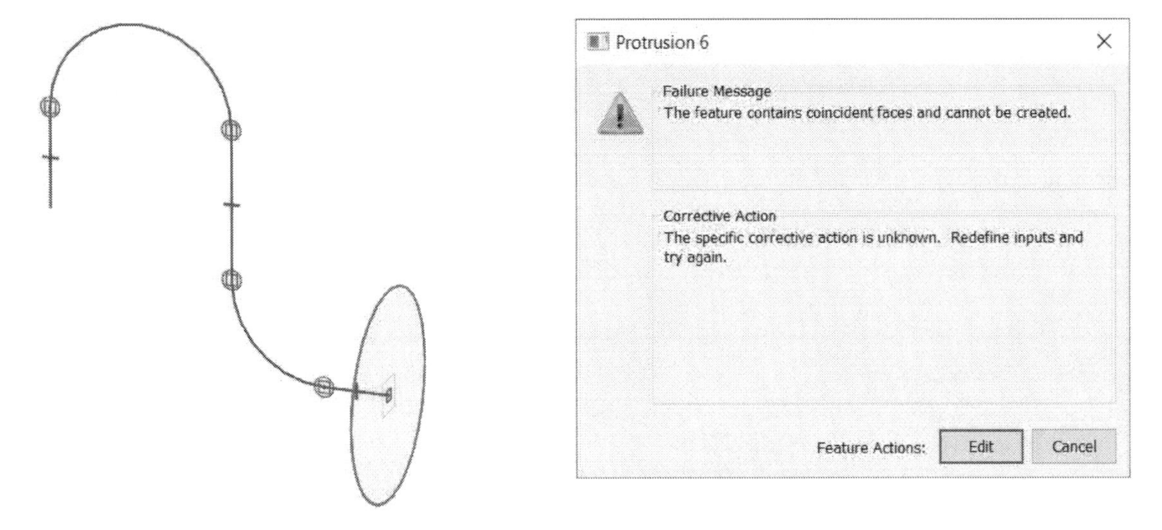

A sweep profile must be created as a sketch. However, a path can be a sketch, curve, or edge. The following illustrations show various types of paths and resultant sweep features.

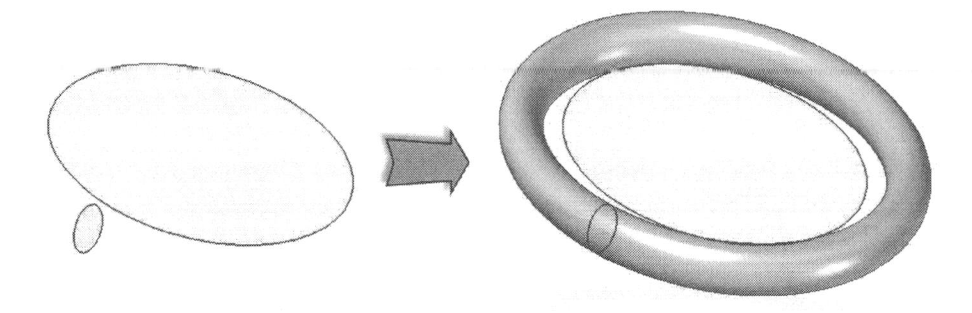

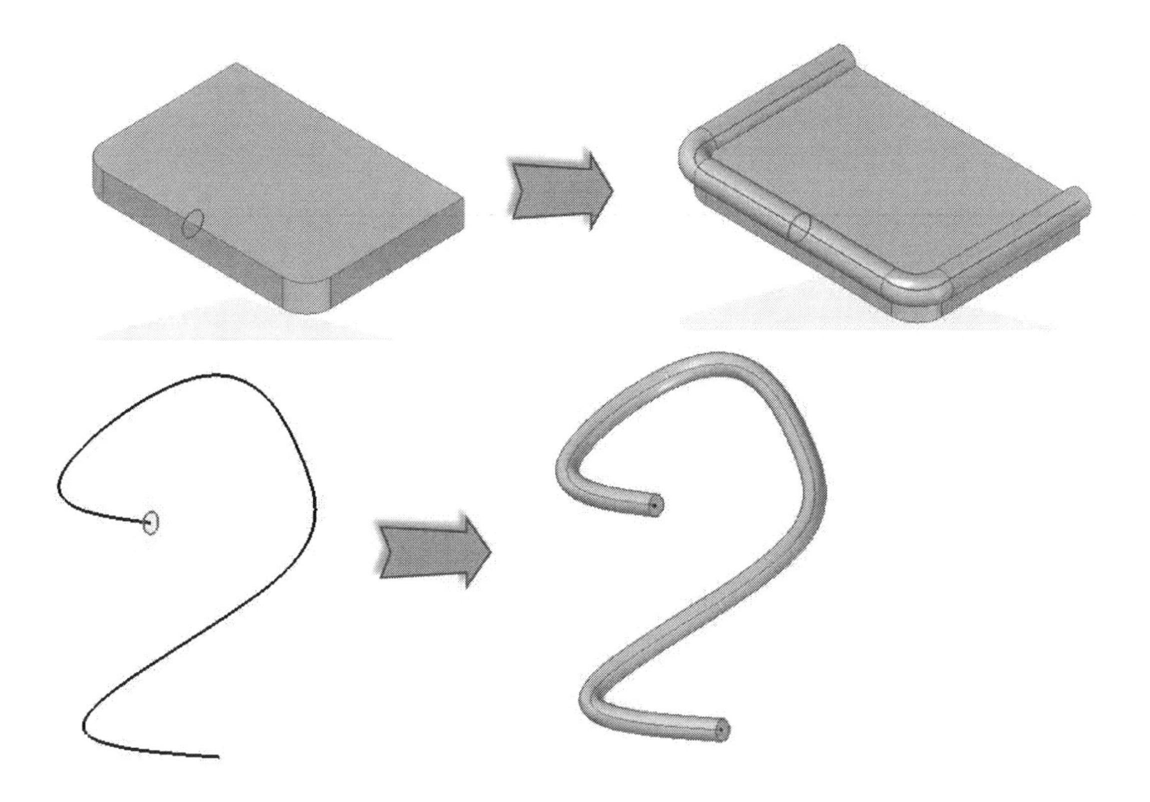

Face Merging

There are three options to merge faces of a sweep feature. These options are available on the **Sweep Options** dialog. The **No Merge** option creates a sweep feature without merging its faces. The **Full Merge** option merges all the faces of a sweep feature. The **Along path** option merges the faces along the direction of the path.

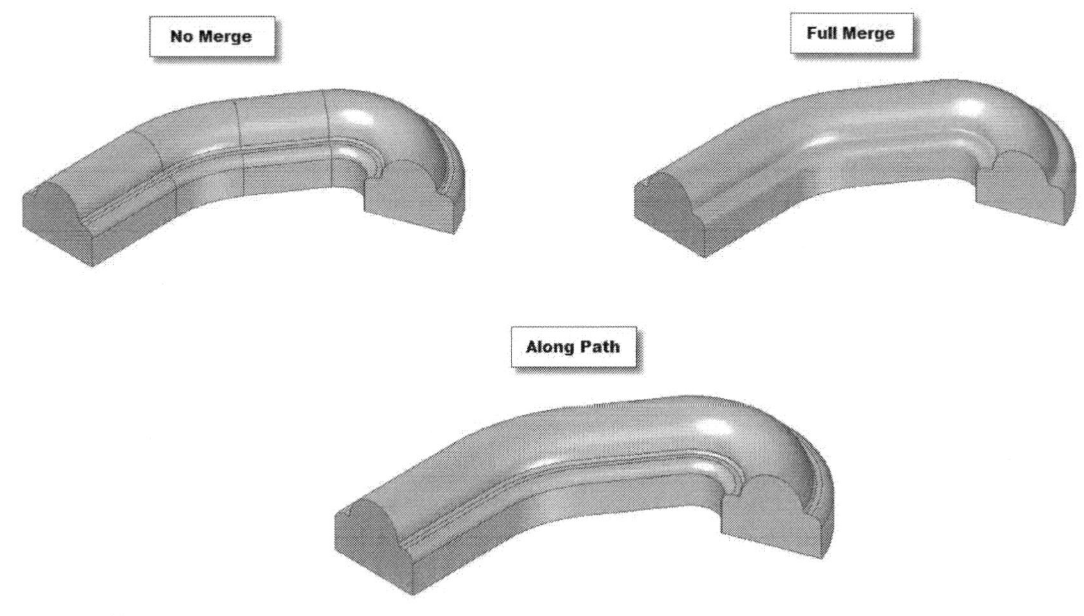

Section Alignment

The section alignment options define the orientation of the resulting geometry. The **Normal** option sweeps the cross-section in the direction normal to the path. The **Parallel** option sweeps the cross-section in the direction parallel to itself.

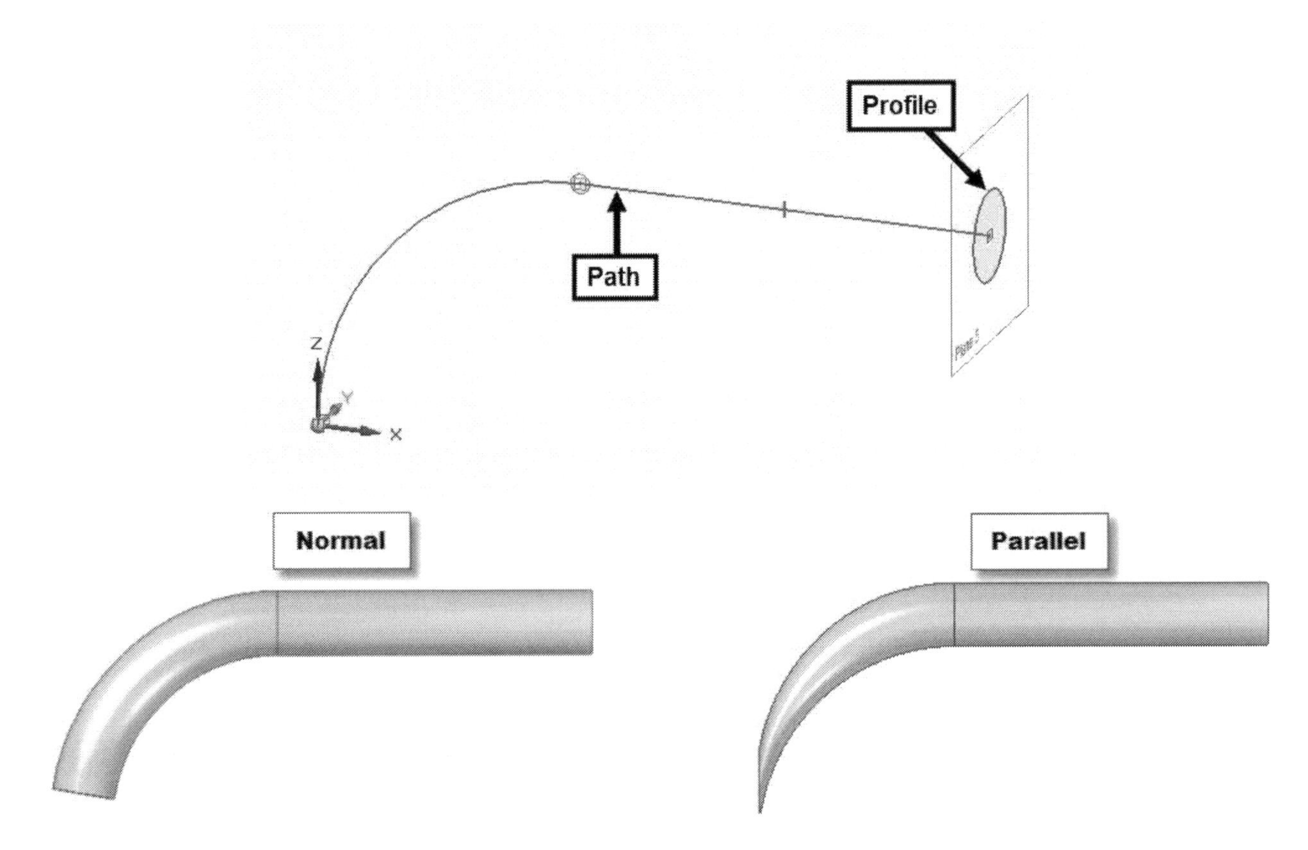

Face Continuity

The **Face Continuity** options define the tangency condition between the faces of a sweep feature. The **Tangent Continuous** option makes two faces tangent and continuous to each other. The **Curvature Continuous** option maintains the tangency as well as radius of curvature between two faces of a sweep feature.

Scale

Solid Edge allows you to scale the sweep along the path. Select the path and cross-section, and then click the **Options** icon on the command bar. Check the **Scale along path** option on the **Sweep Options** dialog, and then type-in the start and end scale factors. Click **OK** and **Finish** creating a scaled sweep feature.

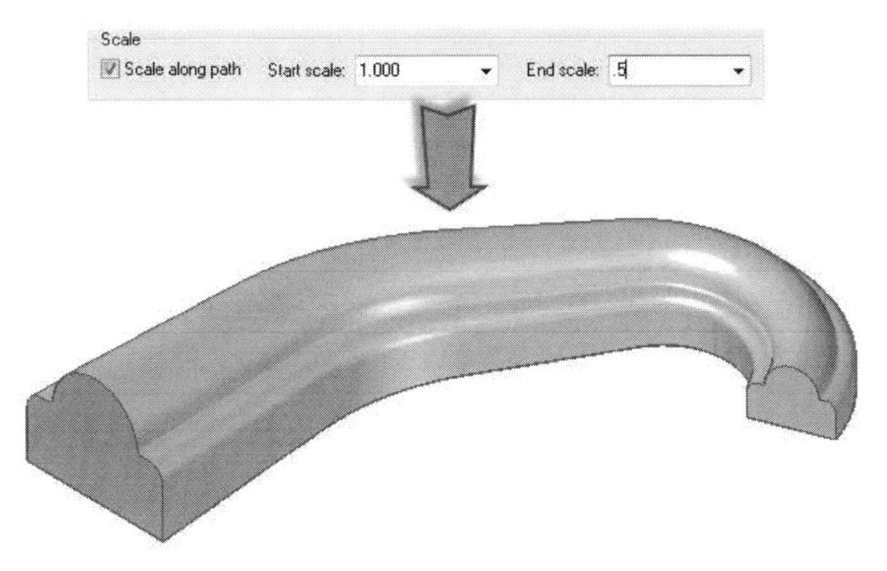

Twist

Solid Edge allows you to twist the cross-section along the path. Define the path and cross-section, and then click the **Options** icon on the command bar. The **Twist** options on the **Sweep Options** dialog help you to apply a twist to the cross-section.

The **Number of Turns** option turns the cross-section by the value you enter in the box.

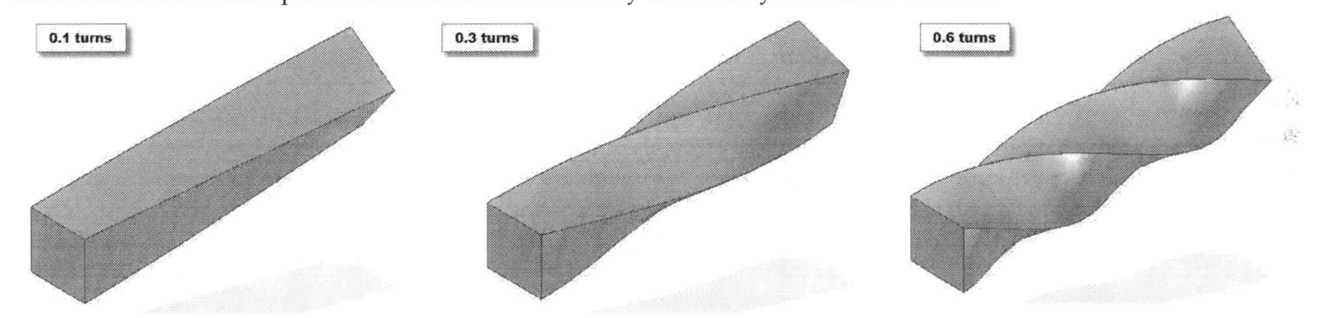

The **Turns per Length** option twists the cross-section by number of turns and length that you enter in the boxes.

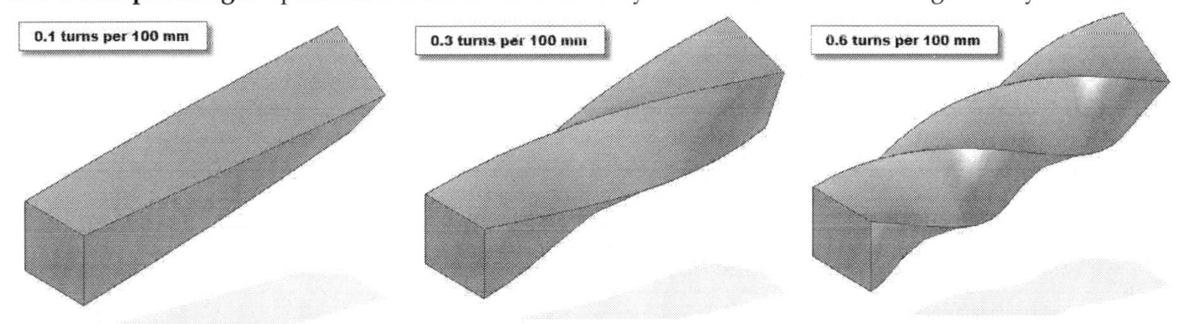

The **Angle** option twists the cross-section by an angle. Select this option and type-in values in the **Start Angle** and **End Angle** boxes.

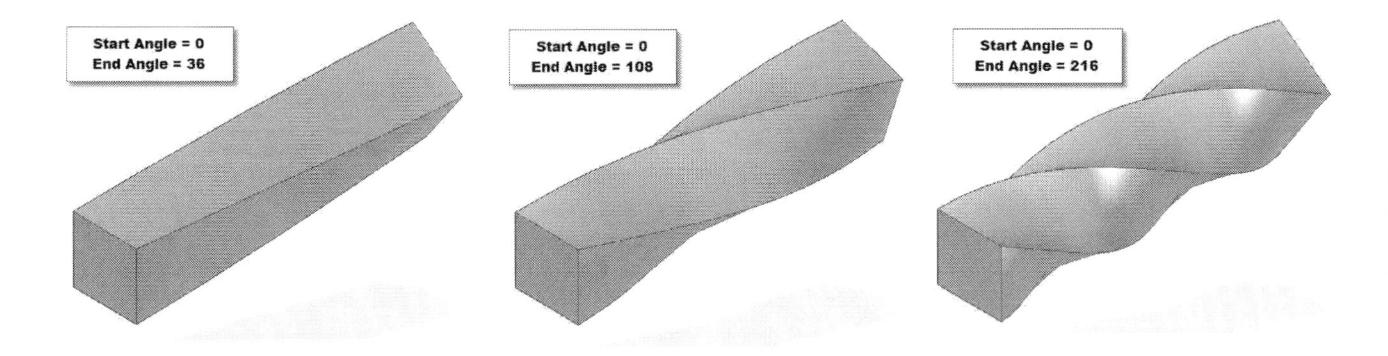

Axis Step

The **Axis Step** option on the command bar will be useful while sweeping a cross-section along a non-planar path. For example, define a path and cross-section similar to the one shown in figure and click the **Axis Step** option on the command bar. Select a line or axis from the Base coordinate system. The cross-section and the axis will be locked in the same plane. As a result, the orientation of the cross-section and axis become same and the cross-section will be swept maintaining the orientation of the axis.

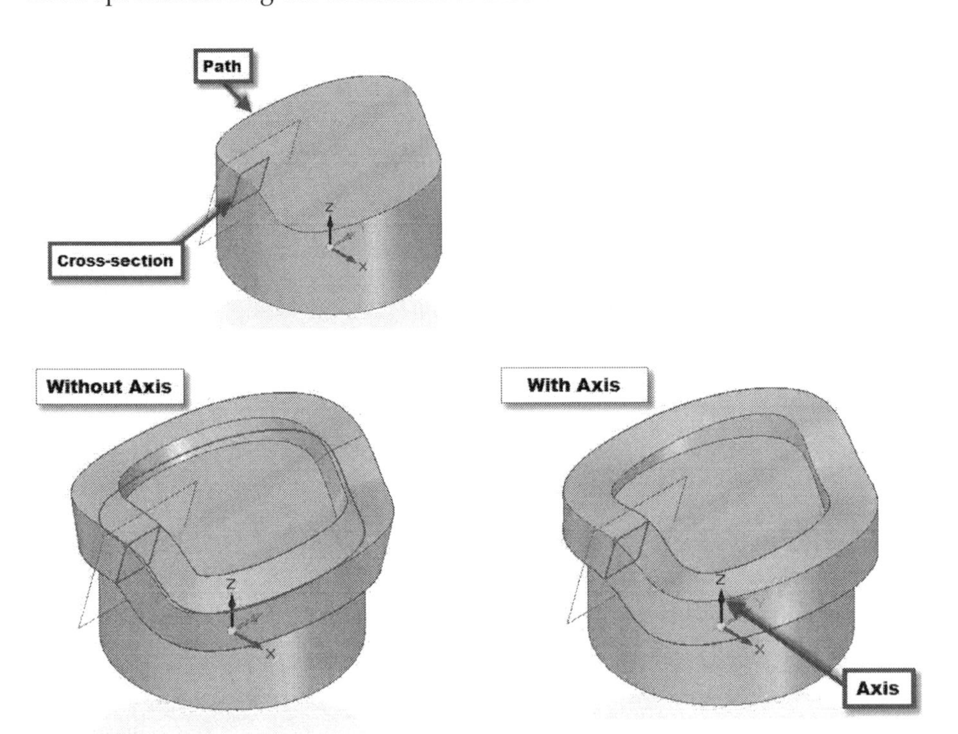

Multiple paths and cross-sections sweeps

Solid Edge allows you to create sweep features with multiple paths and cross-sections. This can be useful while creating complex geometry and shapes. To create this type of sweep feature, first create multiple paths and cross-sections as shown in figure. Activate the **Sweep** command and select **Multiple paths and cross-sections** on the **Sweep Options** dialog. Click **OK** to close the dialog.

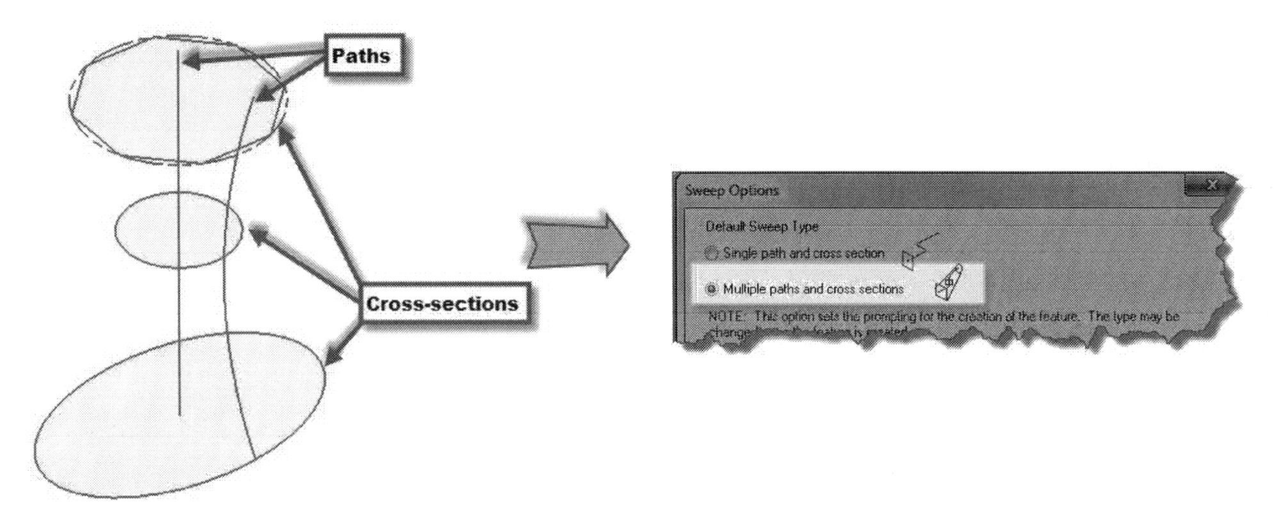

Select the first path and click the green check on the command bar. Select another path and click the green check on the command bar. Select the third path, if available. Otherwise, click **Next** on the command bar. Select the all the cross-sections one-by-one and click **Preview** on the command bar. The preview of the geometry will appear. Click **Finish** to complete the feature.

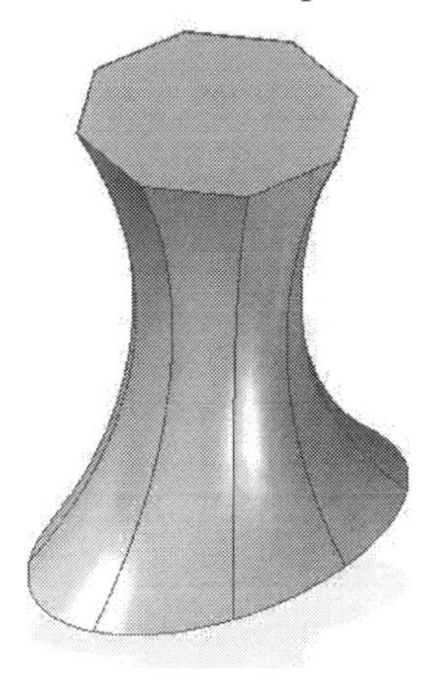

Swept Cutout

In addition to adding swept features, Solid Edge allows you to remove geometry using the **Swept Cutout** command. Activate this command (click **Home > Solids > Swept Cutout** on the ribbon) and select the sweep type from the **Sweep Options** dialog. Click **OK** and select the path. Click the green check on the command bar to accept the path. Select the cross-section and click **Finish** to create the swept cutout.

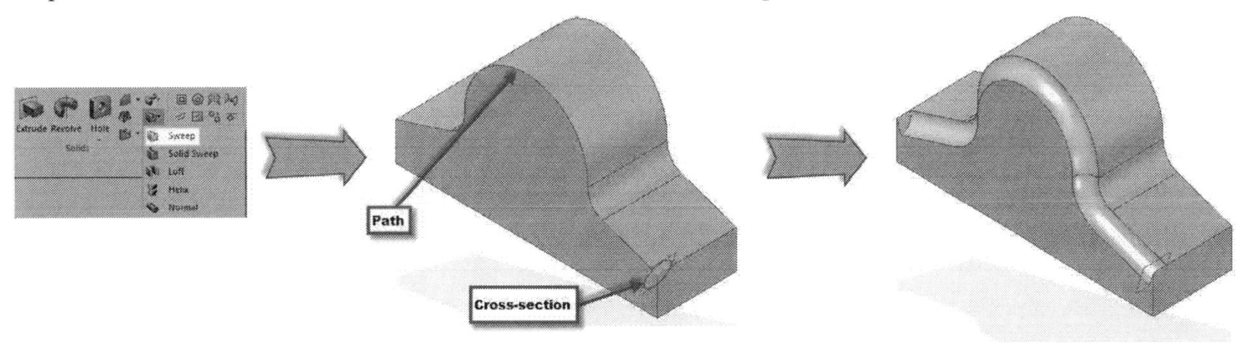

You will notice that the swept cutout is not created throughout the geometry. This is because the cross-section is swept only up to the endpoints of the path. In this case, you must define a new path, which extends beyond the geometry. Delete the swept cutout from the Pathfinder and create two lines, which are continuous and collinear with the path. Activate the **Derived** command (click **Surfacing > Curves > Derived** on the ribbon) and select the edges and lines. Click the green check on the command bar to create a new curve. Now, create a swept cutout by using the curve as path. The resultant swept cutout will be throughout the geometry.

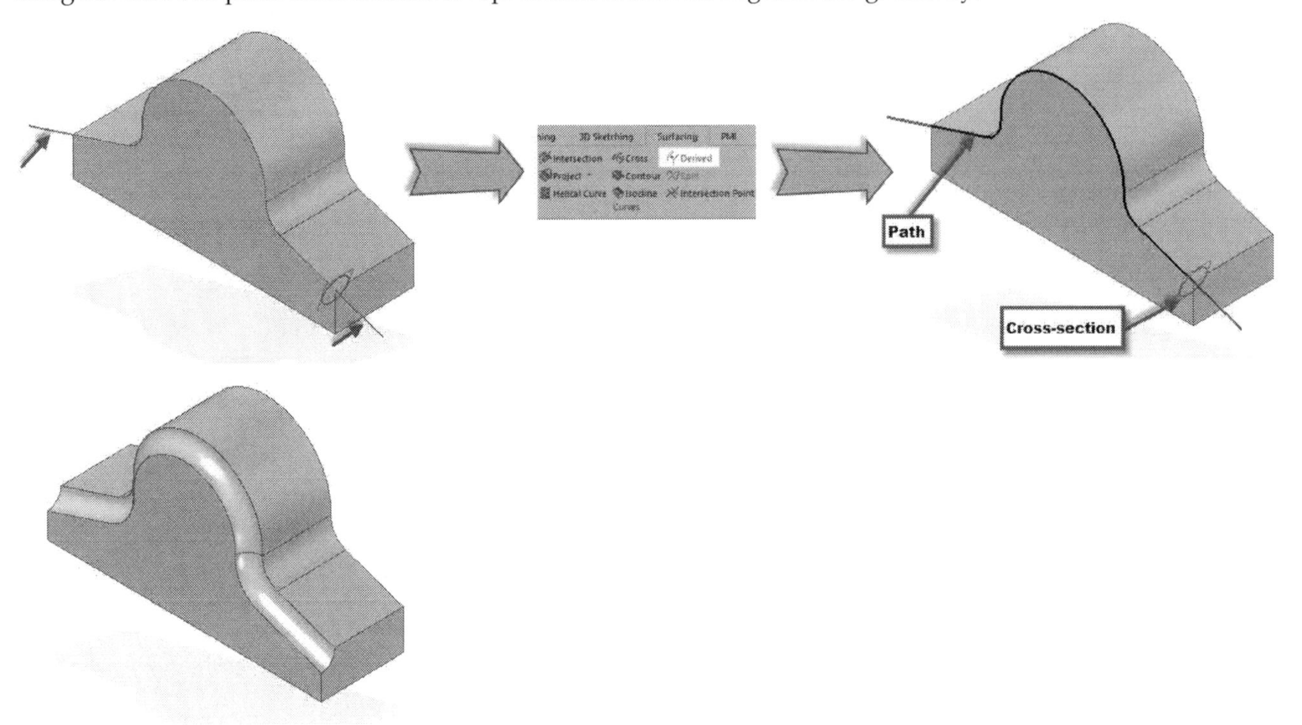

Helix

This command creates are spring shape feature. To create this type of feature, you must have a cross-section and a line (axis). They can be on a same plane or on different planes. Activate the **Helix** command (click **Home > Solids > Helix** on the ribbon), and then select the cross-section and line. Click the green check on the command bar. The preview of the geometry appears on the screen.

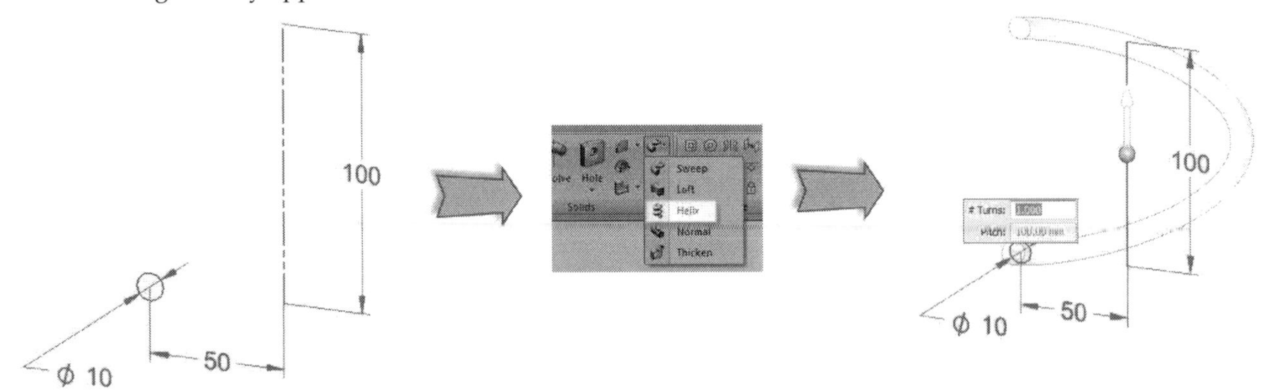

Now, define the **Helix Method** on the command bar. There are three helix methods: **Axis & Pitch**, **Axis & Turns**, **Pitch & Turns**. The **Axis & Pitch** method creates a helix by using the length of the axis and distance between the turns. The **Axis & Turns** method creates a helix by using the axis length and number of turns. The **Pitch & Turns** method uses the pitch and number of turns you specify to create the helix.

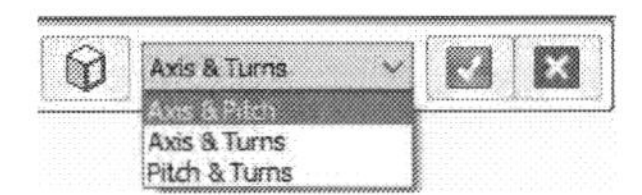

For more helix options, click the **Helix Options** icon on the command bar. The **Helix Options** dialog pops up on the screen. This dialog has many options to define the parameters of the helix feature (such as helix direction, taper, and pitch). Define the helix direction by selecting the **Right-handed** or **Left-handed** option.

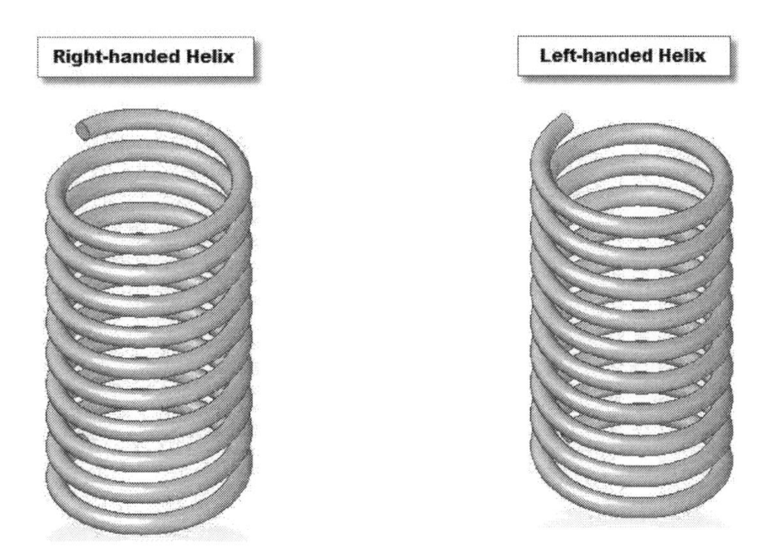

The **Taper** options on the **Helix Options** dialog help you to apply taper to the helix. There are two methods to apply taper to a helix: **By Angle** and **By Radius**. The **By Angle** method applies a taper to the helix by using the taper angle that you enter in the **Angle** box. The **Inward** or **Outward** options define the taper direction. The **By Radius** method applies a taper to the helix by using the start and end radius that you specify.

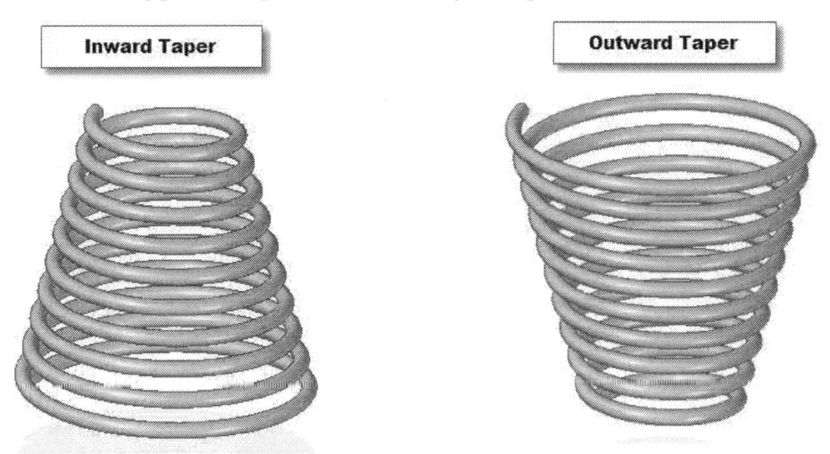

The **Pitch** options on the **Helix Options** dialog help you to create a variable pitch helix. Select the **Variable** option from the drop-down menu and type-in the **Pitch ratio** and **End Pitch** values. For example, if you specify the **Start Pitch =10**, **Turns = 10**, and **End Pitch = 20**, the pitch of the helix varies from 10 to 20. The rate of change in the pitch is calculated by the formula:

$$Rate\ of\ change\ in\ pitch = \frac{End\ Pitch - Start\ Pitch}{No.of\ turns} = \frac{20-10}{10} = 1$$

$$\text{The start pitch} = \text{Start Pitch} + \frac{\text{Rate of change in pitch}}{2}$$

$$\text{The end pitch} = \text{End Pitch} - \frac{\text{Rate of change in pitch}}{2}$$

Therefore, the pitch of the first turn = 10+.5 =10.5
Second turn = 10.5+1 = 11.5
Third turn = 11.5+1= 12.5..............................tenth turn=19.5

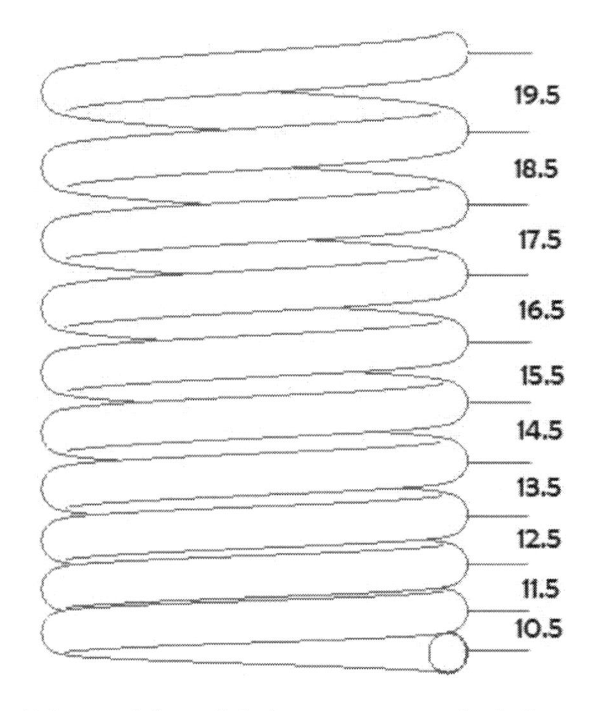

Click **OK** on the **Helix Options** dialog, and then click **Accept** to create the helix.

Helical Cutout

This command removes material from the part geometry by creating a helical feature. To create this feature, first you must have an existing geometry, and the sketches of the cross-section and axis. Activate this command (click **Home > Solids > Swept Cutout > Helical Cutout** on the ribbon) and select the cross-section and axis. Click the green check on the command bar to accept the selection. Define the number of turns and pitch using anyone of the Helix methods described in the previous topic. Next, right-click to create the helical cutout.

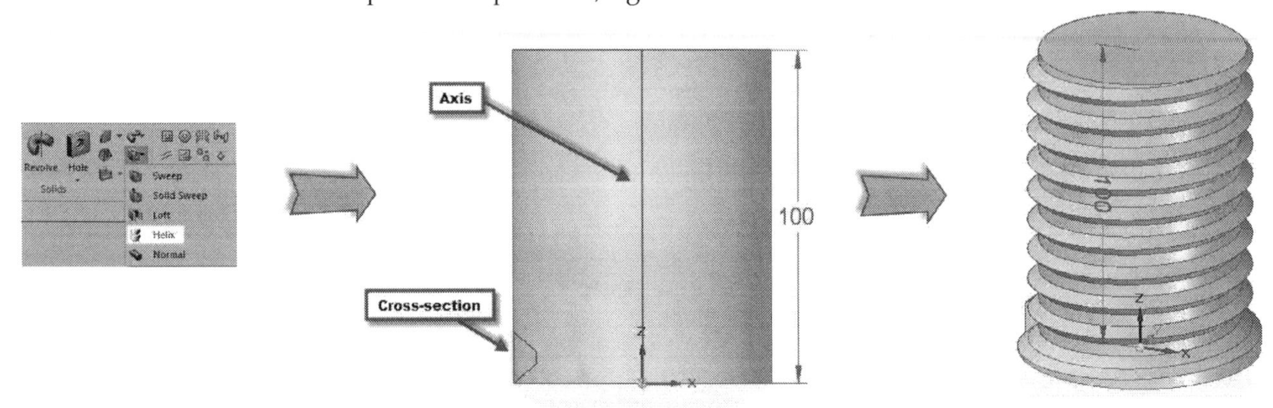

Examples

Example 1 (Inches)

In this example, you will create the part shown below.

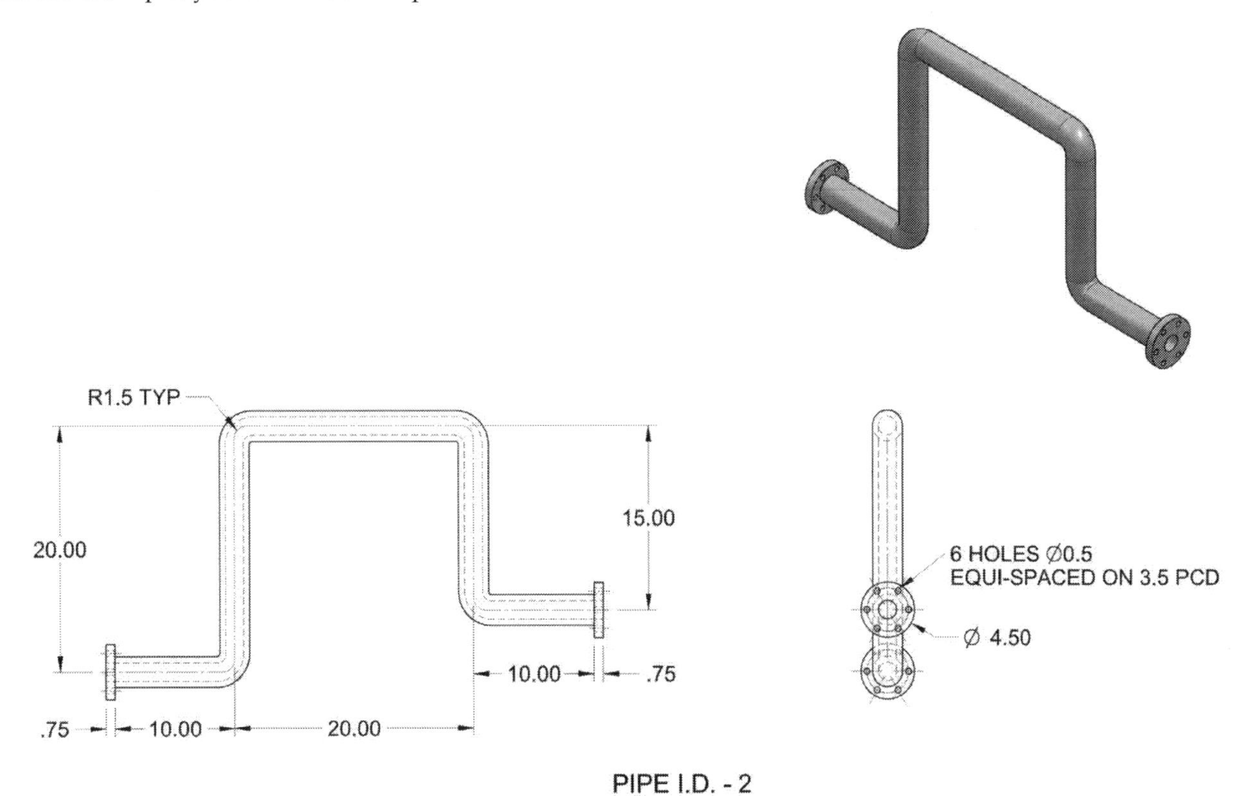

PIPE I.D. - 2
PIPE O.D. - 2.5

1. Start **Solid Edge 2019**.
2. On the **Quick Access Toolbar**, click **New**; the **New** dialog pops up.
3. On this dialog, click **Standard Templates > ANSI Inch**. Select the **ansi inch part.par** template and click **OK**.
4. On the ribbon, click **Home > Draw > Line** and draw the sketch on the XZ plane, as shown below.

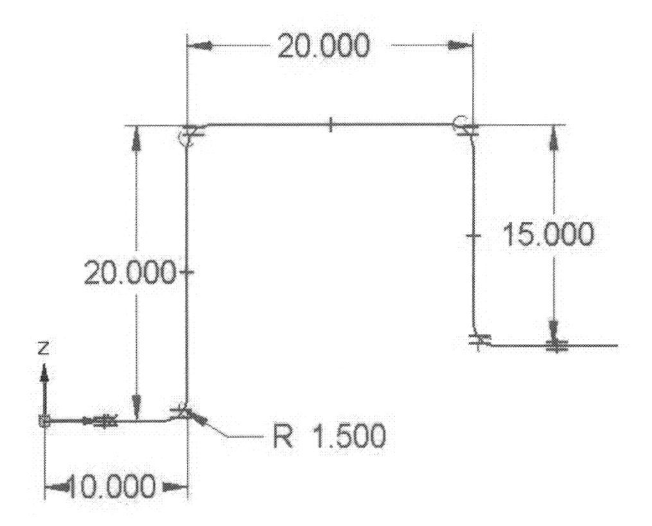

5. Unlock the sketch plane.
6. On the ribbon, click **Home > Planes > More Planes > Normal to Curve** and click on the lower horizontal line.
7. Click on the endpoint of the line to locate the plane.

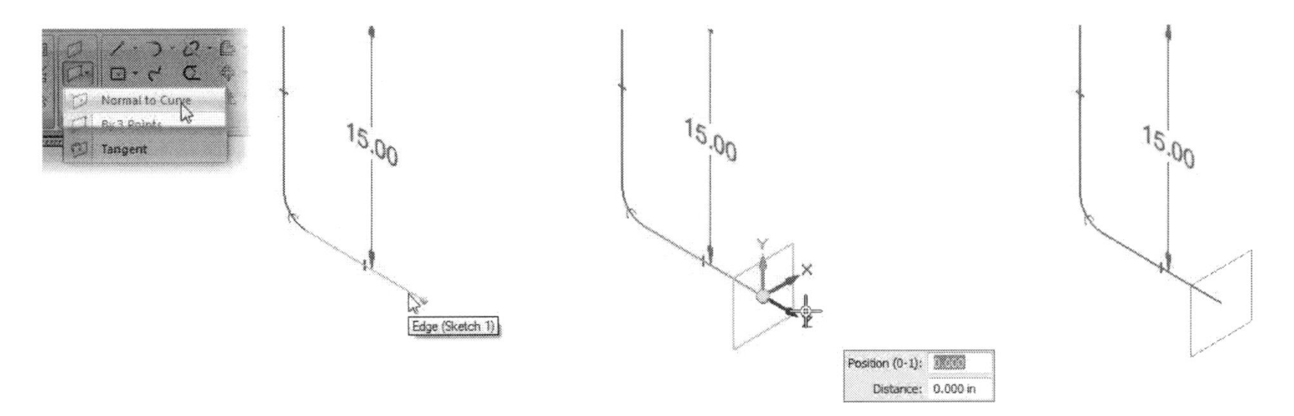

8. On the ribbon, click **Home > Draw > Circle by Center Point** and draw a circle of 2.5-inch diameter on the plane normal to curve.

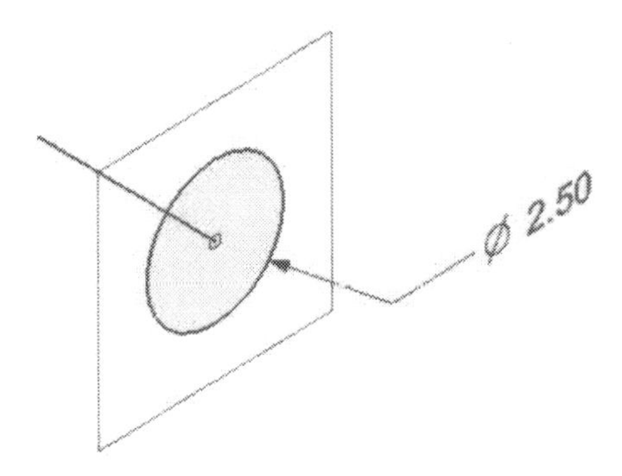

9. On the ribbon, click **Home > Solids > Sweep**; the **Sweep Options** dialog pops up.
10. On this dialog, set the **Default Sweep Type** to **Single path and cross-section**.
11. Set the **Face Merging** option to **No Merge**.
12. Set the **Section Alignment** option to **Normal**. Click **OK** to close the dialog.
13. Click on the first sketch to define the path of the *Sweep* feature. Click the green check on the command bar.
14. Click on the circle to define the cross section.

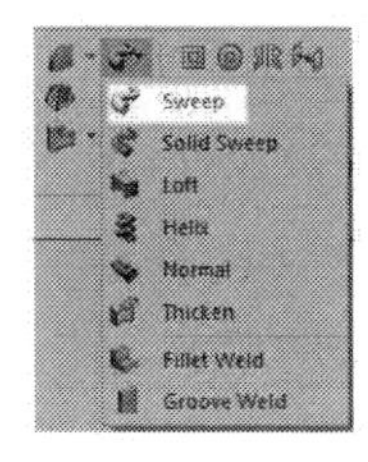

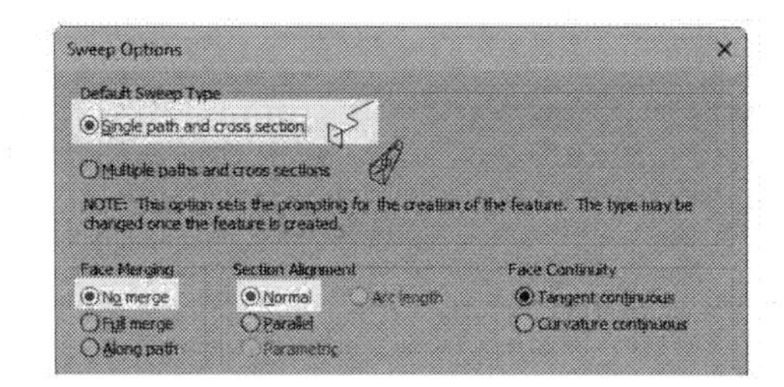

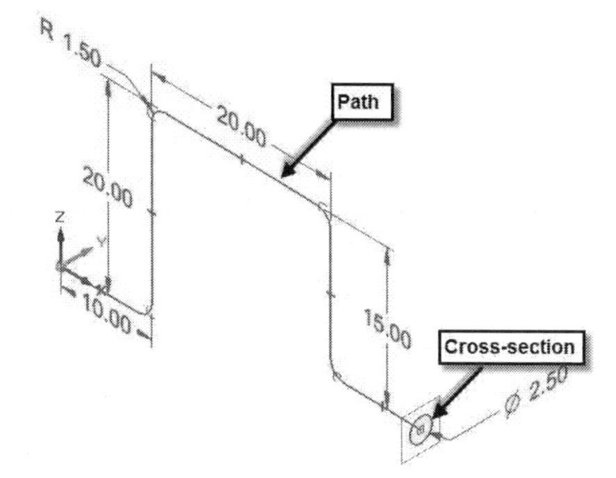

15. Click **Finish** to complete the *Sweep* feature.

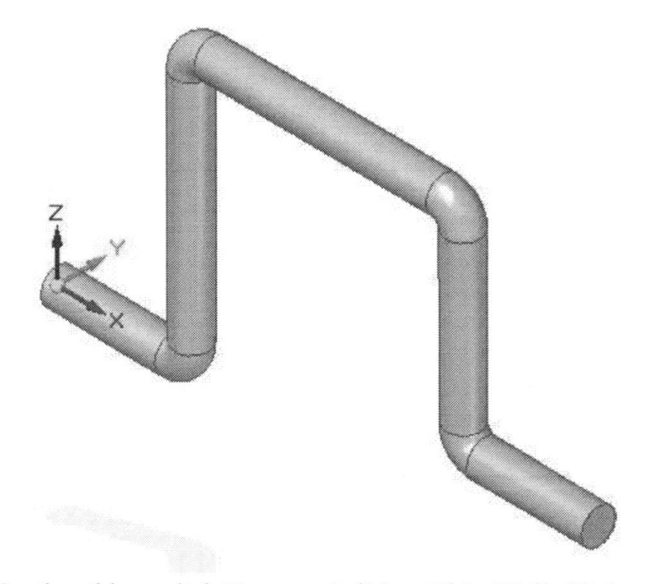

16. On the ribbon, click **Home > Solids > Thin Wall**. Click on the end face of the *Sweep* feature.
17. Rotate the part geometry and click the end face on the other side.
18. Type-in **0.5** in the box that appears on the geometry. Press Enter to shell the *Sweep* feature.

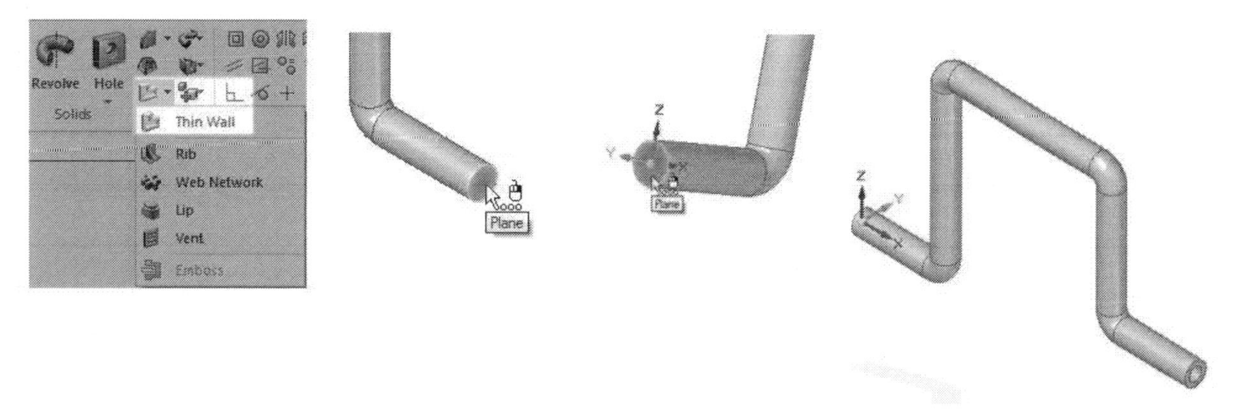

19. On the ribbon, click **Home > Draw > Project to Sketch.** Next, place the cursor on the end face and click on the lock icon.

20. Leave the default options on the **Project to Sketch Options** dialog and click **OK**.
21. Click on the inner edge of the end face to project it.
22. Draw a circle of 4.5 in diameter. Unlock the sketch plane.

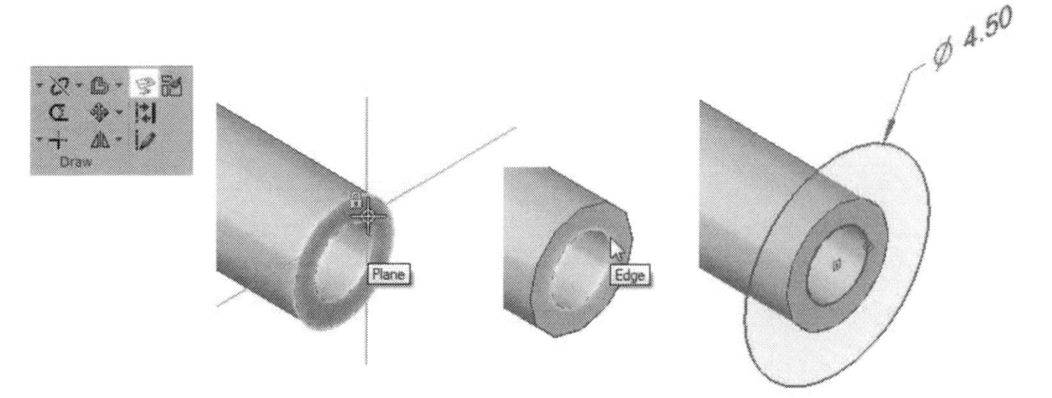

23. Activate the **Extrude** command and click inside the two sketch regions.
24. On the Extrude command bar, select the **Include Internal Loops** option.
25. Right-click to accept the selection and move the mouse pointer. Type-in 0.75 in the box that appears on the geometry.
26. Press Enter to create the flange.

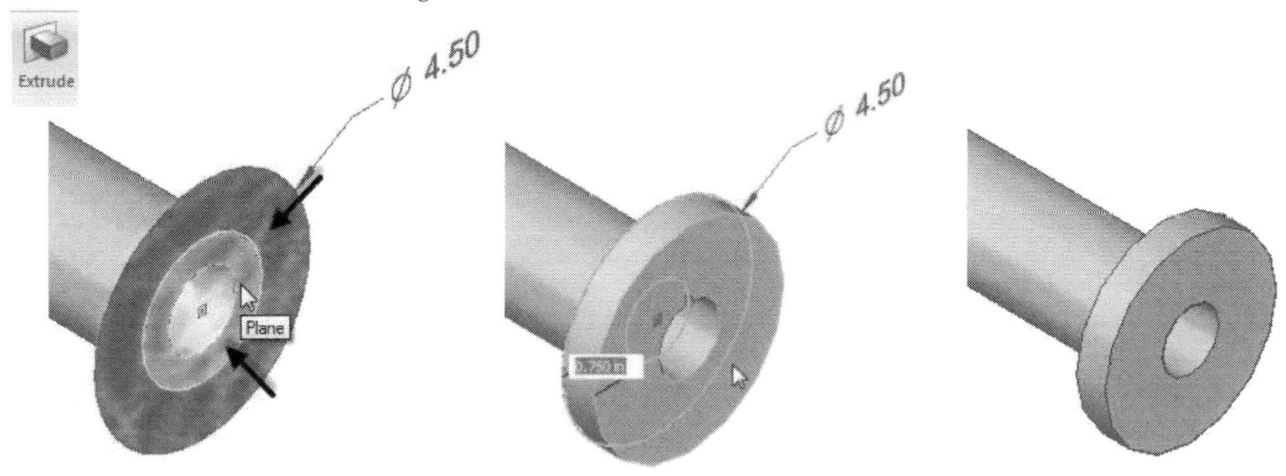

27. Draw a sketch on the flange face and create the *Cutout* feature.

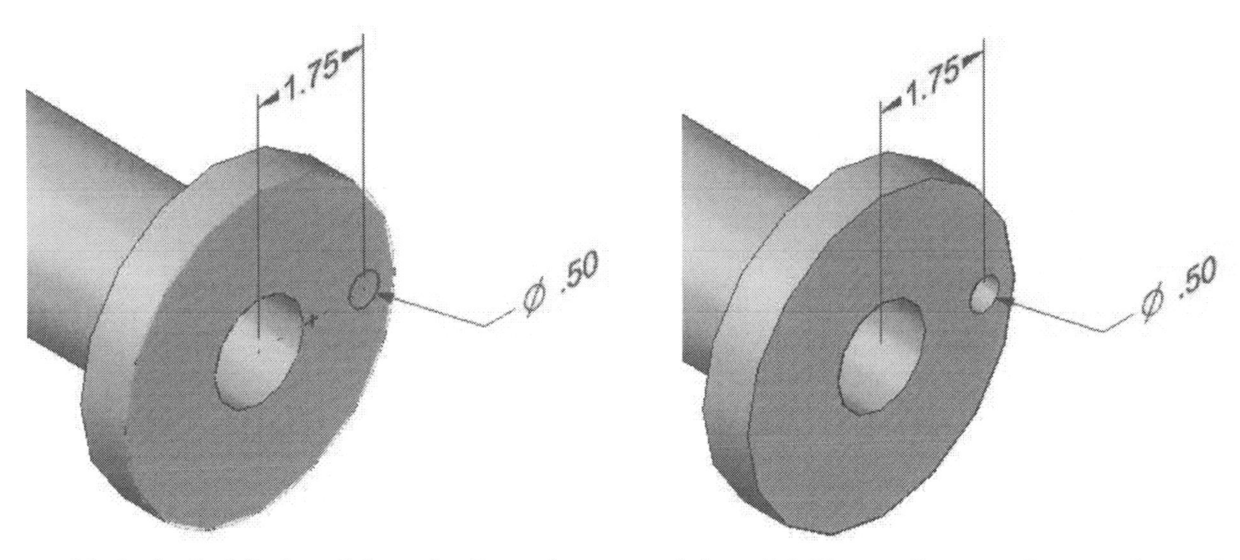

28. In the Pathfinder, click on the *Cutout* feature, and then click **Home > Pattern > Rectangular > Circular** on the ribbon.
29. Place the pointer on the flat face of the flange and click the lock icon. Now, you have to define the axis of the circular pattern.
30. Place the pointer on a circular edge of the flange and click when the center point of the circular edge is selected. This defines the pattern axis.
31. Type-in **6** in the **Count** box and click the green check on the command bar. The cutouts are patterned in a circular fashion.

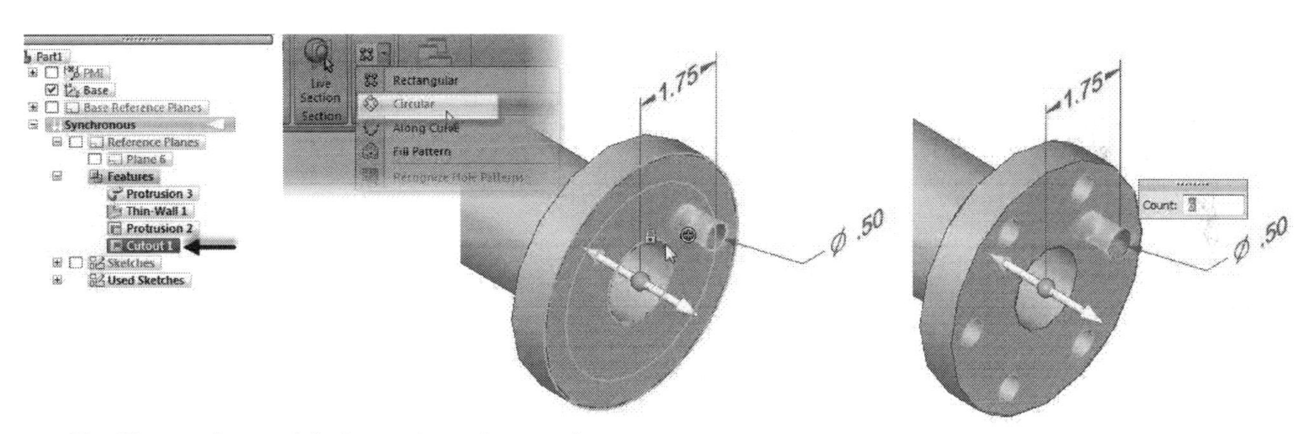

32. Change the model view orientation, as shown.

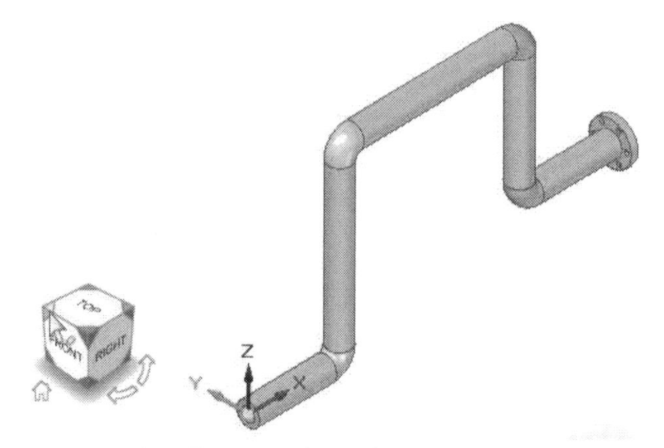

33. Create another flange and circular pattern.

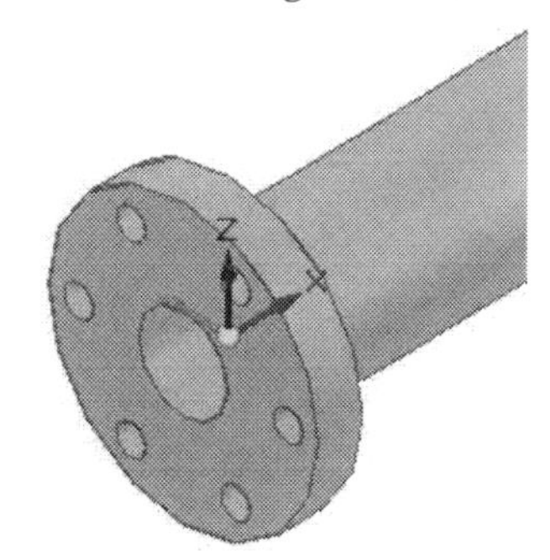

34. Save and close the part file.

Questions

1. List the methods to create the *Sweep* features.
2. How to apply twist and turns to *Sweep* features?
3. Write the formula to calculate the variable pitch of a helical feature.
4. Why do we define the axis of a *Sweep* feature?
5. List any two methods to create helical features.

Exercises

Exercise1

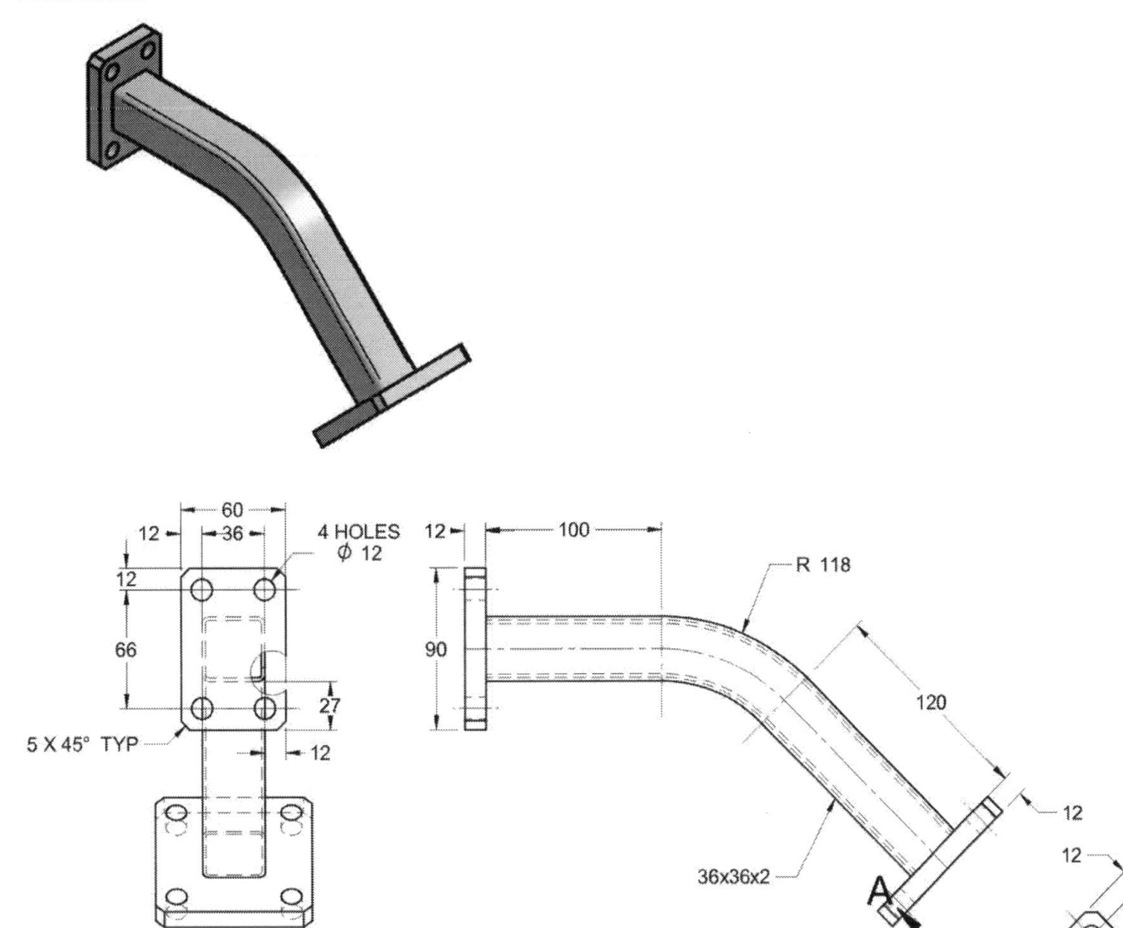
60
12
36
4 HOLES
Ø 12
12
66
27
5 X 45° TYP
12
100
R 118
90
120
12
36x36x2
A
90
66
4 HOLES Ø 12
66
90
12

Chapter 7: Loft Features

The **Loft** command is one of the advanced commands available in Solid Edge that allows you to create simple as well as complex shapes. A basic loft is created by defining two cross-sections and joining them together. For example, if you create a loft feature between a circle and a square, you can easily change the cross-sectional shape of the solid. Whereas the sweep feature allows you to control the cross-section at the start or end points.

The topics covered in this chapter are:

- *Basic Lofts*
- *Loft options*
- *Loft Cutouts*

Loft

This command creates a loft feature between different cross-sections. To create a loft, first create two or more sections on different planes. The planes can be parallel or perpendicular to each other. Activate the **Loft** command (click **Home > Solids > Sweep > Loft** on the ribbon); the **Loft** command bar appears. The **Cross-Section Step** icon is active and you can select the cross-sections that will define the loft. You need to select two or more cross-sections to define a loft. Select the cross-sections from the graphics window. The loft feature is sensitive to the location at which you will click to select the cross-section. For example, select the first cross-section by clicking near the right corner, and then select the second cross-section by clicking at a corresponding location. Click **Preview** on the command bar; the loft preview immediately appears, as shown below.

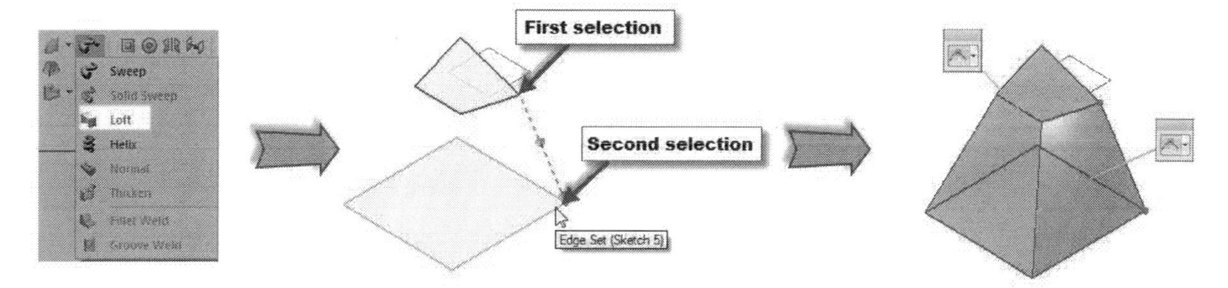

Now, click the **Cross-Section Step** icon on the command bar, and then click **Define Start Point**. Select the opposite corner on the first cross-section, and then click the **Preview** button; you will notice that a different result appears. For this reason, you have to be careful about where you click to select the cross-sections. However, if you do happen to make a mistake, you can use the **Define Start Point** icon to fix any unwanted twisting.

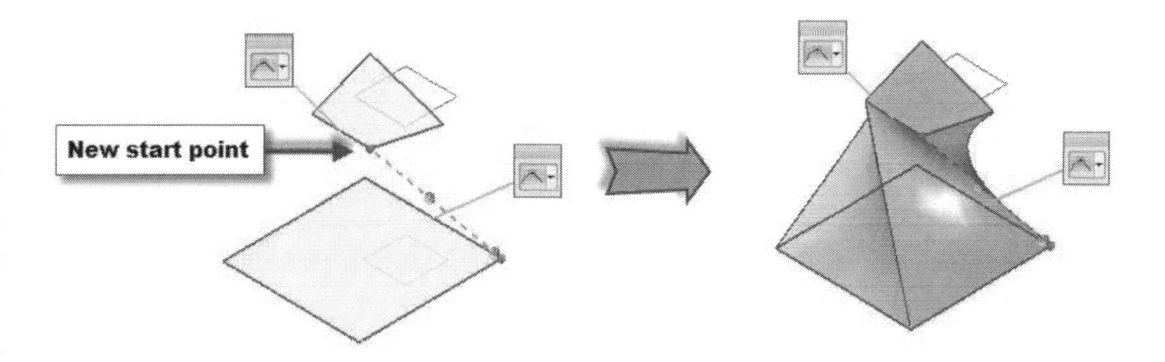

Tangency Controls

The shape of a simple loft is controlled by the cross-sections and the plane location. However, the **Tangency Controls** connected to the cross-sections can control the behaviour of the side faces. If you would like to change the appearance of the side faces, you can use the **Tangency Controls** either at the beginning of the loft, the end of the lofts or both. For instance, click on **Tangency Control** on the beginning of the loft and select **Normal to Section**; the preview of the loft updates. You can notice that the beginning of the loft starts in a direction normal to the cross-section. You can control how much influence the **Normal to Section** option will have by adjusting the parameter in the box attached to the cross-section. A lower value will have lesser effect on the feature. As you increase the value, the more noticeable the effect will be, eventually. If you increase the number high enough, the normal effect will lead to some weird results. You can also click and drag the handle attached to the cross-section to control the normal effect. If you want to change the direction of the Normal to Section effect, enter a negative value in the box attached to section. The same options can also be applied to the end cross-section of the loft.

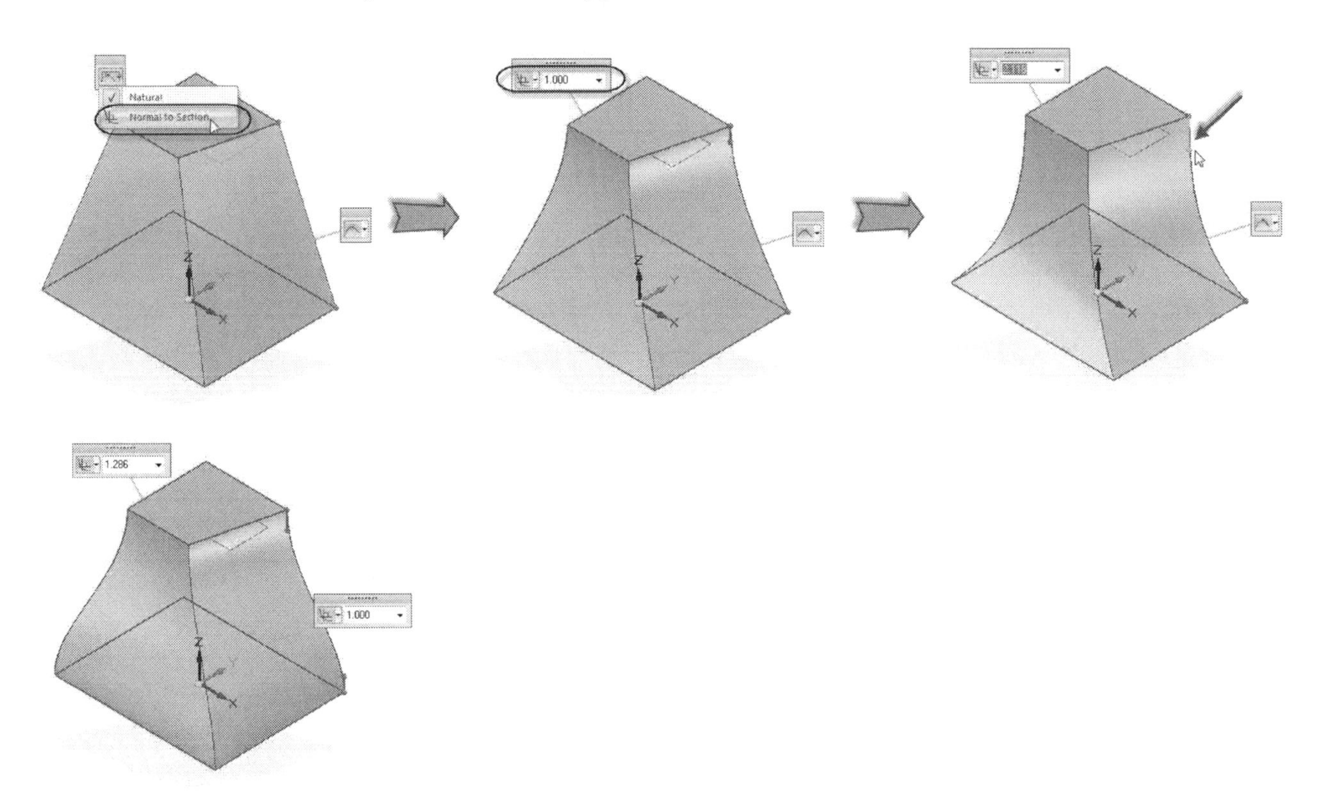

Loft Cross-sections

In addition to 2D sketches, you can also define loft cross-sections by using different element types. For instance, you can use existing model faces, surfaces, curves, and points. The only restriction is that the points can be used at the beginning or end of a loft. Set the appropriate option in the **Select** drop-down menu to select different element types. In the following example, a loft feature is created using a face, single element and point.

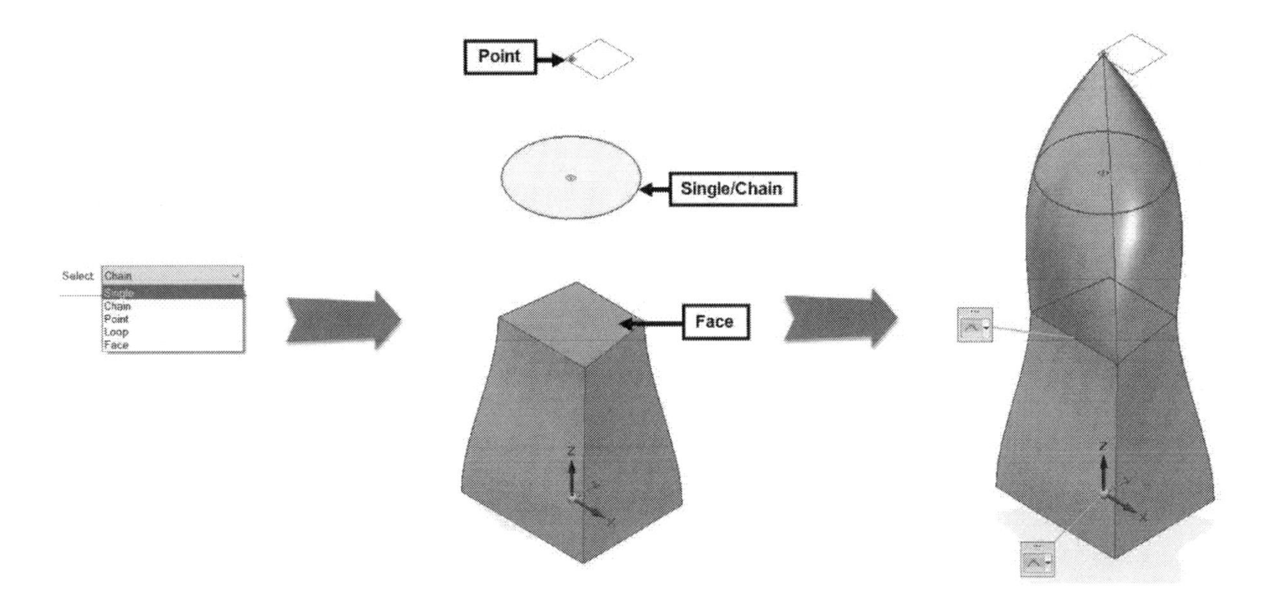

Closed Extent

Solid Edge allows you to create a loft that closes on itself. For example, to create a ring that lofts between each of the shapes, you must select four sketches as shown in figure, and then click the **Extent Step** icon on the command bar. Next, click the **Closed Extent** icon on the command bar, and click **Preview**; this will give you a closed loft.

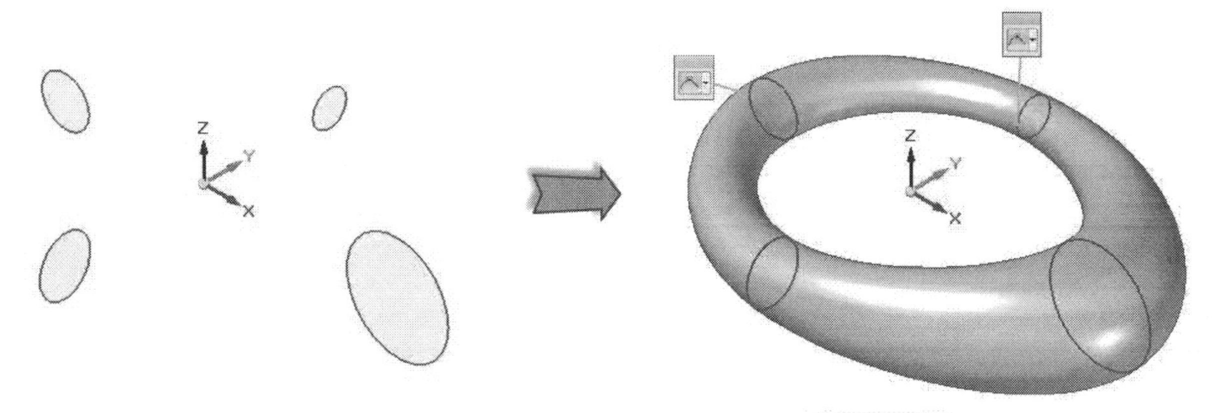

Guide Curves

Similar to **Tangency Controls**, guide curves allow you to control the behaviour of a loft between cross-sections. You can create guide curves by using 2D sketches. You can also use the **Keypoint Curve** command to create guided curves. Activate this command (click **Surfacing > Curves > Keypoint Curve** on the ribbon) and select points to create a curve, as illustrated below. Right-click and click **Finish** to complete the curve. Likewise, create the other curves.

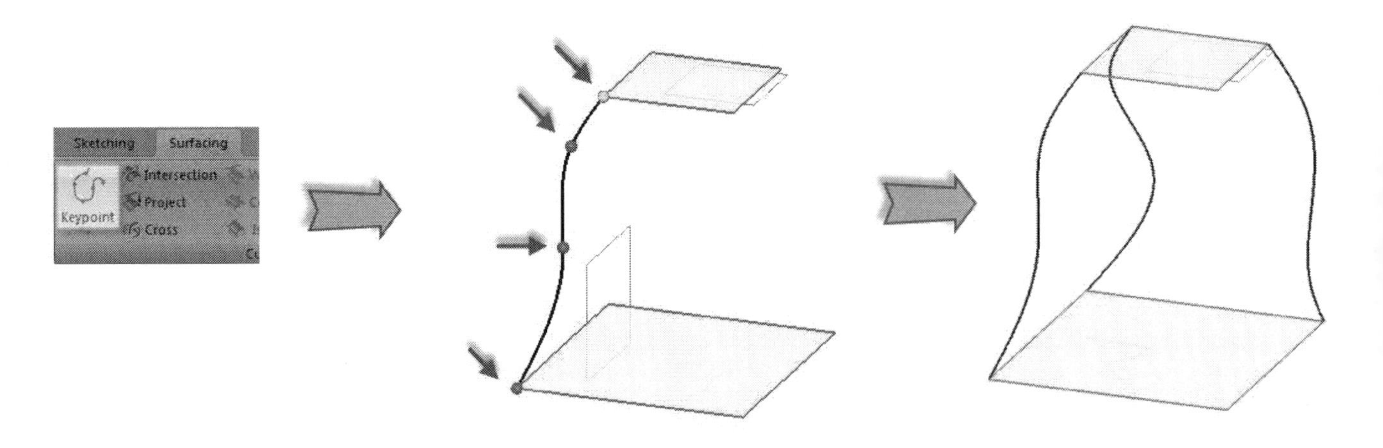

Now, activate the **Loft** command and select the cross-sections. To select guide curves, click the **Guide Curve Step** icon on the command bar and select the first guide curve, and then click the green check on the command bar. In the same way, select the other guide curves and click the **Preview** icon; you will see that the preview updates. Notice that the edges with guide curves are affected. The one without guide curve remains as it is.

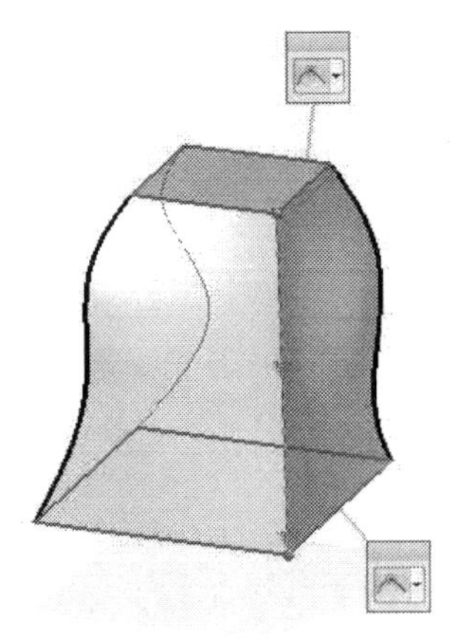

Section Geometry

Sections used for creating lofts should have a matching number of segments. For example, a three sided section will loft nicely to another three sided section despite the differences in the shape of the individual segments. The **Loft** command does a good job of generating smooth faces to join them.

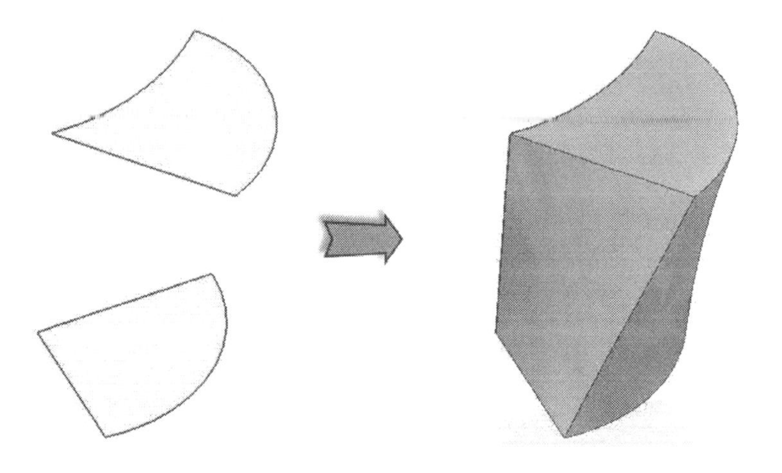

Look at an example for a loft with mismatching segments. Activate the **Loft** command and select the cross-sections. Click the **Extent Step** icon on the command bar, and then click the **Vertex Mapping** icon; the **Vertex Mapping** dialog pops up. You can use this dialog to map vertices of the cross-sections. Click **Set 1** on the dialog and select two vertices. Click **Add** on the dialog and select two vertices; another set of vertices is added to the loft. Similarly, add another set and close the dialog. Click **Finish** to complete the loft.

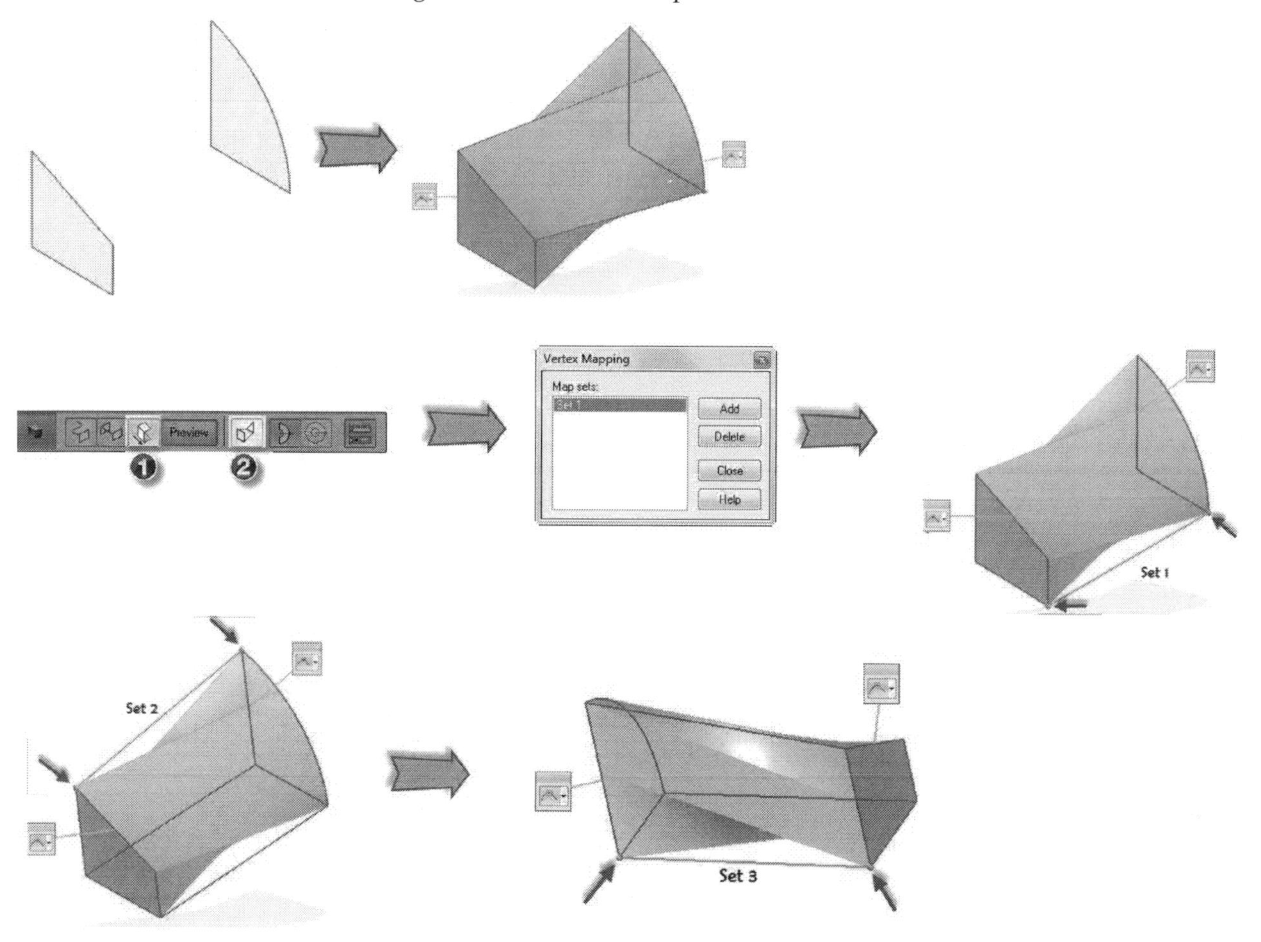

Loft Cutout

Like other standard features such as extrude, revolve and sweep, the loft feature can be used to add material or remove material. You can remove material by using the **Loft Cutout** command. Activate this command (click **Home > Solids > Swept Cutout > Loft Cutout** on the ribbon) and select the cross-sections. Click **Preview** and **Finish** to create the loft cutout.

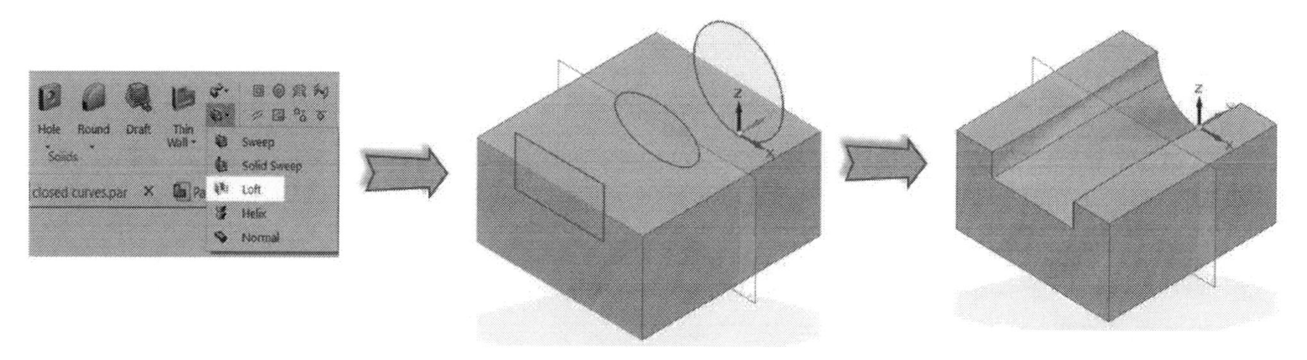

Examples

Example 1 (Millimetres)

In this example, you will create the part shown below.

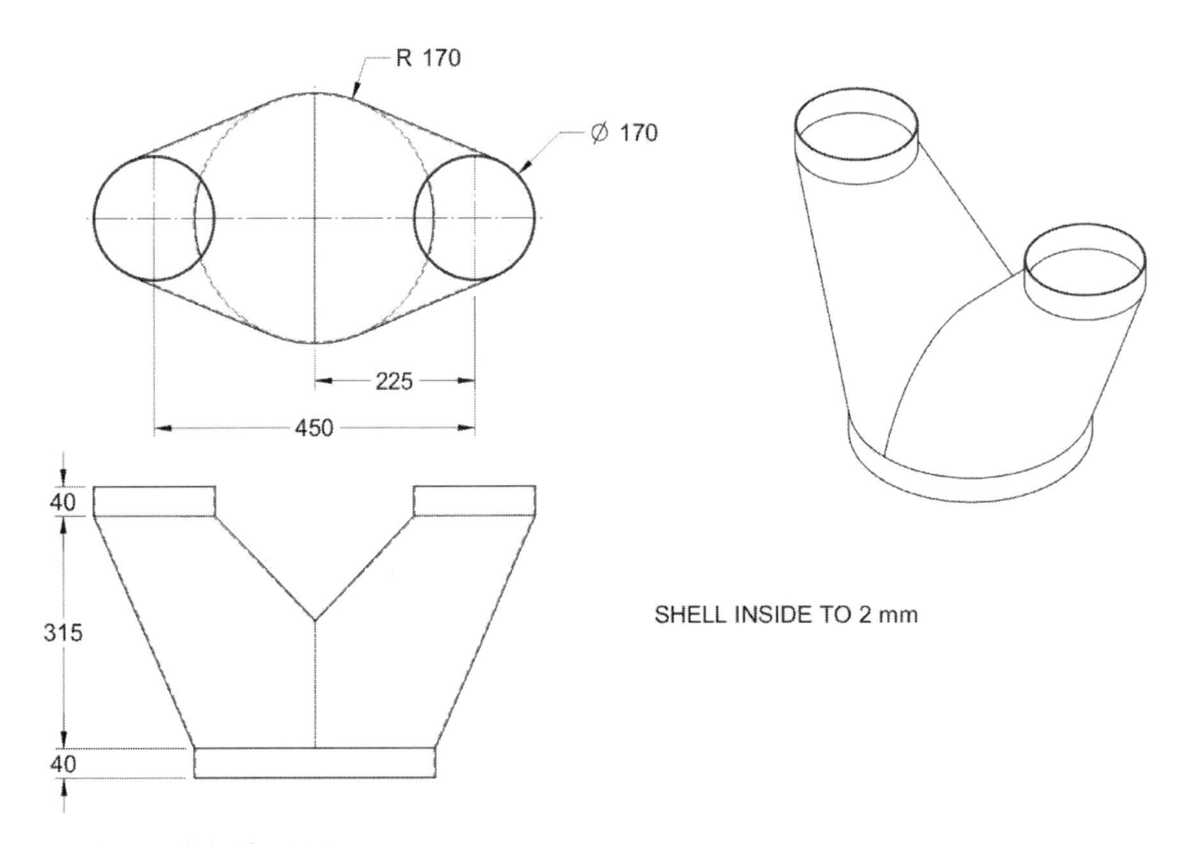

1. Start **Solid Edge 2019**.
2. On the Application Menu, click **New > ISO Metric Part**; a new part file is opened.
3. To start a new sketch, click **Home > Draw > Circle by Center Point** on the ribbon.
4. Lock the XY plane and draw a circle of 340 mm diameter. Unlock the sketch plane.
5. Create the *Extrude* feature with 40 mm thickness.

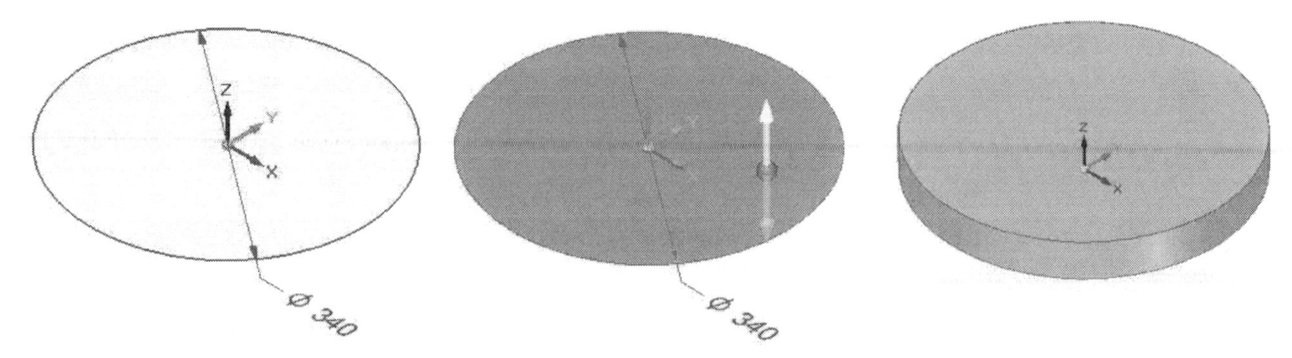

6. On the ribbon, click **Home > Planes > Coincident Plane**.
7. Click on the top face of the geometry to create a coincident plane.
8. Click on the Z-axis of the Steering Wheel tool and move the mouse pointer upward.
9. Type-in 315 mm in the dimension box and press Enter.

10. Activate the **Circle by Center Point** command and draw a circle of 170 mm diameter on the new plane. Also, add dimensions and constraints to the circle, as shown. Unlock the sketch plane.

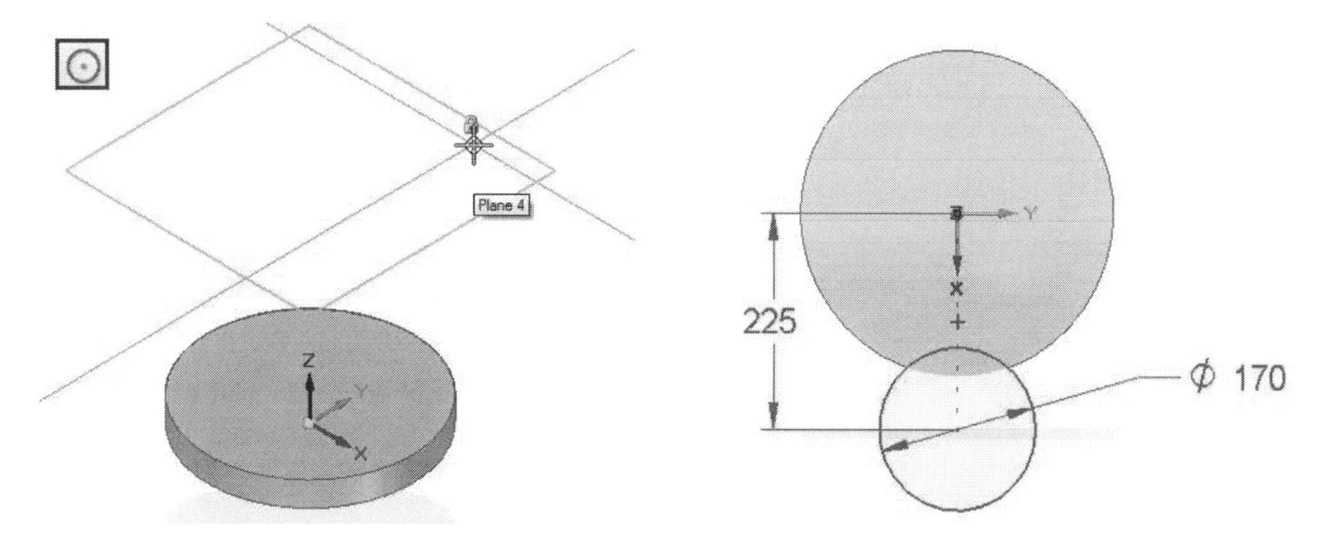

11. Change the model view orientation to ISO View.
12. On the ribbon, click **Home > Solids > Sweep > Loft**.
13. Click on the circle and the top circular edge of the *Extrude* feature.
14. Click **Preview** on the command bar to preview the loft protrusion.
15. Click **Finish** to complete the *Loft* feature.

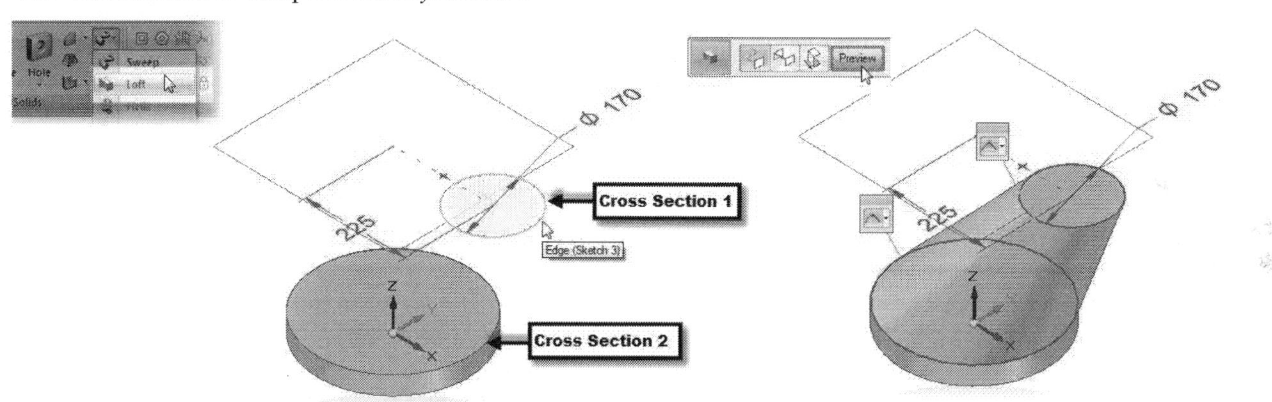

16. Activate the **Extrude** command and click on the top face of the *Loft* feature. Right-click to accept the selection.
17. Move the mouse pointer up and type 40 in the box that appears on the geometry. Press Enter.

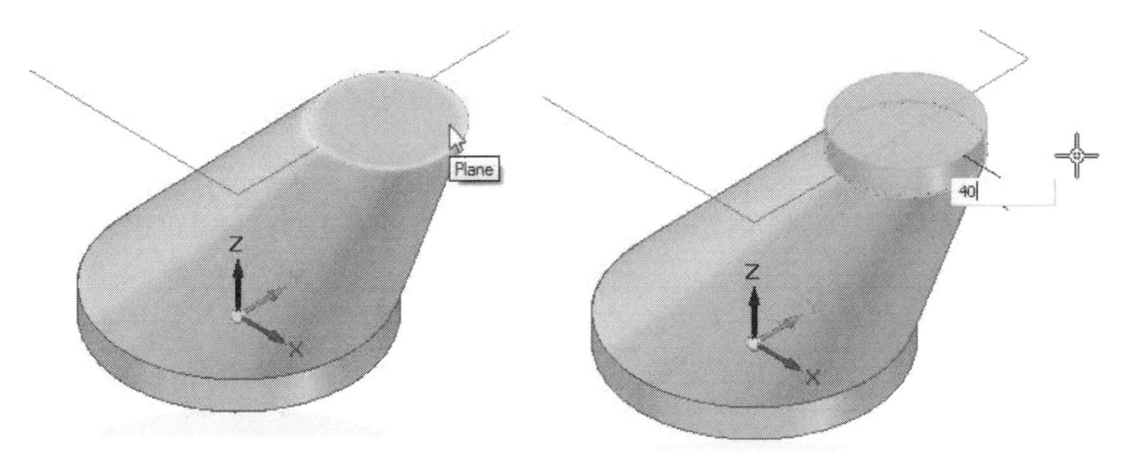

18. In the Pathfinder, press the Shift key and click on the *Loft Protrusion* and two *Extrude Protrusions*. Activate the **Mirror** command.
19. Click on the YZ plane of the coordinate system to mirror the selected features.

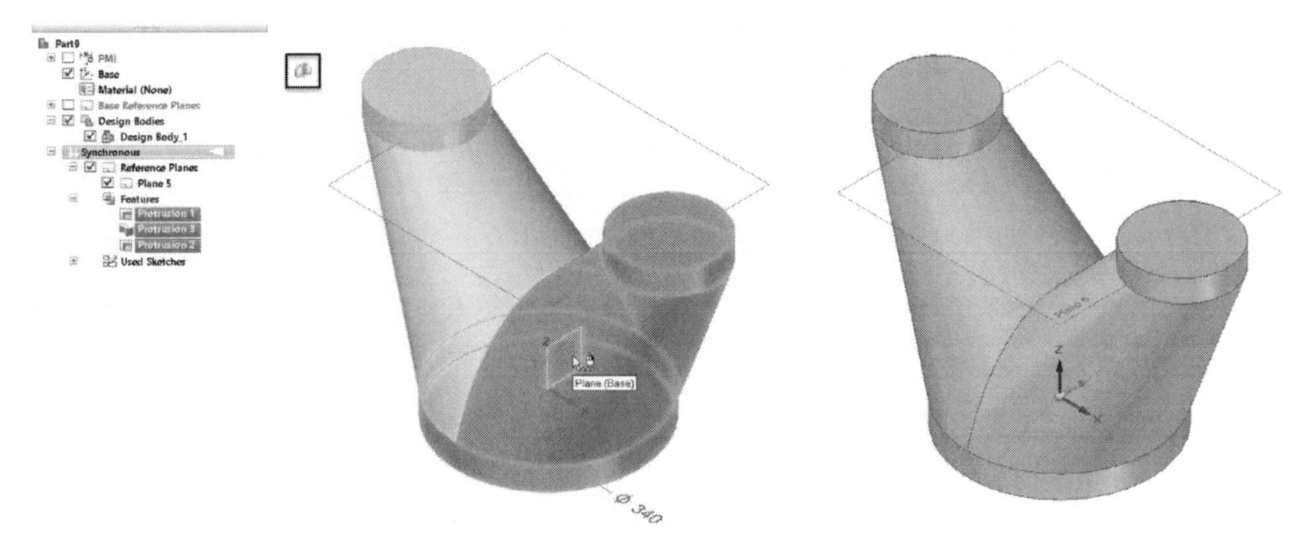

20. On the ribbon, click **Home > Solids > Thin Wall** and click on the flat faces of the part geometry.
21. Type 2 in the box that appears on the geometry and press Enter. The part geometry is shelled.

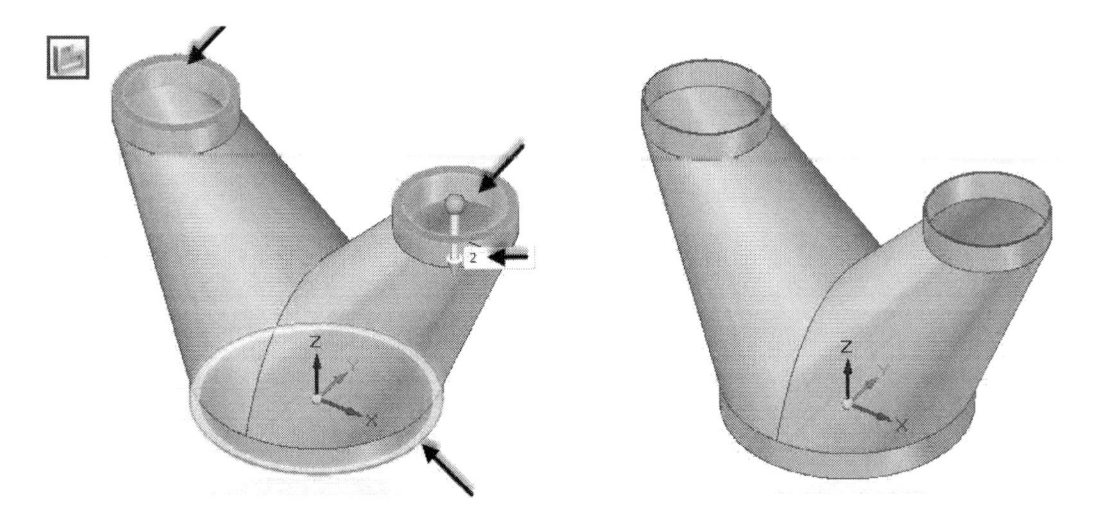

22. Save and close the part file.

Example 2 (Inches)

In this example, you will create the part shown below.

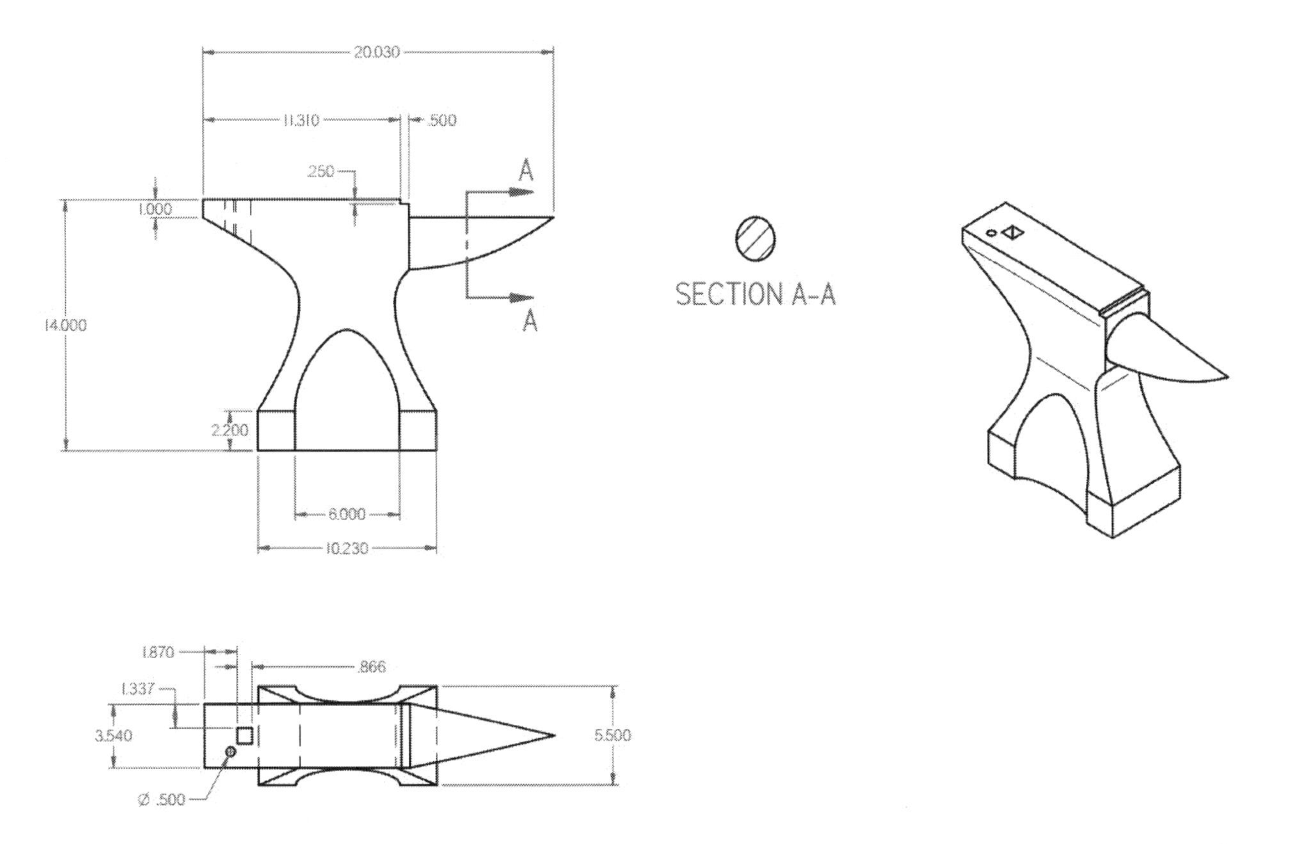

1. Start **Solid Edge 2019**.
2. On the Application Menu, click **New > New**; the **New** dialog appears. On this dialog, click **Standard Templates > ANSI Inch**.
3. Select **ansi inch part.par** from the available templates, and then click **OK**.
4. To start a new sketch, click **Home > Draw > Rectangle by Center** on the ribbon.
5. Place the mouse pointer on the coordinate system; the XZ plane is highlighted and a mouse icon appears.
6. Click the right mouse button to display the **QuickPick** box.
7. Select the XY plane from the **QuickPick** box.

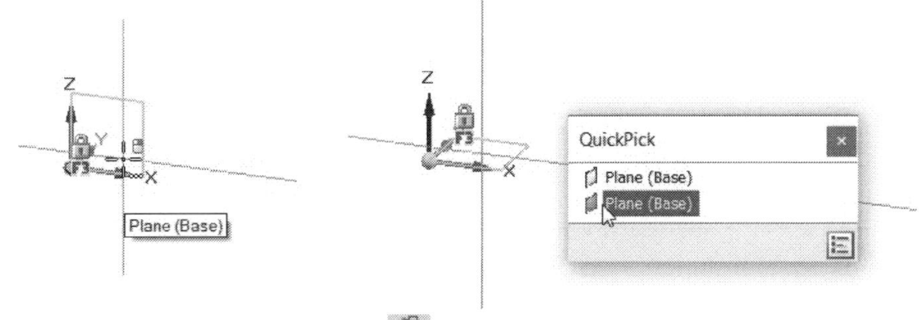

8. Click the **Sketch View** icon located at the bottom of the window. This orients the sketch plane normal to the screen.

9. Click on the origin point to define the center point of the rectangle. Move the mouse pointer diagonally and click to draw a rectangle.
10. Activate the **Smart Dimension** command and apply dimensions to the sketch, as shown below.

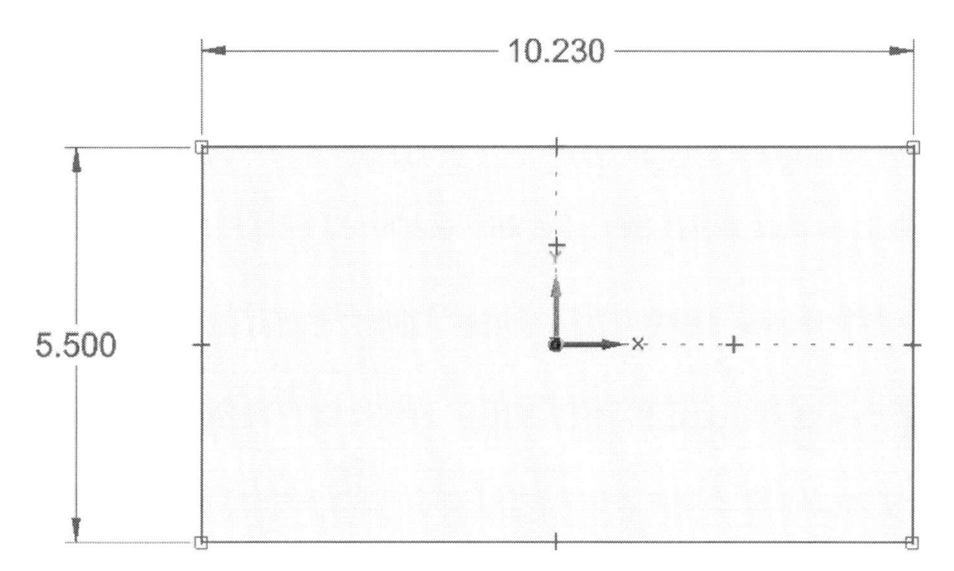

11. Click the **Home** icon located at the bottom left corner of the ViewCube.
12. Click **Home > Solids > Extrude** on the ribbon and click inside the sketch region. Right-click and move the pointer upward. Type 2.2 and press Enter.
13. Click **Home > Planes > Coincident Plane** on the ribbon and click on the top face of the model.
14. Click on the arrow pointing upwards. Next, move the pointer upward, type 6, and then press Enter.

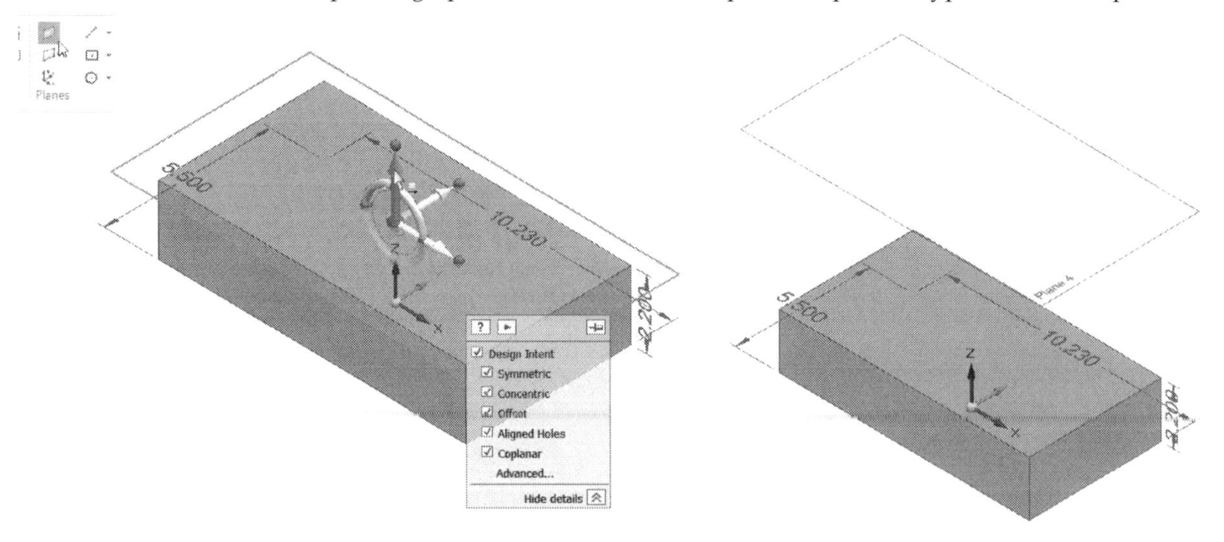

15. Click **Home > Draw > Rectangle by Center** on the ribbon and place the pointer on the newly created plane. Click the lock symbol that appears on the plane to lock it.
16. Click the **Sketch View** icon located at the bottom of the window. This orients the sketch plane normal to the screen.
17. Click on the origin point to define the center point of the rectangle. Move the mouse pointer diagonally and click to draw a rectangle.

18. Activate the **Smart Dimension** command and apply dimensions to the sketch.
19. Click the Unlock (F3) icon located at the top right corner in the graphics window.
20. Click the **Home** icon located at the bottom left corner of the ViewCube.

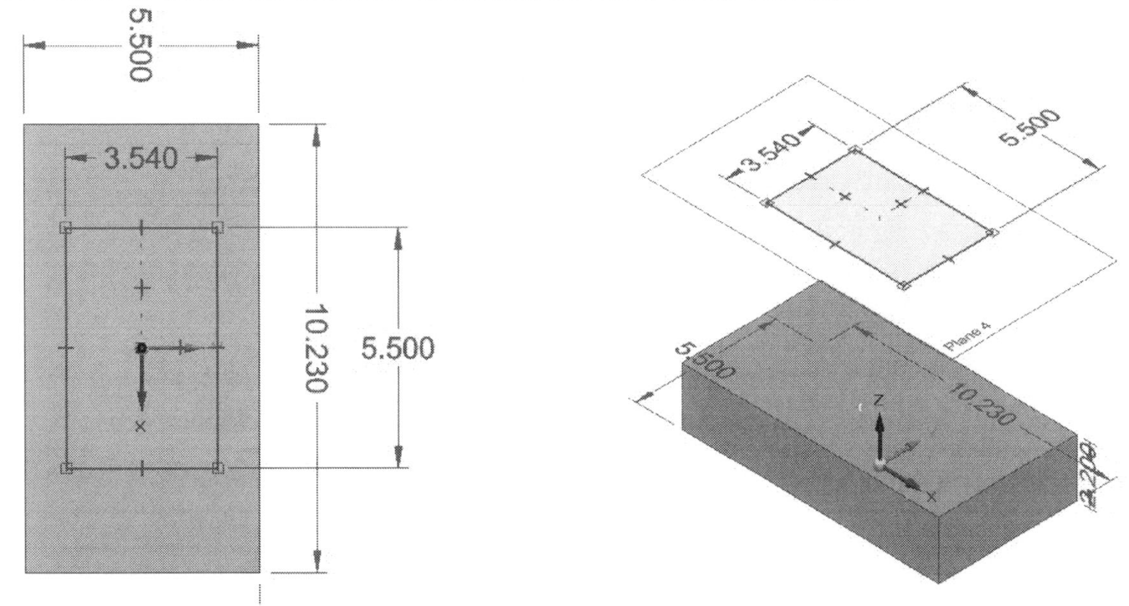

21. Click **Home > Solids > Add drop-down > Loft** on the ribbon.
22. On the command bar, click **Select > Loop**. Next, place the pointer near the lower right corner on the top face of the first feature; the edge loop is highlighted. Click to select the loop; also, the lower right corner point of the top face of the first feature is selected as the start point of the loft feature.
23. Place the pointer at the lower right corner of the rectangle. Next, click to select the rectangular sketch; the lower right corner point of the rectangle is selected as the end point of the loft.

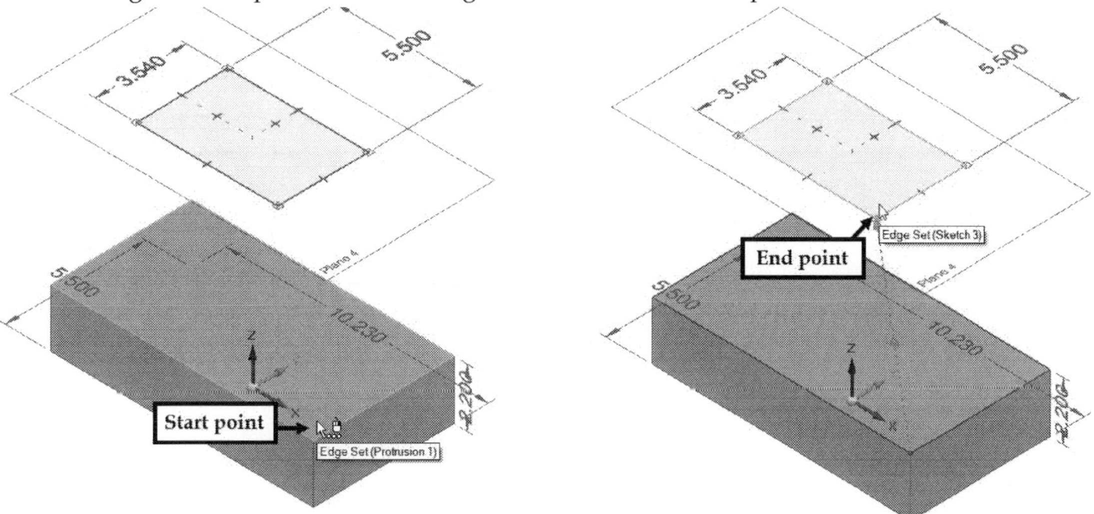

24. Click **Preview** on the command bar. Next, select Normal to section from the Tangency control handle attached to the second section. Type 1 in the value box available on the Tangency control handle.
25. Click **Finish** and **Cancel** on the command bar.

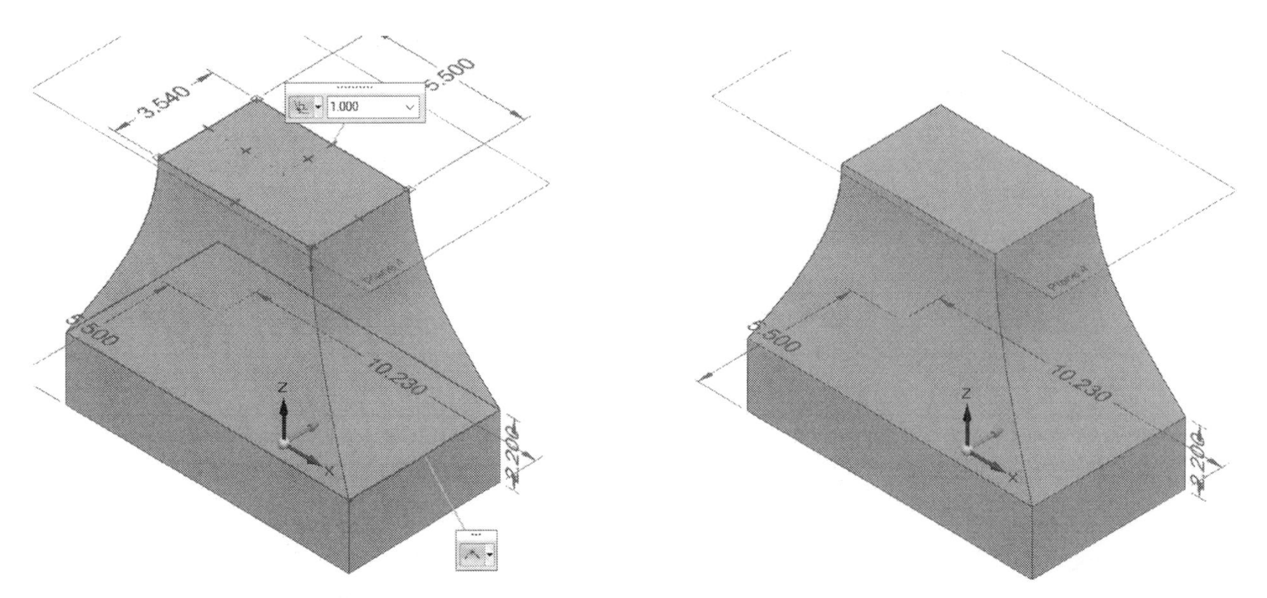

26. Click **Home > Draw > Circle drop-down > Ellipse by Center Point** on the ribbon. Place the pointer on the top plane, and then click on the lock icon that appears on it.
27. Click the **Sketch View** icon located at the bottom of the window. This orients the sketch plane normal to the screen.
28. Place the pointer on the left vertical edge and select the midpoint. Move the mouse pointer horizontally and click to specify the first axis. Next, move the pointer vertically and click to create second axis of the ellipse.

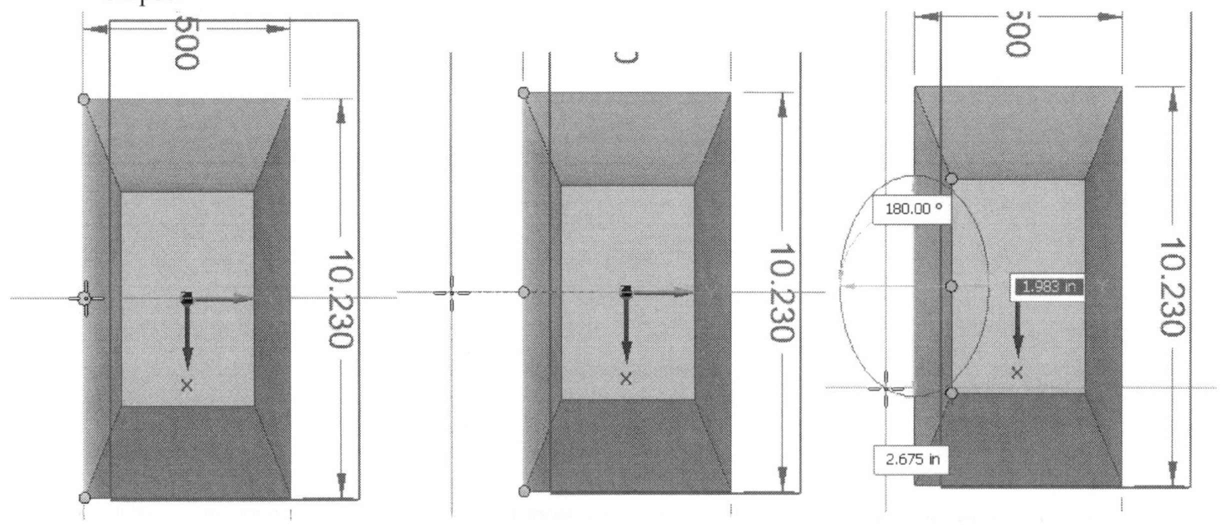

29. Click **Home > Relate > Horizontal/Vertical** on the ribbon and select the two quadrant points, as shown; they are aligned horizontally.
30. Select the remaining two quadrant points to align them vertically.

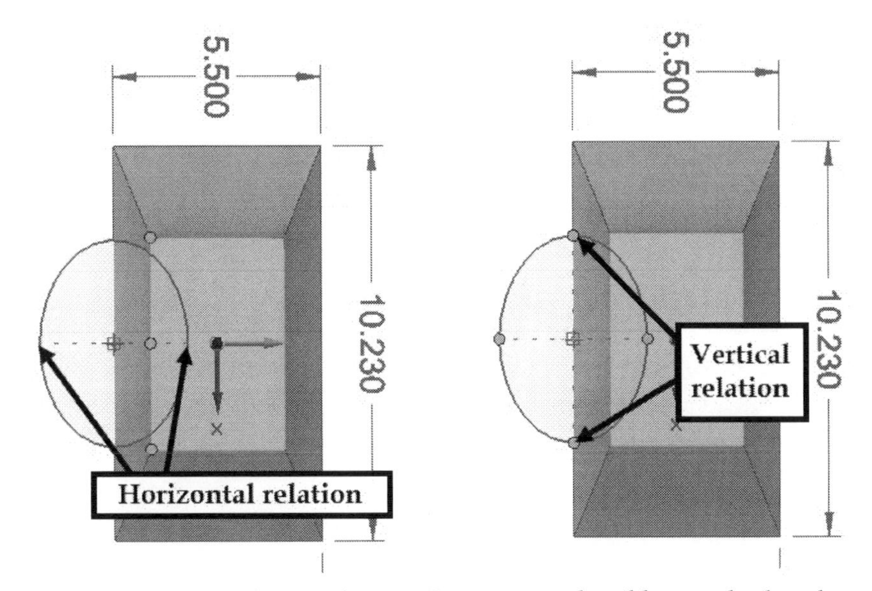

31. Click **Home > Dimension > Distance Between** on the ribbon and select the two quadrant points, as shown. Place the dimension, type 6 in the dimension box, and press Enter.
32. Select the remaining two quadrant points and place the dimension. Type 1.8 in the dimension box and press Enter.

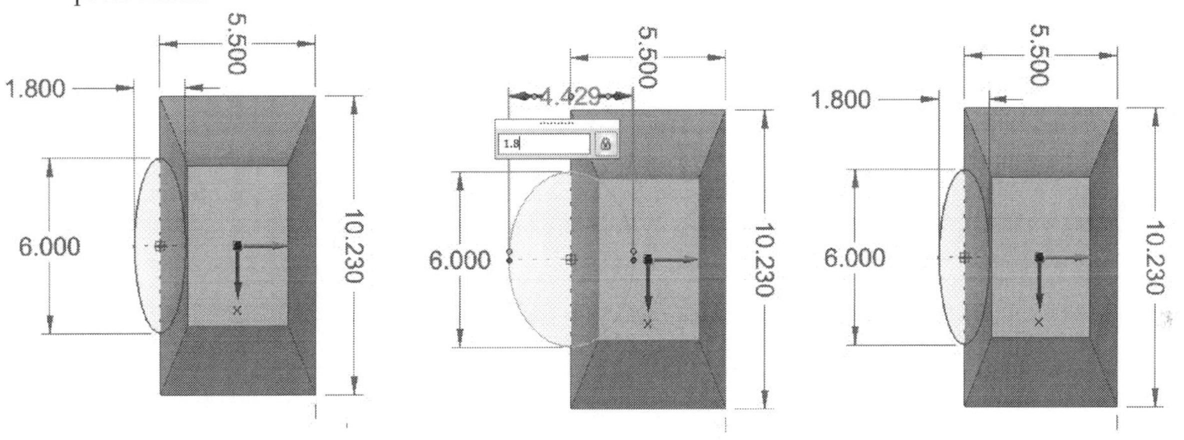

33. Click the Unlock (F3) icon located at the top right corner in the graphics window.
34. Click the **Home** icon located at the bottom left corner of the ViewCube.
35. Click in the sketch region, and then click the arrow pointing downward.
36. On the command bar, click **Add/Cut** drop-down > **Cut**. Next, select **Extent Type > Through All**.
37. Click in the graphics window to create the extruded cut.

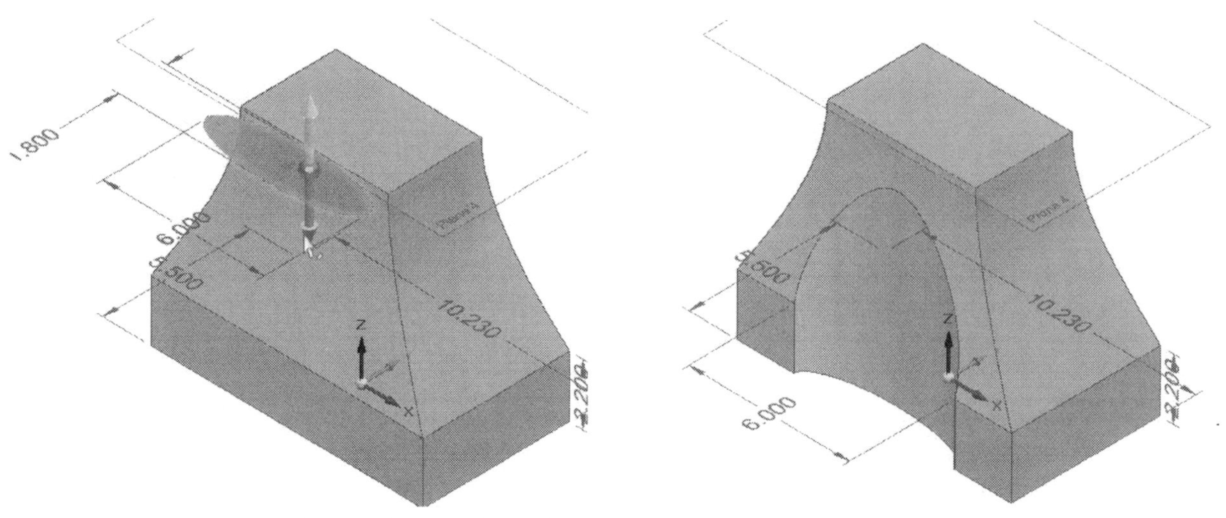

38. Mirror the extruded cut feature about the XZ plane.

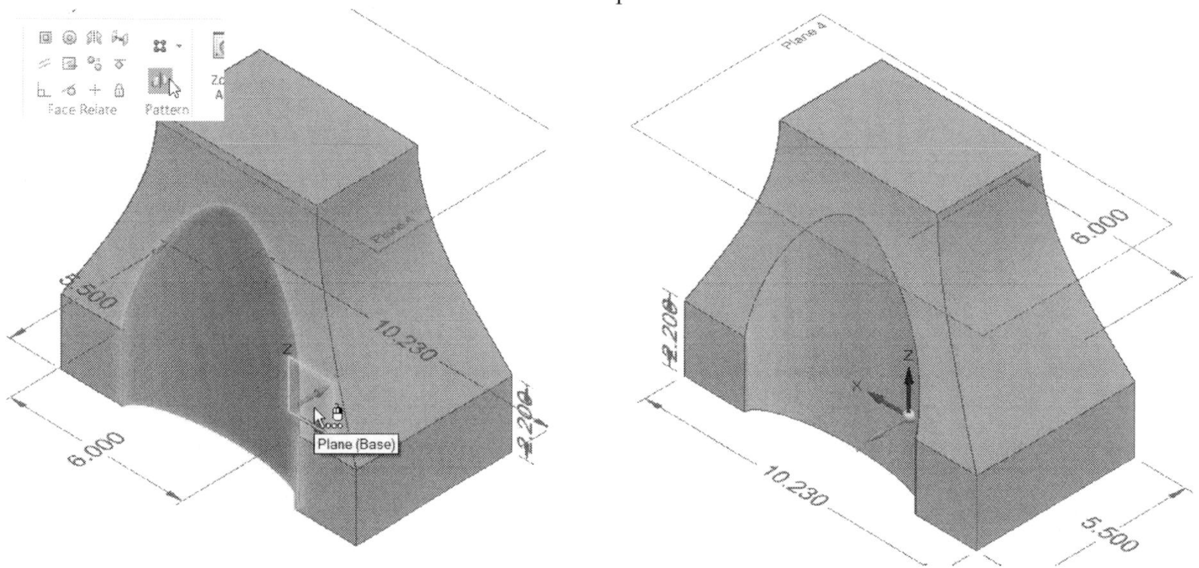

39. Click on the plane displayed on the top face of the loft feature. Select the arrow pointing upwards, move the pointer upward, and enter 4.8.

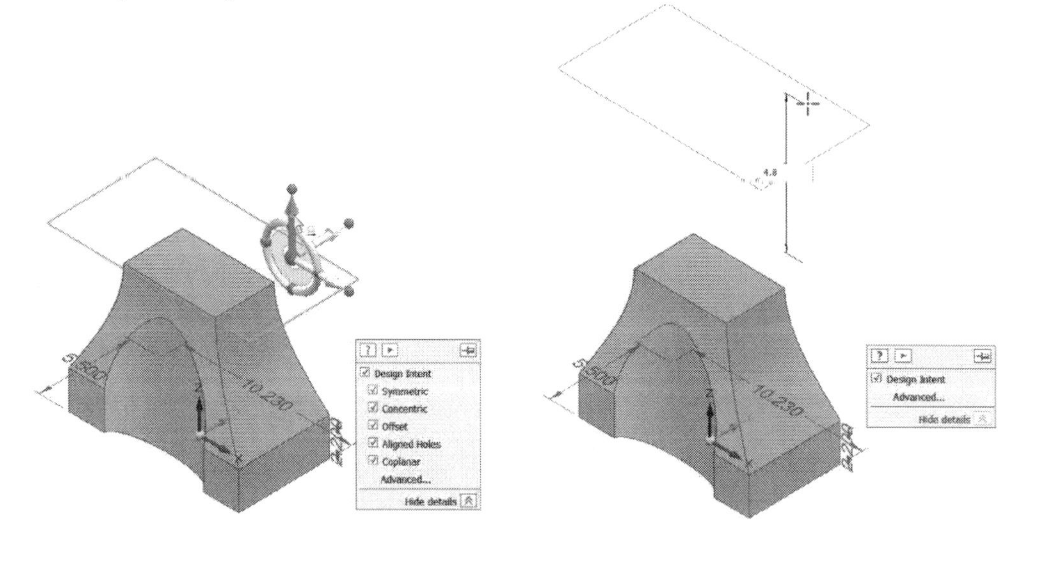

40. Click **Home > Draw > Rectangle by Center** on the ribbon and place the pointer on the plane. Click the lock icon to lock the plane.
41. Click the **Sketch View** icon located at the bottom of the window. This orients the sketch plane normal to the screen.
42. Create a rectangle, as shown. Next, click the Unlock (F3) icon located at the top right corner in the graphics window.

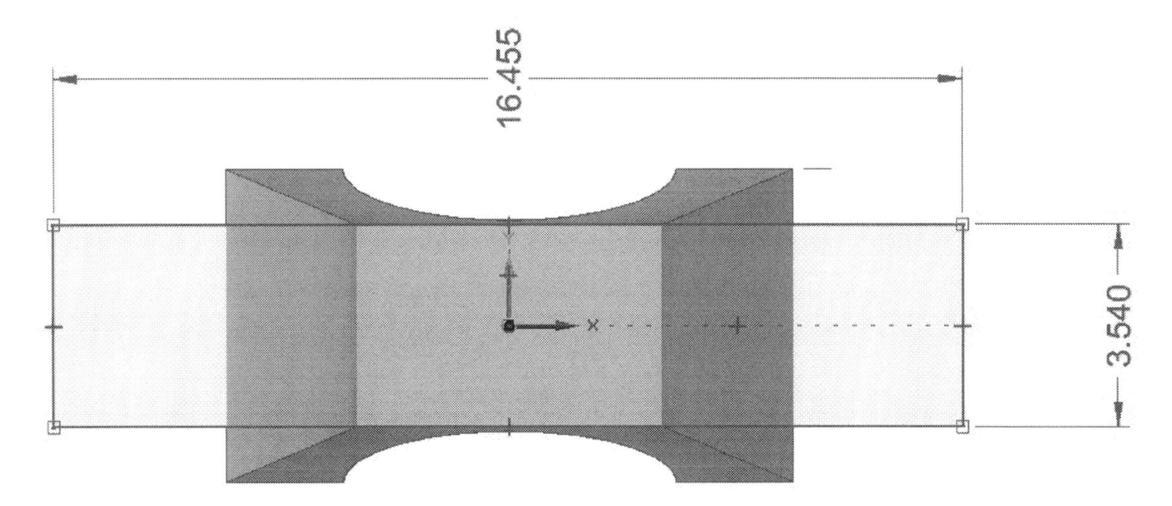

43. Click the **Home** icon located at the bottom left corner of the ViewCube.
44. Click **Home > Solids > Add drop-down > Loft** on the ribbon.
45. On the command bar, click **Select > Loop**. Next, place the pointer near the lower right corner on the top face; the edge loop is highlighted. Click to select the loop; also, the lower right corner point of the top face is selected as the start point.
46. Place the pointer at the lower right corner of the rectangle. Next, click to select the rectangular sketch; the lower right corner point of the rectangle is selected as the end point of the loft.

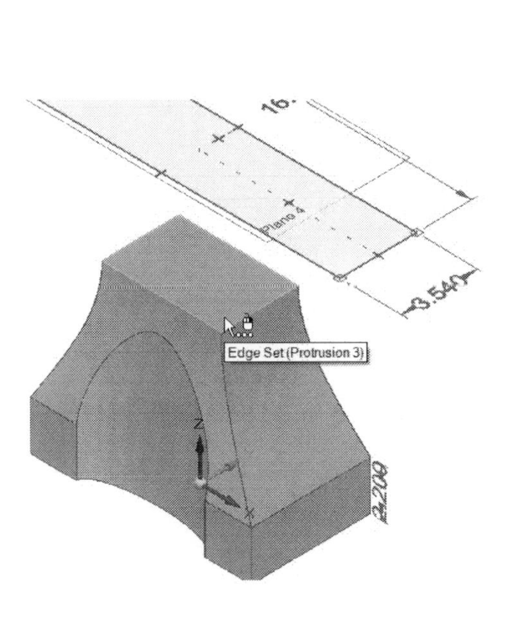

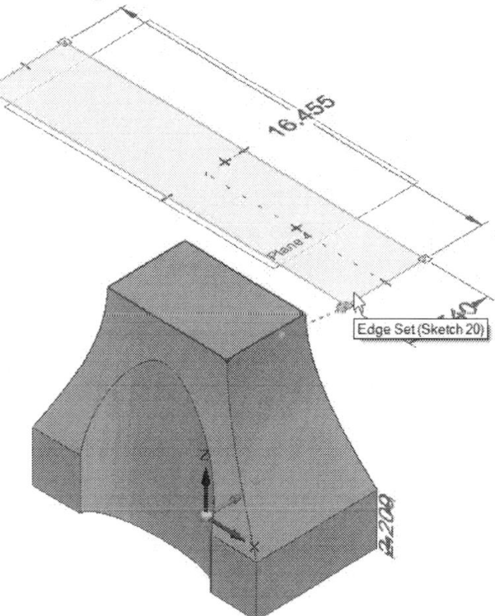

47. Click **Preview** on the command bar. Next, select **Curvature Continuous** from the Tangency control handle attached to the first section. Type 1 in the value box available on the Tangency control handle.
48. Click **Finish** and **Cancel** on the command bar.

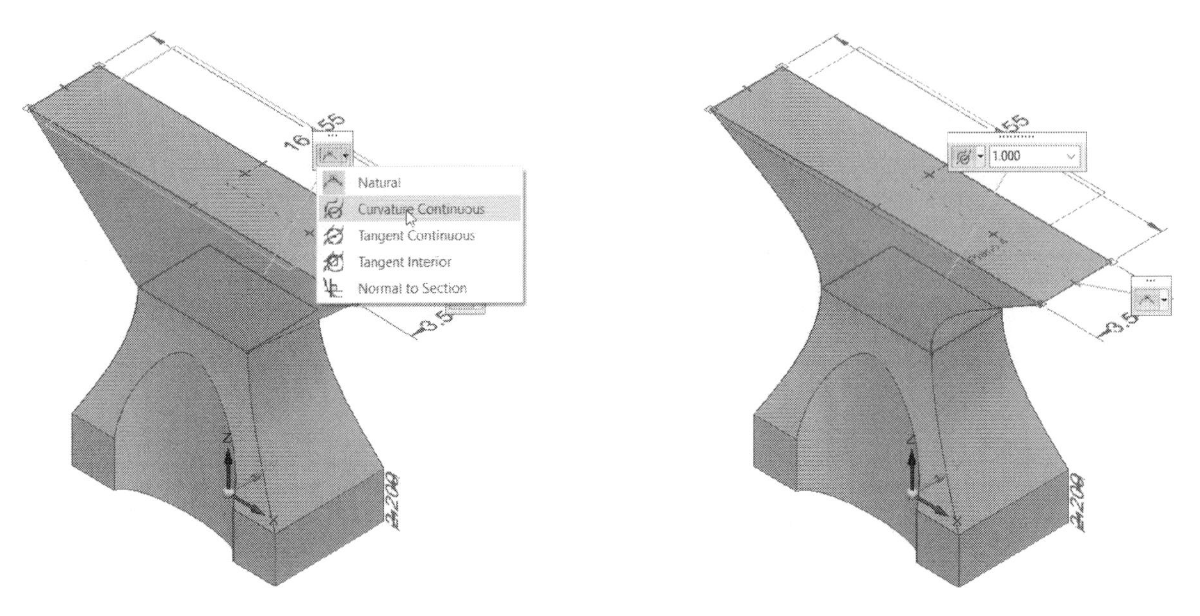

49. Click **Home > Draw > Line** on the ribbon and place the pointer on the plane. Click the lock icon to lock the plane.
50. Click the **Sketch View** icon located at the bottom of the window. This orients the sketch plane normal to the screen.
51. Create a line and add a dimension to it, as shown. Press Esc to deactivate the **Smart Dimension** command.

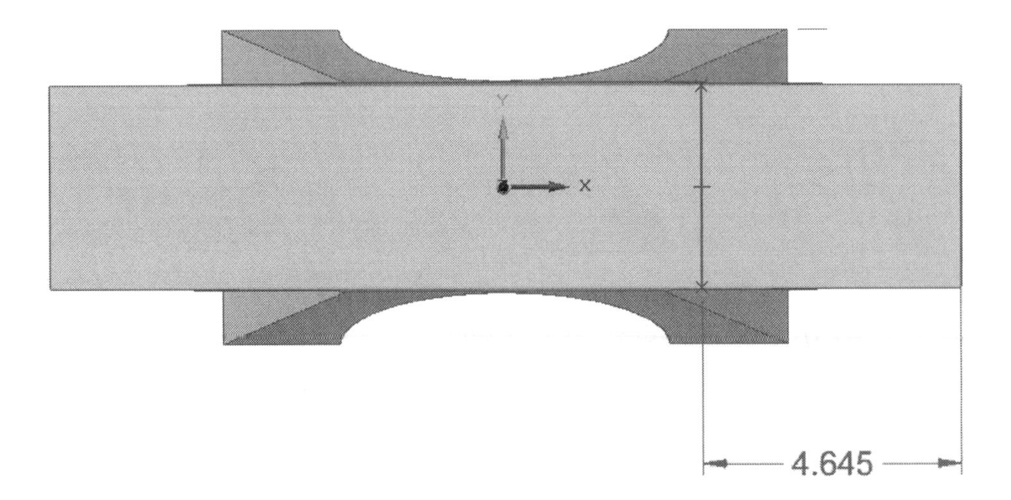

52. Next, click the Unlock (F3) icon located at the top right corner in the graphics window.
53. Click the **Home** icon located at the bottom left corner of the ViewCube.
54. Place the pointer on the right side of the top face; notice that a region is highlighted, as shown. Click to select the region.
55. Click the down arrow, and then select **Extents Type > Through Next** on the command bar.

56. Click in the graphics window to create the cut feature.

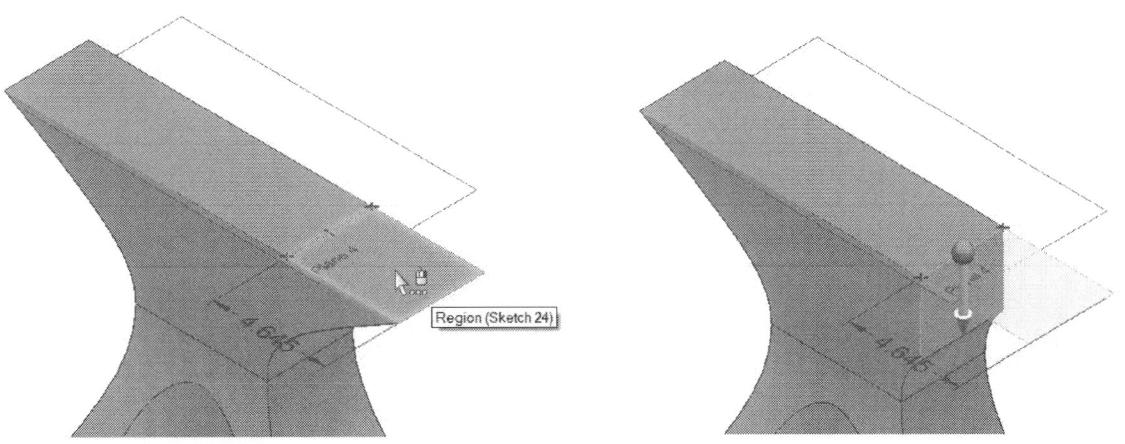

57. Click **Home > Solids > Extrude** on the ribbon, and then click on the top face of the model. Right click to accept the selection.
58. Move the pointer upward, type 1 and press Enter.

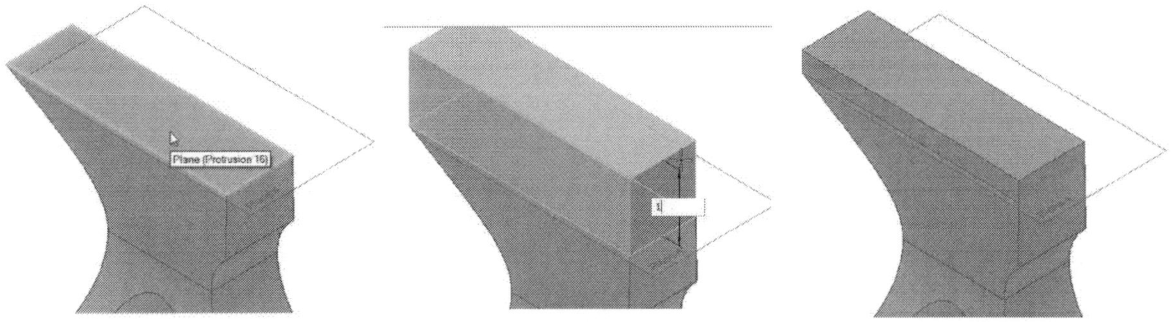

59. Click **Home > Draw > Circle drop-down > Ellipse by 3 Points** on the ribbon and place the pointer on the right face, as shown. Click the lock icon to lock the face.
60. Click the **Sketch View** icon located at the bottom of the window. This orients the sketch plane normal to the screen.
61. Specify the three points of the ellipse, as shown. Next, click **Home > Relate > Connect** on the ribbon.
62. Select the left vertical edge of the sketch plane and the left quadrant of the ellipse.

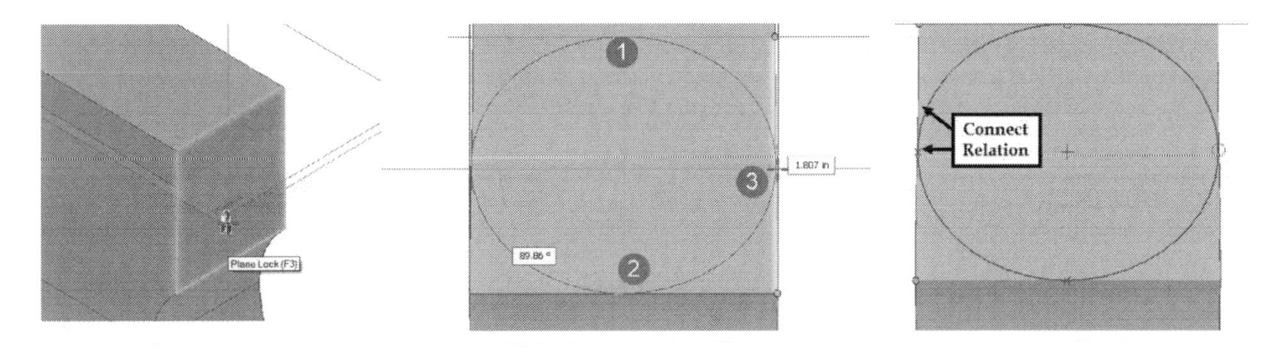

63. Next, click the Unlock (F3) icon located at the top right corner in the graphics window.
64. Click the **Home** icon located at the bottom left corner of the ViewCube.
65. Create a new plane coincident to the right flat face. Next, move it towards right up to a distance of 8.22.

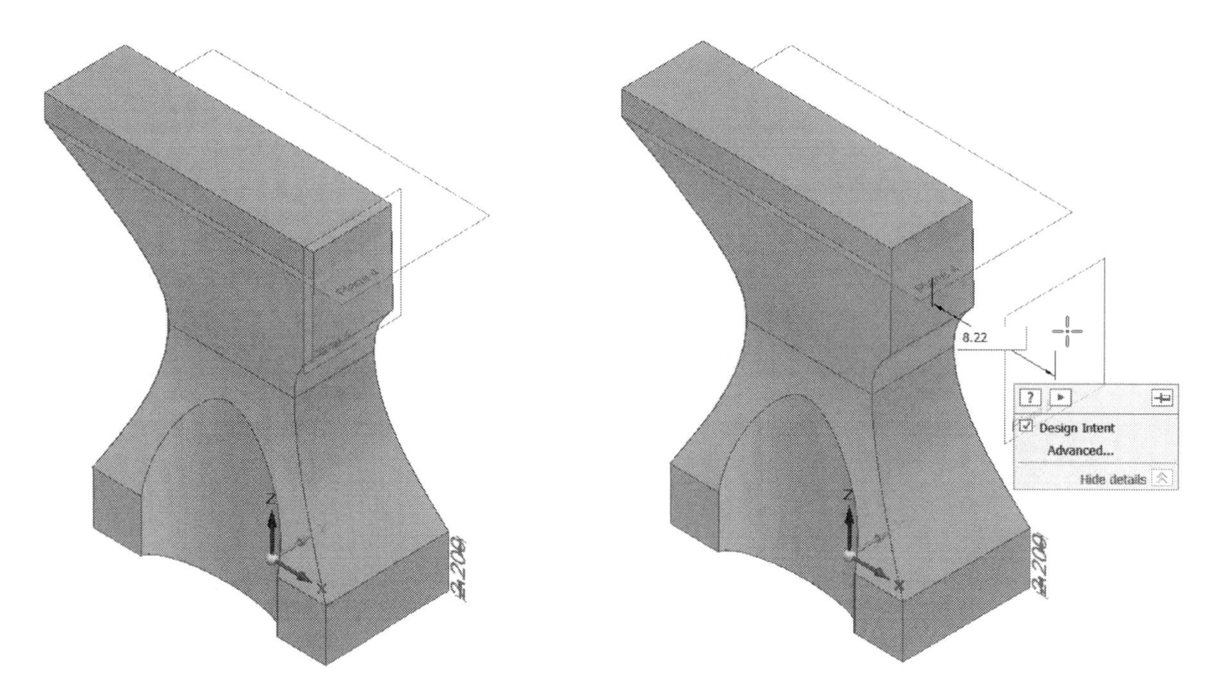

66. Click **Home > Draw > Line drop-down > Point** on the ribbon and place the pointer on the newly created plane, as shown. Click the lock icon to lock the plane.
67. Click the **Sketch View** icon located at the bottom of the window. This orients the sketch plane normal to the screen.
68. Click to specify the location of the point, as shown.

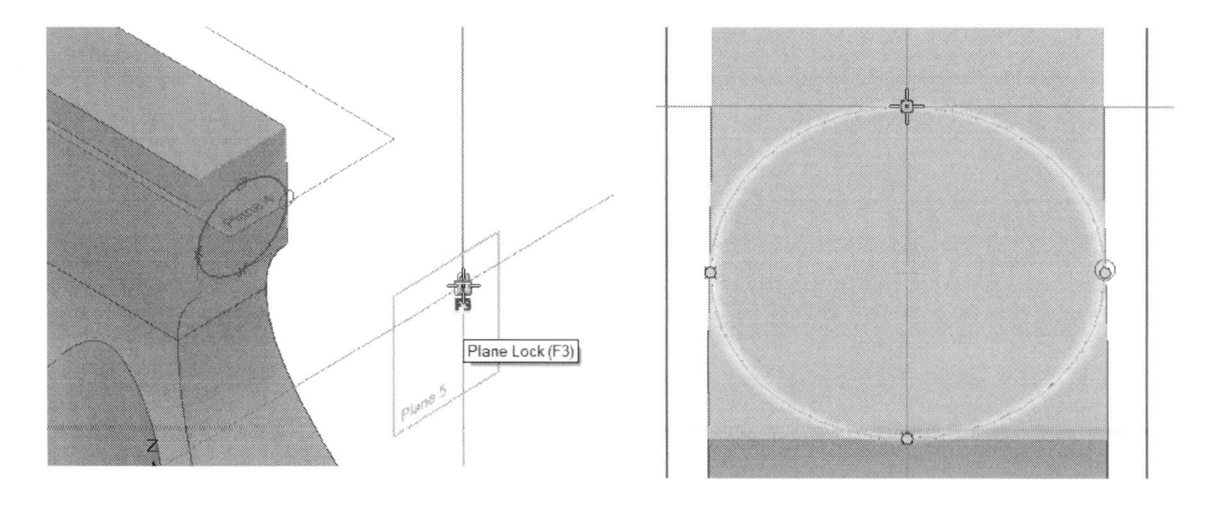

69. Next, click the Unlock (F3) icon located at the top right corner in the graphics window.
70. Click the **Home** icon located at the bottom left corner of the ViewCube.
71. Click **Home > Draw > Curve** on the ribbon and place the pointer on the XZ plane, as shown. Click the lock icon to lock the face.
72. Click the **Sketch View** icon located at the bottom of the window. This orients the sketch plane normal to the screen.

73. Specify the first two points of the curve, as shown. Next, select the sketch point displayed on the plane to specify the third point. Right click to create the curve.

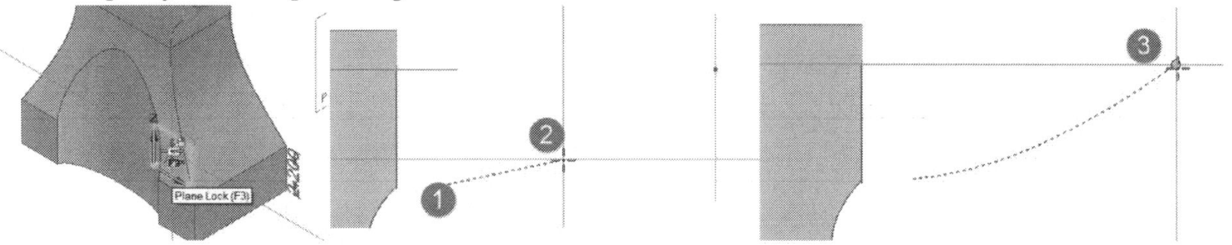

74. Click **Home > Relate > Connect** on the ribbon and select the start point of the curve. Next, select the bottom quadrant point of the ellipse, as shown.

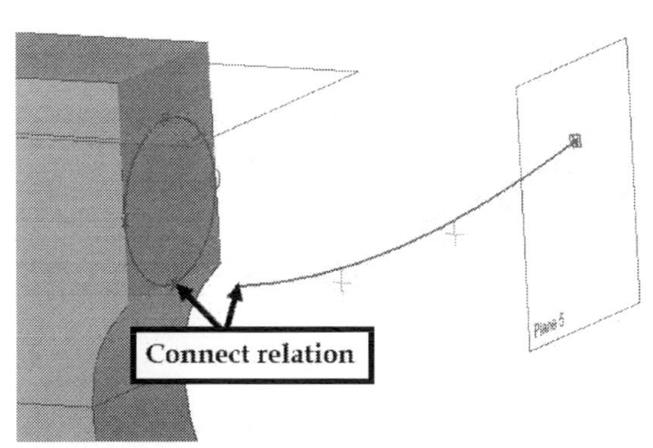

75. Click the Unlock (F3) icon located at the top right corner in the graphics window.
76. Click the **Home** icon located at the bottom left corner of the ViewCube.
77. Click **Home > Draw > Line** on the ribbon and place the pointer on the XZ plane. Click the lock icon to lock the face.
78. Create a line connecting the top quadrant point of the ellipse and the sketch point.

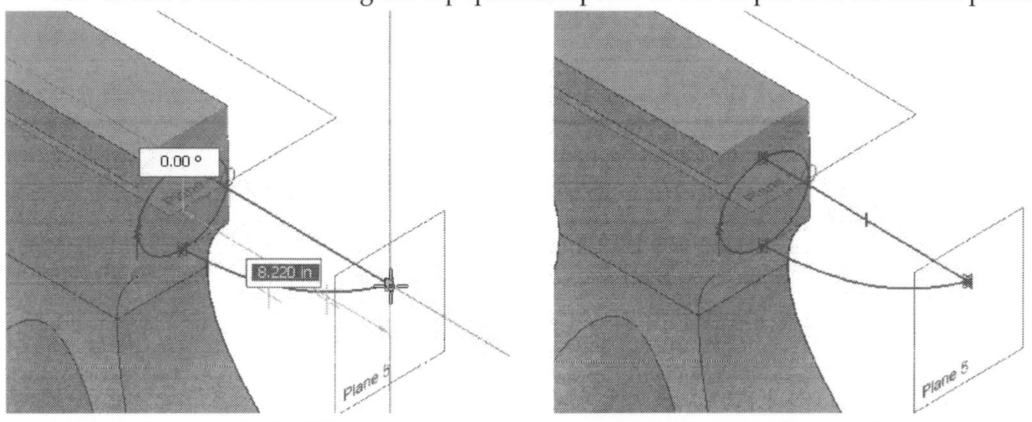

79. Click the Unlock (F3) icon located at the top right corner in the graphics window.
80. Click **Home > Solids > Add drop-down > Loft** on the ribbon and select the ellipse.
81. On the command bar, select **Select > Point**, and then select the sketch point.
82. Click the **Guide Curve Step** icon on the command bar and select **Select > Single**.
83. Select the first guide curve, as shown. Next, click the **Accept** button on the command bar.
84. Select the second guide curve, and then click the **Accept** button.
85. Click the **Preview** button, and then click **Finish** and **Cancel**.

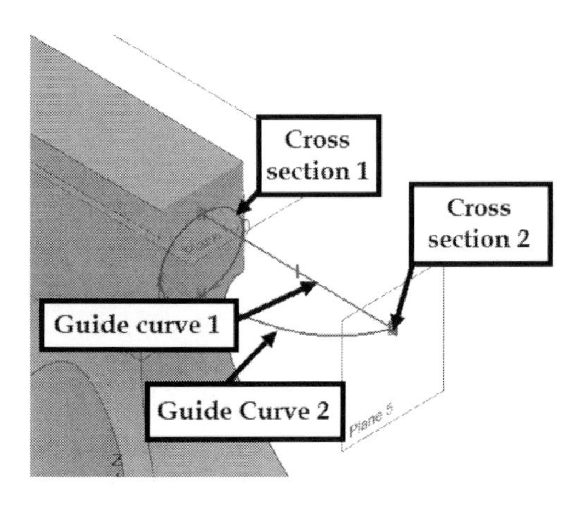

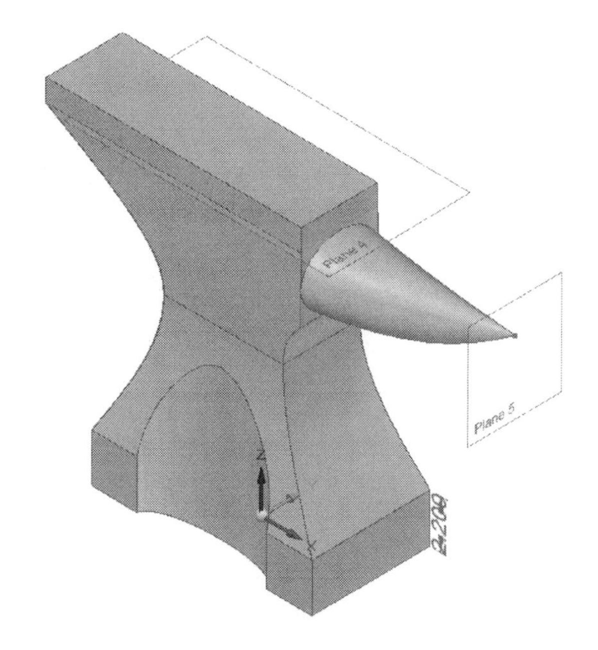

86. Create a cut feature on the side face, as shown.

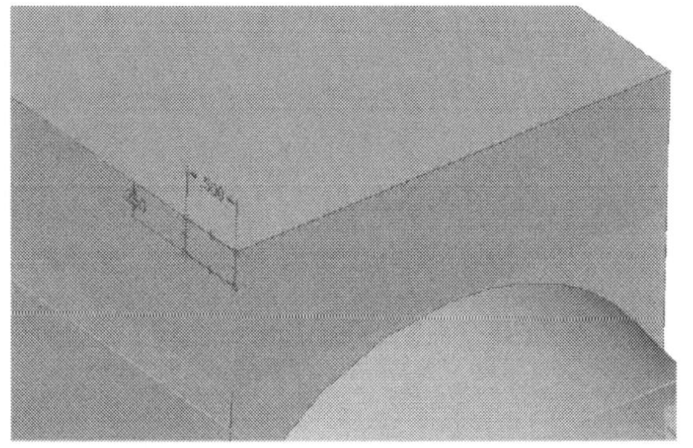

87. Create a hole and the square cut features on the top face, as shown.

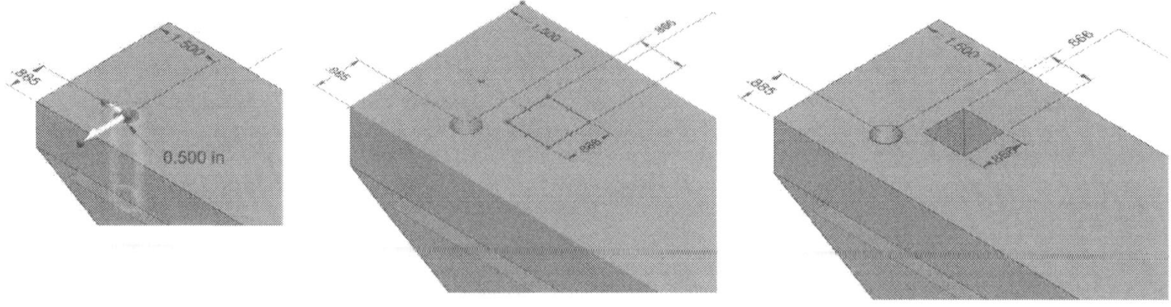

88. Save and close the file.

Questions

1. Describe the procedure to create a *Loft* feature.
2. List any two **Tangency Control** options.
3. List the type of elements that can be selected to create a *Loft* feature.

Exercises

Exercise 1

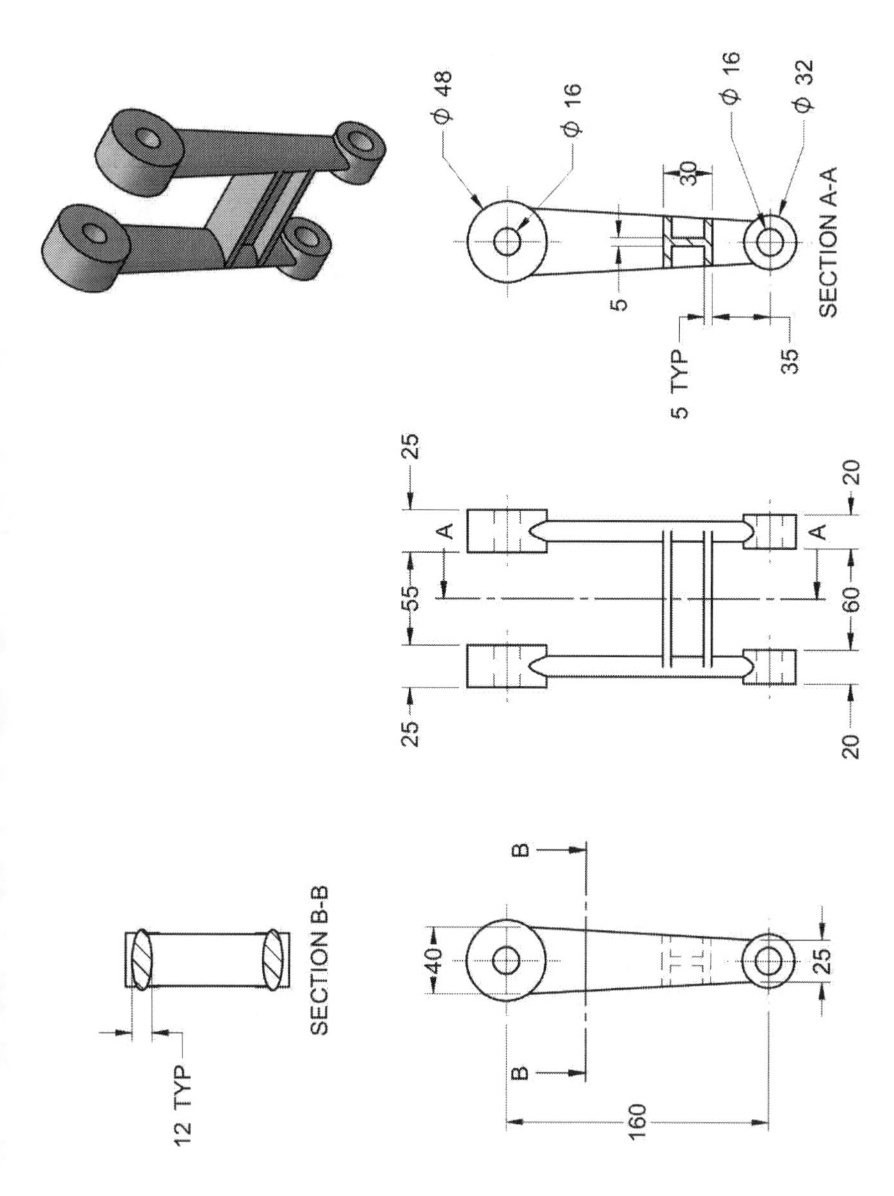

Chapter 8: Additional Features and Multibody Parts

Solid Edge offers you some additional commands and features which will help you to create complex models. These commands are explained in this chapter.

The topics covered in this chapter are:

- *Ribs*
- *Web Network*
- *Mounting bosses*
- *Lips*
- *Vents*
- *Slots*
- *Multi-body parts*
- *Split bodies*
- *Boolean Operations*
- *Emboss features*

Rib

This command creates a rib feature to add structural stability, strength and support to your designs. Just like any other sketch-based feature, a rib requires a two dimensional sketch. Create a sketch, as shown in figure and activate the **Rib** command (click **Home > Solids > Thin Wall > Rib** on the ribbon). Select the sketch and click the green check; the preview of the geometry appears. You can add the rib material to either sides of the sketch line or evenly to both sides. Set the **Alignment** type to **Centered** to add material to both sides of the sketch line. Type-in the thickness value of the rib feature in the box displayed on the model. You can use the steering wheel to change the direction of the rib.

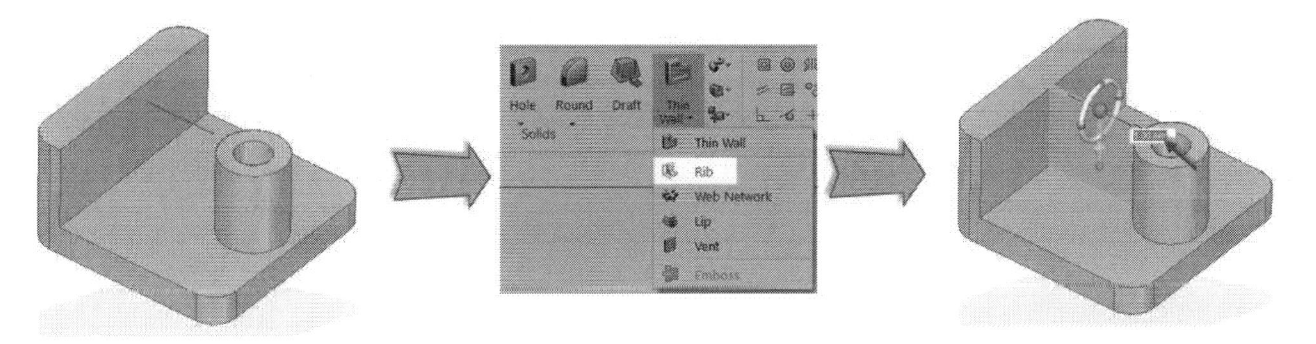

If you activate the **No Extend** option on the command bar, the material will not extend to meet the faces of the surrounding features.

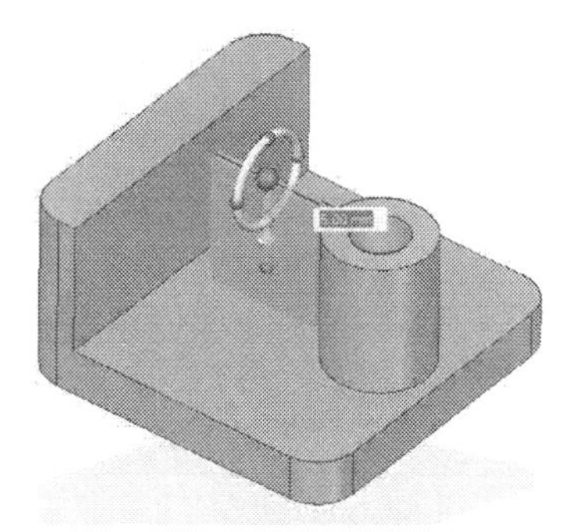

Activate the **Finite Depth** option, if you want to add material only up to some distance. Click the green check to complete the rib feature.

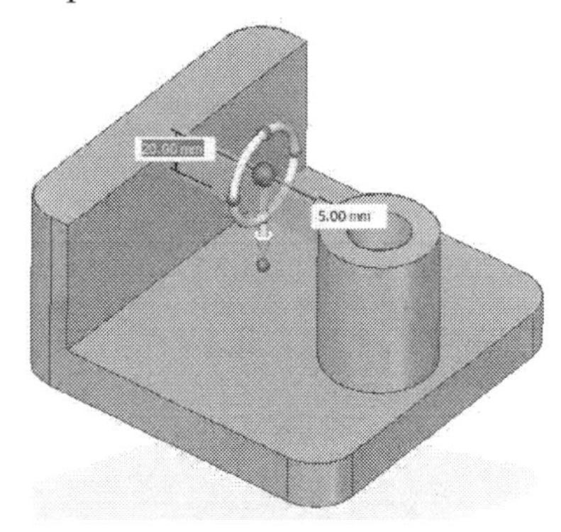

Web Network

This command is similar to the **Rib** command, but creates multiple ribs at a time forming a network. Create a two dimensional sketch, as shown in figure and activate the **Web Network** command (Click **Home > Solids > Thin Wall > Web Networks** on the ribbon). Select the sketch elements one-by-one and click the green check; the preview of the geometry appears.

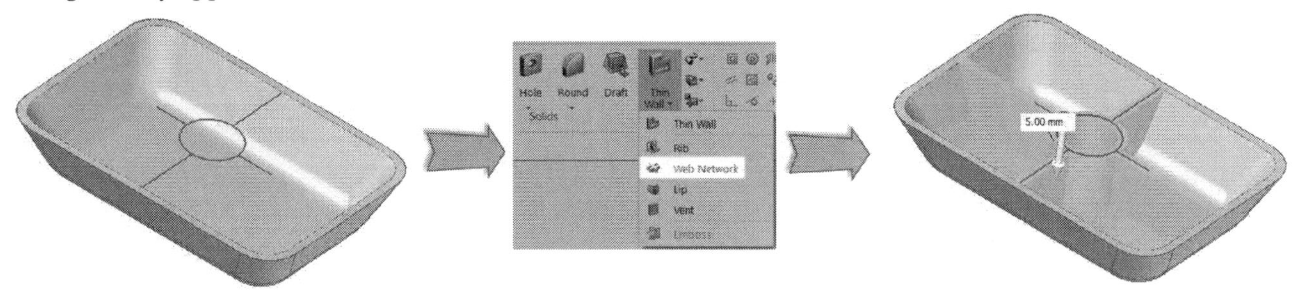

Use the **Draft** option on the command bar to add a draft to the web network feature. Click the green check to complete the feature.

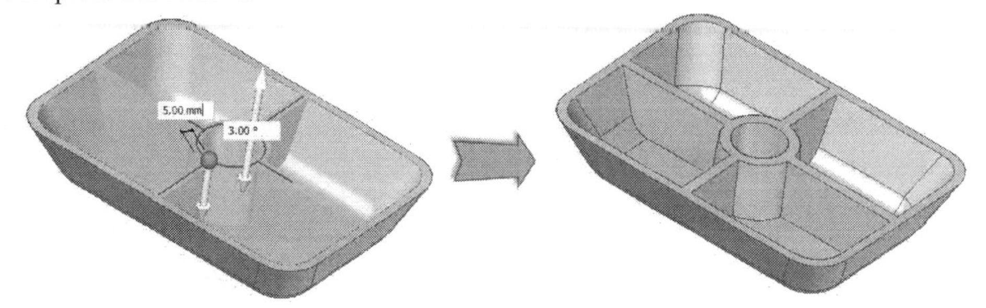

Mounting Boss

The process of creating mounting bosses can be automated using the **Mounting Boss** command. This command is available only in the Ordered mode. Switch to the Ordered mode and activate the **Mounting Boss** command (click **Home >Solids >Thin Wall > Mounting Boss** on the ribbon). On the command bar, select **Coincident Plane** from the **Create from Options** drop-down menu. Select the top face of the model and define the location of the mounting bosses. Click **Close Sketch** on the ribbon and define the side of the bosses.

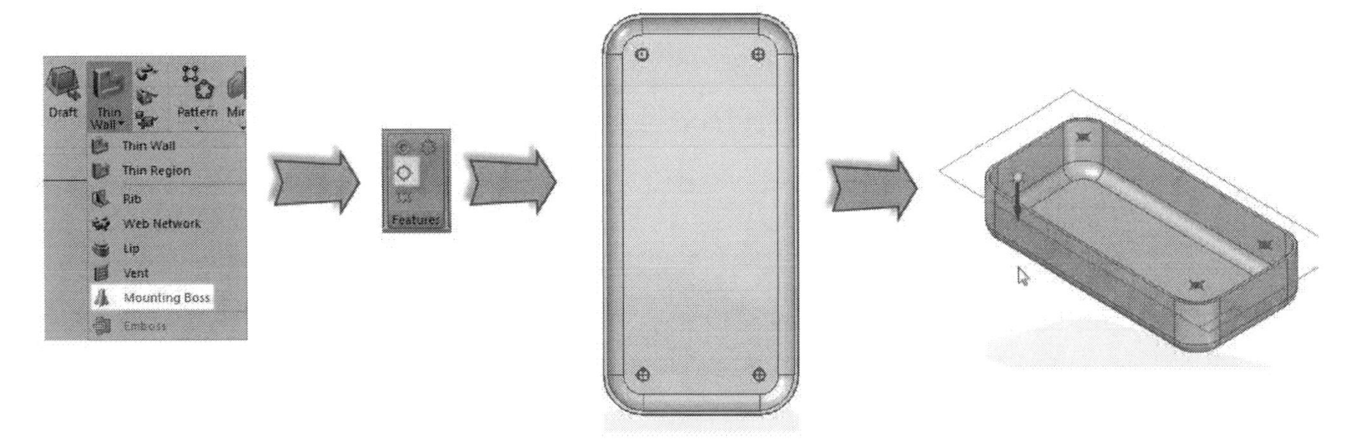

Click the **Options** icon on the command bar; the **Mounting Boss Options** dialog pops up on the screen. Define the parameters of the mounting boss. The parameters are self-explanatory. Click **OK** and then **Finish** completing the feature.

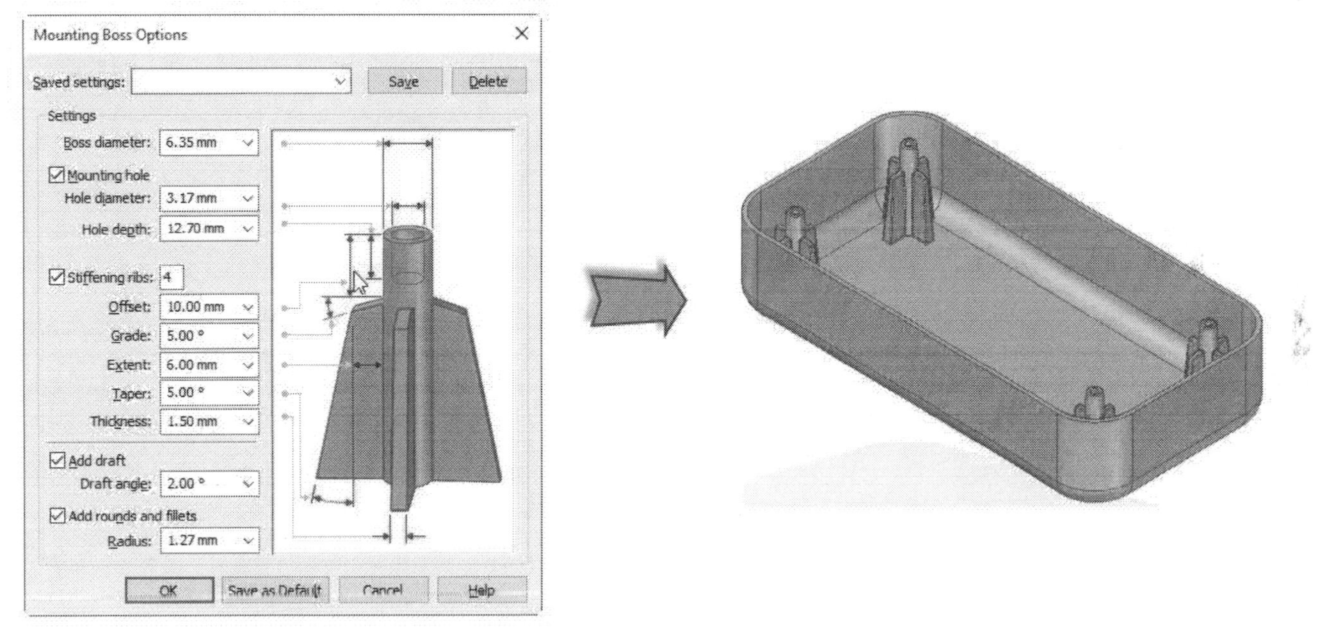

Lip

This command allows you to create lips and grooves on edges of parts, saving you time by not having to create a series of manual cuts. Activate this command (click **Home > Solids > Thin Wall > Lip** on the ribbon) and select a chain of edges. Click the green check on the command bar and define the side of the lip. Click the left mouse button, and then click **Finish** to complete the feature.

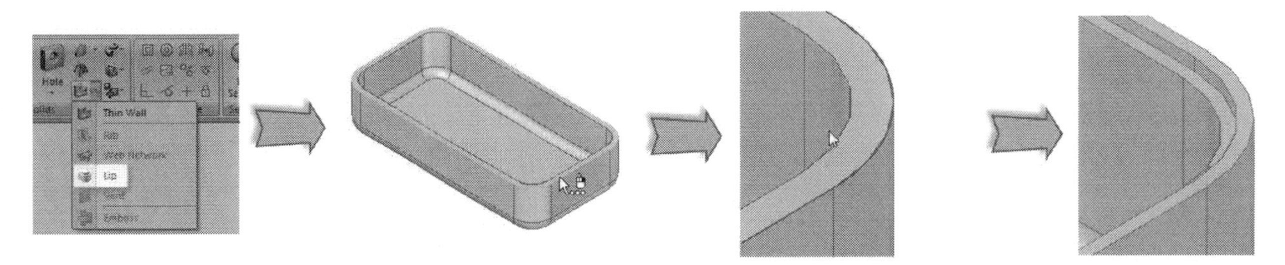

Vent

This command allows you to take a two-dimensional sketch of a vent and convert it into a 3D cutout. To create a vent feature, first create a 2D sketch and activate the **Vent** command (click **Home > Solids > Thin Wall > Vent** on the ribbon). As you activate this command, the **Vent Options** dialog pops up on the screen.

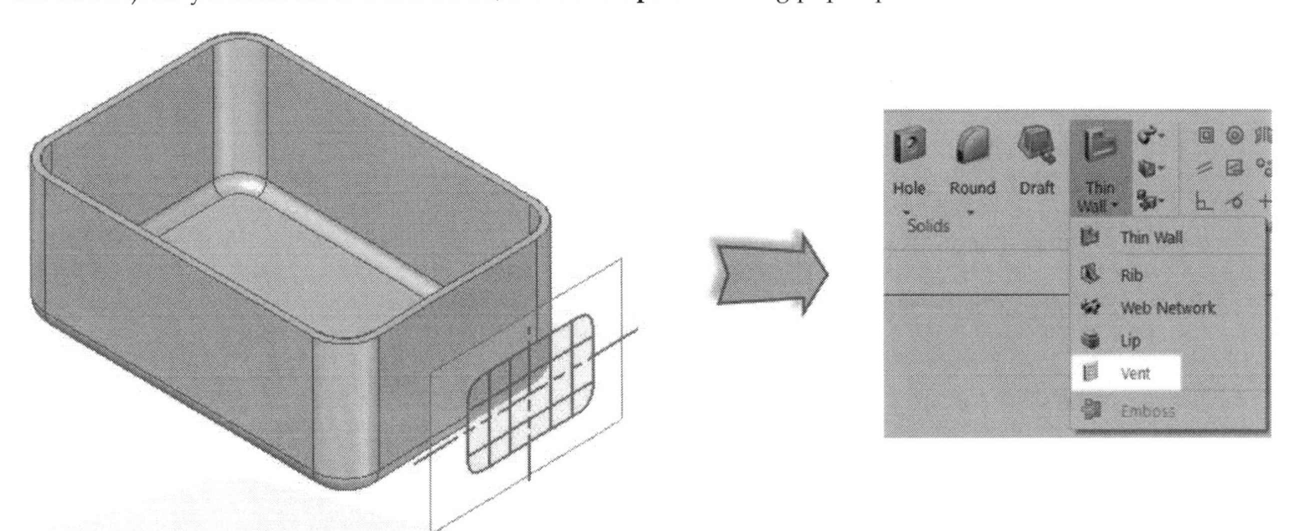

Set the thickness of **Ribs** and **Spars** to 4.

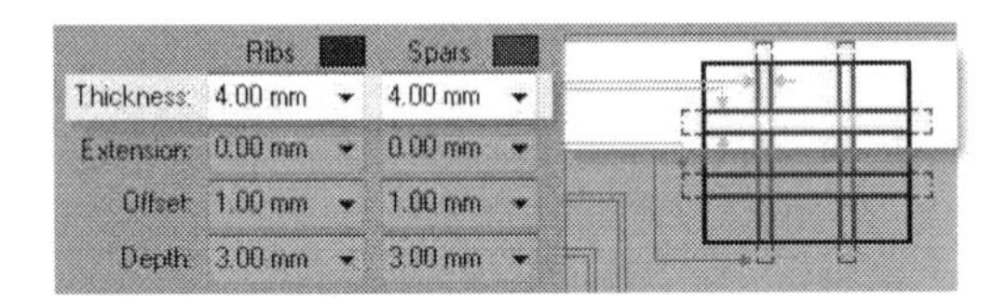

Set the **Offset** values to 1 and **Depth** to 3.

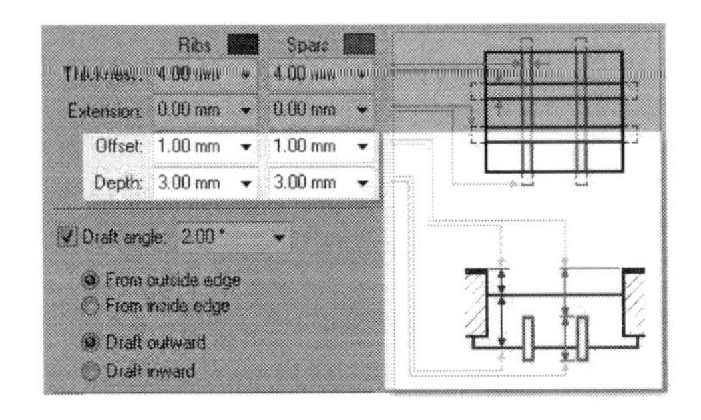

Check the **Draft angle** option and enter 5 in the box.
Check the **Round & fillet radius** option and enter 0.5 in the box. Click **OK** on the dialog.

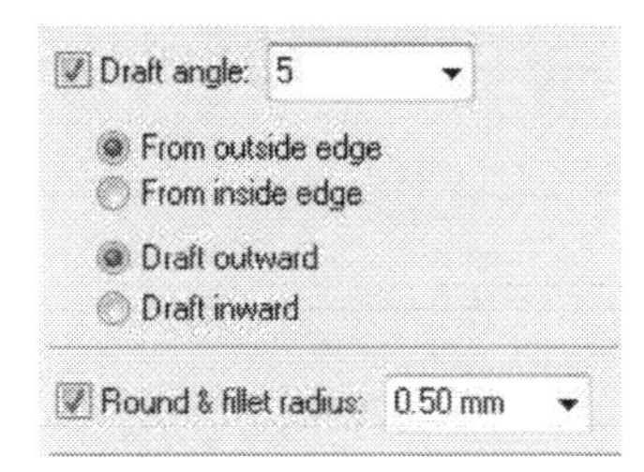

Select the boundary and click the green check on the command bar. Likewise, select the ribs and spars, and then click on the model to define the side of the vent. Next, click **Finish** to complete the feature.

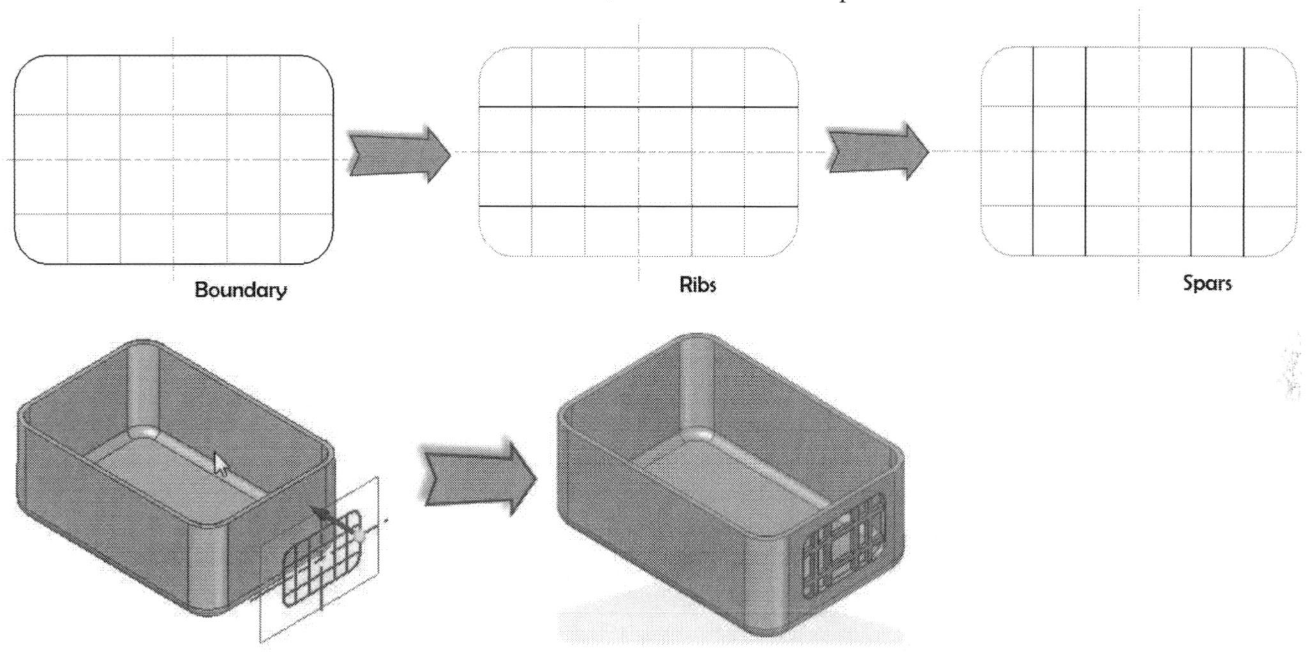

Slot

This command creates a slot by using a 2D sketch. The sketch can have a single or multiple elements. If the sketch is having multiple elements, they should be tangent and continuous to each other. To create a slot, create a 2D sketch on a face, and then activate the **Slot** command (click **Home > Solids > Hole > Slot** on the command); the command bar pops up on the screen. Click the **Options** icon on the command bar to open the **Slot Options** dialog. Type-in a value in the **Slot width** box and select the end type. Click **OK** and select the sketch. Next, define the extent of the slot feature. Click the right mouse button to complete the slot feature.

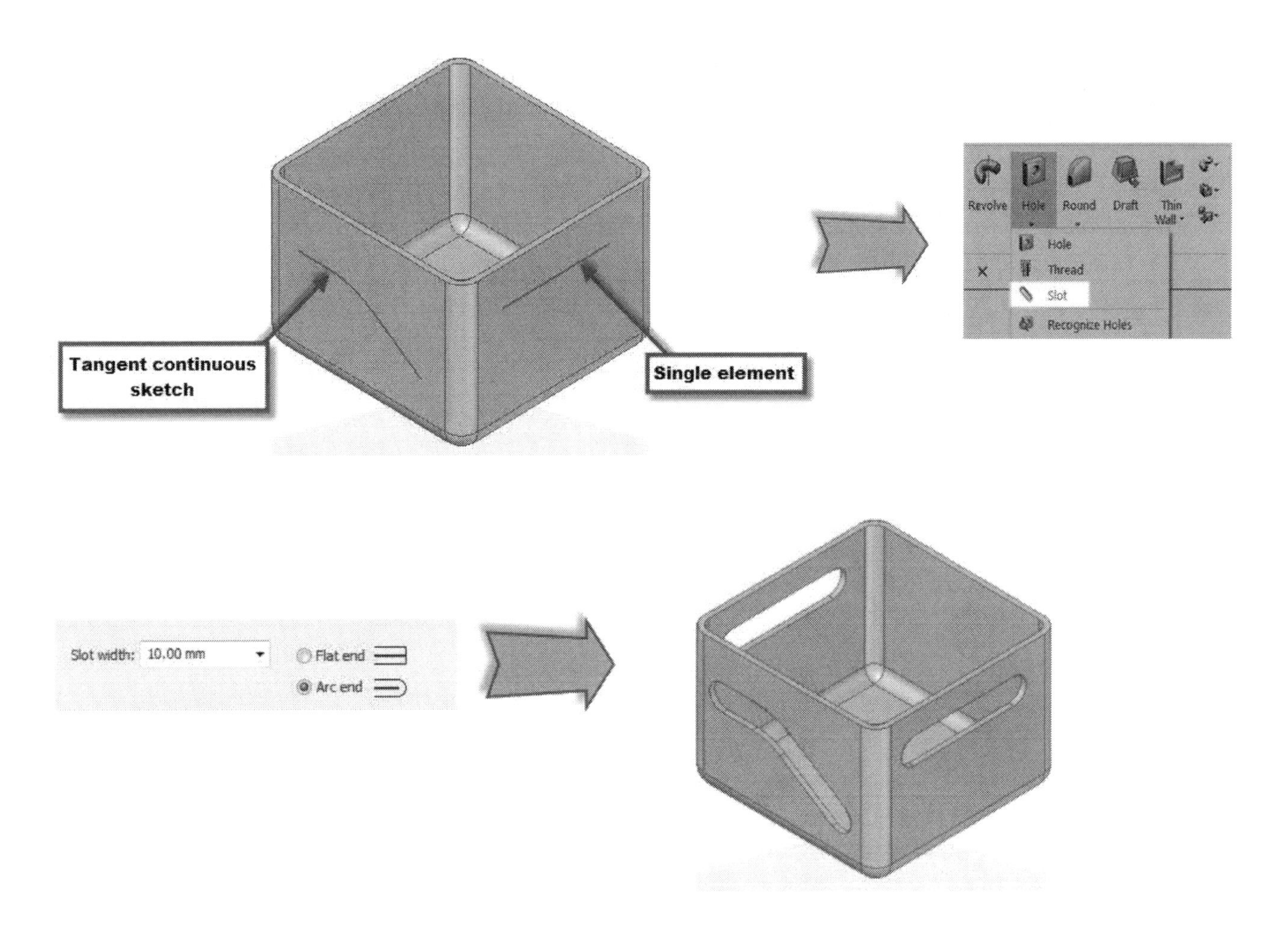

If you want to create a counterbore slot, check the **Counterbore** option on the **Slot Options** dialog and define the **Path Offset** and **Depth Offset** values. You can create two types of the counterbore slots: **Recessed** and **Raised**.

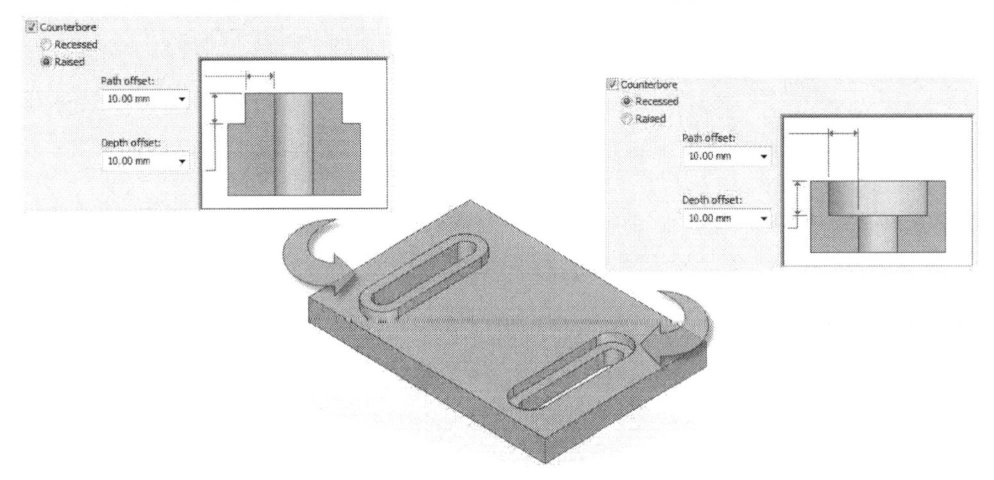

Creating Multiple Slots

In Solid Edge 2019, you can create multiple slots in a single instance by selecting elements with identical parameters. You can only select the sketch elements lying on a same plane or face. Activate the **Slot** command (on the ribbon, click **Home > Solids > Hole > Slot**), and then select the first sketch element. Next, define the extent of the slot using the options available in the **Extents** drop-down. After defining the extents, select the remaining sketch elements one-by-one or by dragging a selection window across them. You can press hold the Shift or Ctrl key, and then click on the selected sketch elements to remove them from the selection. Click **Accept** on the command bar or right-click in the empty space to complete the feature.

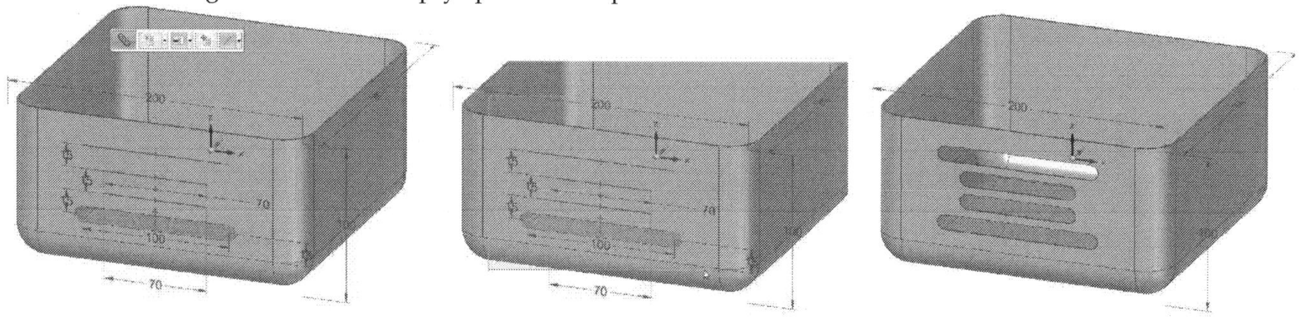

Multi-body Parts

Solid Edge allows the use of multiple bodies when designing parts. This opens the door to several design techniques that would otherwise not be possible. In this section, you will learn some of these techniques.

Creating Multibodies

The number of bodies in a part can change throughout the design process. Solid Edge makes it easy to create multiple bodies, and combine them into a single body.

To create multiple bodies in a part, first create a solid body, and then activate the **Add Body** command (click **Home > Solids > Add Body** on the ribbon); the **Add Body** dialog pops up on the screen. Select the **Add Part body** or **Add Sheet Metal body** option, enter the body name in the **New body name** field, and then click **OK**. Now, create another solid body using anyone of the solid modeling tools; the **Design Bodies** entry will be added to the **Pathfinder** tree.

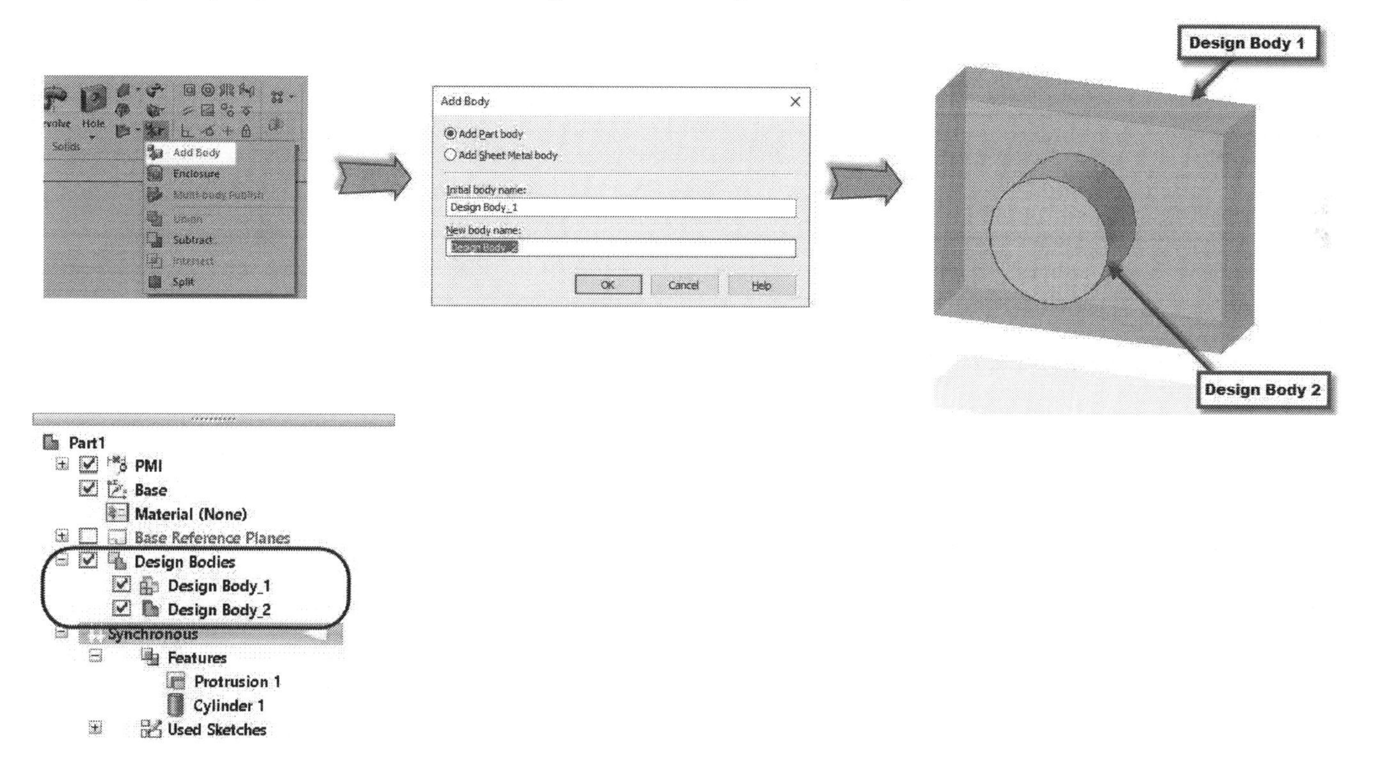

Split

The **Split** command can be used to separate single bodies into multiple bodies. This command can be used to perform local operations. For example, if you apply the shell feature to the front portion of the model shown in figure, the whole model will be shelled. To solve this problem, you must split the solid body into multiple bodies.

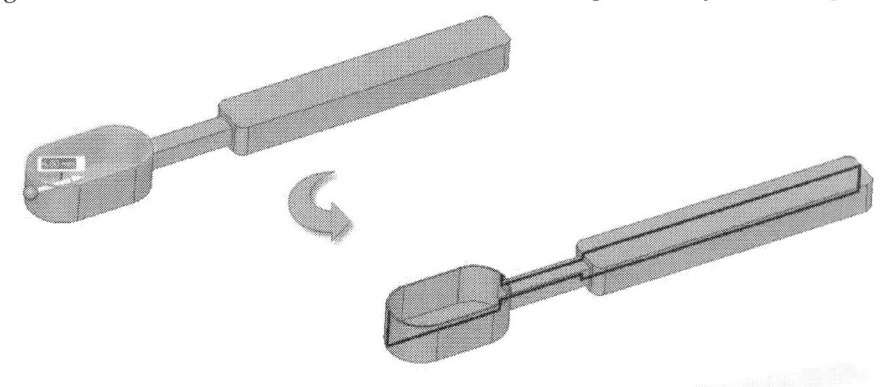

To split a body, you must have a splitting tool such as planes, sketch elements, surface, or bodies. In this case, a surface can be used as a splitting tool. To create a surface, activate the **Bounded** command (click **Surfacing > Surfaces > Bounded** on the ribbon) and set the **Selection Type** on the command bar to **Single**. Now, select the edges, as shown in figure. Click the green check on the command bar and click **Finish** to create the bounded surface.

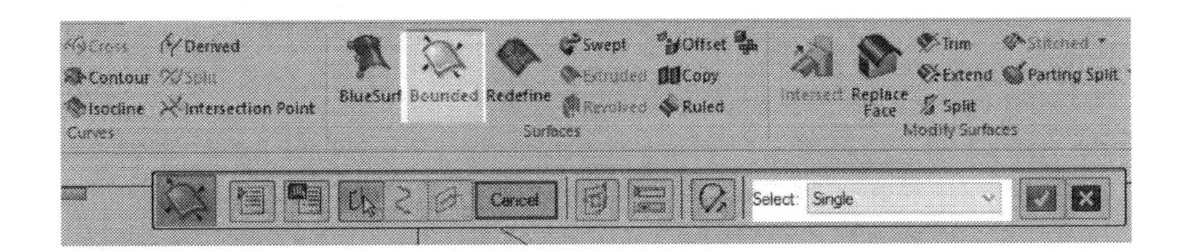

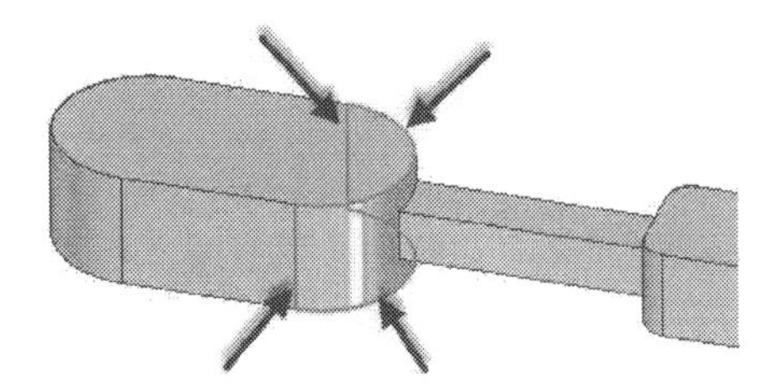

Activate the **Split** command (click **Home > Solids >Add Body > Split** on the ribbon) and select the solid body from the graphics window. Next, select the bounded surface as the splitting tool and click the green check on the command bar. This results in two separate bodies. Press Esc to deactivate the **Split** command.

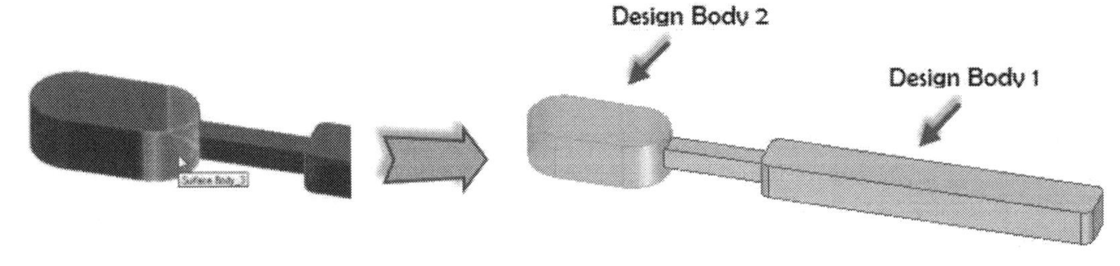

Now, double-click on **Design Body 2** in the **Pathfinder** tree to activate it, and then create the shell feature.

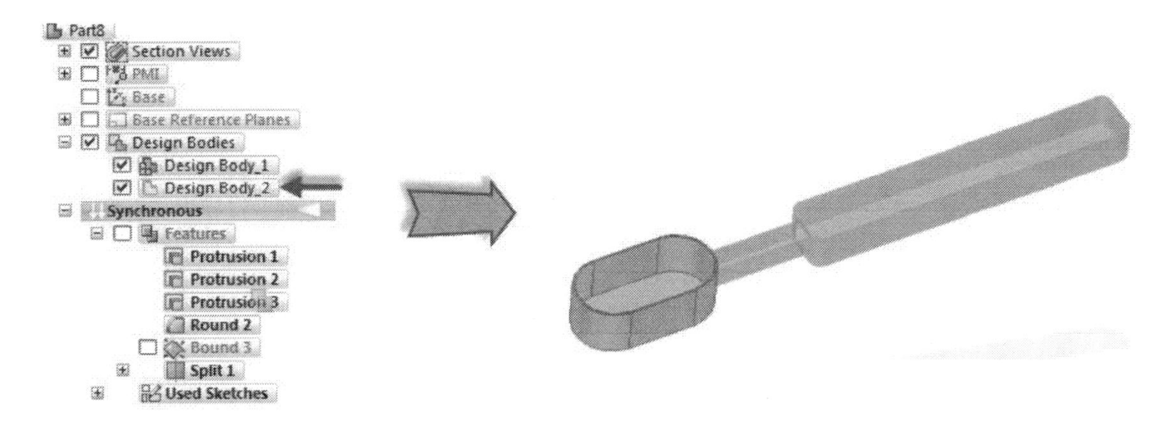

Union

If you apply rounds to edges between two bodies, it will result in a different outcome as shown in figure. To solve this problem, you must combine the two bodies using the **Union** command. Activate this command (click **Home > Solids > Add Body > Union** on the ribbon) and select the bodies. Click the green check on the command bar to combine the bodies. Now, apply rounds to the edges.

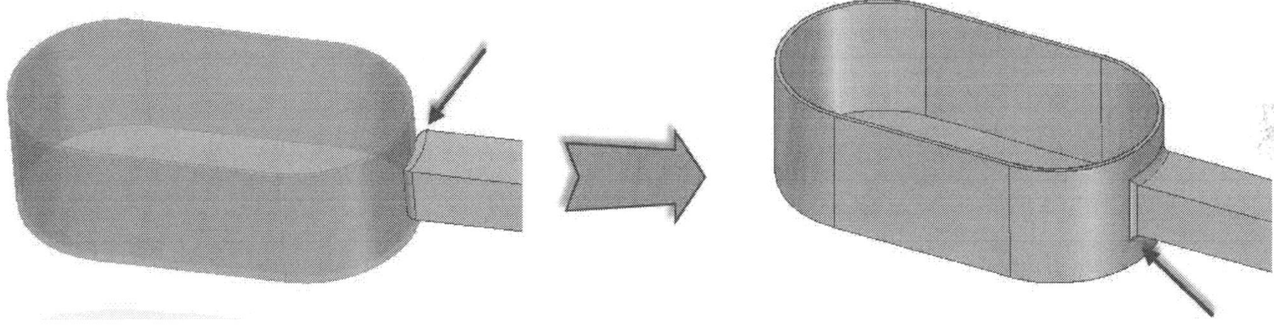

Intersect

By using the **Intersect** command, you can generate bodies defined by the intersecting volume of two bodies. Activate this command (click **Home > Solids > Add Body > Intersect** on the ribbon) and select two bodies. Click the green check to see the resultant single solid body.

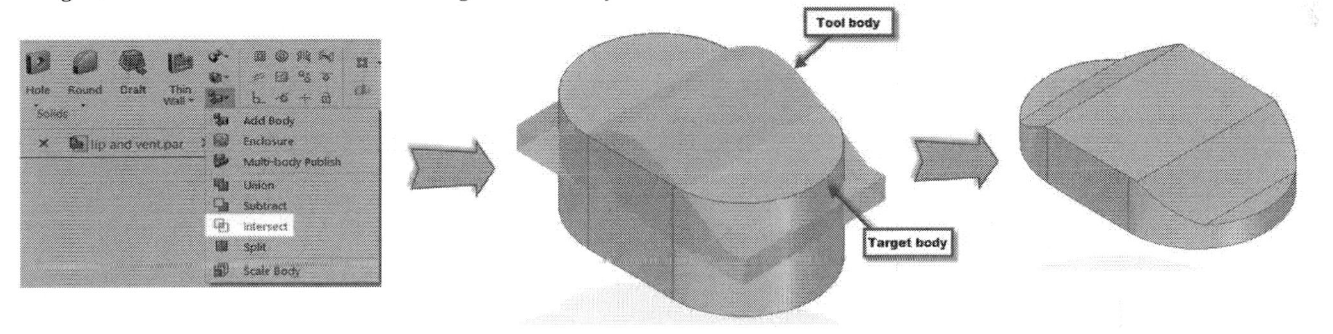

Subtract

This command performs the function of subtracting one solid body from another. Activate this command (click **Home > Solids > Add Body > Subtract** on the ribbon) and select target body. Click the green check, and then select the tool body. Again, click the green check to subtract the tool body from the target. Hide the tool body to see the result.

Multi Body Publish

In addition to creating multiple bodies, Solid Edge also offers an option to generate an assembly from the resulting bodies. For example, create the model shown in figure and split it into two separate bodies. Next, add lip/groove feature to the model, and then save the model.

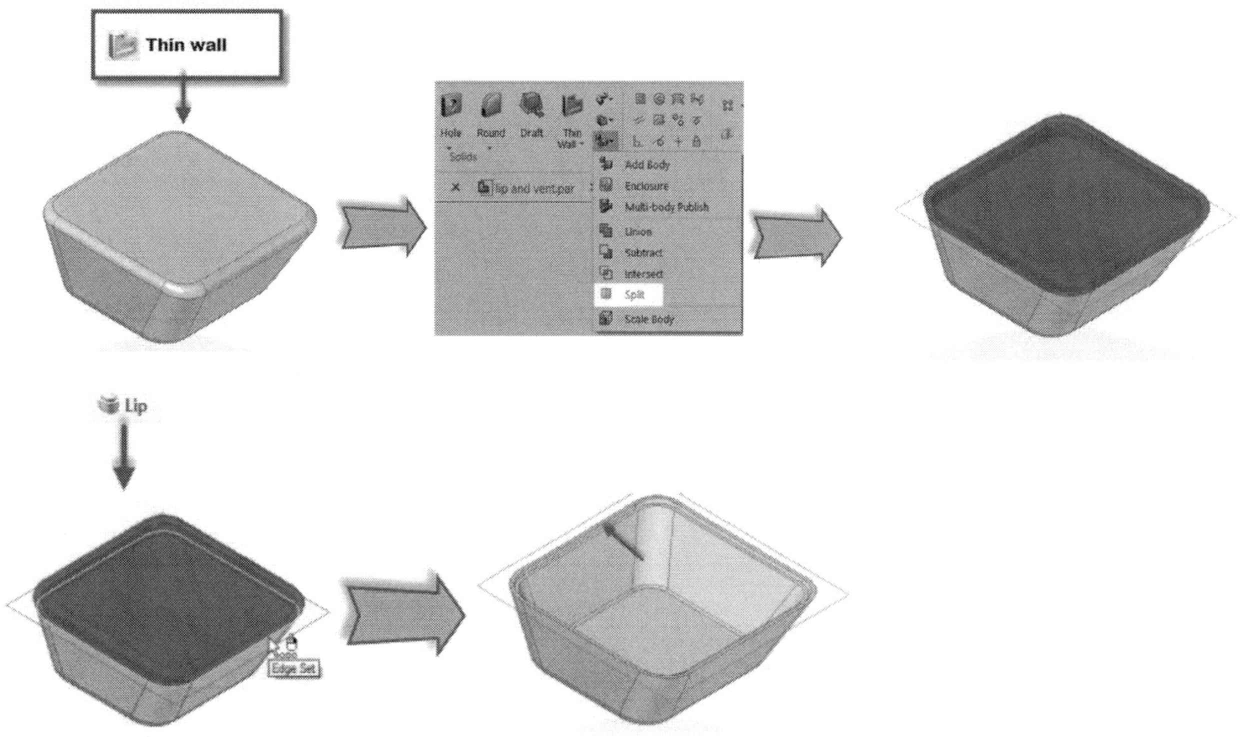

Activate the **Multi Body Publish** command (click **Home > Solids >Add Body > Multi Body Publish** on the ribbon); the **Multi-body Publish** dialog pops up on the screen. Click **Save Files** on the dialog to save the design bodies as individual files and create an assembly. Click the right mouse button on **Create Assembly path** and select **Open**. The newly created assembly is opened. Now, you can apply assembly relationships to the parts.

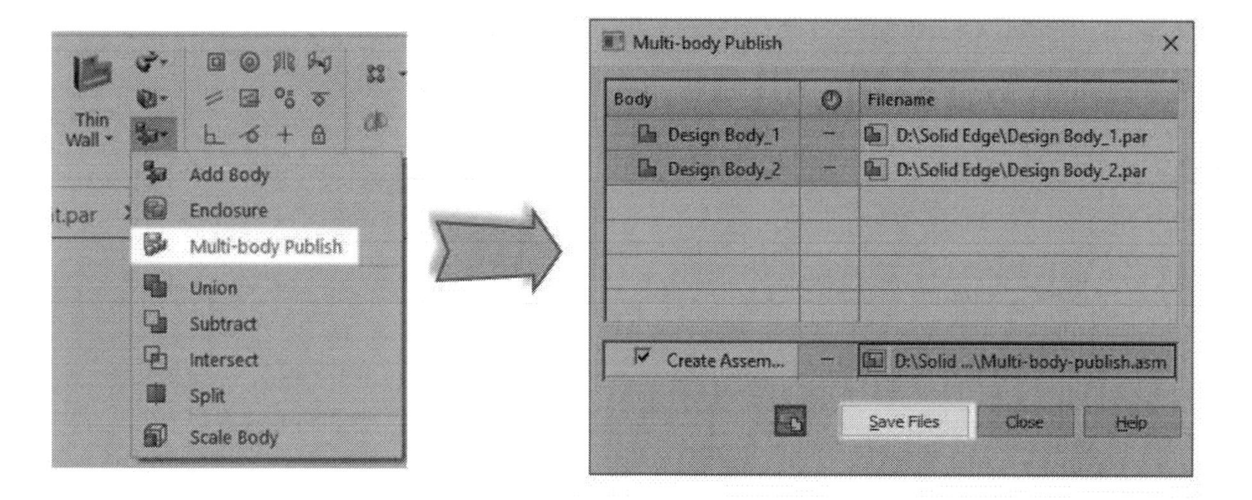

Emboss

This command allows you to change the shape of a solid body by using another solid body. The solid body that is changed is called the target body and the solid body that causes the changes is called the tool body. To create an emboss feature, you must have two solid bodies in a part. Activate the **Emboss** command (click **Home > Solids > Thin Wall > Emboss** on the ribbon) and select the target and tool bodies. Type-in values in the **Clearance** and **Thickness** boxes. Use the **Direction** icon on the command bar to define the side on which the body is embossed. Click the green check on the command bar to complete the emboss feature.

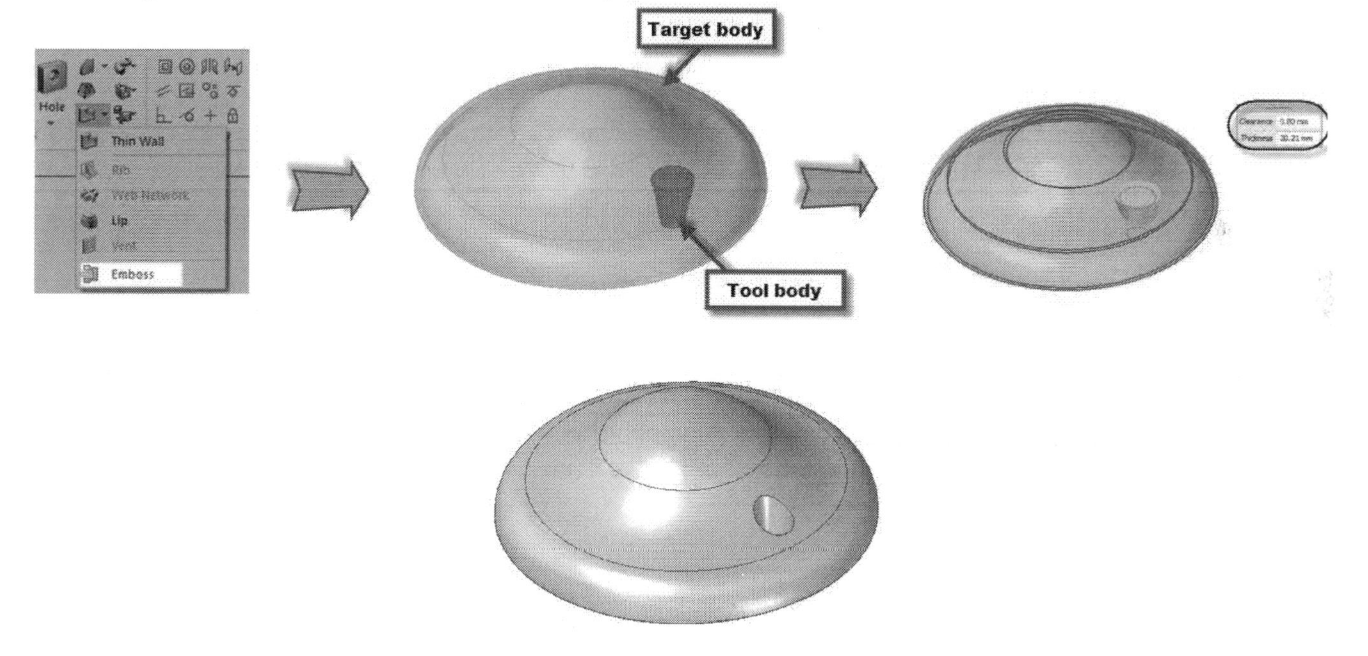

Solid Sweep Cutout

This command removes material by sweeping the volume of another solid body along a planar or non-planar curve. To solid sweep cutout, first create the target body and the tool path, and then create the tool body. Note that you should create the target and tool bodies as separate bodies (Use the **Add body** command to create the tool body). Also, the tool body should be created in such a way that it intersects the target body. The tool path should be created on the target face. The target face can be planar or non-planar.

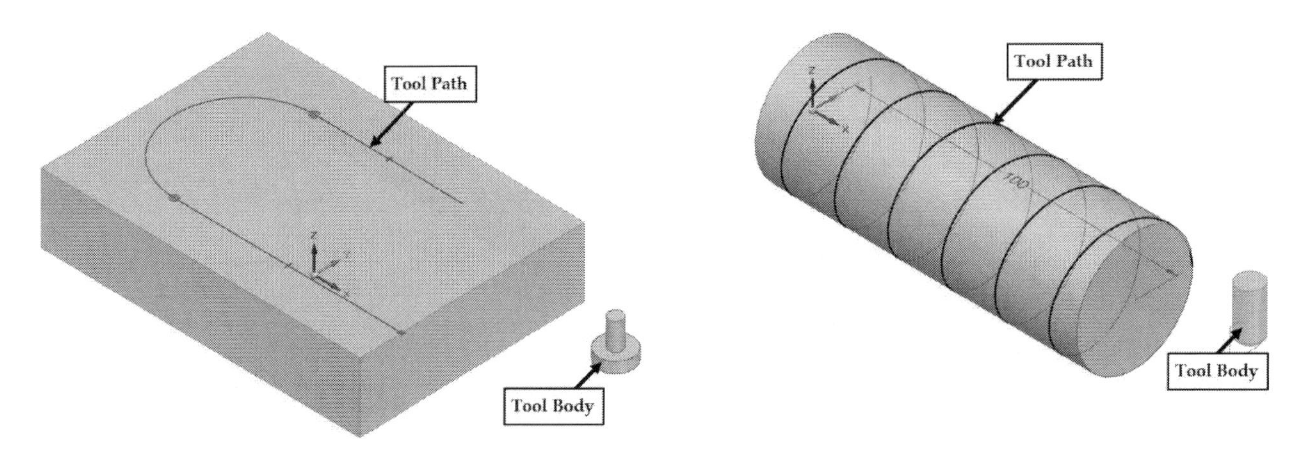

The tool body should be cylindrical or revolved solid. The following figure shows some examples of tool bodies. However, you may need to create the axis manually for the tool similar to the third one.

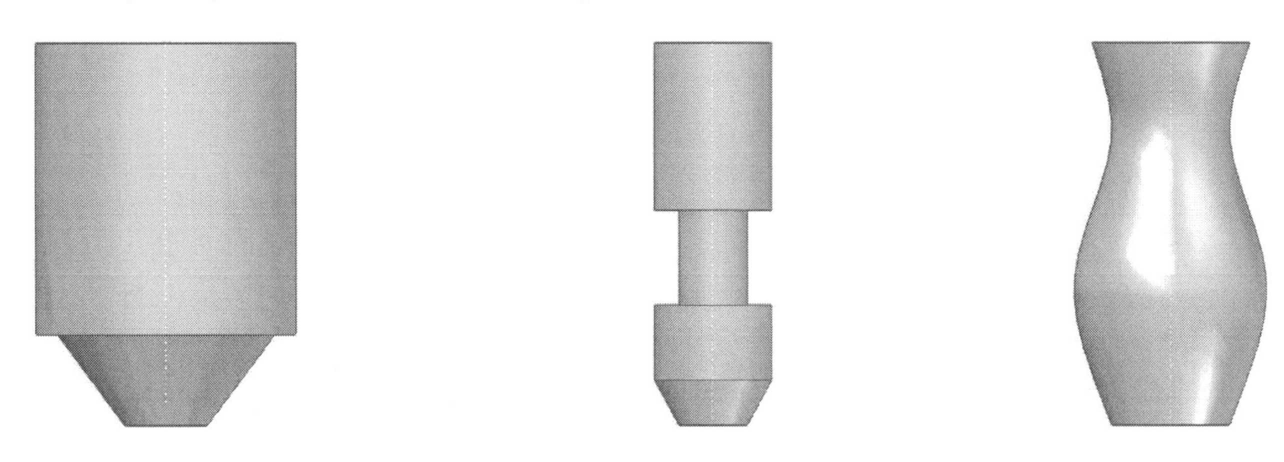

Creating Solid Swept Cutout using 2D Tool Path

Activate the target body by double-clicking on it in the Pathfinder. Activate the **Solid Sweep Cutout** command (on the ribbon, click **Home > Solids > Cut drop-down > Solid Sweep Cutout**). Next, select the tool path and click the **Accept** button on the command bar. Next, select the tool body and make sure that the **Place Tool on Path** icon is selected. This creates a Solid Sweep Cutout even when the tool body is not on the tool path. Click the **Accept** button; the **Axis step** button is activated and the axis of the tool body is selected. Select the axis of the tool body, if it is not selected. Click **Accept** on the command bar; the orientation of the tool body is specified automatically, if the tool body is perpendicular to the tool path. However, you need to specify the orientation using the **Lock Direction Step**, if the tool body is not perpendicular to the tool path. Click **Finish** and **Cancel** on the command bar to create the Solid Swept cutout.

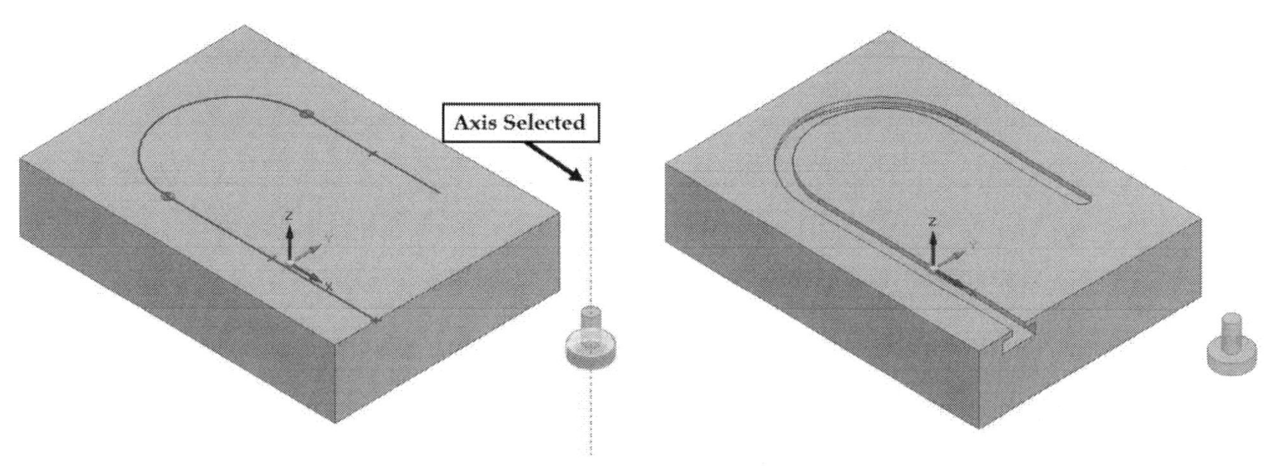

Sweeping Volume using 3D Tool Path

Activate the target body by double-clicking on it in the Pathfinder. Activate the **Solid Sweep Cutout** command (on the ribbon, click **Home > Solids > Cut drop-down > Solid Sweep Cutout**). Next, select the tool path and right-click to accept. Next, select the tool body and right click; the axis is selected automatically. Right click and select the axis perpendicular to the tool body. Click **Finish** and **Cancel** on the command bar to create the Solid Swept cutout.

Note: *Refer to the Helical Curve topic of Chapter 13: Surface Design to learn how to create the helical curve used in the following example.*

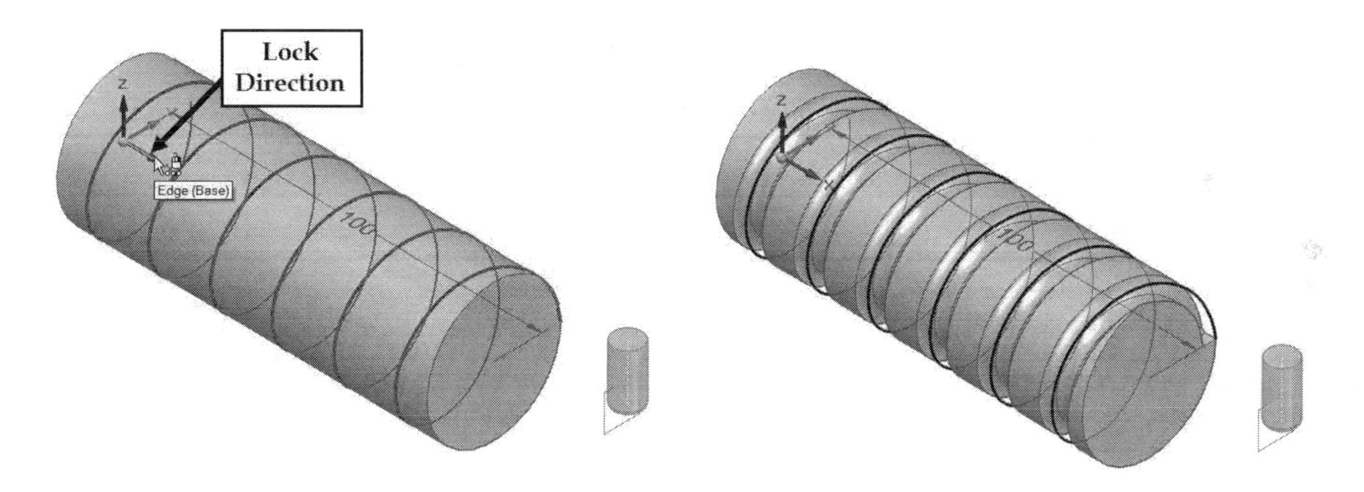

Solid Sweep Protrusion command

This command creates a protrusion by sweeping a solid body along the 2D or 3D path. To create the Solid Sweep

Protrusion, first create the solid body and the path, as shown. Activate the **Solid Sweep** command (on the ribbon, click **Home > Solids > Sweep drop-down > Solid Sweep**). Select the path and right click. Next, select the solid body and right click; the axis of the solid body is selected automatically. Right click to accept the selection. Select an element (planar face, axis, linear edge, or reference plane) to define the orientation of the solid sweep. Right click to accept the selection. The following figure shows the solid sweep when the face perpendicular to the axis of the solid body is selected in the **Lock Direction Step**. Next, click the **Finish** button.

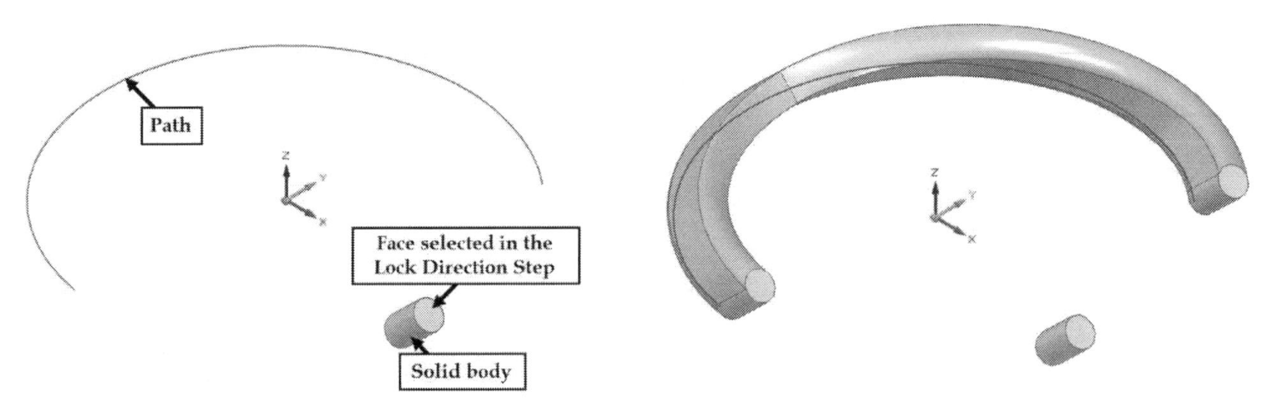

The following figure shows the solid sweep when the Z-axis is selected in the Lock Direction Step.

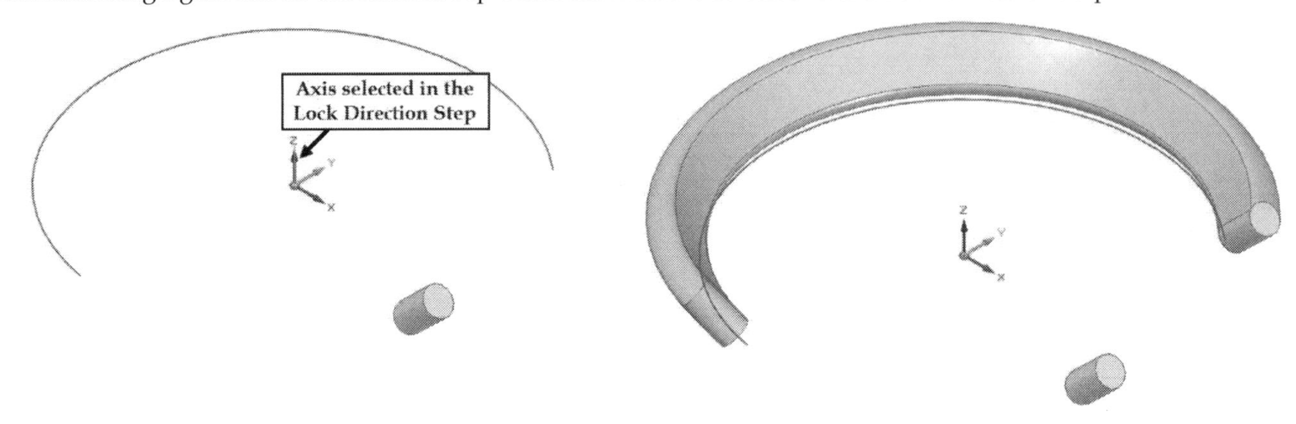

Creating Cut Features in Multi-body Parts

Solid Edge has some additional options (**Cut Active Body** and **Cut Selected Bodies**) while creating cuts in Multi-body parts. These options are available on the **Extrude** and **Revolve** command bars in the **Apply Cut Features** drop-down. However, you can select them only when the **Extent Type** and **Add/Cut** are set to **Finite** and **Cut**, respectively. The options in the **Apply Cut Features** drop-down are explained next.

Cut Active Body

Create three separate bodies in a part file, and then create a sketch on the active body, as shown. Activate the **Extrude** command (on the ribbon, click **Home > Solids > Extrude**). On the **Extrude** command bar, select **Extent Type > Finite** and **Add/Cut > Cut**, respectively. Next, select **Cut Active Body** from the **Apply Cut Features** drop-down. Select sketch region and right click to accept the selection. Move the pointer into the model and click to create the cut feature. Notice the cut is added to the active body only.

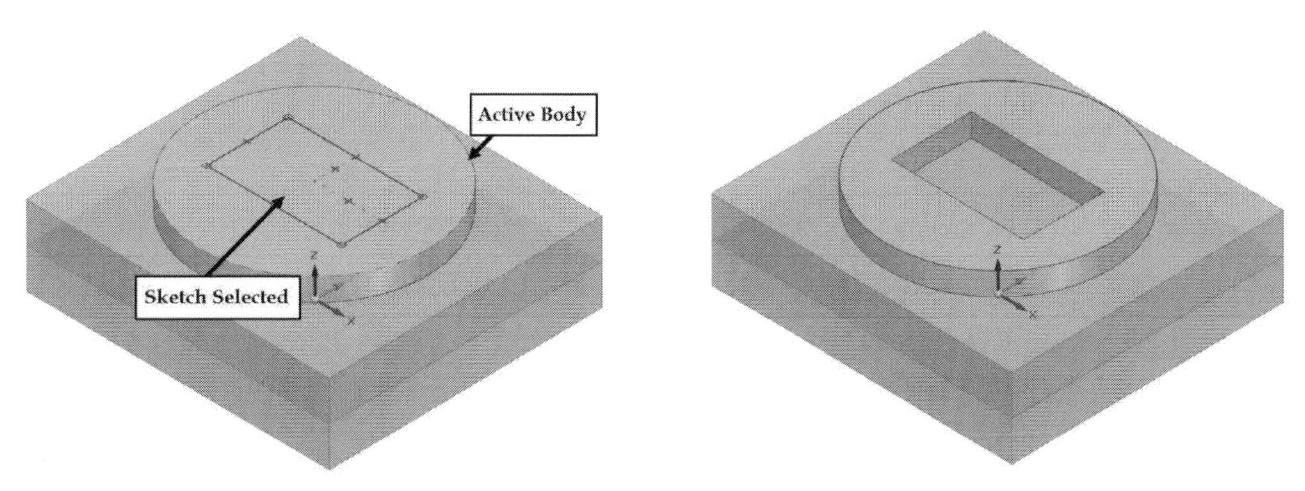

Cut Selected Bodies

Activate the **Extrude** command (on the ribbon, click **Home > Solids > Extrude**). On the **Extrude** command bar, select Extent **Type > Finite** and **Add/Cut > Cut**, respectively. Next, select **Cut Selected bodies** from the **Apply Cut Features** drop-down. Select the sketch region and right click to accept the selection. Move the pointer across all the bodies of the part, and then click. Notice that all the bodies of the part are selected. Press the Shift key and click on the bodies to be removed from the selection. Right click to add the cut to the selected bodies.

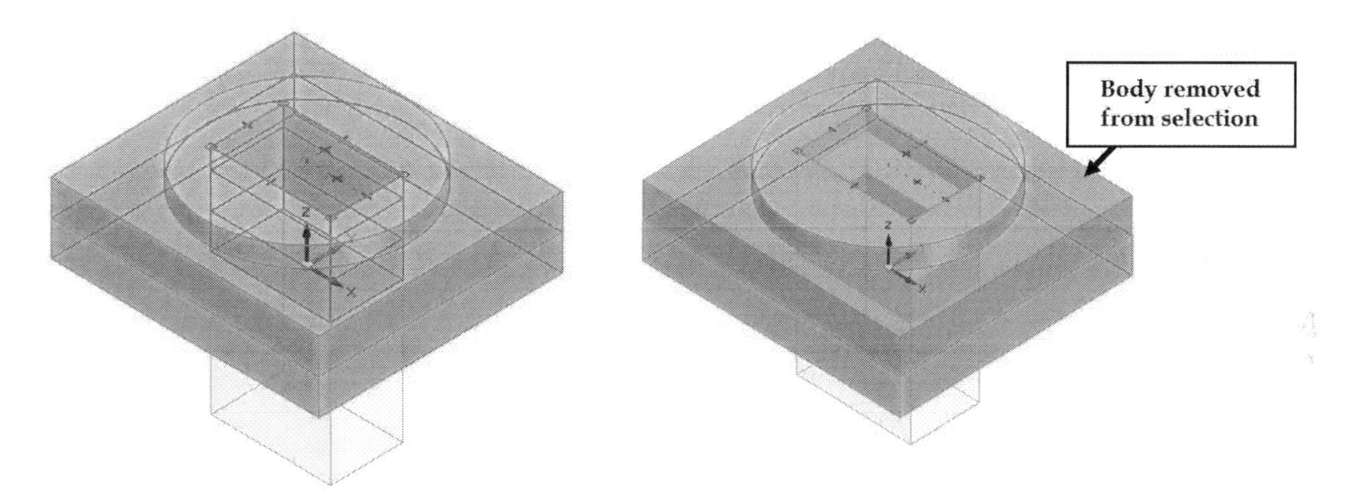

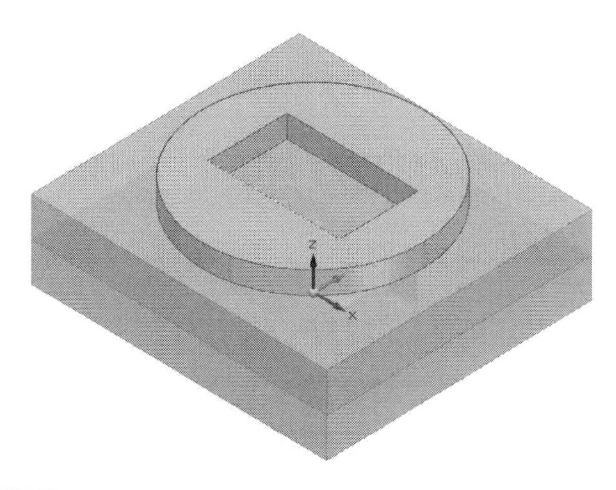

Scale Body

This command scales the part geometry with reference to the Base Coordinate System origin or any other keypoint that you specify. On the **Home** tab of the ribbon, click **Solids** panel > **Add Body** drop-down > **Scale Body**. On the command bar, click the **Options** icon, and then select the **Uniform Scaling** option. Next, click **OK** to close the dialog. Click on the part geometry to scale, and then click **Accept** on the command bar. Next, select a keypoint on the part geometry to define the reference point of the scaling (or) leave the Base Coordinate System as the default reference point. Type-in the scaling factor in the **Scale** box.

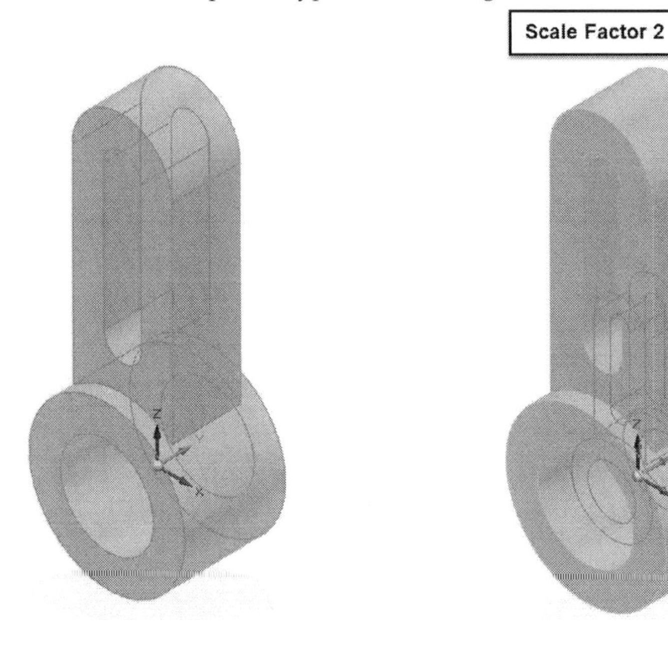

Non-uniform Scaling

This option scales the part geometry along three directions using the coordinate values that you specify. On the **Home** tab of the ribbon, click **Solids** panel > **Add Body** drop-down > **Scale Body**. On the command bar, click the **Options** icon, and then select the **Non-Uniform Scaling** option. Next, click **OK** to close the dialog. On the command bar, type-in values in the X, Y, Z boxes.

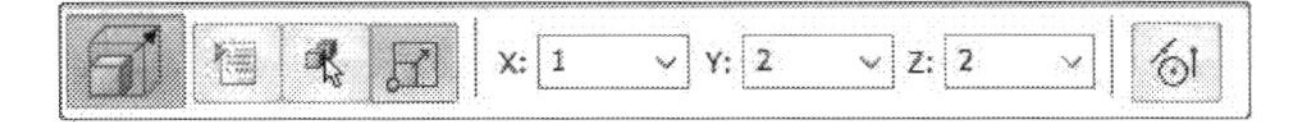

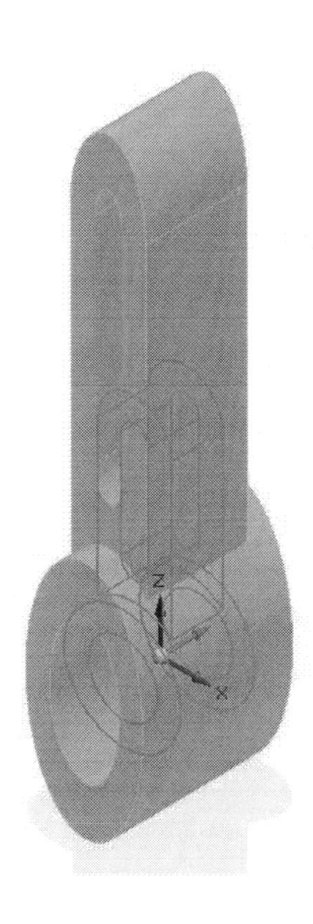

Examples

Example 1 (Millimetres)

In this example, you will create the part shown below.

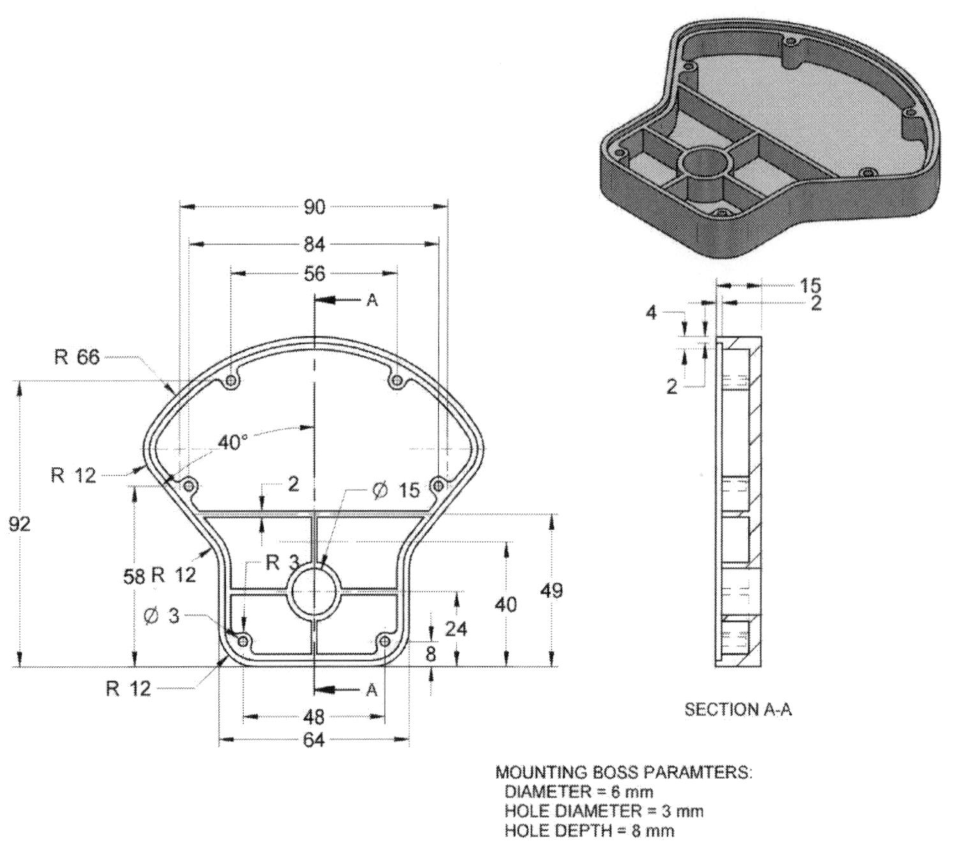

1. Start **Solid Edge 2019**.
2. On the Application Menu, click **New > ISO Metric Part**; a new part file is opened.
3. On the ribbon, click **Home > Draw > Line.**
4. Lock the XY plane in coordinate system. Click the Top face of the Quick View Cube located at the bottom right corner.
5. Create the closed sketch, as shown (refer to the examples of **Chapter 2: Sketch Techniques** to learn how to create a closed sketch using the **Line** tool).

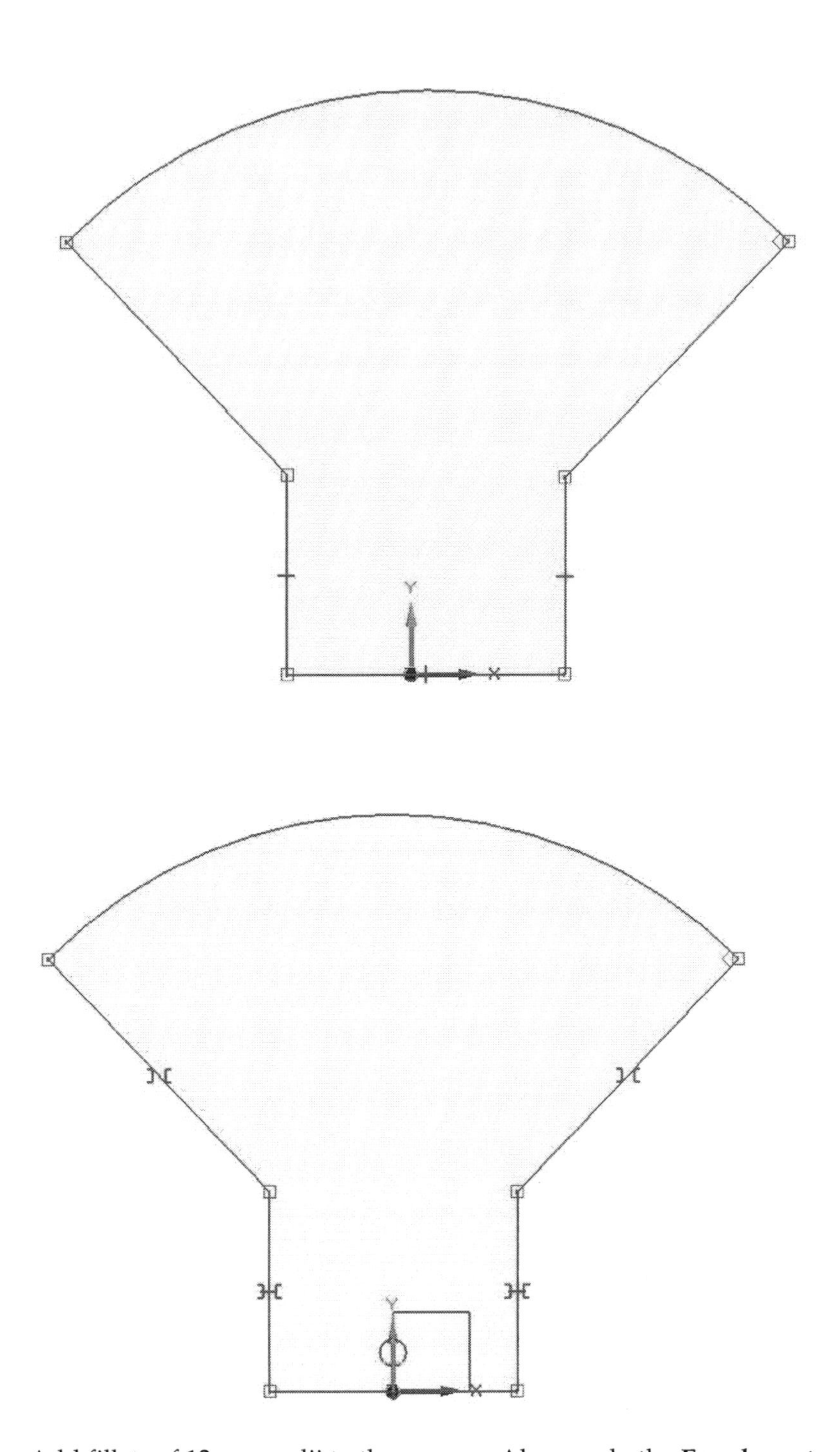

6. Add fillets of 12 mm radii to the corners. Also, apply the **Equal** constraint between the fillet.

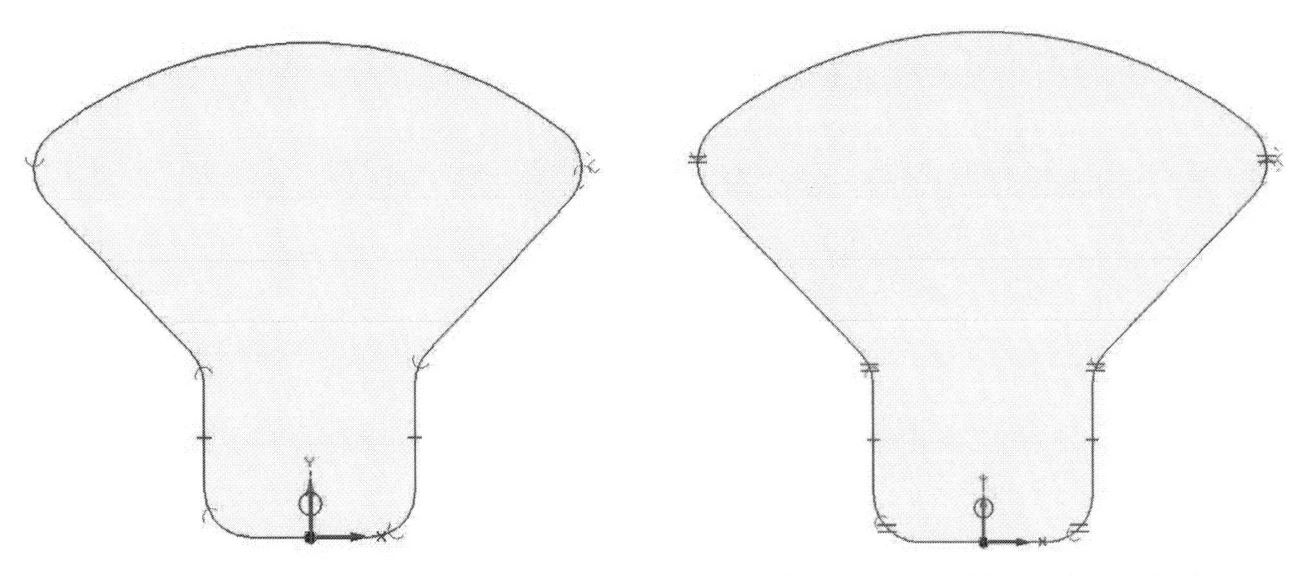

7. Apply the Symmetric relationship between the vertical axis of the two vertical lines. Also, make the two inclined lines symmetric about the vertical axis (refer to the Symmetric topic of **Chapter 2: Sketch Techniques**).

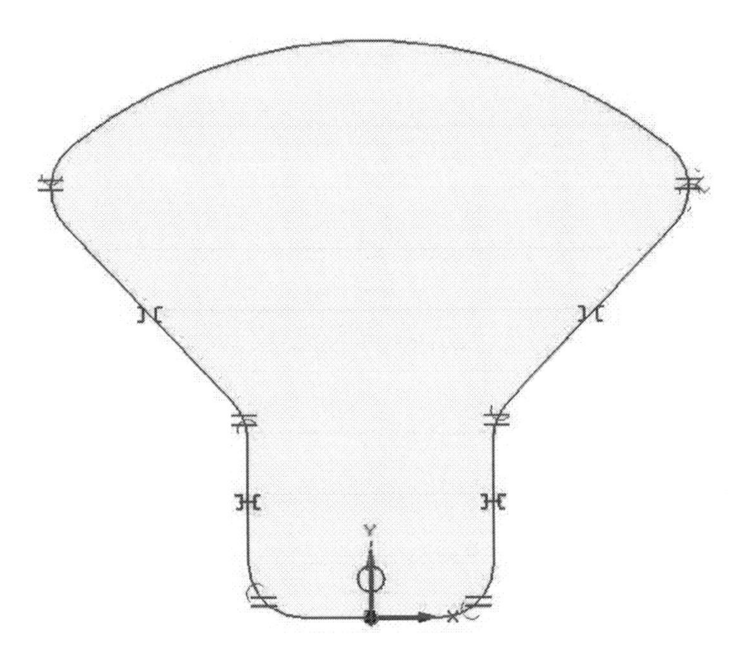

8. Apply dimensions to the sketch, and then unlock the sketch plane.

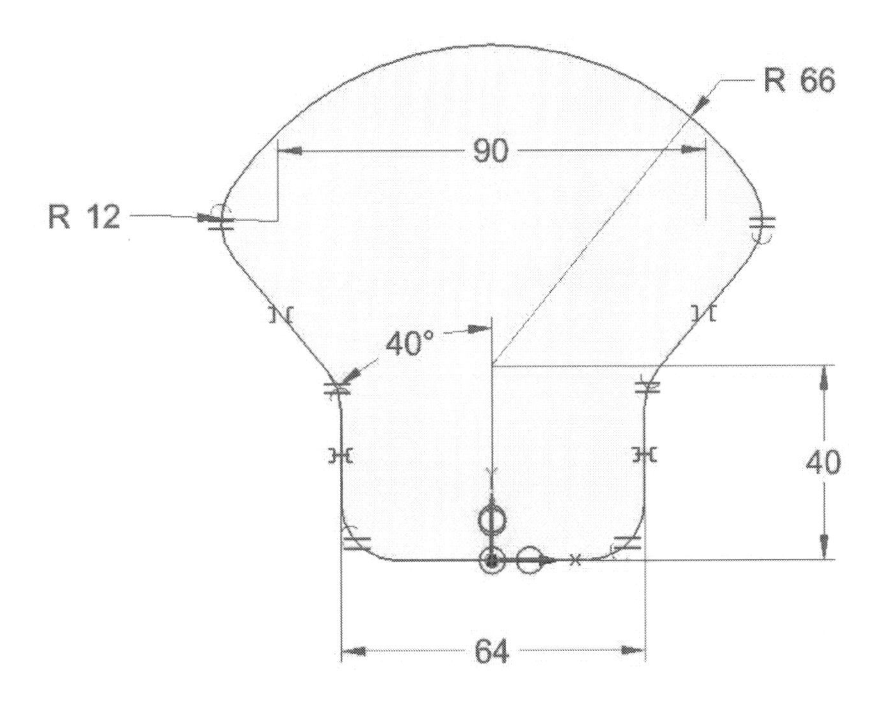

9. Create the *Extrude* feature of 15 mm depth.
10. Create the *Thin Wall* feature of 4 mm depth.

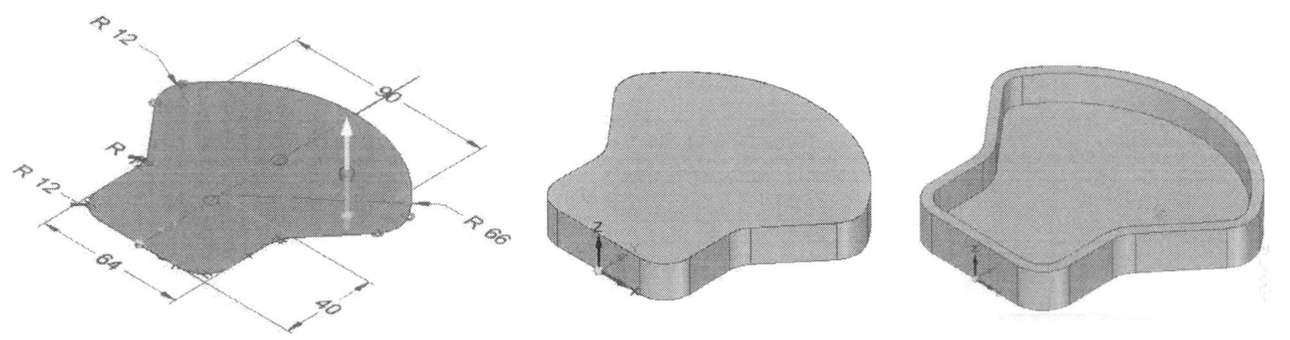

11. On the ribbon, click **Home > Solids > Thin Wall > Lip**.
12. Click on the inner edge of the *Thin Wall* feature and click the green check on the command bar.
13. Type **2** in the **Width** and **Height** boxes, respectively. Click inside the model to define the side of the lip. Click **Finish** and **Cancel** to complete the lip feature.

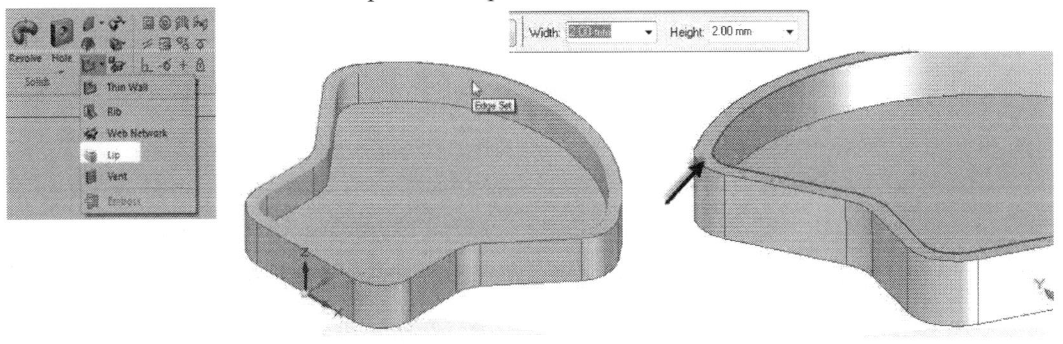

14. Click the right mouse button in the screen and select **Transition to Ordered**.
15. In the Ordered mode, click **Home > Solids > Thin Wall > Mounting Boss** on the ribbon.

16. On the command bar, select **Coincident Plane** from the **Create-From Options** menu.
17. Click on the top face on the lip feature.

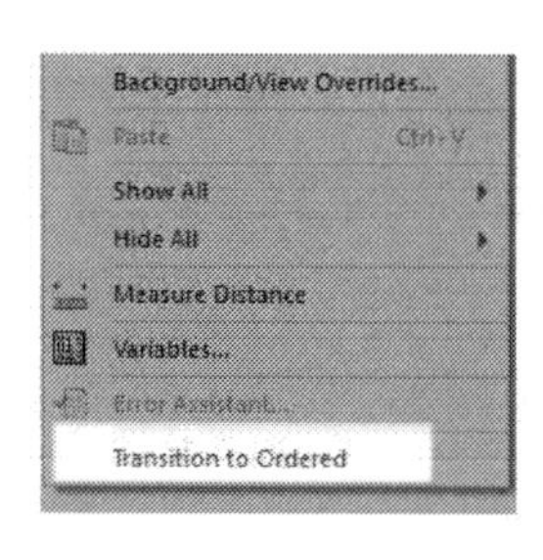

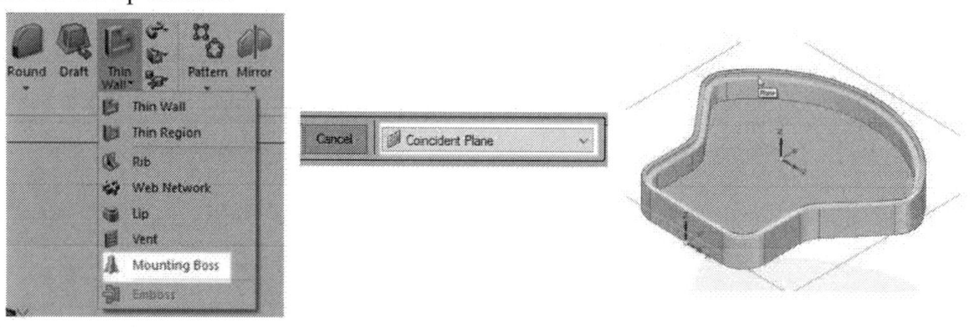

18. Define mounting boss locations and add dimensions. Click **Close Sketch** on ribbon.
19. On the command bar, click **Mounting Boss Options** icon.
20. On the **Mounting Boss Options** dialog, set the **Boss diameter** to 6, check the **Mounting hole** option, and then set the **Hole diameter** to 3 and **Hole depth** to 8. Click **OK** to close the dialog.
21. Move the mouse pointer downward and click to define the side of the mounting boss.
22. Click **Finish** and **Cancel** to complete the mounting boss feature.

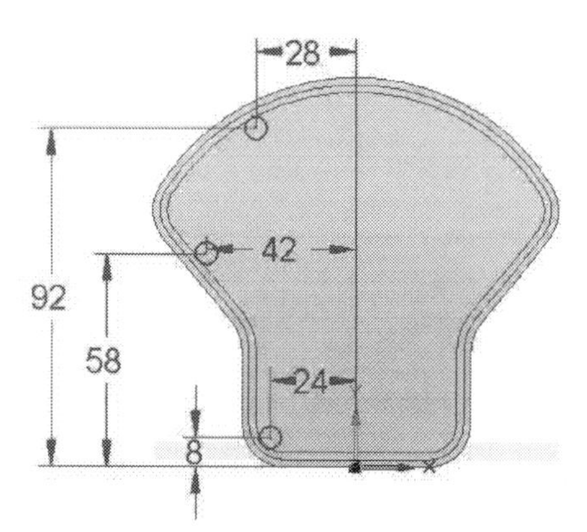

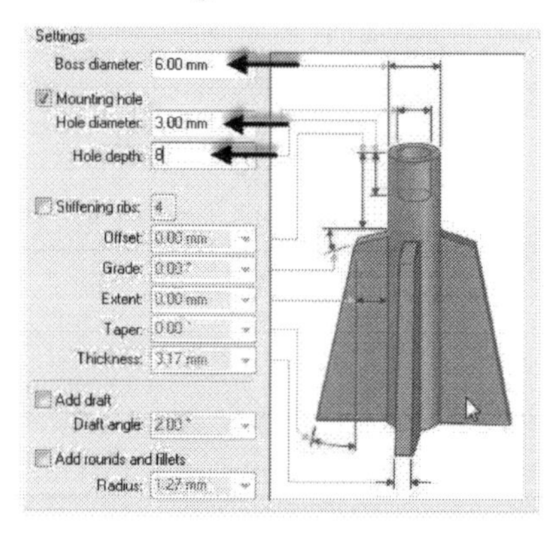

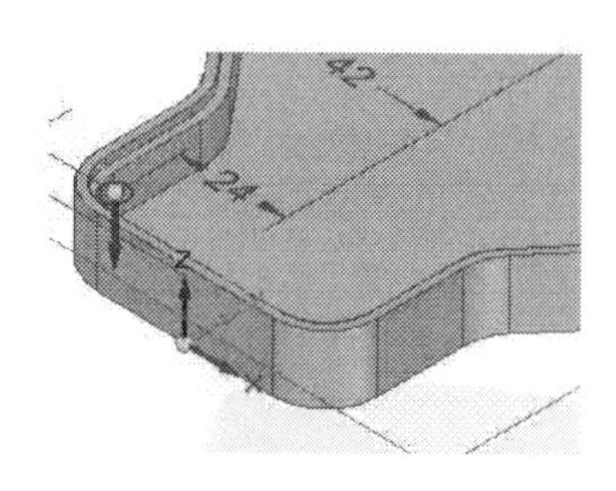

23. In the Pathfinder, click the right mouse button on the *Mounting Boss* and select **Move to Synchronous**.
24. Click the right mouse in the screen and select **Transition to Synchronous**.
25. Press and hold the Shift key and click on the *Lip* and *Mounting Bosses* in the Pathfinder.
26. On the ribbon, click **Home > Pattern > Mirror**.
27. Click on the YZ plane of the base coordinate system. The mounting bosses are mirrored.

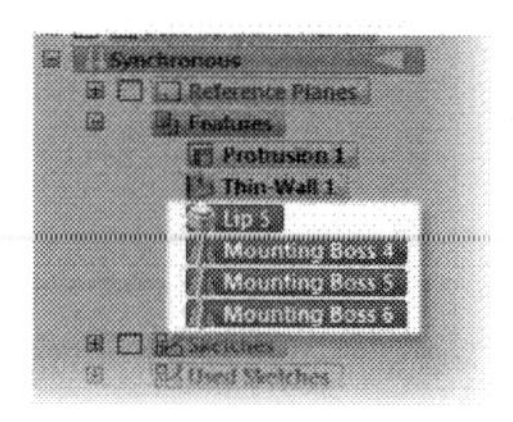

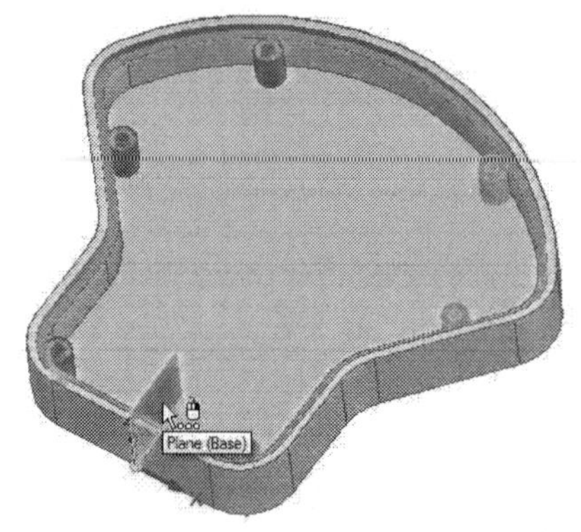

28. On the ribbon, click **Home > Solids > Round** and select the edges where the mounting bosses meet the walls of the geometry.
29. Type **2** in the box displayed on the geometry. Click the right mouse button to round the selected edges.

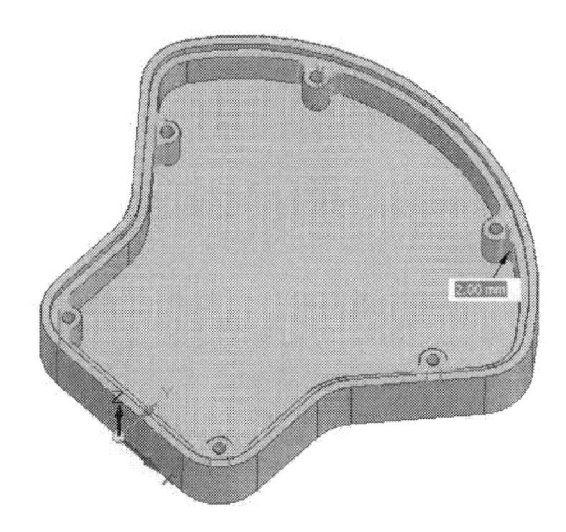

30. Activate the **Hole** command and create a hole on the flat face of the geometry. The hole diameter is 15 mm.

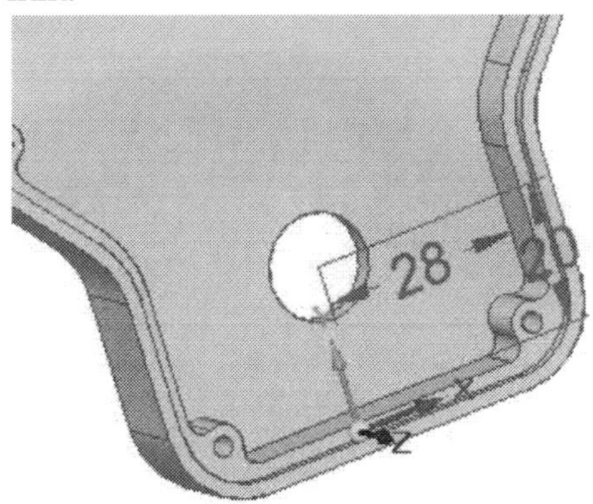

31. On the ribbon, click **Planes > Coincident Plane**. Click on the top face on the lip feature.
32. On the ribbon, click **Draw > Circle by Center Point**. Lock the new plane.
33. Draw the sketch on the locked plane, as shown below. Next, unlock the plane.
34. On the ribbon, click **Home > Solids > Thin Wall > Web Network**.
35. Click on the elements of the sketch, and then click the green check on the command bar.
36. Type **2** in the box that appears on the part geometry, and then click the green check.

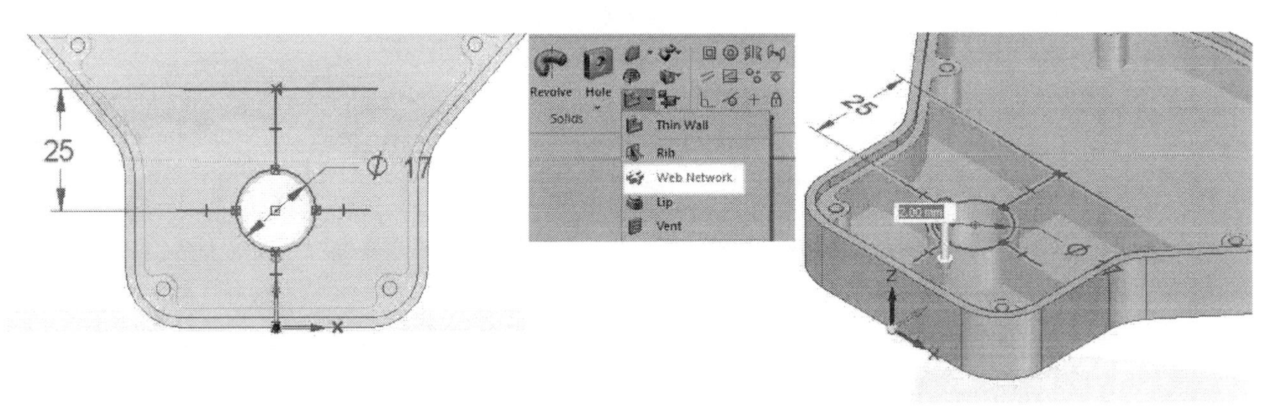

37. Save and close the file.

Questions

1. What is the use of the **Web Network** command?
2. How many types ribs of can be created in Solid Edge?
3. Why do we create multi body parts?
4. Describe the terms 'Rib' and 'Spar' in the *Vent* feature.
5. What is the use of the **Multi Body Publish** command?

Exercises

Exercise 1

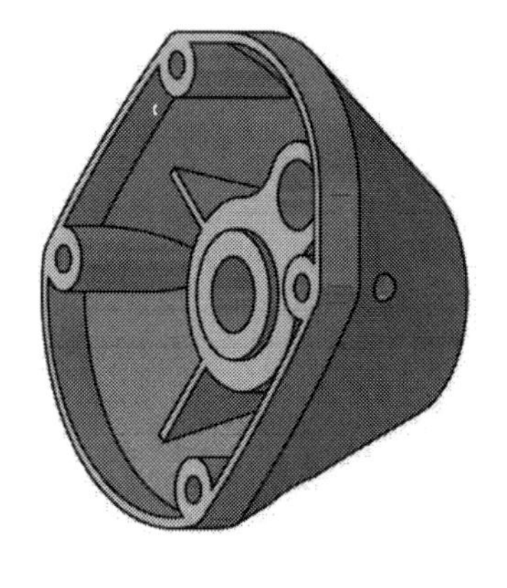

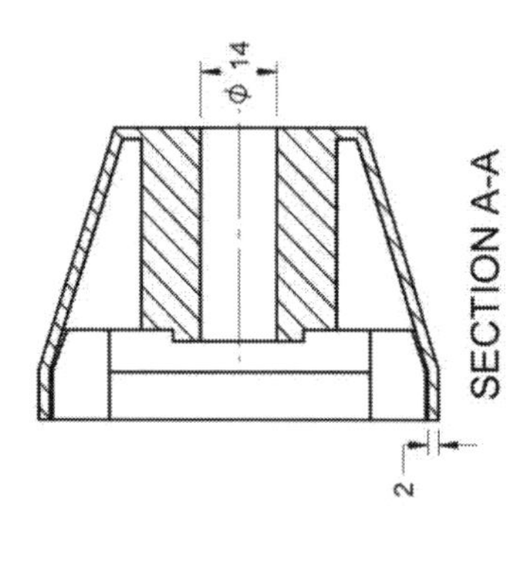

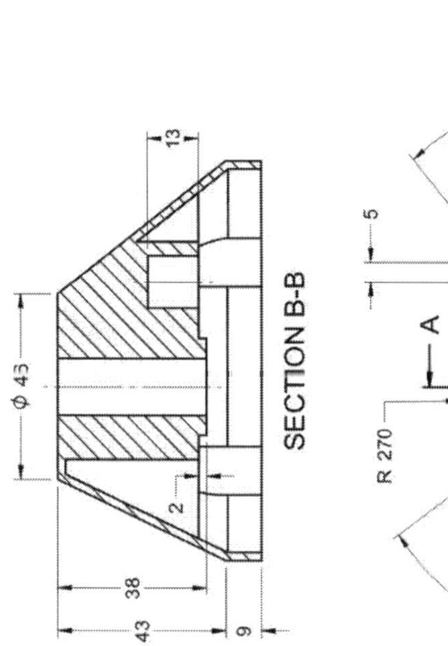

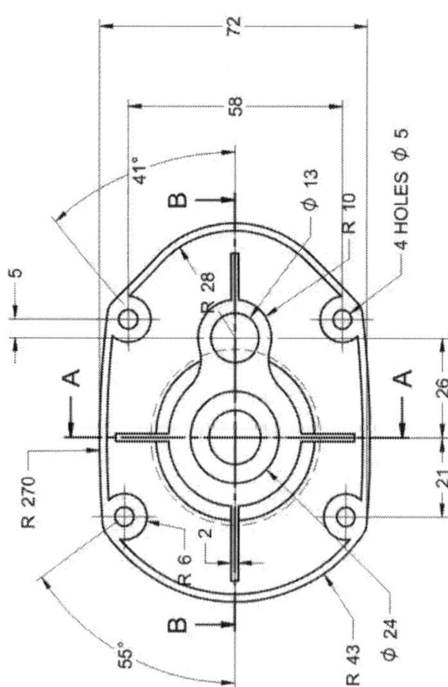

Exercise 2

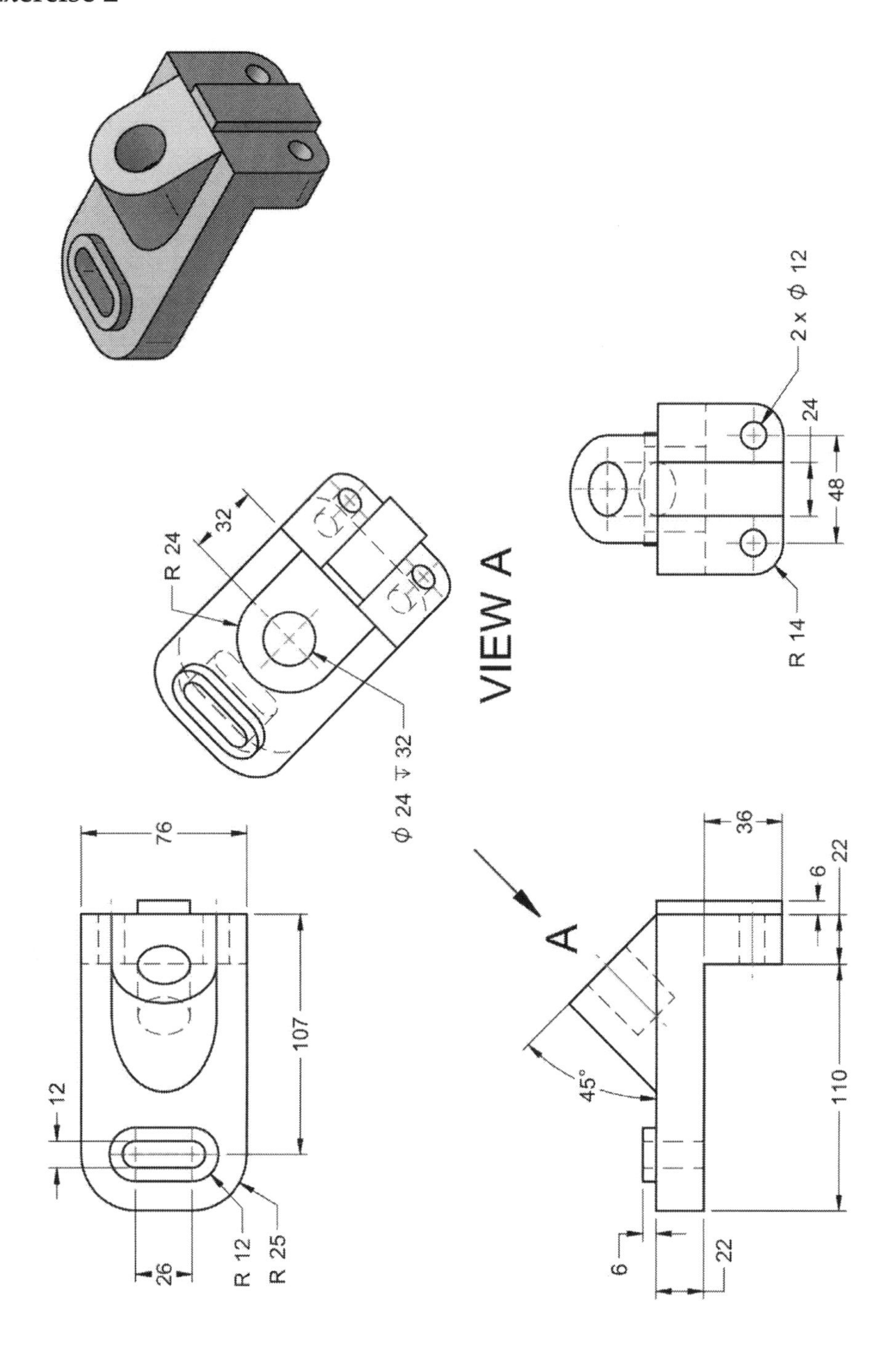
VIEW A
A
2 x Φ 12
R 14
24
48
R 24
32
Φ 24 ⊽ 32
76
107
12
26
R 12
R 25
36
6
22
110
45°

Exercise 3 (Inches)

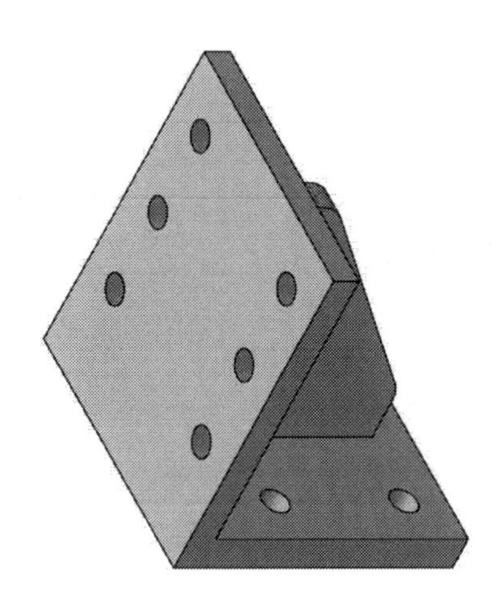

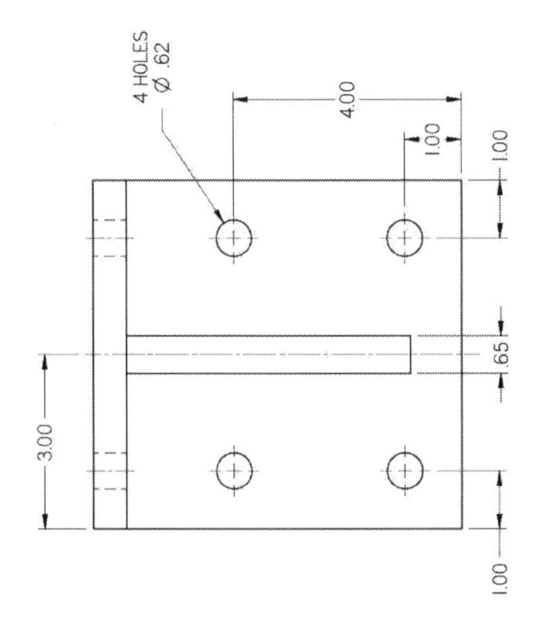

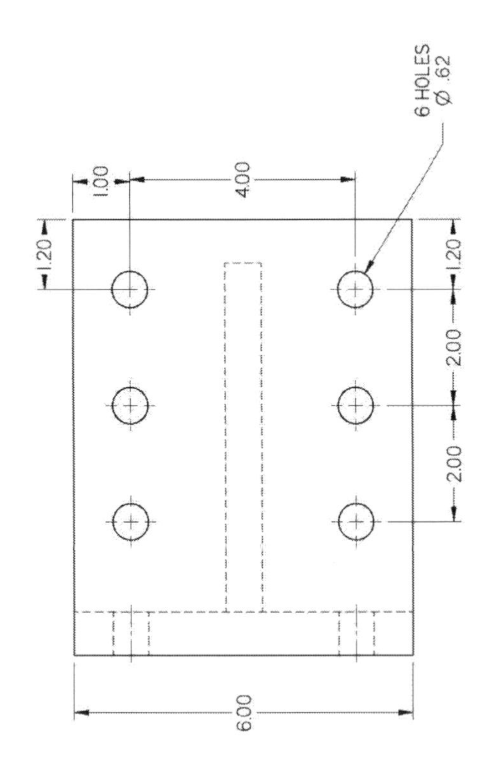

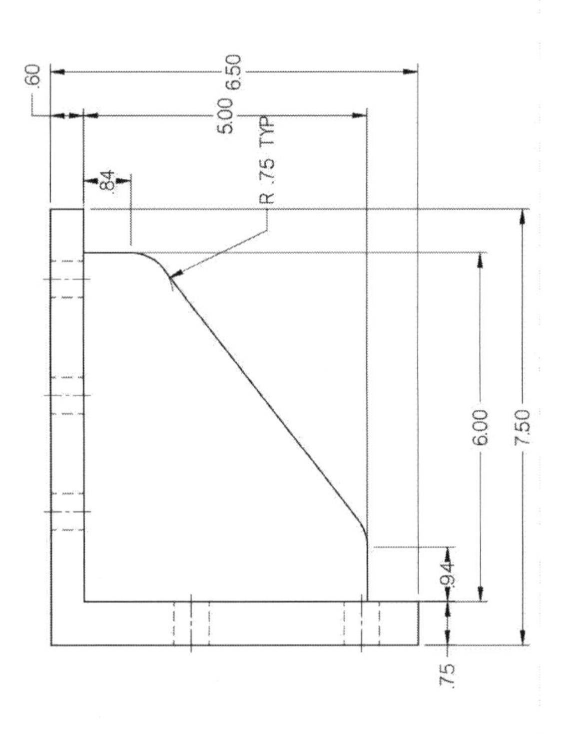

Chapter 9: Modifying Parts

In design process, it is not required to achieve the final model in the first attempt. There is always a need to modify the existing parts to get the desired part geometry. In this chapter, you will learn various commands and techniques to make changes to a part.

The topics covered in this chapter are:

- *Face Relations*
- *Modify models using steering wheel*
- *Live Rules*
- *Change model dimensions*
- *Live sections*

Face Relations

Solid Edge allows you to define relations between faces. This will help you to control the behaviour of the faces when you modify the part geometry. There are different relations that can be applied between faces. These are explained next.

Coplanar

This command brings the selected faces onto one plane. Activate this command (click **Home > Face Relate > Coplanar** on the ribbon) and select the first face. Right-click and select the second face. Again, right-click to make the two faces coplanar.

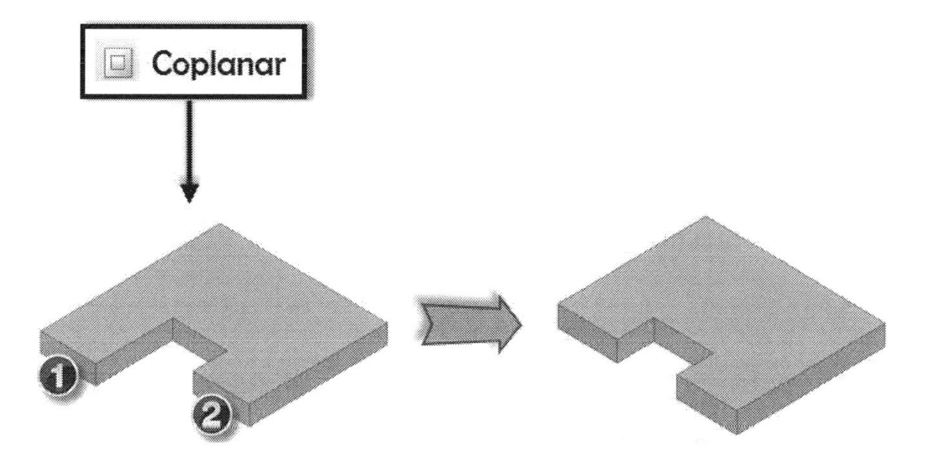

Concentric

This command makes two cylindrical faces share a same centerpoint. Activate this command (click **Home > Face Relate > Concentric** on the ribbon) and select the first cylindrical face. Click the right mouse button and select the second cylindrical face. Click the green check on the command bar to make the first face concentric to the second one.

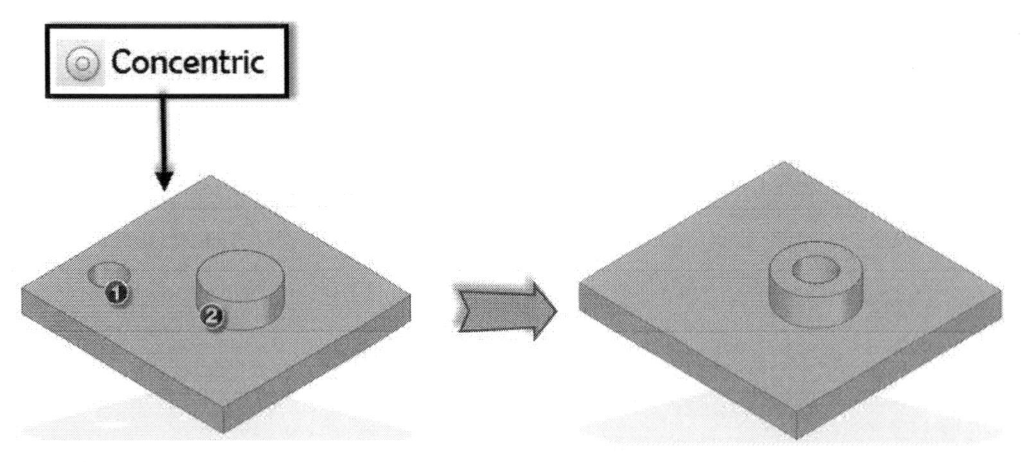

Symmetry

This command makes two faces symmetric about a plane. Activate this command (click **Home > Face Relate > Symmetry** on the ribbon), select the first face and click **Accept** on the command bar. Select the second face and click **Accept** on the command bar. Select the symmetric plane and click **Accept** to make the faces symmetric.

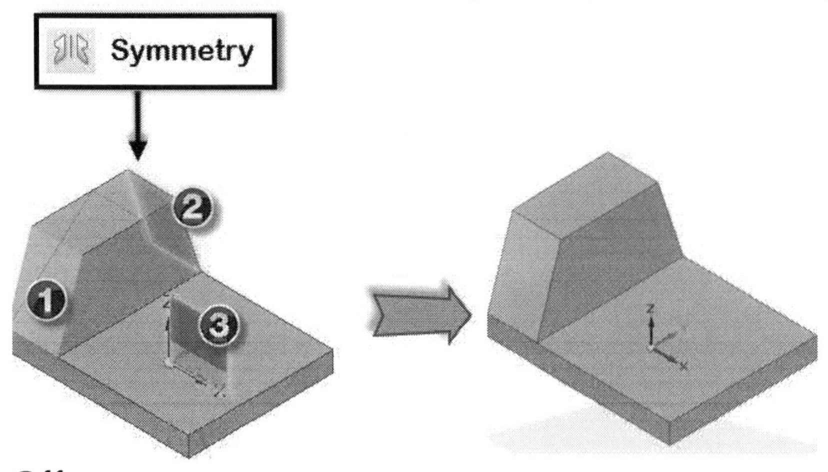

Offset

This command defines an offset distance between two faces. The selected faces should share a common face which is perpendicular to both of them. Activate this command (click **Home > Face Relate > Offset** on the ribbon), select the first face, and then click **Accept** on the command bar. Select the second face and click **Accept**. Type-in an offset value in the box displayed on the model. Click **Accept** on the command bar; the first face will be offset from the second face by the value you specified.

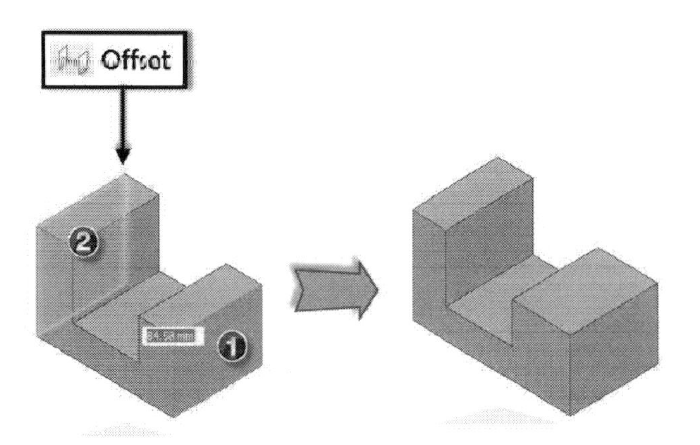

Parallel

This command makes two faces parallel to each other. Activate this command (click **Home > Face Relate > Parallel** on the ribbon), select the first face, and then click **Accept** on the command bar. Select the second face and click **Accept**. The first face will become parallel to the second face.

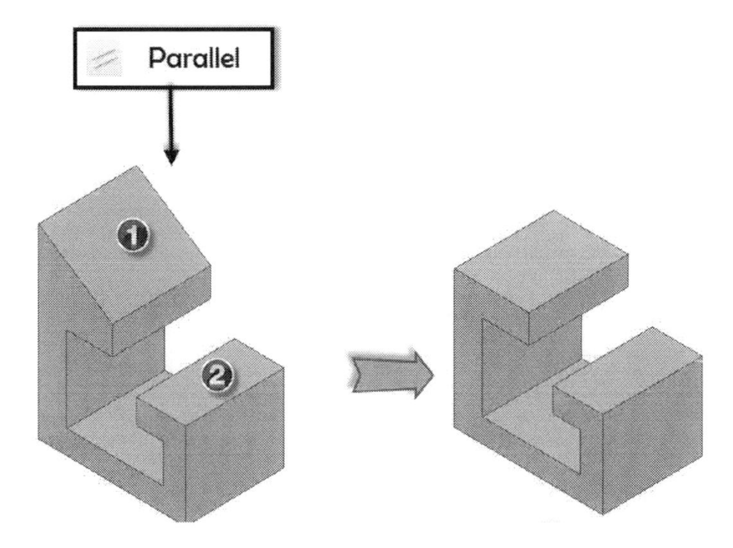

Aligned Holes

This command makes the axes of selected cylindrical or conical faces lie on a same plane. Activate this command and select cylindrical faces from the part geometry. Click **Accept** on the command bar and select a plane or point. Click **Accept** on the command bar; you will notice that the axes of selected faces will be moved onto one plane.

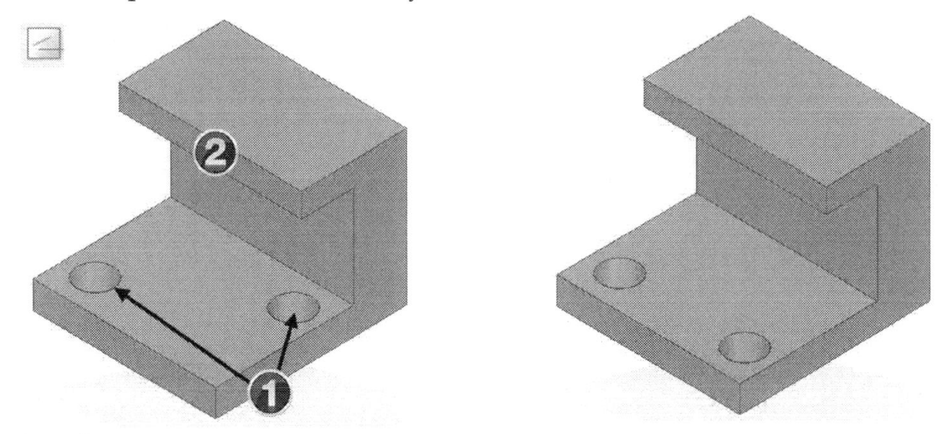

Equal

This command makes the selected cylindrical faces equal in radius. The radius of the first face will be equal to that of the second face.

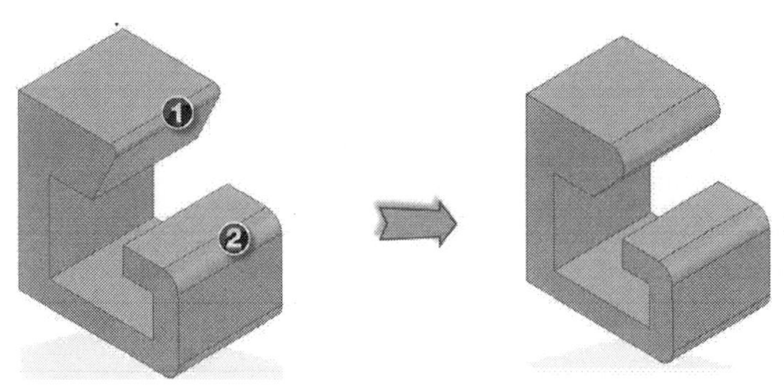

Tangent

This command makes two faces tangent to one another. Select a cylindrical face and right click to accept. Next, select a planar face connected to the selected cylindrical face. Right click to accept the selection; the two faces are made tangent to each other.

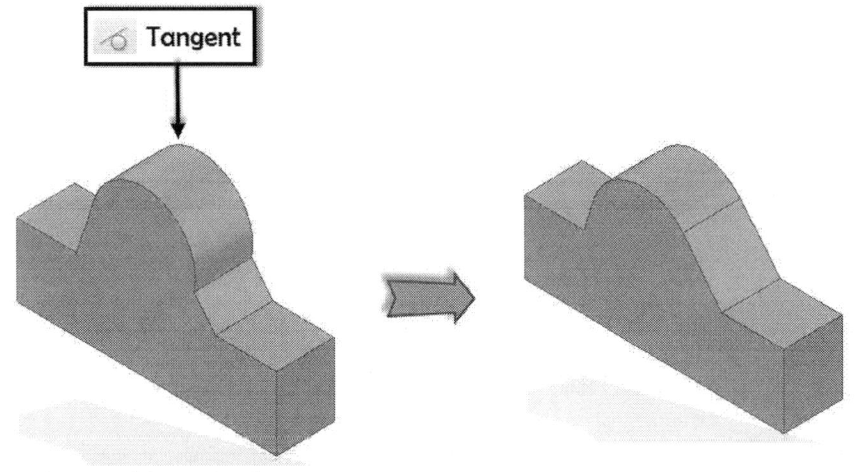

Horizontal/Vertical

This command aligns the selected faces or keypoints vertically/horizontally.

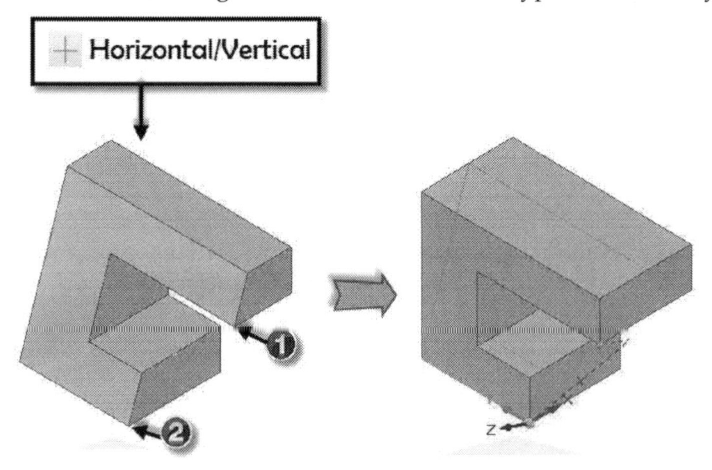

Using the Steering Wheel Tool to Modify Models

Solid Edge provides you with a special tool called Steering Wheel Tool to modify faces and planes of part geometry. You can perform two operations using this tool: **Move** and **Rotate** faces.

Move faces

To move a face, click on it and select the arrow displayed on it. Move the pointer and click to define the distance.

You can also type-in a value in the box displayed on the model. The **Extend/ Trim** option on the command bar extends or trims the adjacent faces to match the new location of the selected face.

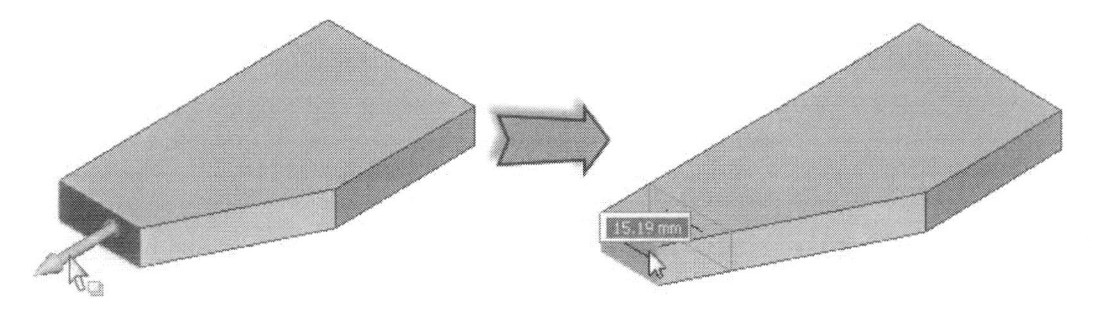

Use the **Tip** option, if you want to adjust the orientation of the faces connected to the selected face.

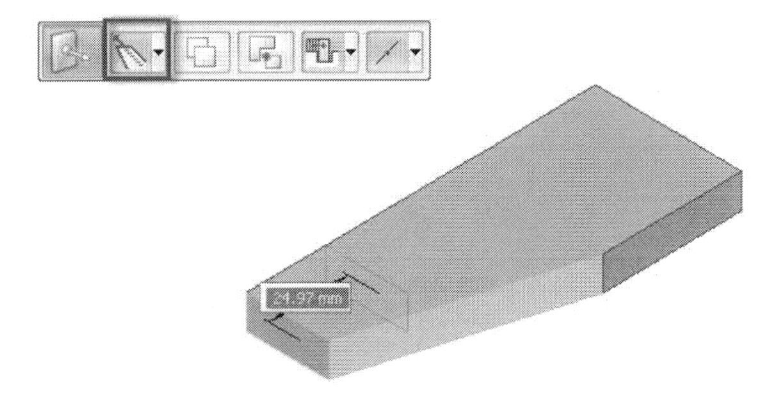

Use the **Lift** option, if you want to lift the selected face and add new faces to model.

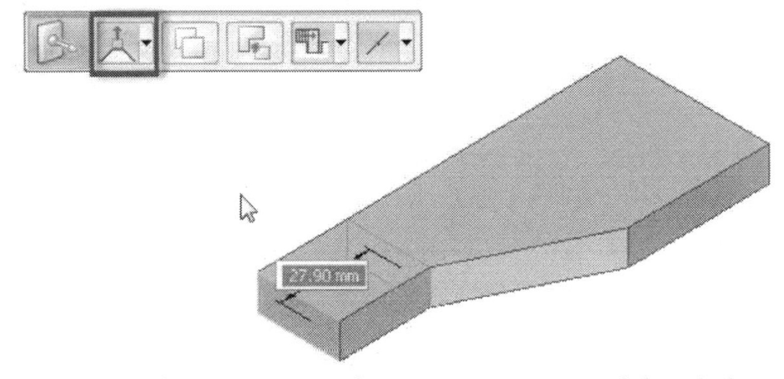

Use the **Detach Faces** option, if you want to move and detach the selected face from the model.

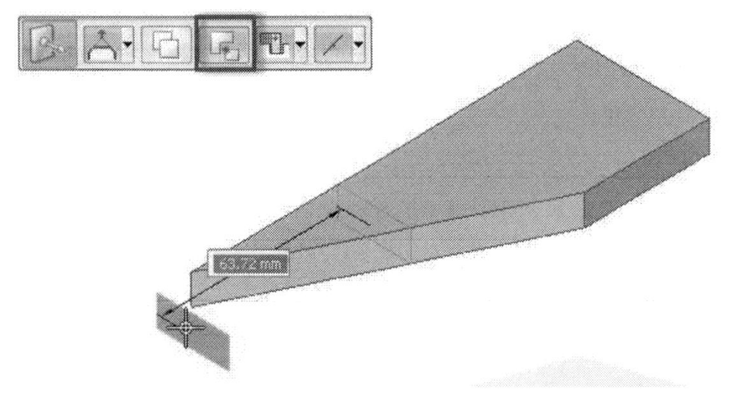

Use the **Copy** option, if you want to move and copy the selected face.

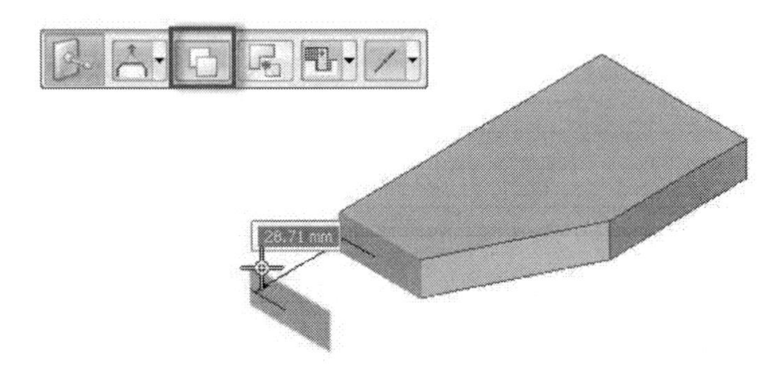

Use the **Model Priority** option, if you want to move the selected face only up to the next face in the model.

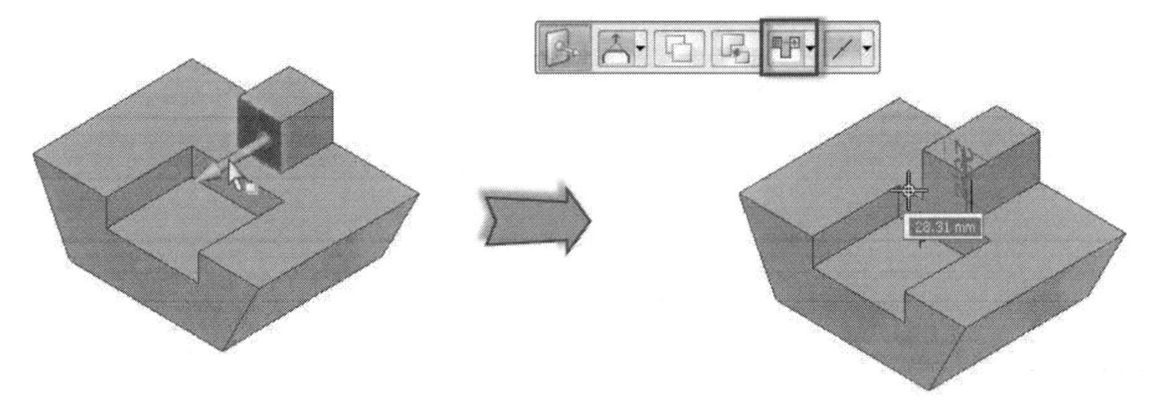

Use the **Select Set Priority** option, if you want to move the selected face beyond any faces in the model.

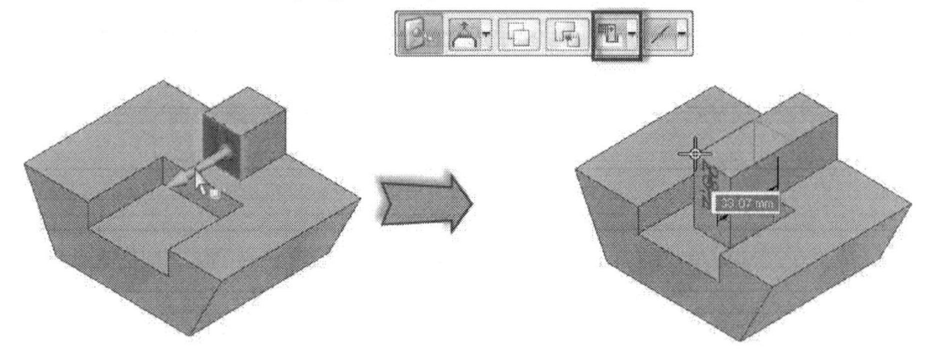

Note that when you move a face adjacent to a chamfer, the chamfer is unaffected.

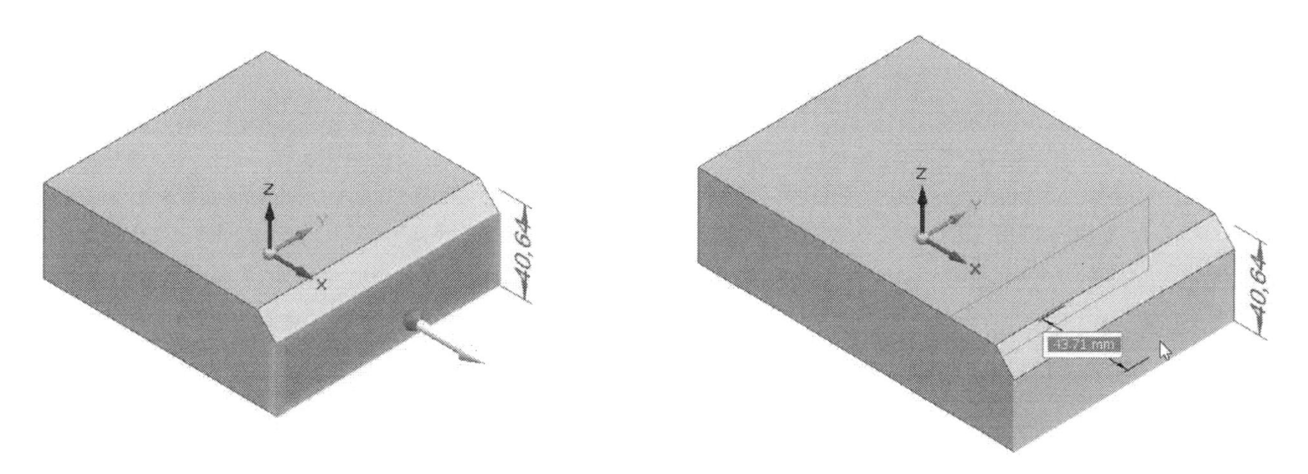

Rotate faces

To rotate a face, you must click on it to display the Steering Wheel arrow. Click and drag the spear attached to the arrow, and then align it to an edge. This defines the axis of rotation. Now, click on the torus of the steering wheel and rotate the face. You can also type-in an angle value in the box displayed on the model.

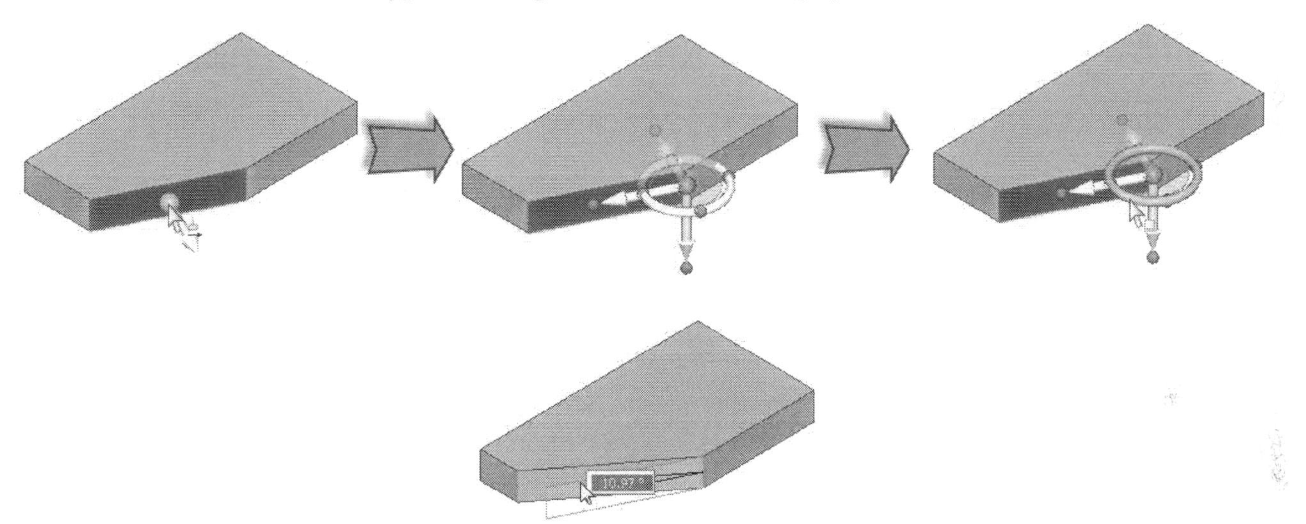

Design Intent Panel

The options on the **Design Intent** panel controls the part geometry when you modify its faces. It appears automatically on selecting any model face. For example, if you move a hole which is concentric to another cylindrical face, the **Concentric** option on the **Design Intent** panel maintains the relationship between the two faces. As a result, the both the faces will be moved. If you turn off this option, only the hole will be moved. Similarly, other options maintain the corresponding relationships while modifying the faces.

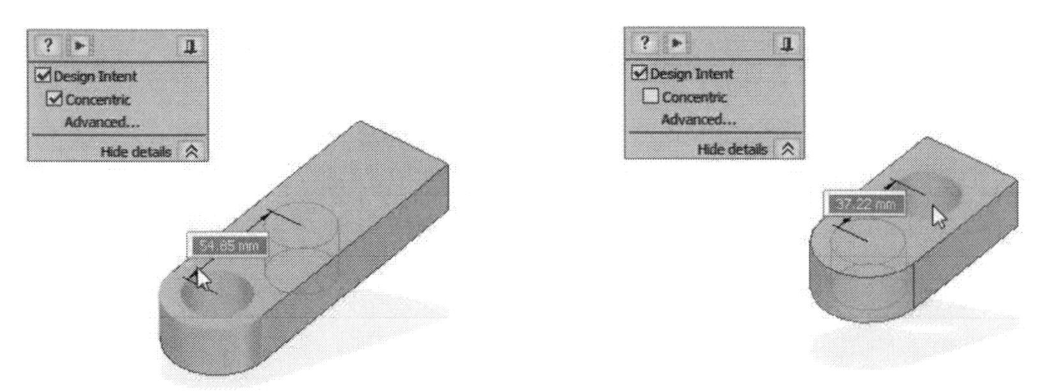

For example, if you move a face, which is symmetric to another face about the YZ plane, the **Maintain Symmetry About Base Planes** option maintains the symmetric relationship. As a result, the other face will also be moved in the opposite direction. If you turn off this option, only the selected face will be moved.

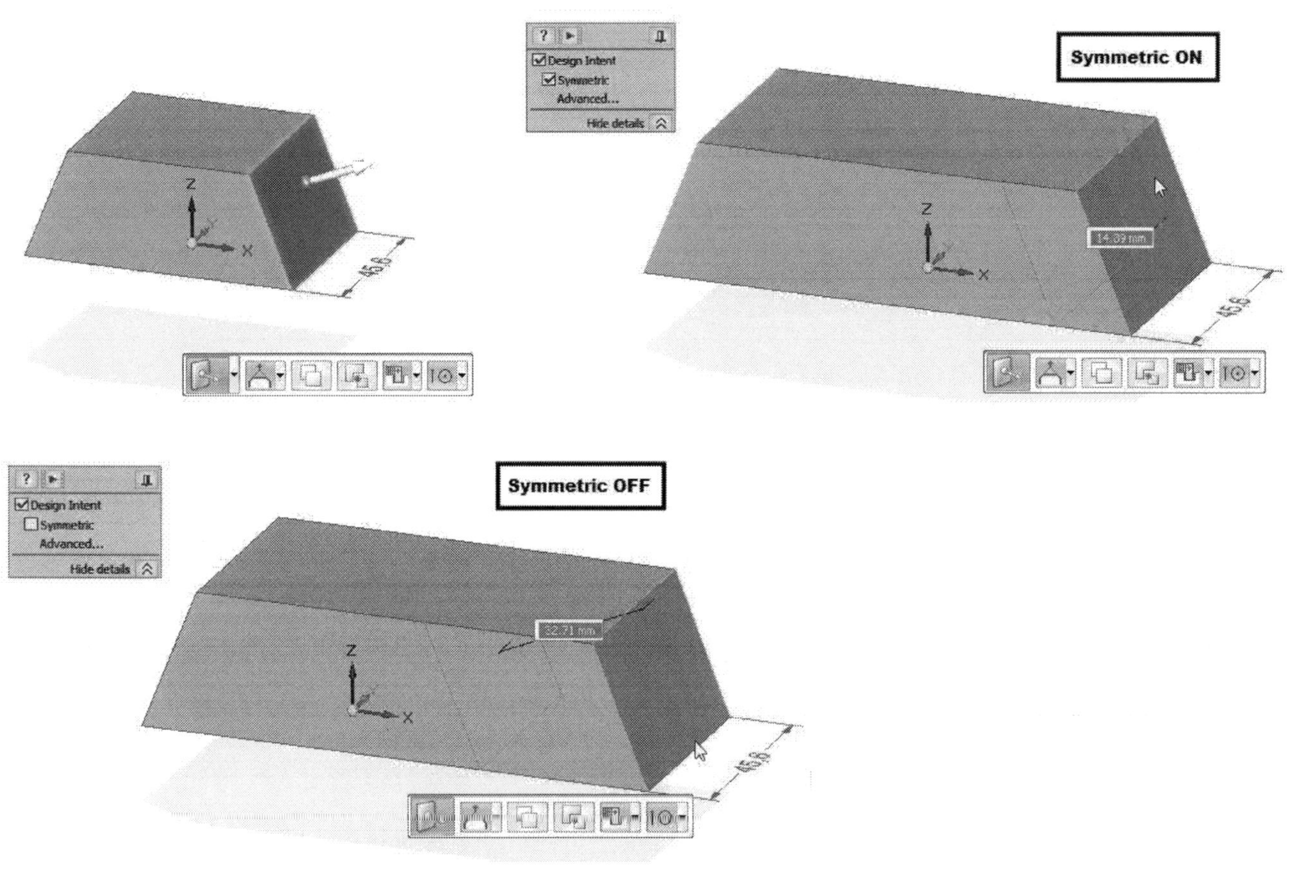

Click the **Advanced** option on the **Design Intent** panel to display all the design intent options on the bottom portion of the graphics window. Uncheck the **Design Intent** option on the **Design Intent** panel, if you want to suspend all the design intent options. You can also use the **Relax Dimensions** and **Relax Persistent Relationships** options to relax the dimensions and face relationships.

Use the **Solution Manager** option to solve the errors while moving or rotating faces of the part geometry. Click this option and you will notice that the model faces are highlighted in different colors. The selected face is highlighted in green color. The error path is highlighted in orange color and the faces which are being solved are highlighted in blue color. Click on the blue faces, if you want to suppress the relationship between the selected face (highlighted in green) and them. Click on the maroon faces, if you want to restore the relationship between them and the selected set (green faces). Move or rotate the green face to get the desired result. Click the right mouse button to accept the result.

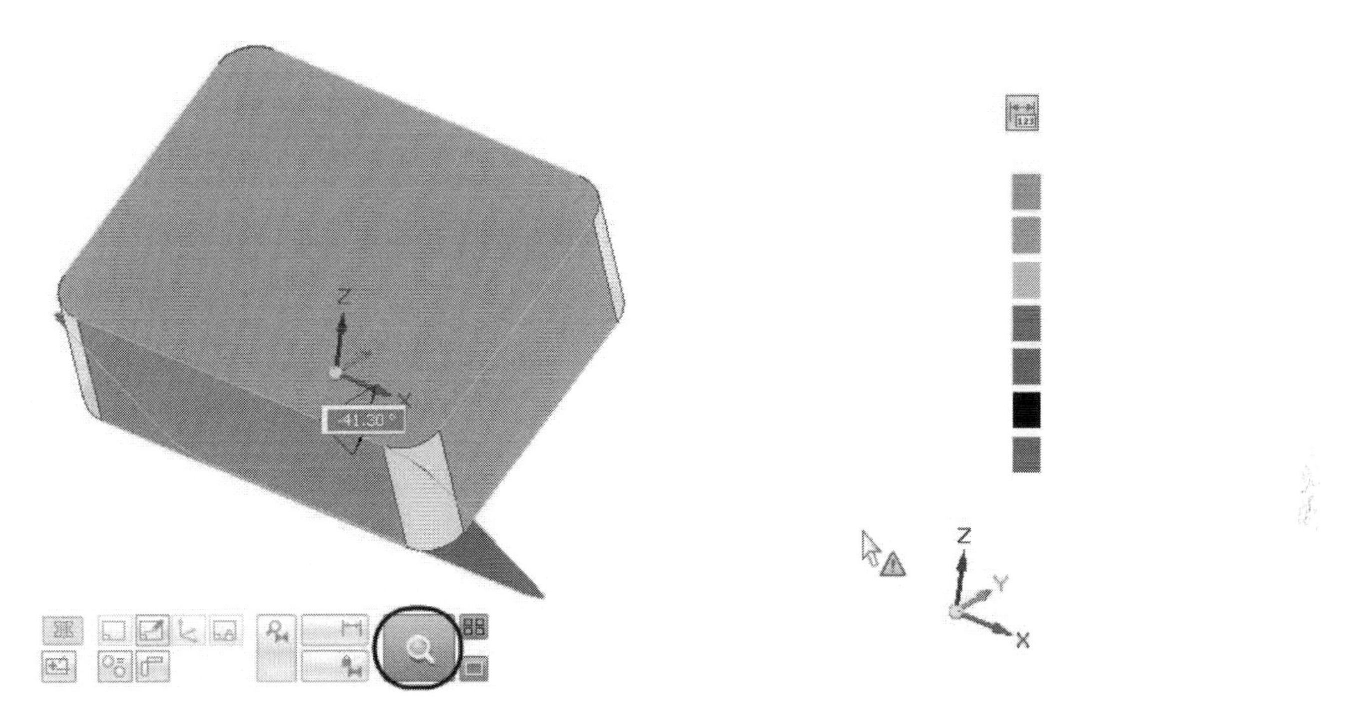

Modify the Part dimensions

Solid Edge allows you to add and modify dimensions between the faces of the part geometry. This will make the modification process much easy. To add a dimension, activate the **Smart Dimension** command and select an edge connecting two faces. Position the dimension and you will notice that a box pops up on the screen. Type-in a value in the box and click the arrows displayed on the box to define the face to be affected. Click the double-arrow button, if you want to modify both the faces. Click the lock button on the box to lock the dimension. The locked dimension will act as a driving dimension.

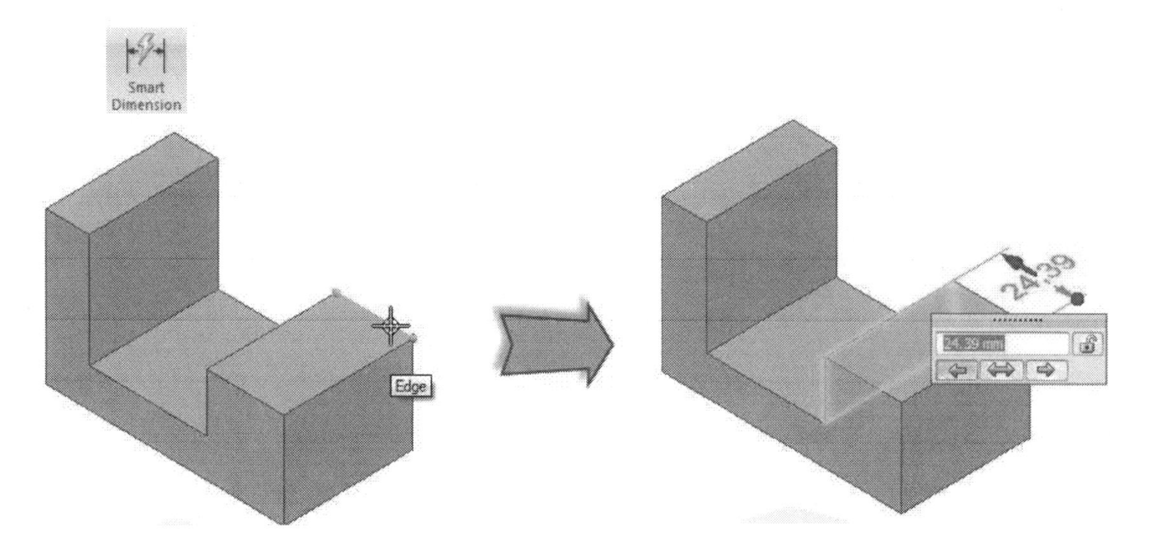

Live Sections

A Live section displays the cross section at a particular location in a part geometry. If you want to create live sections, activate the **Live Section** command (click **Surfacing > Section > Live Section** on the ribbon) and select a plane intersecting with the model. A live section will be created. Use the Steering Wheel Tool to move the live section. You can click and drag the edge of a live section. The part geometry will be modified, automatically.

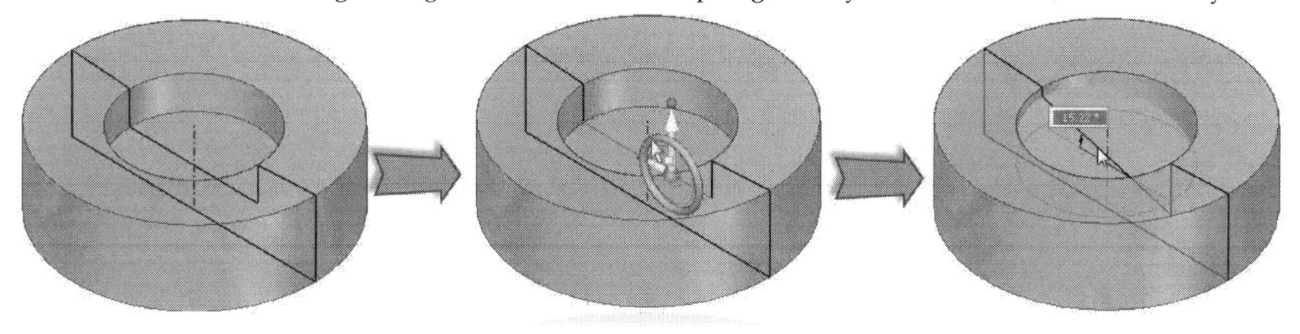

Examples

Example 1 (Millimetres)

In this example, you will create the part shown below, and then modify it using the Synchronous editing tools.

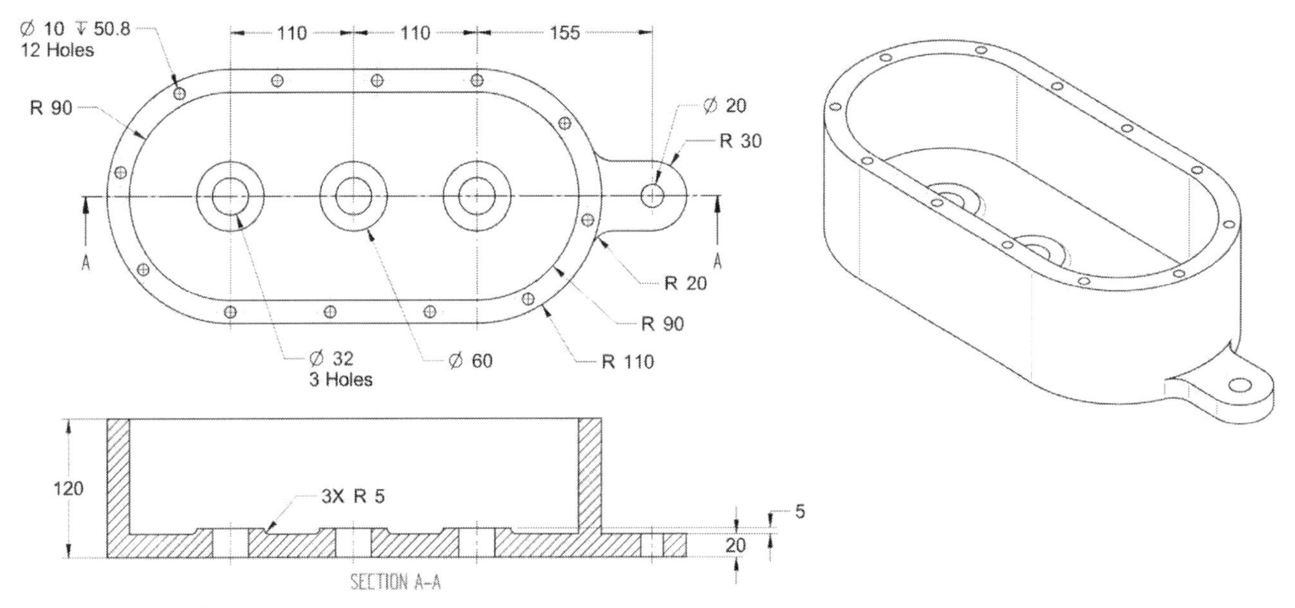

1. Start **Solid Edge 2019**.
2. Click the Application Menu button located at the top left corner. On the **Application Menu**, click **ISO Metric Part**; a new part file is opened.
3. Create the part using the tools and commands in Solid Edge. You can download the part file from our website, if you find it difficult to create it.
4. Click on the 20 mm diameter hole.
5. Click on the dimension value of the hole; the **Hole** command bar pops up on the screen.
6. On the command bar, click the **Hole Options** icon; the **Hole Options** dialog pops up.
7. On the **Hole Options** dialog, set the **Standard** to mm and select the **Counterbore** button.
8. Set the **Counterbore diameter** to 30 and **Counterbore depth** to 10. Click **OK** to close the dialog.
9. Click the right mouse button to accept the changes.

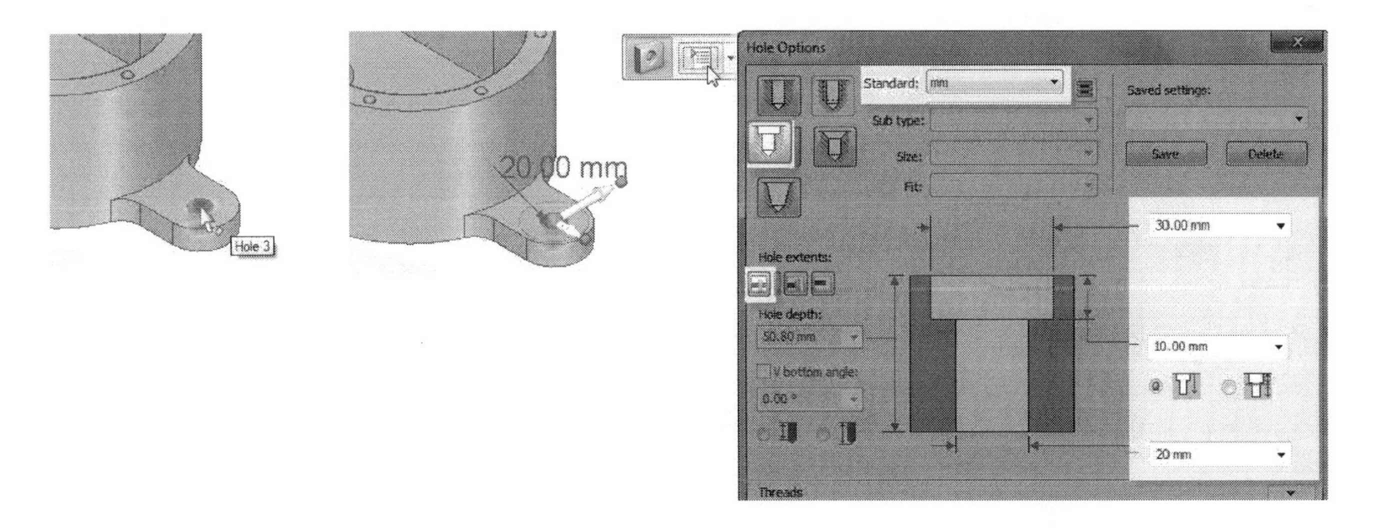

10. Again, click on the counterbore hole and select the small arrow that appears on it.
11. On the **Design Intent** panel, make sure that the **Concentric** option is turned ON.

12. Move the mouse pointer and type 20 in the box that appears on the part geometry. Press Enter to move the hole.

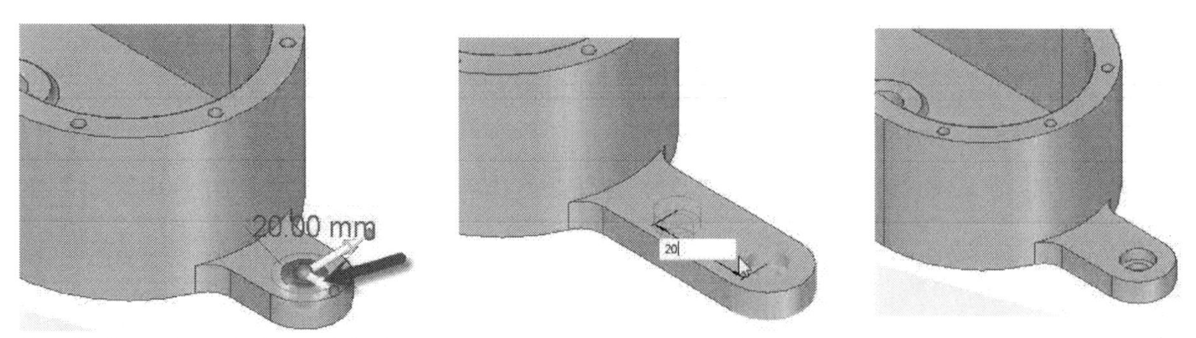

13. Click on the side face of the bottom feature. An arrow handle appears on the face.
14. Click the spear of the handle and drag it; the *Steering Wheel Tool* appears.
15. Align the Z-axis of the *Steering Wheel Tool* to the vertical edge, as shown in figure.

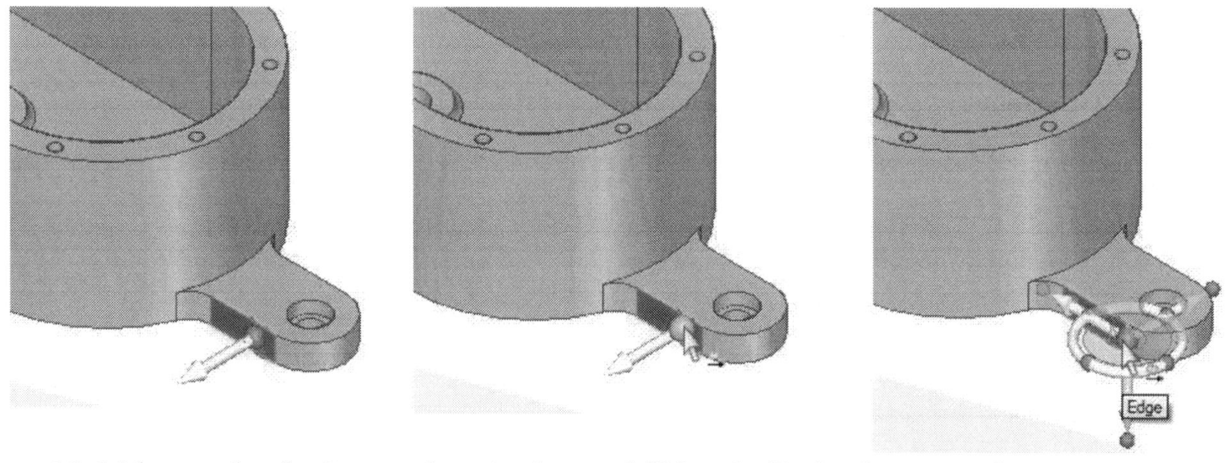

16. Make sure that the **Symmetric** option is turned ON on the **Design Intent** panel.
17. Click on the torus of the *Steering Wheel Tool* and move the mouse pointer outside.
18. Type -20 in the box that appears on the geometry. Press Enter to rotate the faces.

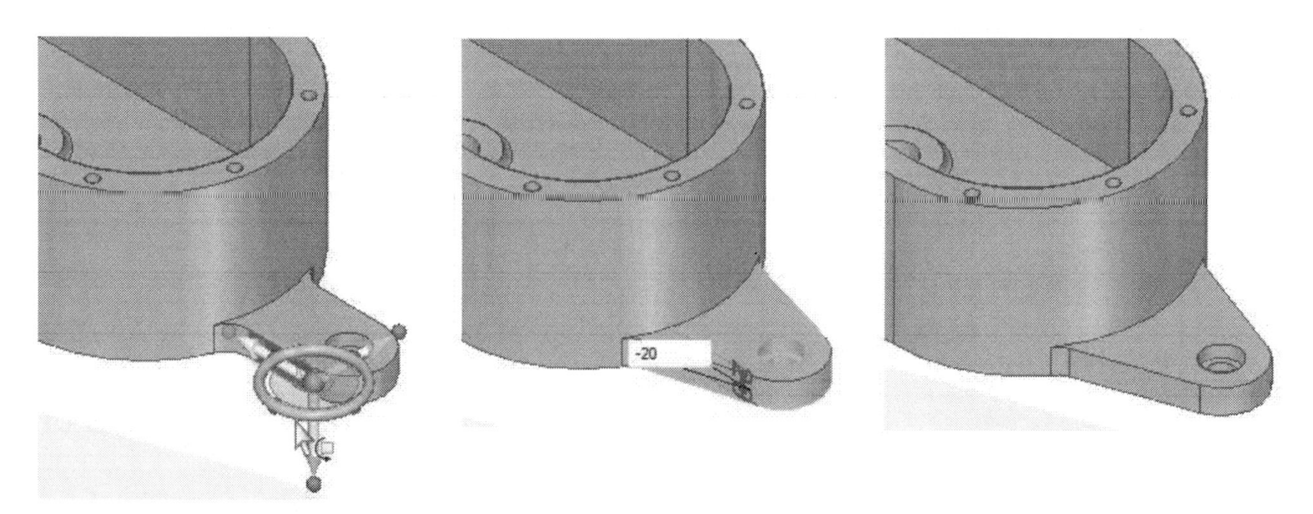

19. Click on anyone of the hole of the *Along Curve* pattern; the whole pattern is selected.

20. Click the *Pattern Handle* that appears on the geometry; the **Along Curve** command bar pops up along with the **Count** box.
21. Type 14 in the **Count** box and press Enter to update the pattern.

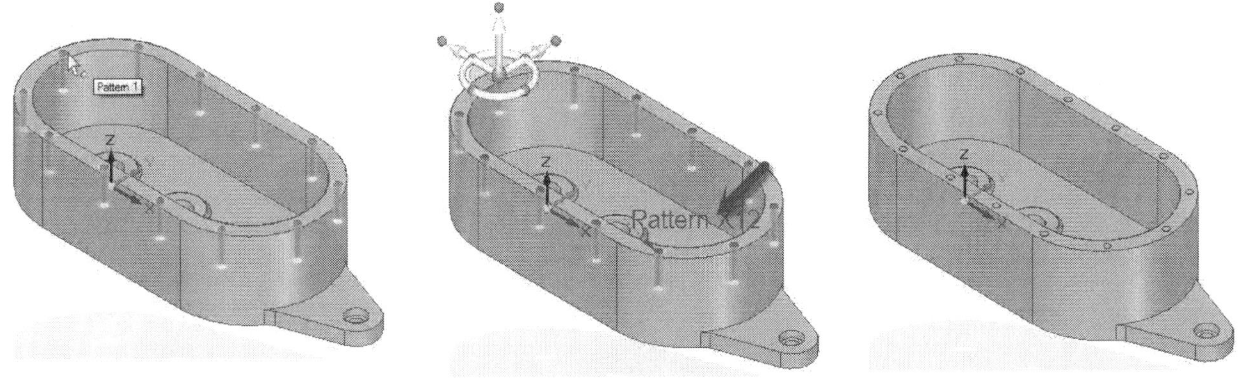

22. Click on the top face of the geometry to display an arrow.
23. Click on the arrow and drag the mouse pointer down. Type 40 in dimension box and press Enter to update the model.

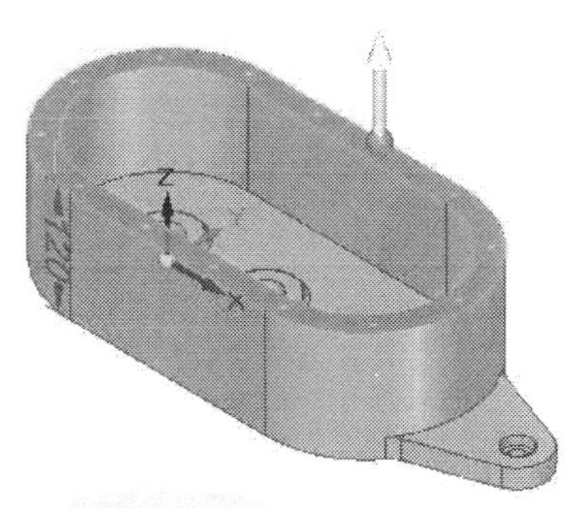

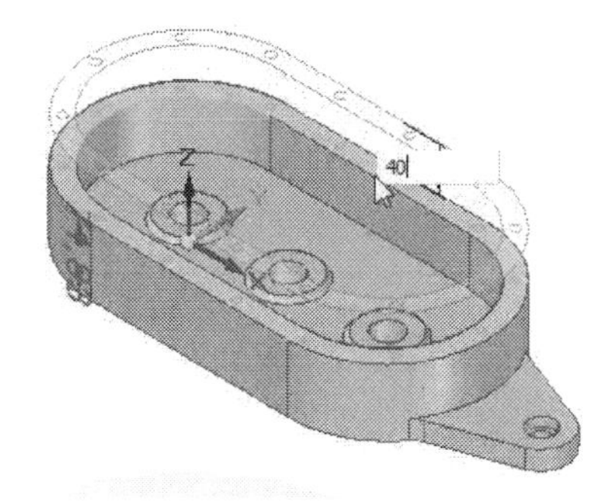

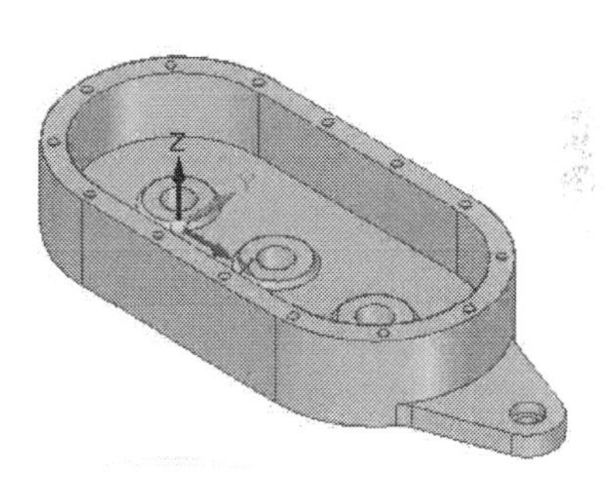

24. Save and close the file.

Questions

1. List any two face relationships.
2. How do you activate the **Move** command?
3. List the three options on the **Move** command bar that help you in moving faces
4. List any two live rules.
5. How do you modify revolved features using Live Sections?
6. What is **Select Set Priority**?

Exercises

Exercise 1

Initial Part

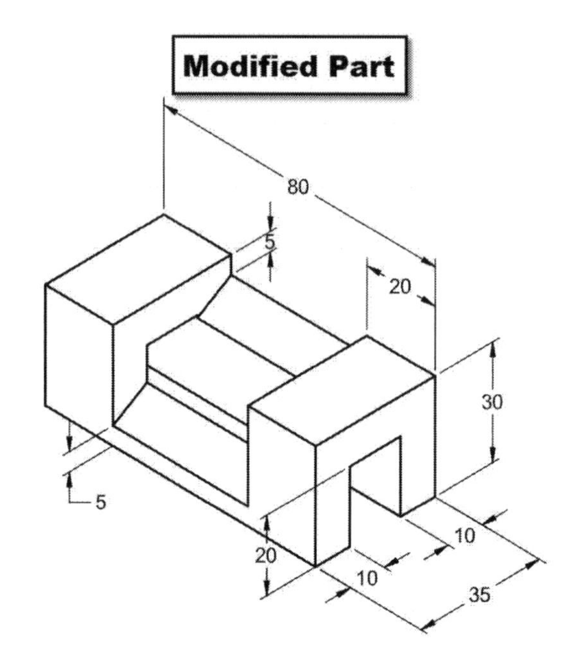

Chapter 10: Assemblies

After creating individual parts, you can bring them together into an assembly. By doing so, it is possible to identify incorrect design problems that may not have been noticeable at the part level. In this chapter, you will learn how to bring parts into the assembly environment and position them.

The topics covered in this chapter are:

- *Starting an assembly*
- *Inserting Parts*
- *Adding Relationships*
- *Dragging and Moving parts*
- *Check Interference*
- *Capture Fit*
- *Editing Assemblies*
- *Replace Parts*
- *Pattern and Mirror Parts*
- *Transfer Parts*
- *Create Subassemblies*
- *Disperse assemblies*
- *Assembly Features*
- *Top-down Assembly Design*
- *Assembly Relationship Assistant*
- *Create Exploded Views*

Starting an Assembly

To begin an assembly file, you can use the **ISO Metric Assembly** option or use the **New** icon and select an assembly template.

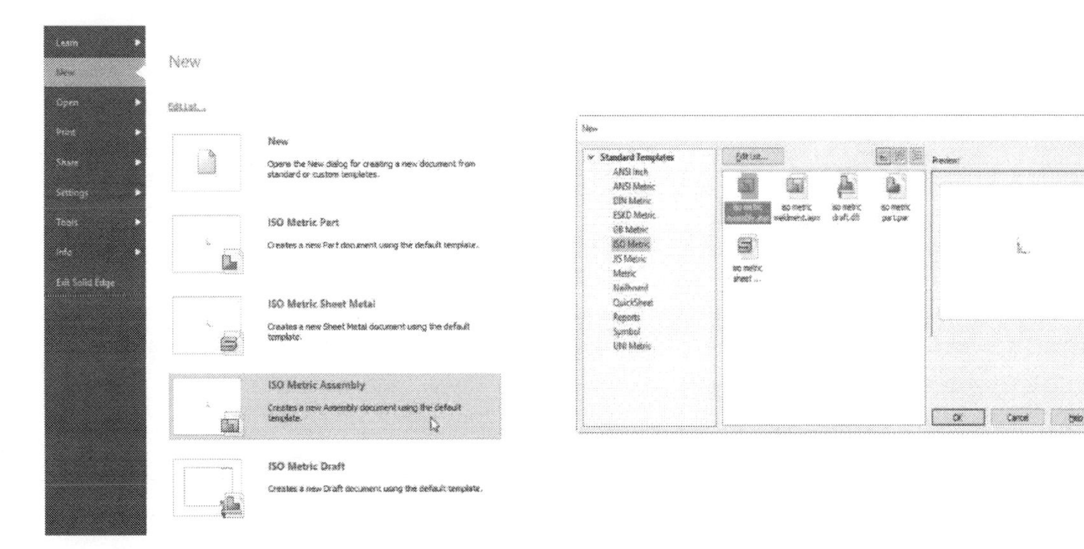

Now, you can insert parts into the assembly by using the **Parts Library** window. On the ribbon, click **Home > Assemble > Insert Component** to open the **Parts Library**. You can browse to the location of the parts by using the drop-down menu on the **Parts Library** window. As you select a component from the list, you can see a preview of the part in the **Preview** box. Now, double-click on the part to drop it into the graphics window.

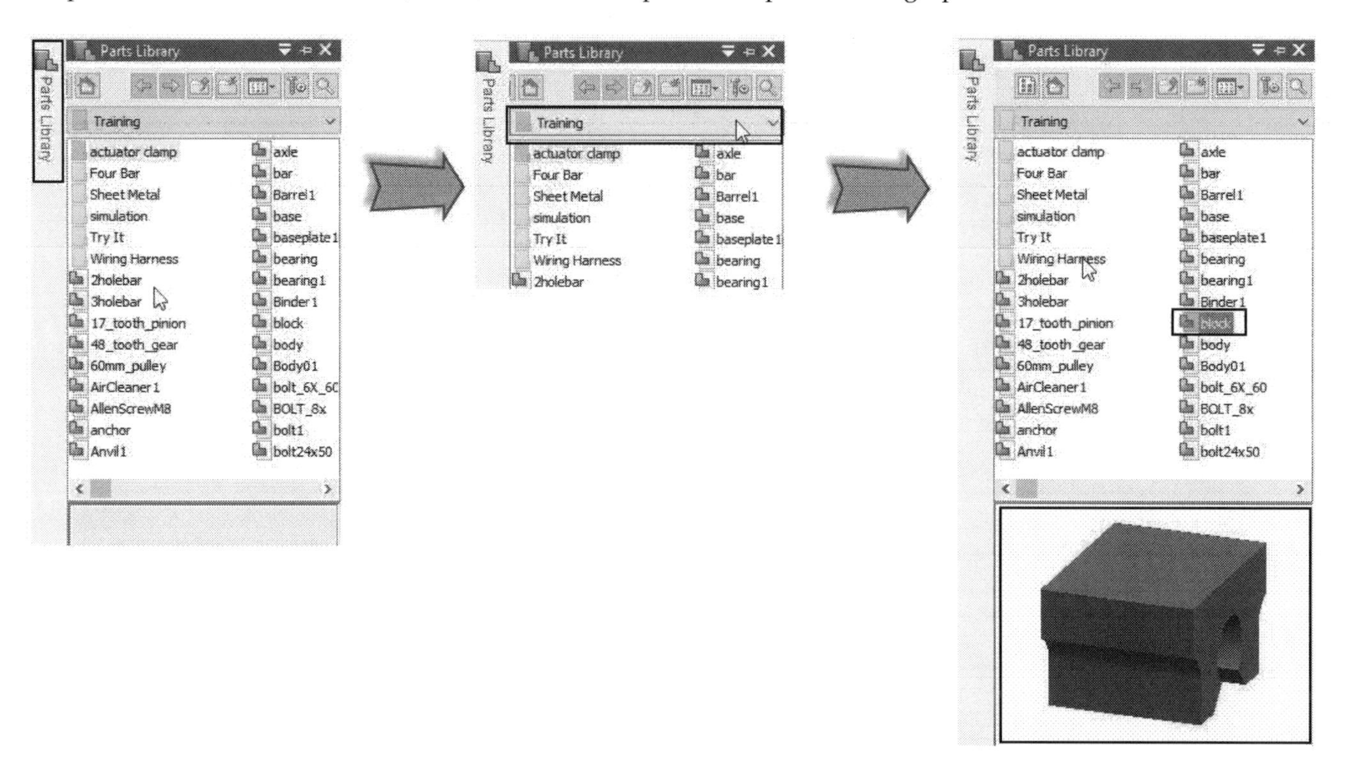

Another way to start an assembly is to create it while a part is open. On the Solid Edge **Application Menu**, click **New > Assembly of Active Model**. The **Create Assembly** dialog pops up on the screen. Click the **Browse** button and select an assembly template from the **New** dialog. Click **OK** twice to start the assembly. You will notice that the part will be placed at the origin. By default, the first part will be grounded at the origin. Also, the ribbon displays the commands related to the assembly environment.

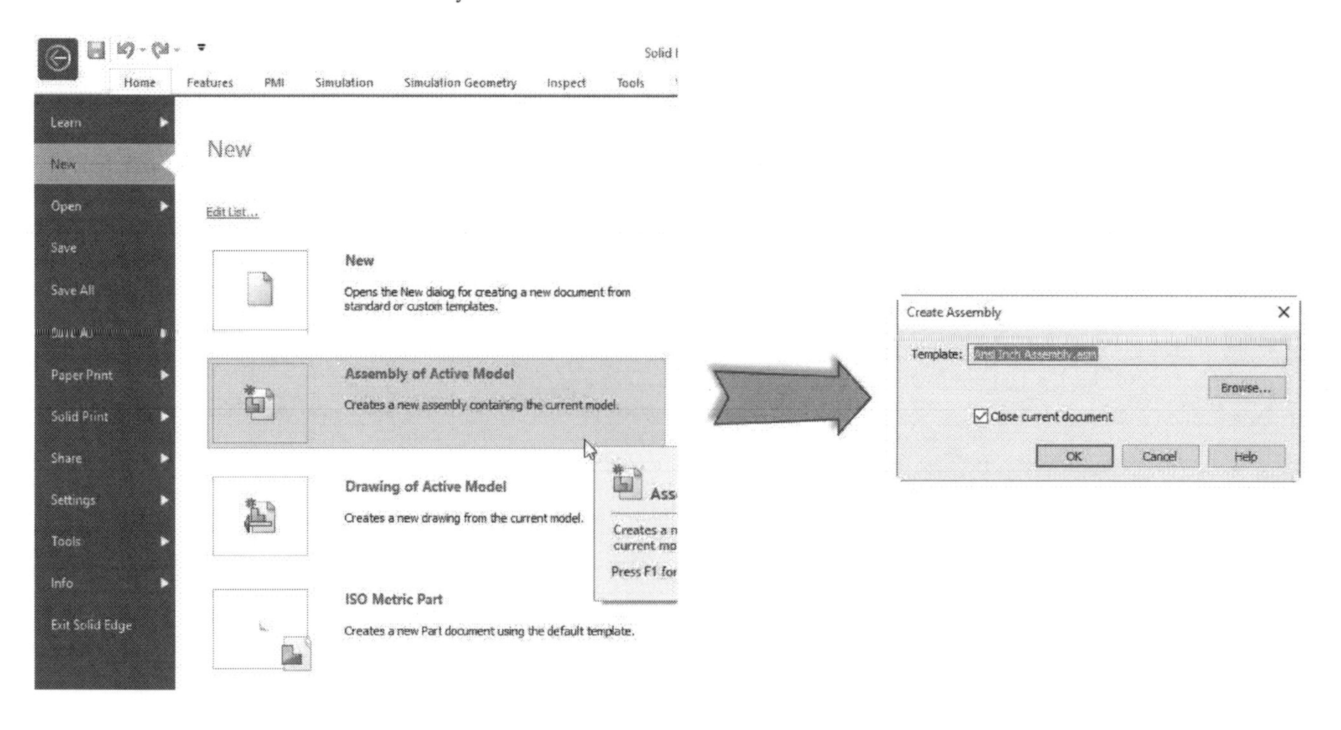

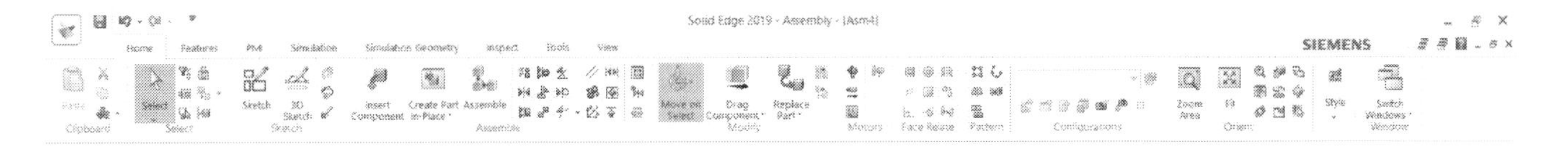

Inserting Parts

There are two different methods to insert an existing part into an assembly. The first one is to drag the part from the **Parts Library** and place it into the graphics window. The second way is to drag it directly from Windows Explorer. In the second method, you are not required to open these parts in Solid Edge. You can simply drag-and-drop the part into the assembly.

Adding Relationships

After inserting parts into an assembly, you have to define relationships between them. By applying relationships, you can make parts to flush with each other or make two cylindrical faces concentric with each other, and so on. As you add relationships between parts, the degrees of freedom will be removed from them. By default, there are six degrees of freedom for a part (three linear and three rotational). Eliminating degrees of freedom will make parts attached and interact with each other as in real life. Now, you will learn to add relationships between parts.

In the **Parts Library**, browse to the folder of the parts to be assembled. Click and drag the first part from the **Parts Library** into the assembly window; it will be fixed at the origin. As a result, all degrees of freedom of the part will be eliminated. Now, drag the second part into the assembly window, the **Assemble** command bar pops up on the screen. Select a face on the newly inserted part, and then click on a face of the fixed part. The two selected faces will mate with each other.

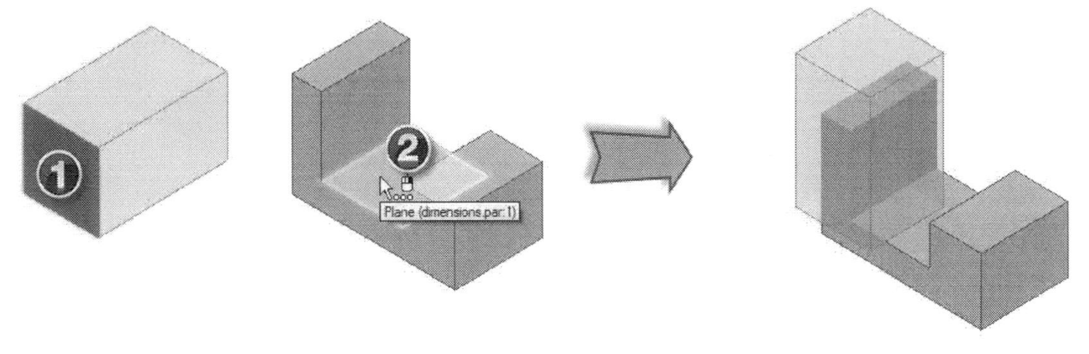

You can use the **Flip** icon on the command bar to flip the part.

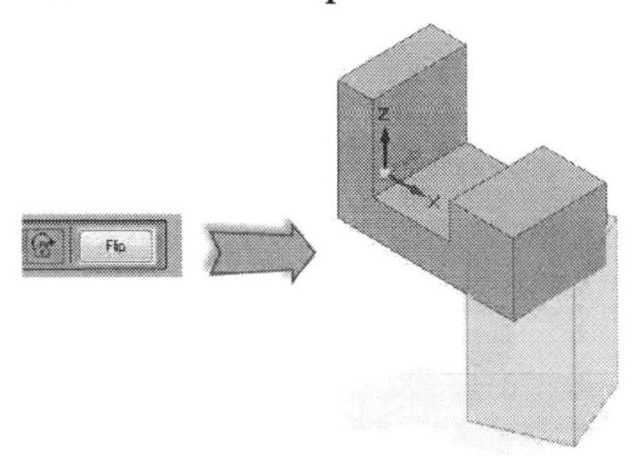

Select the second set of faces.

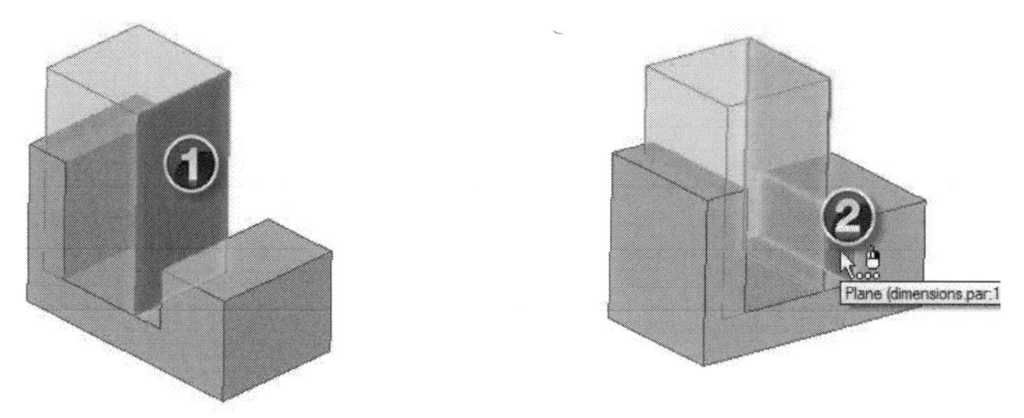

Select the third set of faces; the part will be fully positioned. To confirm this, place the pointer on the corresponding part in the Pathfinder; a message will appear showing that the part is fully positioned.

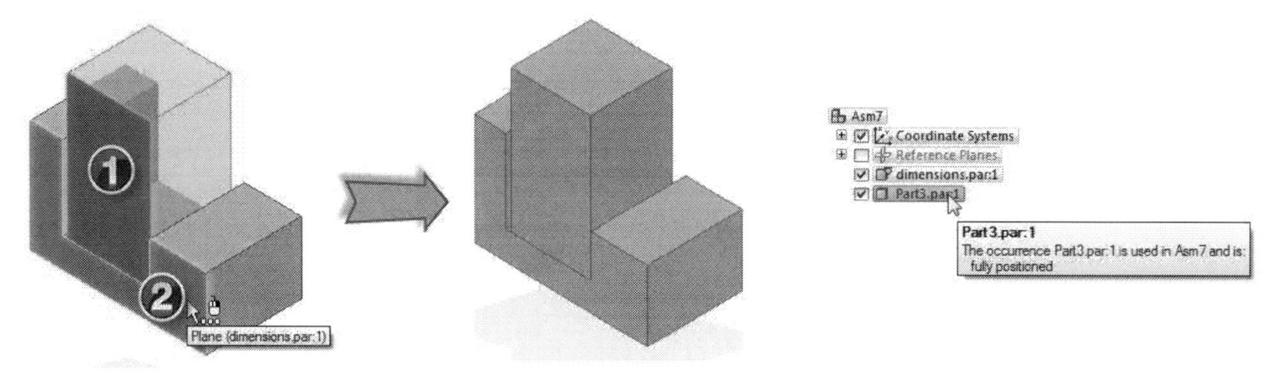

Drag Components

As you insert a part into an assembly, Solid Edge prompts you to define relationships between parts. If you choose not to define any relationships, press the **Esc** key. The part will be under-constrained and free to move and rotate. You can use the **Drag Components** command to move or rotate the under-constrained parts in the assembly window. Activate this command by clicking **Home > Modify > Drag Component** on the ribbon. The **Analysis Options** dialog pops up on the screen. The options on this dialog are self-explanatory. Check the required options on this dialog and click **OK**. Select a part from the assembly window and drag it to a new location.

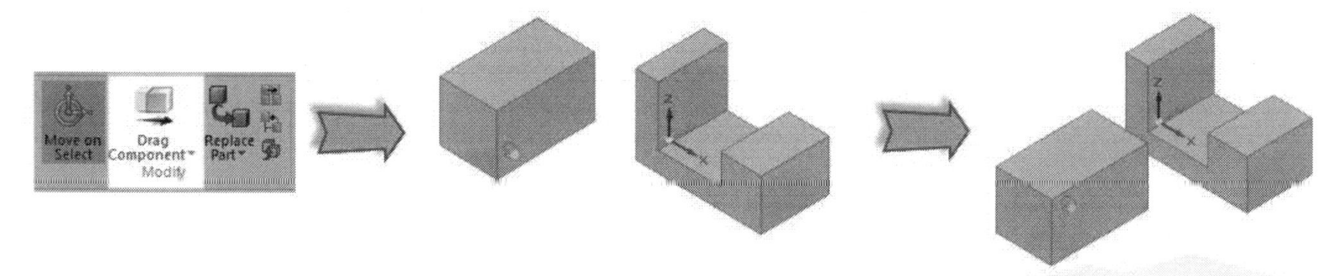

Use the **Move** option on the command bar to move the part in a particular direction. For example, to move the part in the X-direction, select the X-axis and move it (press and hold the left mouse button and drag the pointer).

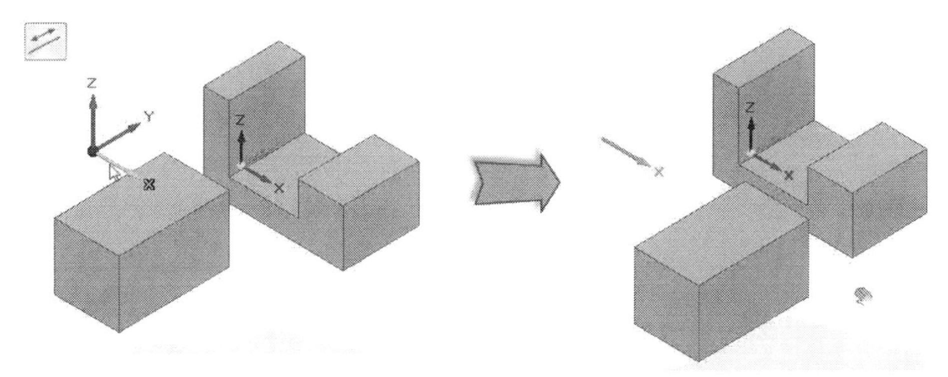

Use the **Rotate** option on the command bar to rotate the part about an axis. For example, to rotate the part about the X-axis, select the X-axis and rotate it (press and hold the left mouse button and drag the pointer).

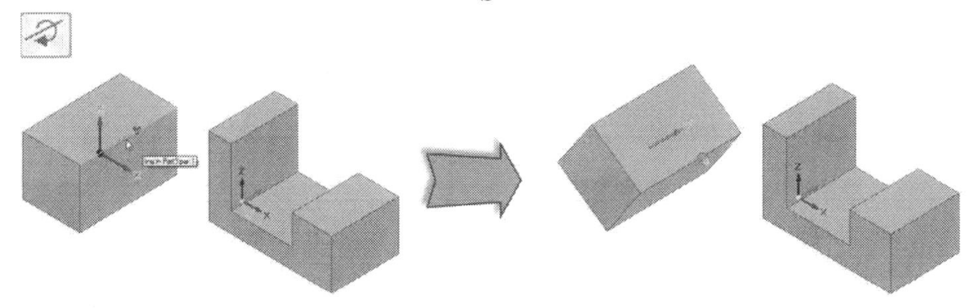

Likewise, use the **Freeform Move** option on the command bar to move or rotate the component randomly.

Use the **Detect Collisions** option on the command bar to detect collisions while moving or rotating the parts.

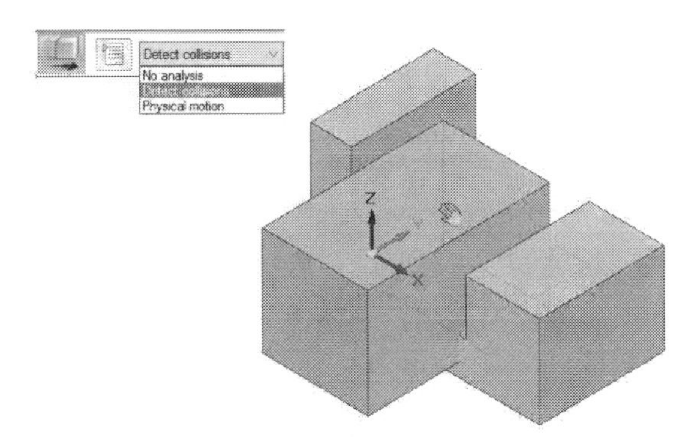

Use the **Physical Motion** option on the command bar to stop the part when it collides with another part.

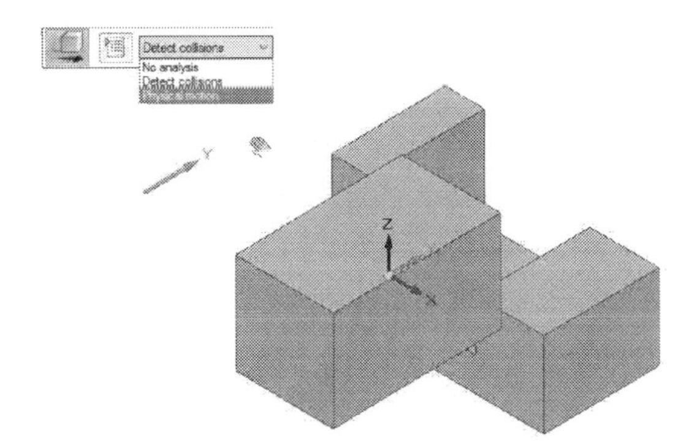

You can also move or rotate grounded parts using the **Drag Component** command. Click the **Options** icon on the command bar and check the **Locate grounded components** option on the **Analysis Options** dialog. Click **OK** on the dialog to close it. Now, select and move (or rotate) the grounded part.

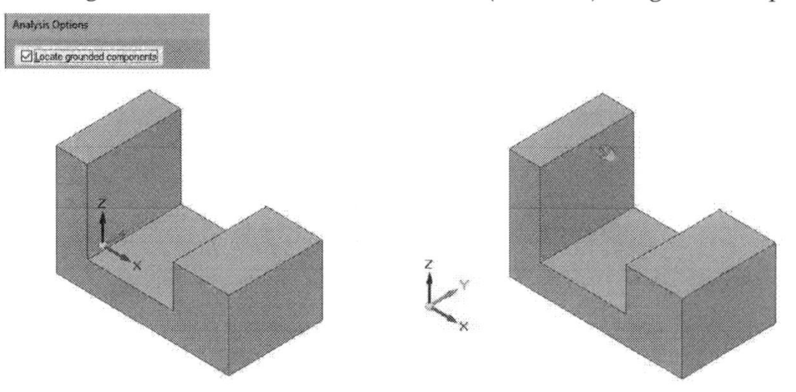

Mate Relationship

The **Mate** relationship makes two faces coincident and opposite to each other. You can define the **Mate** or any relationship between two parts immediately after you insert them. As you click and drag the part from Parts Library into the assembly window, the **Assemble** command bar pops up on the screen. On the command bar, click the **Relationship Types** icon and select **Mate**. Select a face of the inserted part, and then click on a face of the target part. The two selected faces will mate with each other.

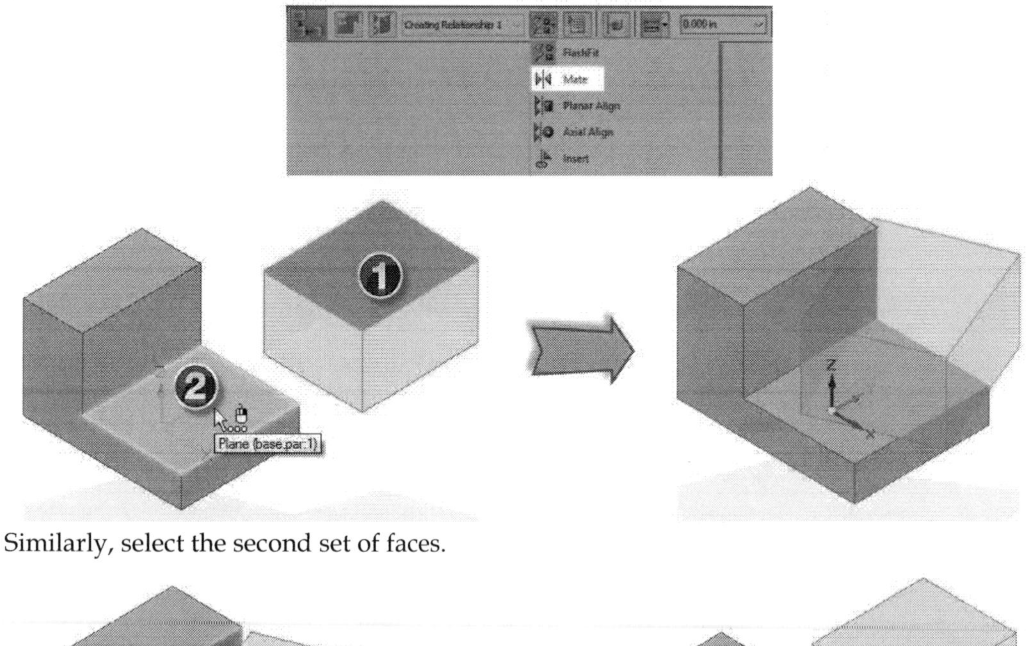

Similarly, select the second set of faces.

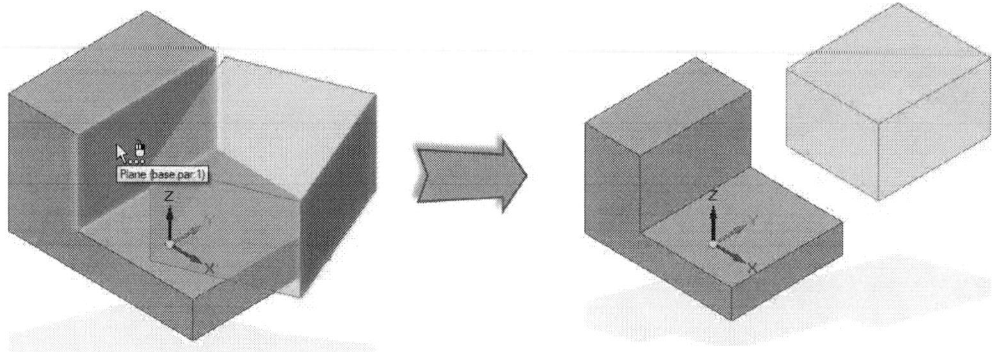

Planar Align Relationship

The **Planar Align** relationship makes two faces flush with each other. To define this relationship, click the **Relationship Types** icon and select **Planar Align** on the **Assembly** command bar. Select a face on the placement part, and then a face on the target part. The two faces will be levelled.

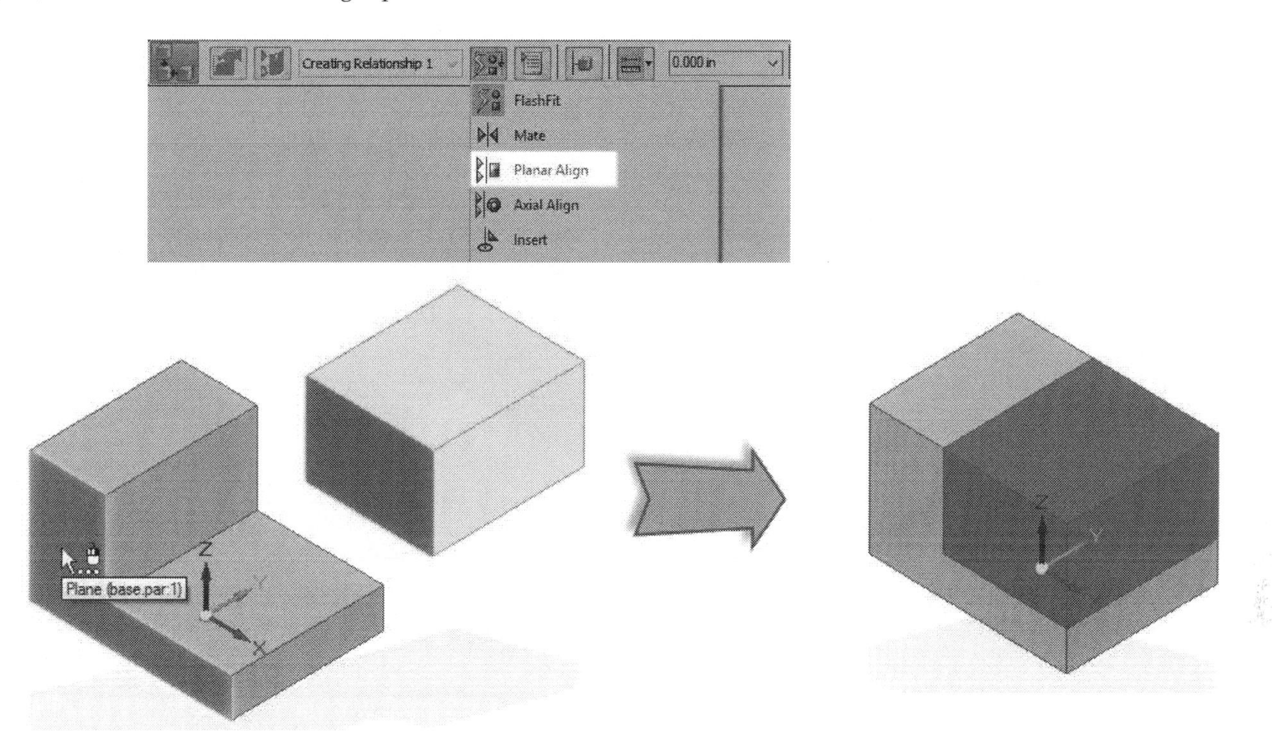

Axial Align Relationship

The **Axial Align** relationship makes the axes of two cylindrical faces coincide with each other. You can activate this command either from the **Assemble** command bar (click **Relationship Types > Axial Align)** or from the ribbon (click **Home > Assemble > Axial Align**). After activating this command, click on a cylindrical face, linear edge, or axis of the placement part. Click the **Lock Rotation** icon on the command bar, if you want to lock the rotation of the part. Next, click on an element on the target part. The two cylindrical axes will be aligned together.

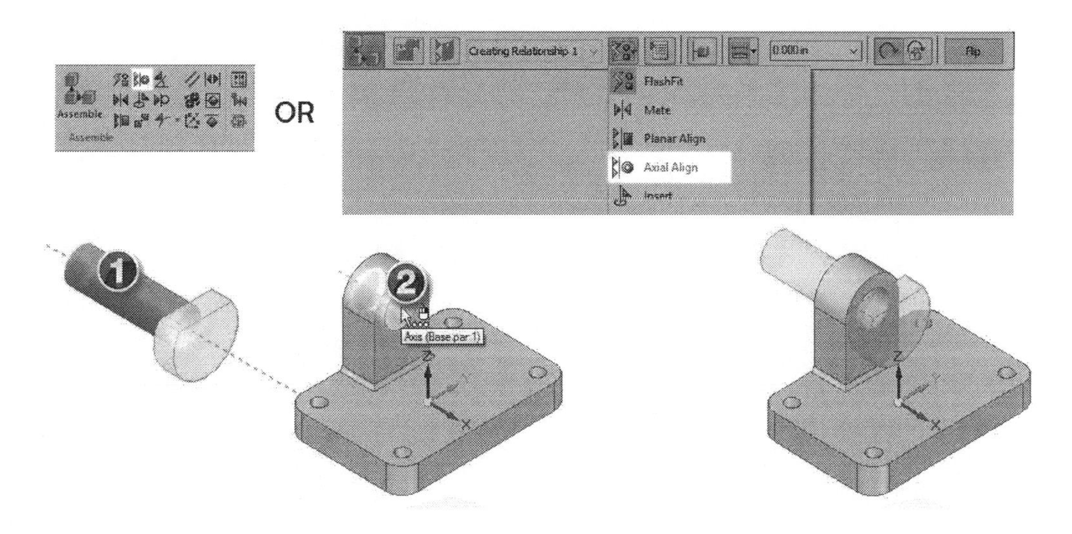

Insert Relationship

The **Insert** relationship helps you to position cylindrical parts into holes. This relationship is a combination of two relationships: **Axial Align** and **Planar Align**. It aligns the cylindrical axes and the end faces of two parts. Activate this command either from the **Assemble** command bar (click **Relationship Types** > **Insert)** or from the ribbon (click **Home > Assemble > Insert**). After activating this command, click on a cylindrical face or axis to align. Next, click on a cylindrical face on the target part. Click on a face to mate on the first part, and then click on a face on the target part. The first part will be inserted into the second part.

Angle Relationship

The **Angle** relationship is used to position faces at a specified angle. Activate this command either from the **Assemble** command bar (click **Relationship Types** > **Angle)** or from the ribbon (click **Home > Assemble > Angle**). After activating this command, type-in a value in the **Angle Value** box on the command bar and click on a plane or linear element of the first part. Next, click on a plane or linear element of the second part. Click on a plane on which the angle will lie. The first part will be positioned at the specified angle.

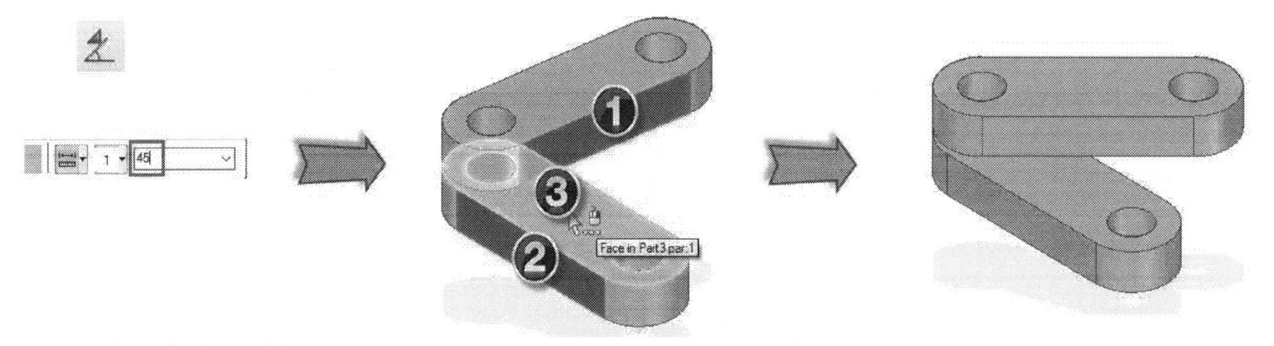

Tangent Relationship

The **Tangent** relationship is often used when working with cylinders and spears. It causes the geometry to maintain contact at a point of tangency. Activate this command either from the **Assemble** command bar (click **Relationship Types** > **Tangent)** or from the ribbon (click **Home > Assemble > Tangent**). After activating this command, click on the face to be made tangent. Next, click on the tangent face on the target part. The first part will be made tangent to the target part.

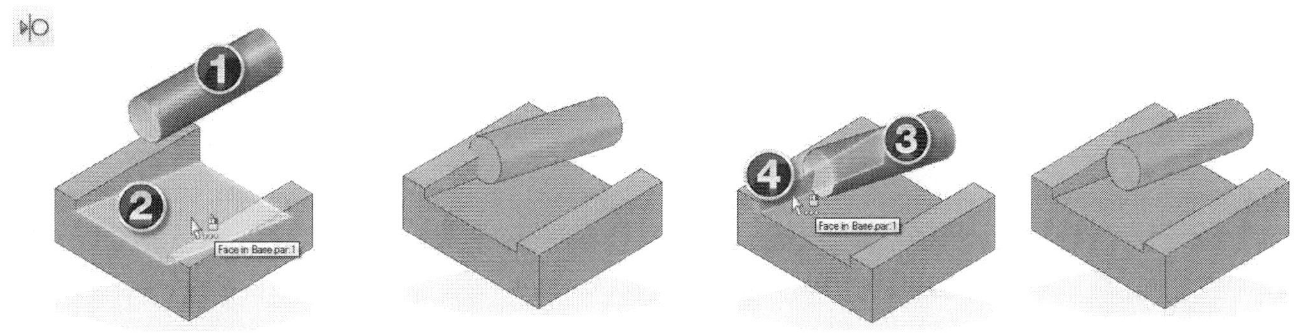

Connect Relationship

The **Connect** relationship connects a keypoint of one part to that of another part. Activate this command either from the **Assemble** command bar (click **Relationship Types** > **Connect)** or from the ribbon (click **Home** > **Assemble** > **Connect**). After activating this command, click on a keypoint on the first part. Next, click on a keypoint, edge, or face to connect to. The first part will be connected to the second part.

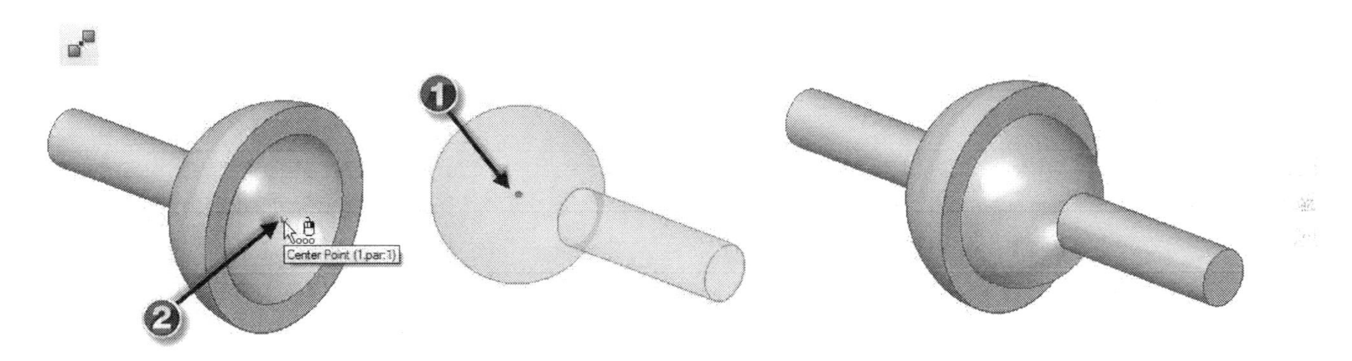

Parallel Relationship

The **Parallel** relationship makes an axis or edge of one part parallel to that of another part. Activate this command either from the **Assemble** command bar (click **Relationship Types** > **Parallel)** or from the ribbon (click **Home** > **Assemble** > **Parallel**). After activating this command, type-in a value in the **Offset Value** box on the command bar and click on a cylindrical face, linear edge, or axis of the first part. Next, click on an element of the second part. The two selected edges or axes will be parallel to each other.

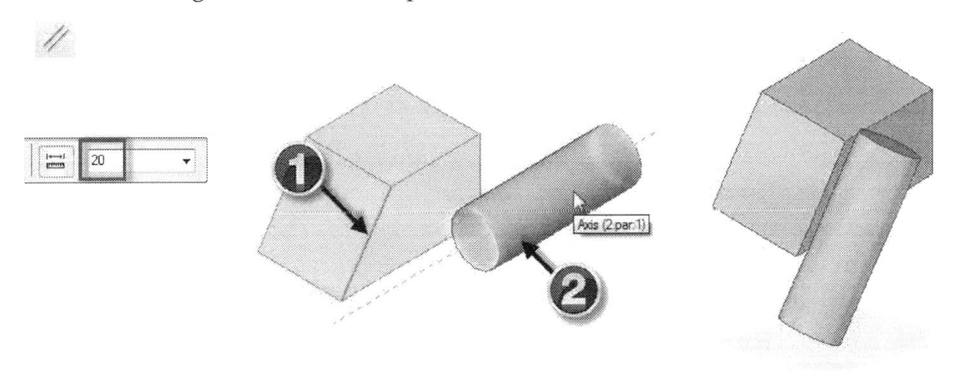

Center-Plane Relationship

The **Center-Plane** relationship allows you to center a part between two faces. Activate this command either from the **Assemble** command bar (click **Relationship Types** > **Center-Plane)** or from ribbon (click **Home** > **Assemble** > **Center-Plane**). After activating this command, you must select the object to be positioned at the center of two planes. Click on a planar face, edge, axis, keypoint, or reference plane on the first part. Next, click on two faces or reference planes on the second part. The first part will be centred between the two planes.

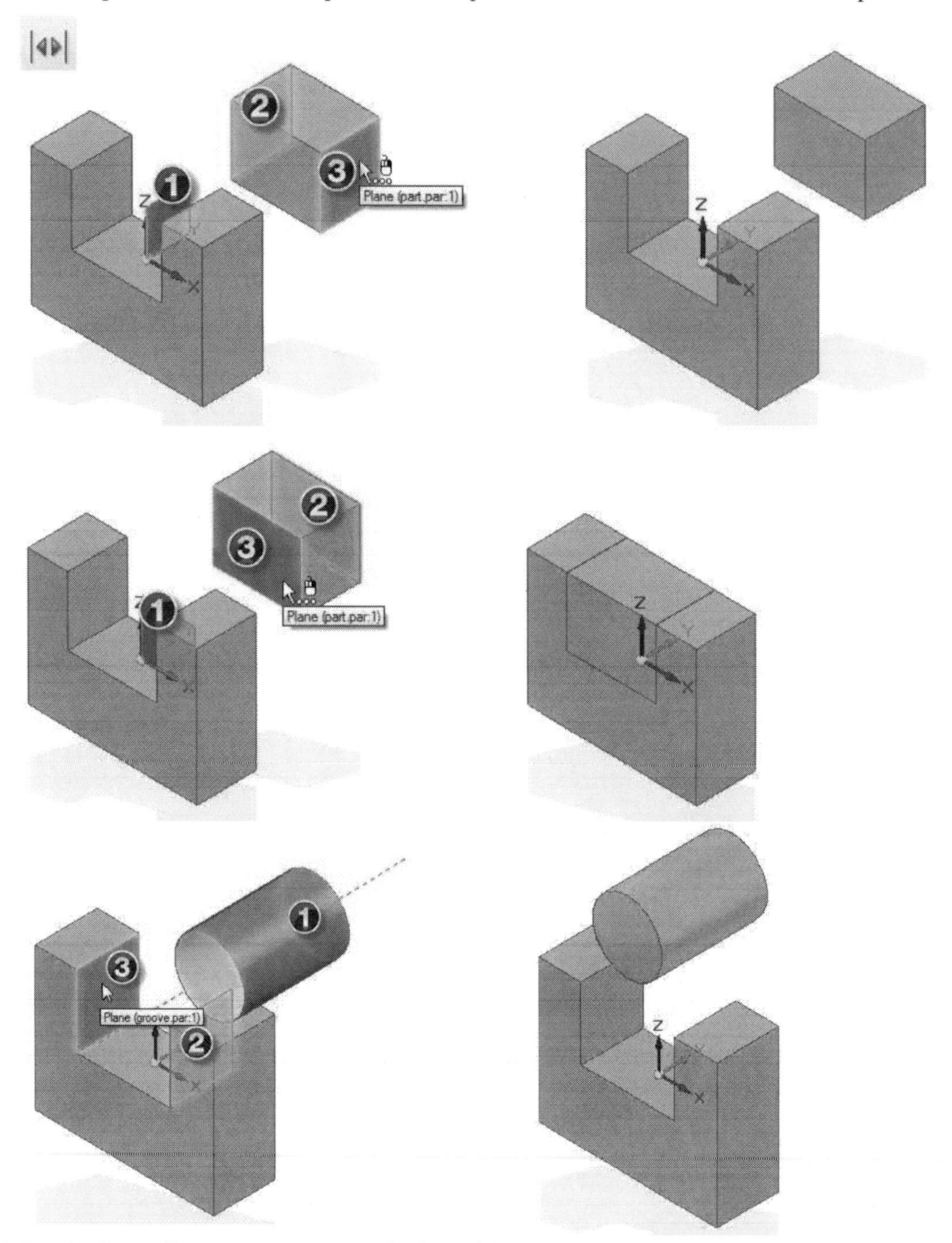

Match Coordinate Systems Relationship

The **Match Coordinate Systems** relationship matches the coordinate systems of two parts. This is the easiest way to constrain parts in an assembly. To apply this relationship, first you must display the coordinate systems of the parts. You can do so by clicking the **Construction Display** icon on the **Assembly** command bar and selecting the **Show Coordinate Systems** option (or) by right-clicking on the part and selecting **Show Hide Component**, and then turning on **Coordinate Systems**.

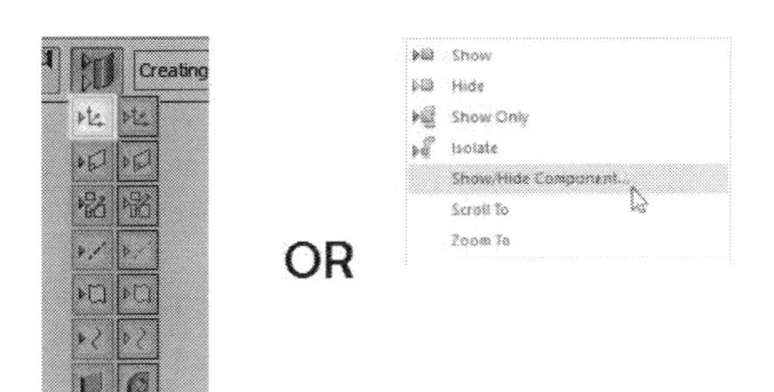

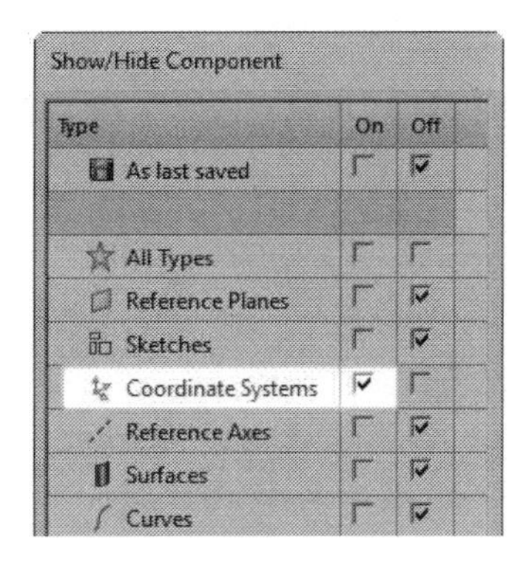

Activate this command either from the **Assemble** command bar (click **Relationship Types** > **Match Coordinate Systems)** or from the ribbon (click **Home** > **Assemble** > **Match Coordinate Systems**). After activating this command, you have to select the coordinate systems of two parts. They will be positioned together.

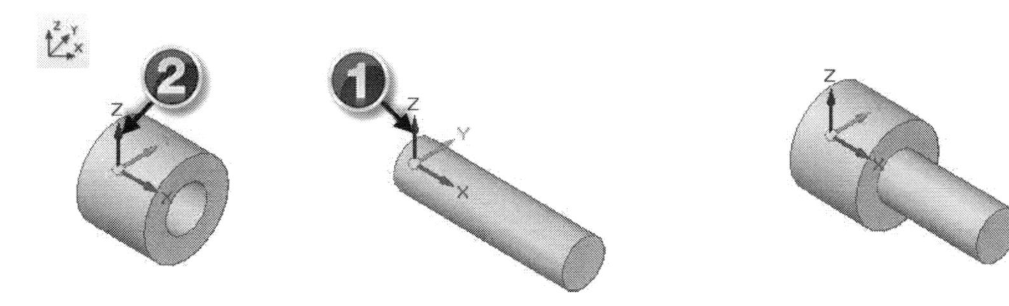

Rigid Set Relationship

The **Rigid Set** relationship makes the selected parts to form a rigid set. As you move a single part of a rigid set, all the other parts will also be moved. Activate this command from the ribbon (click **Home > Assemble > Rigid Set**); a command bar pops up on the screen. On the command bar, select an option from **Shared Relationships** menu. You can select to **Suppress**, **Delete**, or **Ignore** already existing relationships between the parts. Next, select parts from the assembly window and click the green check on the command bar. The selected parts will form a rigid set. Now, if you change the position or orientation of one part, all the other parts of the rigid set will also be affected.

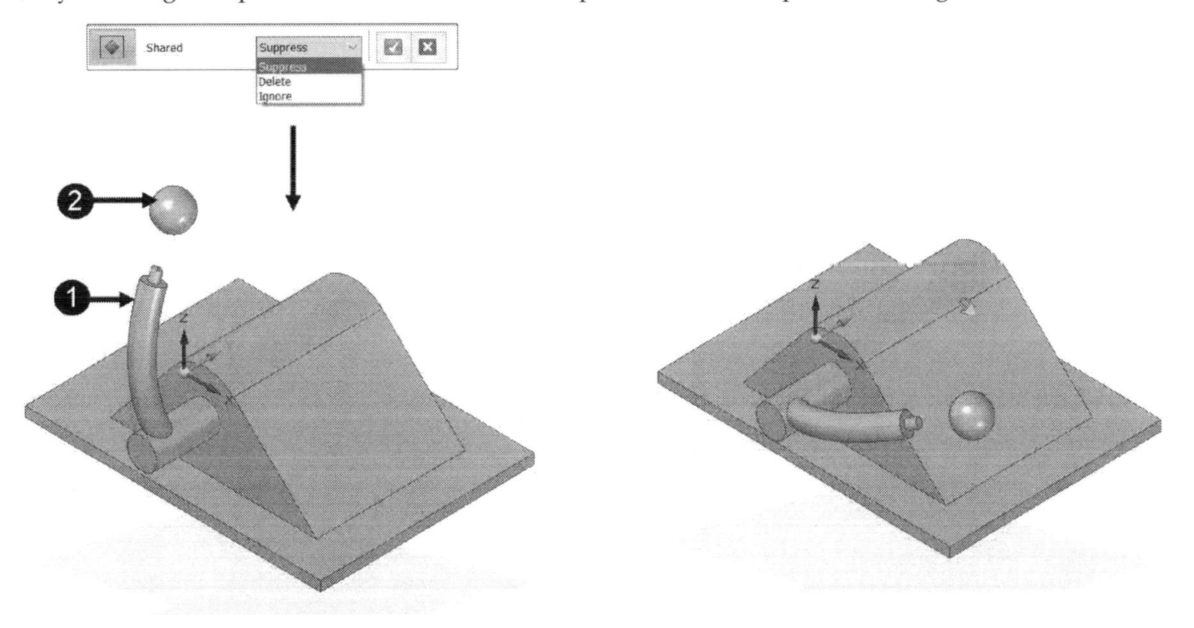

Ground Relationship

By default, the first inserted part in an assembly is grounded or fixed. As a result, all the degrees of freedom of the part are constrained. However, you can make any other part grounded by using the **Ground** command. Activate this command (click **Home > Assemble > Ground** on the ribbon) and select the part to ground. A ground symbol appears on the selected part in the Pathfinder.

Path Relationship

The **Path** relationship is used to constrain a selected point or line along a path. Activate this command either from the **Assemble** command bar (click **Relationship Types** > **Path)** or from the ribbon (click **Home > Assemble > Path > Path**). After activating this command, click on a point or linear edge to define the follower. Next, click on an edge to define the path. Click the green check on the command bar to apply this relationship. Use the **Drag Component** command to drag the follower.

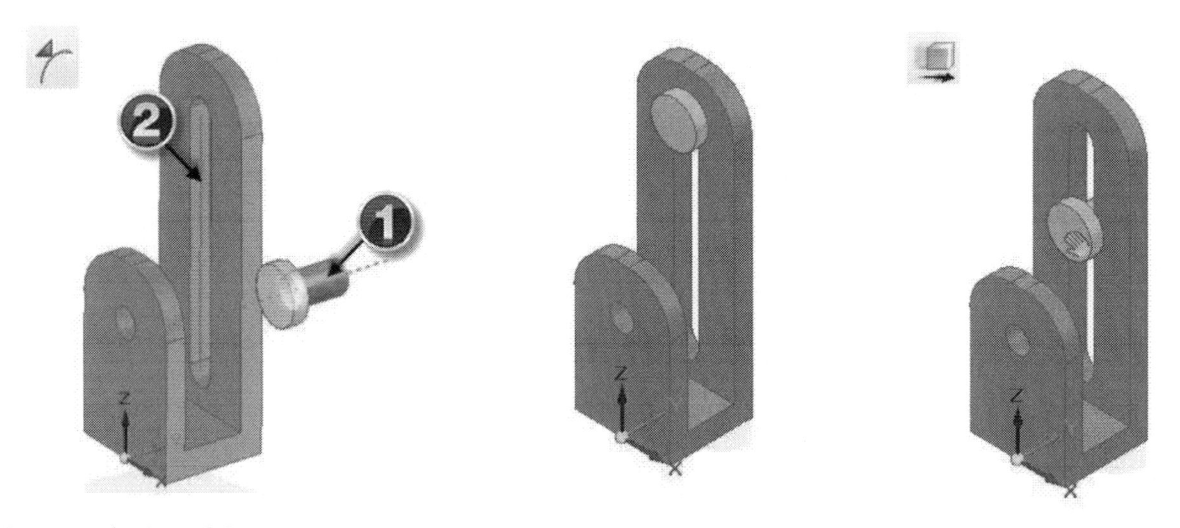

Cam Relationship

The **Cam** relationship is similar to a **Tangent** relationship except that it allows you to mate a cylinder, plane, or point to a series of tangent faces. Activate this command either from the **Assemble** command bar (click **Relationship Types** > **Cam)** or from the ribbon (click **Home > Assemble > Path > Cam**). After activating this command, click on a face or a point to define the follower. Next, click on a face chain to define the cam. Click the green check on the command bar to apply this relationship.

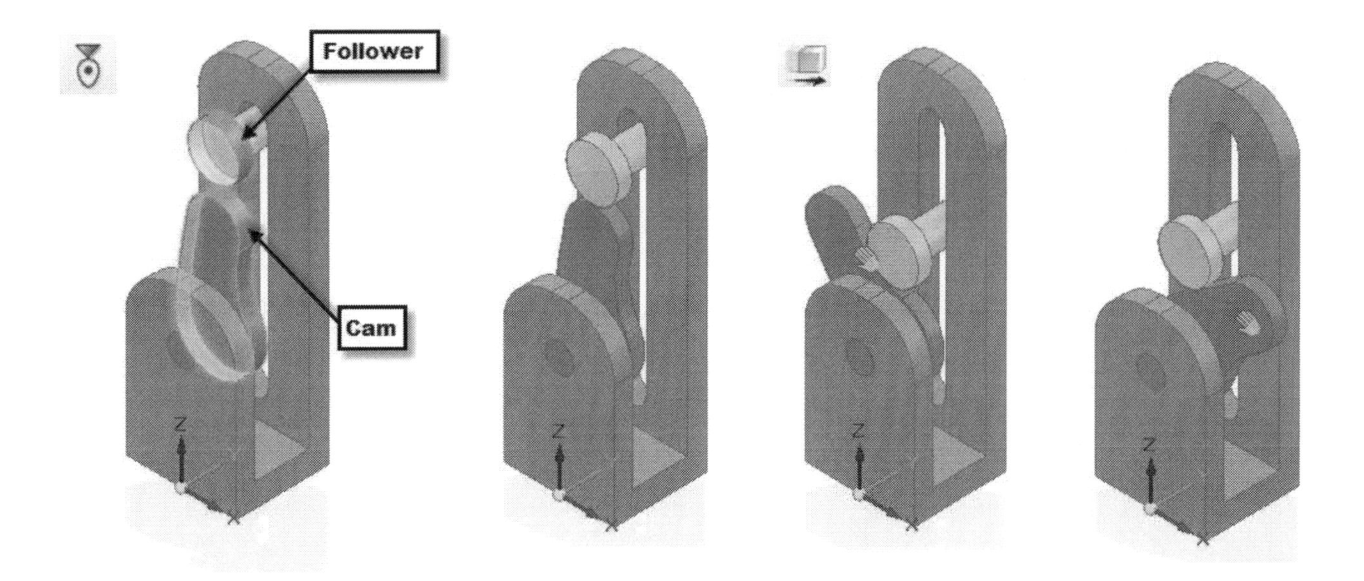

Check Interference

In an assembly, two or more parts can overlap or occupy the same space. However, this would be physically impossible in the real world. When you add relations between parts, Solid Edge develops real-world contacts and movements between them. However, sometimes interferences can occur. To check such errors, Solid Edge provides you with a command called **Check Interference**. Activate this command (click **Inspect > Evaluate > Check Interference** on the ribbon) and select the first set. Click the green check on the command bar and select the second set. Click the **Accept** button, and then click the **Process** icon to show the interference. If there is no interference, a message box appears showing that there are no interferences in the assembly.

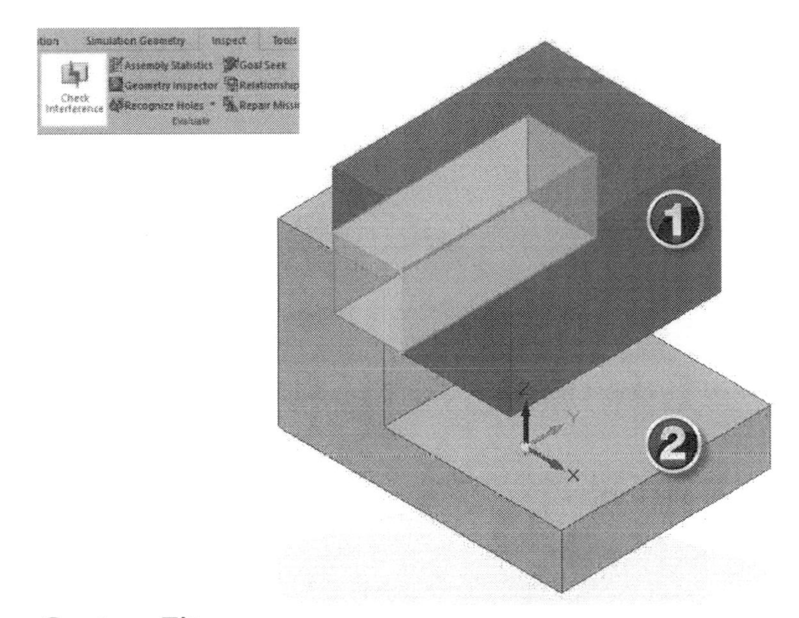

Capture Fit

If you have an assembly in which you need to assemble the same part multiple times, it would be a tedious process. In such cases, the **Capture Fit** command will drastically reduce or even eliminate the time used to assemble commonly used parts. To use this command, first you need to define a relation or set of relations between two parts. For example, define the **Insert** relationship between the screw and the hole.

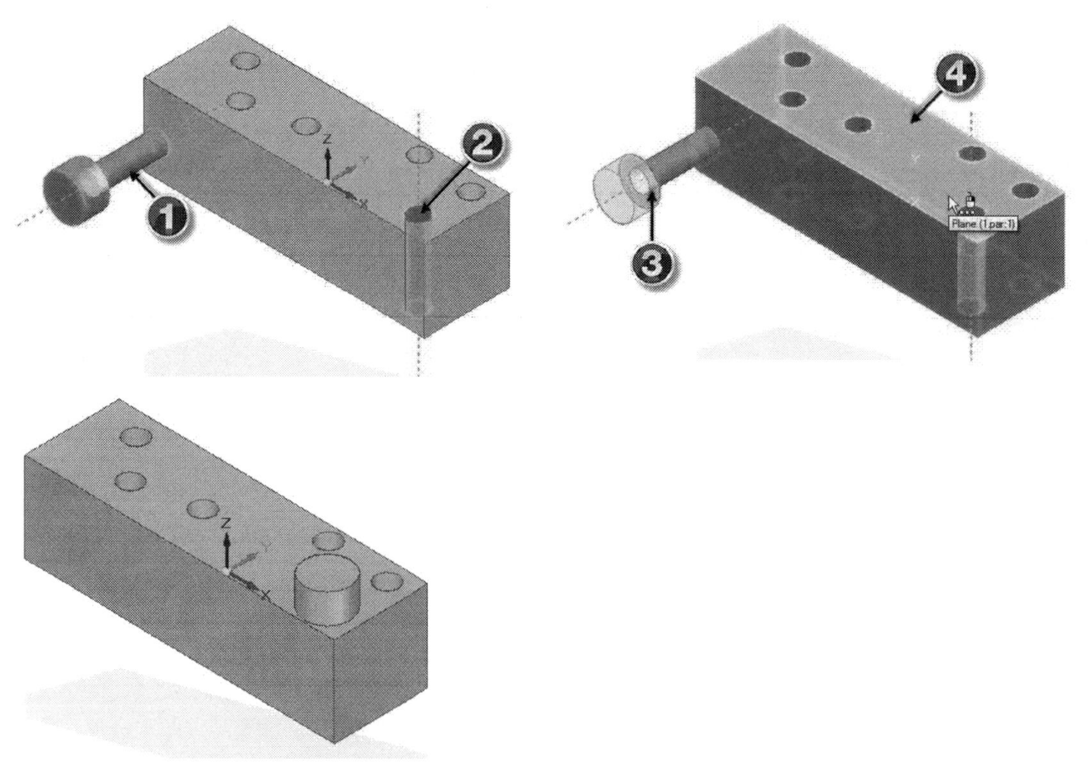

Next, save the assembly and select the screw. Activate the **Capture Fit** command (click **Home > Assemble > Capture Fit** on the ribbon); the **Capture Fit** dialog pops up on the screen. This dialog shows the list of relations that can be captured. If you do not want to capture some relations, select them from the list and click **Remove**. Next, click **OK** on the dialog to capture the relations.

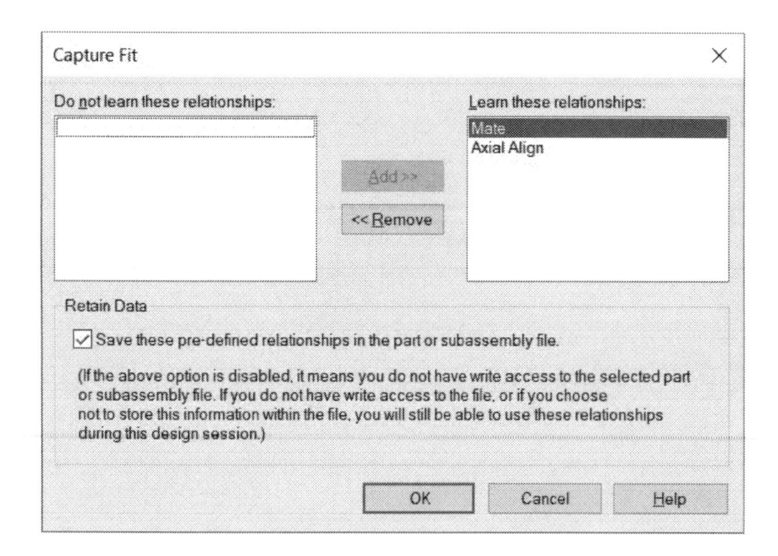

Now, click and drag the screw from the **Parts Library** and place it into the assembly window; you will notice that the flat face on the screw is selected automatically. Select the top face of the block; the axis of the cylinder is selected, automatically. Select the axis of anyone of the holes on the block; the screw is inserted into the hole.

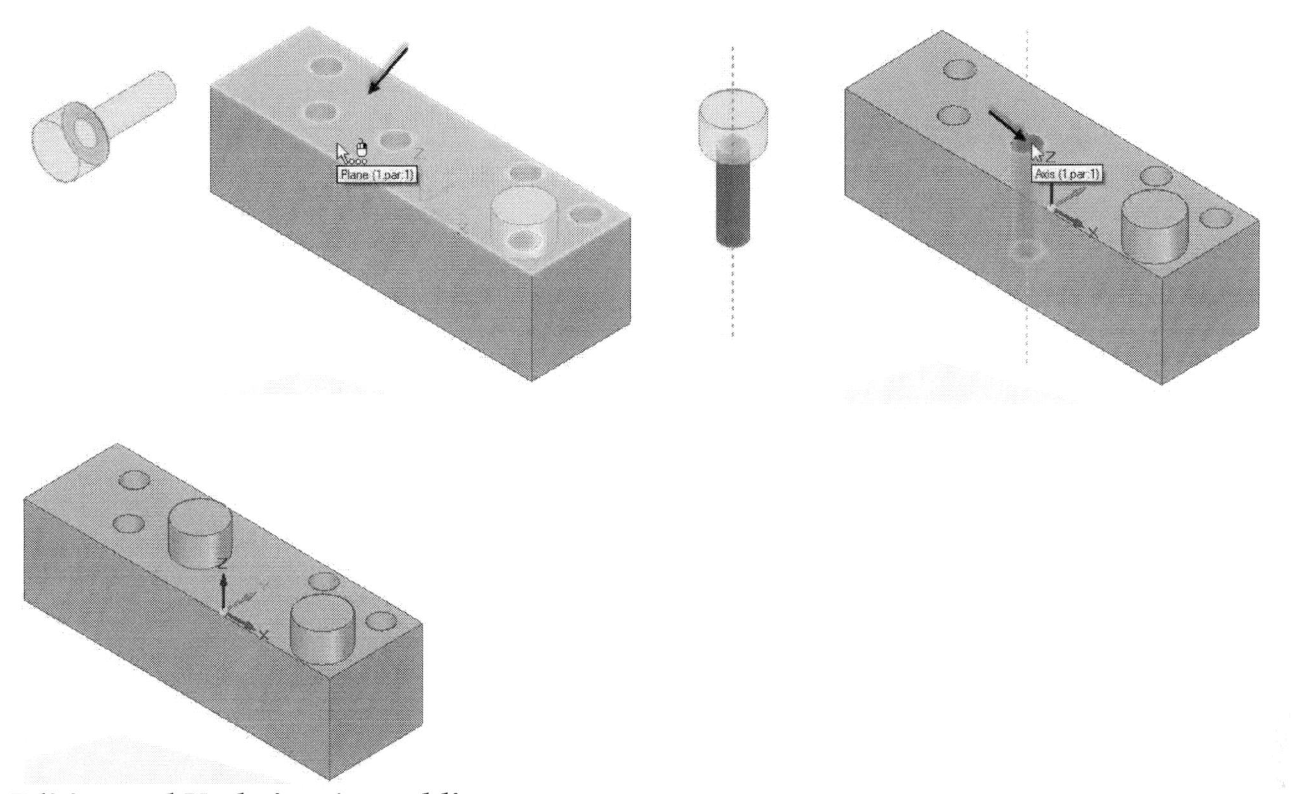

Editing and Updating Assemblies

During the design process, the correct design is not achieved on the first attempt. There is always a need to go back and make modifications. Solid Edge allows you to accomplish this process very easily. To modify a part in an assembly, right-click on it and select **Open**; the part will be opened in a separate window. Make changes to the part and save it. Next, switch to the assembly window. The part will be updated in the assembly automatically. If it is not updated, click **Tools > Update > Update Active Level** on the ribbon.

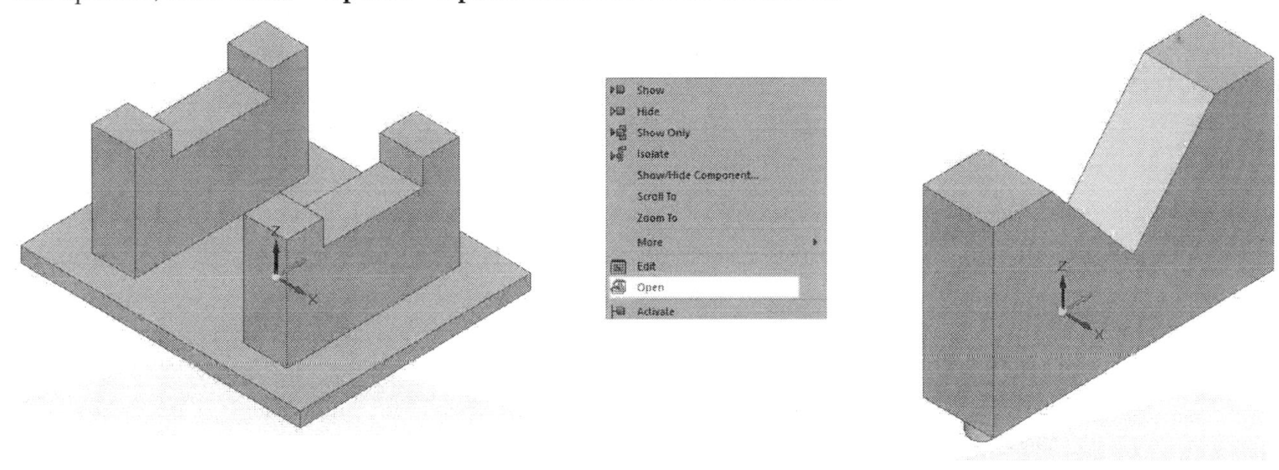

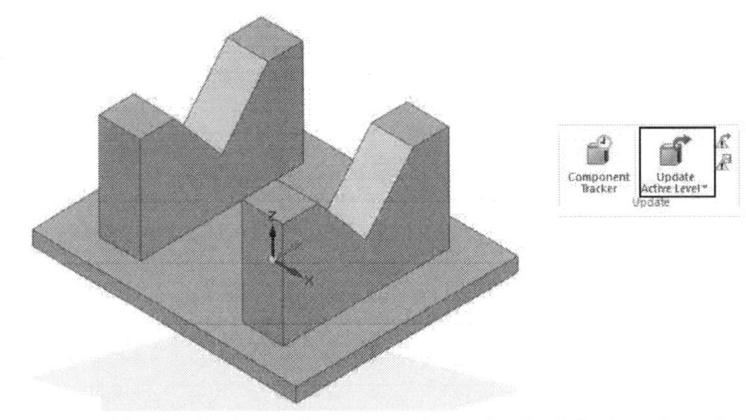

You can also edit relationships in an assembly. To do this, select a part from the Pathfinder; the relationships applied to the part appear at the bottom of the Pathfinder. Click the right mouse button on the relationship to edit; a menu appears with four options: **Delete Relationship**, **Suppress**, **Flip**, and **Edit Definition**. If you select the **Edit Definition** option from the menu, the **Assemble** command bar pops up on the screen. You can reselect the faces or elements between which the relationship is applied. For example, if you want to edit a **Mate** relationship, right-click on it and select **Edit Definition**. On the **Assemble** command bar, click the **Placement Part -Element** icon, and then click on a face of the placement part. Next, click on a face of the target part, and then right-click to apply the relationship.

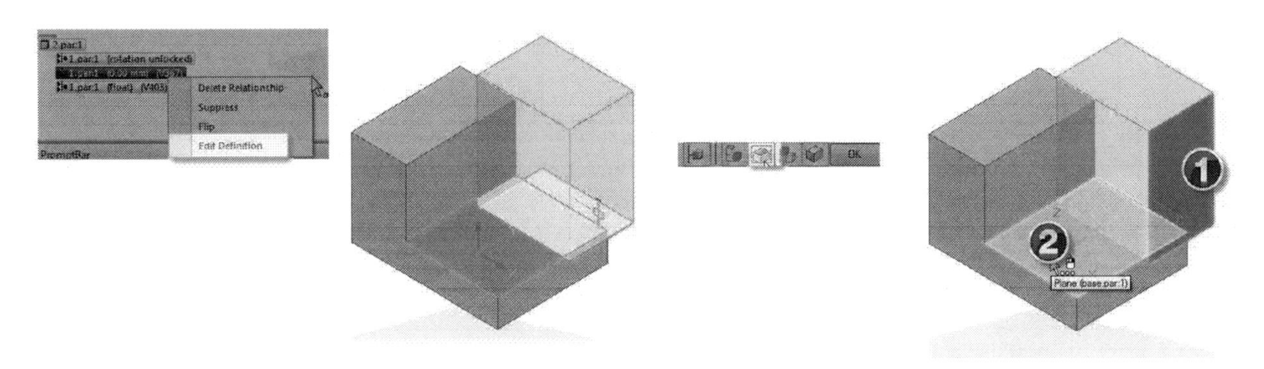

Replace Part

Solid Edge allows you to replace any part in an assembly. Activate the **Replace Part** command (click **Home > Modify > Replace Part** drop-down **> Replace Part** on the ribbon), and then click on parts to replace. Click the green check on the command bar to accept; the **Replacement Part** dialog pops up on the screen. Browse to the location of the replacement part and double-click on it; the **Assembly** message box pops up on the screen. It shows, "The affected Assembly relationships must be either deleted or suppressed to complete the operation". Click **Delete** or **Suppress** on the message box to replace the part. You can suppress or delete relationships based on the differences in the original and replacement part. You can redefine relationships after deleting them.

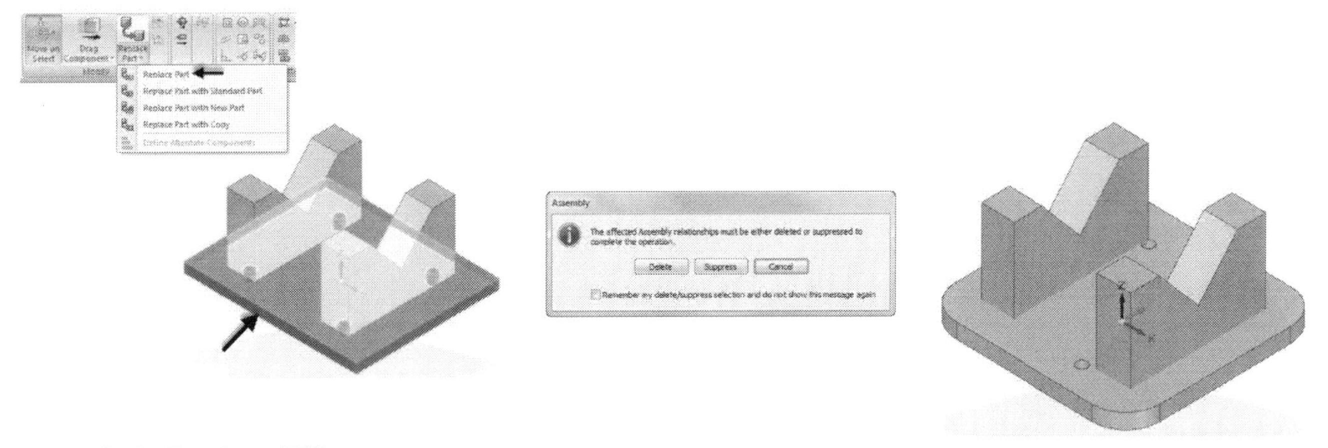

Repair Missing Files

Solid Edge provides a tool to find missing part files of an assembly. Whenever you open an assembly with some missing files, the **Repair Missing Files** dialog appears on the screen. On this dialog, select the missing part and use various options (**Search In**, **Replace Part**, **Rep lace with Standard Parts**, and **Replace with New Part**) to repair the assembly.

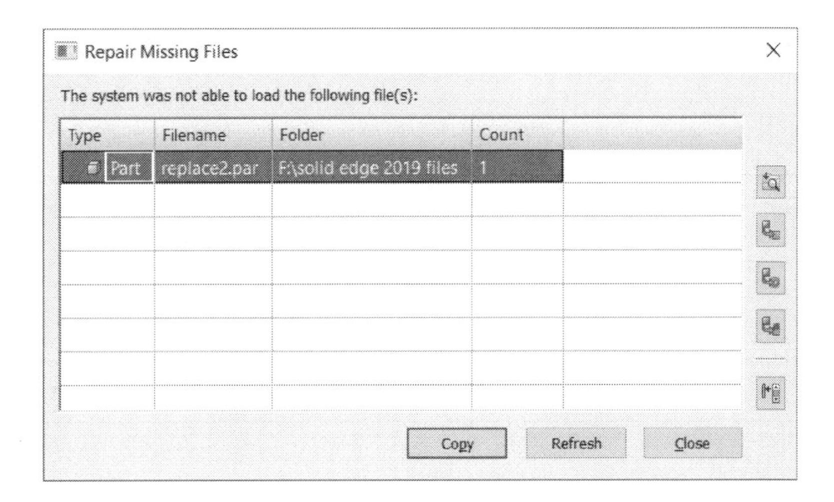

The **Search In** option brings up the **Search In** dialog, which helps you to search various folders on your computer. Click the **Browse** button on the **Search In** dialog, select a folder from the **Browse to Folder** dialog, and click **OK**. On the **Search In** dialog, click the **Add** button to add the selected folder to the **Search in these folders** list. Likewise, add other folders to the list. You can change the order of the folders using the **Move up** and **Move down** buttons. After adding folders to the list, click the **Search and Replace** button to start the search; Solid Edge will search the folder located at the top of the list. If the part file is not found, it will search the next folder.

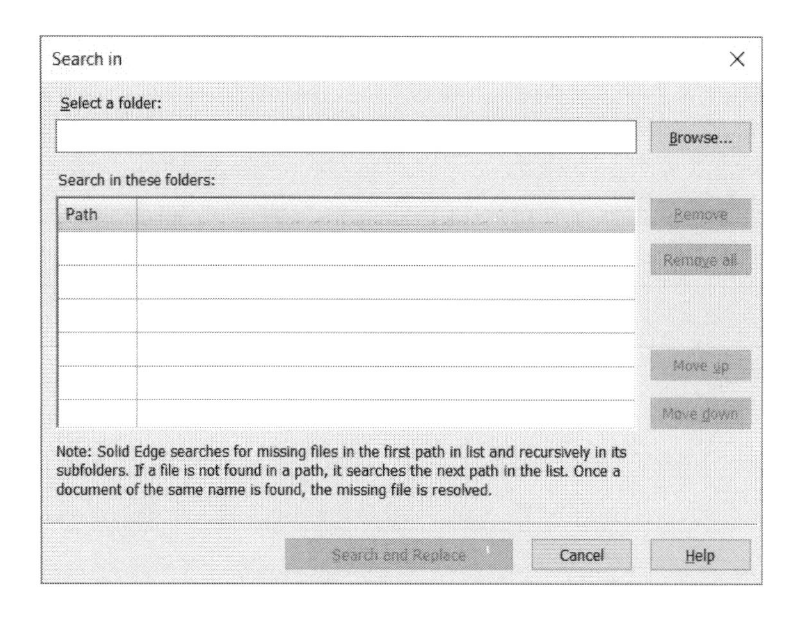

You can also replace the missing parts using the **Replace Part**, **Replace with Standard Parts**, and **Replace with New Part** options.

Pattern

The **Pattern** command allows you to replicate individual parts in an assembly. However, instead of defining layouts of rectangular or circular patterns, you can select an existing pattern as a reference. For example, in the assembly shown in figure, you can position one screw using relationships, and then use the **Pattern** command to place screws in the remaining holes.

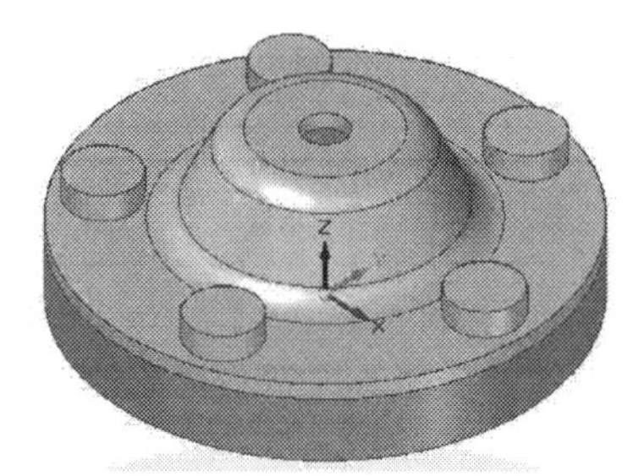

First, position the screw in one hole using the **Insert** relationship. Next, activate the **Pattern** command (click **Home > Pattern > Pattern** on the ribbon) and click on the part to include in the pattern. Click the green check on the command bar to accept the selection. Next, click on the part or sketch which contains the pattern.

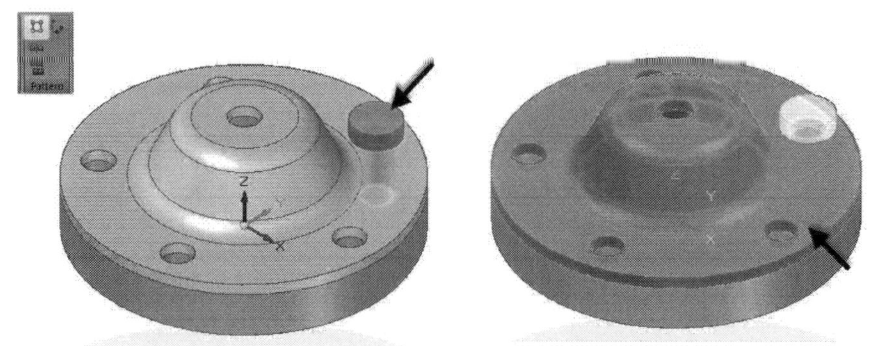

Select the pattern, as shown on the left side in the figure. Next, select the reference feature from the pattern, and then click **Finish** on the command bar to create the pattern.

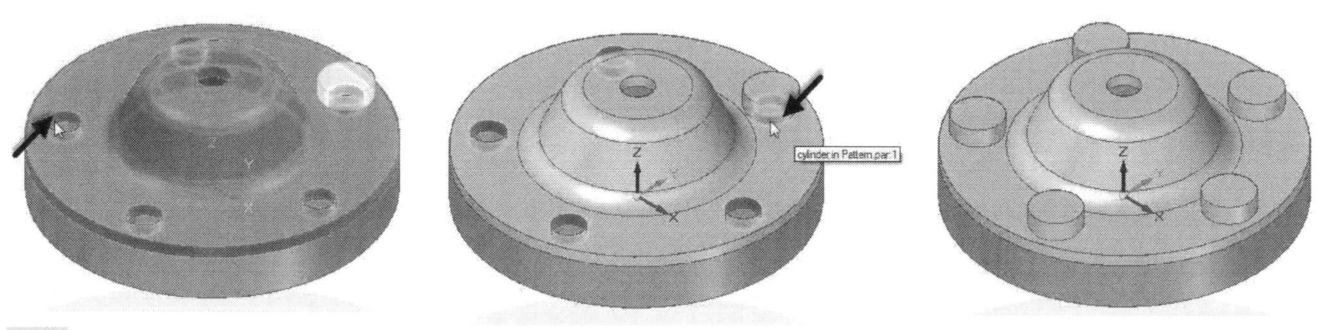

Clone Component

The **Clone Component** command allows you to clone a part or subassembly using the geometry recognition method. Activate this command (on the ribbon, click **Home** tab > **Pattern** panel > **Clone Component**), and then select the part or subassembly to be cloned. Click the **Accept** icon on the command bar. Next, select a reference face to be recognized to place the clones. Right click to accept the selection. Next, select the target component on which the clones are to be placed. Next, select the **Adaptable** or **Exact** button on the command bar. The **Adaptable** option places the clones on locations which are similar but not exactly same as the reference face. The **Exact** option places the clones on locations which are exactly same as the reference face. Click **Accept** on the command bar to accept the selected target component; the occurrence handles (pink dots) appear on the occurrences. Click on an occurrence handle, and then select the **Keep/Remove Occurrence** icon to remove an occurrence. Click the **Flip Orientation** icon, if you want to change the orientation of the occurrence. Next, click the **Finish** button to clone the components.

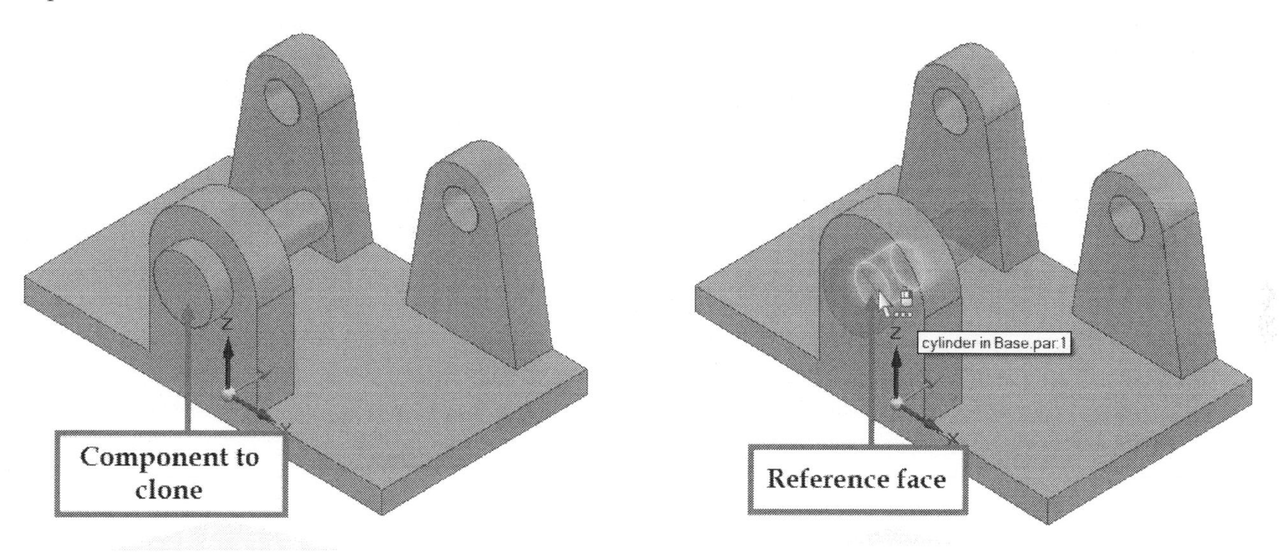

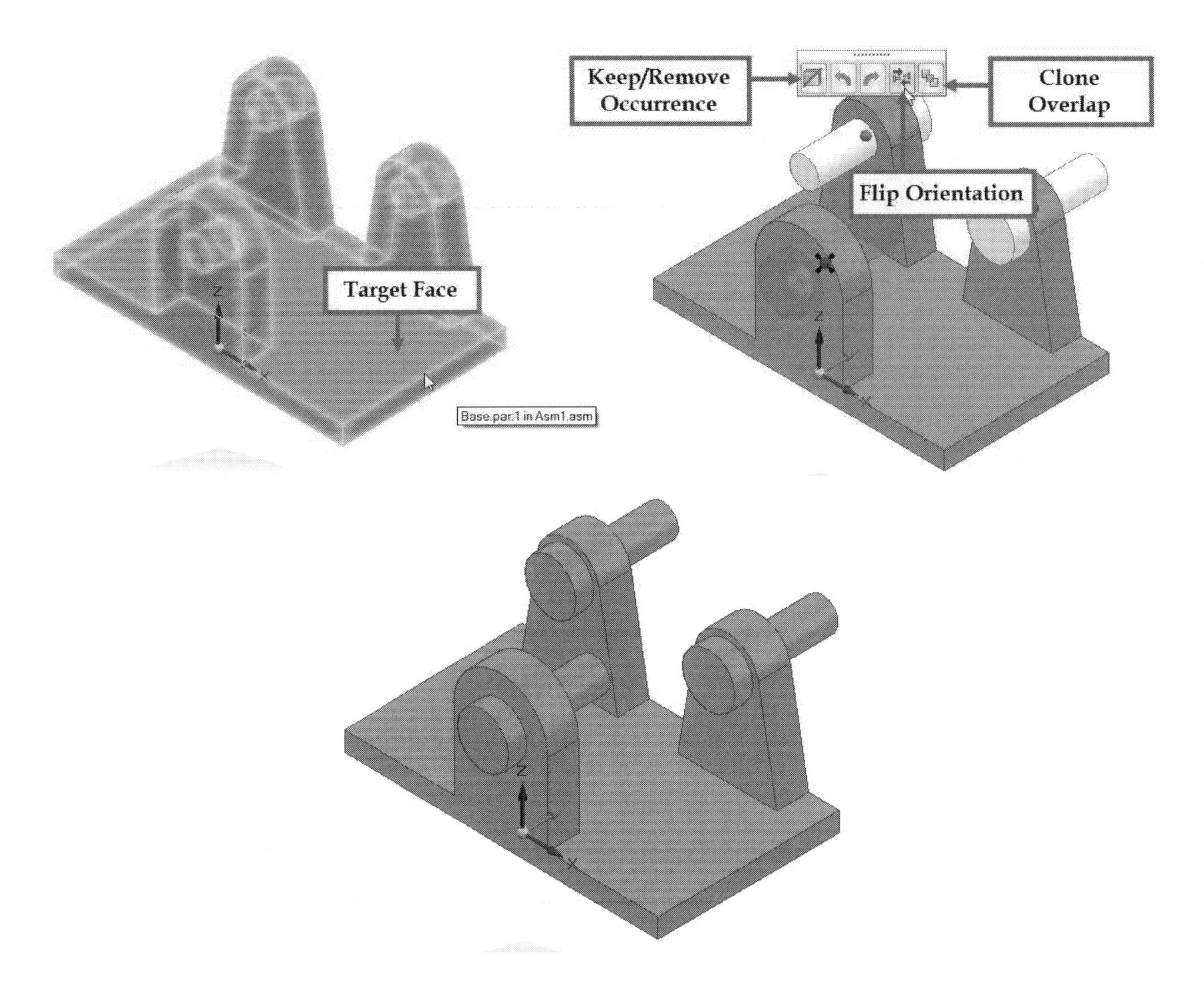

Mirror Components

When designing symmetric assemblies, the **Mirror Components** command will help you in saving time and capture the design intent. Activate this command (click **Home > Pattern > Mirror Components** on the ribbon) and click on the parts to be mirrored. Click the green check on the command bar, and then click on an assembly reference plane to mirror about; the **Mirror Components** dialog pops up on the screen. On this dialog, select the required action from the **Action** drop-down menu. Type-in the output file name in the **Output File** field and click **OK**. Next, click **Finish** to complete the mirroring.

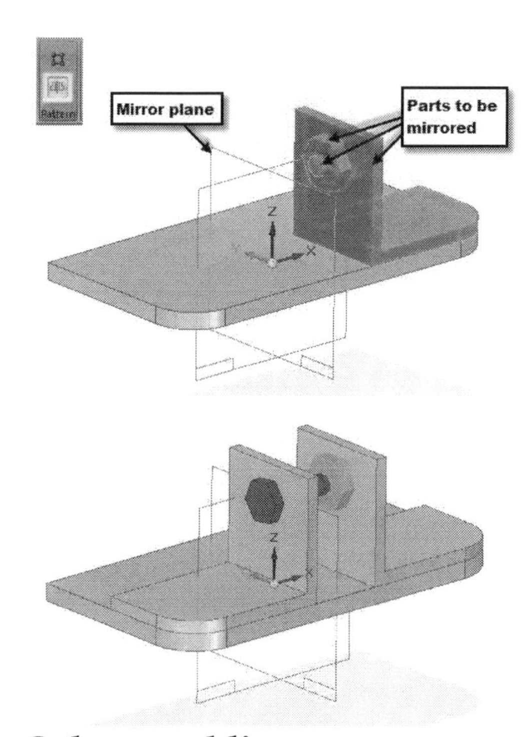

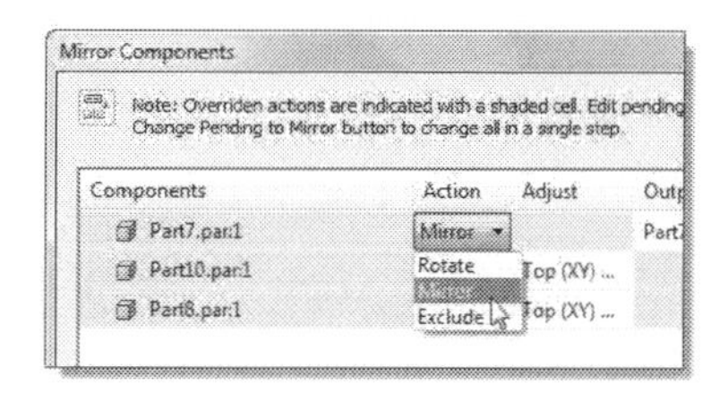

Sub-assemblies

The use of sub-assemblies has many advantages in Solid Edge. Sub-assemblies make large assemblies easier to manage. They make it easy for multiple users to collaborate on a single large assembly design. They can also affect the way you document a large assembly design in 2D drawings. For these reasons, it is important for you to create sub-assemblies in a variety of ways. The easiest way to create a sub-assembly is to insert an existing assembly into another assembly. You need to simply drag and place the assembly from the **Parts Library** window into an existing assembly. Next, apply relationships to constrain the assembly. The process of applying relationships is also simplified. You are required to apply relationships between only one part of a sub-assembly and a part of the main assembly. In addition, you can easily hide a group of parts with the help of sub-assemblies. Click the right mouse button on a sub-assembly and select **Hide**.

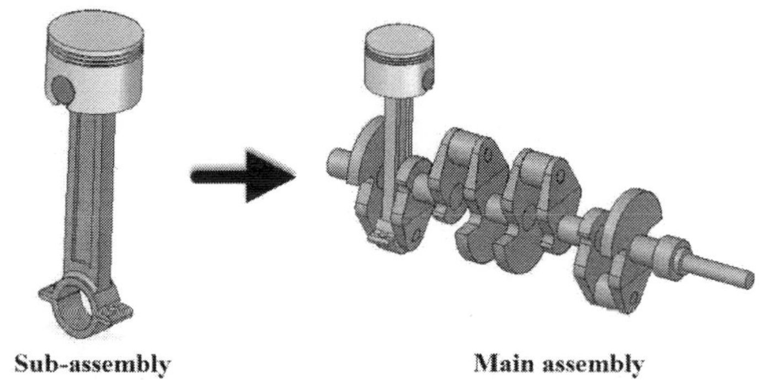

Sub-assembly **Main assembly**

Rigid and Adjustable Sub-Assemblies

By default, Solid Edge makes a sub-assembly as a rigid body. When you move a single part of a sub-assembly, the entire sub-assembly will be moved. If you want the individual parts of a sub-assembly to be moved, you must define the sub-assembly as adjustable. Click the right mouse button on the sub-assembly in the **Pathfinder** and select **Simplified/Adjustable > Adjustable Assembly**. Now, you can move the individual parts of a sub-assembly. In case, if you have multiple occurrences of a sub-assembly, each occurrence can be defined as rigid or adjustable. To help you recognise the difference between the rigid and adjustable assemblies, Solid Edge displays a different icon for each of them in the Pathfinder.

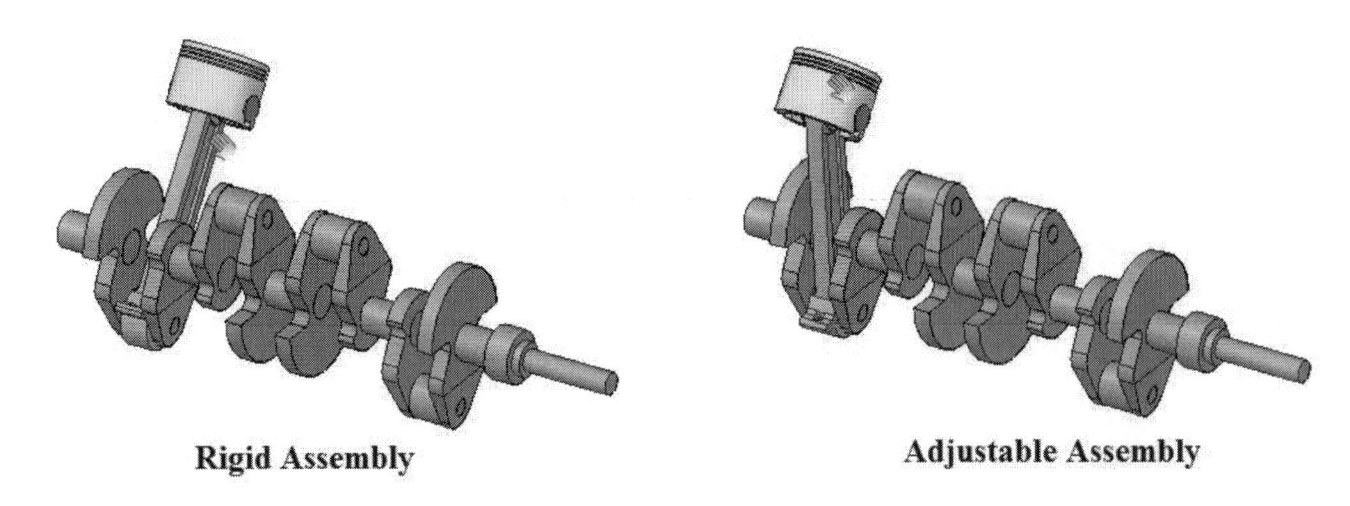

Rigid Assembly Adjustable Assembly

Transfer

In addition to creating sub-assemblies and inserting them into another assembly, you can also take individual parts that already exist in an assembly and make them into a sub-assembly. For example, press and hold the **Shift** key and select the four parts from the assembly. Next, activate the **Transfer** command (click **Home > Modify > Transfer** on the ribbon); the **Transfer to Assembly Level** dialog pops up on the screen.

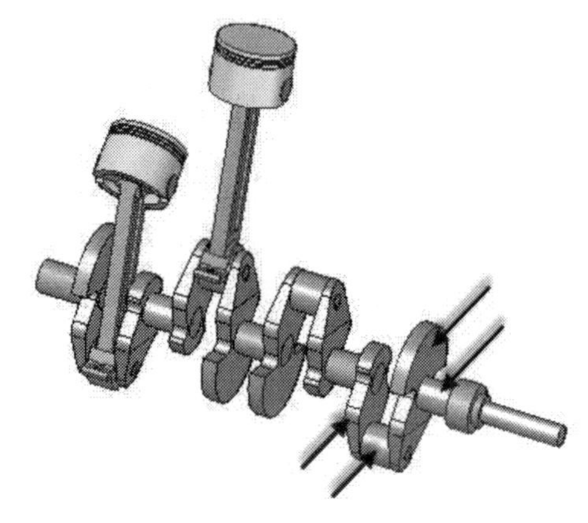

On this dialog, click on the **Assem** option, and then click **New Subassembly**; the **Create New Subassembly** dialog pops up on the screen. On this dialog, select the assembly template, enter file name, specify location, and specify positioning method. Click **OK** twice; the subassembly is created and listed in the Pathfinder.

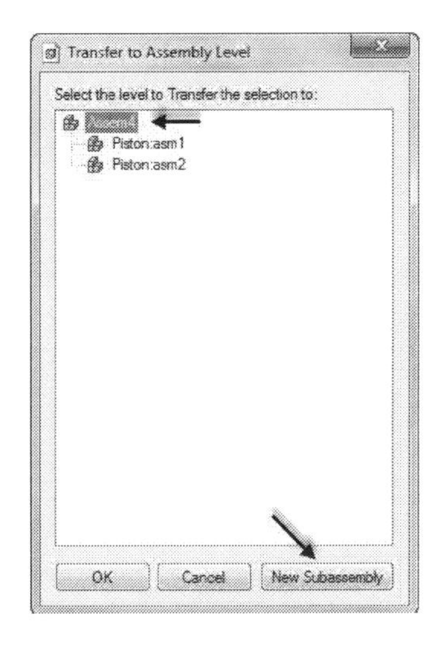

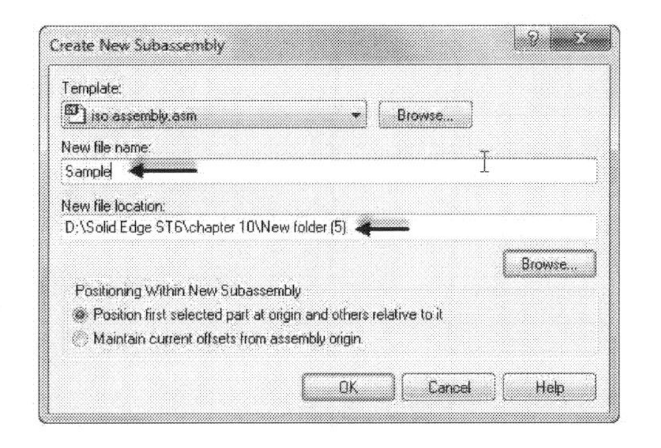

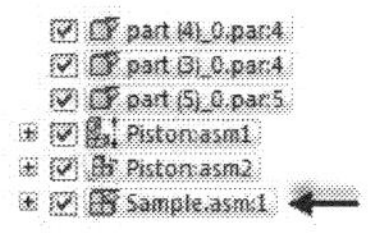

Disperse

After inserting subassemblies, you may require to disperse them into individual parts. Solid Edge provides you with the **Disperse** command to break a subassembly into individual parts. In the Pathfinder, click on the subassembly to disperse, and then activate this command (click **Home > Modify > Disperse** on the ribbon); the **Disperse Assembly** dialog pops up on the screen. On this dialog, click **Disperse Selected Assembly** to disperse the selected assembly (or) click **Disperse All Assemblies** to break down all subassemblies in to individual parts. After clicking the required option, a message box pops up showing, "Transfer the parts in the selected assembly to the next higher level, and delete the selected assembly occurrence". Click **Yes** to transfer the parts to the main assembly.

Assembly Features

Assembly features are the features that exist only in assemblies i.e. instead of creating them at the part level, they are created at the assembly level. Most often the features created at the assembly level are cuts, revolved cuts, holes, and welds. These features are commonly created at the assembly level to represent post assembly machining. For example, to add a cut feature to the assembly shown in figure, activate the **Cut** command (click **Features > Assembly Features > Cut** on the ribbon); the **Assembly Feature Options** dialog pops up on the screen. On this dialog, select **Create Assembly Features** and click **OK**. Select the top face of the block and draw the sketch of the cut feature. Finish the sketch and extrude it using the **Through All** option.

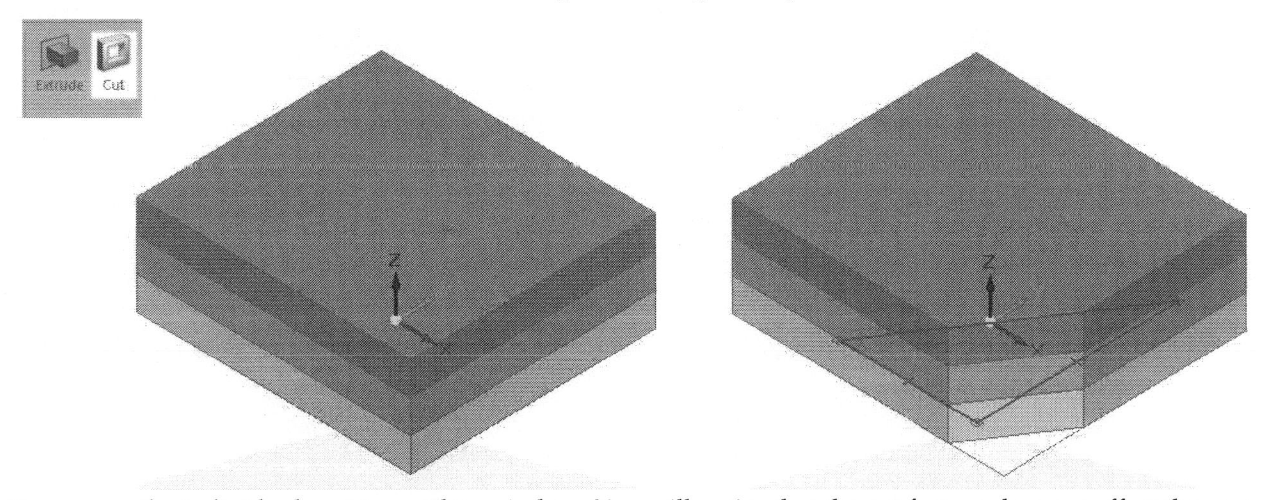

Now, open the individual part in another window. You will notice that the cut feature does not affect the part.

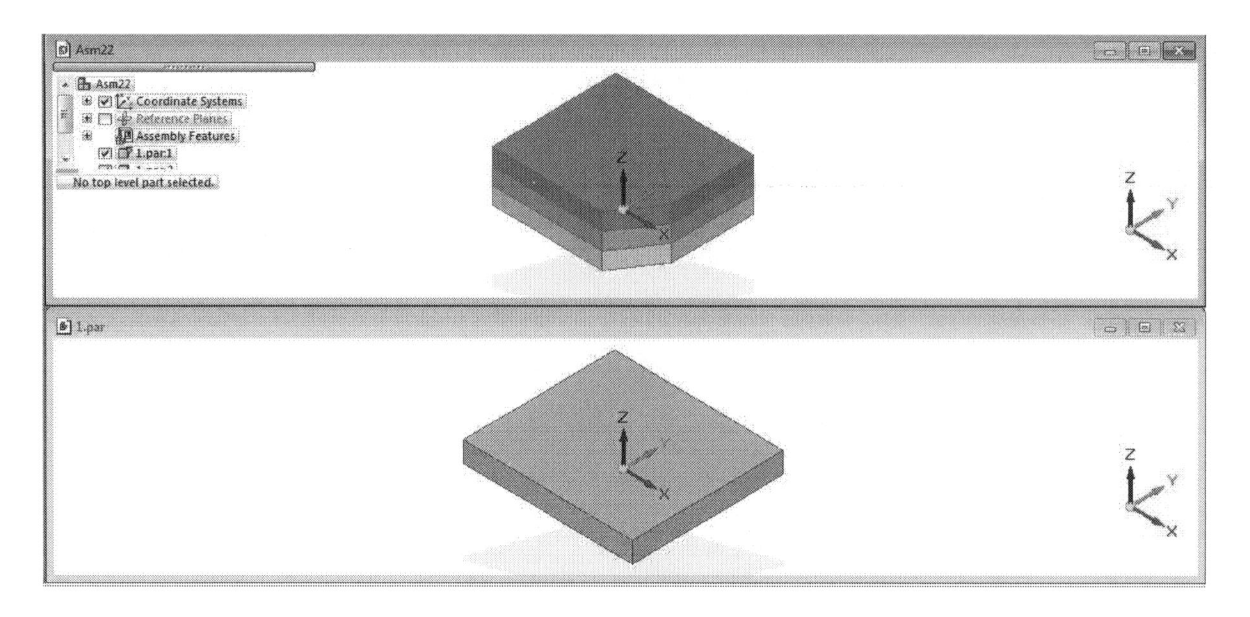

You will also notice that the Cut feature is added to the Pathfinder. You can edit the cut feature by clicking the right mouse button on the **Cut** feature and selecting **Edit Definition**. A command bar pops up on the screen. On the command bar, click the **Edit - Select Parts Step** icon, and then press the Shift key and select the parts to be excluded from the cut feature. Click the green check on the command bar, and then click **Finish**; you can see that the cut feature no longer affects the selected components.

If you add a new part to the assembly, the cut feature will not affect it. Again, you need to edit the cut feature and use the **Edit - Select Part Step** icon to include the part in the cut feature.

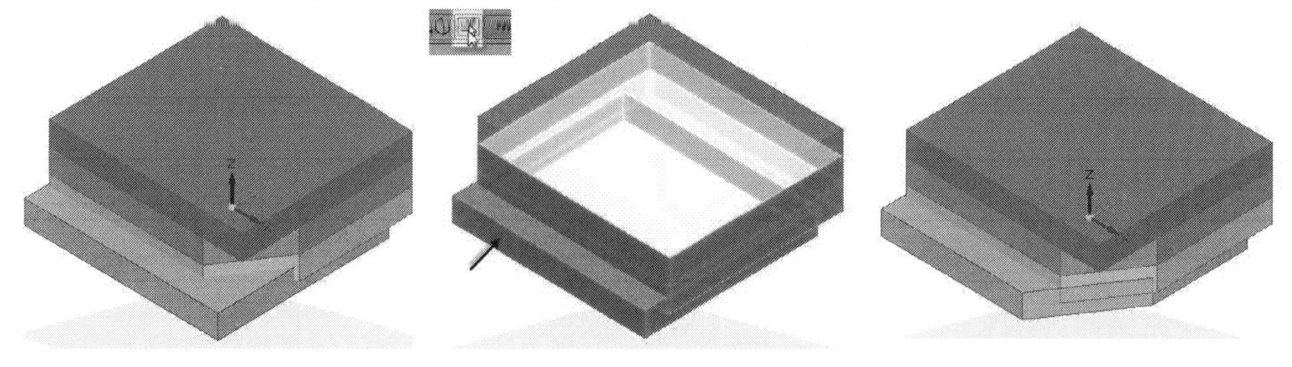

Assembly-Driven Part Features

Assembly-Driven Part features are features, which are created in an assembly and are also reflected in the part documents. To create this type of feature, first save the assembly file, and then activate anyone of the commands available in the **Assembly Features** panel. Next click **Create Assembly-Driven Part Features** on the **Assembly Options** dialog and click **OK**. Create the assembly driven feature and it is listed in the Pathfinder. Now, open the individual part in another window. You will notice that the feature also affects the part. In addition, the feature is listed in the **Ordered** environment of the part file. When you update the feature in the assembly file, it will be reflected in the part file as well.

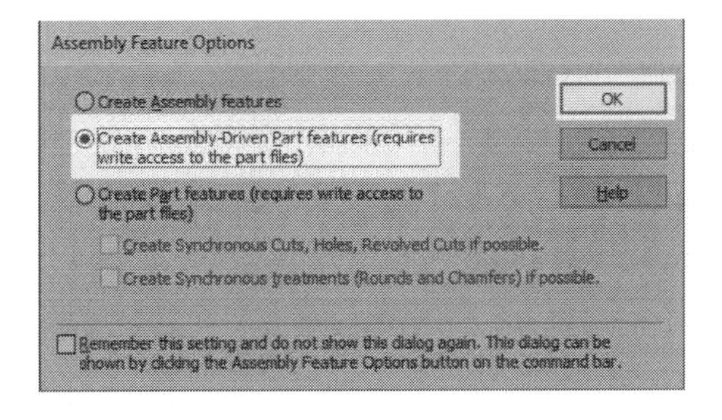

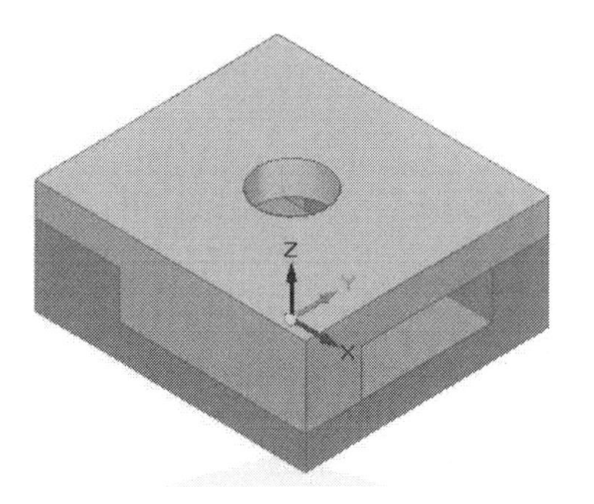

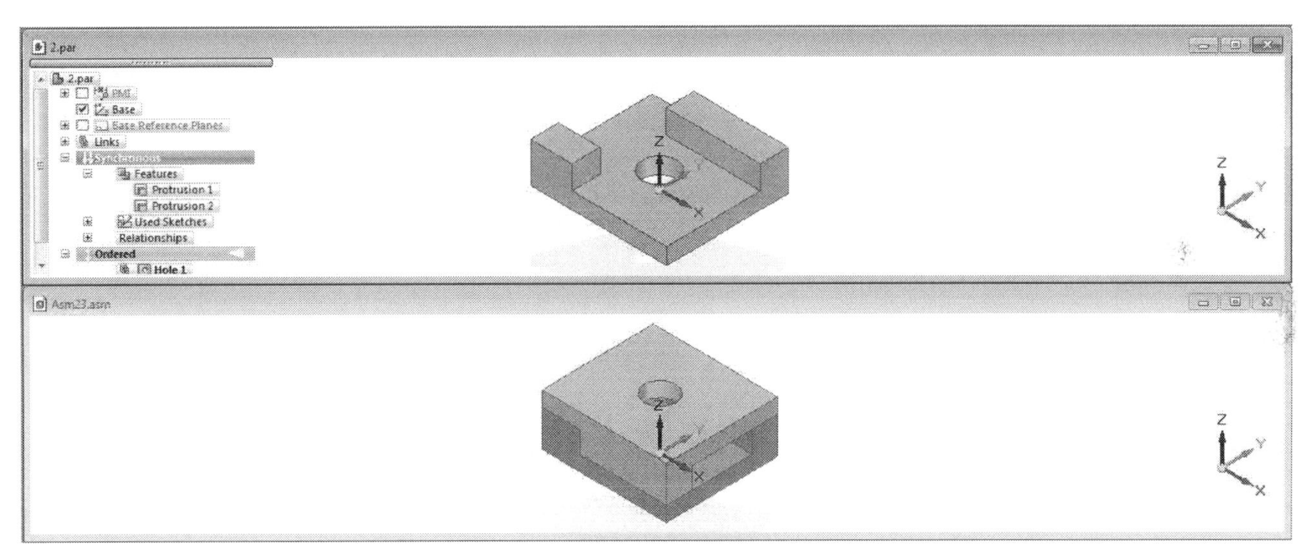

Part Features

Part features are features created in an assembly but are not associated to the assembly. Instead, they are associated to the part file on which they are created. To create a part feature, activate anyone of the commands available in the **Assembly Features** panel and click **Create Part features** on the **Assembly Feature Options** dialog. Click **OK** and create the part feature. Now, open the individual part in another window. You will notice that the feature also affects the part. In addition, the feature is listed in the **Synchronous** environment. If you want to edit a part feature, you must open the part file and make changes to it.

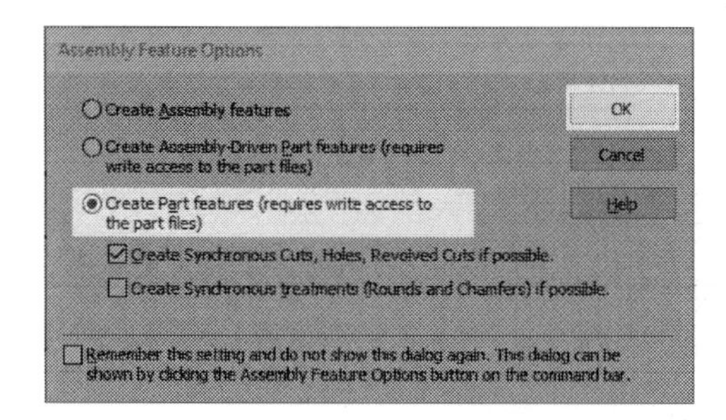

Top Down Assembly Design

In Solid Edge, there are two methods to create an assembly. The method you are probably familiar with is to create individual parts, and then insert them into an assembly. This method is known as Bottom-Up Assembly Design. The second method is called Top Down Assembly Design. In this method, you will create individual parts within the assembly environment. This allows you to design an individual part while taking into account how it will interact with other parts in an assembly. There are several advantages in Top-Down Assembly Design. As you design a part within the assembly, you can be sure that it will fit properly. You can also use reference geometry from the other parts.

Create Part In-Place

Top-down assembly design can be used to add new parts to an already existing assembly. You can also use it to create assemblies that are entirely new. To create a part using the Top Down Design approach, first you must save the assembly file, and then activate the **Create Part In-Place** command (click **Home > Assemble > Create Part In-Place** on the ribbon); the **Create Part In-Place Options** dialog pops up on the screen. The options available on this dialog are self-explanatory. Set the options on this dialog and click **OK** to close it.

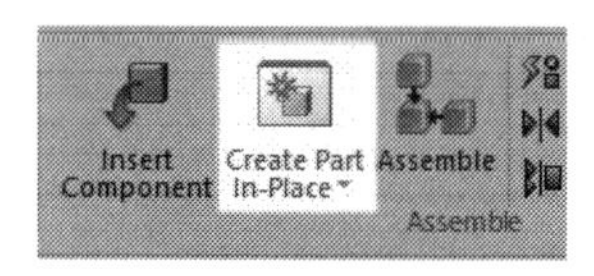

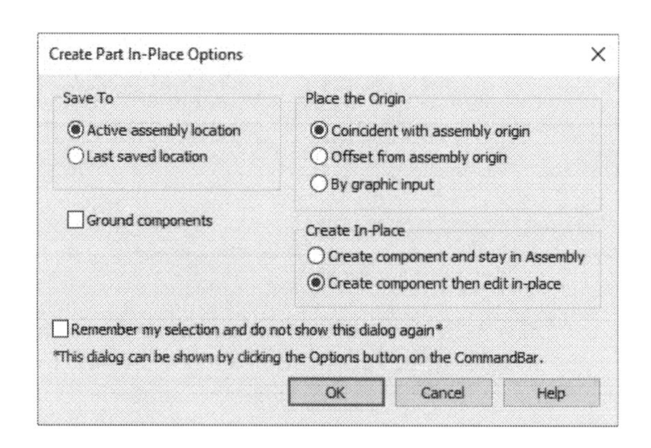

On the command bar, select a part template from the **Template** drop-down menu. If you want to access more templates, click the **Browse for Template** icon to display the **New** dialog. On this dialog, select the template you need and click **OK**.

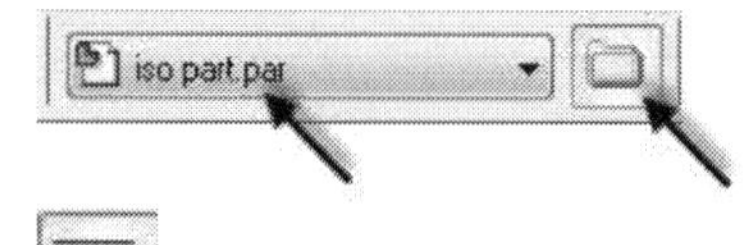

On the command bar, click the **Ground** icon, if you want make the part grounded at the origin.

Use the **Origin** drop-down menu to specify the origin location.

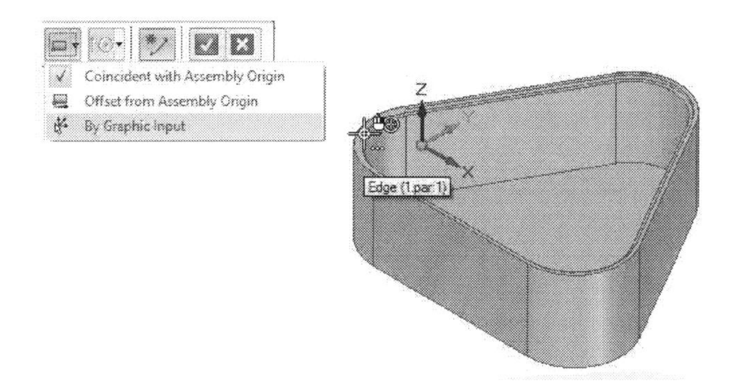

Activate the **Edit In Place** icon to directly switch to the part environment to create the part. Click the green check on the command bar; the **Save As** dialog pops up on the screen. On this dialog, specify the part name and location on the drive, and then click **Save** to create the part. Now, create the features of the part, and then close and return to the assembly. In the example given in figure, the **Project to Sketch** command is used to project the edges of the existing part to create a sketch. The projected sketch is then extruded. This makes it easy to create a part using the edges of the existing part.

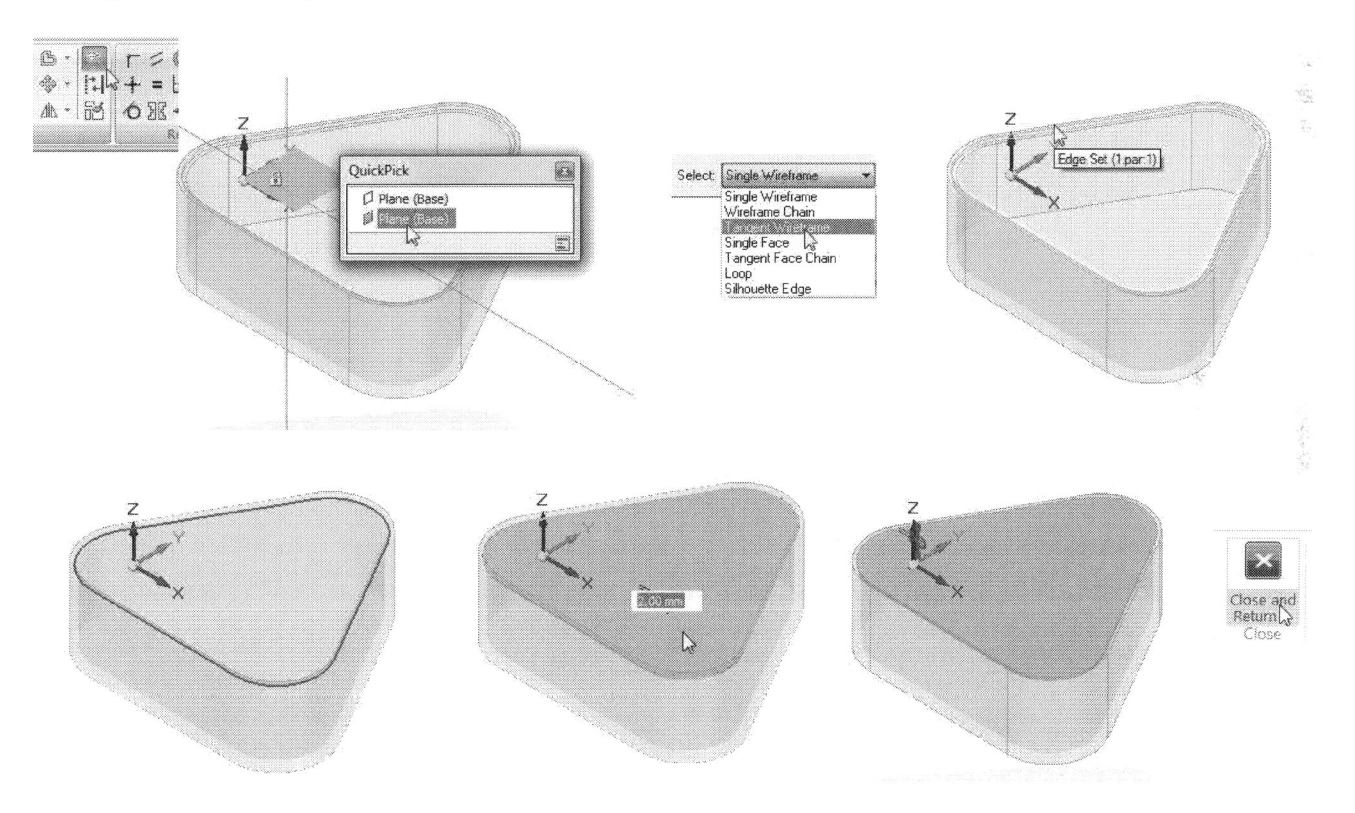

Tip: *You can use the* ***Thicken*** *command create the part features by thickening the faces of the adjacent parts. You can learn about the* ***Thicken*** *command in Chapter 13: Surface Design.*

Assembly Relationship Assistant

Assembly Relationship Assistant is the command provided by Solid Edge intelligent technology. This command helps you to create relationships automatically based on the position and interaction between the parts. This command is very helpful while creating relationships in a top down assembly design. Activate this command (click **Home > Assemble > Assembly Relationship Assistant** on the ribbon); the **Relationship Assistant Options** dialog pops up on the screen. Set the options on this dialog and click **OK**. Click on parts to define the first set, and then click the green check on the command bar. Click on parts to define the second set, and then click the green check on the command bar; the **Relationship Assistant Settings** dialog pops up on the screen. On this dialog, check the **Allowable Relationship Types**, and then click **Process**. This will analyze the position and interaction between the parts, and then apply relationships between them. The possible relationships are listed on the **Relationship Assistant Settings** dialog. Check the required relationships, and click **Accept**. Close the dialog and click **Finish** to apply the relationships.

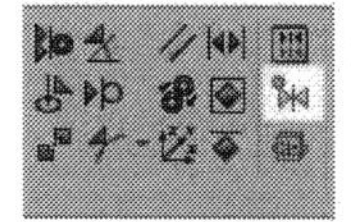

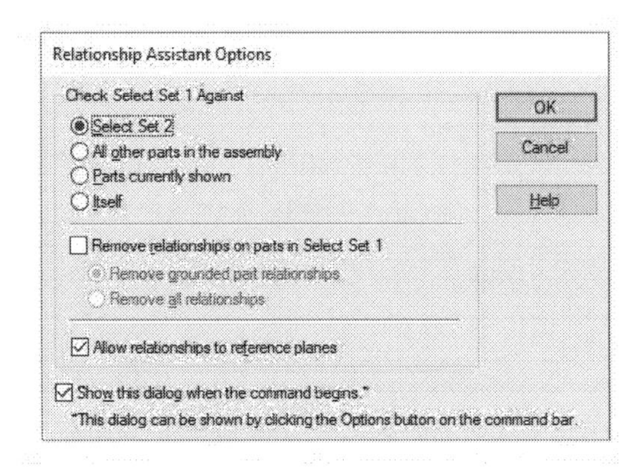

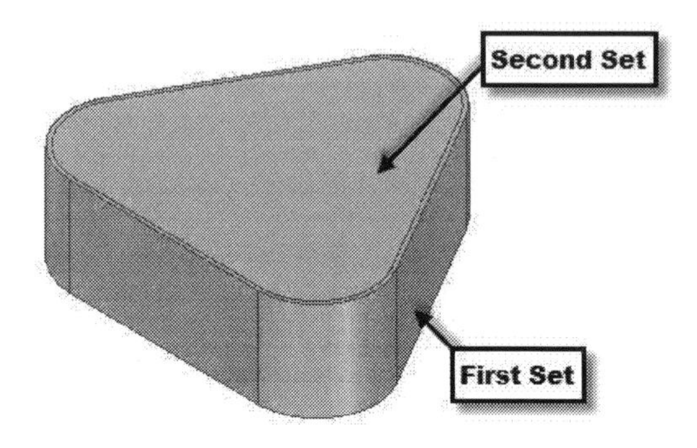

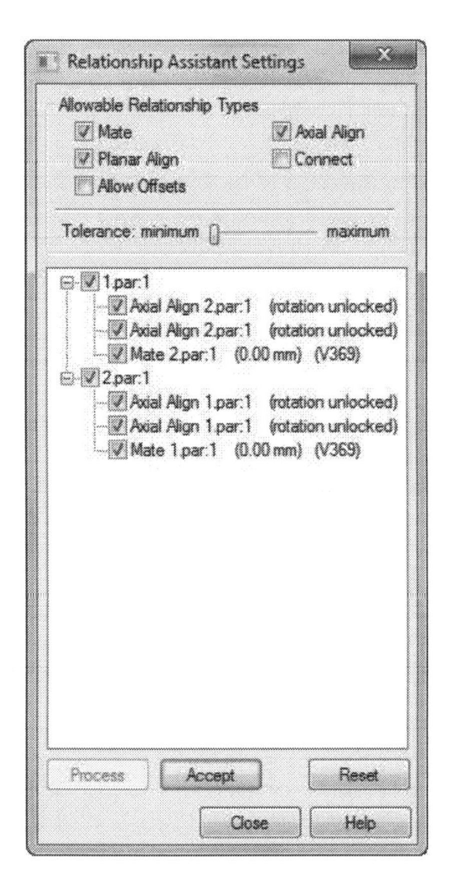

Assembly Relationships Manager

The **Assembly Relationships Manager** command (on the ribbon, click **Home** tab > **Assemble** group > **Assembly Relationships Manager**) displays a table of relationships created in an assembly. You can view and modify the assembly relationships from the table. All the assembly relationships are grouped into different categories. For example, the all the Axial relationships are grouped into the **Axial** category. Expand the **Axial** category to view all the **Axial** relationships in the assembly. You can further expand each relationship to view its status and components associated with it. The **Assembly Relationship Manager** dialog provides you with many options when you right click on a relationship or component associated with a relationship. These options help you to solve the relationship error or edit the component. To do this, right click on a relationship and select **Supress**, **Unsupress**, **Flip**, **Edit Definition,** or **Delete** from the shortcut menu. Close the dialog after making changes to the relationships.

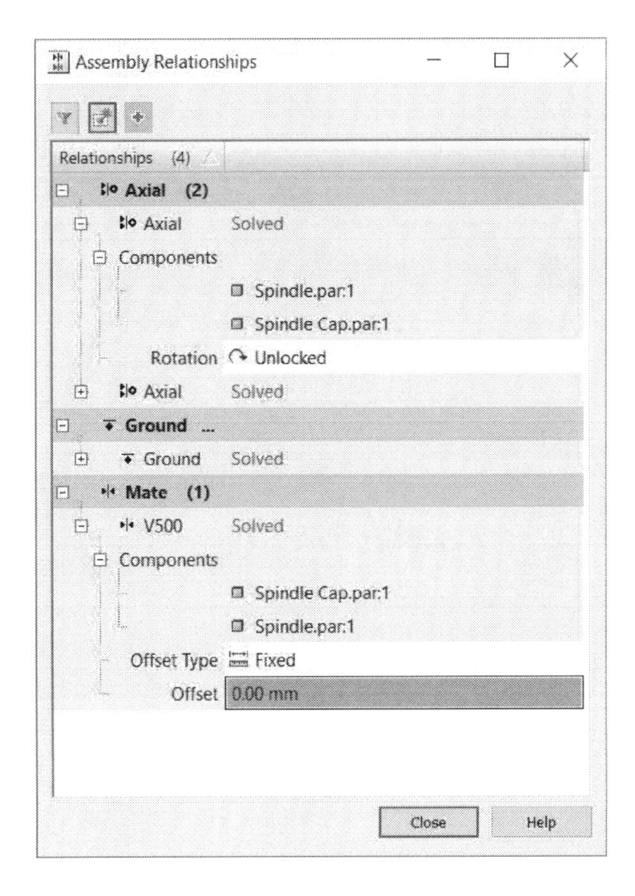

Exploding Assemblies

To document an assembly design properly, it is very common to create an exploded view. In an exploded view, the parts of an assembly are pulled apart to show how they were assembled. To create an exploded view, activate the **ERA** command (click **Tools > Environs > ERA** on the ribbon); the **Explode - Render - Animate** environment is activated.

Use the **Auto Explode** command to explode the assembly, automatically. On activating this command, the **Auto Explode** command bar pops up on the screen. On the command bar, select **Top-level Assembly** if you want to explode the complete assembly. If you want to explode only selected subassemblies, then select the **Subassembly** option. Next, click the green check on the command bar. Deactivate the **Automatic Spread Distance** icon on command bar and type-in the spread distance in the **Distance** box. Click **Explode** and **Finish** to explode the assembly.

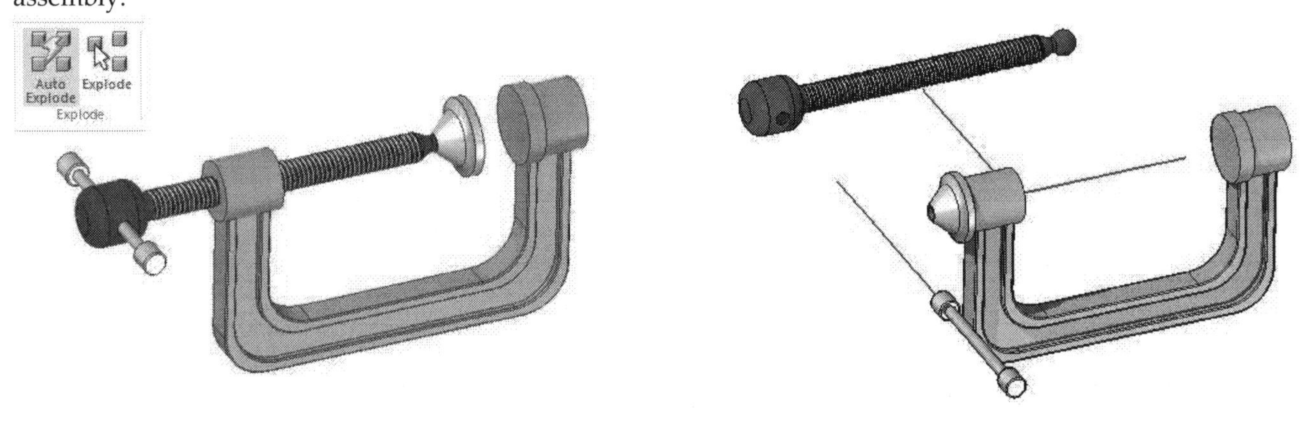

You will notice that the parts are not exploded properly. To get a desired explosion, you need to use the **Explode** command. First, unexplode this assembly using the **Unexplode** command. On clicking this button, the **Solid Edge** message box pops up showing, "This action will delete the current explosion". Click **Yes** to explode the assembly.

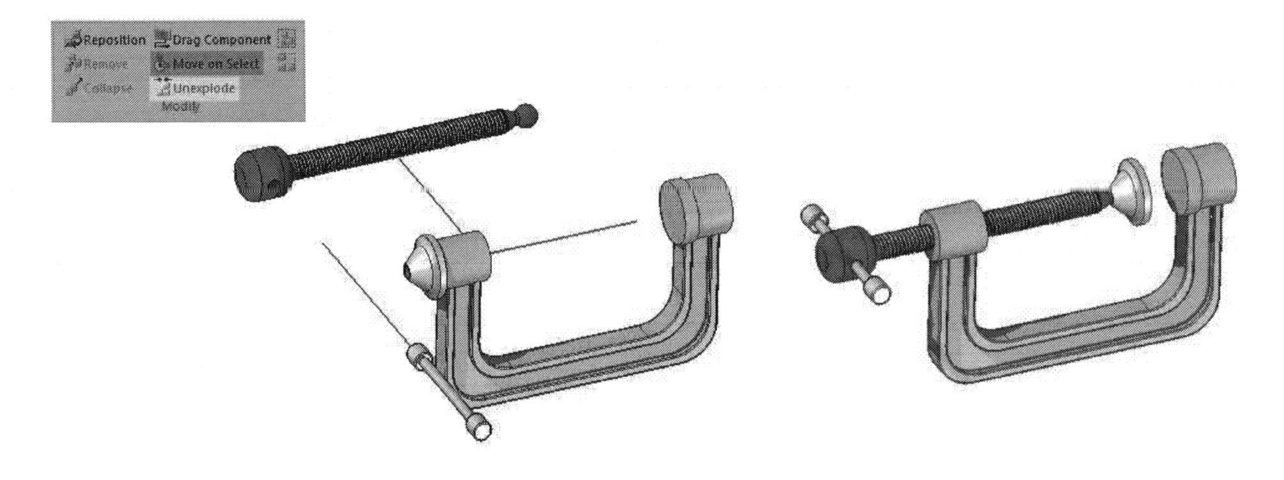

To manually explode an assembly, activate the **Explode** command; the **Explode** command bar pops up. Click on the parts to be exploded, and then click the green check on the command bar. Click on the part to be remained stationary in the explosion.

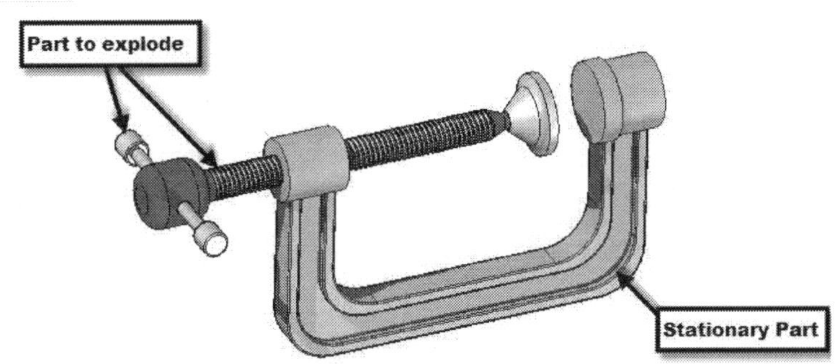

Click on the stationary part face from which you want to explode; an arrow appears on it. Move the pointer and click to define the direction of explosion; the **Explode Options** dialog pops up.

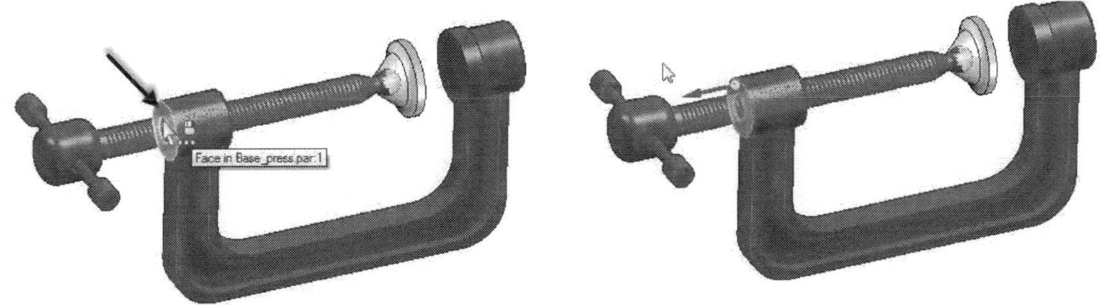

On this dialog, select an option to specify the **Explode Technique**. You can select **Move components as a unit** or **Spread components evenly**. Next, specify the **Explode order** by selecting the parts listed and using the **Move Up** and **Move Down** buttons. Note that you can specify the explode order only when you select the **Spread components evenly** option. Click **OK** to close this dialog.

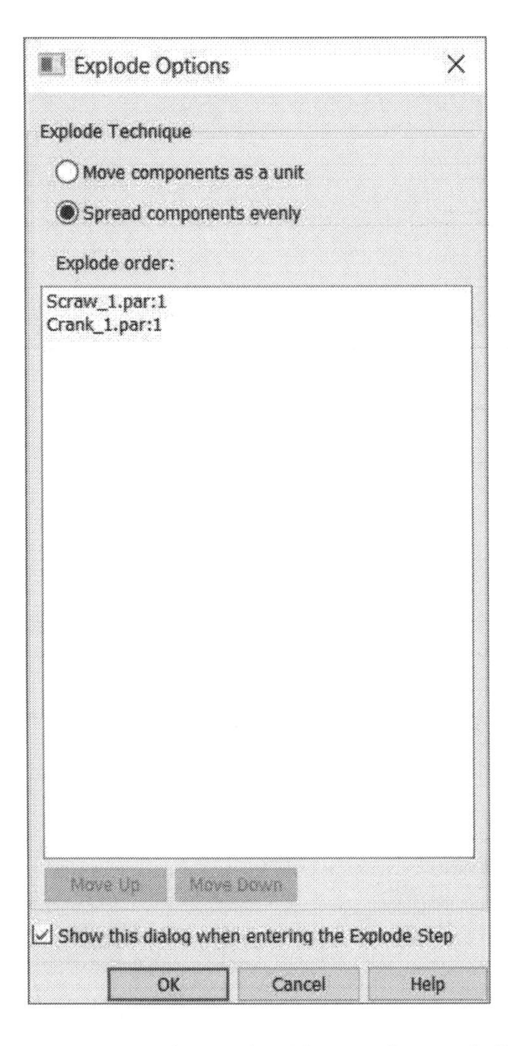

Type-in a value in the **Distance** box and click **Explode** to explode the parts. Click **Finish** to complete the explosion.

If the distance between the exploded parts is less or more, you can use the **Drag Component** command to adjust the spacing between them.

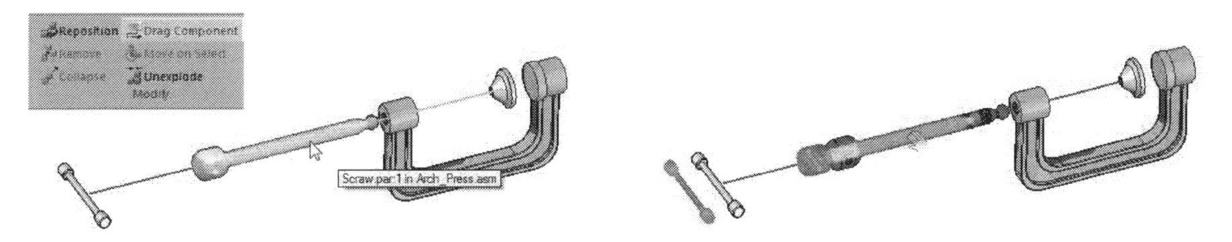

You can also select individual parts and type-in a new explode distance in the **Distance** box.

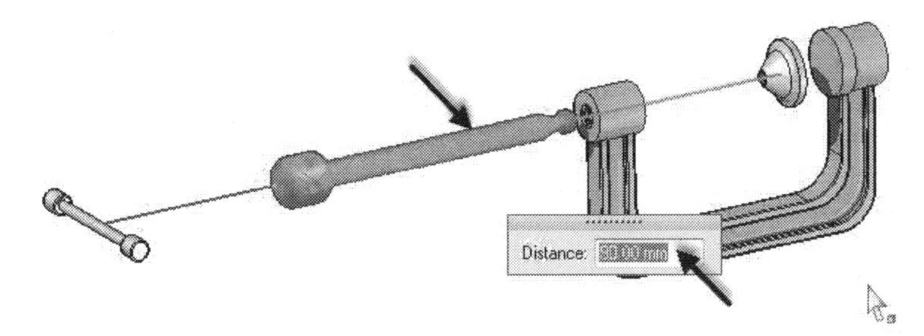

If you want to reorder an exploded component, click **Reposition** on the **Modify** panel and select the component to reorder. Select the component next to it. Click to define the side in which the component will be repositioned.

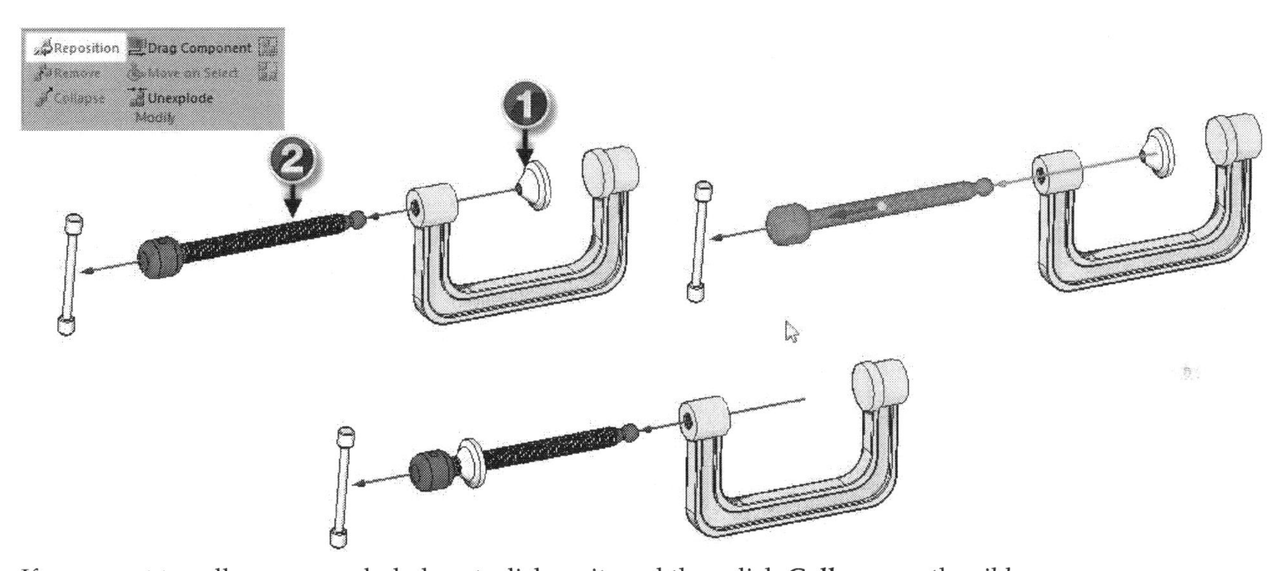

If you want to collapse an exploded part, click on it, and then click **Collapse** on the ribbon.

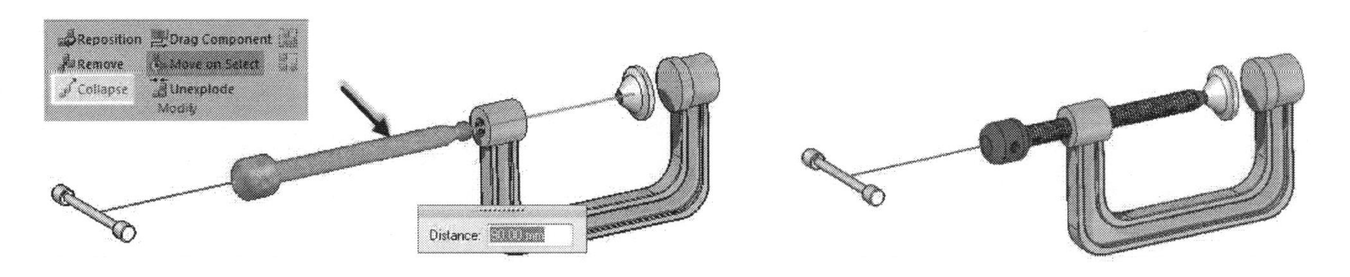

If you want to remove a part from the explosion, click on it, and then click **Remove** on the ribbon.

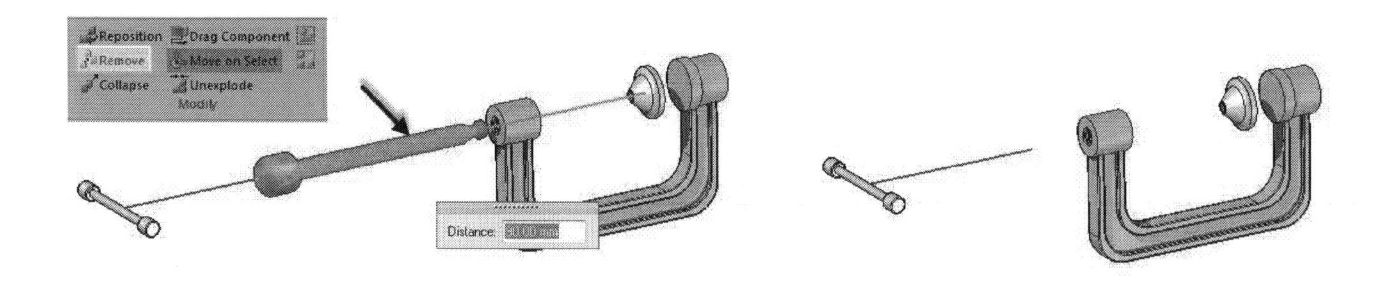

If you want to convert the explosion into a 'move component' operation, then click **Drop** on the **Flow Lines** panel. The flow lines are also converted into annotation flow lines.

If you want to modify a flow line, click **Modify** on the **Flow Lines** panel and select the flow line; two handles appear at start and end points of the flow line. Click on the start point handle, and then redefine the start point of the flow line. Similarly, redefine the end point of the flow line.

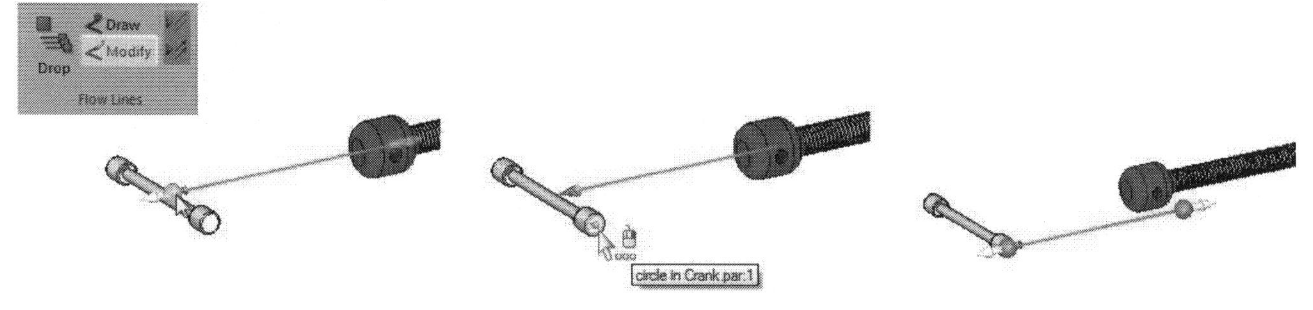

If you want to draw a new flow line, click **Draw** on the **Flow Lines** panel and select the start and end points; a flow line appears between the selected points. On the **Draw** command bar, click the **Next** icon to see different paths of the flow lines. Click **Finish** to complete the flow line creation.

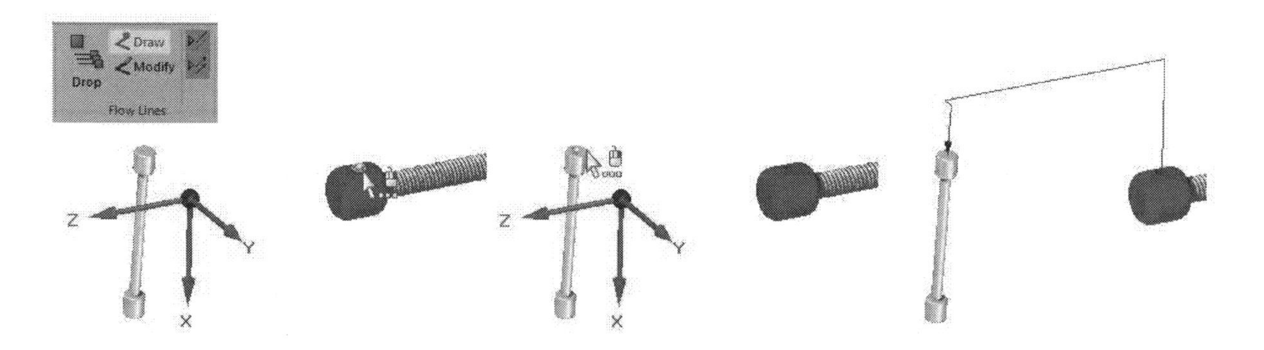

After exploding the assembly, click the **Save** icon on the **Quick Access Toolbar**, specify the location of the assembly file, and click **OK**. The **explode, Solid Edge** configuration is created. You can use this configuration to display the exploded view. Now, click **Close ERA** on the ribbon; the assembly environment appears.

Examples

Example 1 (Bottom Up Assembly)

In this example, you will create the assembly shown below.

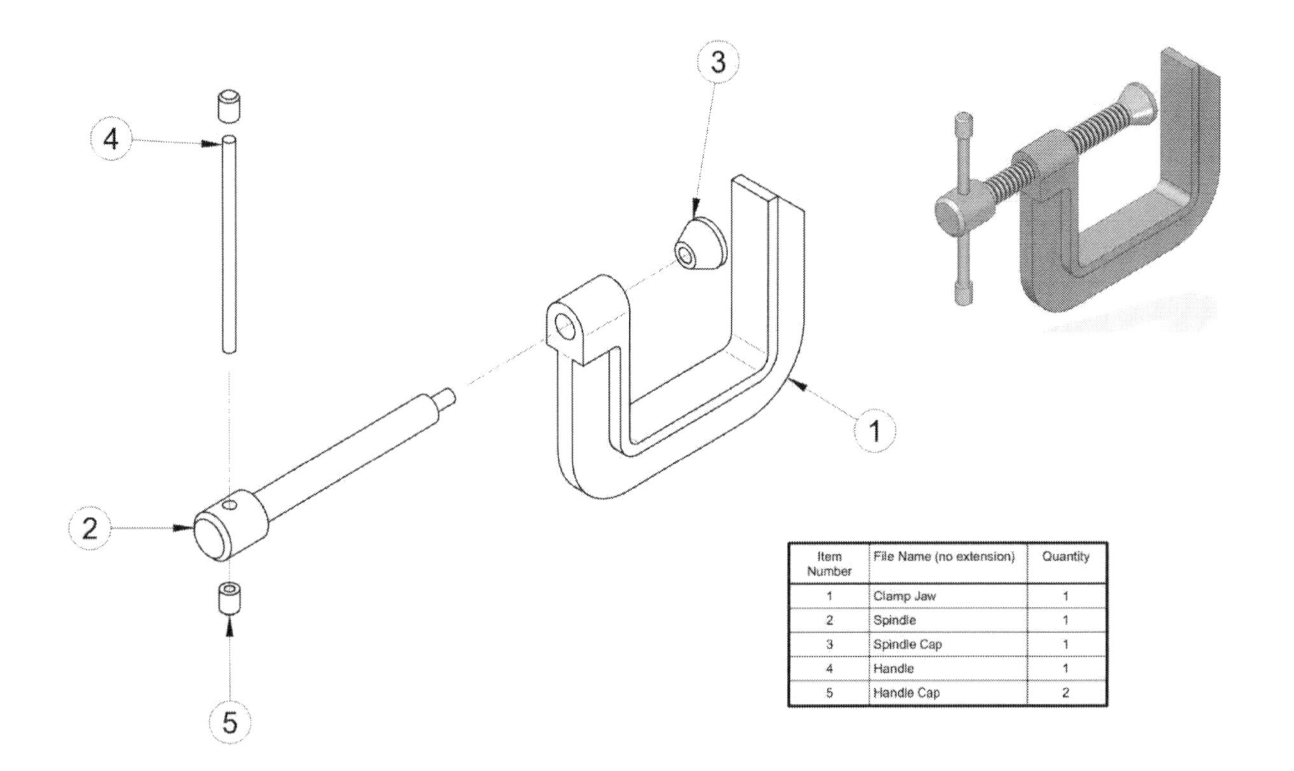

Item Number	File Name (no extension)	Quantity
1	Clamp Jaw	1
2	Spindle	1
3	Spindle Cap	1
4	Handle	1
5	Handle Cap	2

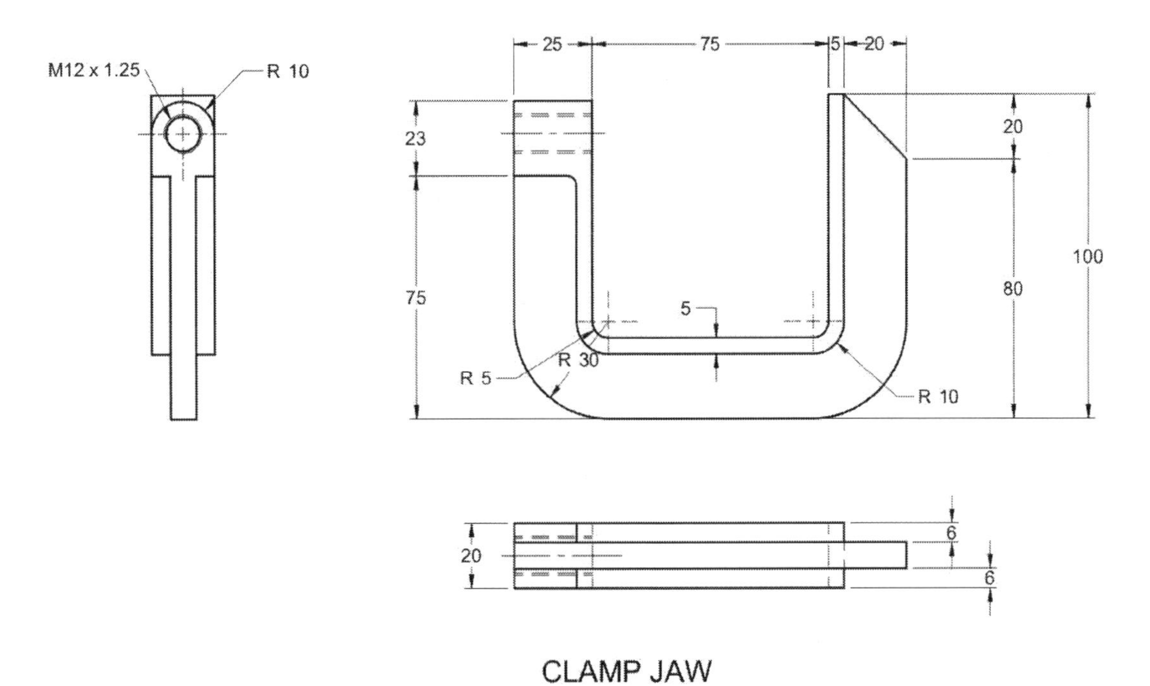

CLAMP JAW

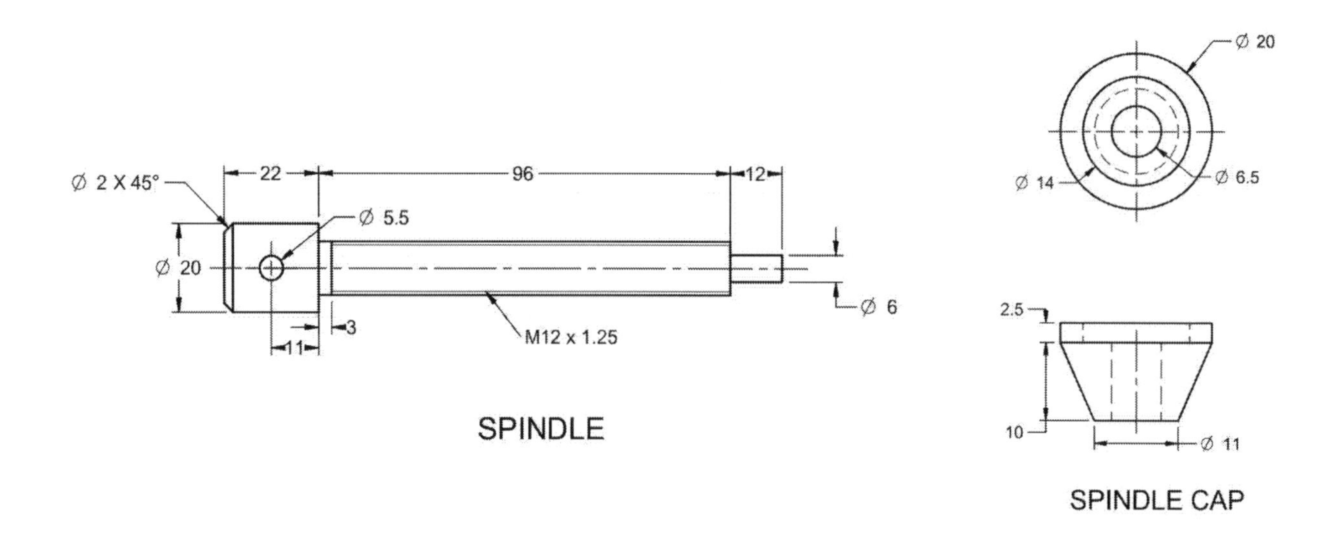

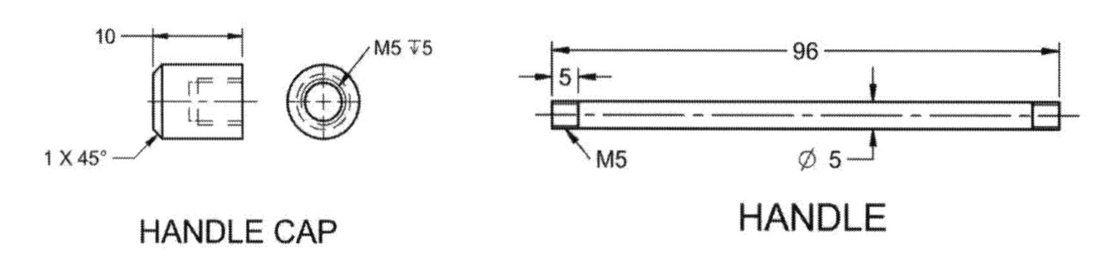

1. Start **Solid Edge 2019**.
2. Create and save all the parts of the assembly in a single folder. Name this folder as *G-Clamp*.
3. On the **Application Menu**, click **New > ISO Metric Assembly** to start an assembly file.
4. Click the **Insert Component** button on the ribbon to display the **Parts Library** window.
5. In the **Parts Library** window, use the drop-down menu and go to the *G-Clamp* folder.
6. In the **Parts Library** window, click *Clamp Jaw* and drag it into the assembly window.

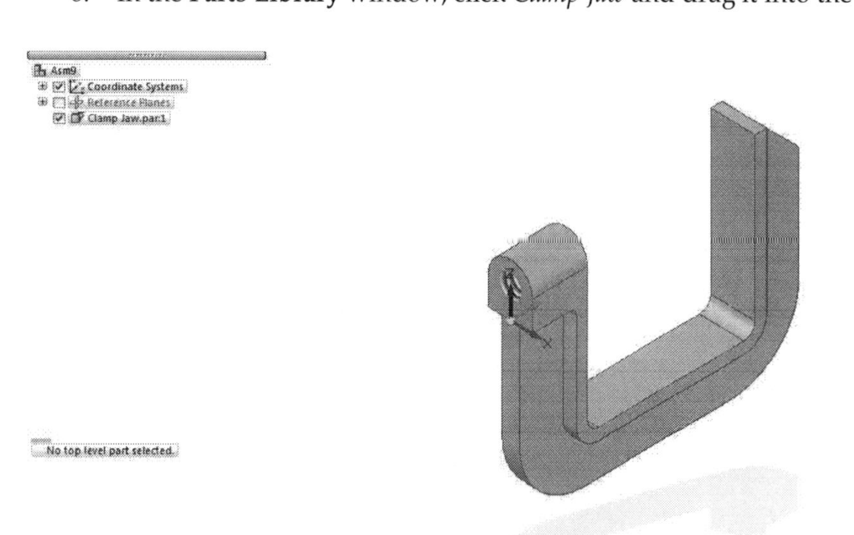

7. In the **Parts Library** window, click *Spindle* and drag it into the assembly window.

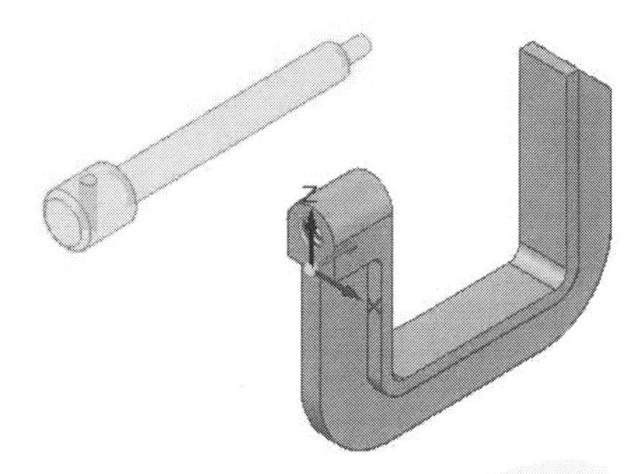

8. On the command bar, click **Relationship Types > Axial Align,** and then click the **Lock Rotation** icon.
9. Click on the cylindrical face of the *Spindle* and hole of the *Clamp Jaw*.

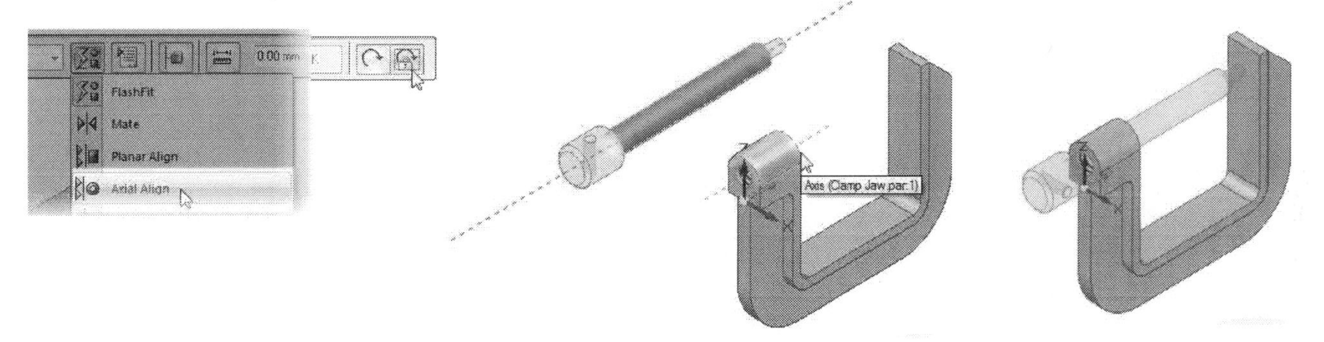

10. On the command bar, click **Relationship Types > Planar Align**, and then type **-40** in the **Offset Value** box.
11. Click on the back face of the *Spindle* and rotate the view.
12. Click on the flat face of the *Clamp Jaw*, as shown in figure.

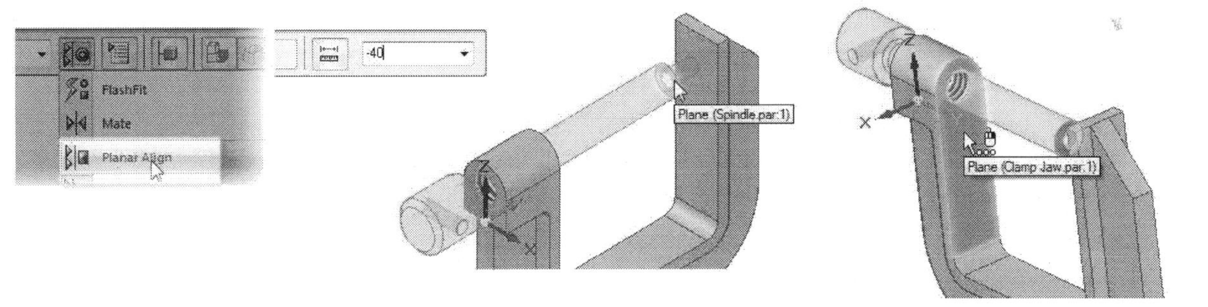

13. In the **Parts Library**, click *Spindle Cap* and drag it into the assembly window.
14. On the command bar, click the **Lock Rotation** icon, and then click on the cylindrical face of the *Spindle Cap* hole.
15. Click on the small cylindrical face of the *Spindle*. The *Spindle* and *Spindle Cap* are axially aligned.

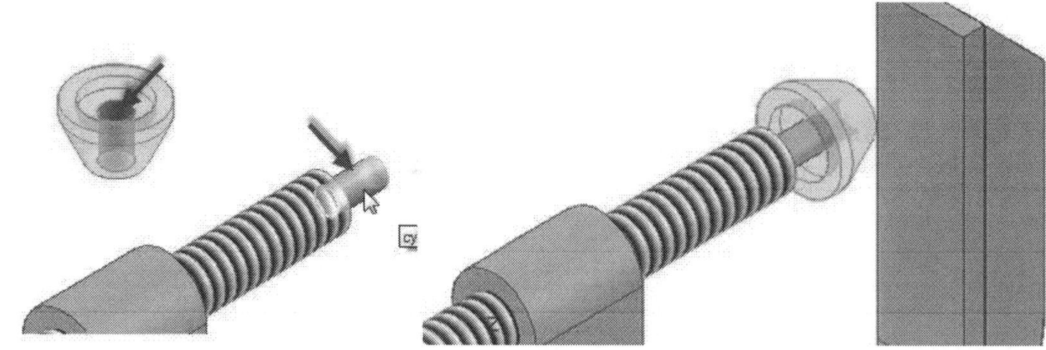

16. Rotate the model view and click on the flat face of the *Spindle Cap*, as shown in figure.
17. Click on the flat face of the *Spindle*. The *Spindle Cap* is assembled and fully constrained. However, you will notice that the part is oriented in reverse direction.

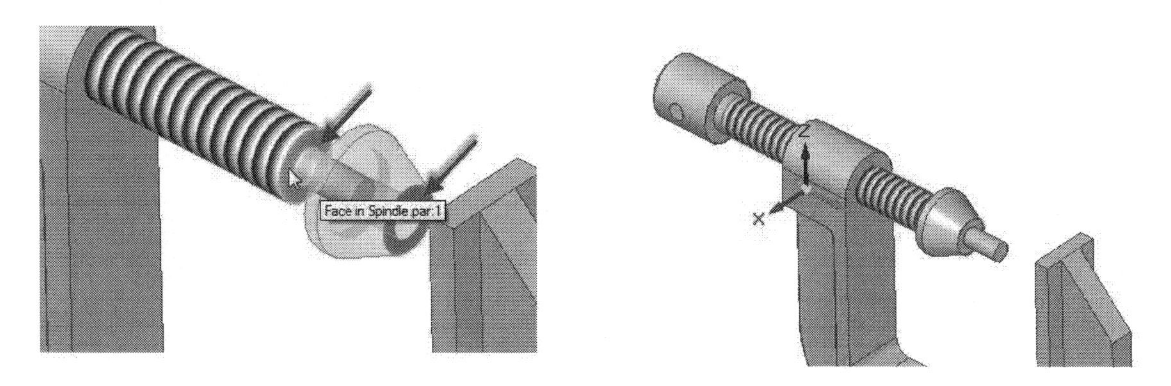

Note: *Skip step 18 and 19, if the Spindle Cap is oriented properly.*

18. In the **Pathfinder**, click *Spindle Cap*. The relations associated with the part appear at the bottom of the **Pathfinder**.
19. Click on the planar align relation, and then click the **Flip** button at the bottom of the screen. The *Spindle Cap* is reversed.

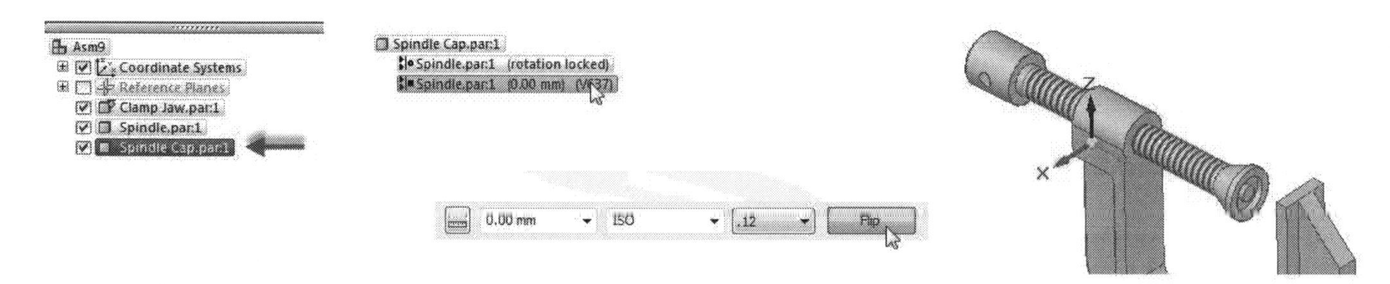

20. In **Parts Library**, click *Handle* and drag it into the assembly window.
21. On the command bar, click **Relationship Types > Center-Plane** and select the axis of the *Spindle*.
22. Click on the front face and back face of the *Handle*.

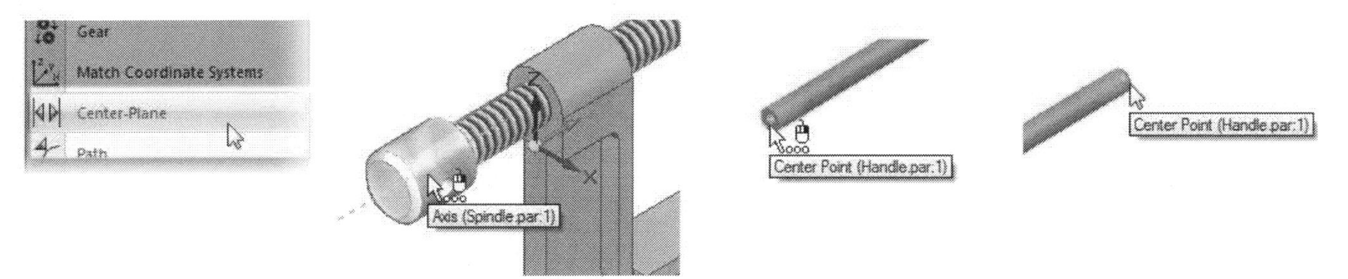

23. On the command bar, click **Relationship Types > Axial Align**. Click the **Lock Rotation** icon, and then click on the cylindrical face of the *Handle*.
24. Click on the hole of the *Spindle*. The *Handle* is axially aligned with the hole.

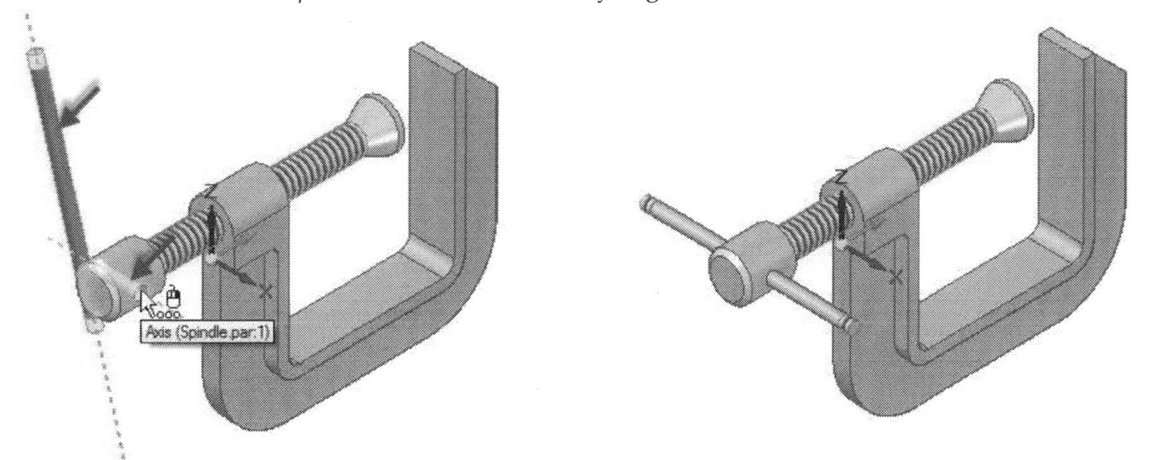

25. In the **Parts Library**, click *Handle Cap* and drag it into the assembly window.
26. On the command bar, click **Relationship Types > Insert**, and then click on a cylindrical face of the *Handle Cap*.
27. Click on the cylindrical face of the *Handle*.
28. On the command bar, type 1 in the **Offset Value** box, and then click on the flat face of the hole of the *Handle Cap*.
29. Click the end face of the *Handle*. The *Handle Cap* is inserted into the *Handle*.

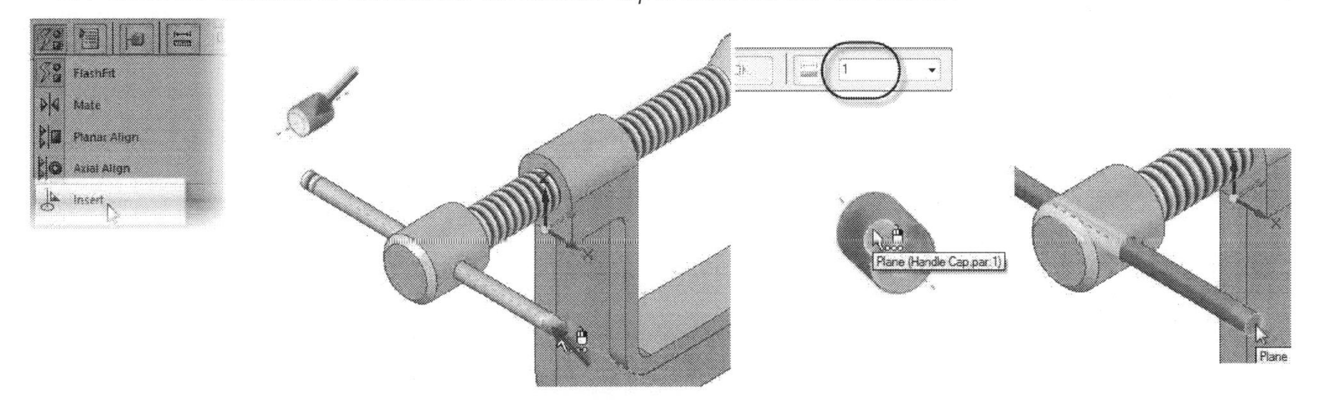

30. Save the assembly with the name **G-Clamp.asm**.
31. In the **Pathfinder**, click on the *Handle Cap*, and then click **Home > Relate > Capture Fit**. The **Capture Fit** dialog appears. Click **OK** to close the dialog.
32. In the **Parts Library**, click *Handle Cap* and drag it to the assembly window. The **Mate** command is activated and the flat face of the *Handle Cap* is selected.

33. Type 1 in the **Offset value** box and click on the end face of the *Handle*. The **Axial Align** command is activated and axis of the H*andle Cap* is selected.
34. Click on the axis of the *Handle* to complete the assembly.

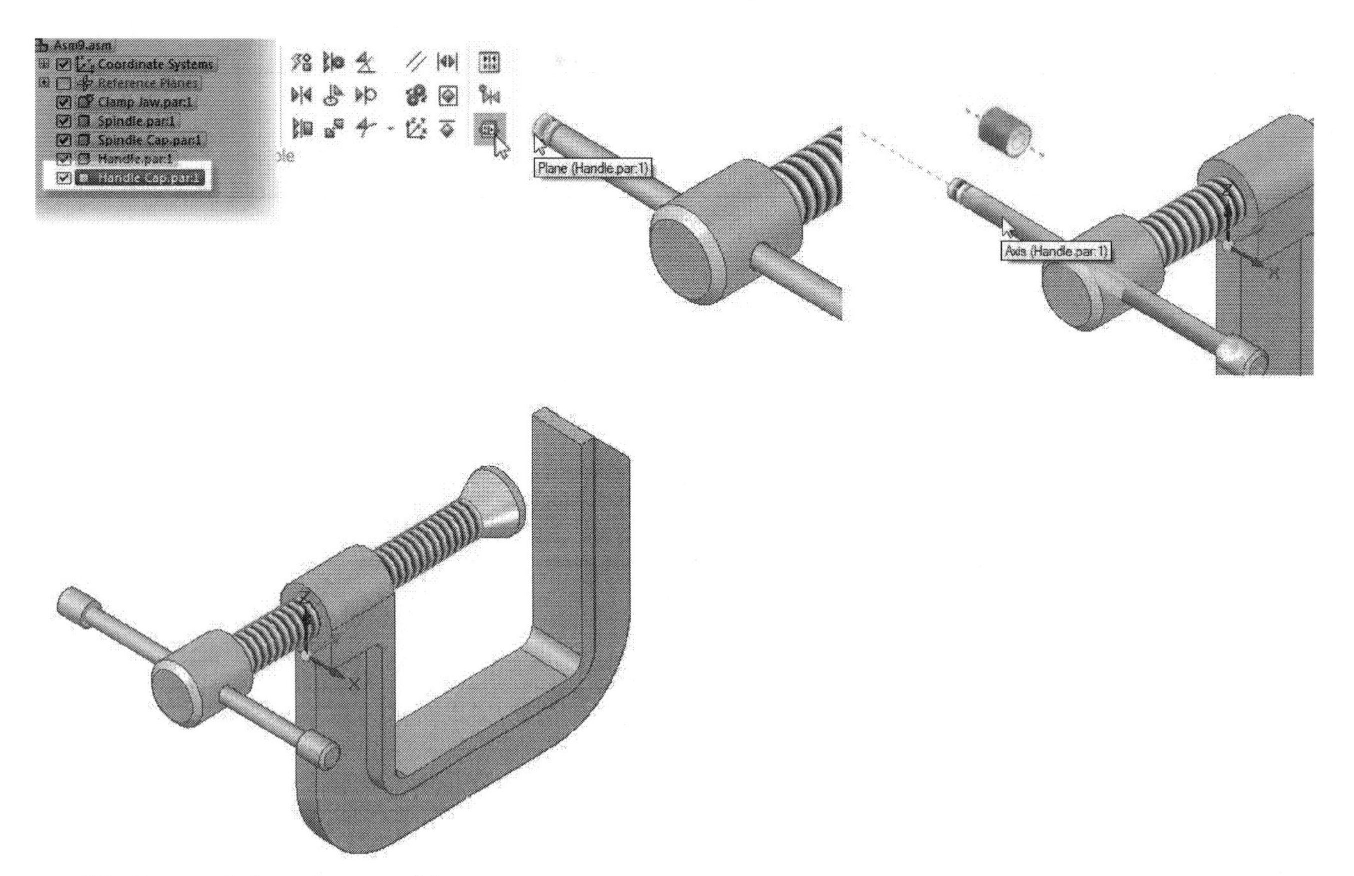

35. Save and close the assembly.

Example 2 (Top Down Assembly)

In this example, you will create the assembly shown below.

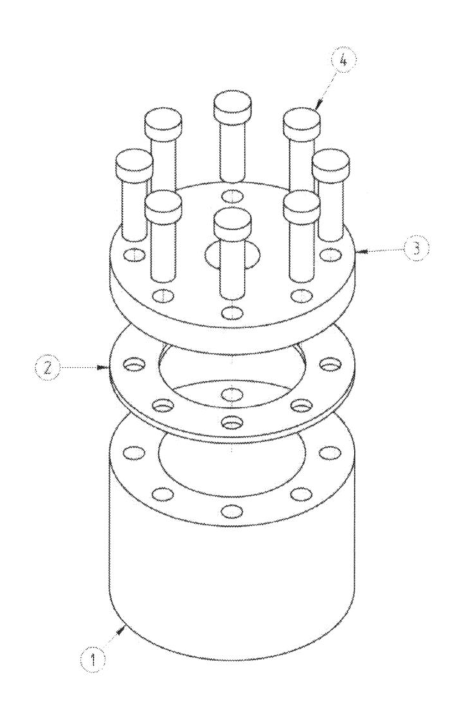

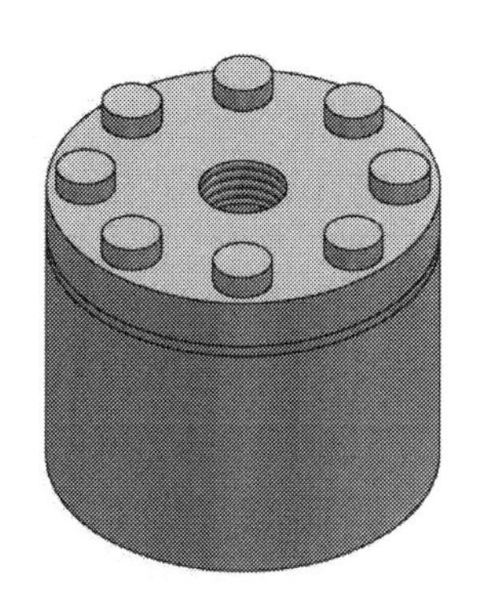

Item Number	File Name (no extension)	Quantity
1	Cylinder base	1
2	Gasket	1
3	Cover plate	1
4	Screw	8

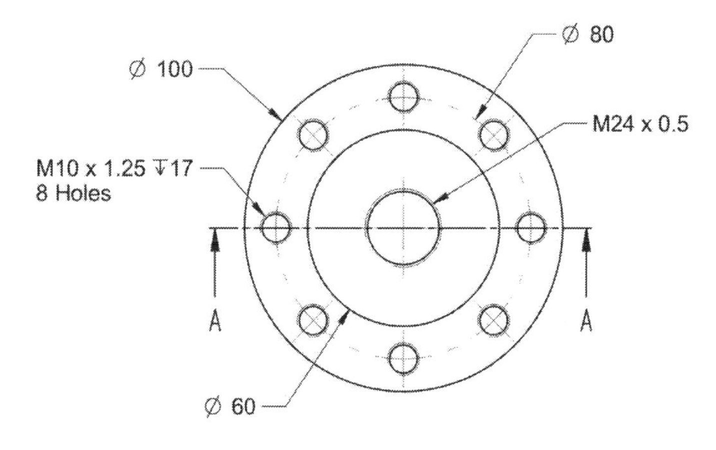

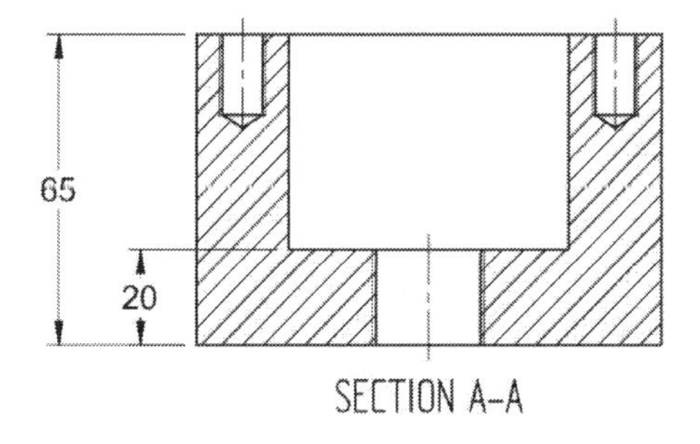

Cylinder Base

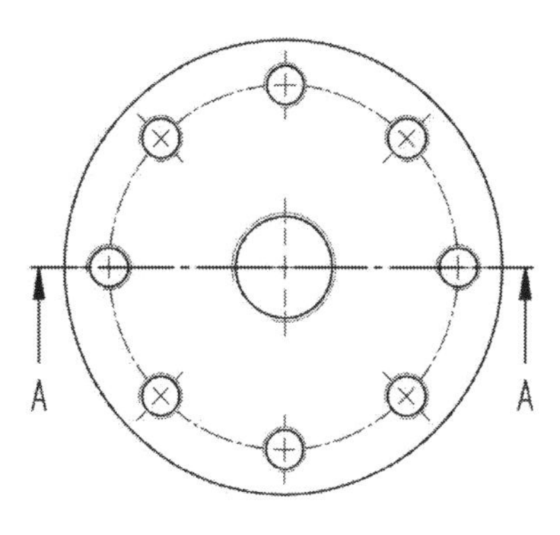

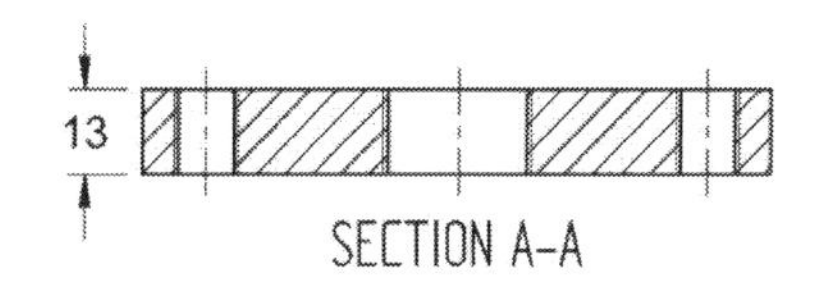

Cover Plate

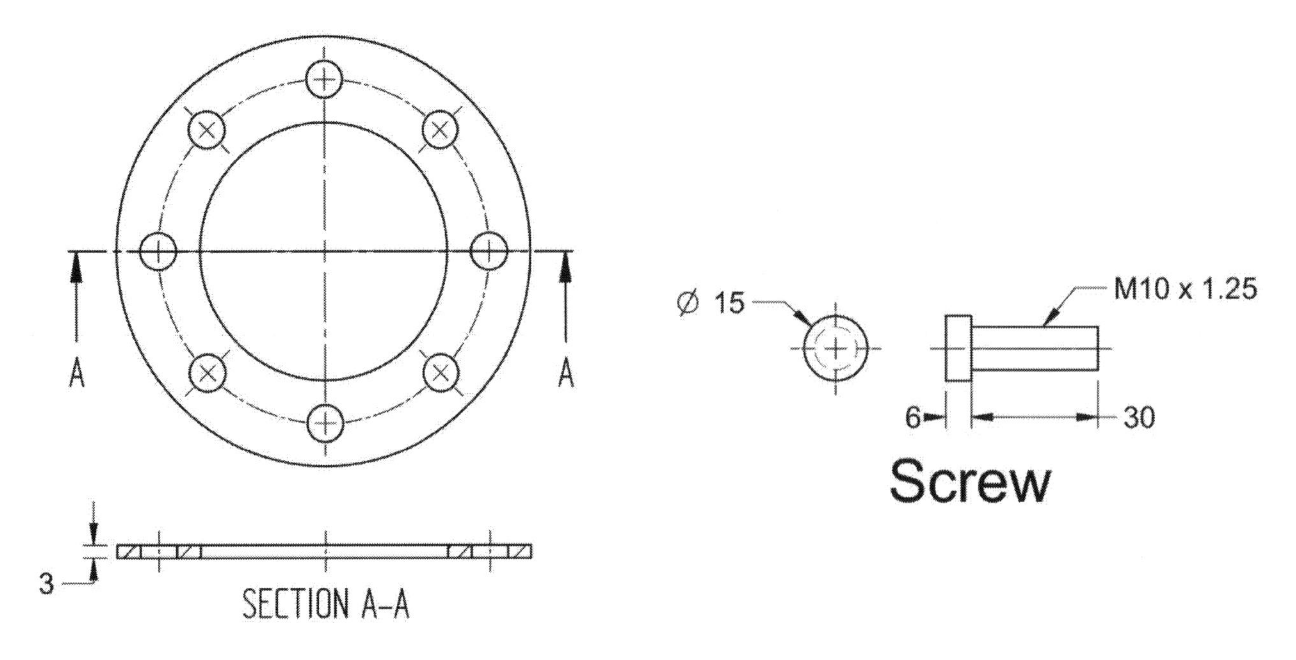

1. Start **Solid Edge 2019**.
2. Start a new part file and create the Cylinder base. Do not create the center hole.

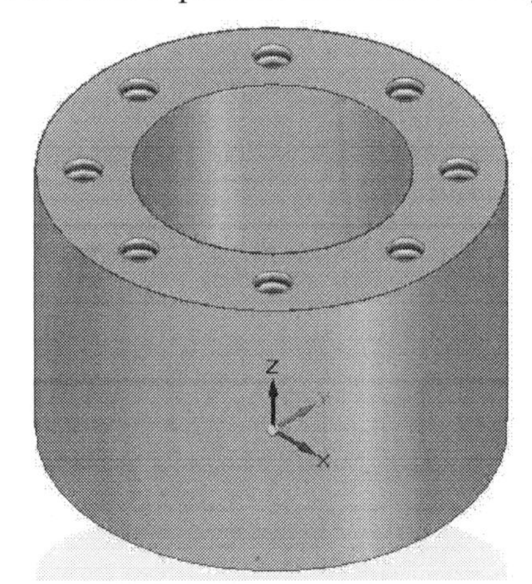

3. Create a new folder with name *Pressure Cylinder.*
4. Save the file with the name *Cylinder base.*
5. On the **Application Menu**, click **New > Assembly of Active Model**. The **Create Assembly** dialog appears.
6. On this dialog, click **OK** to start a new assembly file. The *Cylinder base* is automatically placed at the origin.
7. Save the assembly file in the *Pressure Cylinder* folder.
8. On the ribbon, click **Home > Assemble > Create Part In-Place**. The **Create Part In-Place Options** dialog pops up on the screen.
9. On this dialog, under the **Place the Origin** section, select the **By graphic input** option.
10. Leave the other default options on this dialog and click **OK**. The origin of the new part is attached to the mouse pointer.
11. Place the mouse pointer on the circular edge of the Cylinder base.
12. Press **T** on your keyboard to toggle the orientation of the origin.

13. Click when the orientation of the part origin is same as that of the assembly origin.

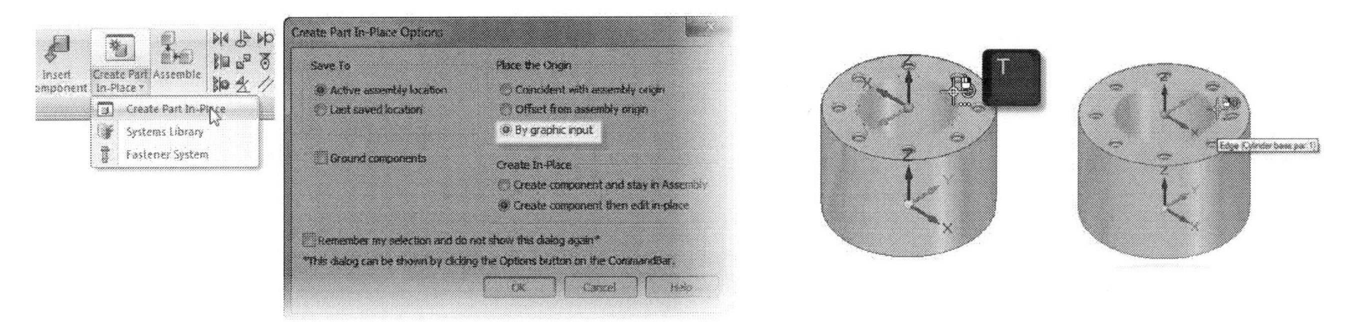

14. On the command bar, click the green check. The **Save As** dialog pops up.
15. Type *Gasket* in the **File name** field and click **Save**. The part file is created and Part environment is activated.
16. On the ribbon, click **Home > Draw > Project to Sketch** and lock the XY plane.
17. Leave the default settings on the **Project to Sketch Options** dialog, and then click **OK**.
18. Click on the circular edges on the top face of the Cylinder base. The edges are projected to the locked plane.
19. Activate the **Extrude** command and click in the region enclosed by the sketch.
20. Right-click to accept the selection, and then set the **Extent Type** to **Finite**.
21. Move the mouse pointer upward.
22. Type 3 in the dimension box and press Enter to create the *Extrude* feature.

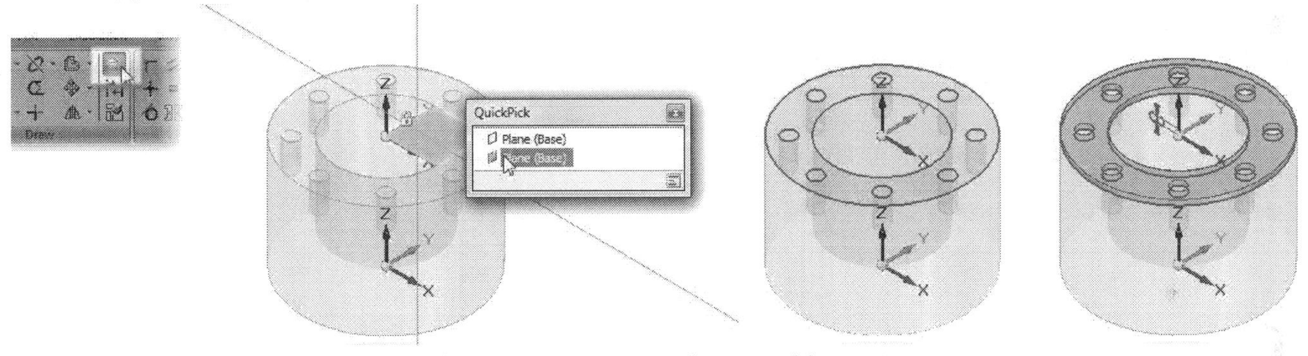

23. On the ribbon, click **Close and Return** to return to the assembly session.
24. On the ribbon, click **Home > Assemble > Create Part In-Place**.
25. On the **Create Part In-Place** dialog, under the **Place the origin** section, select the **Offset from assembly origin** option. Click **OK** to close the dialog. Now, you have to enter X, Y, and Z values (or) select a keypoint to specify the origin of the new part.

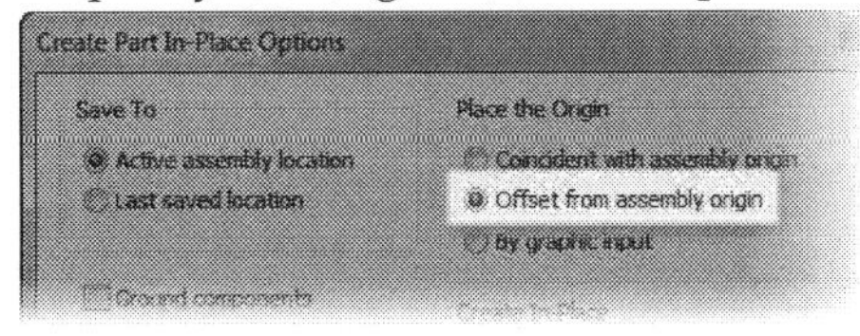

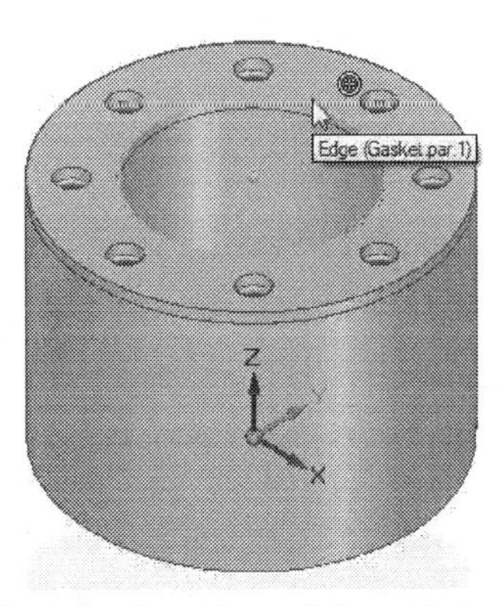

26. Click on the circular edge of the *Gasket* to define the location of the origin.

27. On the command bar, click the green check.
28. On the **Save As** dialog, type *Cover plate* in the **File name** field and click **Save**.
29. In the **Part** environment, activate the **Project to Sketch** command and lock the XY plane.
30. Project the outer and small circular edges.
31. Use the sketch and create an *Extrude* feature. The depth of the extrusion is 13 mm.

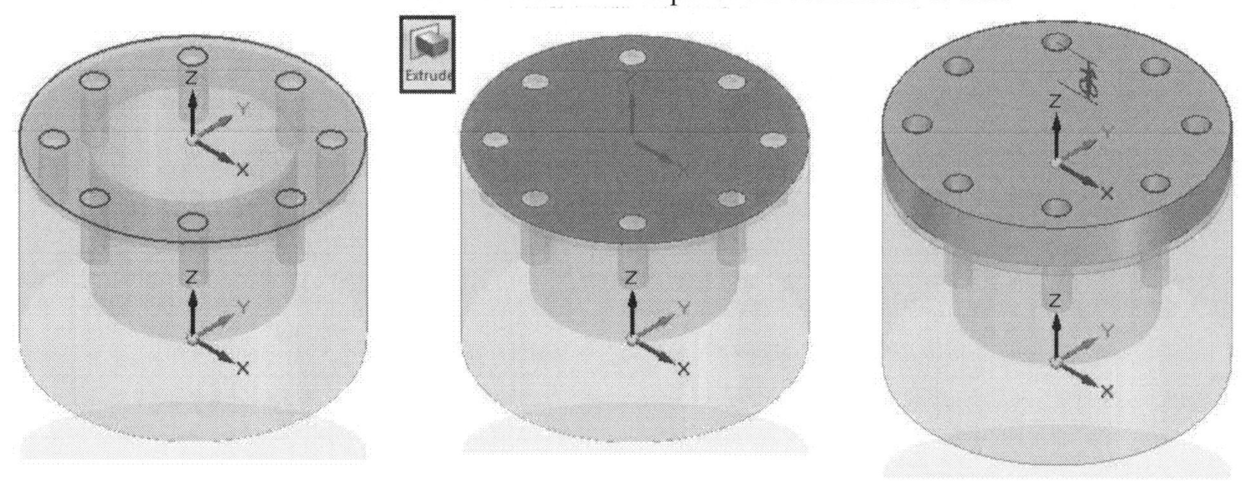

32. Activate the **Thread** command and add M10 x 1.25 to threads to the holes.

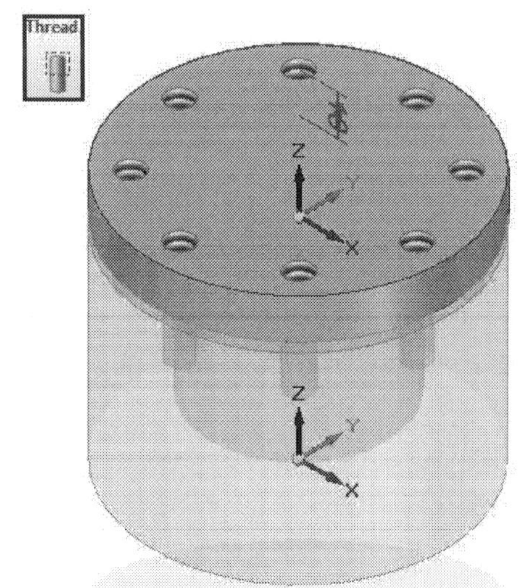

33. On the ribbon, click **Close and Return** to return to the assembly environment.
34. Activate the **Create Part In-Place** command and create the *Screw* on the top face of the *Cover plate*.
35. In the Part environment, activate the **Project to Sketch** command and lock the XY plane.
36. Project the circular edge of the hole.
37. Use the sketch and create an *Extrude* feature of 30 mm depth. The direction of extrusion should be downward.

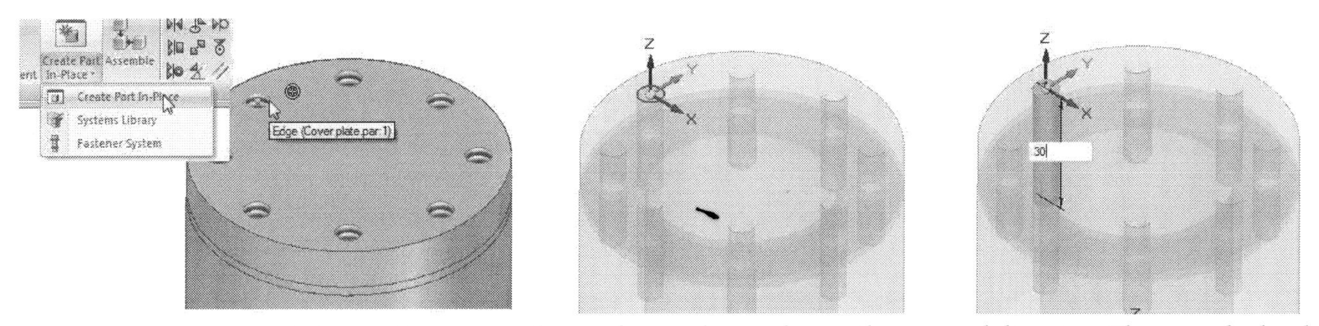

38. Create a circle of 15 mm diameter on the top face and extrude it in the upward direction. The extrude depth is 6 mm.
39. Activate the **Thread** command and add thread to the lower cylindrical face of the part. The thread size is M10 x 1.25.

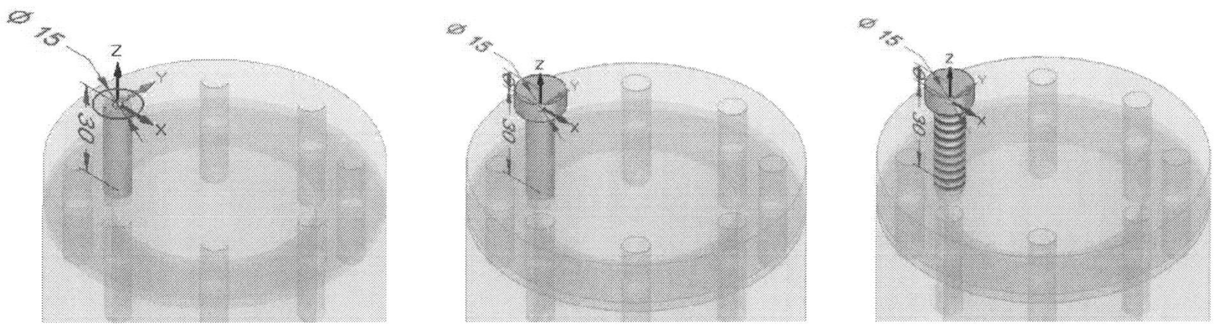

40. On the ribbon, click **Close and Return** to return to the assembly environment. Now, you have to add relationships between parts.
41. On the ribbon, click **Home > Assemble > Assembly Relationship Assistant**. The **Relationship Assistance Options** dialog appears.
42. On this dialog, select the **Select Set 2** option and click **OK**.
43. Click on the *Cylinder base*, and then click the green check on the command bar.
44. Click on the *Gasket*, and then click the green check on the command bar. The **Relationship Assistant Settings** dialog pops up.

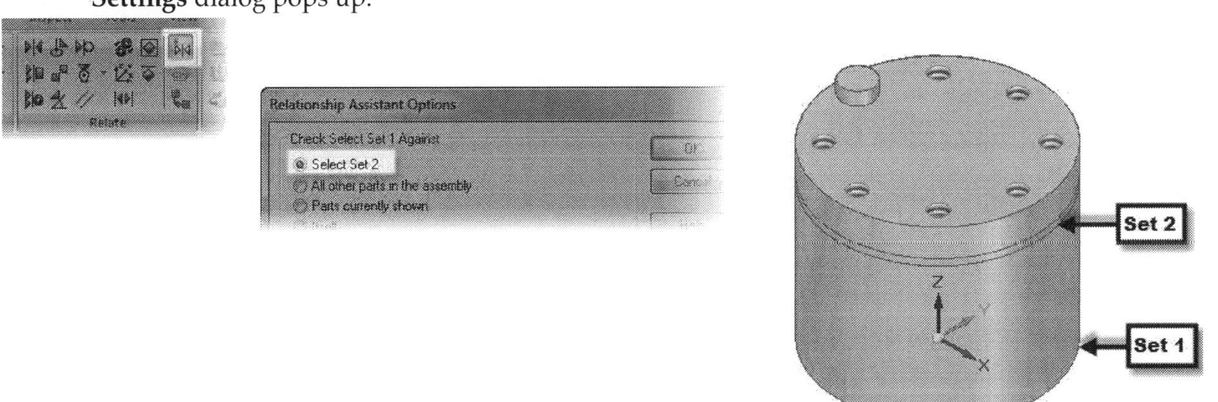

45. On this dialog, make sure that the Mate and Axial Align relationships are turned on. Click the **Process** button to create relationships between the selected parts automatically.

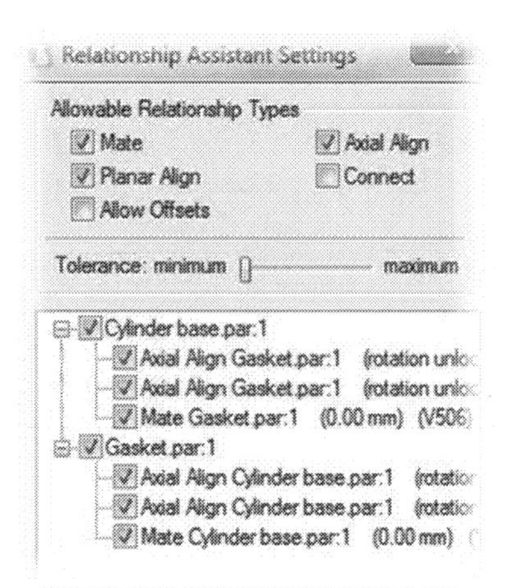

46. Click on the relationships to highlight the faces associated with them.
47. Click **Accept** to create the relationships.
48. Close the dialog and click **Finish** to complete creating the relationships. Click **Cancel** to deactivate the command.
49. In the **Pathfinder**, click on the *Gasket*, and then click on the Axial Align relationship at the bottom.
50. At the bottom of the screen, click the **Lock Rotation** icon to arrest the rotation of the *Gasket*.

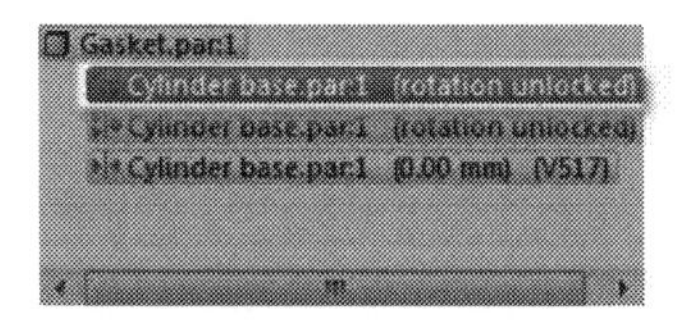

51. Use the **Assembly Relationship Assistant** command and create relationships between the other parts of the assembly.
52. On the ribbon, click **Home > Pattern > Pattern,** and then click on the *Screw*. Click the green check on the command bar to accept the selection.
53. Click on the *Cylinder base* and select the circular pattern.
54. On the command bar, click **Finish** to complete the pattern.

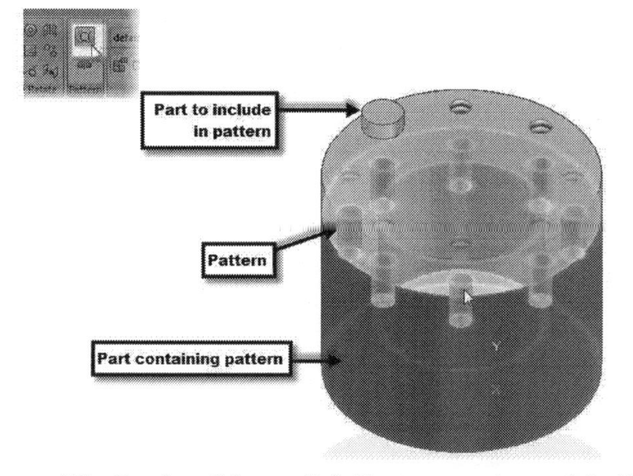

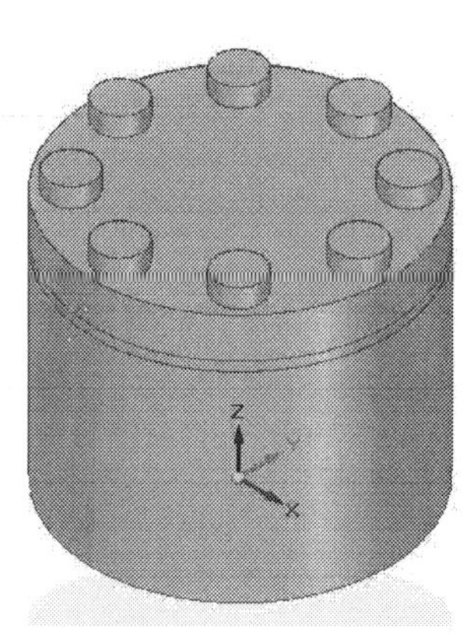

55. On the ribbon, click **Features > Assembly Features > Hole**. The **Assembly Feature Options** dialog pops up.
56. On this dialog, select the **Create Part features** option, and then check the **Create Synchronous Cuts, Holes, Revolved Cuts if possible** option. Click **OK** to close the dialog.

57. On the command bar, click the **Hole Options** dialog and set the options, as shown below. Click **OK** to close the dialog.

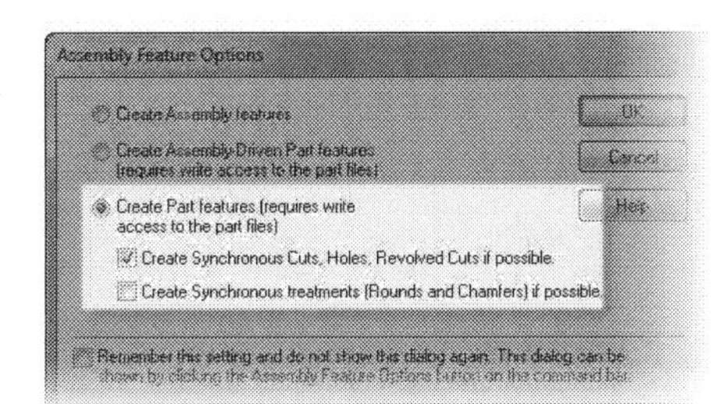

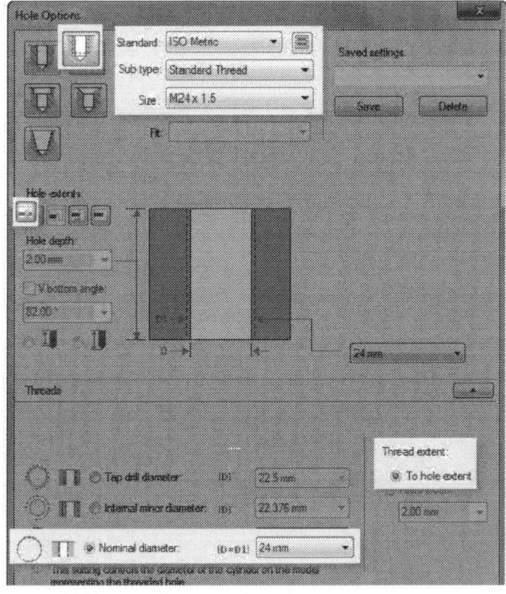

58. Click on the top face of the cover plate and place a hole circle at the center.
59. Click **Close Sketch** to exit the sketch.
60. Move the mouse pointer downwards and click to define the side of the hole.
61. On the command bar, click the green check to create the threaded hole.

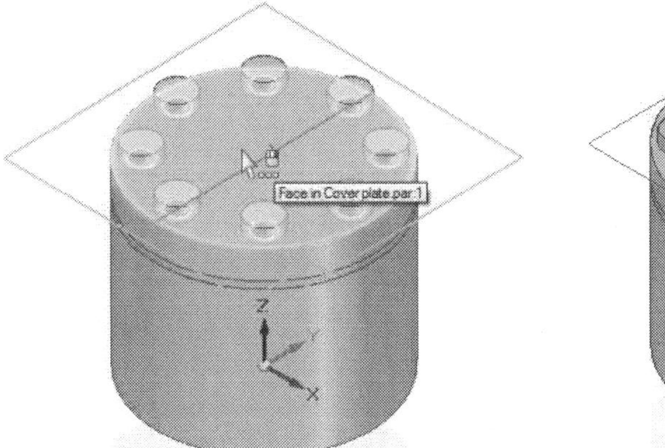

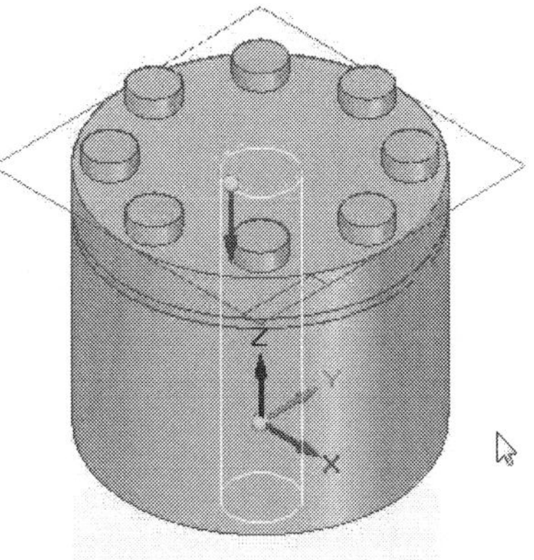

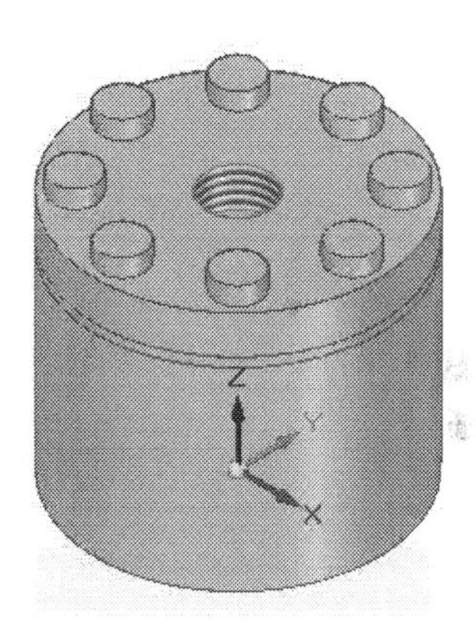

62. On the ribbon, click **PMI > Model Views > Section**.
63. Click on the XZ plane and draw the sketch, as shown in figure.

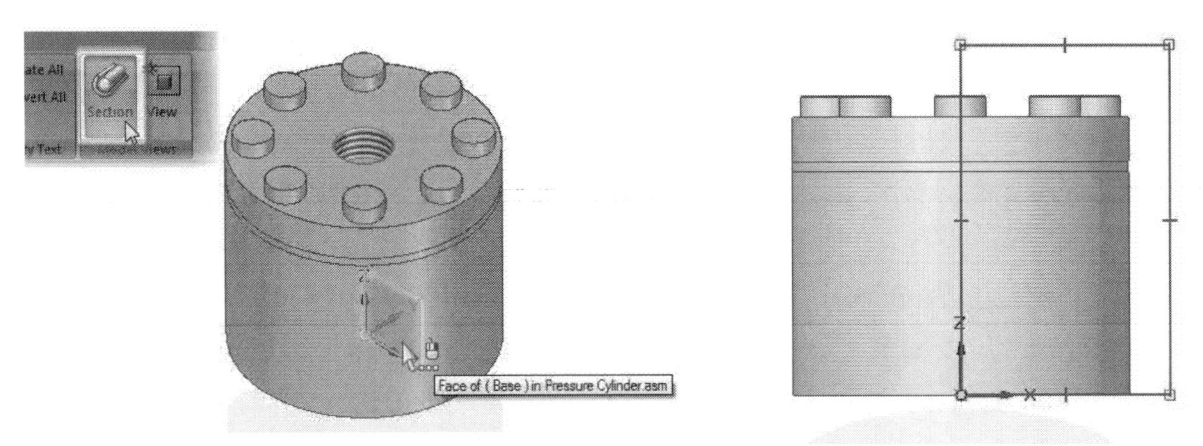

64. On the ribbon, click **Close Sketch** to close the sketch.
65. Move mouse pointer such that the arrow points inside the sketch. Click to define the side of the section cut.
66. Extrude the sketch in the forward direction to create the section cut. Click **Accept** and **Finish** to create the section cut.

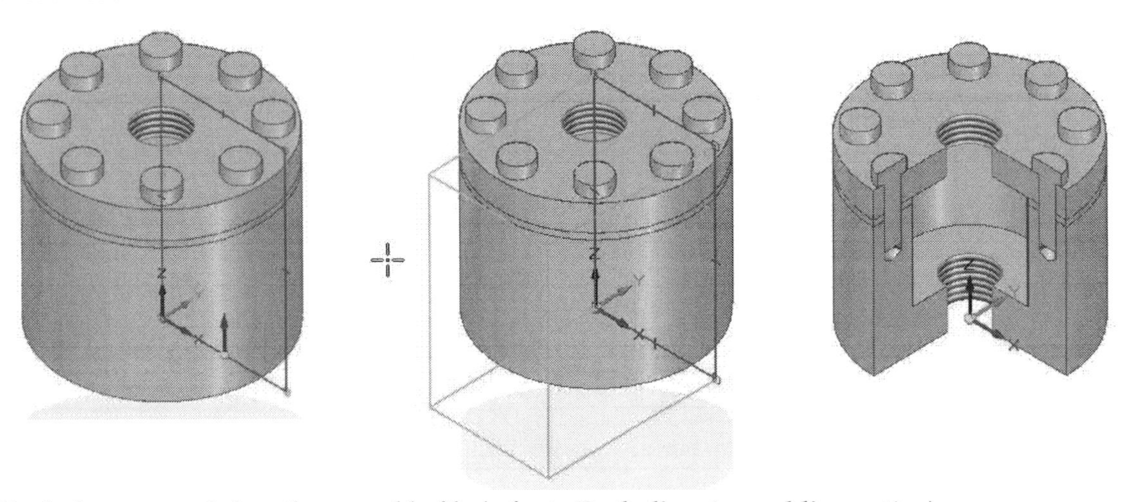

67. Explode, save, and close the assembly file (refer to **Exploding Assemblies** section).

Questions

1. How do you start an assembly from an already opened part?
2. What is the use of the **Capture Fit** command?
3. List the advantages of Top-down assembly approach.
4. What is a grounded part?
5. What is the use of the **Assembly Relationship Assistant** command?
6. How do you create a sub-assembly in the assembly environment?
7. Briefly explain the **Edit-In Place** command.
8. Why do we prefer the **Explode** command to the **Auto Explode** command?
9. What is the difference between rigid and adjustable subassemblies?
10. How to show or hide reference planes of a part?

Exercise 1

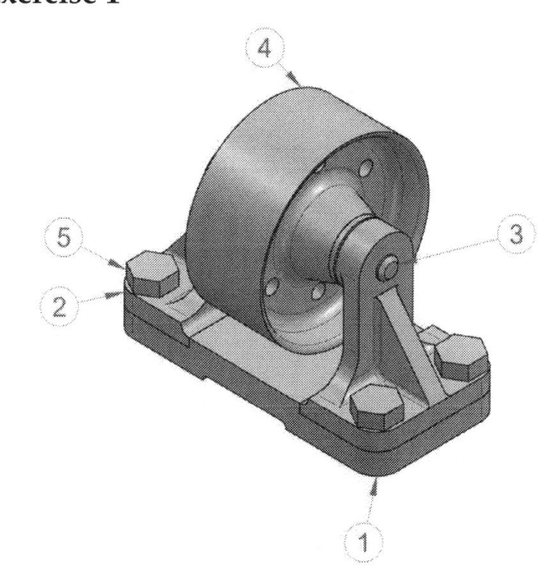

Item Number	File Name (no extension)	Quantity
1	Base	1
2	Bracket	2
3	Spindle	1
4	Roller-Bush assembly	1
5	Bolt	4

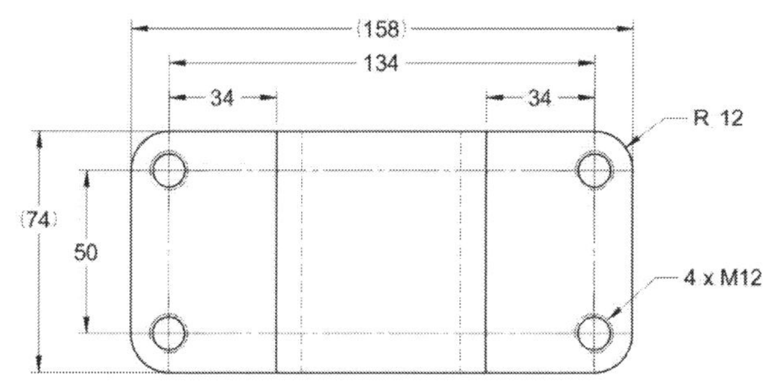

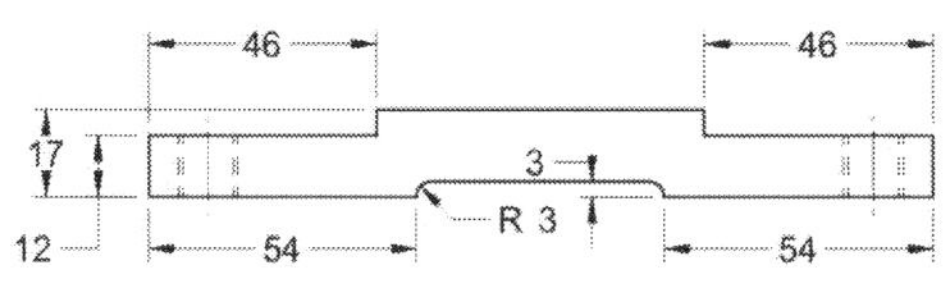

Base

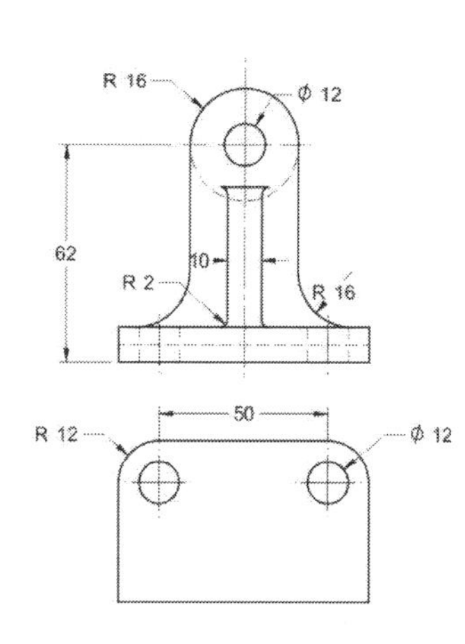

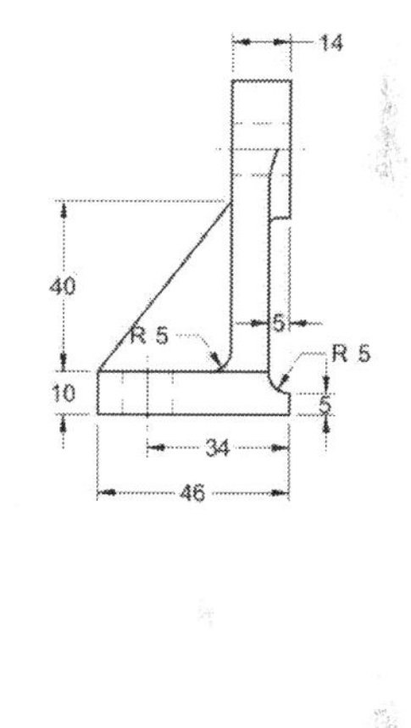

Bracket

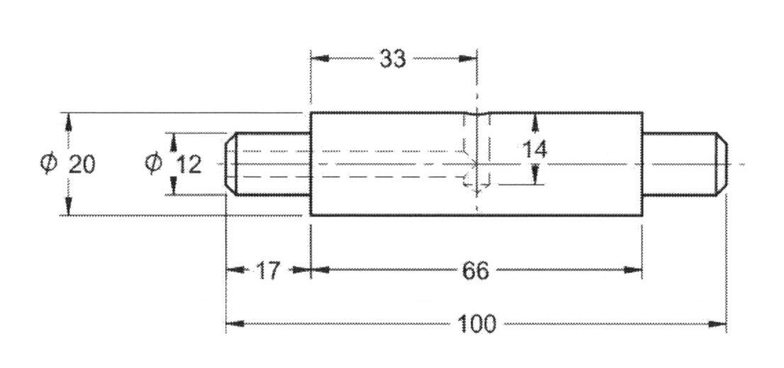

SPINDLE

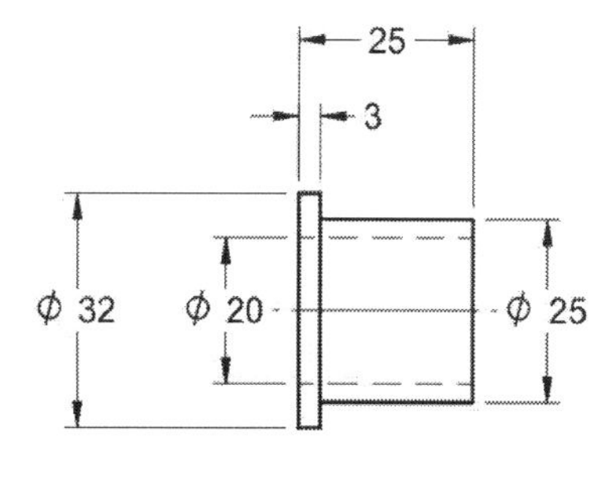

BUSH

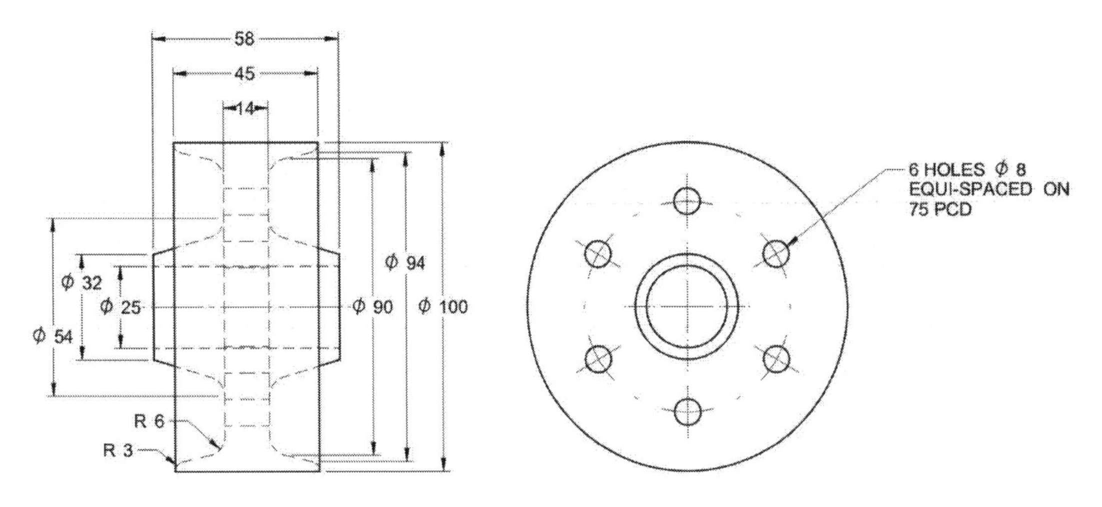
58
45
14
6 HOLES Ø 8
EQUI-SPACED ON
75 PCD
Ø 32
Ø 25
Ø 54
Ø 94
Ø 90
Ø 100
R 6
R 3

Roller

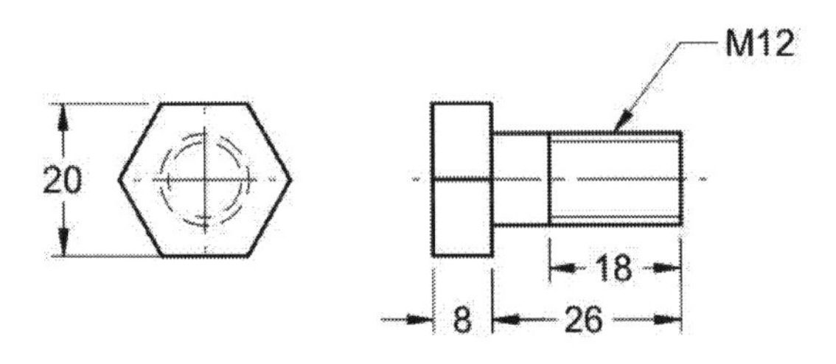
M12
20
18
8
26

Bolt

Chapter 11: Drawings

Drawings are used to document your 3D models in the traditional 2D format including dimensions and other instructions useful for manufacturing purpose. In Solid Edge, you first create 3D models and assemblies, and then use them to generate drawings. There is a direct association between the 3D model and the drawing. When changes are made to the model, every view in the drawing will be updated. This relationship between 3D model and the drawing makes the drawing process fast and accurate. Because of the mainstream adoption of 2D drawings of the mechanical industry, drawings are one of the three main file types you can create in Solid Edge.

The topics covered in this chapter are:

- *Create model views*
- *Projected views*
- *Auxiliary views*
- *Section views*
- *Detail views*
- *Broken-Out views*
- *Break Lines*
- *Display Options*
- *View Alignment*
- *Parts List and Balloons*
- *Retrieve Dimensions*
- *Arrange Dimensions*
- *Maintain Alignment*
- *Remove Alignment*
- *Line Up Text*
- *Ordinate Dimensions*
- *Chamfer Dimension*
- *Center Marks*
- *Centerlines*
- *Automatic Centerlines*
- *Bolt Hole Circles*
- *Callouts and Leaders*
- *Notes*

Starting a Drawing

To start a new drawing, click the **Application Menu** icon on the initial screen, and select **New > ISO Metric Draft** (or) click the **New** icon on the **Quick Access Toolbar**, and then double-click on the **iso metric draft.dft** template on the **New** dialog. If you want to start the drawing in any other standard, select the standard from the **Standard Templates** section and select the required template.

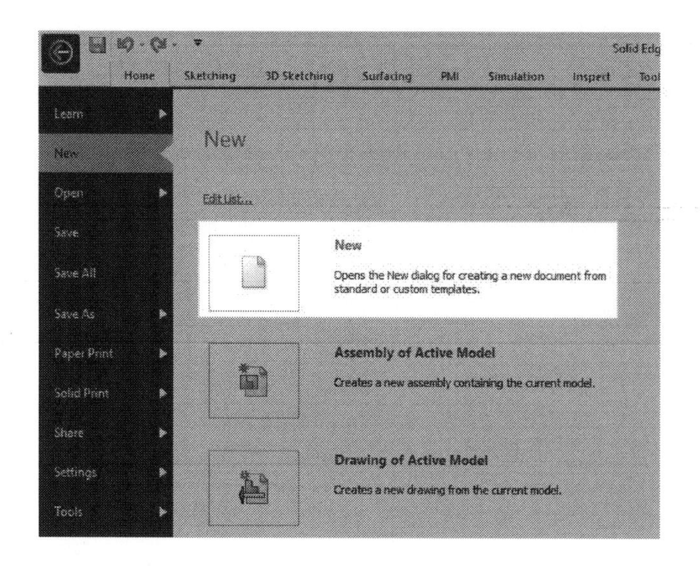

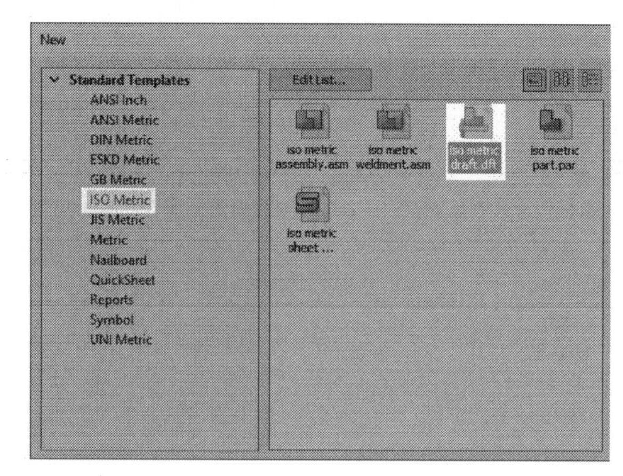

If you already have a part or assembly opened, you can click **Application Menu > New > Drawing of Active Model**; the **Create Drawing** dialog appears. On this dialog, click the **Browse** button to access different sheet templates. Select anyone of the sheet templates and click **OK**. On the **Create Drawing** dialog, check the **Run Drawing View Creation Wizard** option to start creating drawing views. If you uncheck this option, the drawing views will be created automatically. Click **OK** on the **Create Drawing** dialog to start a new drawing.

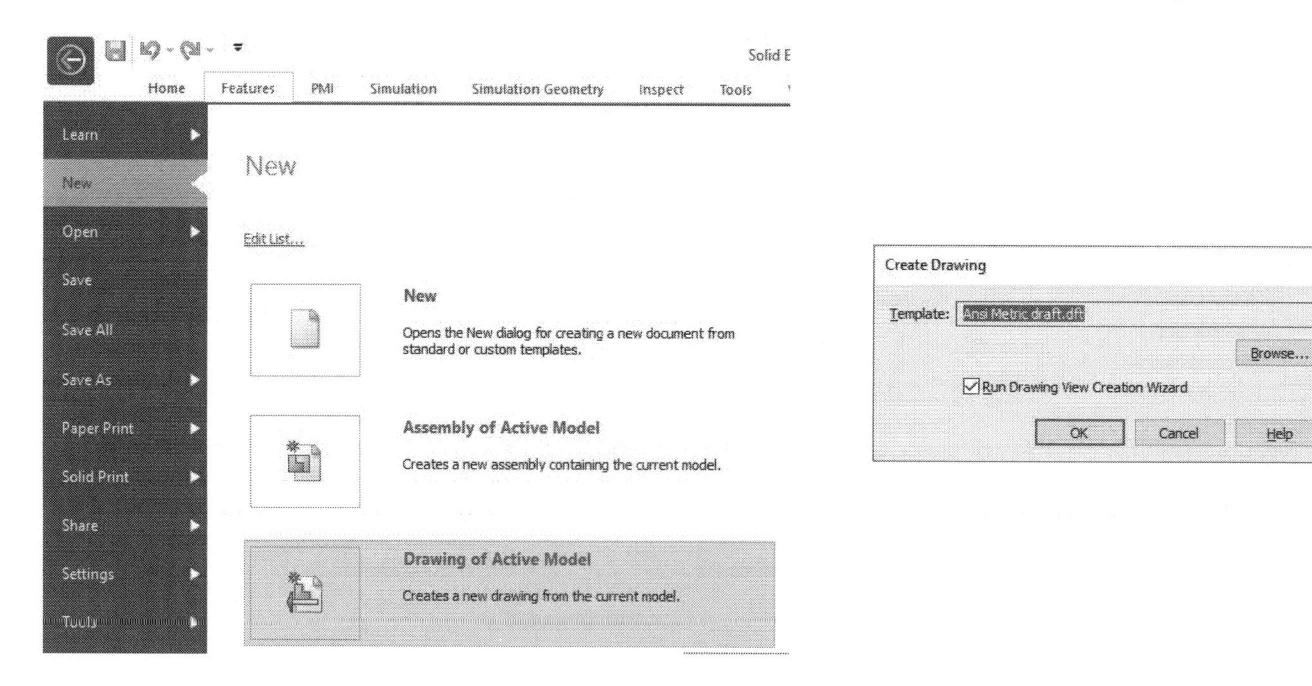

View Creation

There are different standard views available in a 3D part such as front, right, top, and isometric. In Solid Edge, you can create these views using the **View Wizard** command. This command is activated automatically, if you have created a drawing from an already opened part. If it is not activated, click **Home > Drawing Views > View Wizard** on the ribbon. The **Select Model** window appears. Browse to the location of the part or assembly and double-click on it; a model view will be attached to the pointer. In addition, the **View Wizard** command bar pops up on the screen.

Drawings

Click the **Drawing View Layout** icon on the command bar; the **Drawing View Wizard** dialog pops up on the screen. On this dialog, select the first view from the **Primary View** list. Next, click on the icons that represent the standard views that are to be created. After selecting the standard views, click **OK** on the **Drawing View Wizard** dialog. Click the **Set View Scale** icon to adjust the sizes of the views to sheet size. Click on the sheet to create views. Click and drag the views to position them.

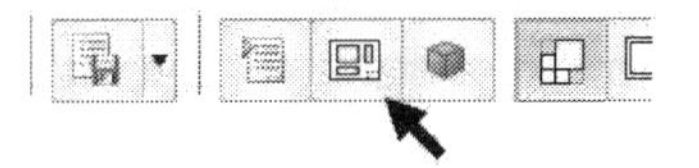

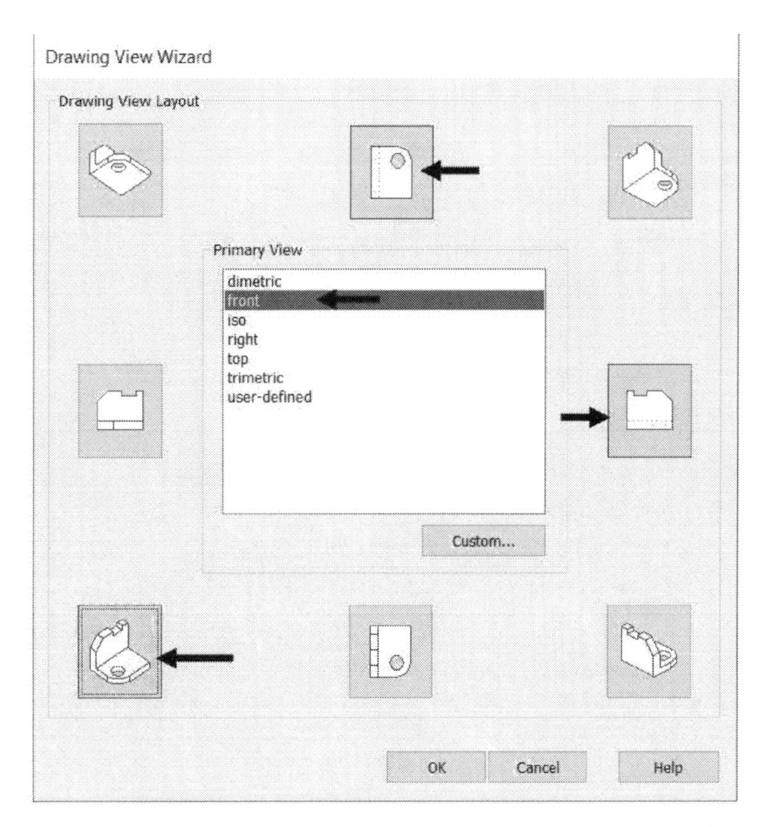

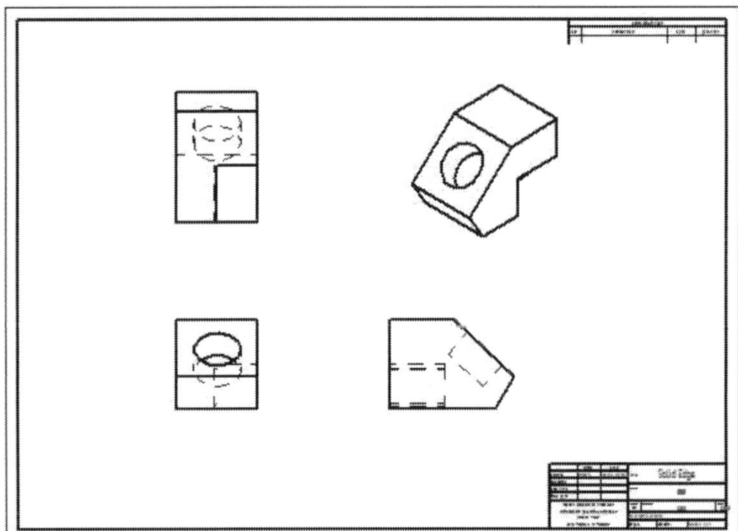

You can change the orientation of a drawing view even after creating all the views associated with it. To do this, select the drawing view and click the **View Orientation** icon on the command bar. Next, select the required orientation from the flyout; the selected view and all its associated views are changed.

Principal View

After you have created the first view in your drawing, a principal view is one of the simplest views to create. Activate the **Principal View** command (click **Home > Drawing Views > Principal View** on the ribbon). After activating the command, select a view you wish to project from. Next, move the pointer in the direction you wish to have the view to be projected. Next, click on the sheet to specify the location; the projected view will be created. Click the right mouse button to deactivate this command.

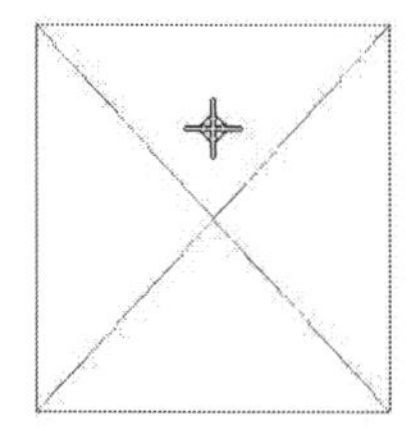

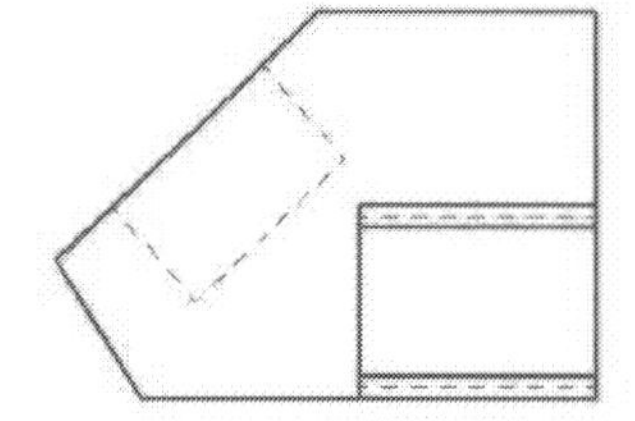

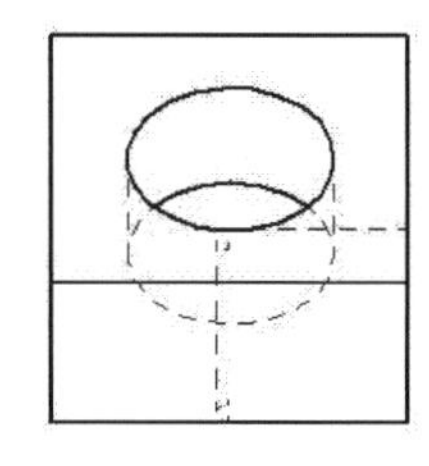

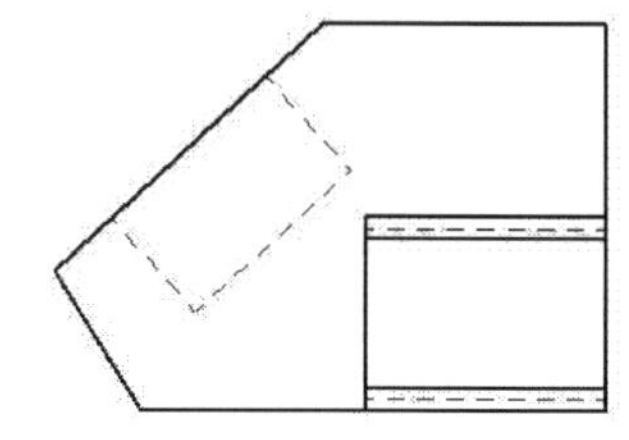

Auxiliary View

Most of the parts are represented by using orthographic views (front, top and/or side views). However, many parts have features located on inclined faces. You cannot get the true shape and size for these features by using the orthographic views. To see an accurate size and shape of the inclined features, you need to create an auxiliary view. An auxiliary view is created by projecting the part onto a plane other than horizontal, front or side planes. To create an auxiliary view, activate the **Auxiliary** command (click **Home > Drawing Views > Auxiliary** on the ribbon). Click the angled edge of the model to establish the direction of the auxiliary view. Next, move the pointer to the desired location and click to locate the view.

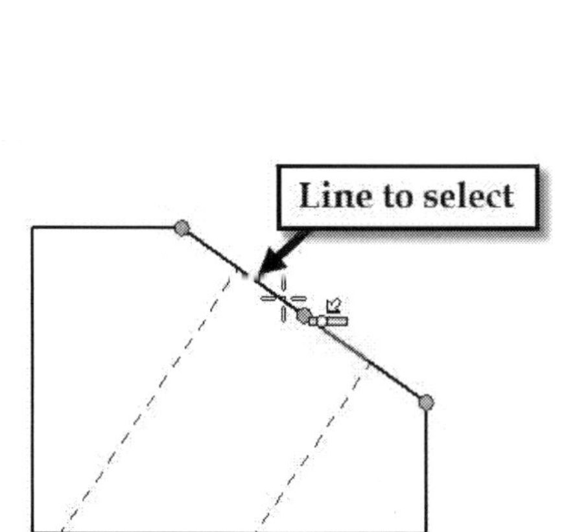

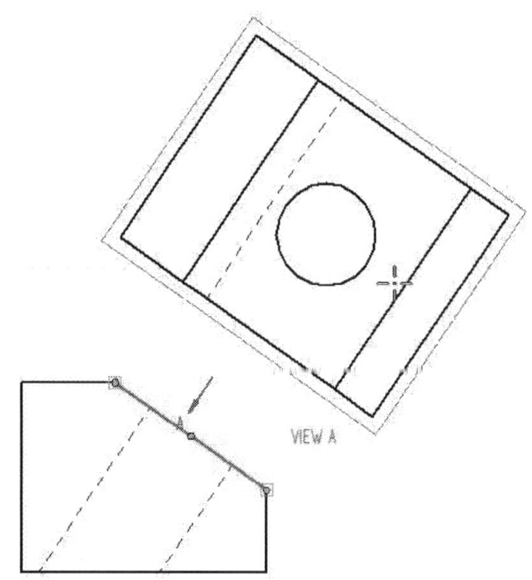

Section View

One of the more common views used in 2D drawings is the section view. Creating a section view in Solid Edge is very simple. Once a view is placed on the drawing sheet, you need to draw a line where you want to section the drawing view. Activate the **Cutting Plane** command (click **Home > Drawing Views > Cutting Plane** on the ribbon) and click on a drawing view. Now, you have to draw a line to define the cutting plane. You can use the geometry of the drawing view to draw the line. After drawing a line, click **Close Cutting Plane** on the ribbon. Next, click on either side of the cutting plane to indicate the view direction.

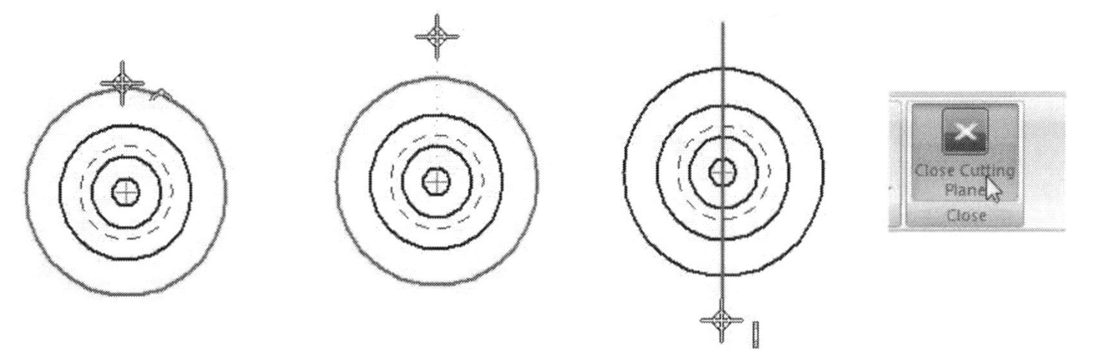

Activate the **Section** command (click **Home > Drawing Views > Section** on the ribbon) and click on a cutting plane. Move the pointer and click to position the section view.

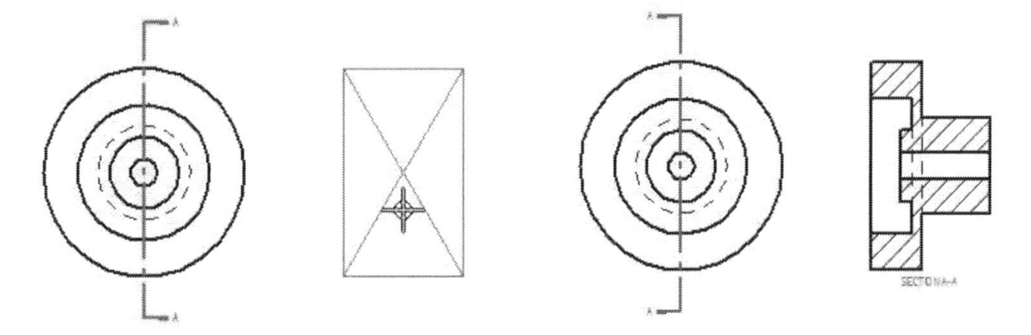

You can also use a multi-segment cutting line to create a section view.

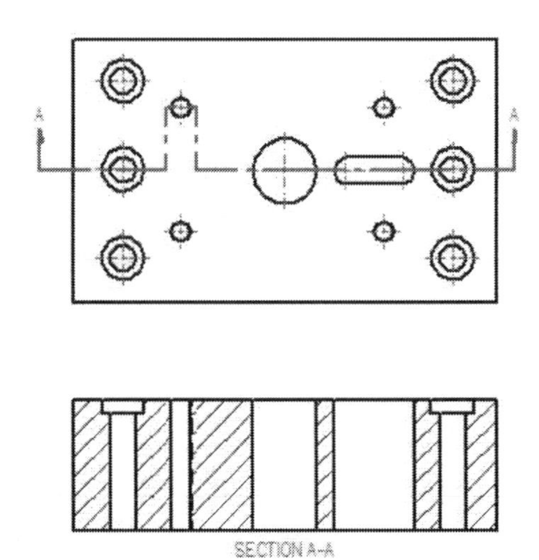

Use the **Section Only** option to display only the geometry on the cutting plane.

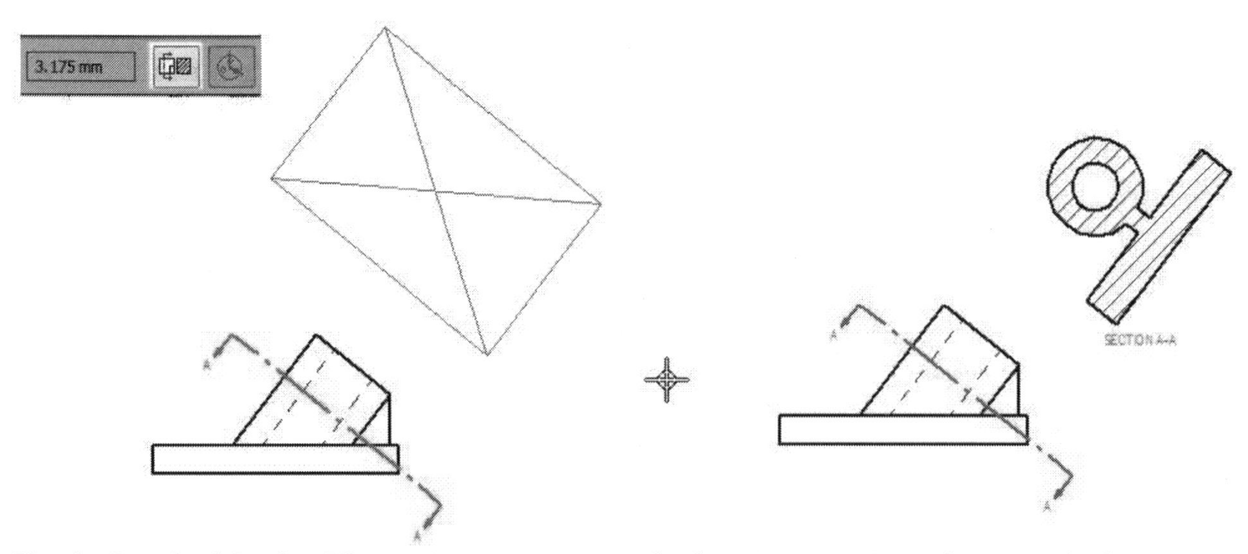

Use the **Revolved Section View** option to create a revolved section view. First, draw a multiple segment cutting plane using the **Cutting Plane** command. Next, activate the **Section** command and select the multi-segment cutting plane. Click on a segment to define the fold angle of the section view. Click the **Revolved Section View** icon on the command bar. Move the pointer and click to position the revolved section view.

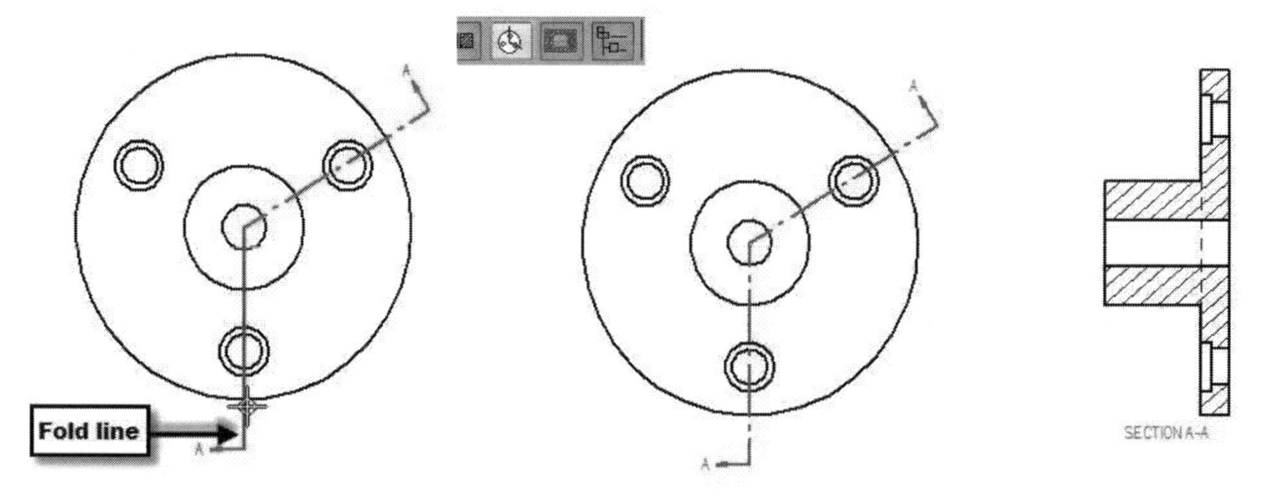

When creating a section view of an assembly, you can choose to exclude one or more components from the section cut. For example, to exclude the piston of a pneumatic cylinder, click the **Model Display Settings** icon on the command bar; the **Drawing View Properties** dialog pops up on the screen. On this dialog, select **piston** from the **Parts list** and uncheck the **Section** option. Click **OK** and locate the section view. You will notice that the piston is not cut.

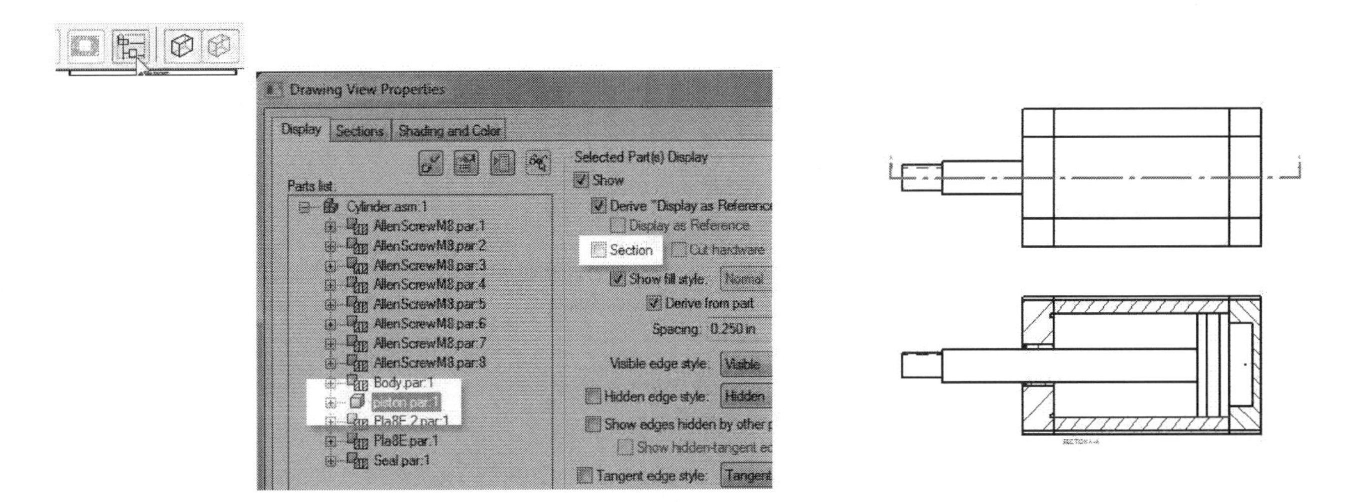

Solid Edge allows you to change the section view type even after creating it.

Detail View

If a drawing view contains small features that are difficult to see, a detailed view can be used to zoom in and make things clear. To create a detailed view, activate the **Detail** command (click **Home > Drawing Views > Detail** on the ribbon); this automatically activates the circle tool. Draw a circle to identify the area that you wish to zoom into. Once the circle is drawn, type in a value in the **Multiplier** box available on the command bar; the detail view is scaled by the number that you enter in the **Multiplier** box. Next, move the pointer and click to locate the view; the detail view will appear with a label.

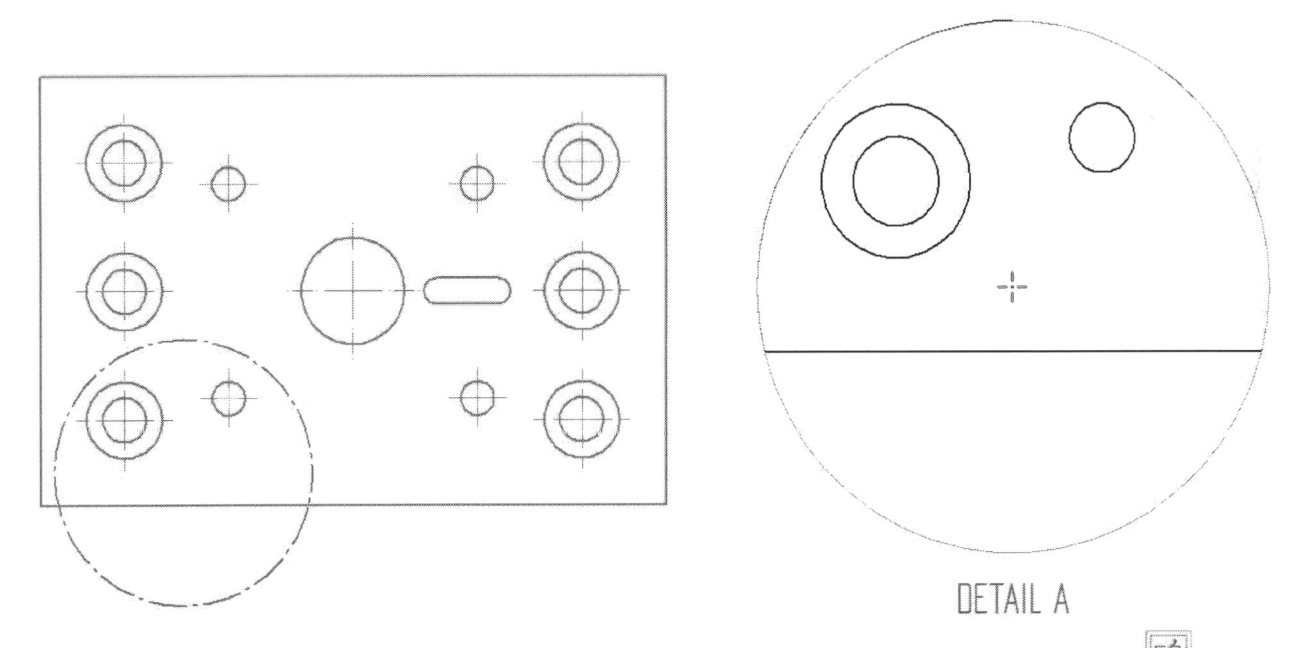

Solid Edge allows you to create custom profile of the detail view. To do this, click the **Define Profile** icon on the command bar, and then select the drawing view; the **Detail Profile** environment is activated. Draw the profile for the detail view using the commands available in the **Draw** panel of the ribbon. Next, click the **Close Detail Envelope** icon on the ribbon. Move the pointer and click to position the detail view.

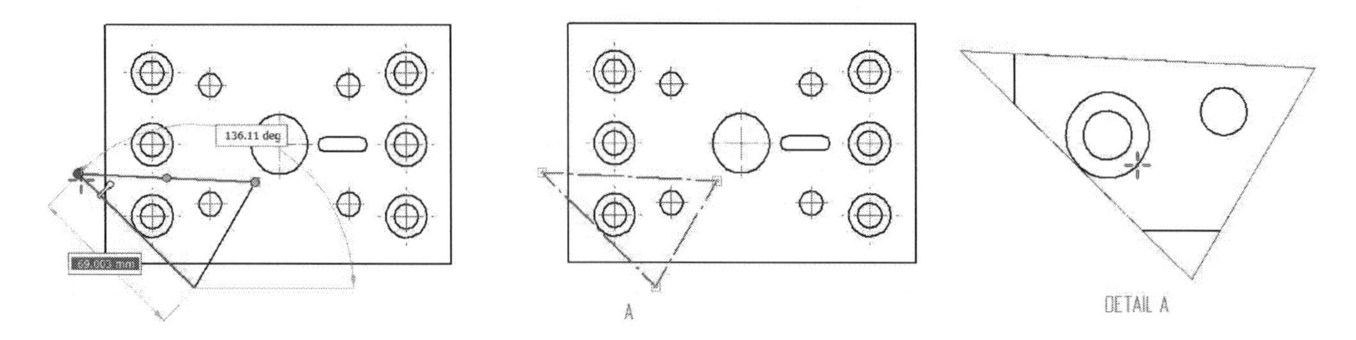

Broken View

Break lines are added to a drawing view, which is too large to fit on the drawing sheet. They break the view so that only important details are shown. To add break lines, select the view and click **Home > Drawing Views > Broken View** on the ribbon; the **Broken** command bar pops up. On this command bar, click the **Vertical Break** or **Horizontal Break** icon and define the **Break Line Type**. Type-in the desired value in the **Break gap** box and move your pointer to the area of the view where you would like to start the break. Click once to locate the beginning of the break. Move the pointer, and click again to locate the end of the break. You can select the keypoint of the view to create associative break lines. The resultant break lines update when the model is updated. Click **Finish** on the command bar; the view is automatically broken.

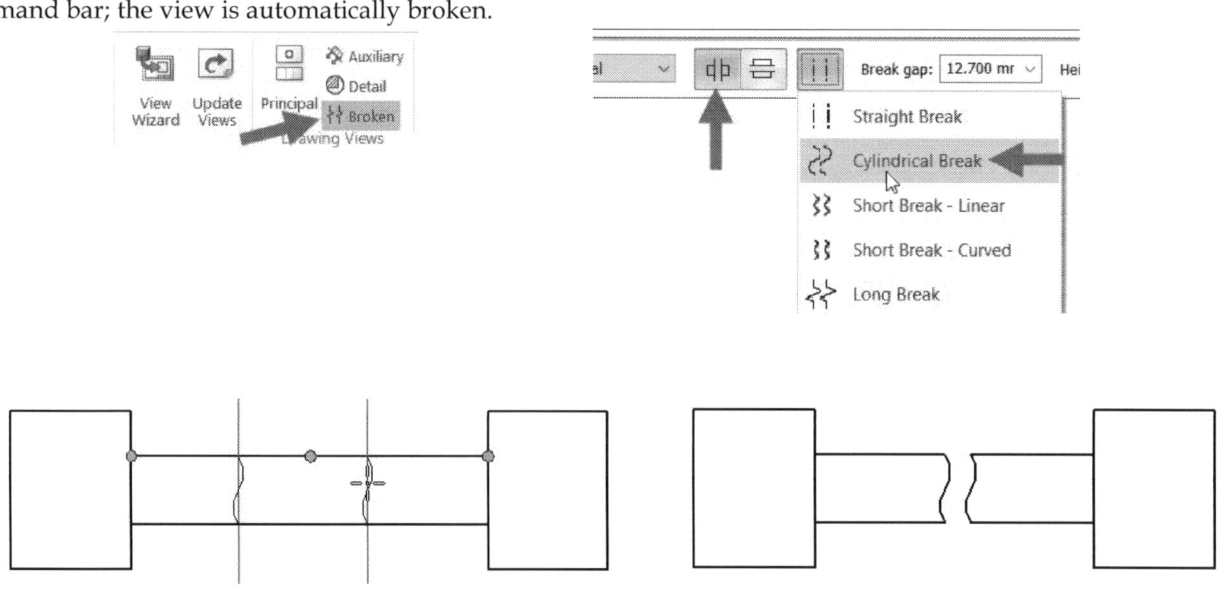

***Tip.** If you want to add a break line to another view in the drawing, then select the view, right click and select **Inherit Break Lines** from the shortcut menu. Next, select the view with break lines.*

Broken Out

The **Broken-Out** command alters an existing view to show the hidden portion of a part or assembly. This command is very useful to show the parts which are hidden inside an assembly view. You need to have a closed profile to break-out a view. For example, if you want to show the piston inside a pneumatic cylinder, activate the **Broken-Out** command (click **Home > Drawing Views > Broken-Out** on the ribbon) and select a drawing view to draw the profile. Draw a closed profile on the selected drawing view, and click **Close Broken Out Section** on the ribbon.

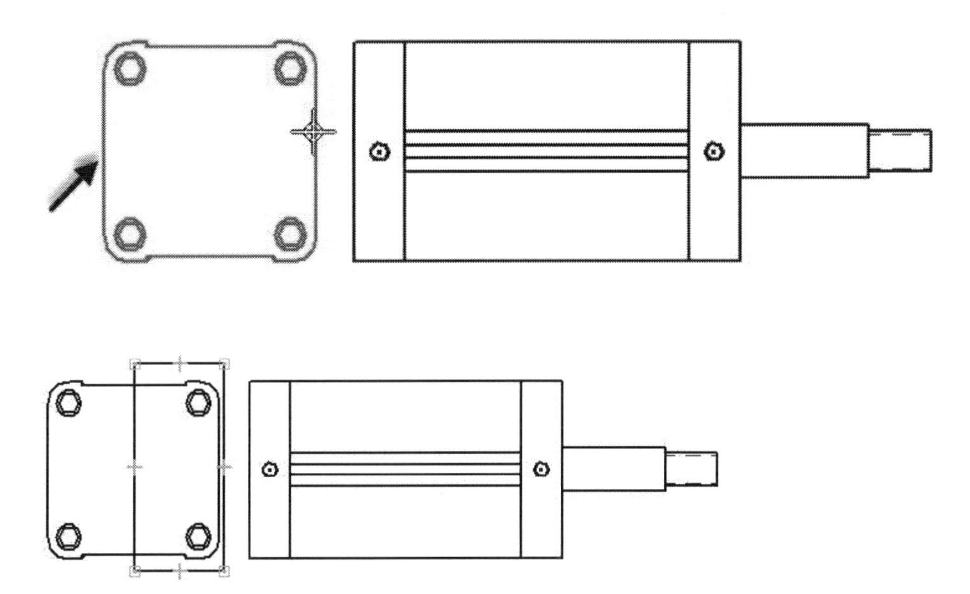

Now, move the pointer and click to specify the depth of the cutout. Select the drawing view to apply the cutout.

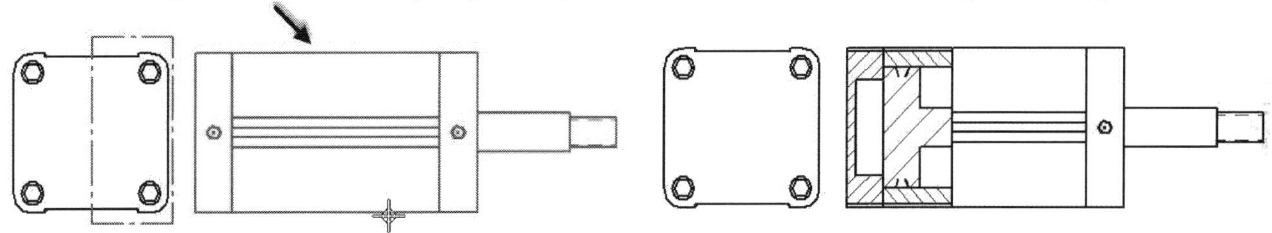

Exploded View

You can display an assembly in an exploded state as long as the assembly already has an exploded view defined. If you want to add an isometric exploded view, activate the **View Wizard** command and select the assembly from the **Select Model** dialog. On the command bar, click the **Drawing View Wizard Options** icon; the **Drawing View Creation Wizard** dialog pops up. On this dialog, select **Explode** from the **.cfg, PMI model view, or Zone** drop-down menu and click **OK**. Click on the drawing sheet to locate the exploded view.

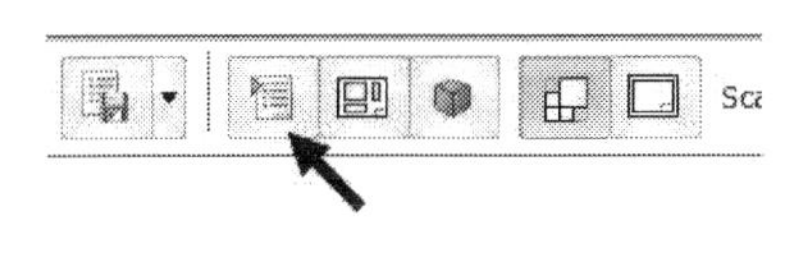

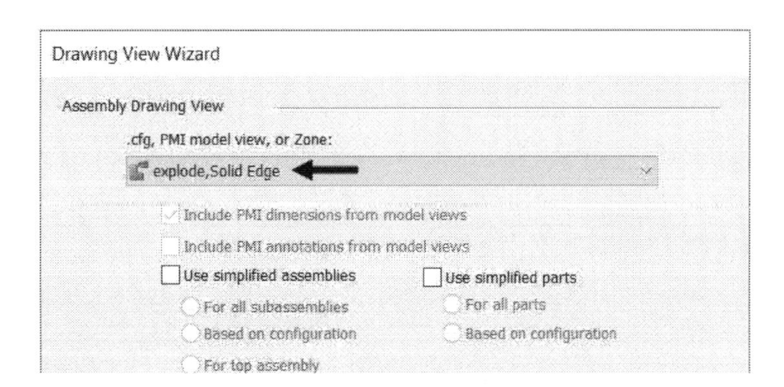

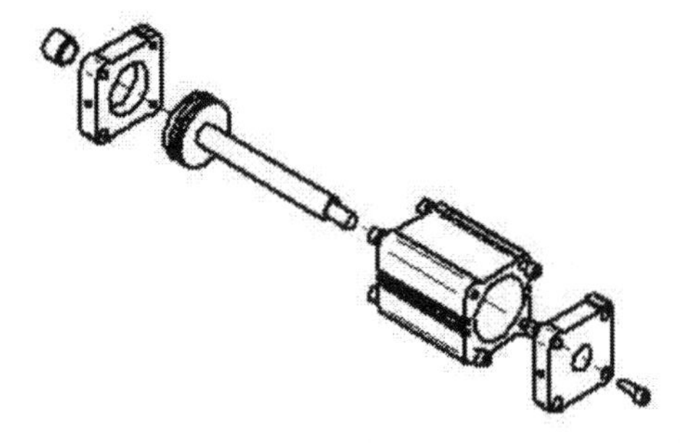

If you want to show an already existing isometric view in an exploded state, all you have to do is right-click the view and select **Properties**; the **High Quality View Properties** dialog pops up. On this dialog, click the **Display** tab and select the explode configuration file from the **.cfg, PMI model view, or Zone** drop-down menu and click **OK**. Next, click **Update Views** on the ribbon; the view will be updated.

Display Options

When working with Solid Edge drawings, you can control the way a model view is displayed by using the display options. Select a view from the drawing sheet and click the **Shading Options** icon on the **Edit Definition** command bar; a menu appears. On this menu, select the desired shading type and click **Update Views** on the ribbon. The shading type of the view will be changed.

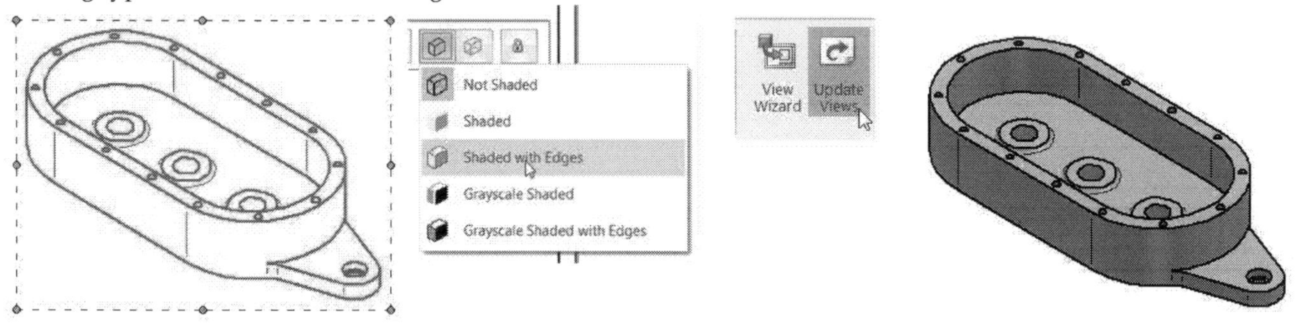

If you want to hide the hidden lines of multiple views, select them and click **Properties** on the command bar; the **High Quality View Properties** dialog pops up. On this dialog, uncheck the **Hidden edge style** option on the **Display** tab, and click **OK**. The hidden lines will disappear from the model view.

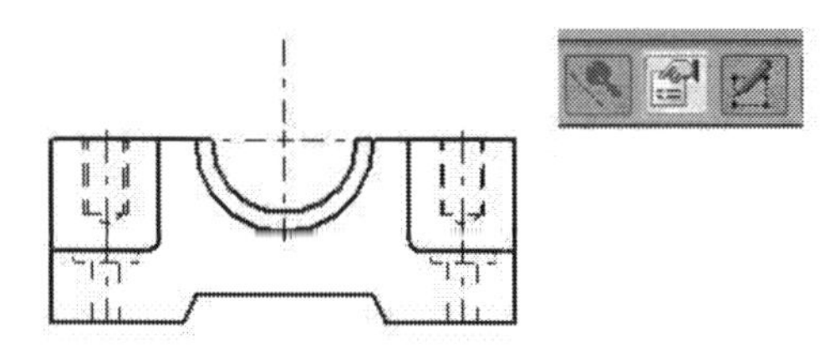

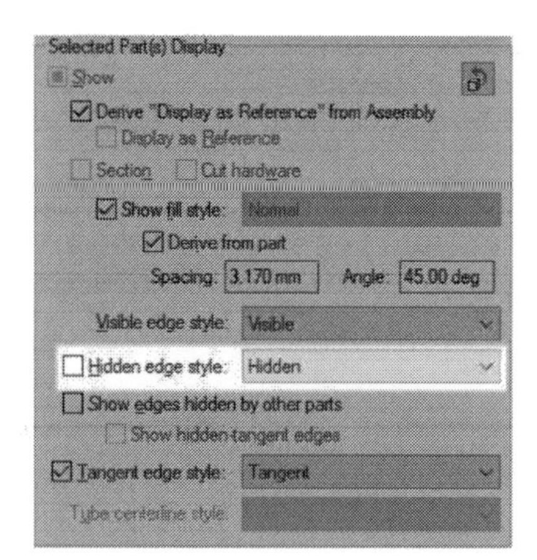

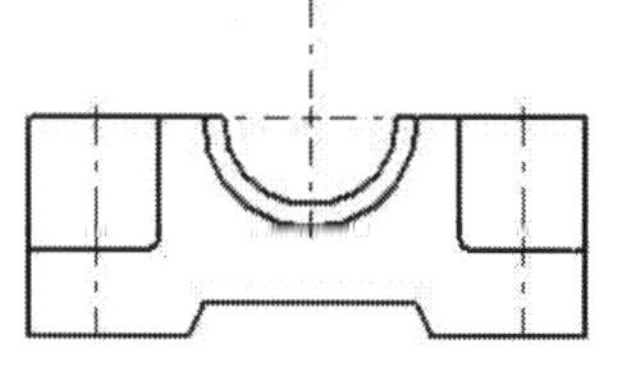

View Alignment

There are several types of views that are automatically aligned to a parent view. These include section views, auxiliary views, and projective views. If you move down a view, the parent view associated with it will also move.

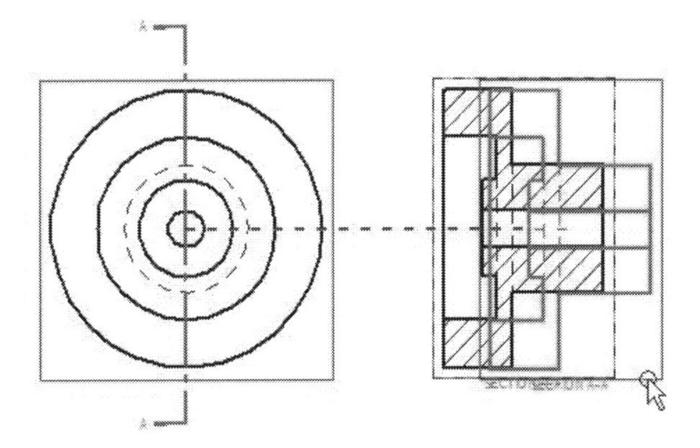

You need to break the alignment between them to move the view separately. Click the right mouse button on the view and select **Delete Alignment**. Now, click on the alignment line that appears between the two views.

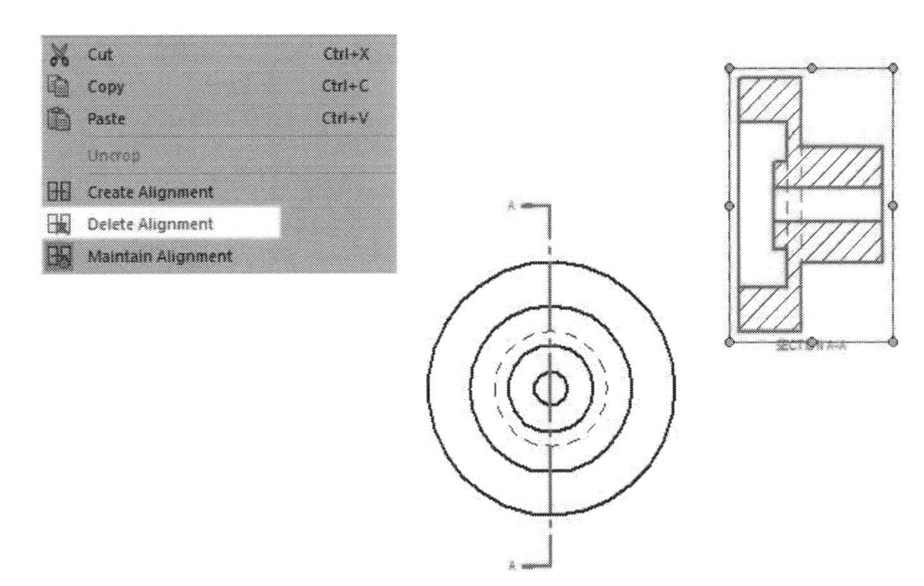

If you want to create alignment between the views, click the right mouse button on the parent view and select **Create Alignment**. On the command bar, select the required alignment option and click on the view to be aligned.

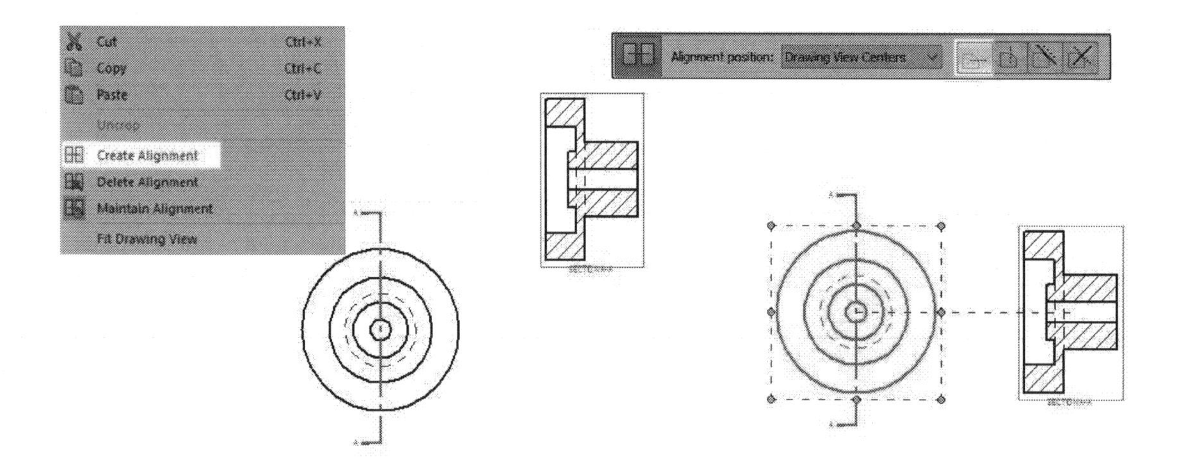

If you want to temporarily delete the alignment between the views, click the right mouse button on the view and deactivate the **Maintain Alignment** option. Now, drag the view to a new location without affecting the position of the parent view.

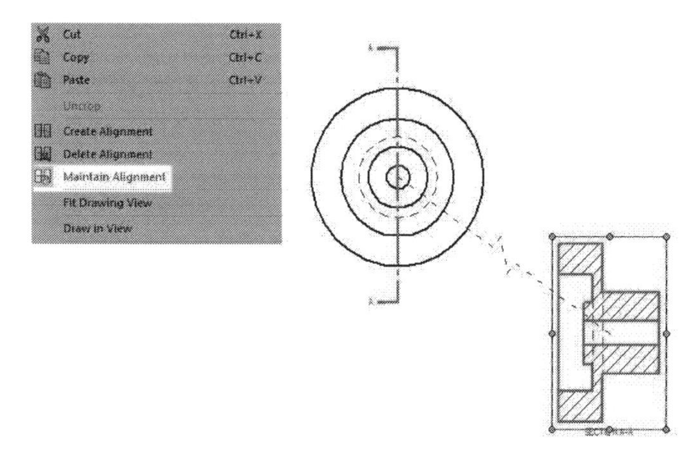

Parts List and Balloons

Creating an assembly drawing is very similar to creating a part drawing. However, there are few things unique in an assembly drawing. One of them is creating parts list. A parts list identifies the different components in an assembly. Generating a parts list is very easy in Solid Edge. First, you need to have a view of the assembly. Next, click **Home > Tables > Parts List** on the ribbon, and then click on the drawing view. On the command bar, click the **Properties** icon to open the **Parts List Properties** dialog. On this dialog, click the **List Control** tab and select an option from the **Global** section. You can select the **Top-level list**, **Atomic list**, or **Exploded list** option. Next, select the required configuration and click the **Columns** tab.

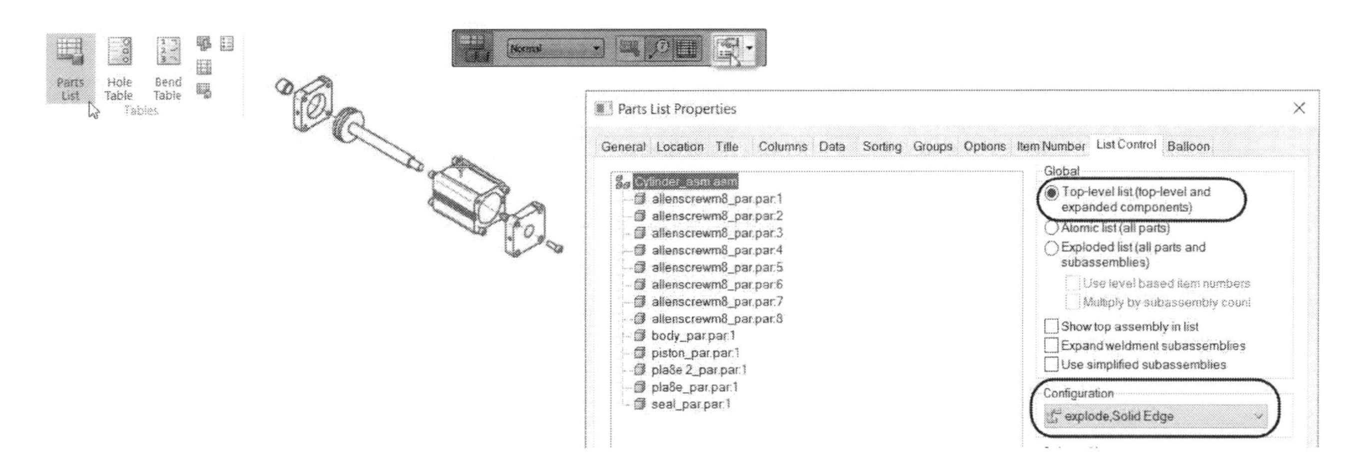

In this tab, select the column names from the **Columns** section and arrange them using the **Move Up** and **Move Down** buttons. To add a new column, select the column name from the **Properties** section and click **Add Column**. To remove a column, select the column name from the **Columns** section and click **Delete Column**. Type-in a value in the **Column width** box.

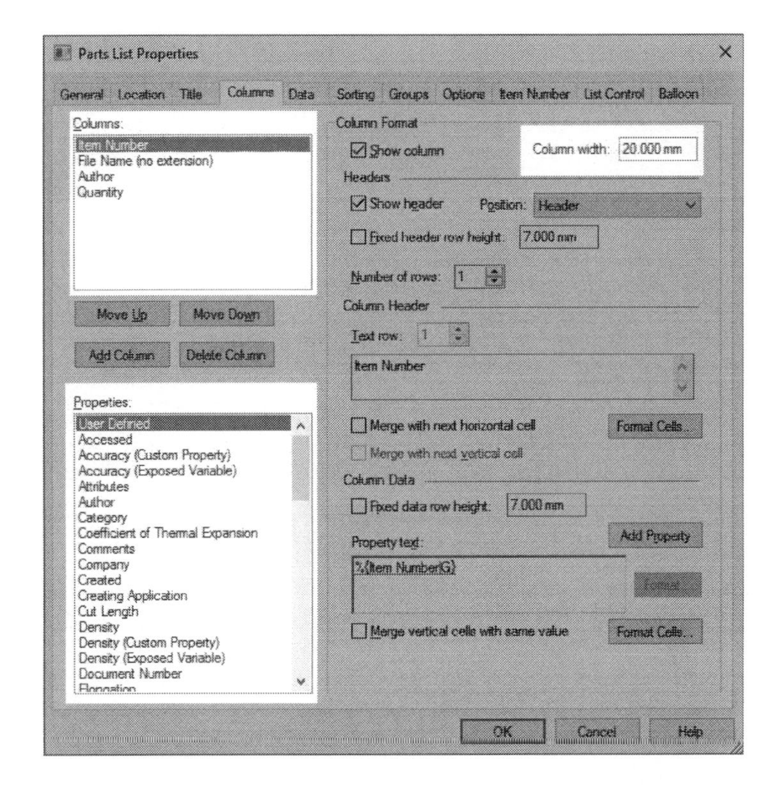

Click the **Balloon** tab and type-in a value in the **Text size** box. Click the **Shape** icon and select the desired balloon shape. If you want to hide the item count inside the balloon, uncheck the **Use Item Count for lower text** option. Under the **Auto-Balloon** section, check the **Create alignment shape** option to create magnetic lines aligning the balloons. Click on the **Pattern** button and select the alignment shape from the menu. Click on the **Order** button to change the direction in which the balloons are created. Click **OK** on the dialog to close it.

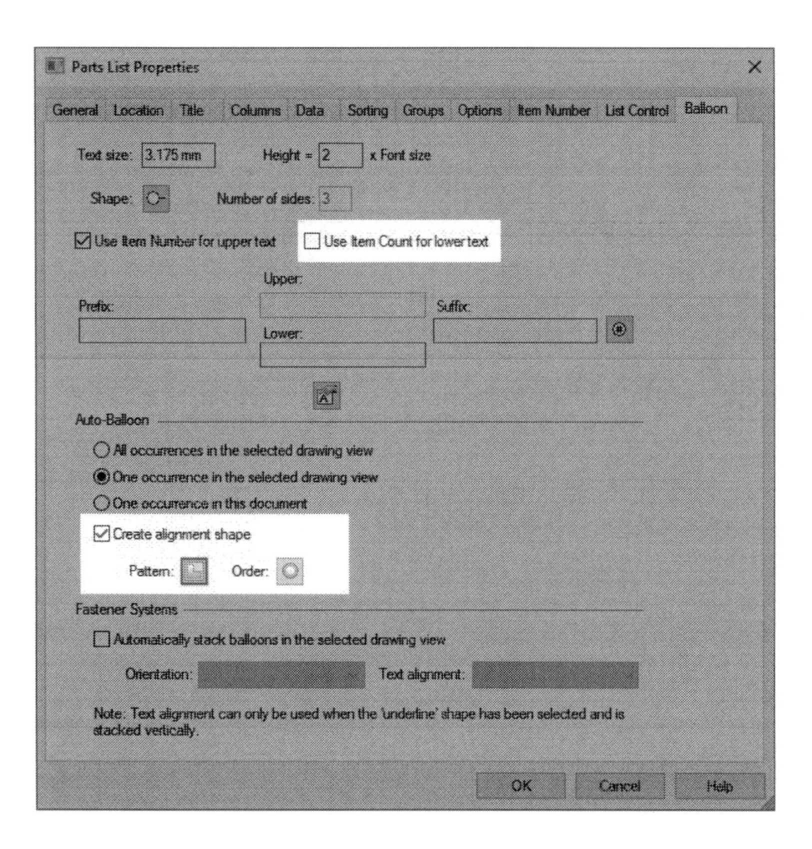

Click on the drawing sheet to place the parts list. The balloons are created automatically.

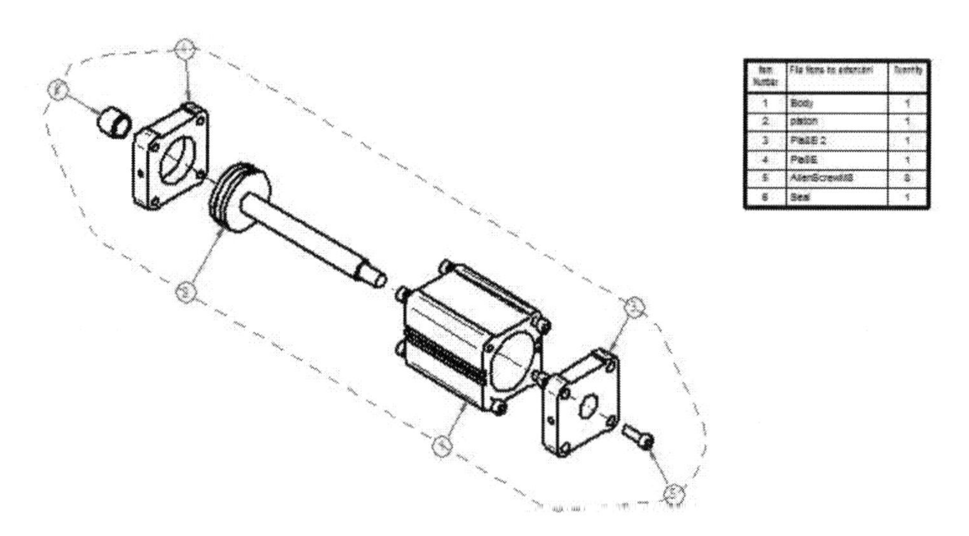

After creating the table, you can change the font type, size, orientation, and alignment by using the direct edit. To do this, double click on the table and select the icon that appears at the top left corner. Next, change the options on the direct edit and click anywhere in the drawing sheet.

Solid Edge ISO 3.500 mm		Author	Quantity
1	Break Lines	anand	1

Creating a Subassembly Parts List and Balloons

Solid Edge allows you to create parts list of a subassembly, which is part of the main assembly. Activate the **Parts List** command (**Home > Tables > Part List** on the ribbon). Select the assembly view from the drawing sheet and click the **Part List - From Selected Subassembly** icon on the command bar. On the **Select Assembly** dialog, select the subassembly from the Assemblies list and click **OK**. Click on the drawing sheet to position the part list table; the balloons are attached to the parts of the selected subassembly.

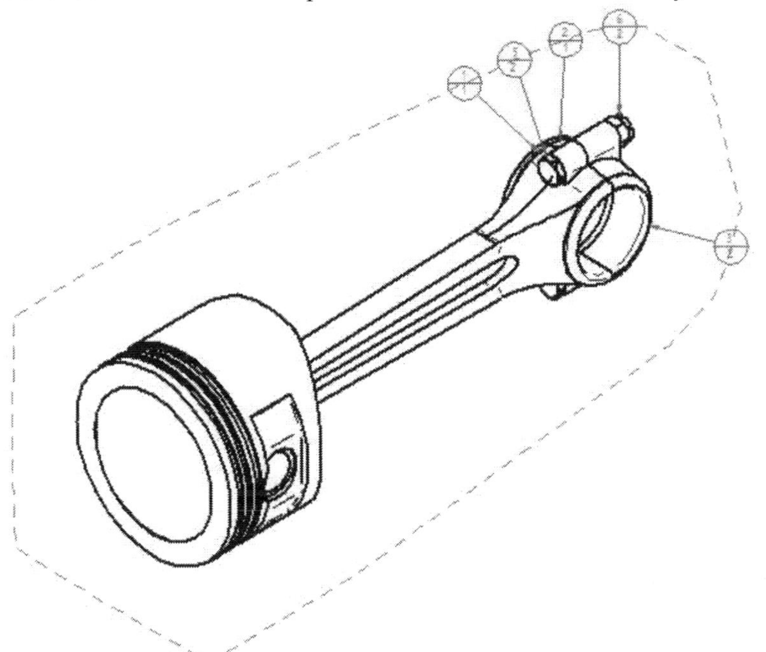

Item Number	File Name (no extension)	Quantity
1	Rod	1
2	Cap	1
3	Bearing_Brass	2
4	Bearing_Bush	1
5	Bolt	2
6	Nut	2

Dimensions

Solid Edge provides you with different ways to add dimensions to the drawing. One of the methods is to retrieve the dimensions that are already contained in the 3D part file. Click **Home > Dimension > Retrieve Dimensions** on the ribbon. On the command bar, select the dimension types that you want to retrieve. You can click the **Multiple Views** icon on the command bar, if you want to retrieve the linear dimensions into multiple views of the drawing view. Click on the drawing view where you want to display the dimensions.

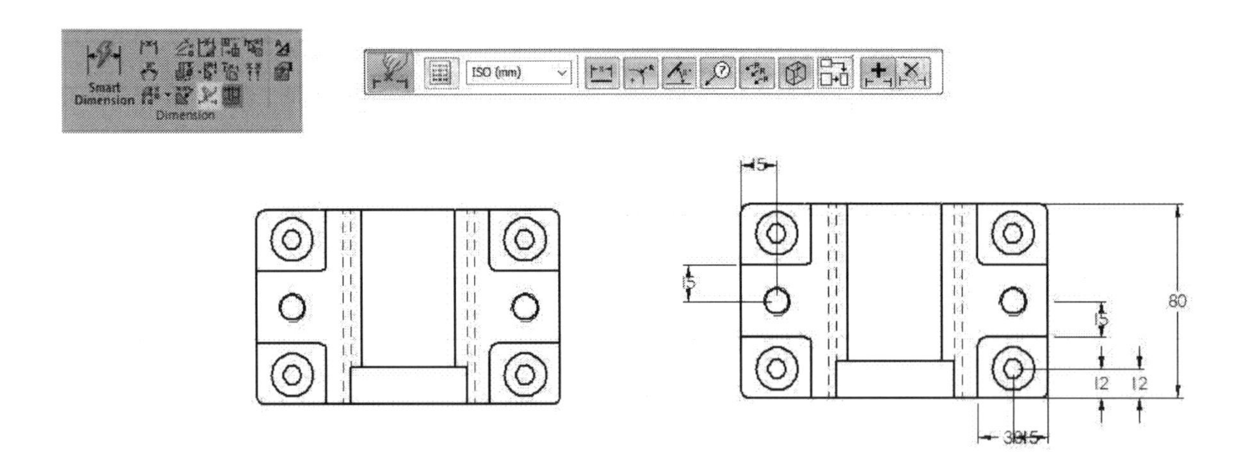

You may notice that there are some unwanted dimensions. Simply select them and press Delete to remove them. In addition, the dimensions may not be positioned properly. To arrange them properly, activate the **Arrange Dimensions** command (click **Home > Dimension > Arrange Dimension** on the ribbon). Click on the dimensions, and then click the green check on the command bar. The dimensions will be arranged properly.

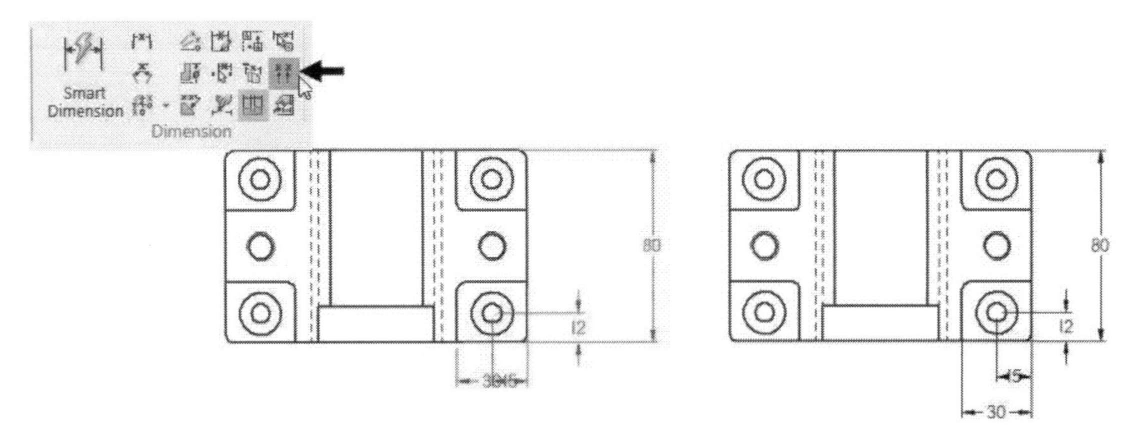

If you want to add some more dimensions, which are necessary to manufacture a part, activate the **Smart Dimension** command and add them to the view. You can also use the **Distance Between** command to add linear dimensions.

Note: *You can use the dimension handles to modify the position of the dimension and size of the dimension and extension lines. The dimension handles are displayed on selecting a dimension.*

Concentric diameter Dimensions

Solid Edge has a new option to create concentric diameter dimensions. Activate the **Smart Dimensions** command (on the ribbon, click **Home > Dimension > Smart Dimension**), and then select the circular edge of the drawing view. On the **Smart Dimension** command bar, click the **Concentric Dimension** icon. Place the diameter dimension, and then select a concentric circle; the second dimension is placed, automatically. Likewise, select other concentric circles to add dimensions to them.

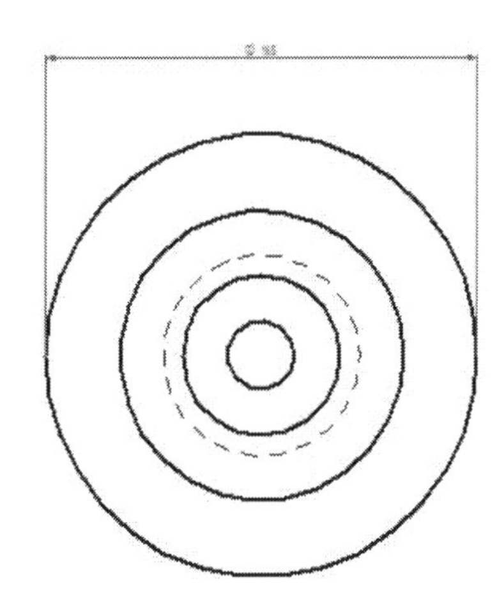

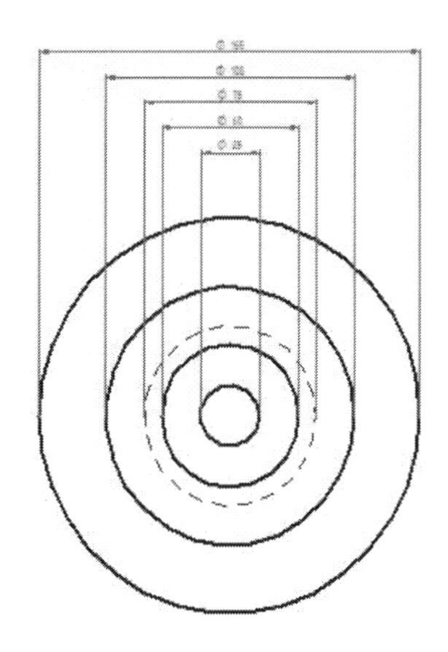

Coordinate Dimensions

Coordinate dimensions are another type of dimensions that can be added to a drawing. To create them, activate the **Coordinate Dimension** command (click **Home > Dimension > Coordinate Dimensions** drop-down **> Coordinate Dimension** on the ribbon), and then click on any edge of the drawing view to define the ordinate or zero reference. Now, click on the points or edges of the drawing view and place the coordinate dimensions.

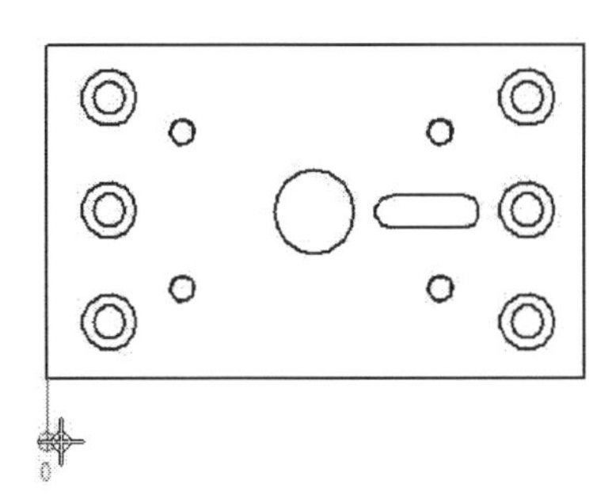

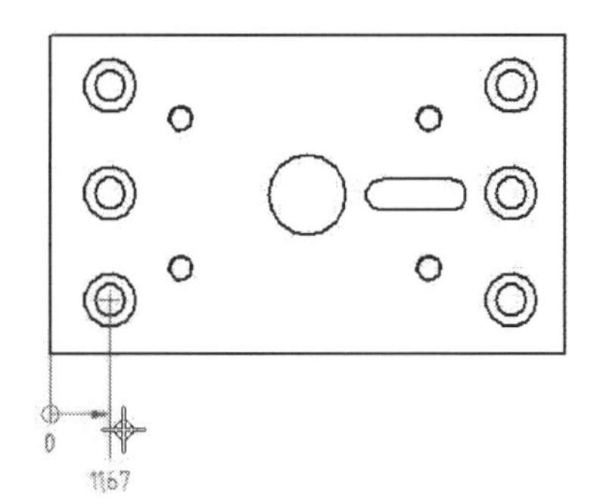

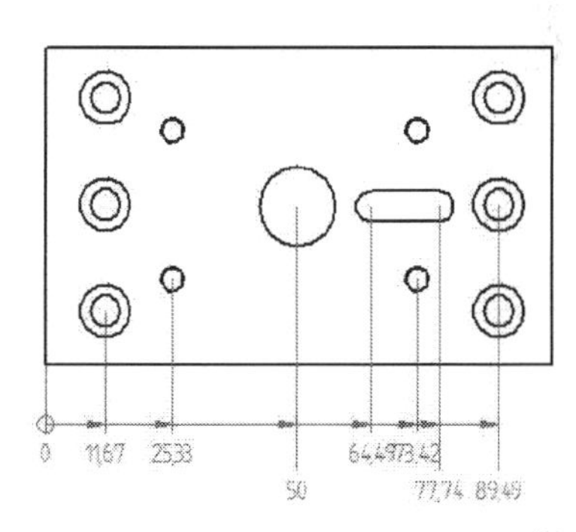

Automatic Coordinate Dimensions

The **Automatic Coordinate Dimensions** command creates coordinate dimensions automatically. On the ribbon, click **Home > Dimension > Coordinate Dimensions** drop-down **> Automatic Coordinate Dimensions**, and click the **Keypoint Options** button on the command bar. On the **Keypoint Options** dialog, select the type of the points that should be selected to create the coordinate dimensions. Click **OK** and select the drawing view. Click **Accept** and select a point on the drawing view to define the origin. Move the pointer vertically or horizontally and click to position the coordinate dimensions.

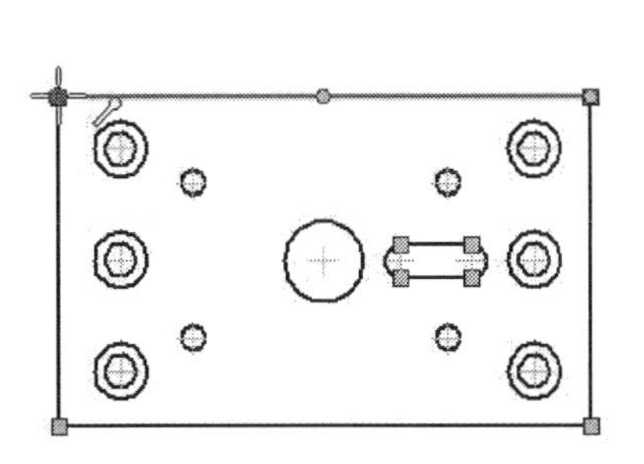

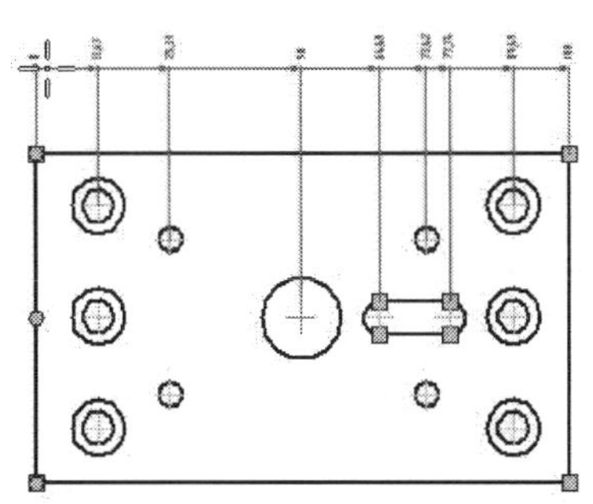

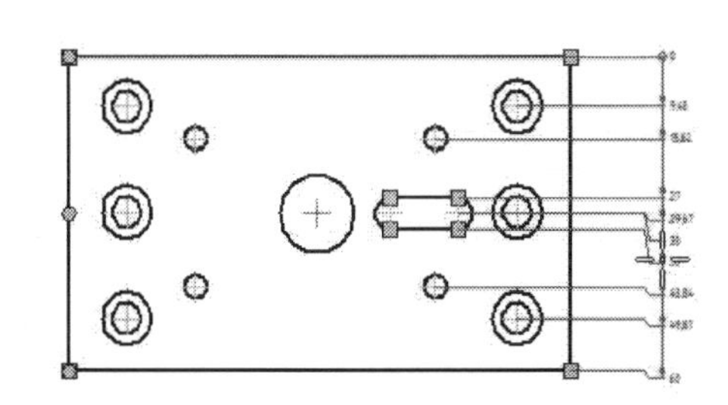

Change Coordinate Origin

The **Change Coordinate Origin** command is used to change the origin of the coordinate dimensions. Activate is command (On the ribbon, click **Home > Dimension > Coordinate Dimensions** drop-down **> Change Coordinate Origin**), and click on a coordinate dimension to change it as the origin.

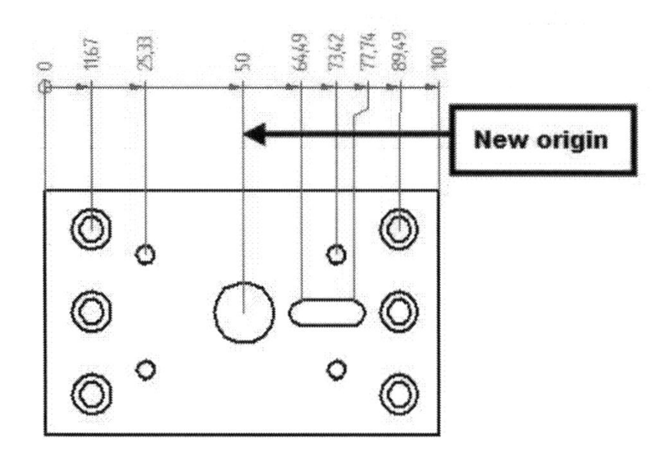

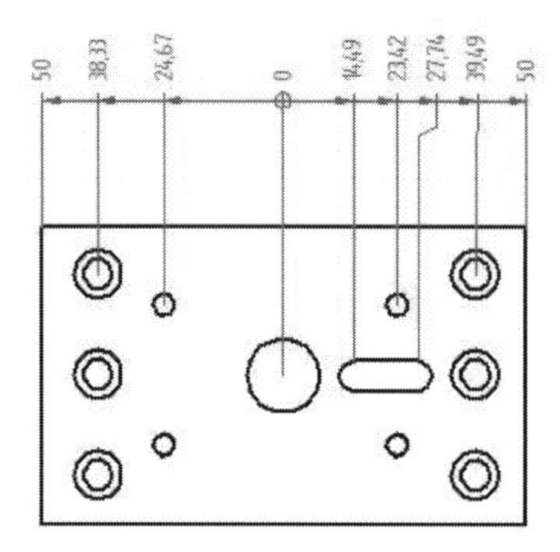

Hole Table

The **Hole Table** command creates a table showing the X, Y, and Z coordinates of the hole, sizes, and other properties. Activate the **Hole Table** command (on the ribbon, click **Home** tab > **Tables** panel > **Hole Table**), and then click on a point of the drawing view to define the origin of the coordinate system. Next, move the pointer vertically and click to place the X1 axis. Again, click on the origin point, move the pointer horizontally, and then click to place the Y1 axis. On the command bar, click the **Properties** icon to open the **Hole Table Properties** dialog. On this dialog, specify the options for table columns, callout, data, and so on. Next, click **OK** on the dialog. On the command bar, click **Selection Method** > **By User Selection**, and then select the hole features one-by-one. The numbering of the holes will be generated based on the order in which you select the holes. You can also drag a selection box across all the hole feature if the numbering order does not matter you.

If you set the **Selection Method** to **By Drawing View**, then you need to define the origin and X1 and Y1 axes. Next, click on the drawing view; all the hole features in the selected drawing view are selected automatically.

Click the **Accept** icon on the command bar after selecting the holes. Click in the drawing sheet to place the hole table.

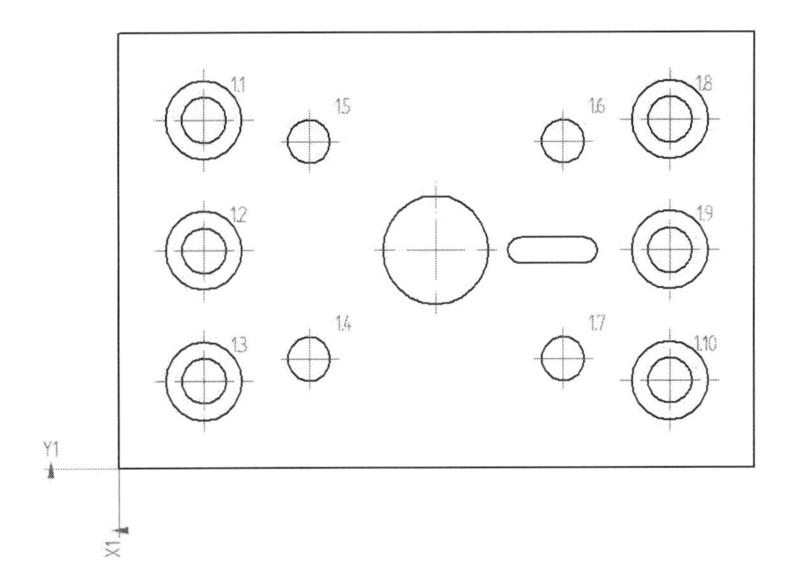

Hole	X	Y	Size
1.1	20 mm	80 mm	10.5 mm
1.2	20 mm	50 mm	10.5 mm
1.3	20 mm	20 mm	10.5 mm
1.4	45 mm	25 mm	10 mm
1.5	45 mm	75 mm	10 mm
1.6	105 mm	75 mm	10 mm
1.7	105 mm	25 mm	10 mm
1.8	130 mm	80 mm	10.5 mm
1.9	130 mm	50 mm	10.5 mm
1.10	130 mm	20 mm	10.5 mm

You can add more holes to the hole table by using the **Add or Remove Holes Step** icon. To do this, select the hole table and click the **Add or Remove Holes Step** icon on the command bar. Next, select the hole feature to be added to the table, and then click the **Accept** icon.

Update Views

The **Update Views** command updates the drawing view to reflect the changes made to the 3D model. If you make any changes to the 3D model, the drawing view related to it is not updated automatically. You need to use the **Update Views** command to update the changes. On the ribbon, click **Home** tab > **Drawing Views** panel > **Update Views**; the **Dimension Tracker** dialog appears if any dimension value is changed in the 3D model. Click the **Clear All** button, and then close the dialog; the dimension value is updated in the drawing view.

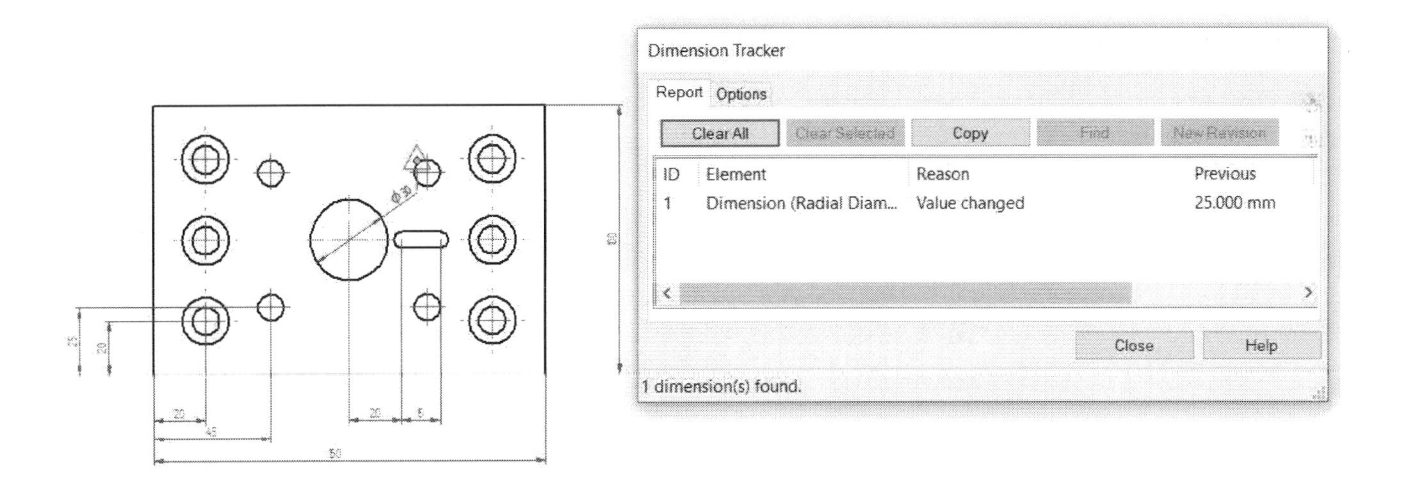

Center Marks and Centerlines

Centerlines and Centermarks are used in engineering drawings to denote hole centers and lines. To add center marks to the drawing, activate the **Center Mark** command (click **Home > Annotation > Center Mark** on the ribbon) and click on the hole circles. On the Command bar, click the **Projection Lines** icon, if you want to add projected centermarks to the circle.

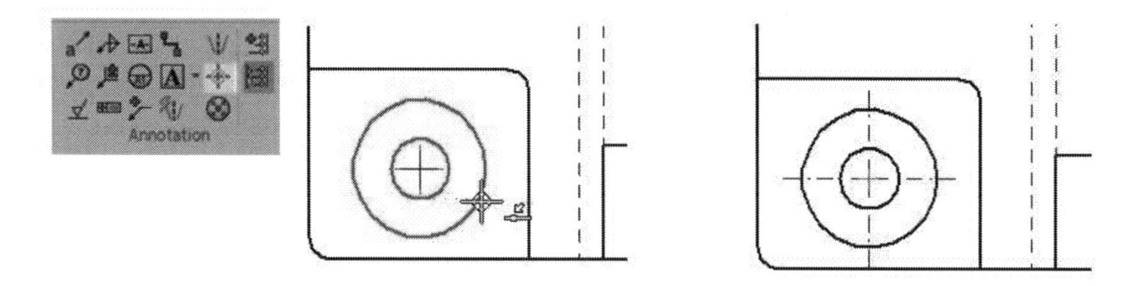

To add centerlines, activate the **Centerline** command (click **Home > Annotation > Centerline** on the ribbon). On the command bar, select **Placement Options > By 2 Lines**, and then click on two parallel edges of the drawing view. A centerline will be created between the two lines.

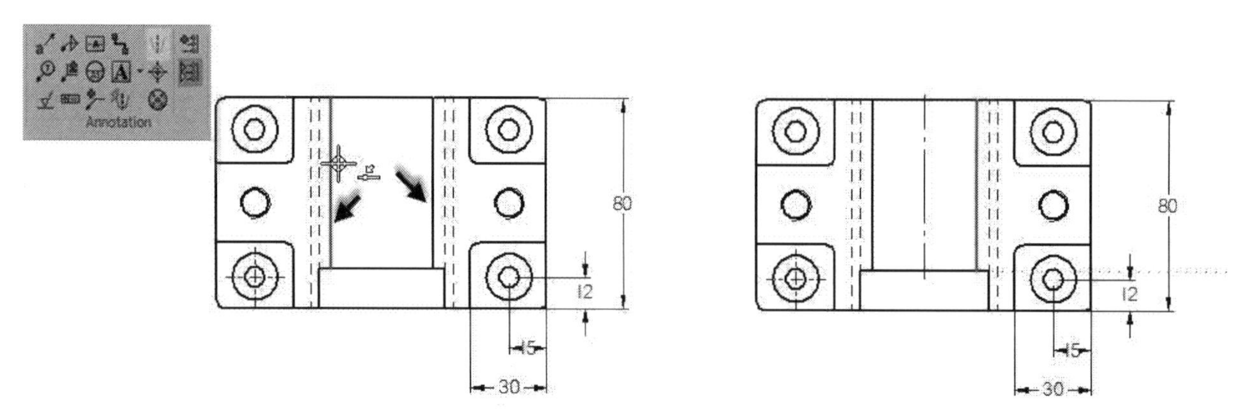

If you want to add centrelines automatically, activate the **Automatic Centerlines** command (click **Home > Annotation > Automatic Centerlines** on the ribbon). The **Automatic Centerlines** command bar pops up. On the command bar, click the **Options** icon to open the **Center Line and Center Mark Options** dialog. On the dialog, select the element to which the centerlines and center marks are to be added. Click **OK** to close the dialog. Click the drawing view to add centerlines and center marks.

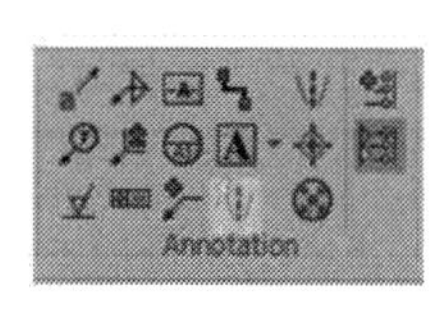

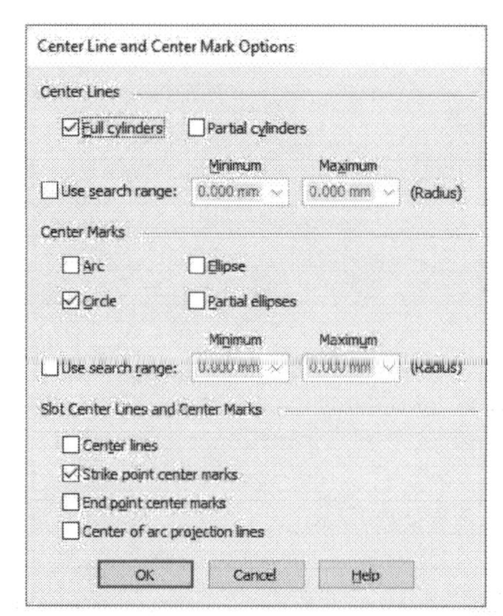

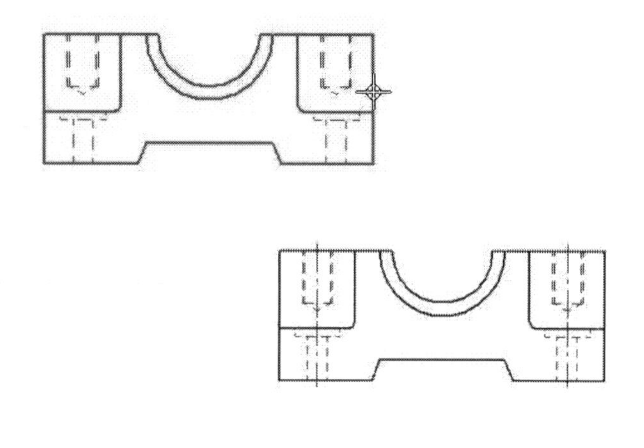

Bolt Hole Circle

The **Bolt Hole Circle** command (click **Home > Annotation > Bolt Hole Circle** on the ribbon) allows you to add center marks to the holes arranged in a circular fashion. Activate this command and click for the center of the bolt hole circle. Drag the pointer and click for the radius point of the bolt hole circle. A bolt circle will be created.

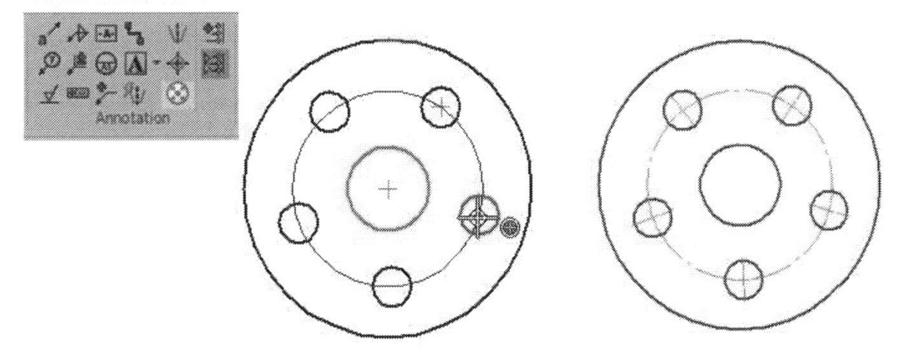

Callouts and Leaders

Callouts and leaders are an essential element in creating drawings. In this section, you will learn to add callouts and leaders to a drawing. For example, to add a counterbore hole callout, activate the **Callout** command (click **Home > Annotation > Callout** on the ribbon). On the **Callout Properties** dialog, type-in values in the **Callout text** and **Callout text 2** boxes. You can use the **Special Character** icons available on the dialog. Click the **OK** button and click on the hole. Drag the mouse pointer and click to place the callout.

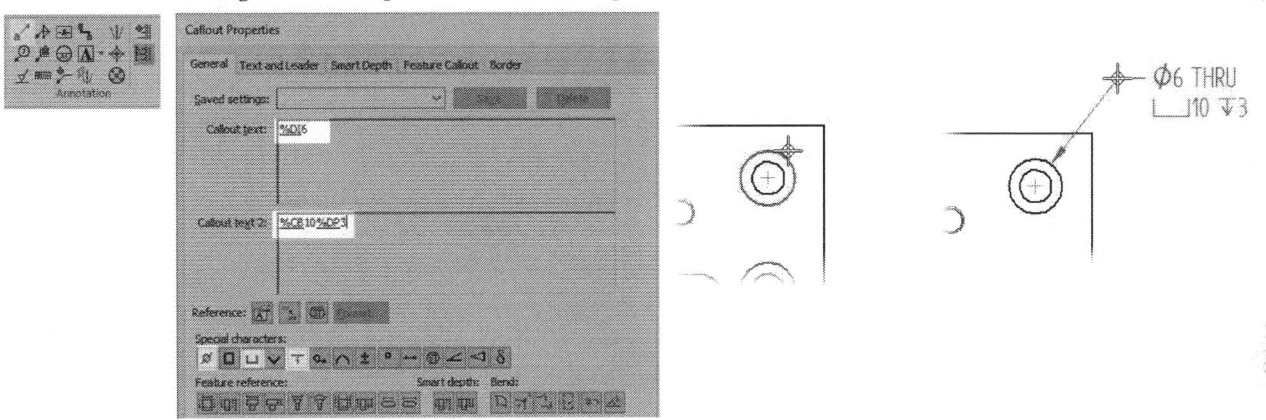

If you have multiple elements in a drawing with a same callout value, you can use leaders to connect them to an existing callout. Activate the **Leader** command (click **Home > Annotation > Leader** on the ribbon) and click on an element. Drag the mouse pointer and click on an already existing callout.

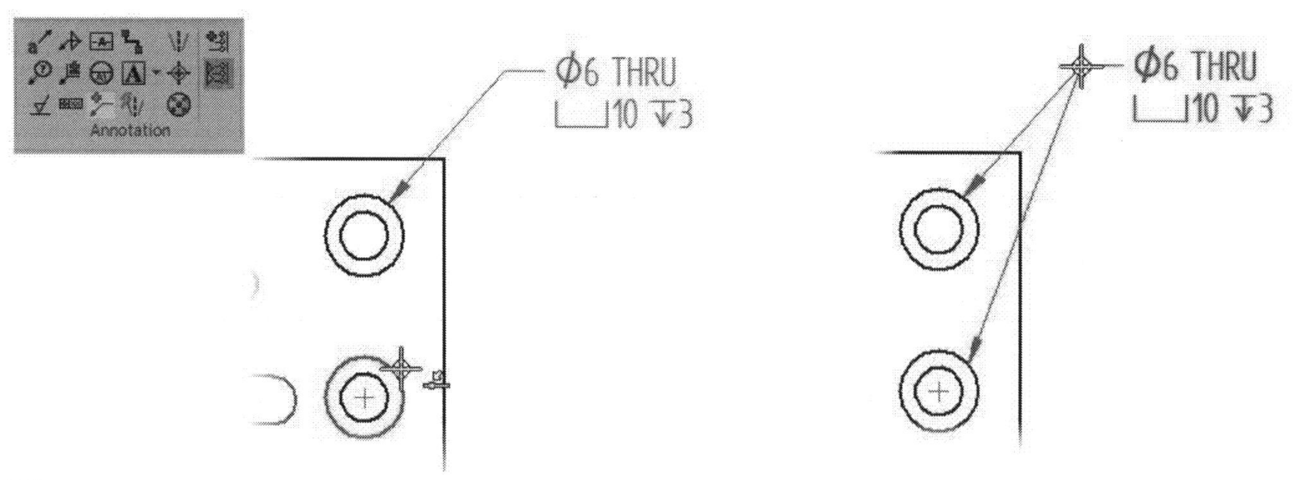

Notes

Notes are important part of a drawing. You add notes to provide additional details, which cannot be done using dimensions and annotations. To add a note or text, activate the **Text** command (click **Home > Annotation > Text** on the ribbon). On the command bar, select the font and font size. Create a box and type text inside it.

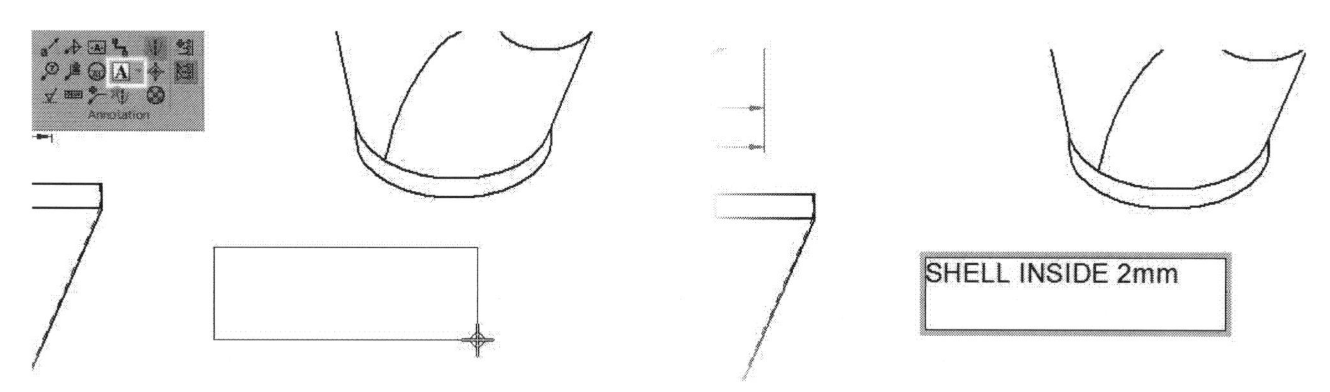

You can use the **Text** command to insert associative text (reference and property texts). To insert an associative text in the drawing, activate the **Text** command and click to define the text position. On the **Text** command bar, click the **Insert Symbols, Characters or Property Text** icon. Select **Insert Property Text** from the flyout. On the **Insert Property Text** dialog, click the **Property Text** icon to open the **Select Property Text** dialog. On this dialog, select a property from the **Properties** list and click **Select**. Click **OK** on the **Select Property Text and Insert Property Text** dialogs; the property text will be inserted on the drawing sheet.

Adding Technical Requirements

Solid Edge has an option to add technical requirements to the drawing. On the ribbon, click **Home > Annotation > Text** drop-down **> Technical Requirements**. On the **Technical Requirements Properties** dialog, click the **General** tab and type-in the technical requirement in the box available at the top. Click **Insert** to add the technical requirement. Likewise, type-in another technical requirement and insert it into the table available at the bottom. Select the bullet style and format from the **Style** and **Format** drop-downs.

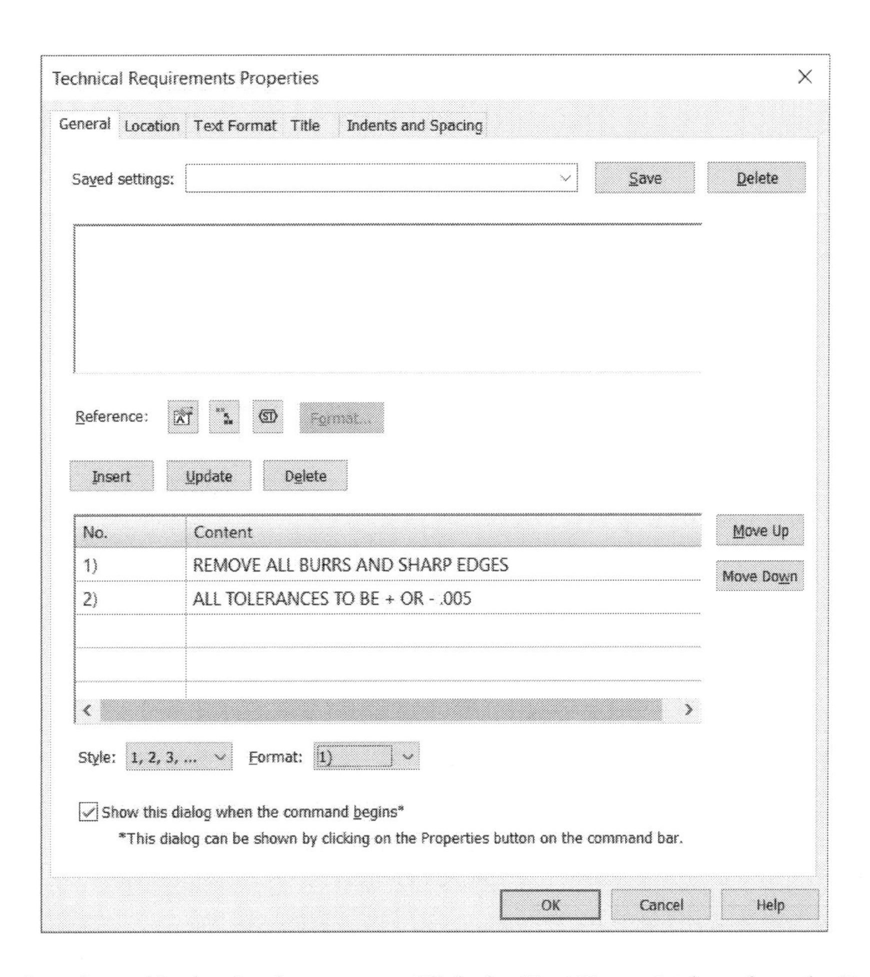

Click the **Location** tab and specify the **Anchor corner**. Click the **Text Format** tab and set the Font type, size, color, aspect ratio, and fit width. Click the **Title** tab and type-in the title of the technical requirements in the **Title text** box. Likewise, specify the Indentation and spacing on the **Indenting and Spacing** tab. Click **OK** on the Technical **Requirement Properties** dialog, and then position the technical requirements on the drawing sheet.

GENERAL NOTES: UNLESS OTHERWISE SPECIFIED
1) REMOVE ALL BURRS AND SHARP EDGES
2) ALL TOLERANCES TO BE + OR - .005

Compare Drawings

Solid Edge allows you to compare two different versions of a drawing file. It is very useful when a drawing file is shared between different members of a team. To compare two drawings, click **Application Menu > Tools > Compare Drawings**. On the **Compare Drawings** dialog, under the **Configure New Comparison** section, click the **Browse** button next to the **File 1** drop-down. On the **Open File** dialog, select the first version of a drawing file, and then click **Open**. Likewise, click the **Browse** button next to the **File 1** drop-down and select the second version of the drawing. Select the sheets to be compared from the **Sheet** drop-downs, and then click **Compare**. The differences between the two versions of the drawing are displayed in the **Differences** section. You can use the **Zoom Area**, **Fit**, **Pan**, and **Zoom** buttons under the **Display** section to view various portions of the drawing. You can also save the comparison file for future use with the help of the **Save** button. You can load an existing comparison file using the **Browse** button in the **Open Existing Comparison** section.

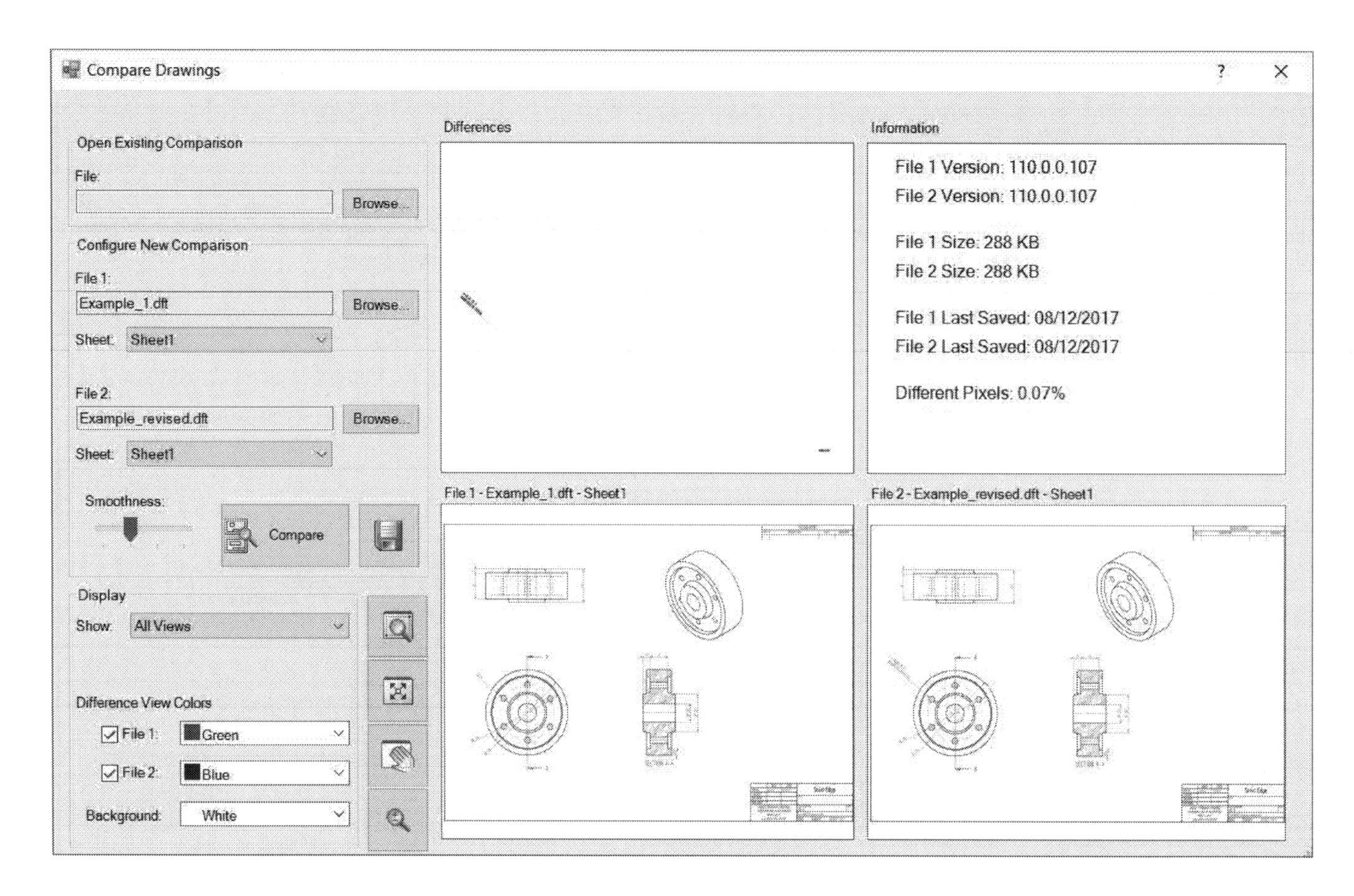

Examples

Example 1

In this example, you will create the 2D drawing of the part shown below.

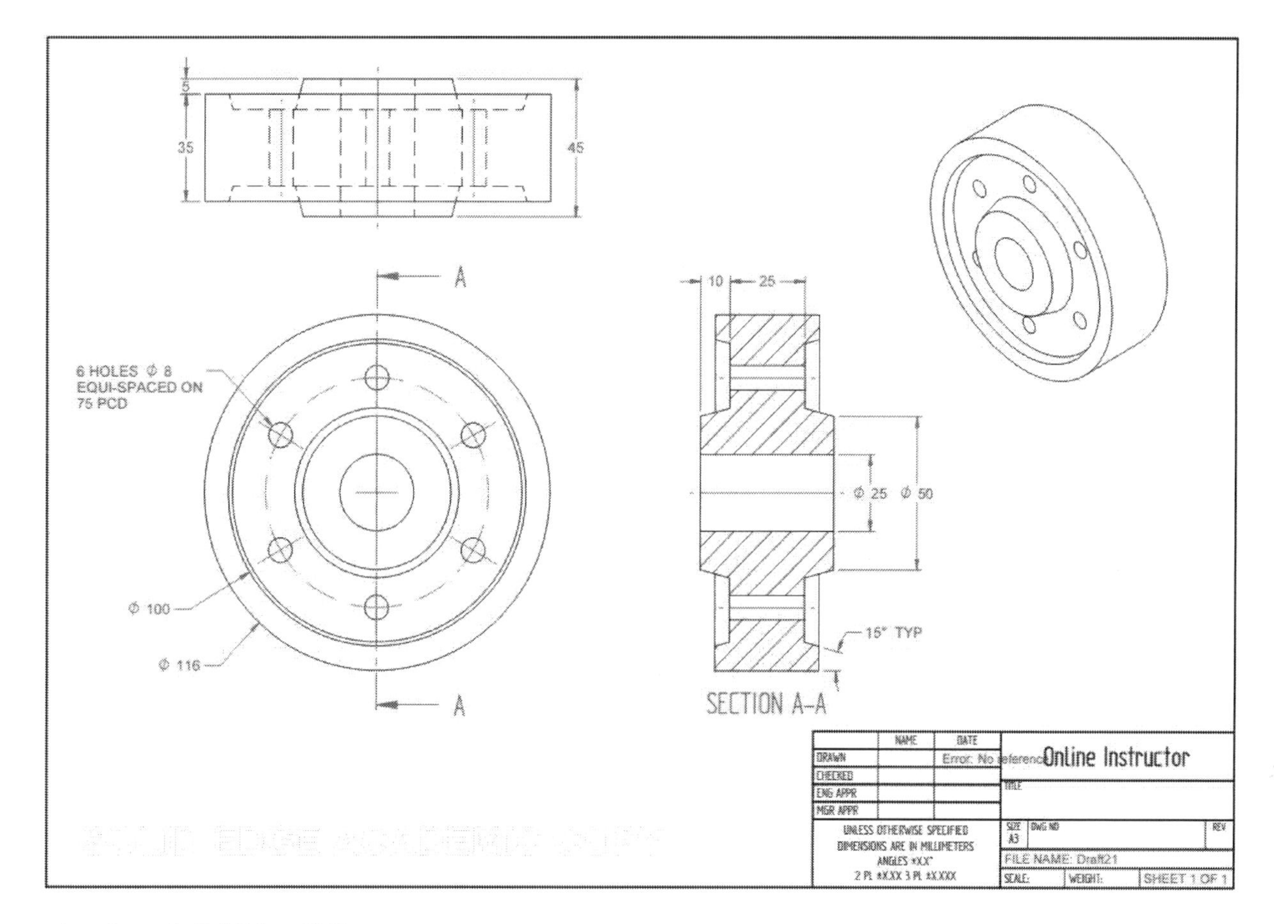

1. Start **Solid Edge 2019**.
2. On the **Application Menu**, click **New > ISO Metric Draft** to start a new drawing.
3. At the bottom of the window, right-click on the **Sheet 1** tab and select **Sheet Setup**.
4. On the **Sheet Setup** dialog, click the **Size** tab and select the **Standard** option. Set the sheet size to **A3 Wide (420mm x 297mm)**.
5. Click the **Background** tab and set the **Background sheet** to **A3-Sheet**. Click **OK** to close the dialog.

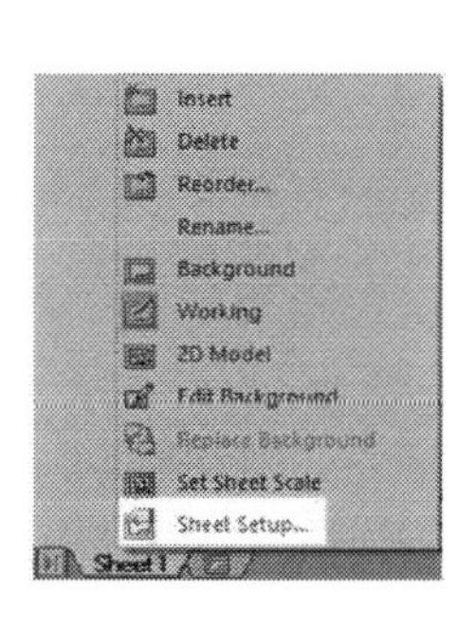

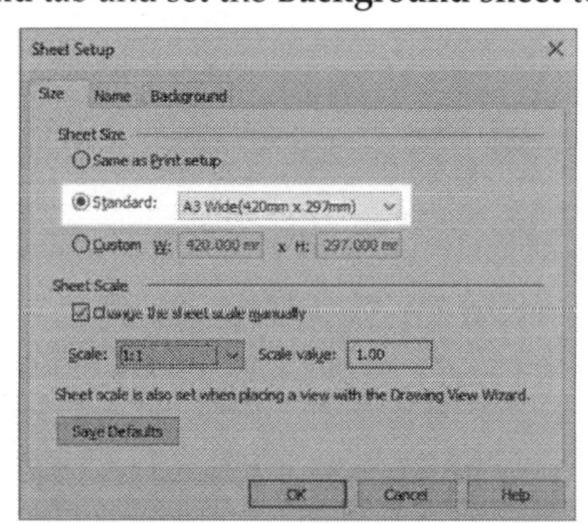

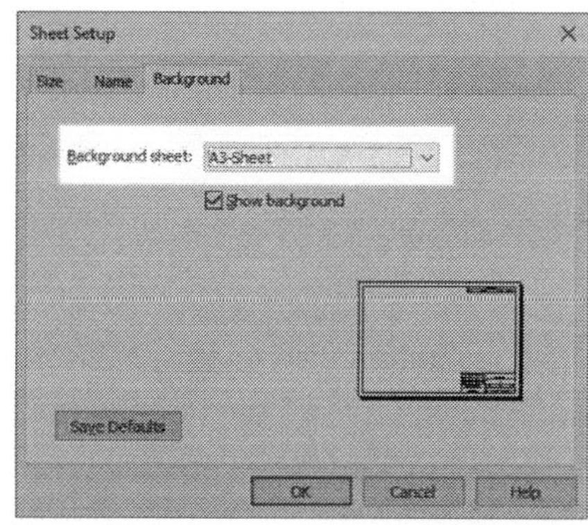

6. On the ribbon, click **View > Sheet Views > Background** to activate the background. Deactivate the **Working** icon located below the **Background** icon.

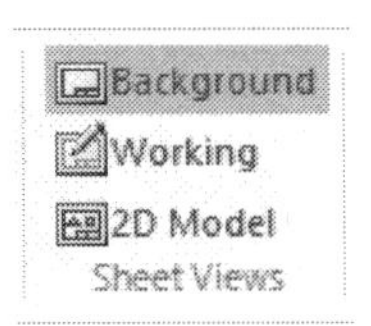

7. At the bottom of the sheet, click **A3-Sheet**.
8. Select the revision table and press Delete on your keyboard.

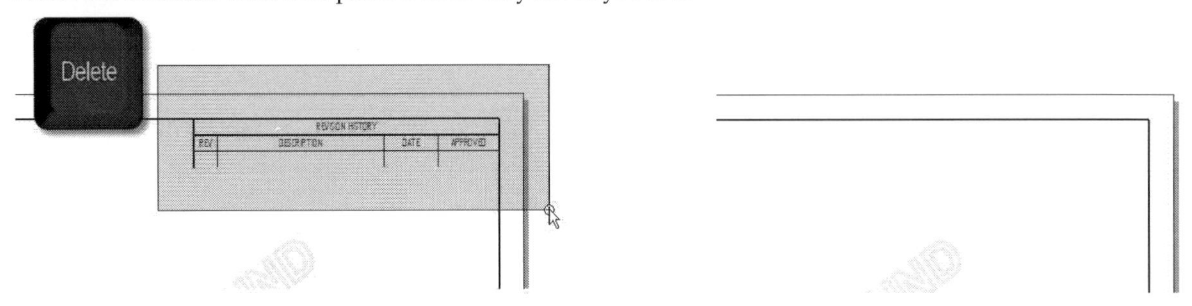

9. In the Title Block, change the company name to Online Instructor.

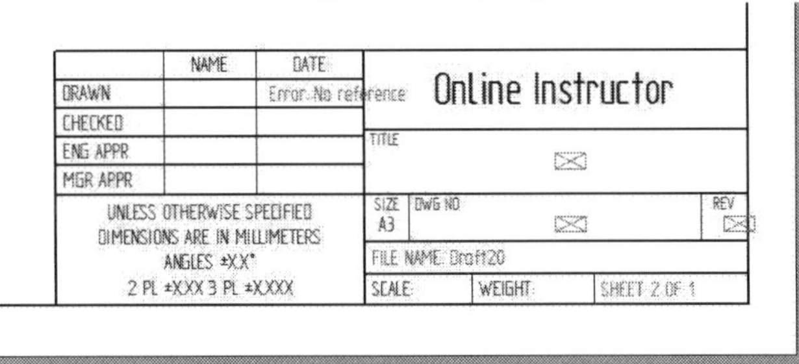

10. Activate the **Working** icon on the **Sheet views** panel, and then deactivate the **Background** icon.
11. On the **Application Menu**, click **Settings > Options** button. On the **Solid Edge Options** dialog, click the **Drawing Standards** tab. Set the **Projection Angle** to **Third** and click the **OK** button.

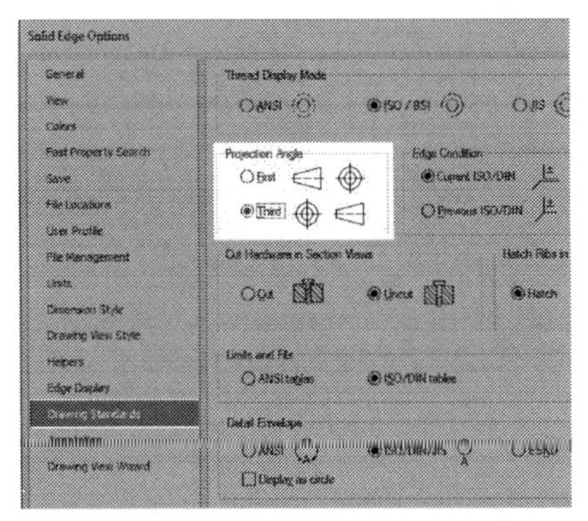

12. Activate the **Styles** command (click **Home > Dimension > Styles** on the ribbon). On the **Style** dialog, set the **Style type** to **Dimension**. Select **ISO (mm)** from the **Styles** box and click the **Modify** button.

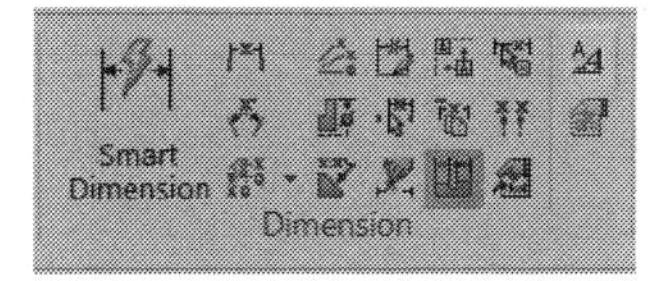

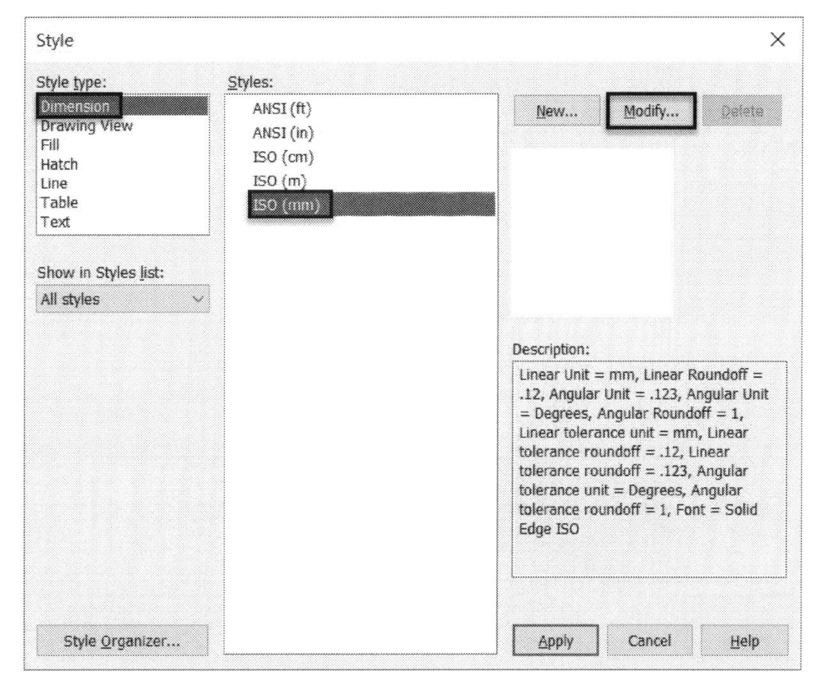

13. On the **Modify Dimension Style** dialog, click the **Text** tab and set the **Font** type to **Arial**. Set the **Orientation** to **Horizontal** and **Position** to **Embedded**.

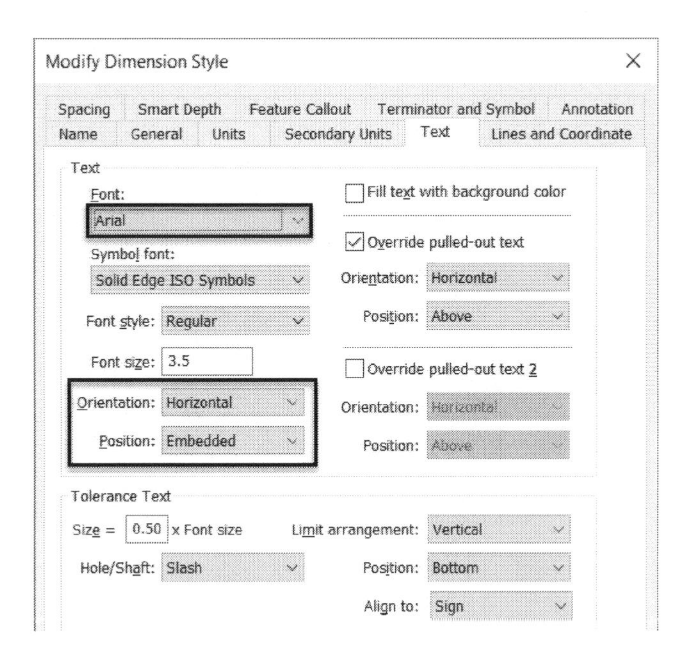

14. Click the **Units** tab and set the **Round-off** value to **1**.

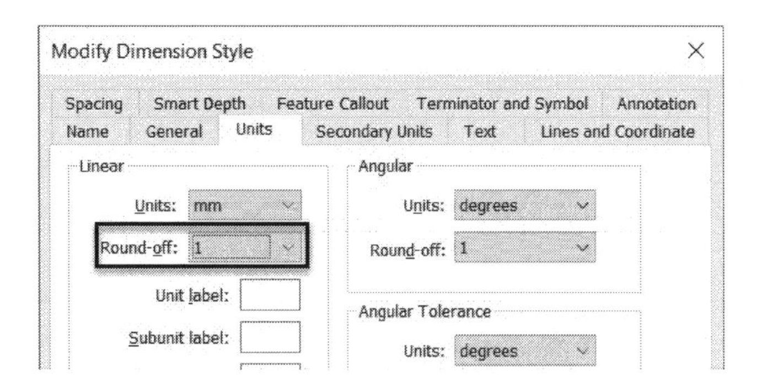

15. Click the **Lines and Coordinate** tab and set the **Element gap** to 0.5 x Font Size. Under **Dimension Lines**, uncheck the **Connect** option. Click **OK** and then **Apply** to make changes to the dimension style.

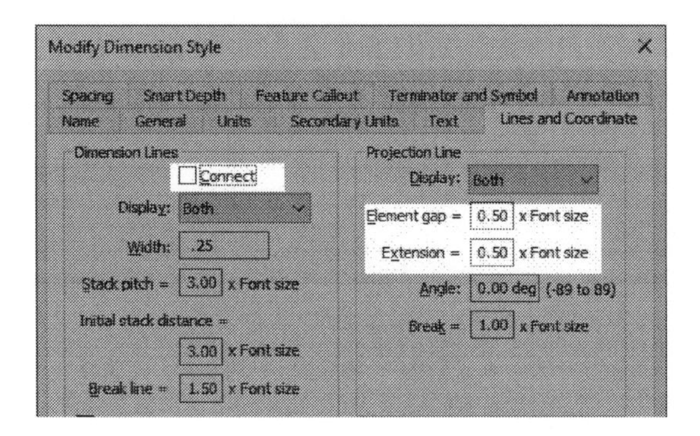

16. On the **Quick Access Toolbar**, click the **Save** icon and browse to the location C:\Program files\Solid Edge 2019\Template\ISO Metric. Type **Online Instructor** in the **File name** box and click **Save**. Close the file.
17. On the **Application Menu**, click the **New** icon to open the **New** dialog. On this dialog, click **Standard Templates > ISO Metric** and select **Online Instructor.dft**. Click **OK** to start a new drawing file.

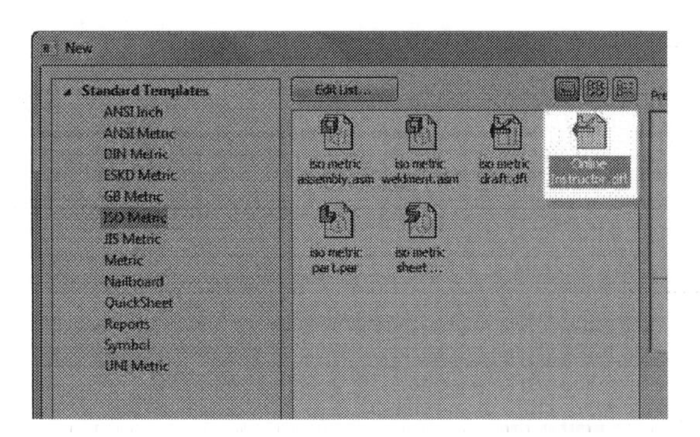

18. Activate the **View Wizard** command (click **Home > Drawing Views > View Wizard** on the ribbon).
19. Browse to the location of Exercise 1 of Chapter 5 and click on the part file. Click the **Open** button.
20. On the command bar, click the **Drawing View Layout** icon.
21. On the **Drawing View Wizard** dialog, set the **Primary View** to **front**. Click on the top view and isometric view icons. Click the **OK** button to close the dialog.

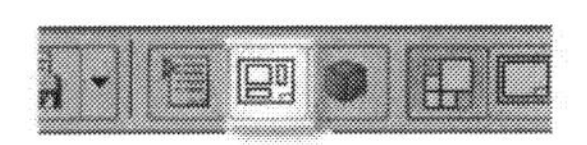

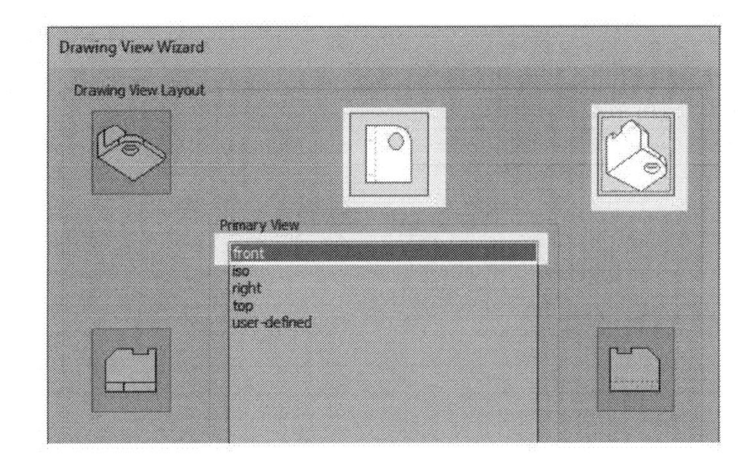

22. On the command bar, set the **Scale** to **1:1**
23. Click on the left portion of the sheet to place the drawing views. Drag the isometric view and position it at the top right corner.

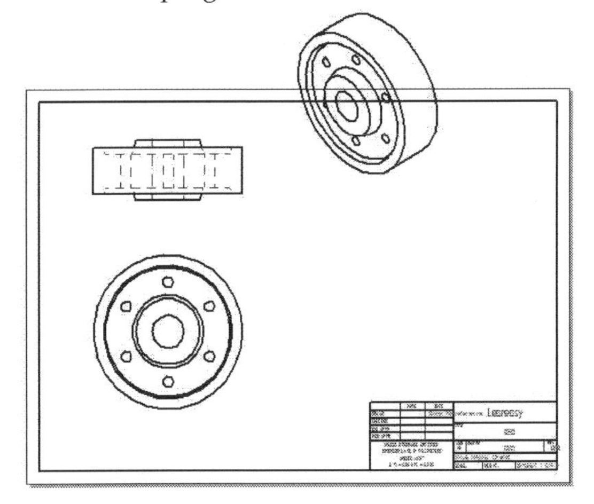

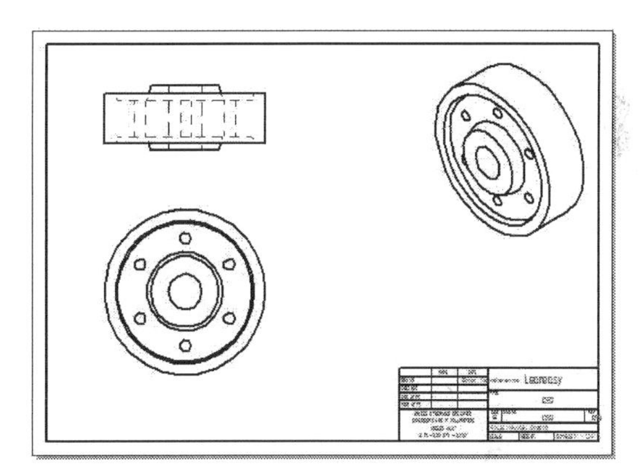

24. Click on the isometric view to activate the command bar. On the command bar, type-in **0.75** in the **Scale** box and press Enter.

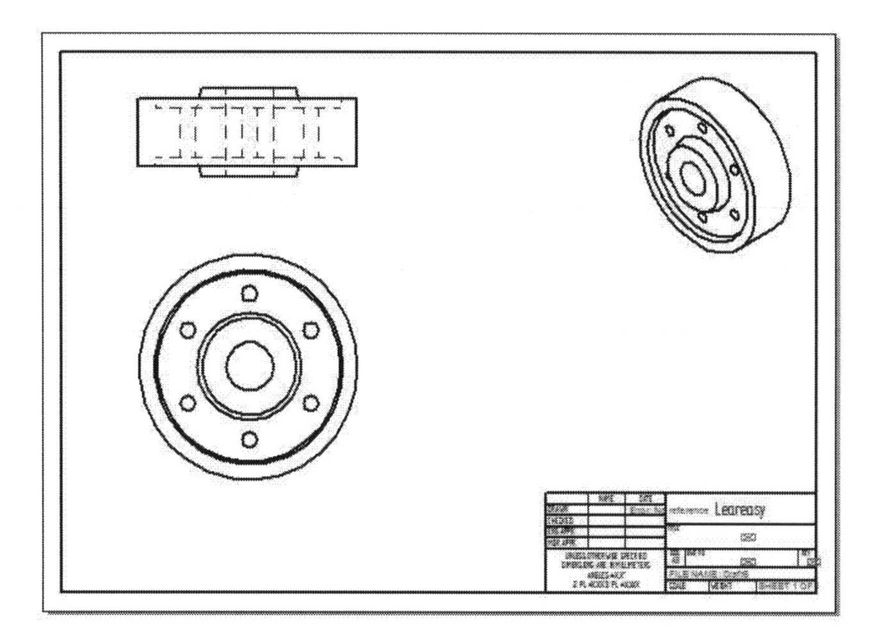

25. Activate the **Cutting Plane** command (click **Home > Drawing Views > Cutting Plane** on the ribbon) and select the front view. Create a cutting plane passing through the center of the front view.

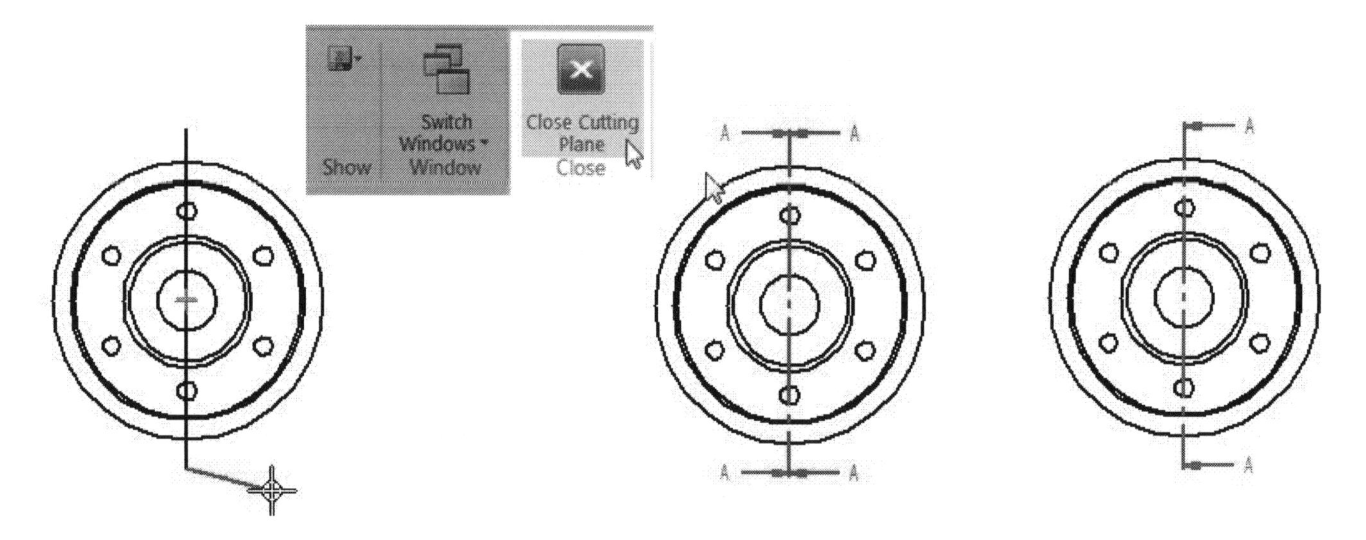

26. Activate the **Section** command (click **Home > Drawing views > Section** on the ribbon) and click on the cutting plane.
27. On the command bar, click the **Model Display Settings** icon. On the **Drawing View Properties** dialog, uncheck the **Hidden edge style** option and click **OK** twice.

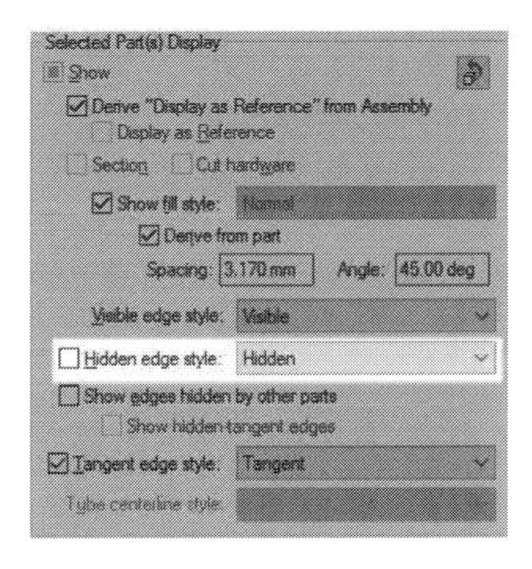

28. Drag the mouse pointer toward right and click to position the view.

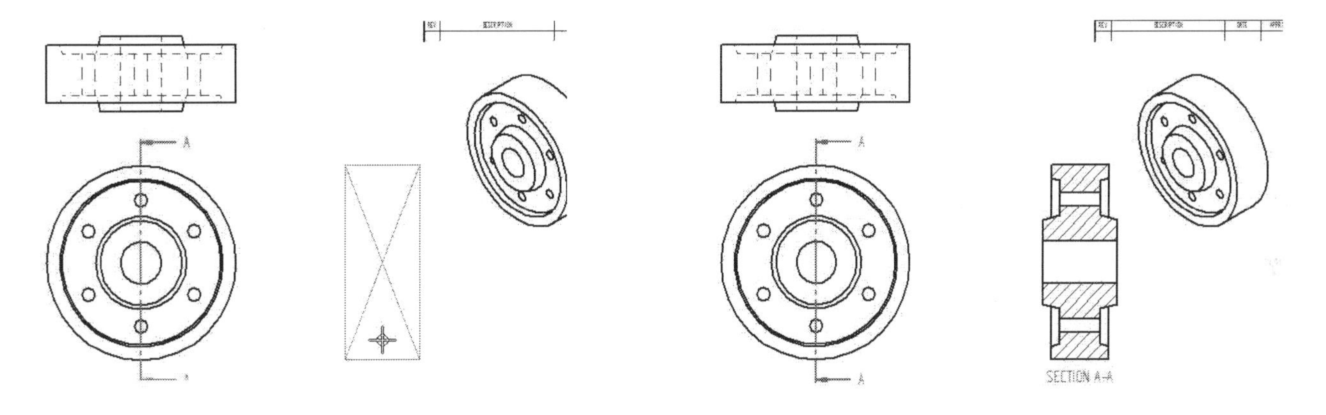

29. Activate the **Automatic Centerlines** command (click **Home > Annotation > Automatic Centerlines** on the ribbon). Click on the top view to apply centerlines.

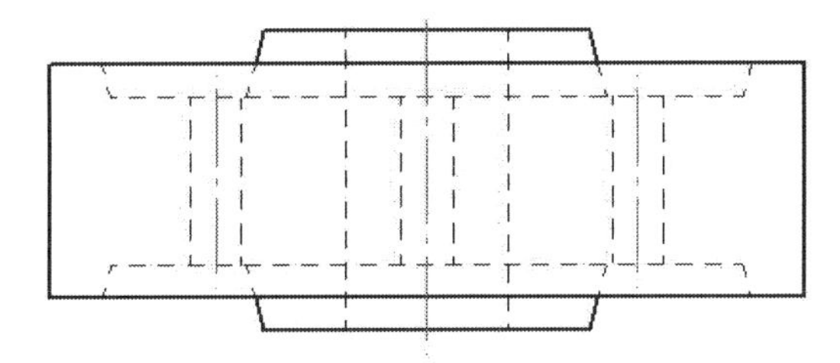

30. Activate the **Centerline** command (click **Home > Annotation > Centerline** on the ribbon). On the command bar, select **By 2 Lines** from the **Placement Options** drop-down menu.
31. Click on the horizontal lines on the section view corresponding to holes. The centerlines are created between the hole lines.

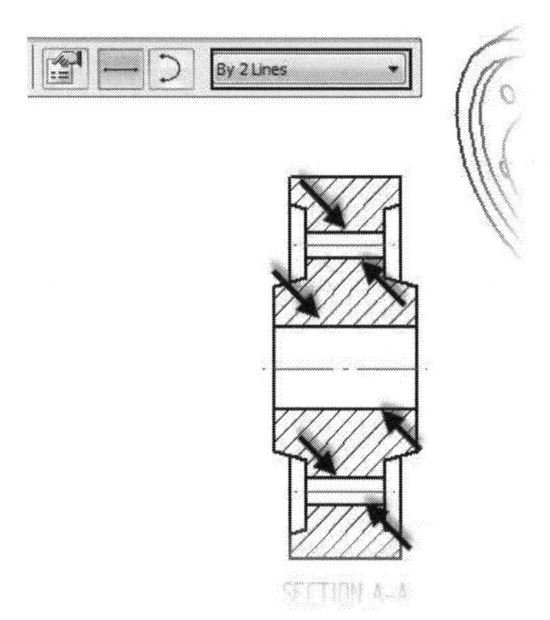

32. Activate the **Center Mark** command (click **Home > Annotation > Center Mark** on the ribbon)
33. On the command bar, set the **Orientation** to **Horizontal/Vertical** and click on the hole located at the center of the front view.
34. Activate the **Bolt Hole Circle** command (click **Home > Annotation > Bolt Hole Circle** on the ribbon) and click on the hole located at the center of the front view. Drag the mouse pointer and click on anyone of the small holes. A bolt hole circle is created.

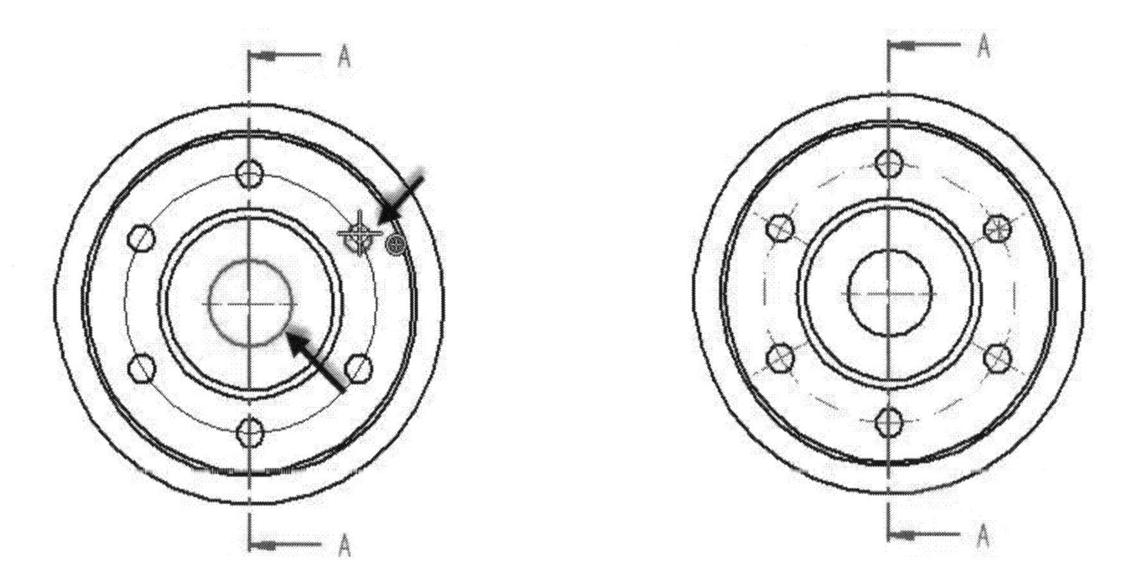

35. Activate the **Smart Dimension** command and apply dimensions to the top view.

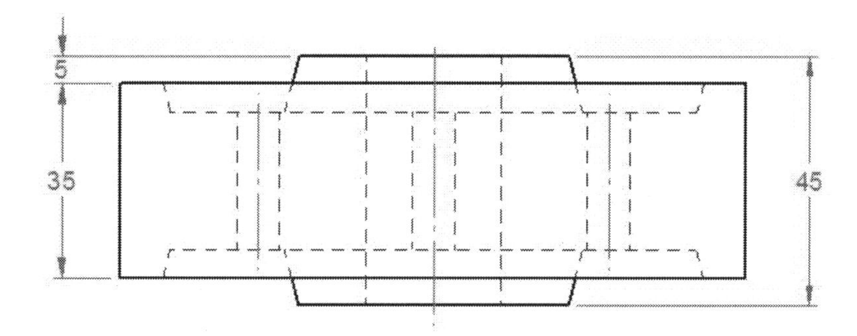

36. Activate the **Symmetric Diameter** command (click **Home > Dimension > Symmetric Diameter** on the ribbon). On the command bar, activate the **Diameter-Half/Full** icon.
37. Click the centerline of the section view and horizontal line of the large hole. Drag the mouse pointer toward right, and position the diameter dimension.

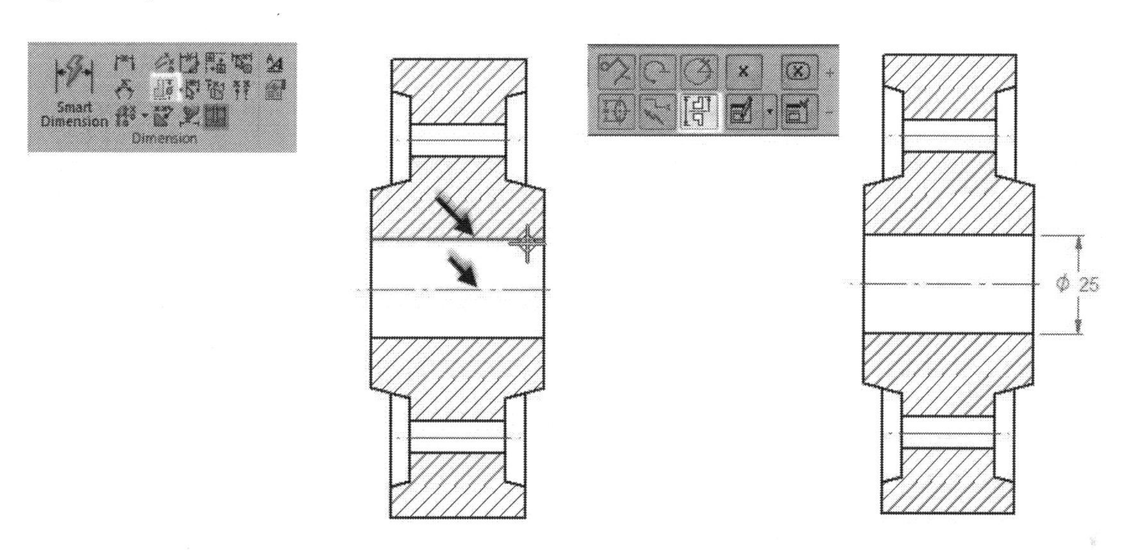

38. Click on the angled edge of the section view and create another diameter dimension.

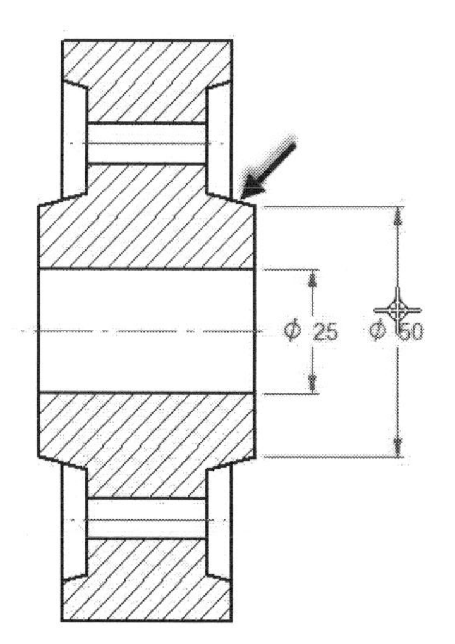

39. Activate the **Smart Dimension** command and click on the lower horizontal edge of the section view. On the command bar, activate the **Angle** icon and click on the inclined edge. Drag the pointer and click to position the angle dimension. Press Esc to deactivate the **Smart Dimension** command.

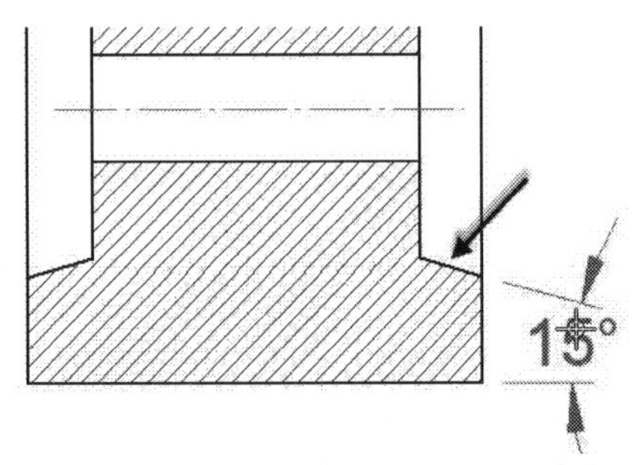

40. Click on the angle dimension and drag upward. On the command bar, click the **Prefix** icon to open the **Dimension Prefix** dialog.
41. On the **Dimension Prefix** dialog, type-in **TYP** in **Suffix** box and click **OK**.

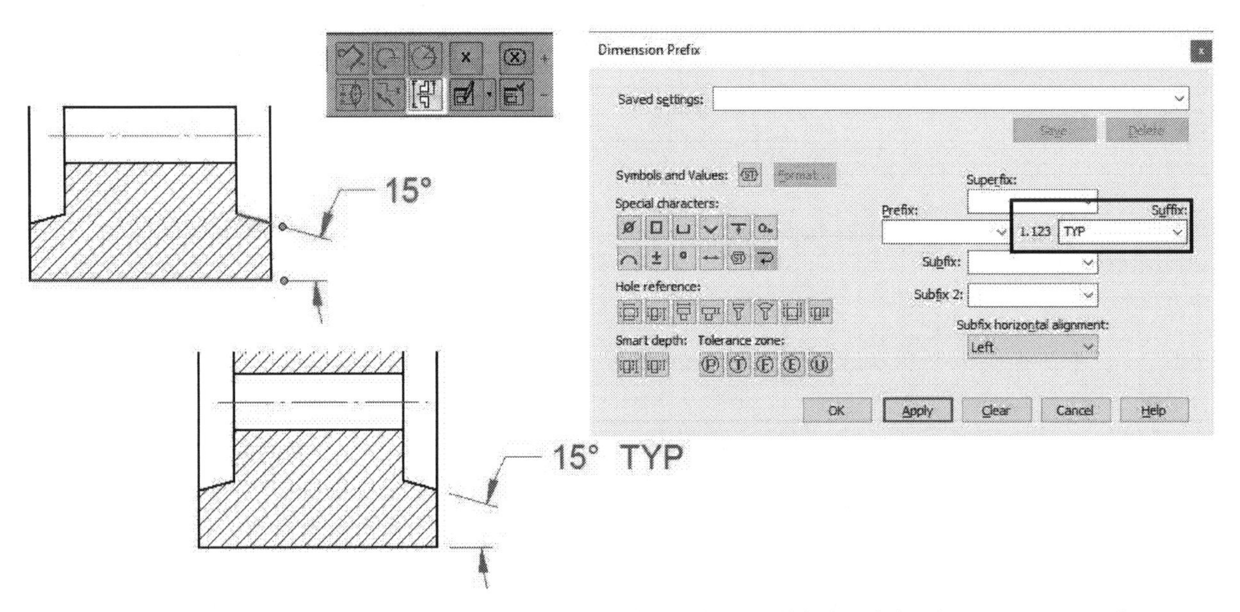

42. Activate the **Smart Dimension** command and click on the small hole of the front view. On the command bar, click the **Prefix** icon.
43. On the **Dimension Prefix** dialog, type-in values in the **Prefix**, **Subfix**, and **Subfix 2** boxes, as shown. Click **OK** and position the hole dimension.

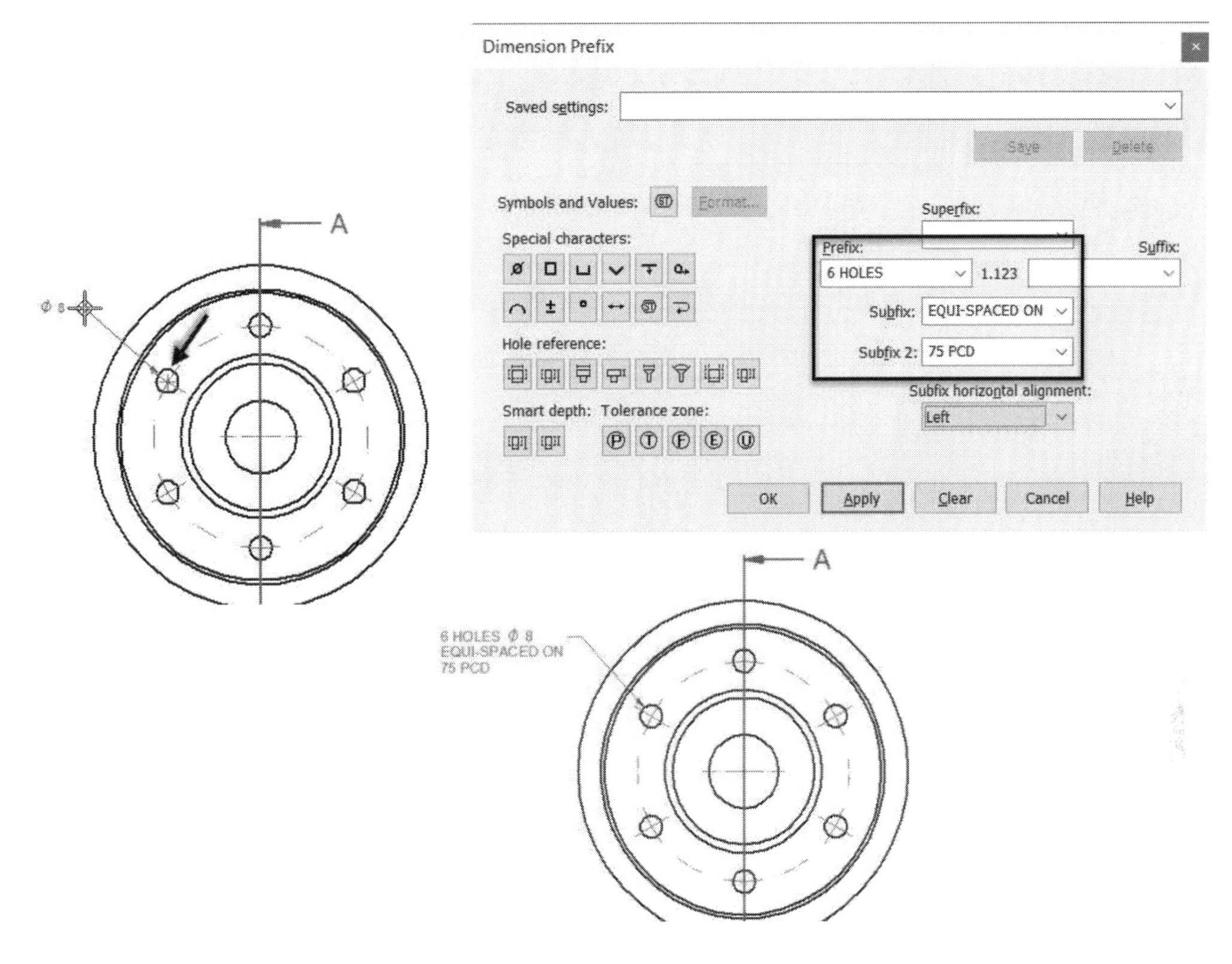

44. Activate the **Smart Dimension** command and open the **Dimension Prefix** dialog. On this dialog, empty the **Prefix**, **Subfix**, and **Subfix 2** boxes and click **OK**.
45. Create the other dimensions in the drawing.

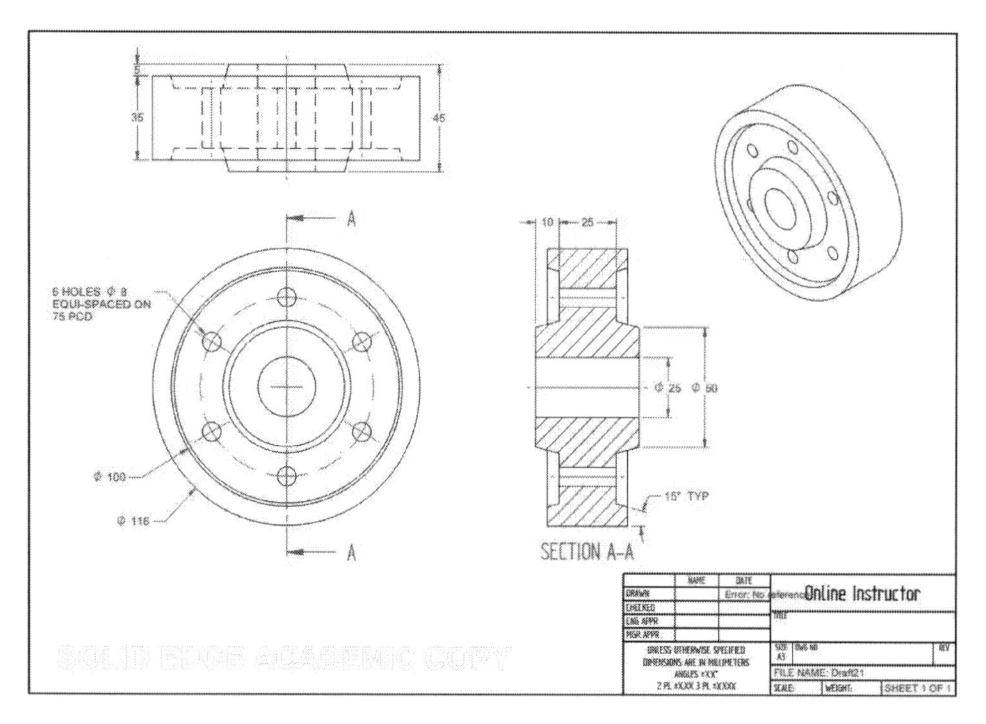

46. Save and close the drawing.

Example 2

In this example, you will create an assembly drawing shown below.

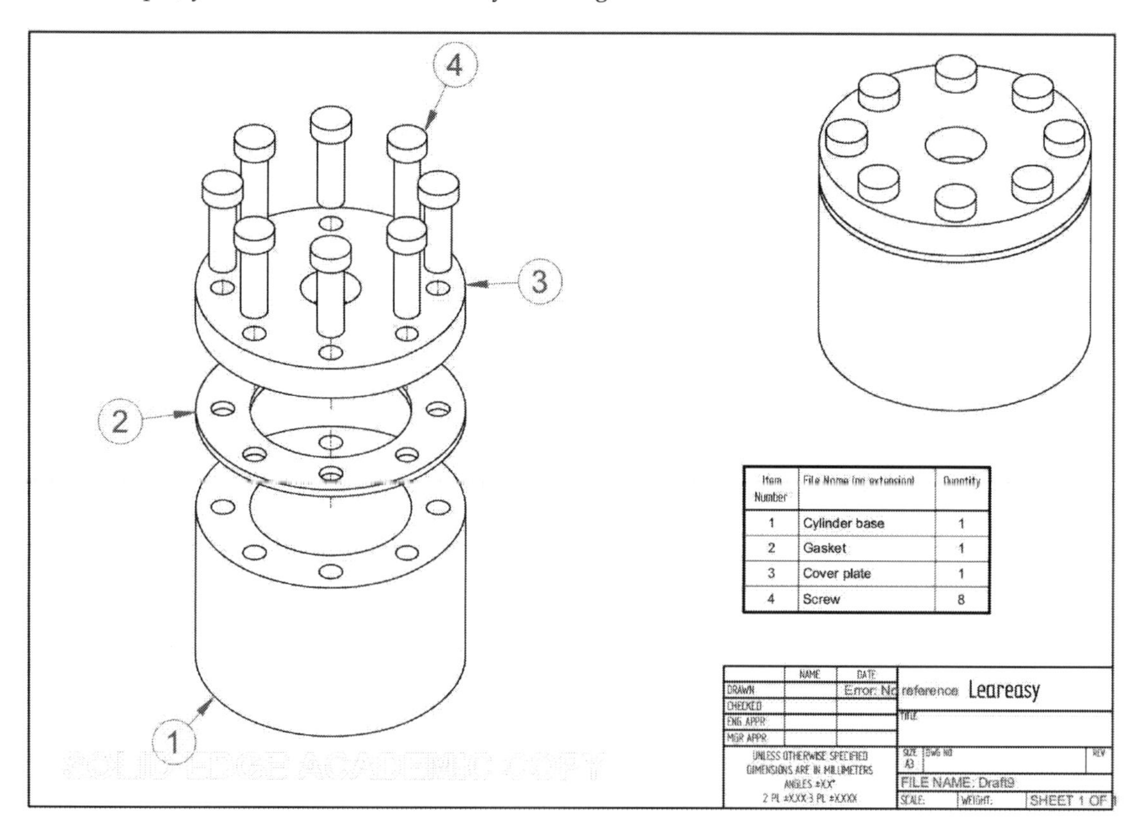

Drawings

1. Start **Solid Edge 2019**.
2. On the **Quick Access Toolbar**, click the **New** icon. On the **New** dialog, click on the **Online Instructor.dft**, and then click **OK**.
3. Activate the **View Wizard** command (click **Home > Drawing Views > View Wizard** on the ribbon).
4. Browse to the location of Example 2 of Chapter 10 and click on the assembly file. Click the **Open** button.
5. Click on the top right corner to place the isometric view of the assembly. Press Esc to stop view projection.
6. Again, activate the **View Wizard** command. On the **Select Attachment** dialog, check the **Create drawing view independent of assembly** option and set the **Configuration** to **explode, Solid Edge**. Click **OK**.
7. On the command bar, set the **Scale** to 1:1. Click on the drawing sheet to position the exploded view.

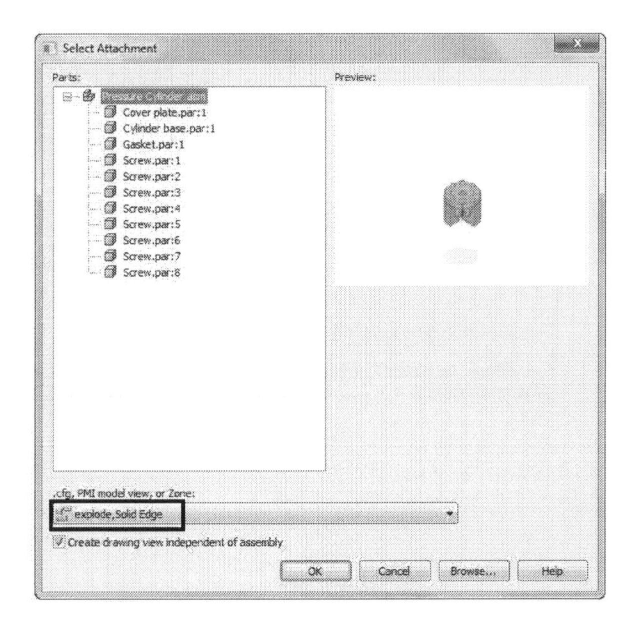

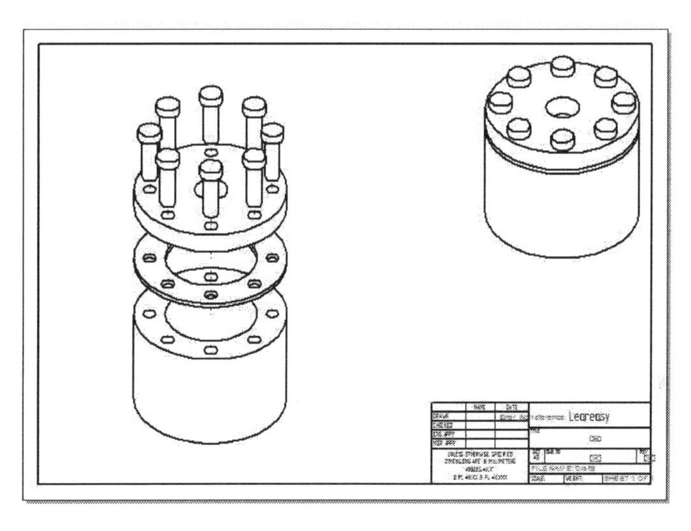

8. Activate the **Parts List** command (click **Home > Tables > Parts List** on the ribbon) and click on the exploded view.
9. On the command bar, click the **Properties** icon to open the **Parts List Properties** dialog.
10. On this dialog, click the **Columns** tab. In the **Columns** box, click on the **Author** option, and then click the **Delete Column** button.
11. On the **Data** tab, press Shift key and select all the cells of the table. Change the **Font** type to **Arial**.

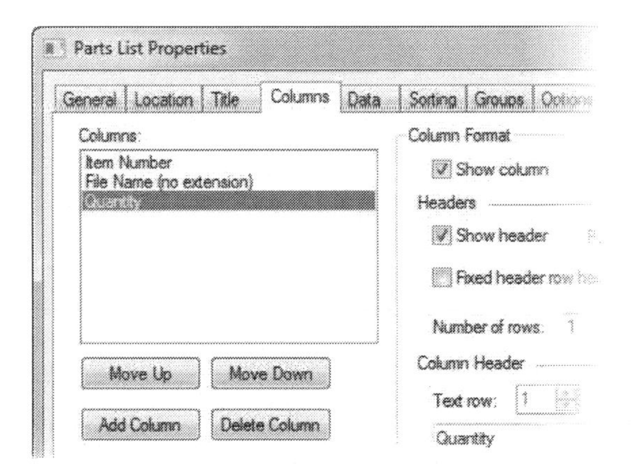

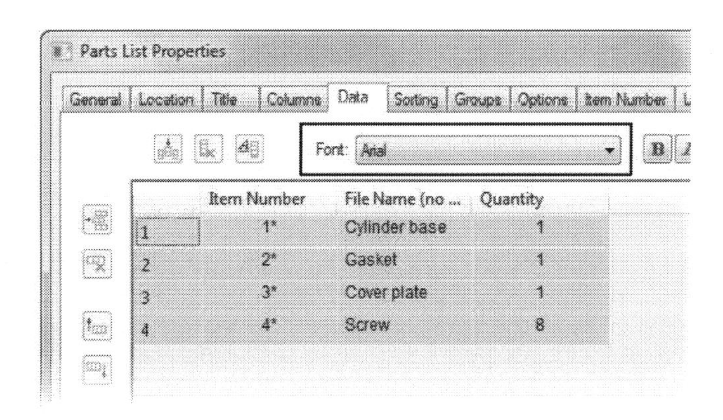

12. On the **Balloon** tab, set the **Text Size** to 8 and uncheck the **Use Item Count for lower text** option. Click **OK**.
13. Position the parts list below the isometric view. You will notice that some balloons are placed outside the sheet.
14. Click on the alignment line connecting the balloons. Square and circle grips appear on it.
15. Click on a square grip and reduce the size of the alignment shape.

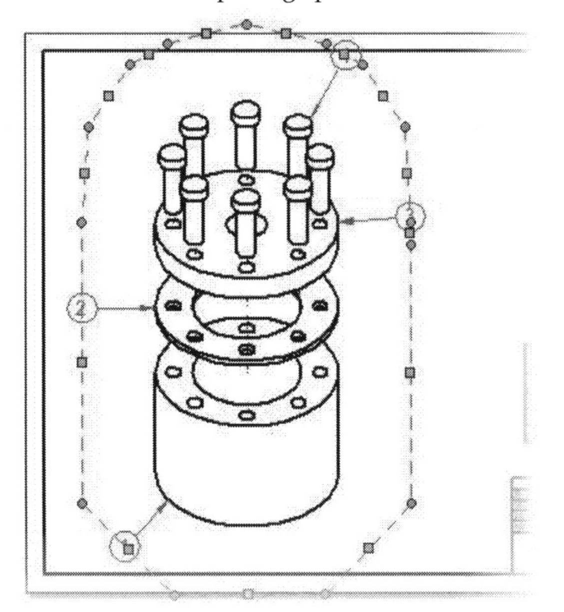

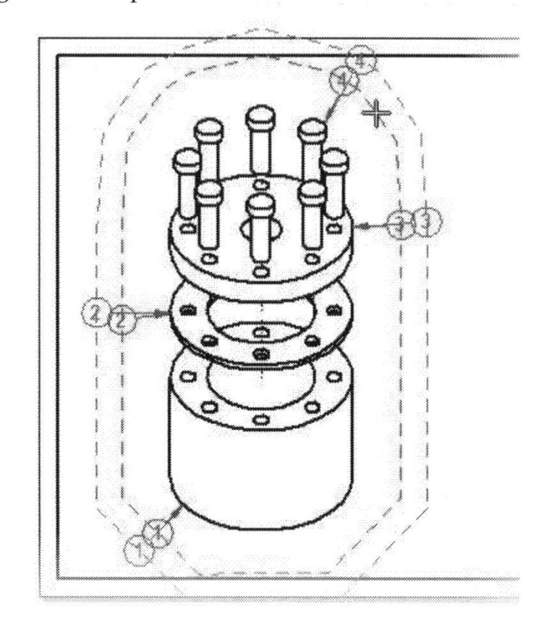

16. At the bottom of the sheet, click the **New Sheet** icon to add a new sheet to the drawing.
17. Right-click on **Sheet2** and select **Sheet Setup**. On the **Sheet Setup** dialog, click the **Background** tab and set the **Background sheet** to **A3-Sheet**. Click **OK**.
18. Activate the **View Wizard** command and uncheck the **Create drawing view independent of assembly** option.
19. From the Parts list, click the Cylinder Base.par file. Click **OK** and place the drawing view on the sheet.
20. Likewise, use the **View Wizard** command and place other part views, as shown below.

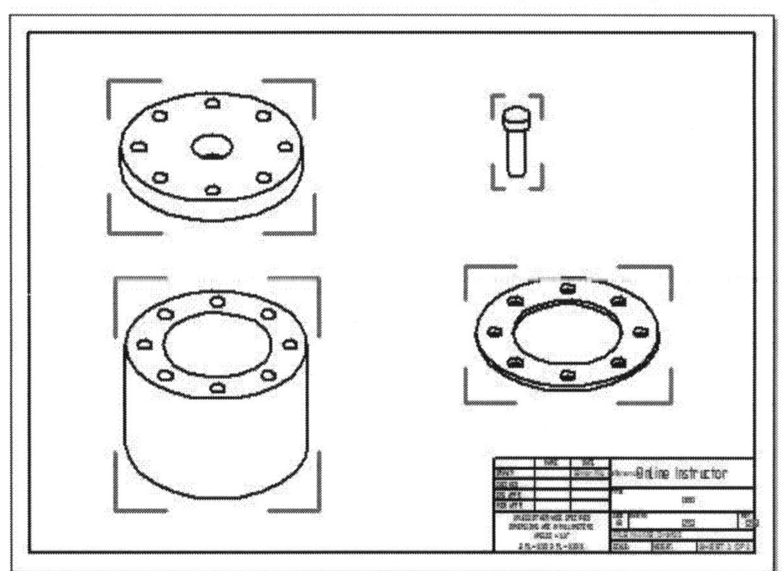

21. On the ribbon, click **Home > Annotation > Balloon**.

22. On the command bar, type-in 2 in the **Text Scale** and **Height** boxes.
23. On the command bar, click the **Link to Parts List** icon, and then activate the **Item Number** icon.
24. Select a point on the cover plate to attach a balloon to it. Move the pointer and click to define the location of the balloon.

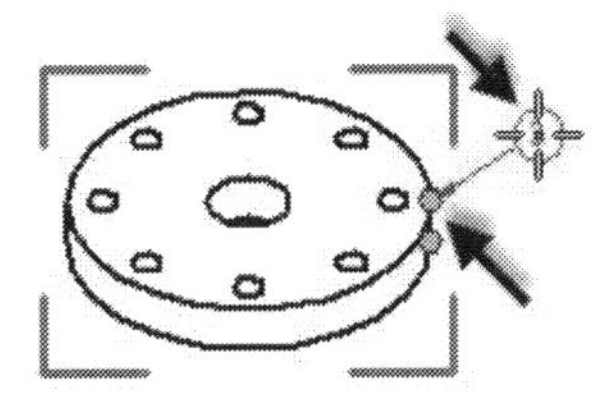

25. Likewise, add balloons to other views.

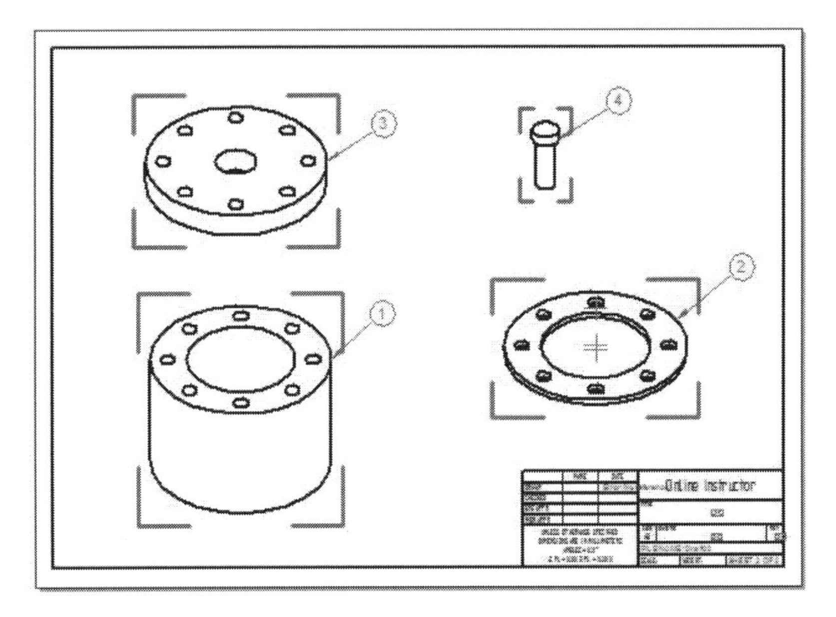

26. Save and close the drawing.

Questions

1. How to create drawing views using the **View Wizard** command?
2. How do you hide hidden edges of a drawing view?
3. How do you change the display style of a drawing view?
4. How do you update drawing views when the part is edited?
5. How do you control the properties of dimensions and annotations?
6. List the commands used to create centerlines and center marks.
7. How do you add symbols and texts to a dimension?
8. How do you add break lines to a drawing view?
9. How do you create revolved section views?
10. How do you create exploded view of an assembly?

Exercises

Exercise 1

Create orthographic views of the part model shown below. Add dimensions and annotations to the drawing.

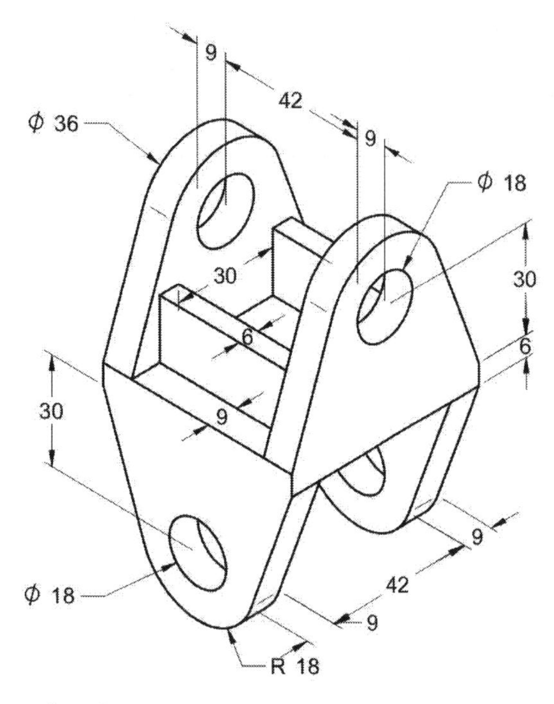

Exercise 2

Create orthographic views and an auxiliary view of the part model shown below. Add dimensions and annotations to the drawing.

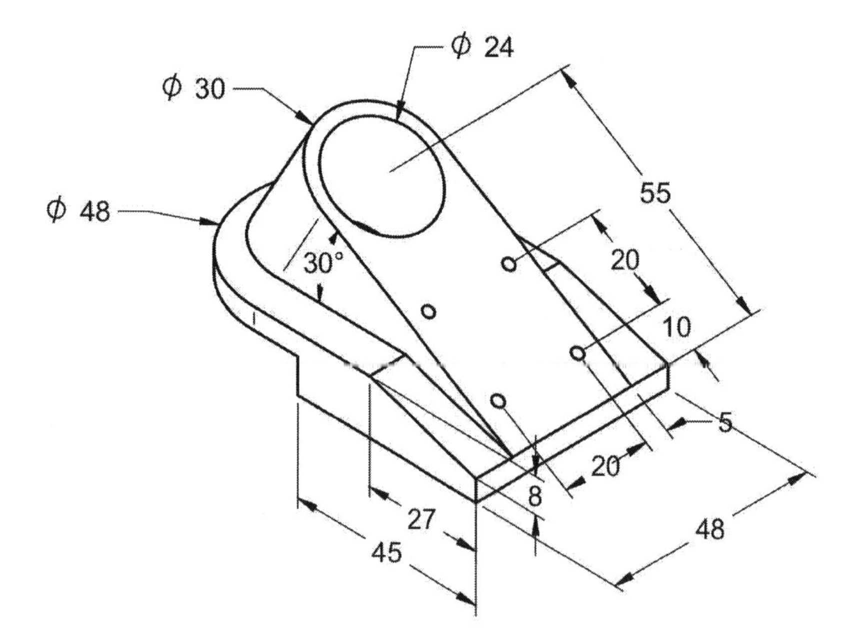

Chapter 12: Sheet Metal Design

Sheet metal parts are made by bending and forming flat sheets of metal. In Solid Edge, sheet-metal parts can be folded and unfolded enabling you to show them in the flat pattern as well as their bent-up state. There are two ways to design sheet-metal parts in Solid Edge. You either start the sheet-metal part from scratch using sheet-metal features throughout the design process, or design it as a regular solid part and convert it to a sheet-metal part. Most commonly, sheet-metal parts are designed in Sheet Metal environment from the beginning. In this chapter, you will learn both the approaches.

The topics covered in this chapter are:

- *Tabs*
- *Flanges*
- *Bend Allowance*
- *Bend Tables*
- *Counter Flanges*
- *Hems*
- *Close 2-Bend Corners*
- *Bends*
- *Jogs*
- *Dimples*
- *Louvers*
- *Drawn Cutouts*
- *Beads*
- *Gussets*
- *Etches*
- *Embosses*
- *Cuts*
- *Convert to Sheet Metal*
- *Rip Corners*
- *Flat Pattern*
- *Export to DXF or DWG*

Starting a Sheet Metal part

To start a new sheet metal part, click the **Application Menu** icon on the top left corner. On the **Application Menu,** click **New** tab > **ISO Metric Sheet Metal** option (or) click the **New** icon on the **Quick Access Toolbar**, and then double-click on the **iso sheet metal.psm** template on the **New** dialog. If you want to start the sheet metal part using any other template, select a standard from the **Standard Templates** section, and then select the sheetmetal template corresponding to the selected standard.

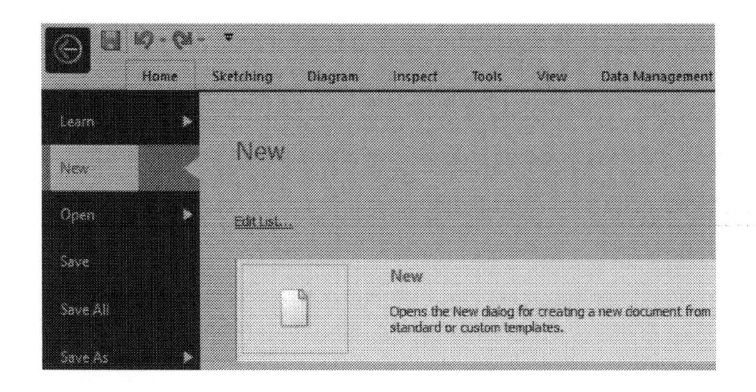

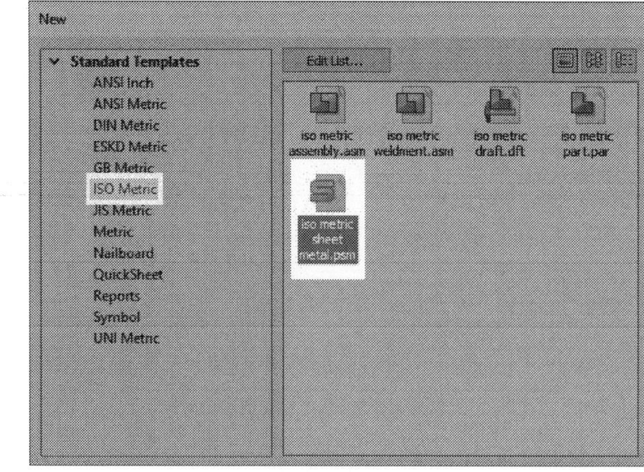

Tab

The tab is a basic type of sheet metal feature. To create a tab, create a closed sketch on a plane and click inside it. An arrow handle appears along with the command bar. On the command bar, click the **Material Table** icon to open the **Material Table** dialog. On this dialog, select a material from the Materials tree; its properties are displayed in the **Material Properties** tab. You can change the properties of the material in the **Properties** table.

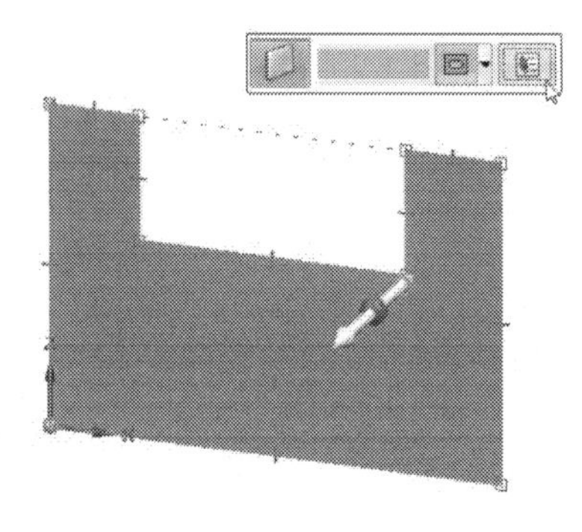

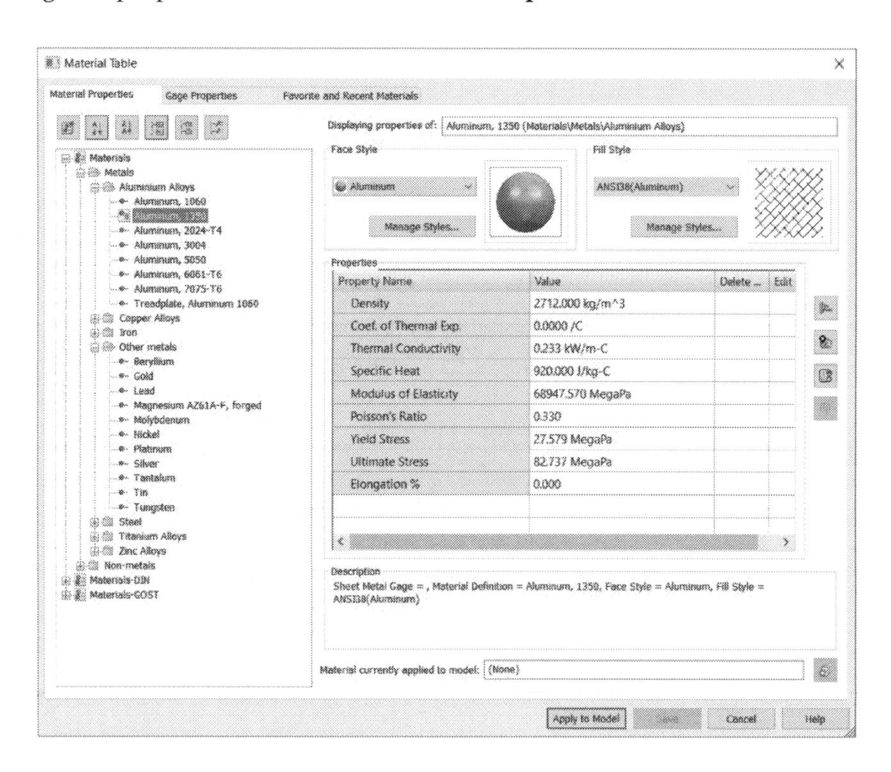

Open the **Gage Properties** tab and define the gage properties of the sheet metal part. Type-in values in the **Material thickness**, **Bend radius**, **Relief depth**, and **Relief width** boxes.

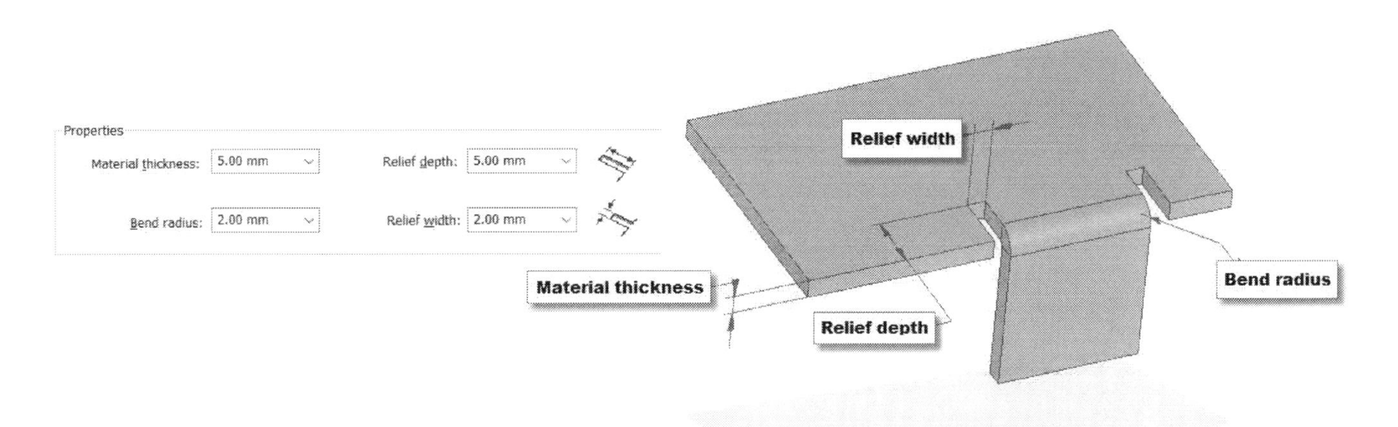

You can also use a spreadsheet to define these values. Check the **Use Excel file** option and select a gage table from the **Use Gage Table** drop-down menu. You can edit the gage table values by clicking the **Edit** button. In the spread sheet, modify the values, and then save and close the file. You can also define the sheet metal properties by selecting anyone of the sheet metal gages available in the **Sheet metal gage** drop-down menu.

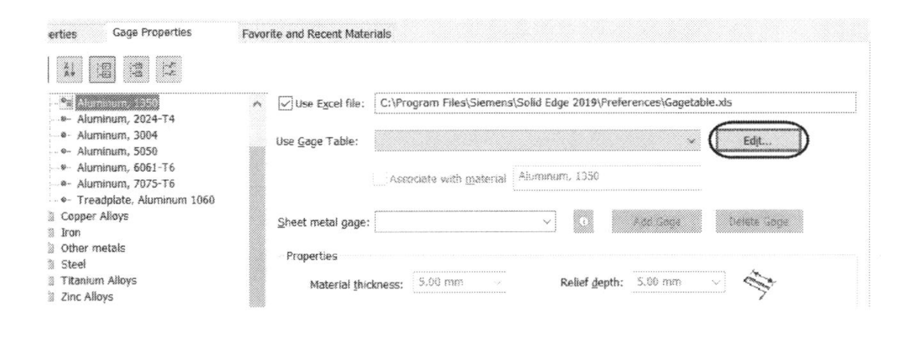

	A	B	C	D	E
1	Process Type				
2	Units	inches			
3	*100				
4	Gage Name	10 Gage			
5	Material Thickness	0.1345			
6	Bend Radius	1			
7	Relief Depth	0.263			
8	Relief Width	0.14			
9	*200	Radius			
10	Angle	0.5	0.75	1	1.125
11	5	0.01	0.05	0.09	0.12
12	15	0.05	0.08	0.13	0.2
13	30	0.08	0.13	0.2	0.28
14	45	0.11	0.18	0.27	0.36
15	60	0.15	0.22	0.34	0.43
16	75	0.2	0.22	0.33	0.44
17	90	0.23	0.34	0.42	0.52
18	105	0.3	0.4	0.5	0.6
19	120	0.35	0.43	0.51	0.62
20	135	0.38	0.49	0.57	0.71
21	150	0.4	0.55	0.66	0.77
22	165	0.47	0.58	0.69	0.8
23	180	0.55	0.65	0.75	0.86

Next, type-in a value in the **Neutral Factor** box. The **Neutral Factor** is the ratio that represents the location of neutral sheet measured from the inside face with respect to the thickness of the sheet-metal. It defines the bend allowance of the sheet metal part. The standard formula that calculates the bend allowance is given below.

$$BA = \frac{\pi(R + KT)A}{180}$$

BA = Bend Allowance

R = Bend Radius

K = Neutral Factor = t/T

T = Material Thickness

t = Distance from inside face to the neutral sheet

A = Bend Angle

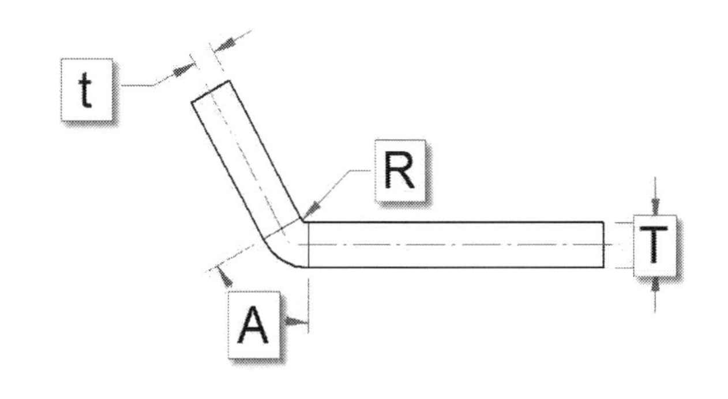

You can also define the bend allowance by using your own formula. Select the **Custom formula** option and type-in a value in the **ProgramID.ClassName** box. Click the **Apply to Model** button to apply the material and gage properties to the model. Now, click on the arrow handle to define the side of the tab feature. Click the right mouse button to create the tab feature.

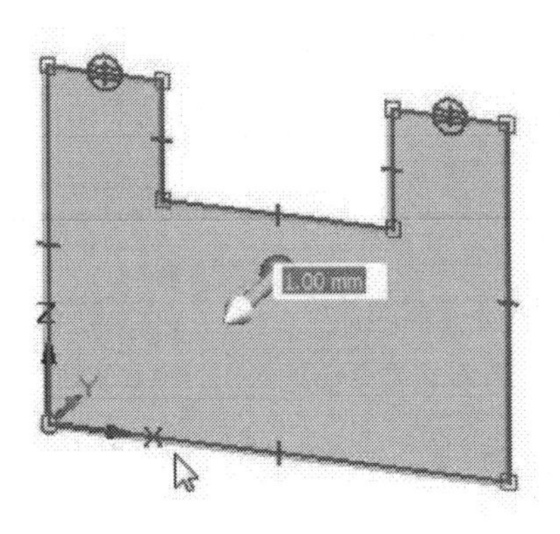

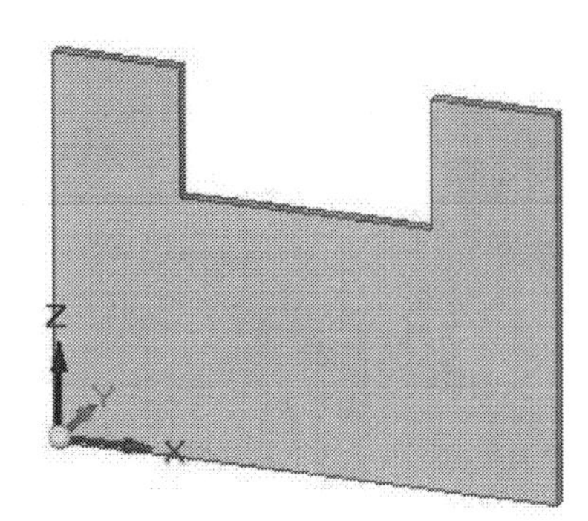

Flange

The second feature after creating a tab is flange. This feature can be created along an edge or multiple edges of a sheet metal part. In order to create a flange, all you need is to click an end face of the tab feature. The flange handle appears on the selected face. Click the small arrow and drag the pointer. A flange feature appears attached to the mouse pointer.

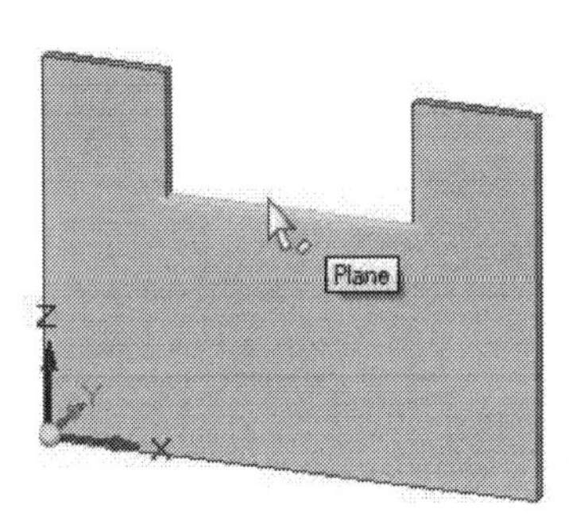

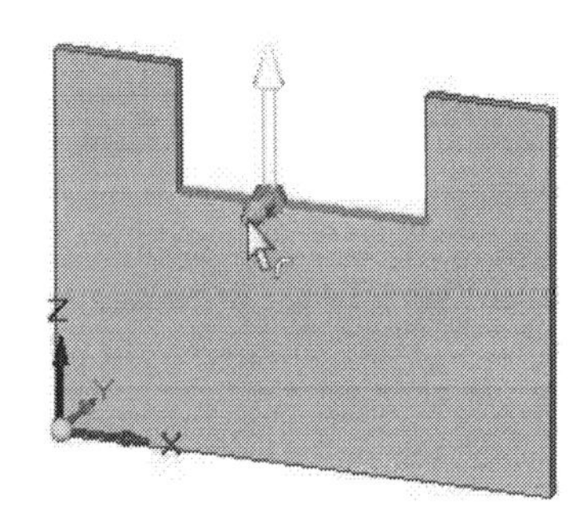

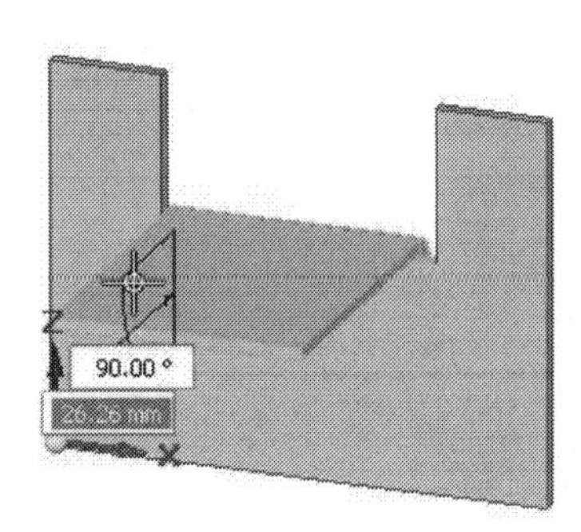

On the command bar, click the **Flange Options** icon to open the **Flange Options** dialog. On this dialog, you can override the gage properties by checking the **Override global value** options available next to each of the gage property.

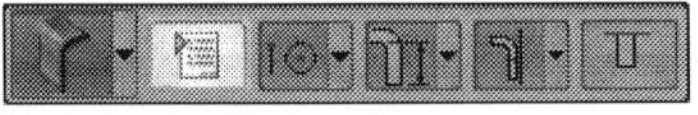

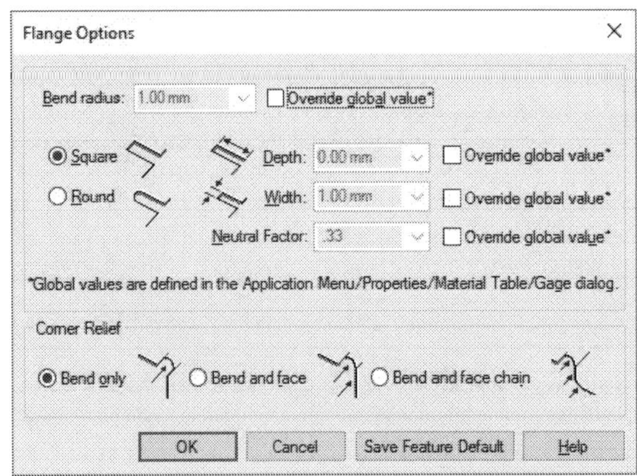

Under the **Corner Relief** section, select an option to define the type of corner relief. The three types of corner reliefs are shown below. Click **OK** to close the dialog.

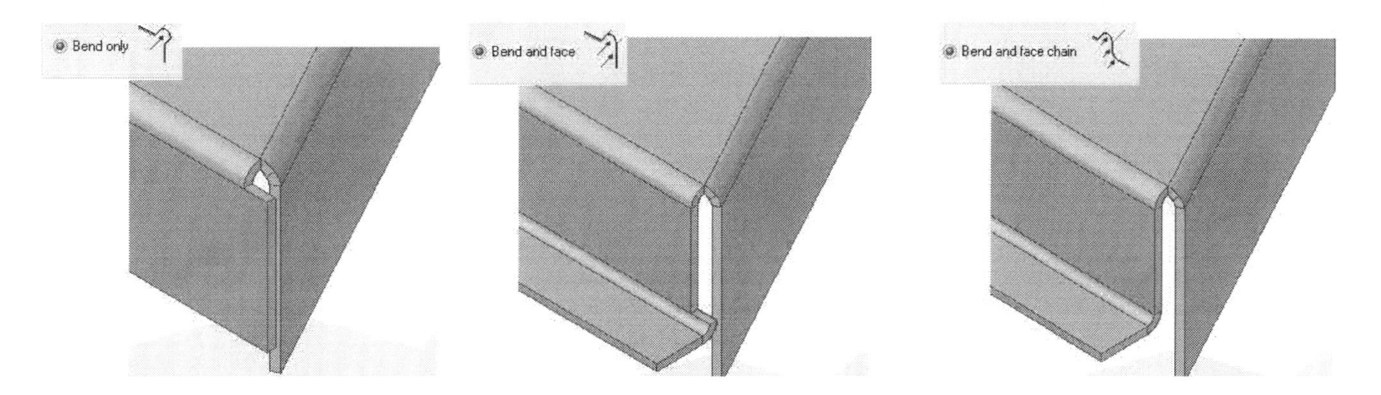

On the command bar, select an option from the **Measurement Point** drop-down menu. Both the measurement points are explained in the illustration below.

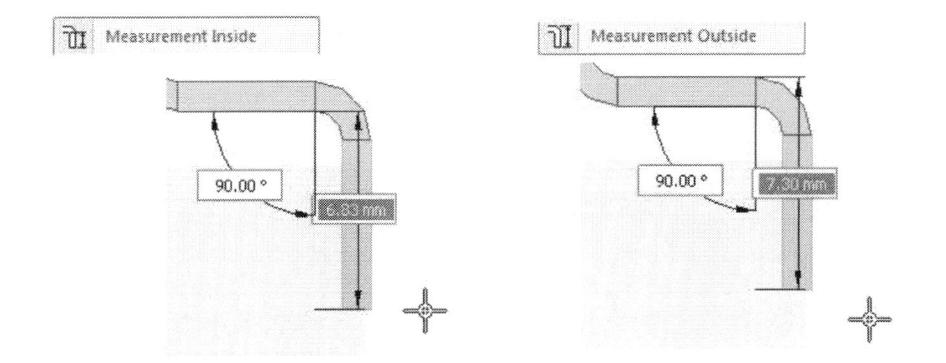

Define the material side using the **Material Side** drop-down menu. The three types of material sides are shown below.

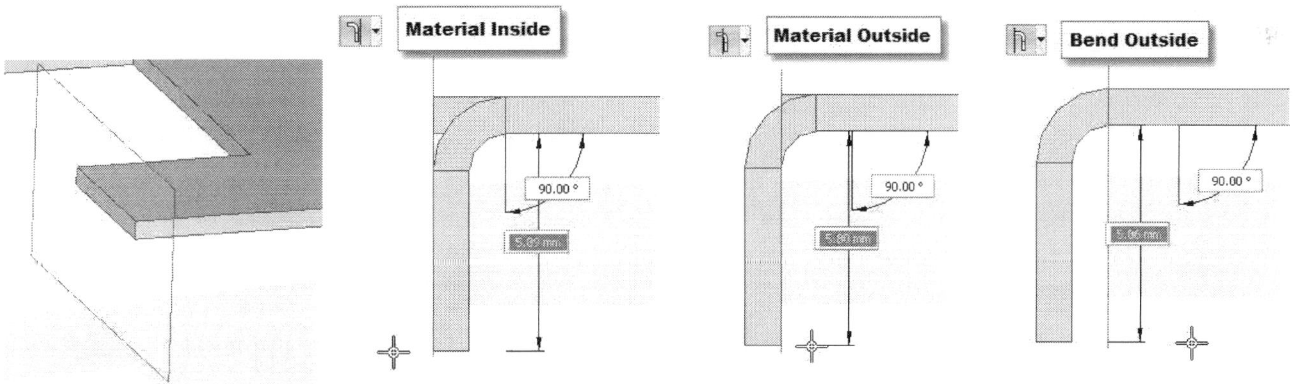

Click the **Partial Flange** icon to create the flange at the middle of the selected edge.

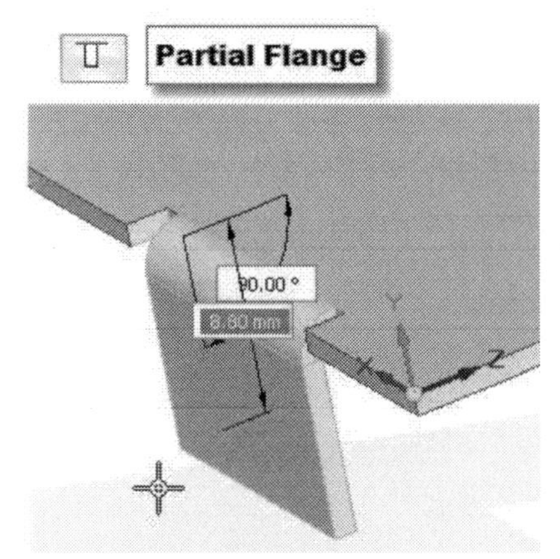

Type-in values in the distance and angle boxes that are attached to the flange. Click the right mouse button to create the flange.

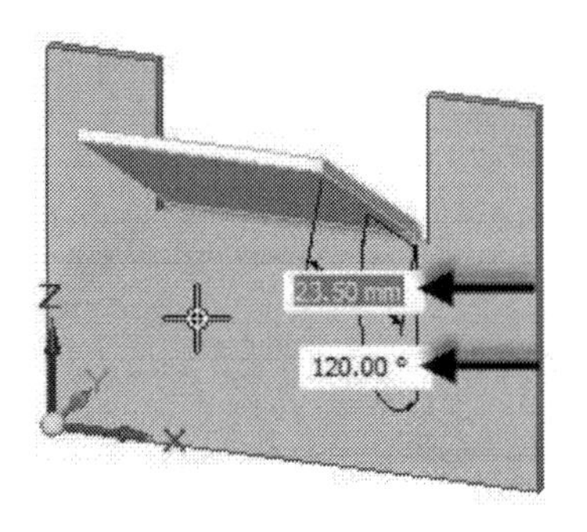

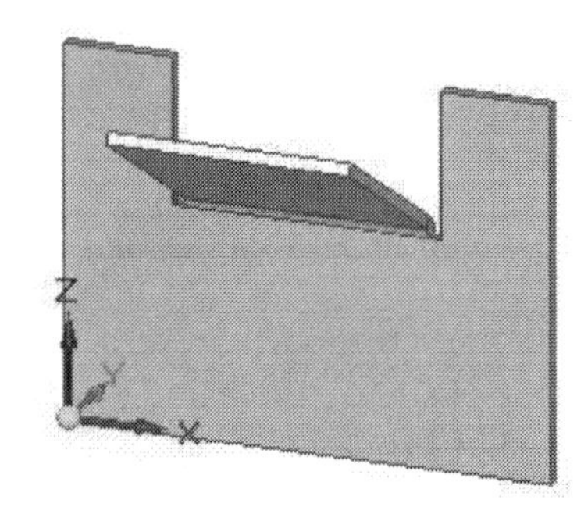

Close 2-Bend Corner

The **Close 2-Bend Corner** command allows you to control the appearance of sheet metal seams. For example, when two flanges meet at a corner, this command allows you to close the gap between them. In addition to that, it applies a corner treatment. Activate this command (click **Home > Sheet Metal > Close 2-Bend Corner** on the ribbon) and click on two bends that meet at a corner. On the command bar, select the required corner treatment.

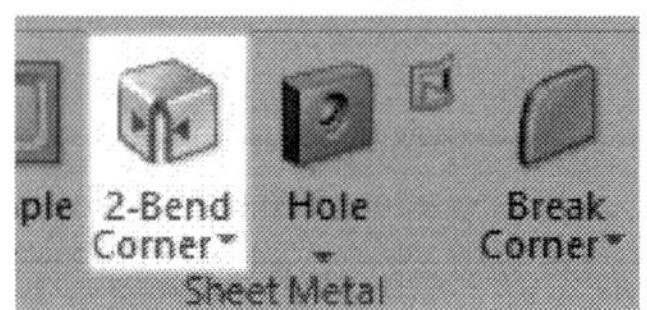

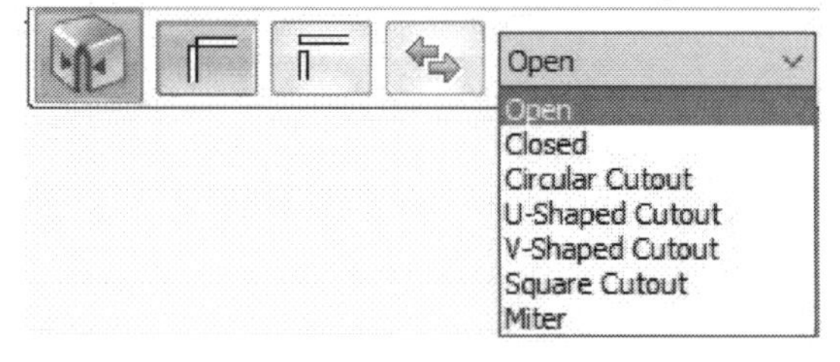

There are seven types of corner treatments available in the **Corner Treatment** drop-down menu, as shown below.

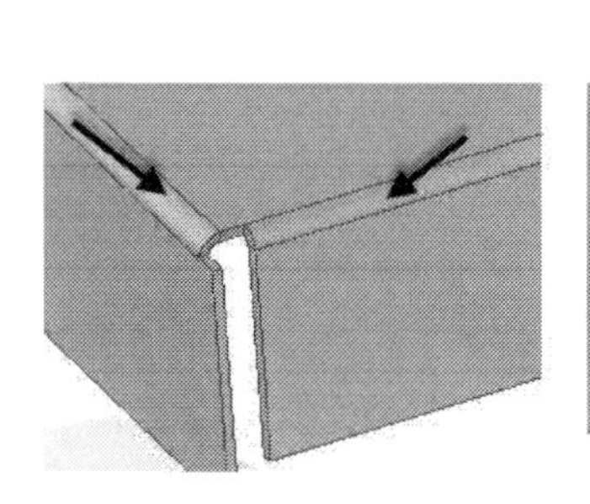

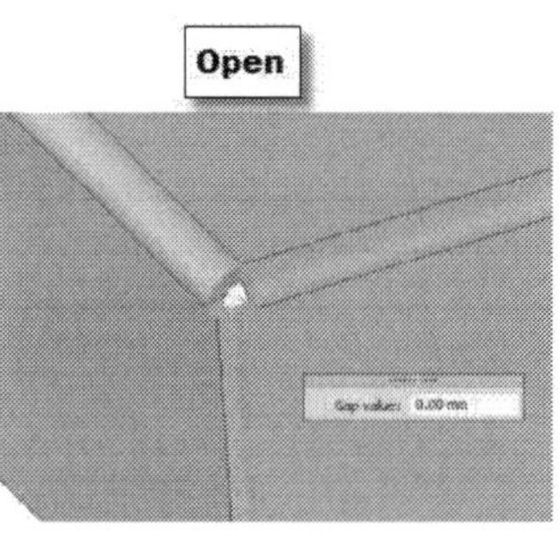

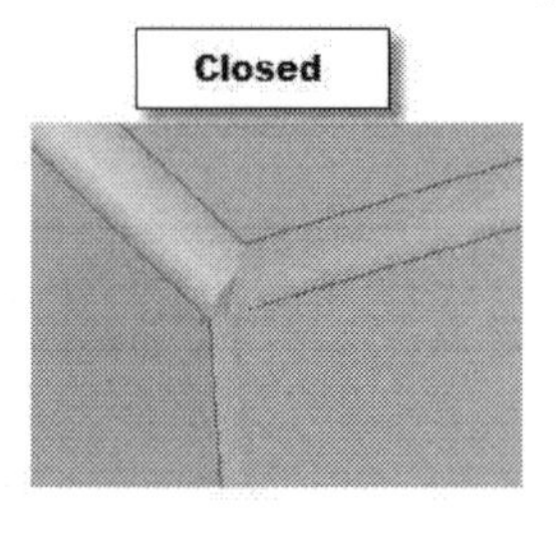

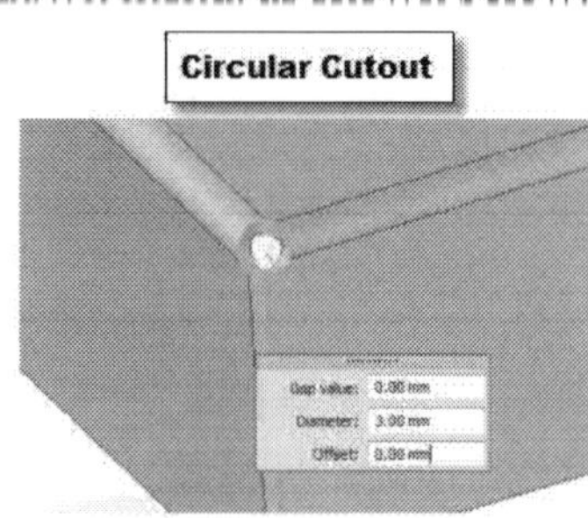

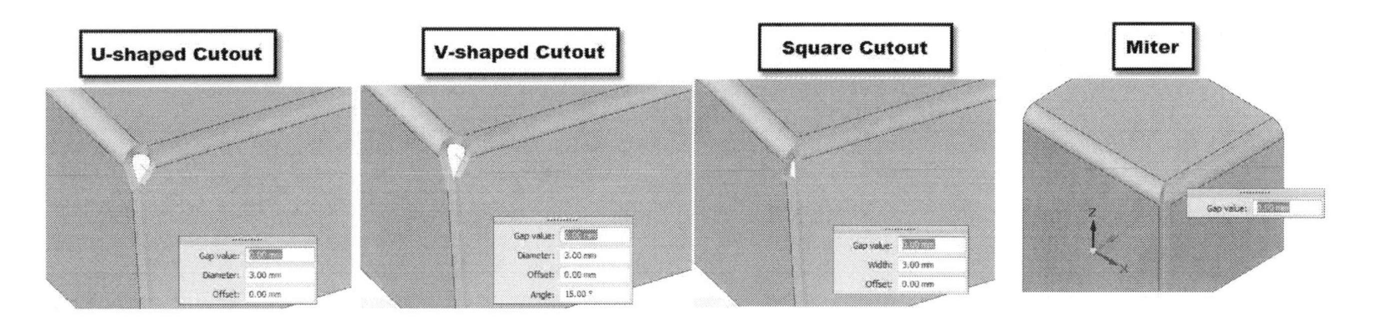

Note: The **Miter** corner treatment can be created only when the flanges are similar and perpendicular to each other.

On the command bar, click the **Overlapping Corner** icon to overlap one flange on the other. Next, type-in a value in the **Overlap ratio** box. Click the **Flip** icon to change the overlapping side.

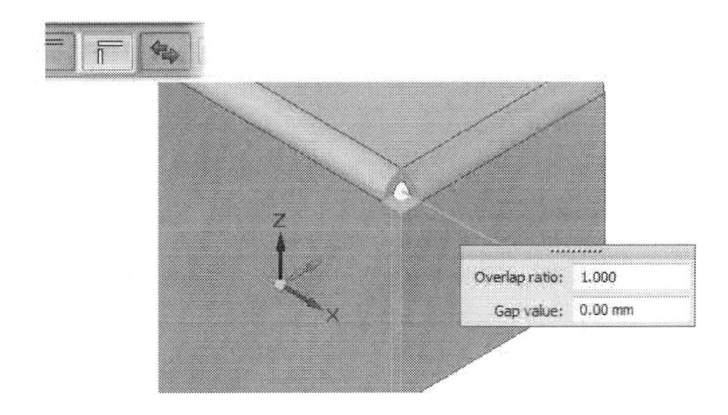

Contour Flange

The contour flange is another basic type of sheet metal feature. To create a contour flange, you need to have an open sketch. Activate the **Contour Flange** command (click **Home > Sheet Metal > Contour Flange** on the ribbon) and click on the open sketch. Drag the mouse pointer and type-in a value in the distance box that is attached to the preview. Press Enter to create the contour flange feature.

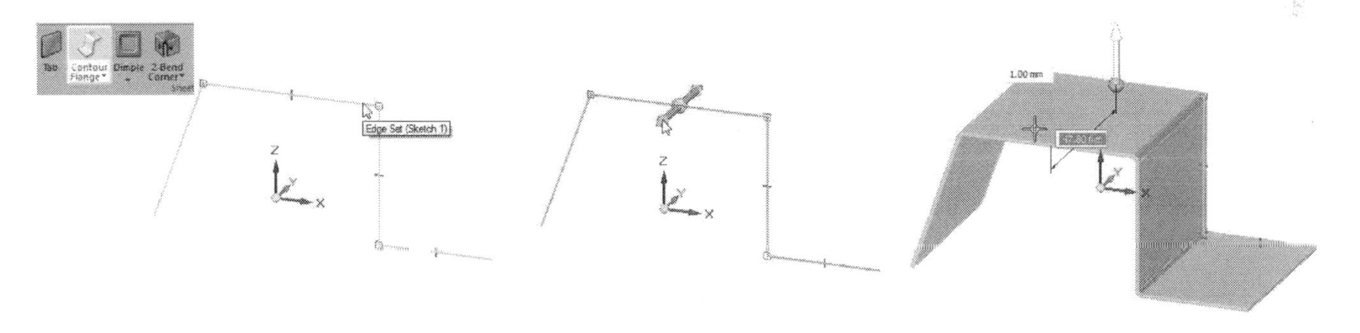

You can also add contour flanges to a base tab. Activate the **Line** command and lock the end face of the tab feature. On the locked face, draw an open sketch, and then activate the **Contour Flange** command. Click on the sketch, and then click on the arrow pointing towards the model. The contour flange preview appears. You will notice that the contour flange is created along face perpendicular to the sketch. You can click on multiple faces to add contour flanges to them. You can also use the **Chain** option to select multiple faces at a time.

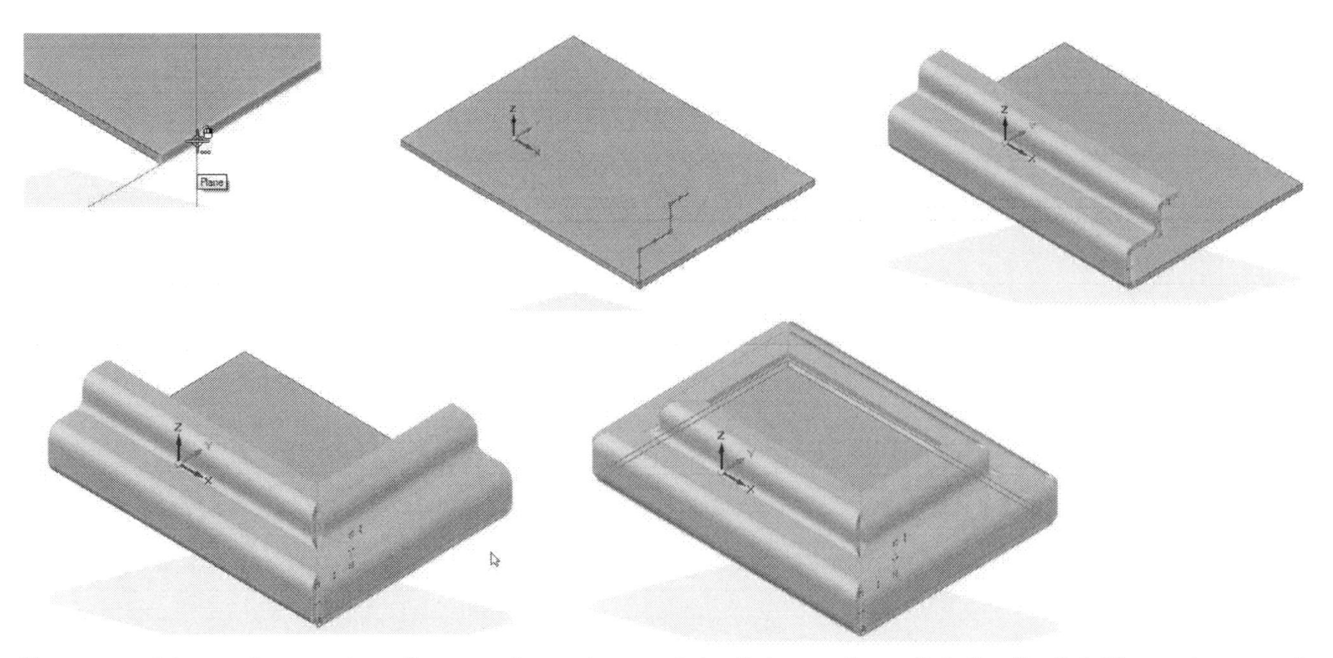

If you want to create a contour flange only up to a certain distance, then click the **Partial Flange** icon on the command bar and type-in the distance value.

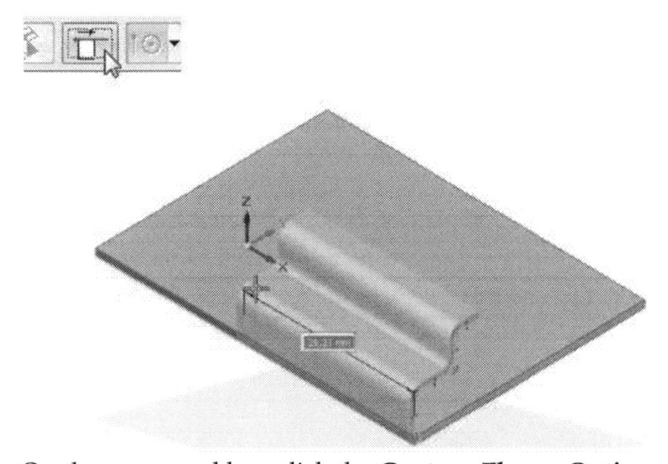

On the command bar, click the **Contour Flange Options** icon to open the **Contour Flange Options** dialog. On this dialog, click the **Miters and Corners** tab and check the **Miter** option to apply miter to the ends of the contour flange. Under the **Interior Corners** section, check the **Close Corner** option to apply treatment to the corners.

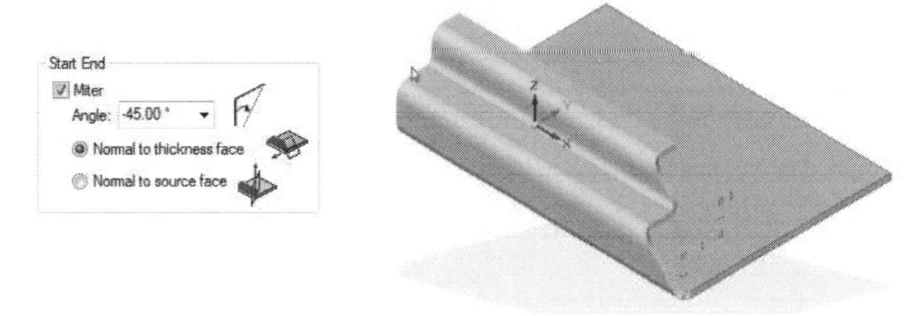

Click **OK** on the dialog to close it. Right click to create the counter flange.

Hem

The **Hem** command is used to fold an edge of a sheet metal part. To add a hem, activate the **Hem** command (click **Home > Sheet Metal > Contour Flange > Hem** on the ribbon) and select the edge you need to fold over. On the command bar, the **Material Setback** drop-down menu controls whether the material is added to inside or outside the existing edge.

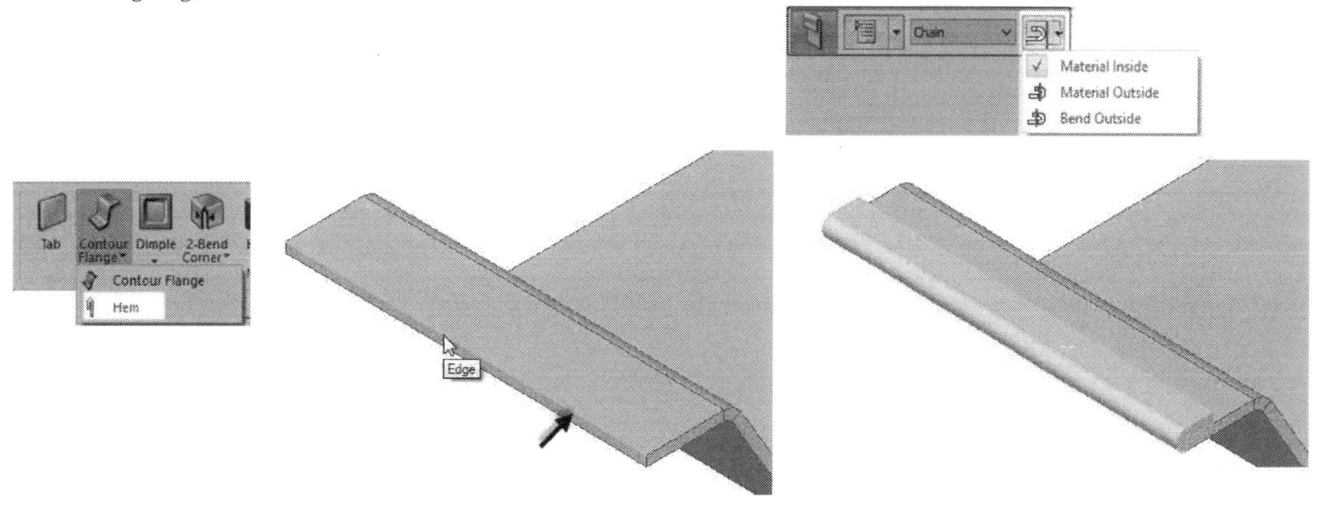

On the command bar, click the **Hem Options** icon to open the **Hem Option** dialog.

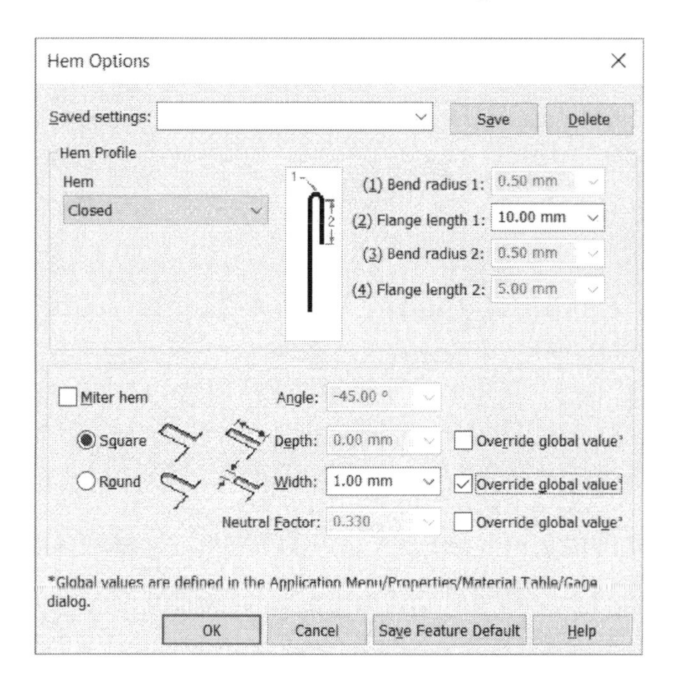

On this dialog, select a hem type from the **Hem type** drop-down menu and define its parameters. Different hem types are shown below.

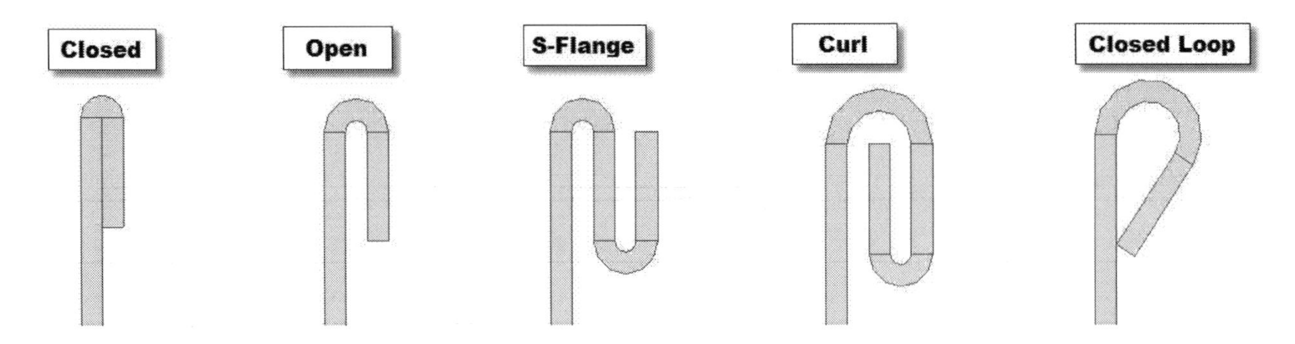

If you want to bevel the end faces of the hem, check the **Miter hem** option and type-in a value in the **Angl**e box. Click the **OK** button to close the dialog, and then right-click to complete the hem feature.

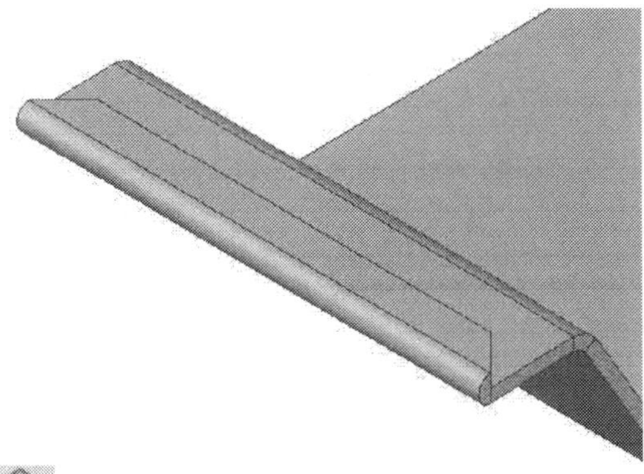

Bend

In addition to adding flanges and contour flanges, you can also bend a flat sheet using the **Bend** command. First, draw a sketch line on the flat sheet. Activate the **Bend** command (click **Home > Sheet Metal > Bend** on the ribbon) and click on the sketched line. A two-sided arrow appears on the line. Click on the either side of the arrow to define the side to be folded. Type-in a value in the angle box to change the folding angle. Click on the arrow attached to the folded face to reverse the folding direction.

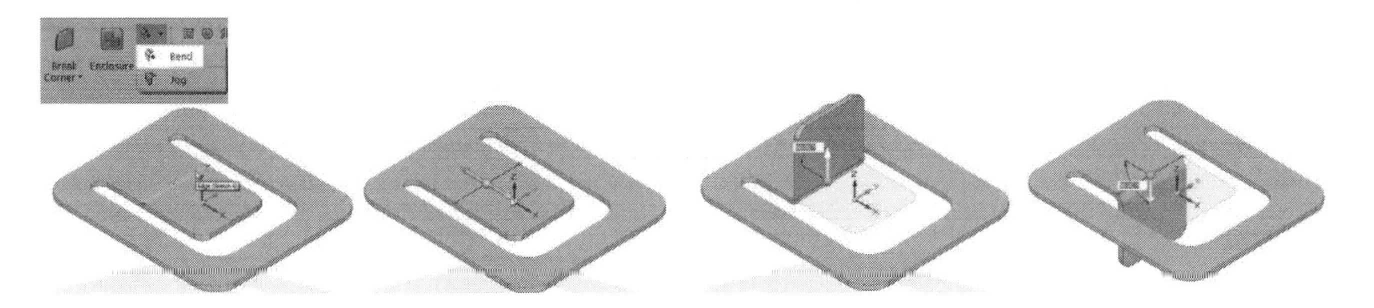

On the Command bar, click **Material Side > Center of Mold Lines** option to position the center of the bend on the sketched line. Click **Material Side > Mold Lines** option to position the sketch line on left or right edge of the bend.

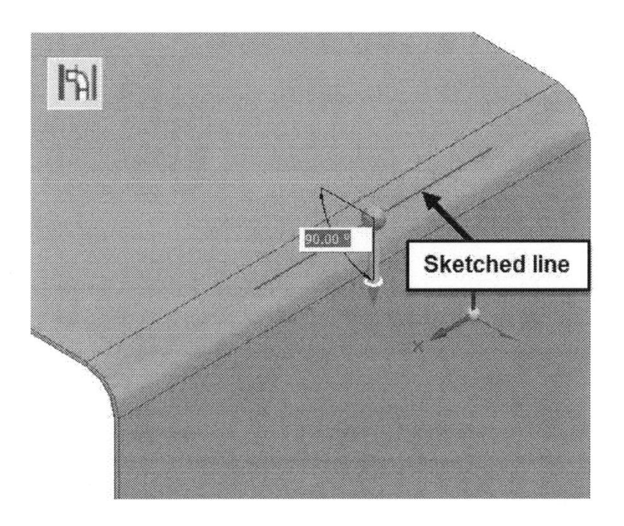

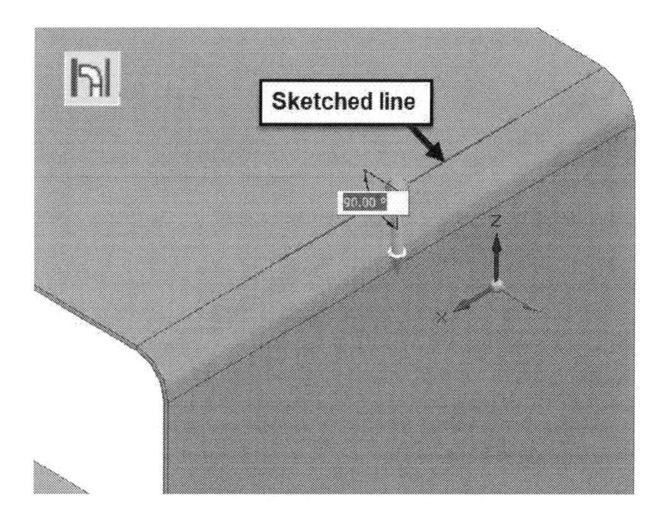

Click anywhere in the graphics window to complete the bend creation.

Jog

The **Jog** command is used to add a jog or offset to a flat sheet. To add a jog to sheet metal part, first you must define its location. You can do this by drawing a sketch line. Next, activate the **Jog** command (click **Home > Sheet metal > Jog** on the ribbon) and click on the sketched line. A two-sided arrow appears on the selected line. Click on either side of the arrow to define the side of bend.

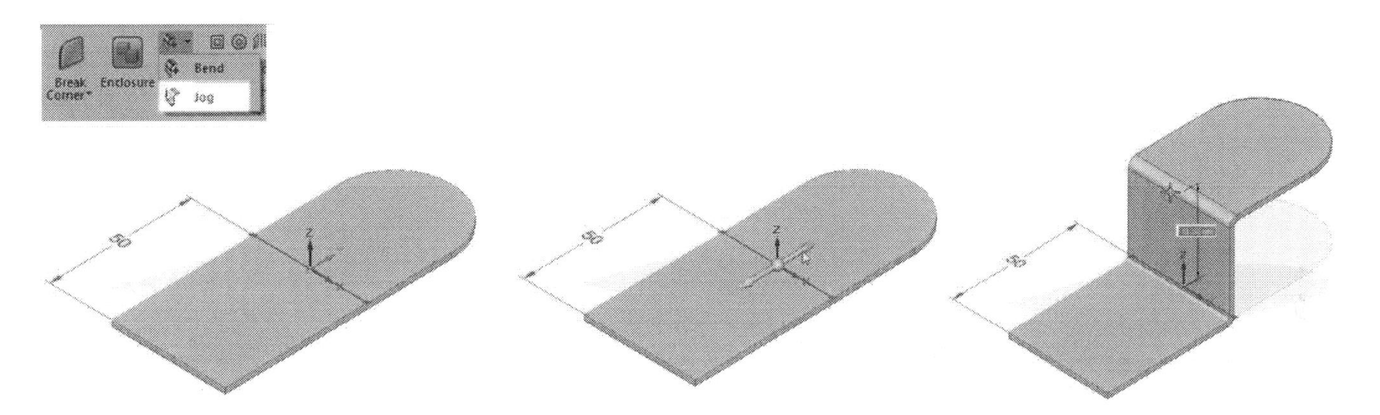

On the command bar, select a measurement point from the **Measurement Point** drop-down menu. Both the measurement points are illustrated below.

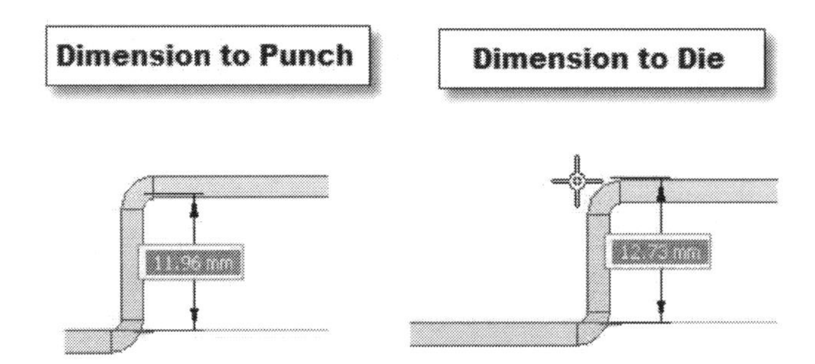

Type-in a value in the distance box and press Enter to add a jog to the sheet metal part.

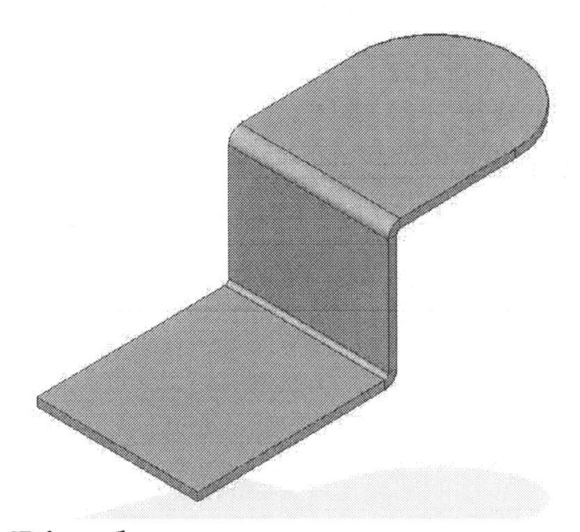

Dimple

The **Dimple** command is used to add a dimple to a flat sheet by deforming it. To add a dimple to a sheet metal part, first you must define its shape, size, and location. You can do this by drawing a closed sketched. After creating a closed sketch, activate the **Dimple** command (click **Home > Sheet Metal > Dimple** on the ribbon) and click in the sketch region. The sketch will be converted into a dimple shape. Click the arrow that appears on the dimple to change its direction.

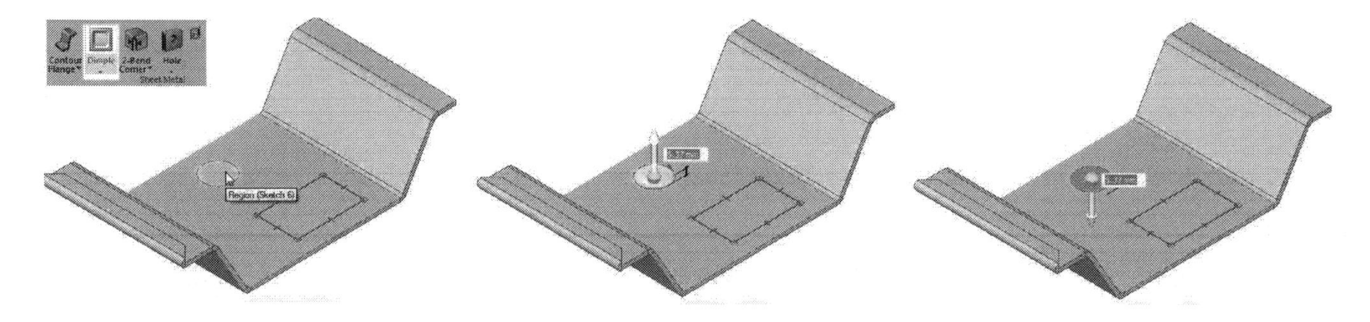

On the command bar, click the **Dimple Options** icon to open the **Dimple Options** dialog. On this dialog, type-in the values of taper angle, punch radius, die radius, and corner radius. Click **OK** to close the dialog.

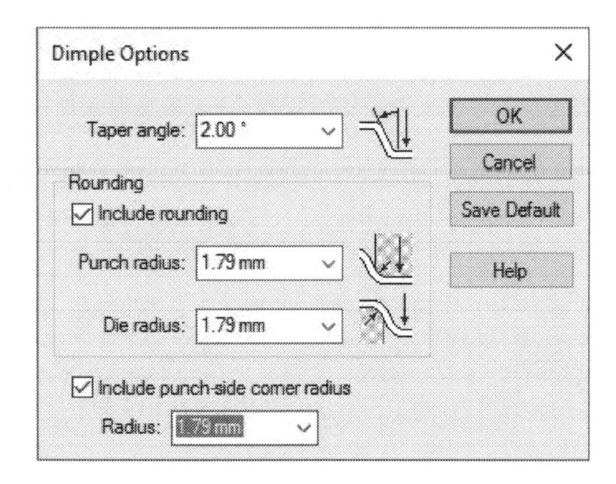

On the command bar, define the representation of the profile. You can select **Profile Represents Die** or **Profile Represents Punch**. Type-in a value in the distance box that is attached to the feature, and then press Enter to create the dimple.

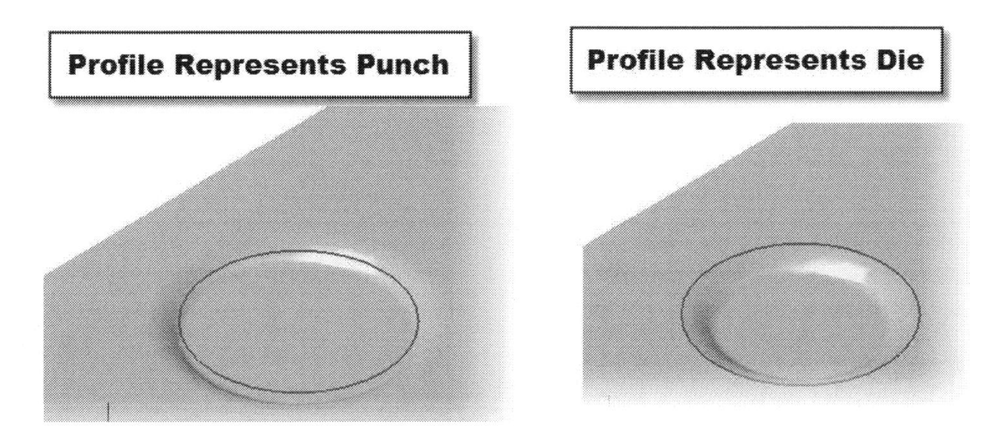

Drawn Cutout

The drawn cutout and dimple feature are almost alike, except that an opening is created in case of a drawn cutout. In order to create a drawn cutout, first you must have a closed sketch. Next, activate the **Drawn Cutout** command (click **Home > Sheet Metal > Drawn Cutout** on the ribbon) and click inside the sketch region. Click on the arrow that appears on the drawn cutout feature to change its direction.

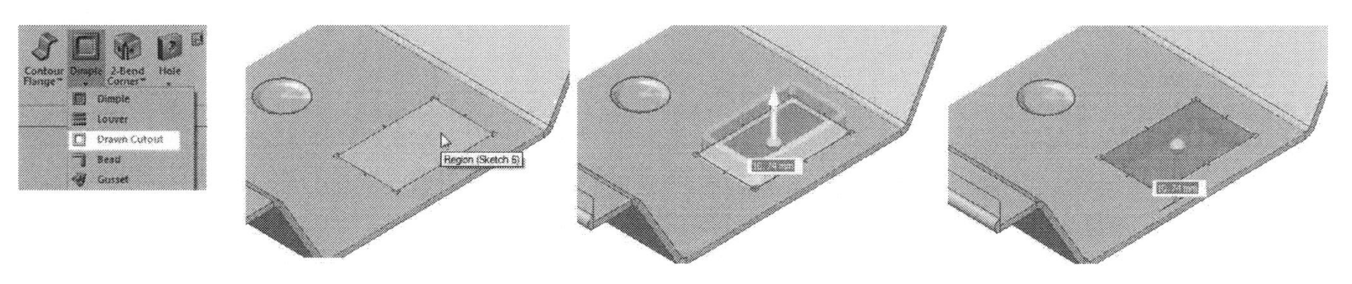

On the command bar, click the **Drawn Cutout Options** icon to open the **Drawn Cutout Options** dialog. Type-in values of taper angle, die radius, and corner radius. Click **OK** to close the dialog. Next, on the command bar, click the **Profile Represents Die** or **Profile Represents Punch** icon. This determines whether the sidewalls are placed inside or outside the sketch profile. Next, type-in a value in the distance box attached to the feature and press Enter.

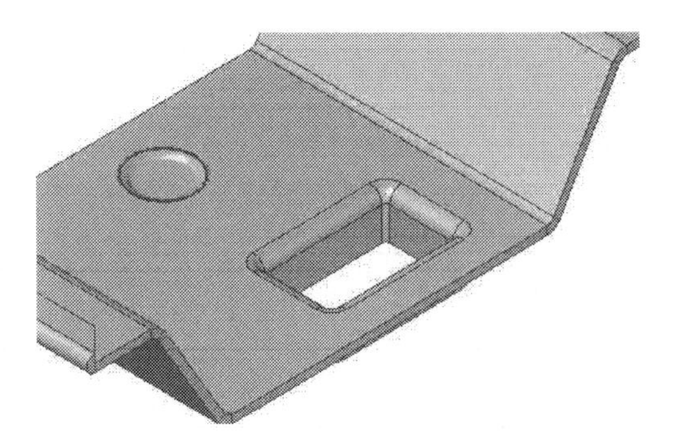

Bead

The **Bead** command creates a bead feature, which stiffens the sheet metal part. To create a bead feature, first you must have a sketch, which defines its size and shape. If the sketch is having curved edges, then ensure that they are tangent continuous. Next, activate the **Bead** command (click **Home > Sheet Metal > Dimple > Bead** on the ribbon) and click on the sketch. Click on the arrow that appears on the bead feature to change its direction.

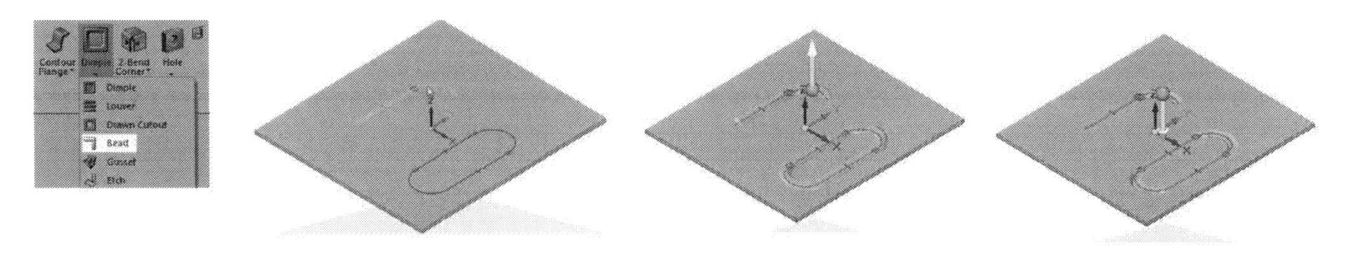

On the command bar, click the **Bead Options** icon to open the **Bead Options** dialog. On this dialog, under **Cross Section**, select the cross section type and define the size parameters. Check the **Include Rounding** option to apply rounds to the edges of the bead feature. Under the **End Conditions** section, select the desired option and click **OK** to close the dialog. Click the right mouse button to complete the bead feature.

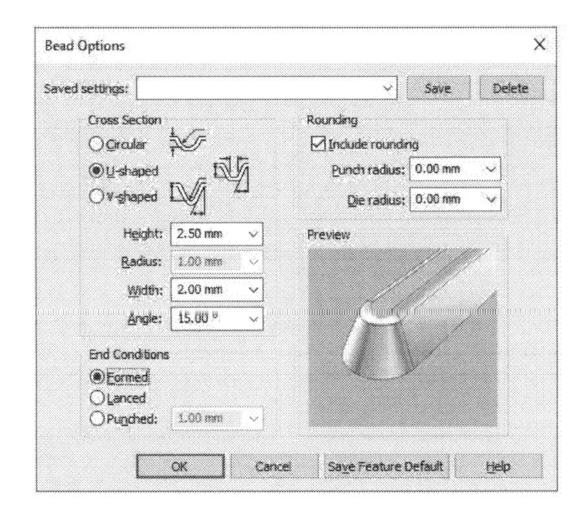

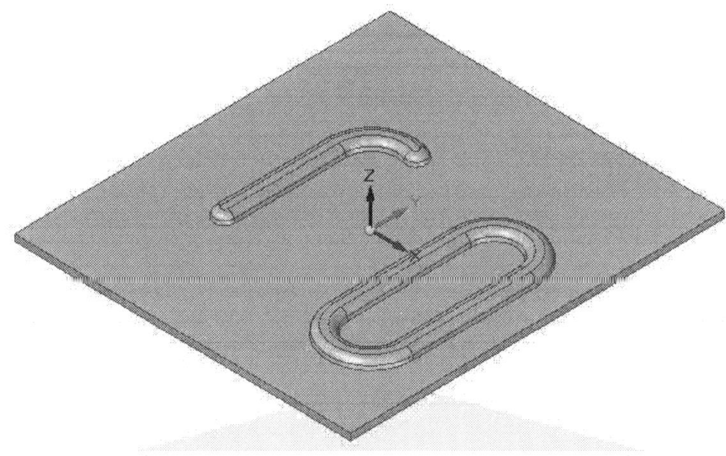

Louver

Solid Edge provides you with the **Louver** command, which makes it easy to create louvers. Activate this command (click **Home > Sheet Metal > Dimple > Louver** on the ribbon) and place the mouse pointer on a face. You will notice that a louver appears parallel to an edge. Press N or B on your keyboard to change the orientation of the louver. Press F3 on your keyboard to lock the face, and then place the mouse pointer on an edge and press E to add location dimension.

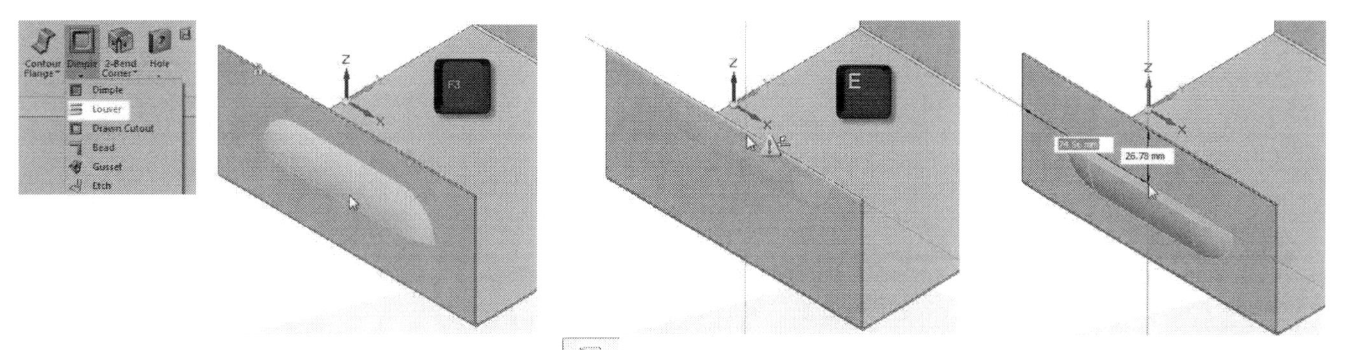

On the command bar, click the **Louver Options** icon to open the **Louver Options** dialog. On this dialog, select the end condition of the louver. The two types of end conditions are shown in figure.

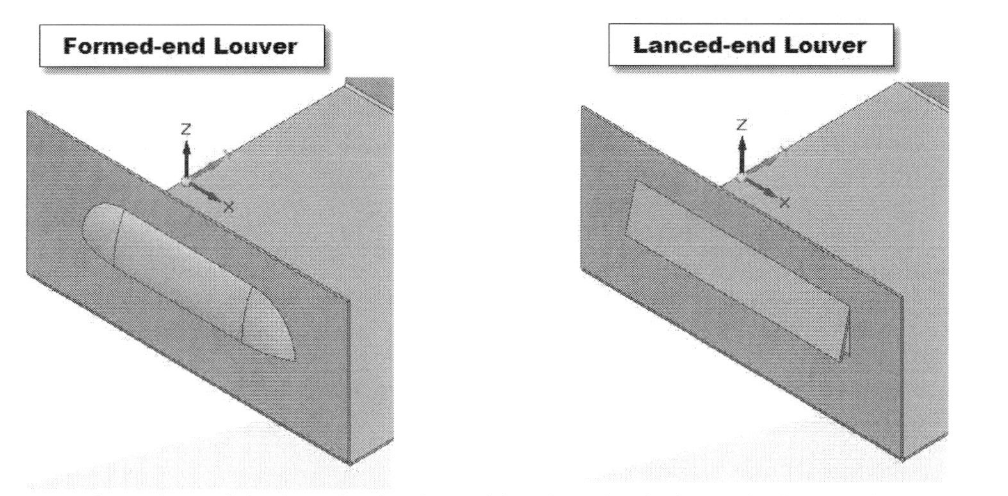

Type-in the values of the length, depth, and height. Check the **Include rounding** option to round the edges of the louver. Type-in values in the **X** and **Y** boxes to shift the default origin of the louver. Click **OK** to close the dialog.

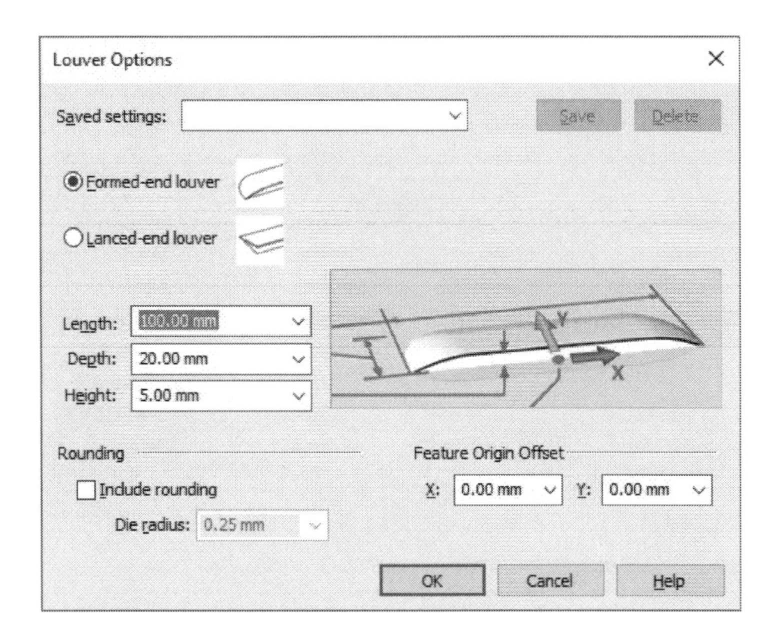

Type-in values in dimension boxes that are attached to the louver, and press Enter to complete the louver feature.

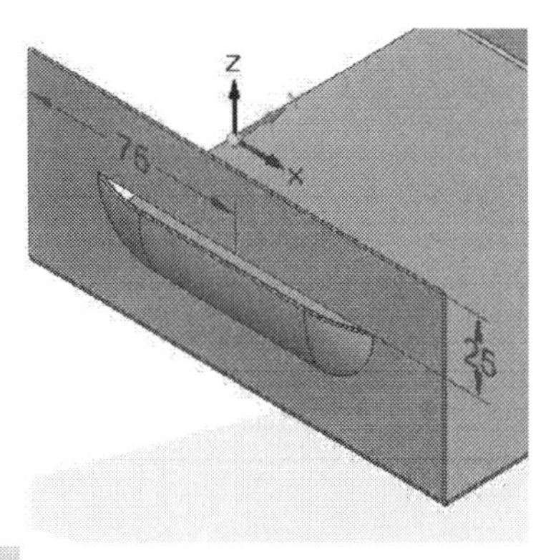

Gusset

Gussets are stiffening features created across a bend to reinforce the sheet metal part. To create a gusset, activate the **Gusset** command (click **Home > Sheet Metal > Dimple > Gusset** on the ribbon) and click on a bend face. A gusset feature appears along with a dimension box attached to it.

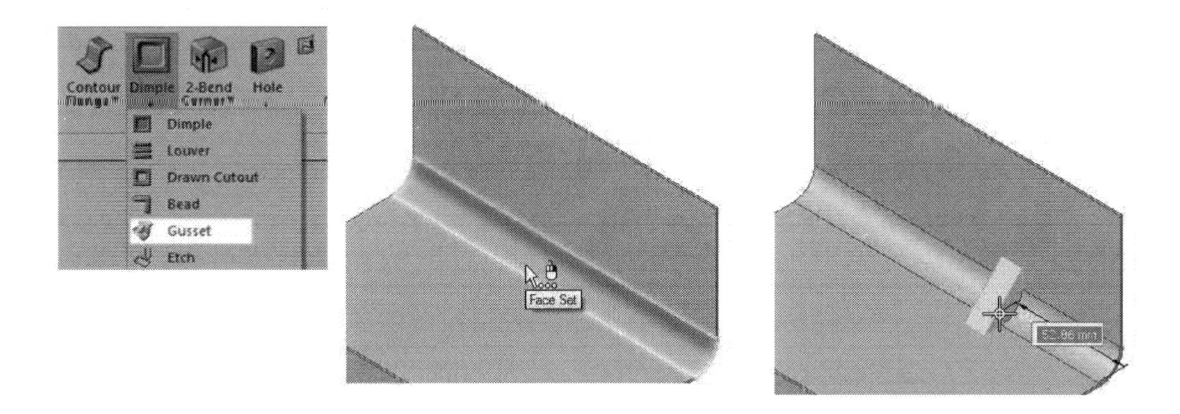

On the command bar click the **Gusset Options** icon to open the **Gusset Options** dialog. On this dialog, select the gusset shape and type-in a value in the **Depth** box. Type-in values of taper angle, width, and radius. Check the **Include rounding** option to round the edges of the gusset, and then click **OK** to close the dialog.

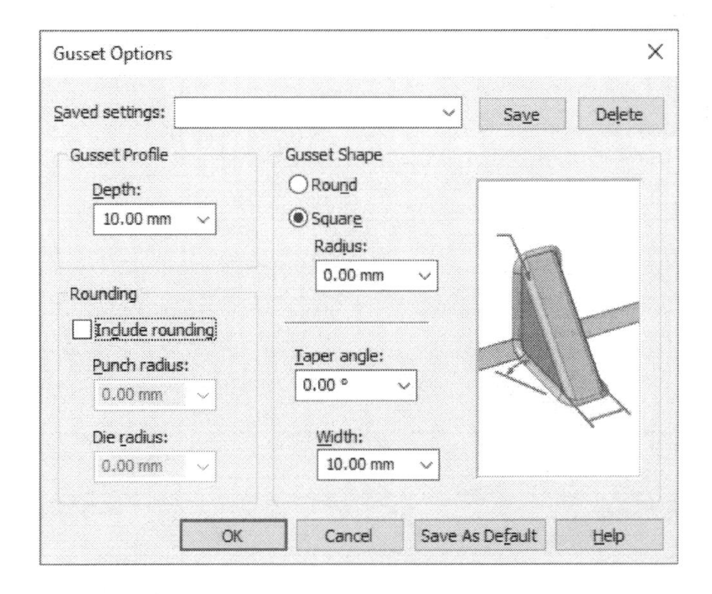

On the command bar, select a patterning option from the **Pattern** drop-down menu. The **Single** option creates a single gusset. The **Fit** option creates pattern along the total length of the bend by using the count value that you specify. The **Fill** option creates a pattern along the total length of the bend by using the spacing value that you specify. The **Fixed** option creates a pattern by using the spacing and the count values that you specify.

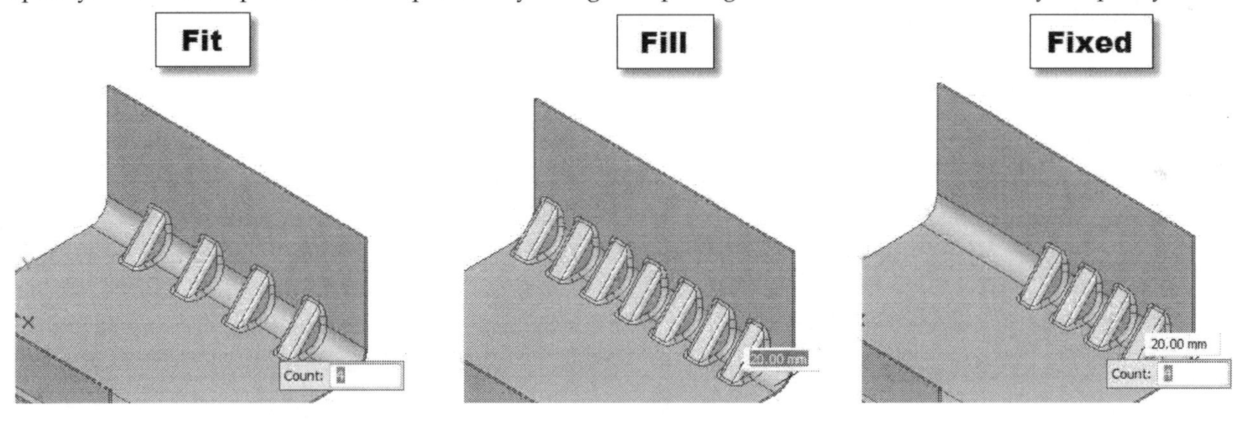

 Cut

When it is necessary to remove material from a sheet metal part, you must use the **Cut** command. First, draw a sketch and click inside it; a two sided arrow appears. Click on this arrow and drag the mouse pointer into the geometry. On the command bar, the select the extent type from the **Extents** drop-down menu. Click the right mouse button to create the cut.

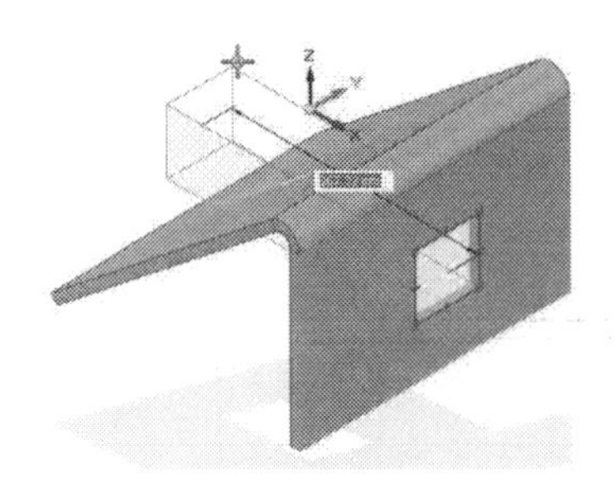

Creating Cut across Bends

If you need to create a cut across a bend, you must use the **Wrapped Cut** option. First, you must create a closed sketch across a bend. Press the Shift key and select all portions of the sketch region. On the command bar, activate the **Wrapped Cut** icon, and then click on the arrow handle pointing toward the model. You will notice that the sheet metal part is flattened and cut is created across the bend. Click the right mouse button to complete the cut.

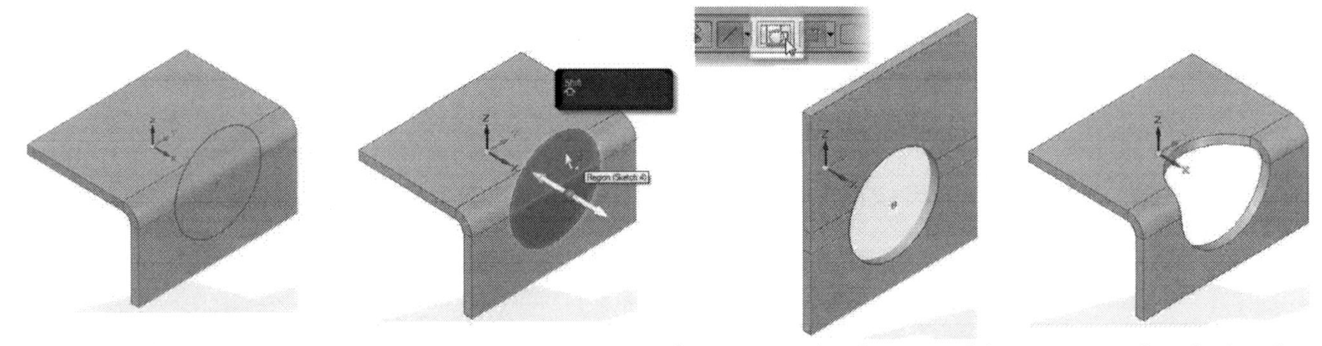

The shape of the wrapped cut-out remains constant even if you move the sheetmetal face connected to the bend.

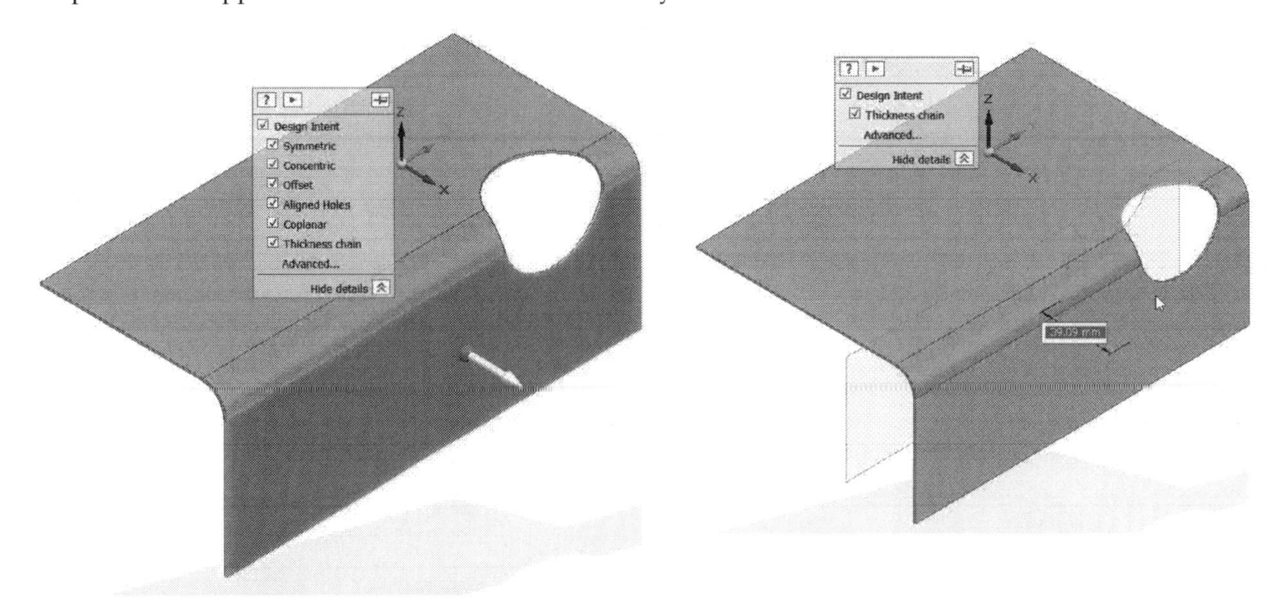

Also, the cut-out maintains its shape when you move its face. However, it displays an error message when you try to move the face over the bend.

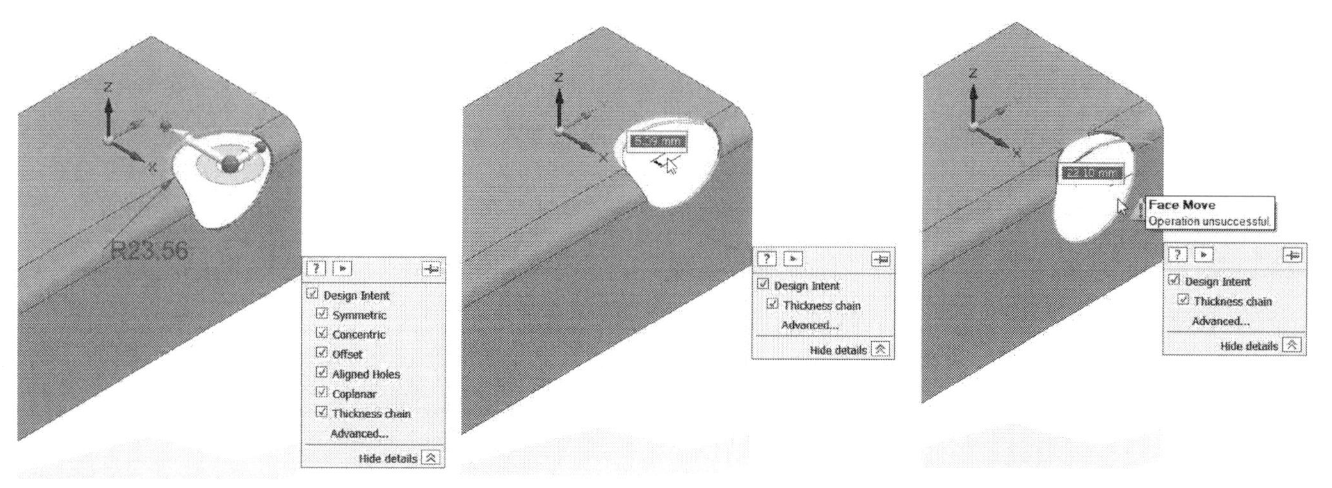

Break Corner

The **Break Corner** command is used to round or chamfer the sharp corner of a sheet metal part. Activate this command (click **Home > Sheet Metal > Break Corner** drop-down **> Break Corner** on the ribbon) and click on the corner edges of the sheet metal part. If you want to break all the corners of the sheet metal part, then drag a window across the geometry. All the corners of the sheet metal part will be selected.

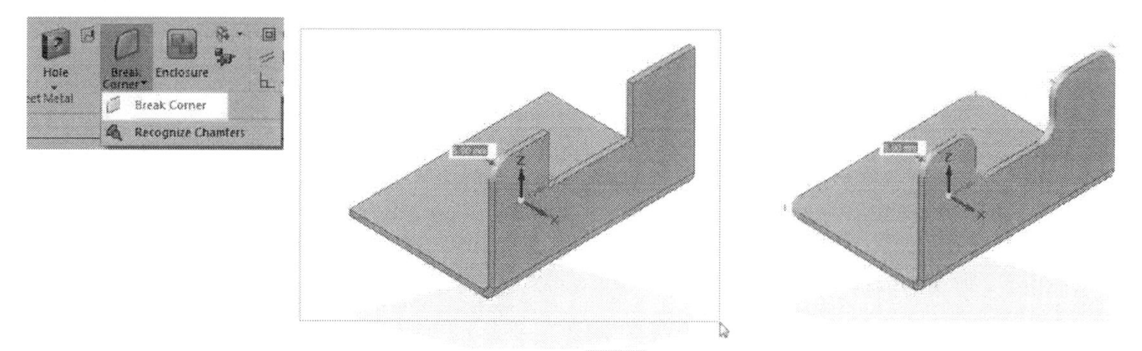

On the command bar, click the **Chamfer Corner** icon to apply chamfers to the corner edges. Type-in a value in the box that is attached to the round or chamfer. Press Enter to complete the break corner feature.

Flat Pattern

The **Flat Pattern** command flattens the part so that the manufacturing information can be displayed easily. To create a flat pattern, activate the **Flat Pattern** command (click **Tools > Model > Flatten** on the ribbon) and click on a base sheet. Next, click on an edge to define the x-axis of the flat pattern. Click the right mouse button to create the flat pattern.

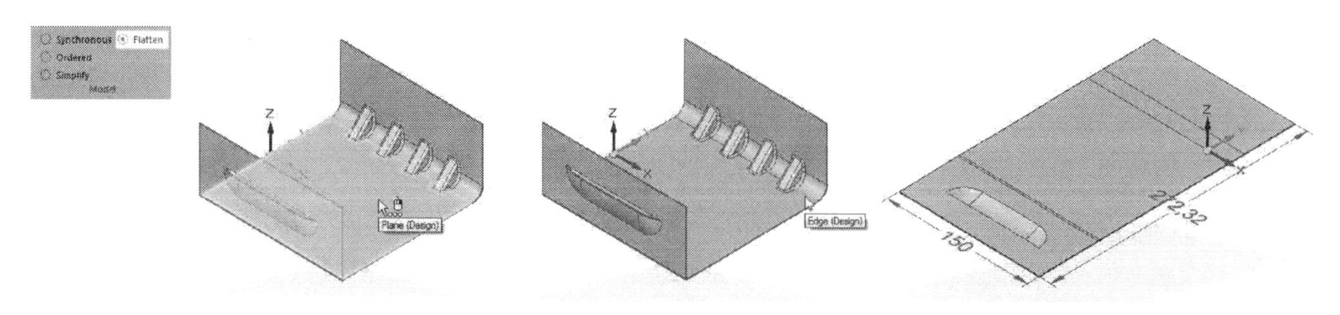

You will notice that a new entry 'Flat Pattern' is created in Pathfinder. You can switch back to the modeling mode by clicking **Tools > Model > Synchronous** on the ribbon.

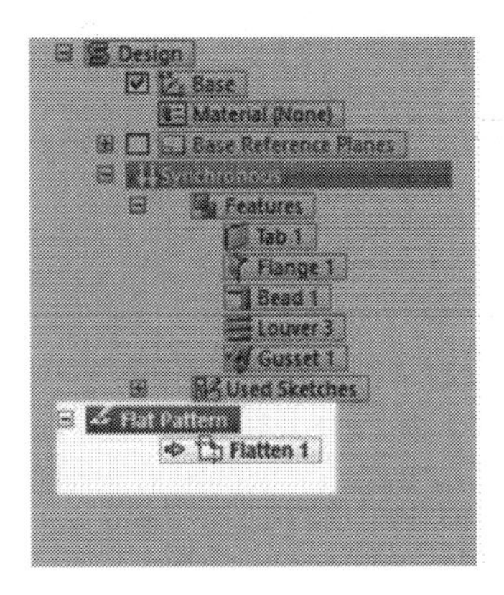

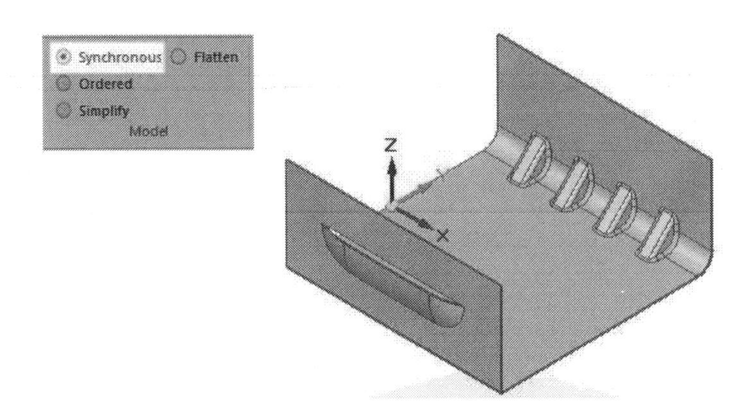

Lofted Flange

The **Lofted Flange** command allows you to create a lofted flange that can be unfolded into flat pattern. In Solid Edge, the **Lofted Flange** command is available only in the **Ordered** environment. Transit to the **Ordered** environment and create two sketches on planes parallel to each other. Ensure that the sketches are not closed. In addition, the openings should be in the same direction.

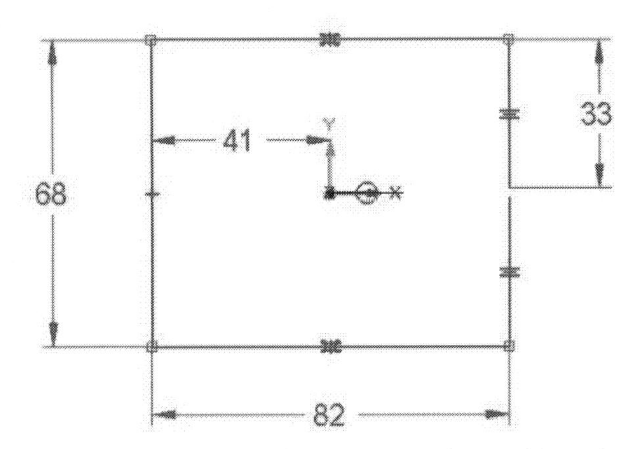

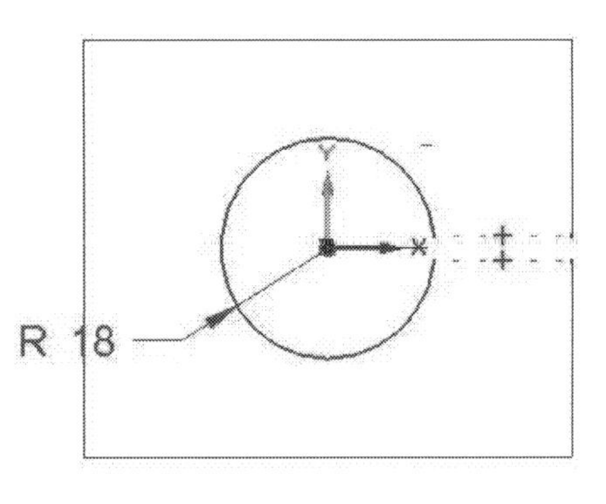

Activate the **Lofted Flange** command (click **Home > Sheet Metal > Contour Flange > Lofted Flange** on the ribbon) click on the first cross section. Click the green check on the command bar to accept the selection. Click on the second cross section and click the green check.

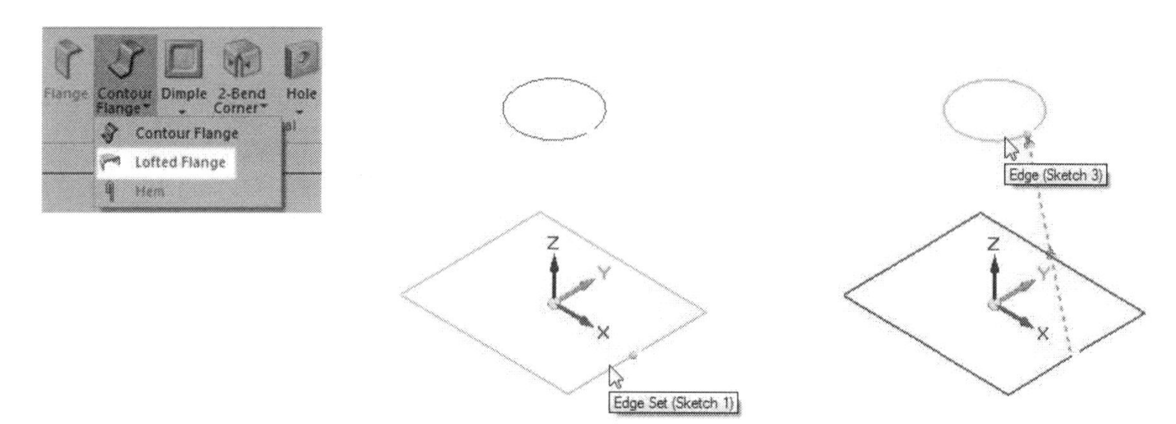

On the command bar, type-in a value in the **Thickness** box. Click inside or outside the sketch to define the side of sheet metal. Click **Finish** to complete the lofted flange feature.

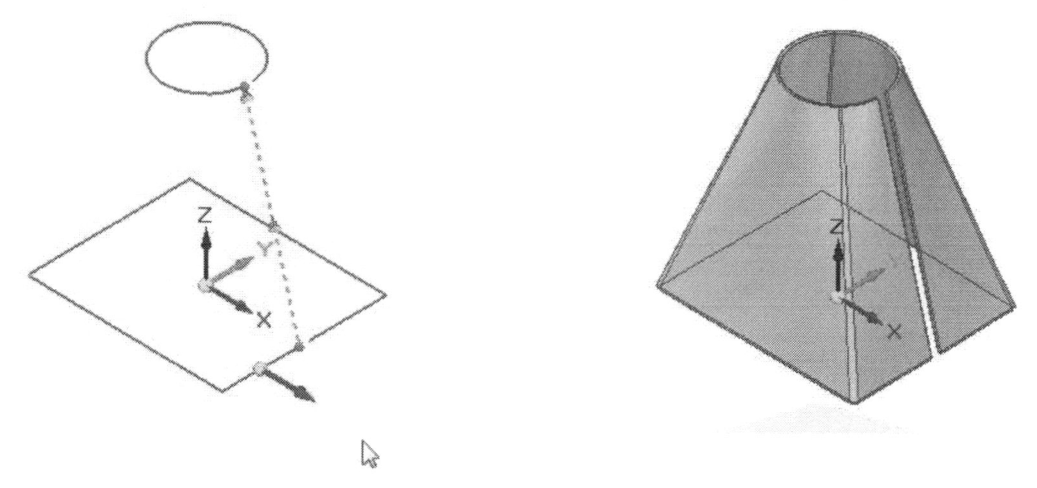

Create the flat pattern of the sheet metal part.

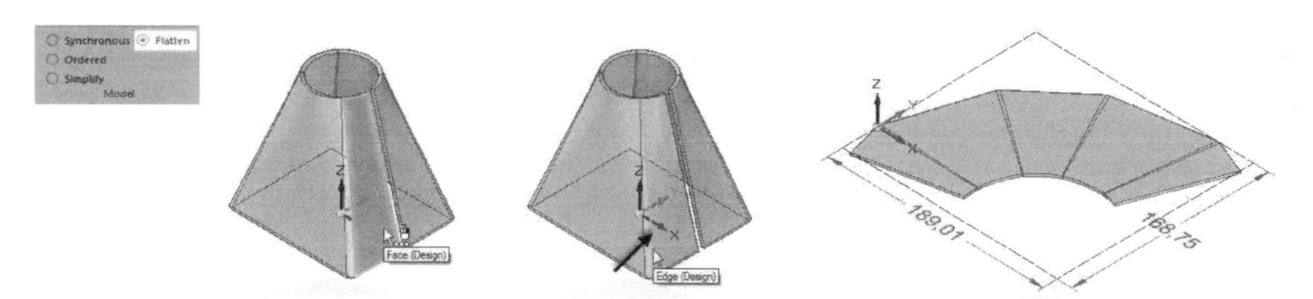

Thin Part to Synchronous Sheet Metal

Solid Edge has a special command, which converts an already existing part into a sheet metal part. This command is called **Thin Part to Synchronous Sheet Metal**. First, create a part in the Synchronous environment, and then shell it using the **Thin Wall** command. Next, click **Tools > Transform > Thin Part to Synchronous Sheet Metal** on the ribbon.

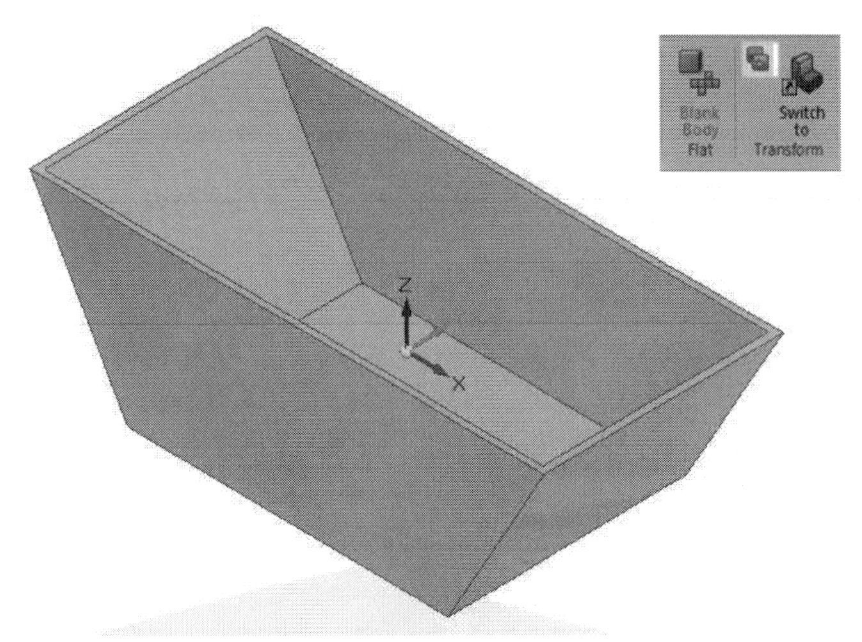

On the command bar, click the **Options** icon to open the **Transform to Sheet Metal Options** dialog. On this dialog, set the relief type: **Square** or **Round**. Next, specify the **Depth**, **Width**, **Bend radius**, and **Neutral Factor** values. Click **OK** to close the dialog.

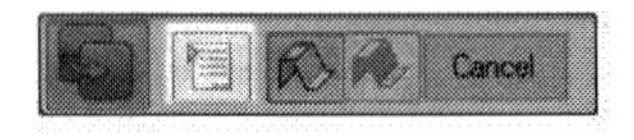

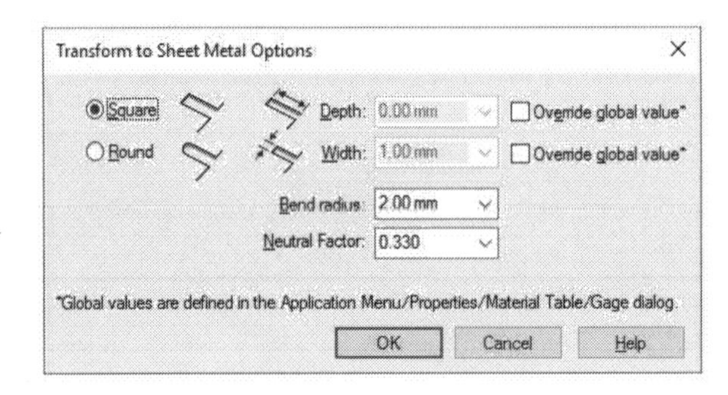

Click on a face of the part geometry to define the base face. A message pops up asking you to rip the edges of the part. Click **OK** to close the message.

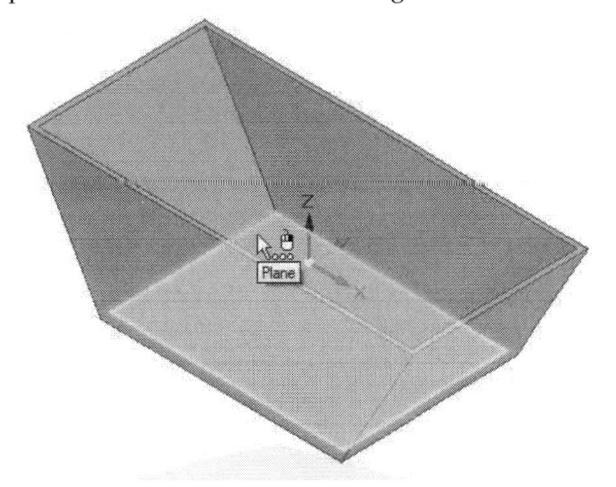

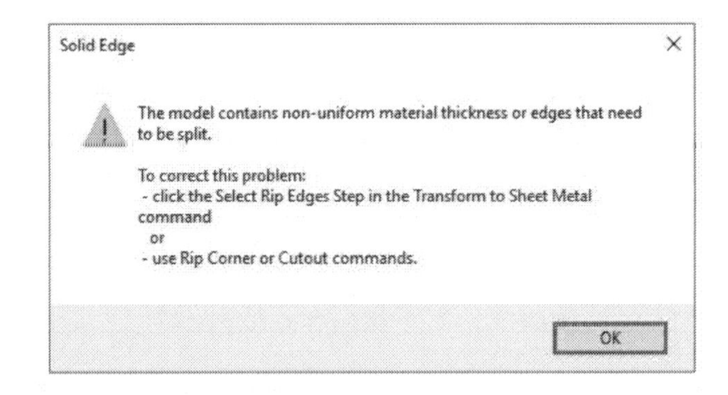

On the command bar, click the **Select Rip Edges Step** icon and click on the side edges of the part. Click the green check to complete the conversion process. Now, you can save and close the file.

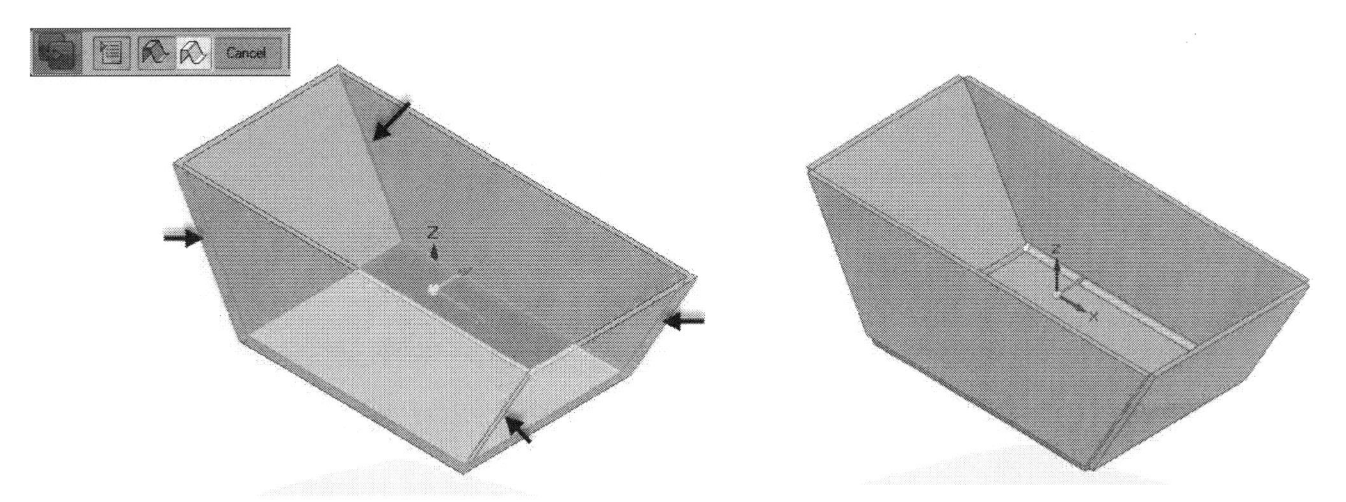

Part to Sheet Metal

The **Part to Sheet Metal** command creates a sheet metal part from a set of planar faces of a solid body. This command is available in the **Ordered** environment only. First, create a solid body using the solid modeling commands, and then activate the **Part to Sheet Metal** command (On the ribbon, click **Tools > Transform > Part to Sheet Metal**). On the **Part to Sheet Metal Options** dialog, specify the sheet metal properties (refer to the Tab section) and click **OK**. Click on linear edges of the solid body. The faces connected to the selected edge are highlighted.

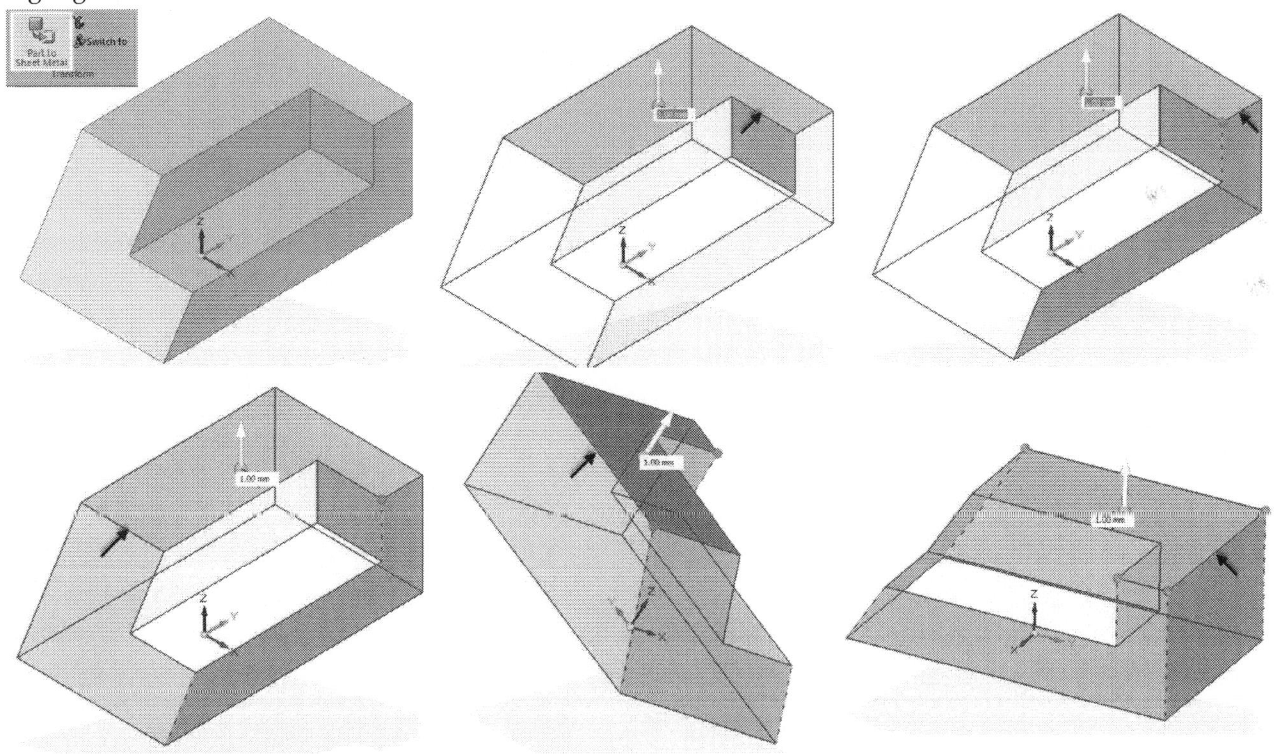

Click the arrow to change the side of the sheet metal. Type-in the sheet metal thickness and right-click to convert the solid to sheet metal.

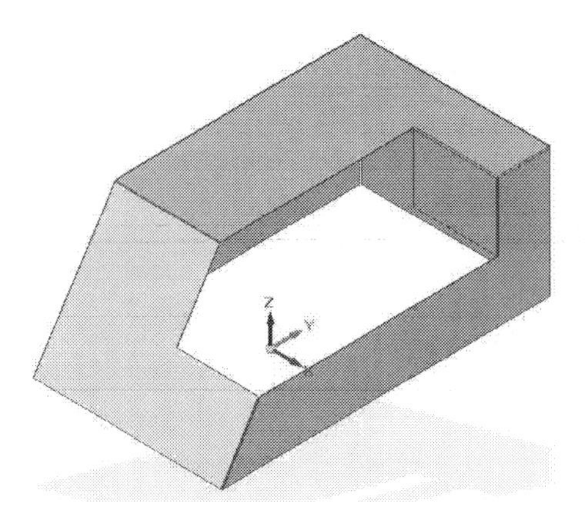

Sheet Metal Drawings

Creating drawings of a sheet metal part is same as any other drawing. However, there are some settings specific to sheet metal flat pattern. You can access these settings in the **Annotation** tab of the **Solid Edge Options** dialog. Note that these settings are available only in the **Solid Edge Options** dialog of the **Drawing** file.

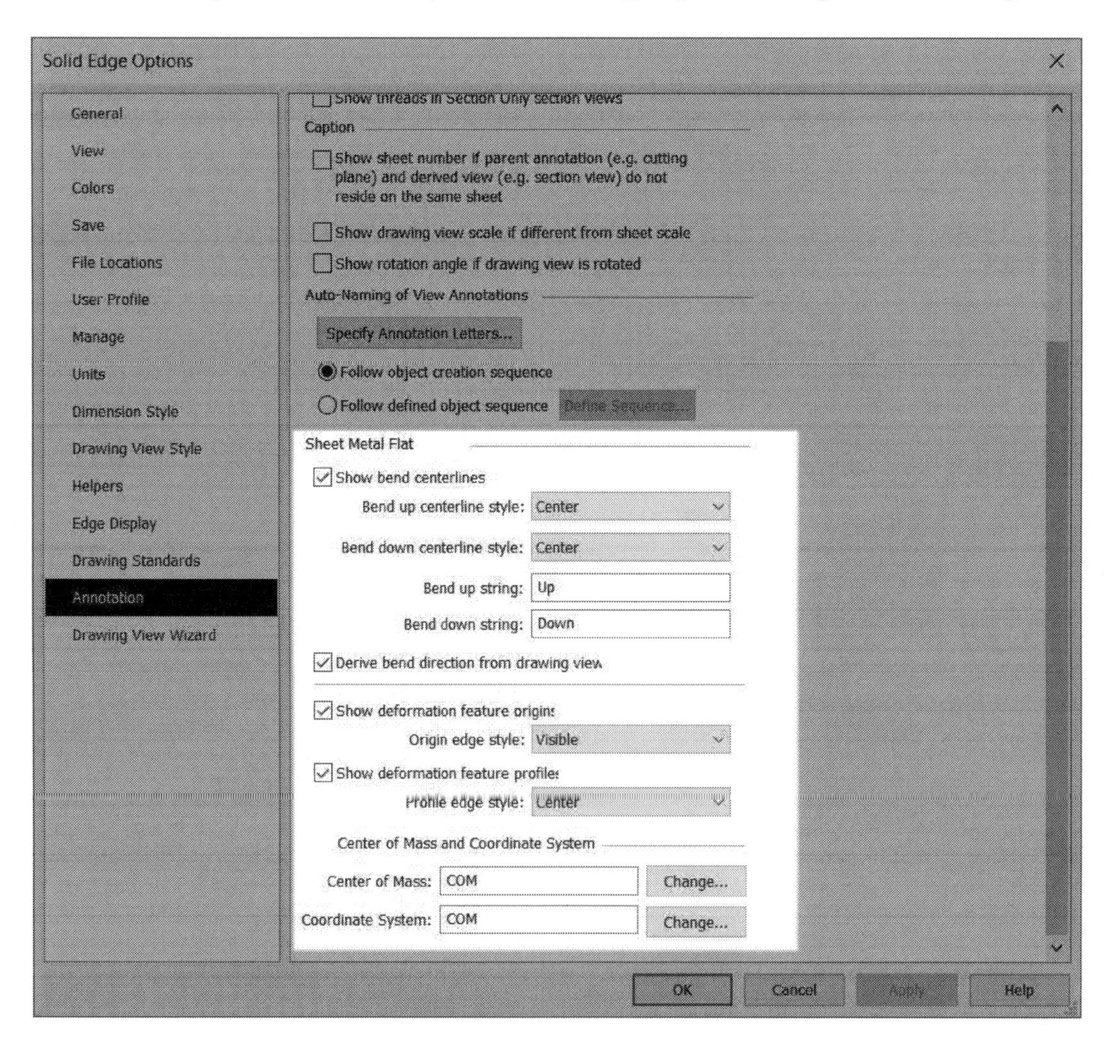

To create a flat pattern view, activate the **View Wizard** command and select the sheet metal part. On the command bar, click the **Drawing View Wizard Options** icon to open the **Drawing View Creation Wizard** dialog. On this dialog, select the **Flat pattern** option and click **OK**.

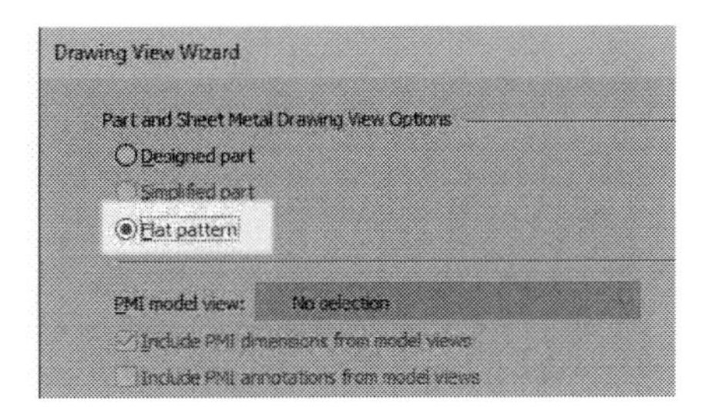

On the command bar, set the **Scale** value and click to place the view. You will notice that the bends are represented by centrelines.

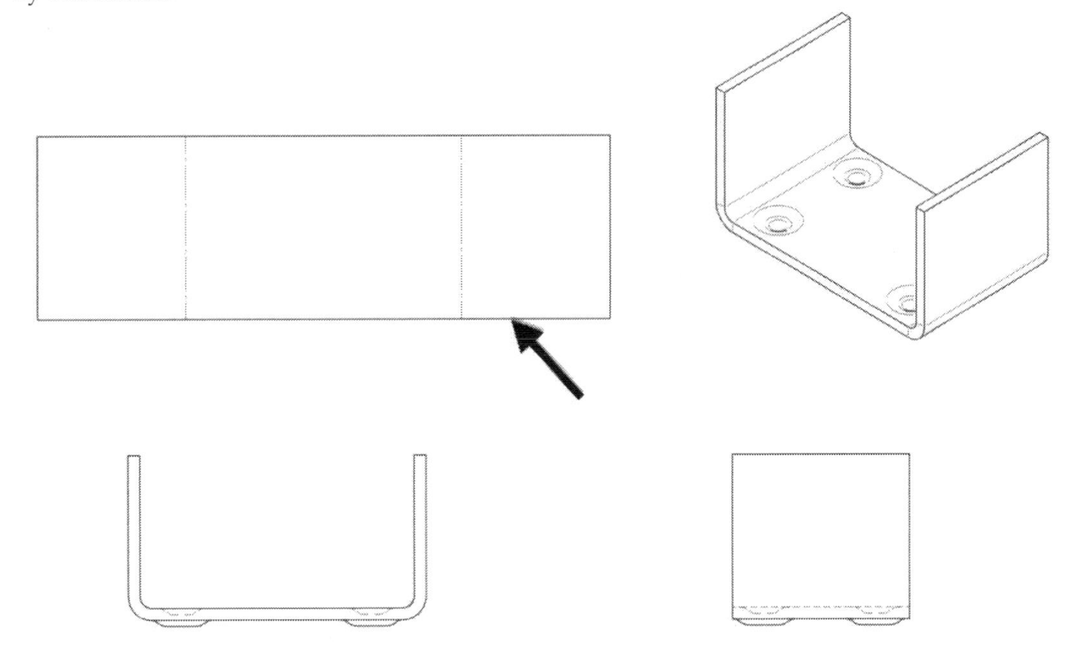

To add a bend table, click **Home > Table > Bend Table** on the ribbon, and then click on the flat pattern view. Click on the sheet to position the bend table.

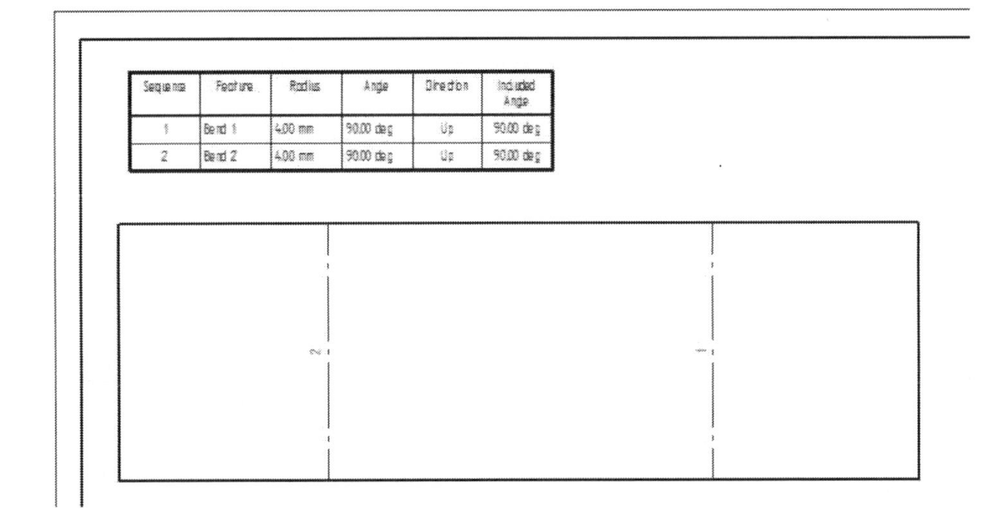

Sequence	Feature	Radius	Angle	Direction	Included Angle
1	Bend 1	4.00 mm	90.00 deg	Up	90.00 deg
2	Bend 2	4.00 mm	90.00 deg	Up	90.00 deg

Export to DWF

In addition to creating drawings, you can directly export a sheet metal part to DWF format which can be opened in AutoCAD. All you have to do is click **Application Menu > Save As > Save As Flat** . Select the sheet metal face that will be orientated upwards. Click on an edge of the selected face to define the x-axis of the DWF file. On the **Save As Flat** dialog, click the **Options** button to open the **Save As Flat DXF Options** dialog. On this dialog, set the layer properties and bend data, and then click **OK**. Type-in a name in the **File name** box and click **Save**. Now, you can open the DWF file in AutoCAD.

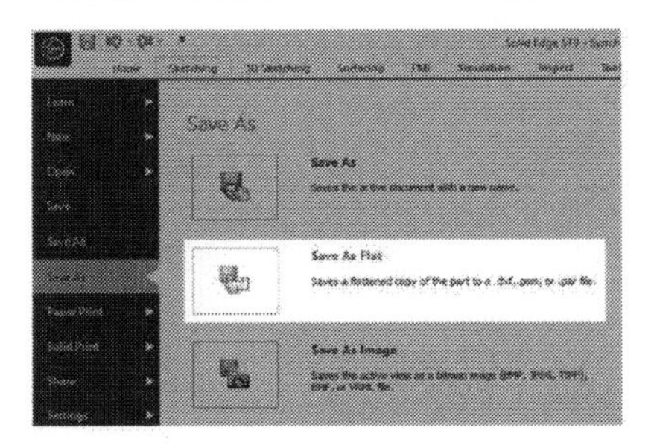

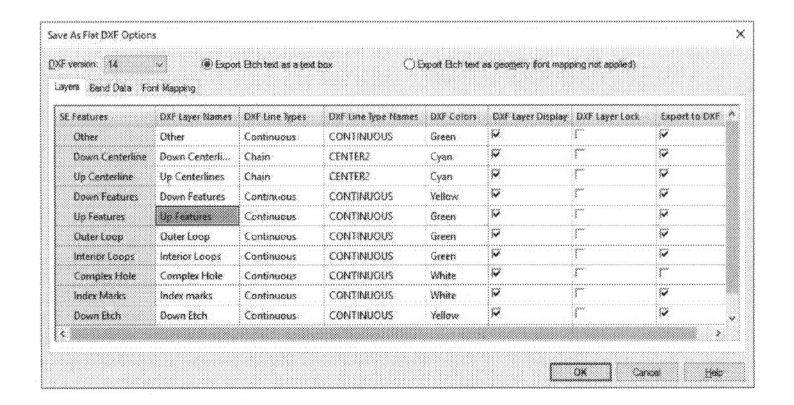

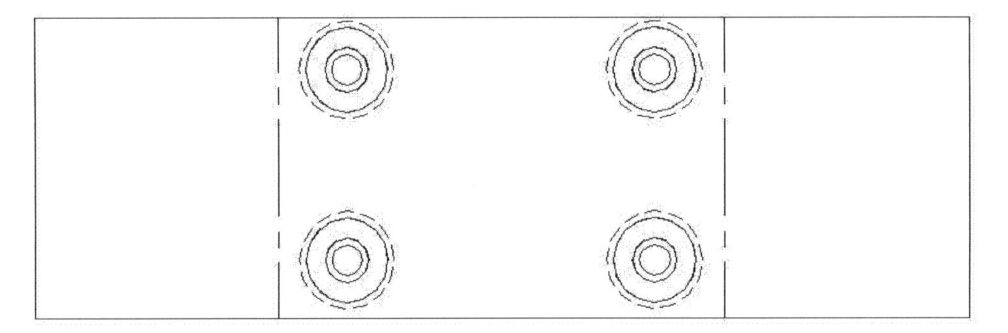

Examples

Example 1

In this example, you will design the sheet metal part shown below.

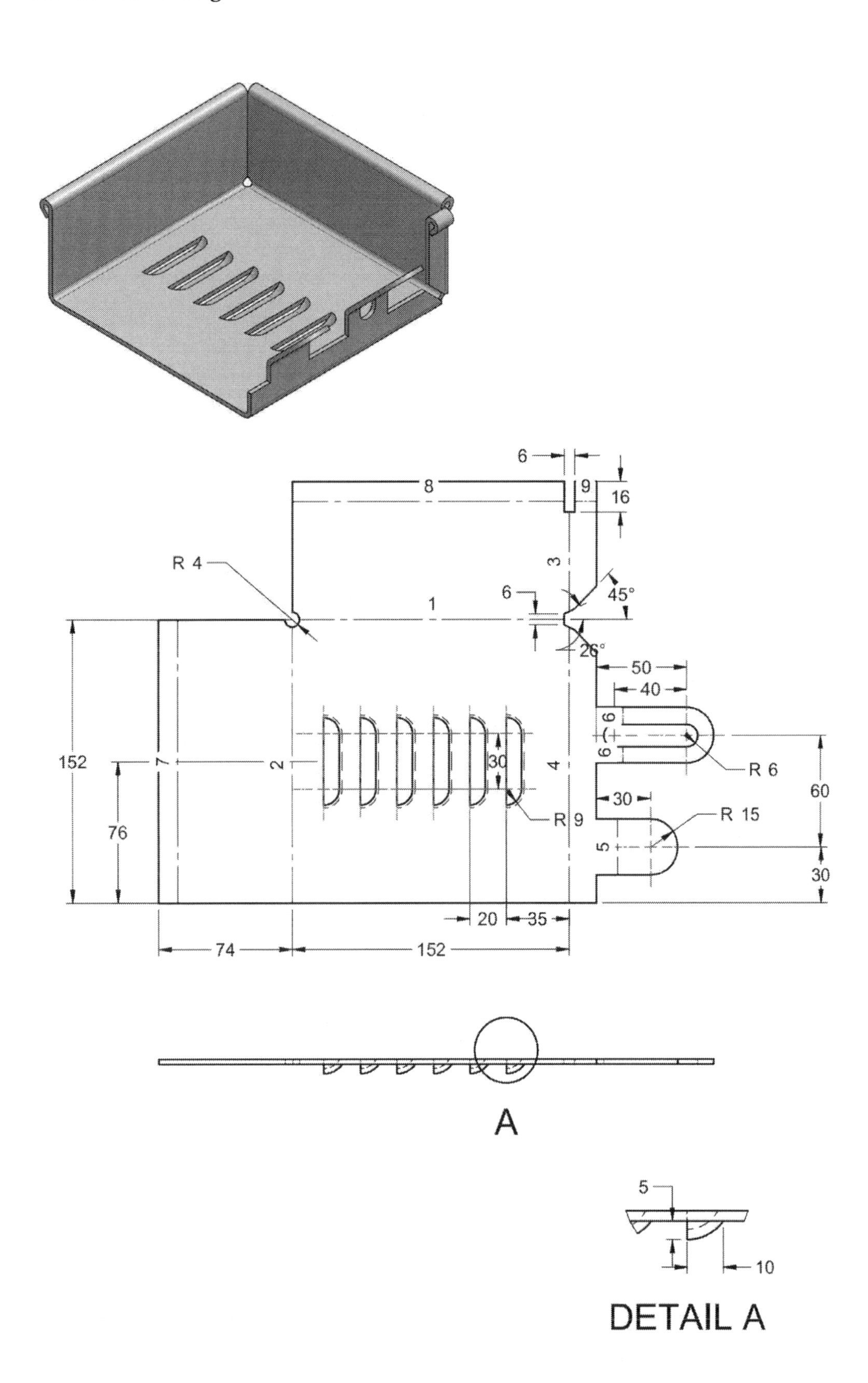
6
8
9
16
R 4
3
6
1
45°
26°
50
40
152
7
2
30
4
6
6
R 6
60
30
R 9
R 15
76
5
30
20
35
74
152
A
5
10
DETAIL A

Sequence	Feature	Radius	Angle	Direction	Included Angle
1	Bend 1	2.77 mm	90.00 deg	Up	90.00 deg
2	Bend 2	2.77 mm	90.00 deg	Up	90.00 deg
3	Bend 4	2.77 mm	90.00 deg	Up	90.00 deg
4	Bend 3	2.77 mm	90.00 deg	Up	90.00 deg
5	Bend 5	2.77 mm	45.00 deg	Down	135.00 deg
6	Bend 9	2.77 mm	45.00 deg	Down	135.00 deg
7	Bend 12	2.00 mm	136.44 deg	Down	43.56 deg
8	Bend 11	2.00 mm	136.44 deg	Down	43.56 deg
9	Bend 10	2.00 mm	136.44 deg	Down	43.56 deg

1. Start **Solid Edge 2019**.
2. On the **Application Menu**, click **New > ISO Metric Sheet Metal** to start a new sheet metal file.
3. Create a sketch on the top (XY) plane. Change the orientation of the model to the ISO View.

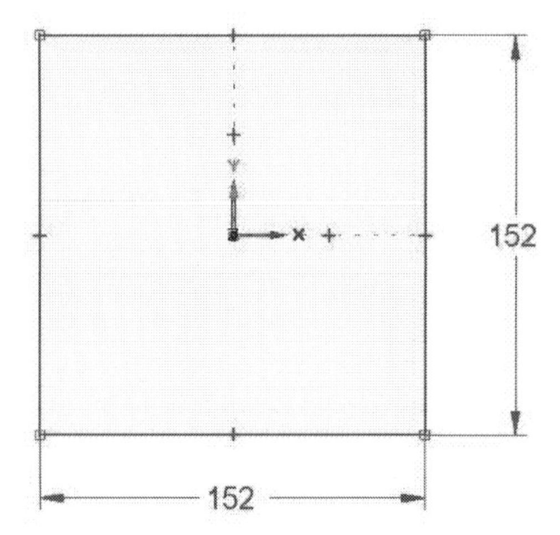

4. Click inside the region enclosed by the sketch. On the command bar, click the **Material Table** icon to open the **Material Table** dialog. On this dialog, open the **Gage Properties** tab and set the **Sheet metal gage** to **12 gage**. Set the **Neutral Factor** to 0.5. Click **Apply to Model** and then click **No** on the Save Changes message box. Next, close the dialog.
5. Click on the arrow handle to make it point upwards. Click the right mouse button to complete the tab feature.

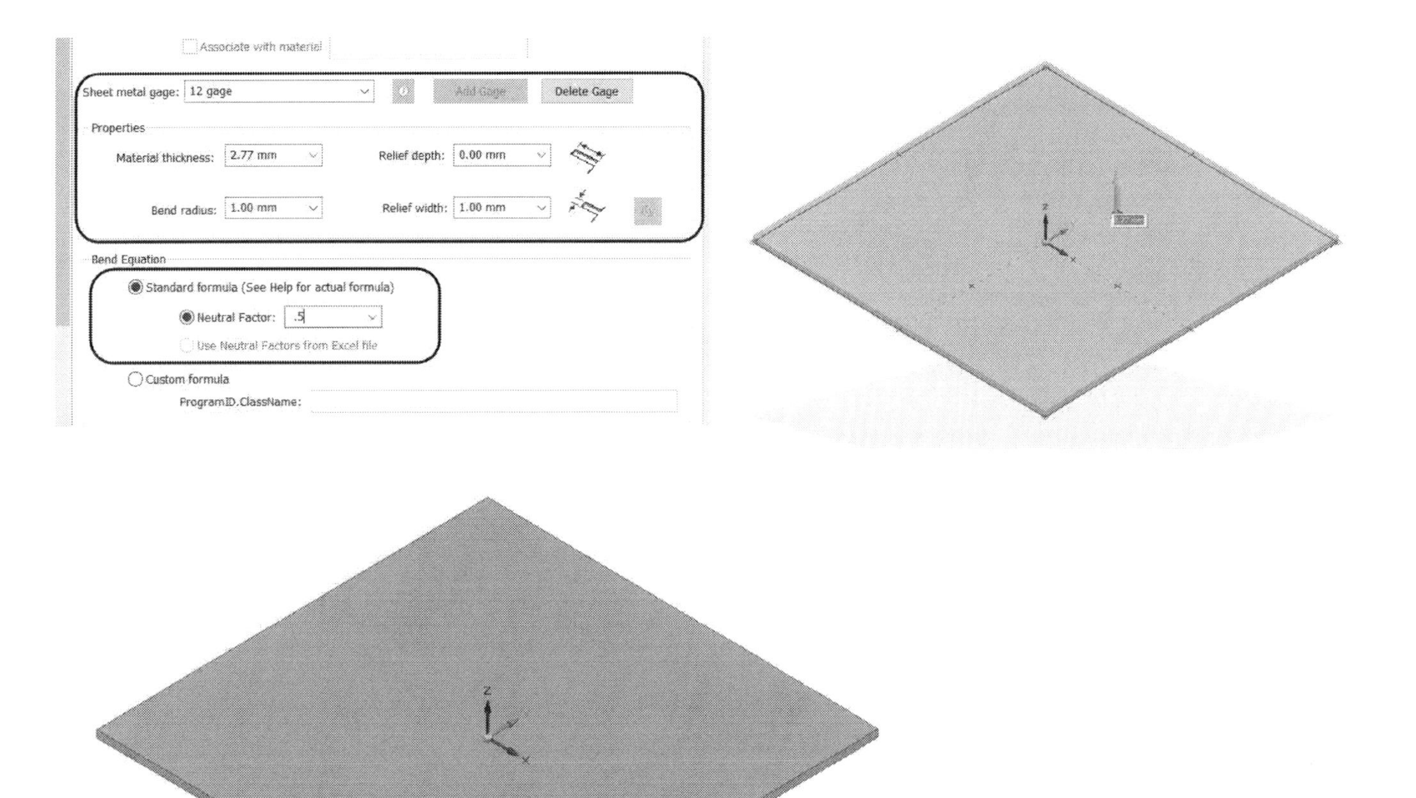

6. Click on the back end face to display the flange handle on it. On the flange handle, click the arrow pointing upwards, and then drag the mouse pointer.
7. On the command bar, set the **Measurement Point** to **Measurement Outside**. Set the **Material Side** to **Material Outside**.
8. Move the mouse pointer up and type-in **65** in the distance box. Press Enter to create the flange.

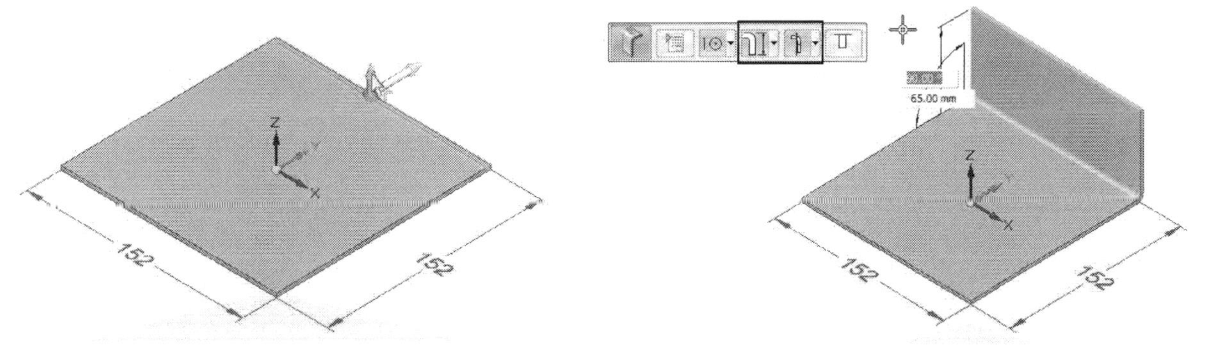

9. Create another flange on the left side. The flange length is 65 mm.

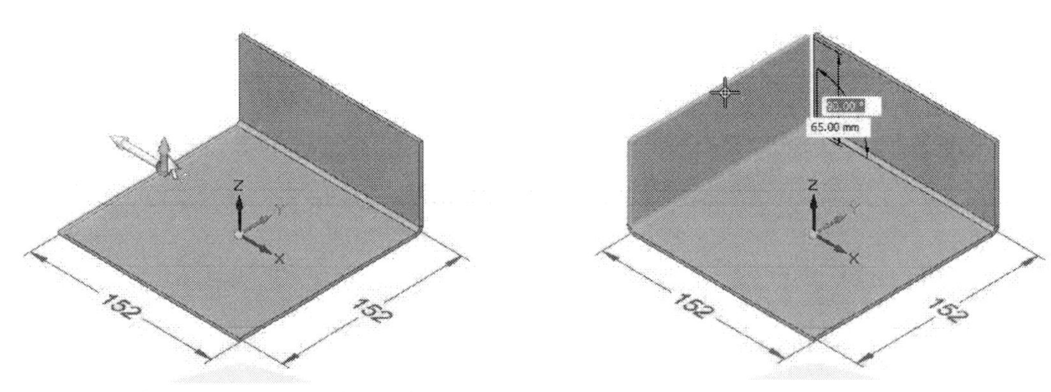

10. Activate the **Line** command (Click **Home > Draw > Line** on the ribbon).
11. Lock the front-end face and draw a vertical line of **15** mm length. Apply the Connect relation between the end point of the line and the top vertex of the corner.

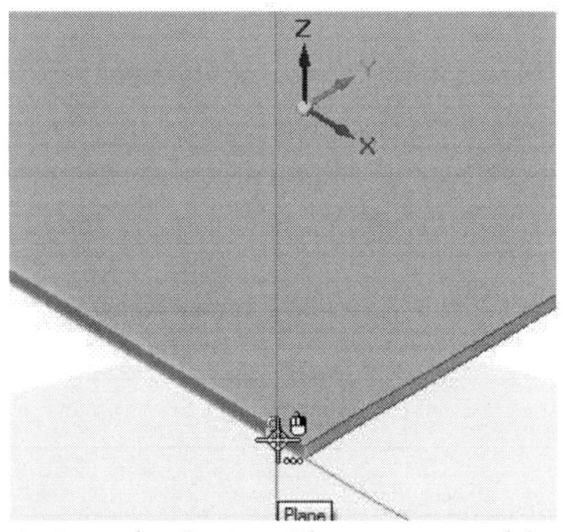

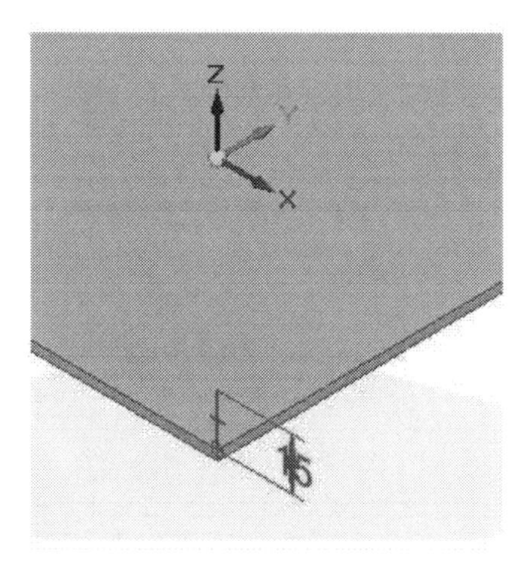

12. Activate the **Contour Flange** command (on the ribbon, click **Home > Sheet Metal > Contour Flange**) and click on the line. Click the arrow pointing toward right.

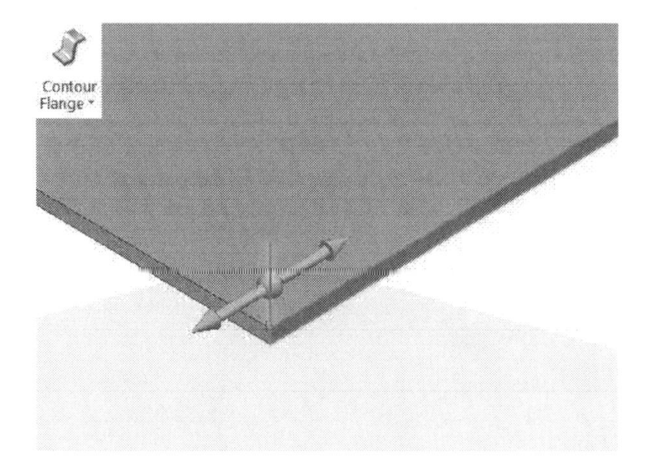

13. Select the end face of the flange perpendicular to the tab feature. Click the **Contour Flange Options** icon on the Command Bar. On the **Contour Flange Options** dialog, click the **Miters and Corners** tab, and then type in 2.77 in the **Gap** box of the **Interior Corners** section. Click **OK** to close the dialog. Click the right mouse button to create the contour flange.

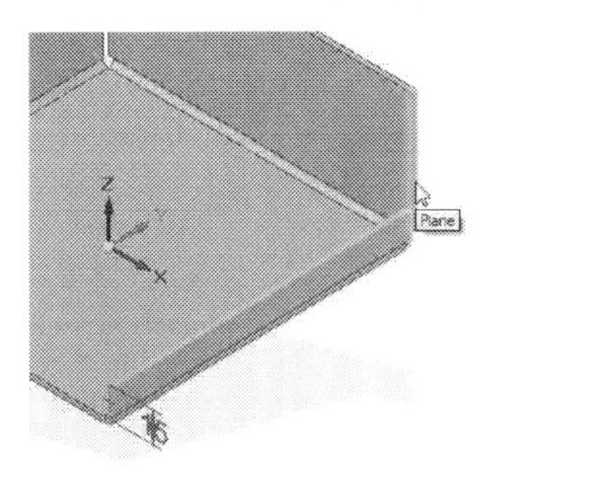

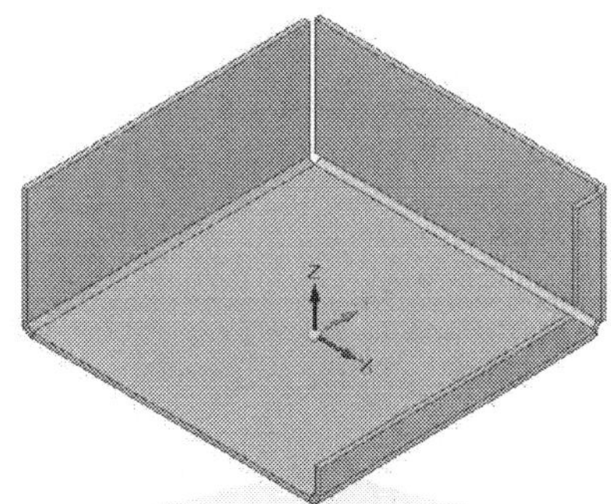

14. Activate the **Line** command (Click **Home > Draw > Line** on the ribbon).
15. Lock the outer face of the contour flange and draw the sketch shown below. Create a tab feature using the sketch.

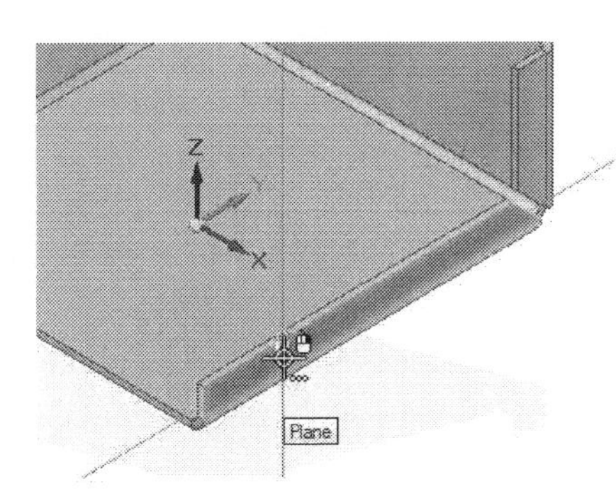

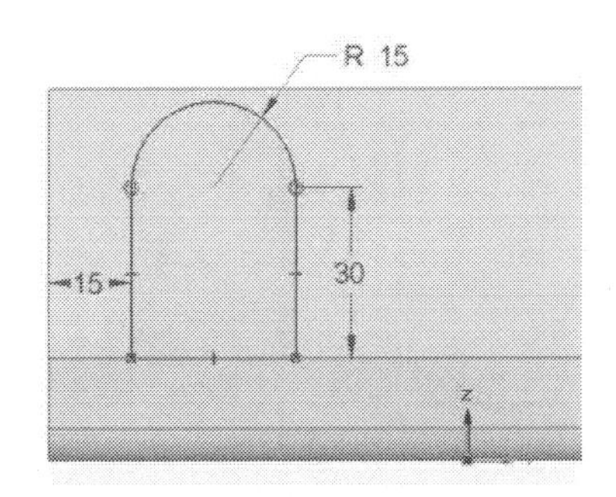

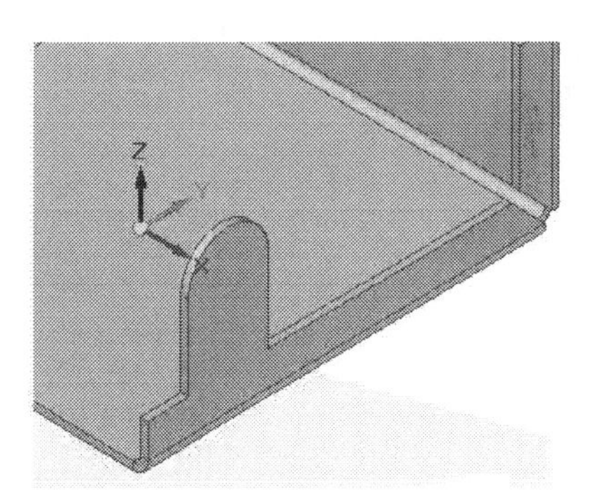

16. Draw a horizontal line on the outer face of the tab. Activate the **Bend** command (click **Home > Sheet Metal > Bend** on the ribbon) and click on the line.
17. Click on the arrow pointing upwards. Type-in **135** in the angle box and press Enter to bend the tab feature.

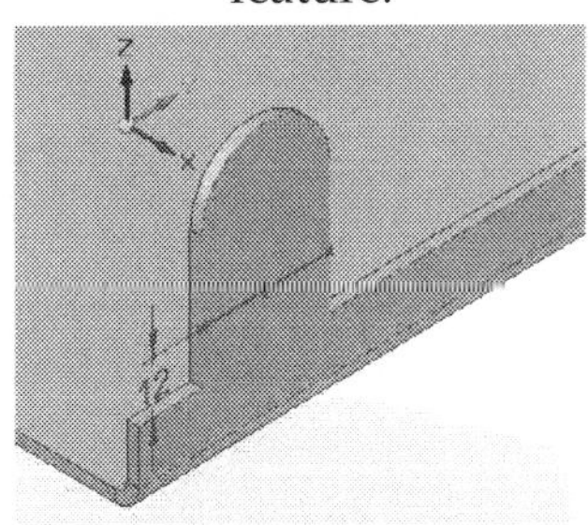

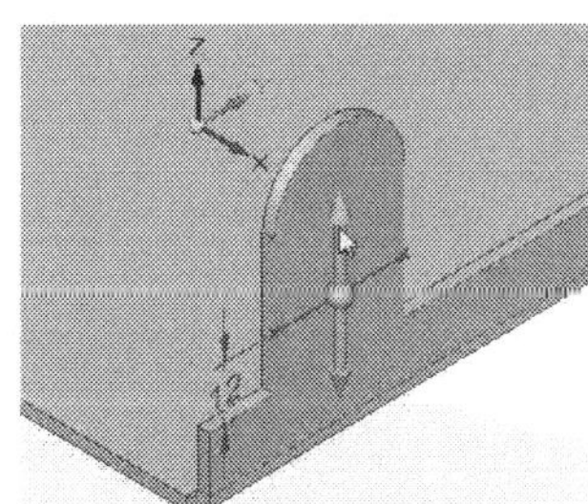

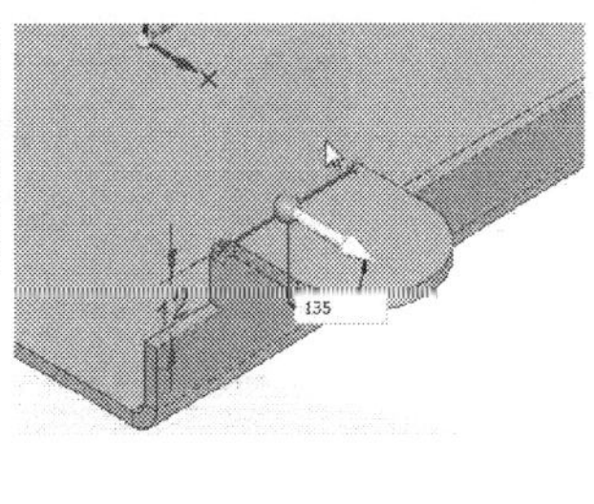

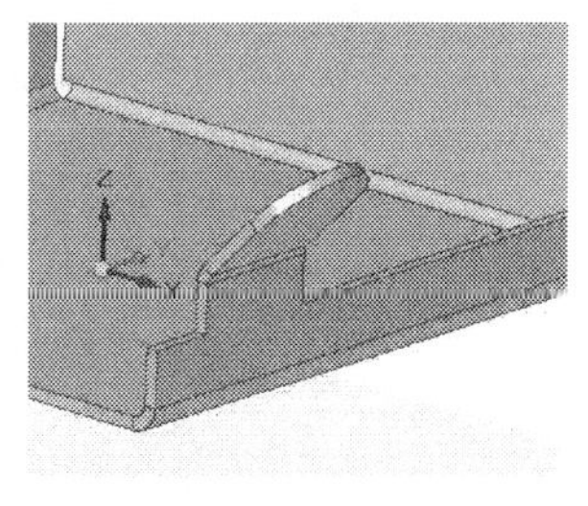

18. Draw another sketch on the outer face of the contour flange.
19. Activate the **Tab** command and create a tab feature using the sketch.

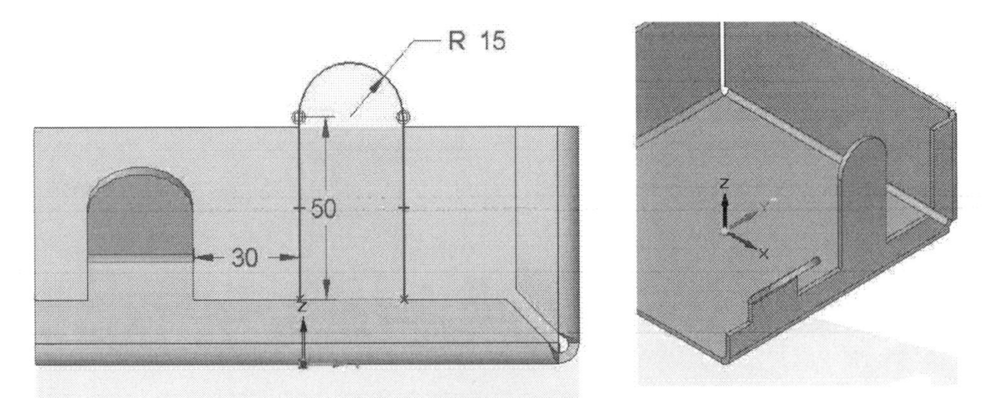

20. Draw a horizontal line on the outer face of the tab feature. Activate the **Bend** command (click **Home > Sheet Metal > Bend** on the ribbon) and click on the sketched line.

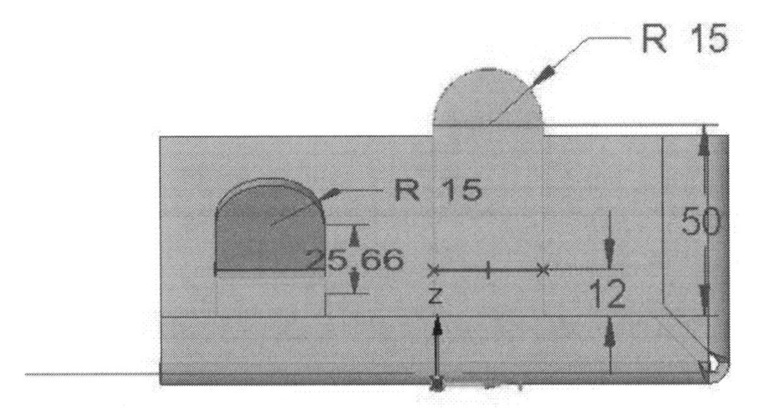

21. Click on the arrow pointing upwards. Type-in **135** in the angle box and press Enter to bend the tab feature.

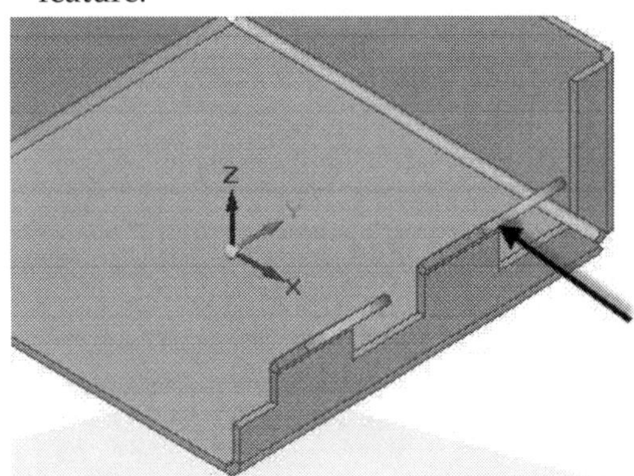

22. Create the sketch on the vertical face of the bend feature (use the **Symmetric Offset** command), as shown in figure.

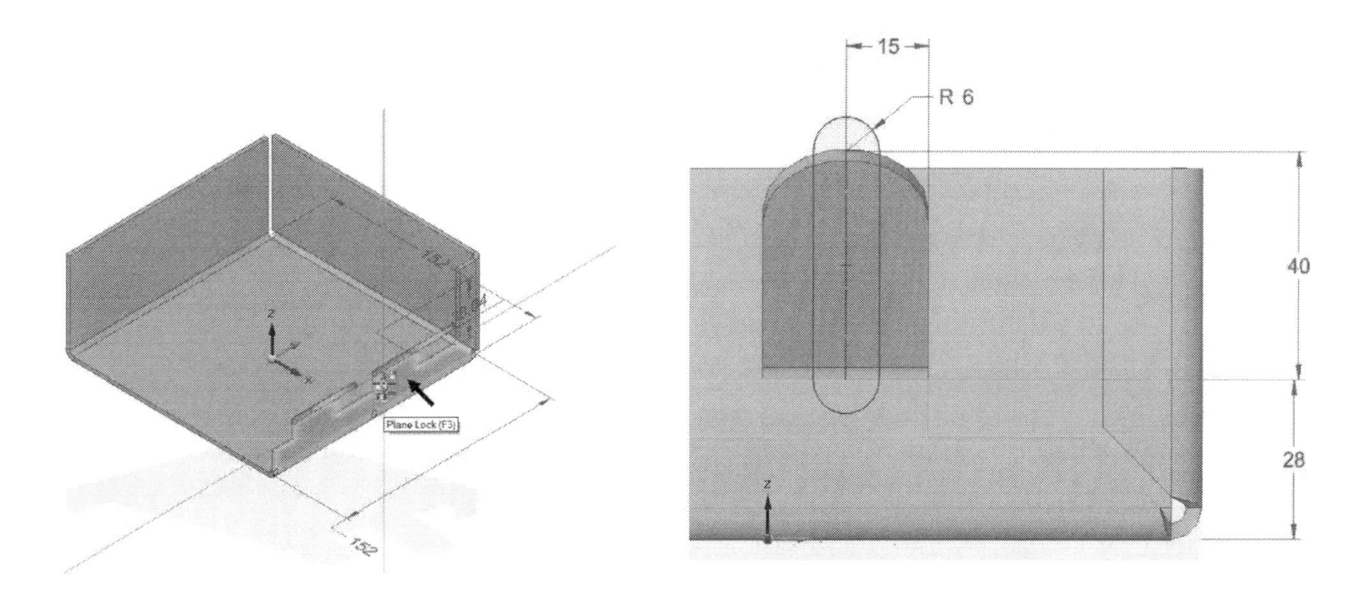

23. On ribbon, click **Home > Sheet Metal > Hole > Cut** .
24. Press Ctrl key and click inside the regions enclosed by the sketch. Click the right mouse button to accept the selection.
25. On the command bar, click the **Wrapped Cut** icon.

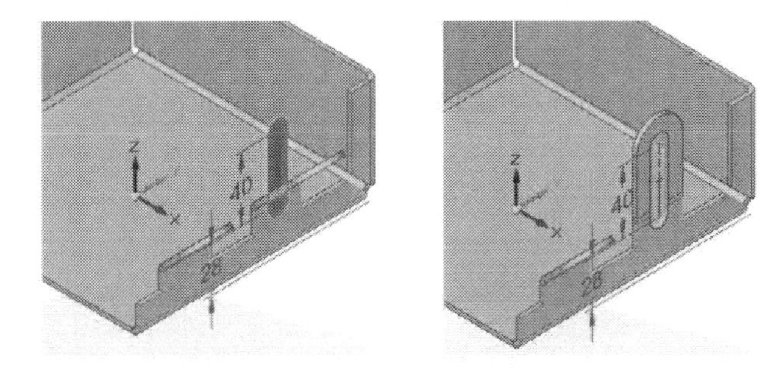

26. Click the right mouse button to complete the cut feature.

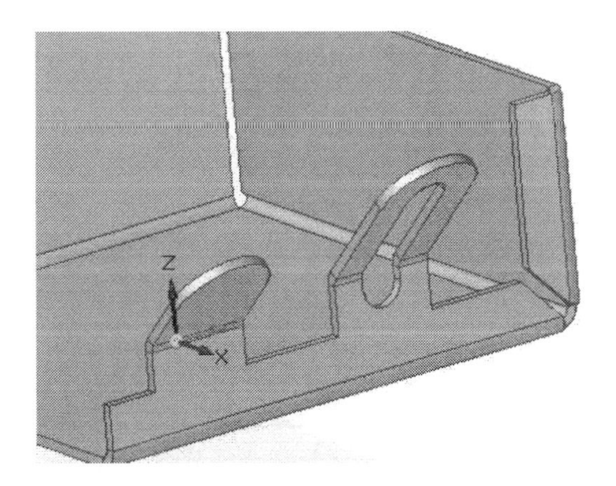

27. Activate the **Close 2-Bend Corner** command (click **Home > Sheet Metal > Close 2 Bend Corner** on the ribbon) and click on the bends of the flange features.

28. On the command bar, set the **Corner Treatment** to **Circular Cutout**. Click the **Closed Corner** icon on the command bar. Set the **Diameter** value to 8 mm. Click the right mouse button to close the bends.

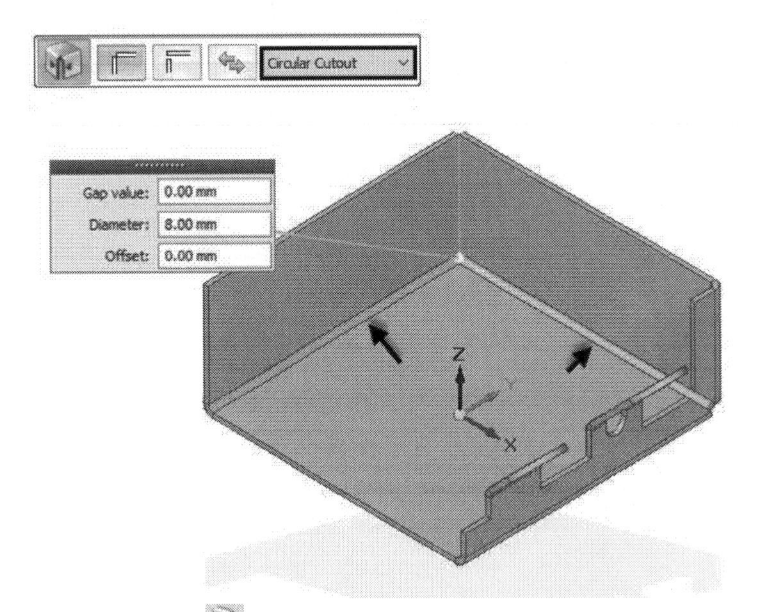

29. Activate the **Hem** command (click **Home > Sheet Metal > Contour Flange > Hem** on the ribbon).

30. On the command bar, click the **Hem Options** icon to open the **Hem Options** dialog. On this dialog, set the **Hem type** to **Closed Loop**. Set the **Bend radius1** to 2 and **Flange length1** to 8. Click **OK** to close the dialog. On the command bar, select **Material Setback > Material Outside** .

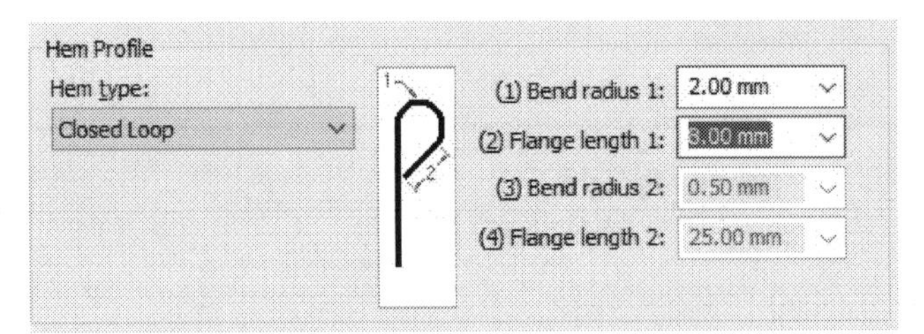

31. Click on the outer edges of the flange features. Click the right mouse button to create the hem features.

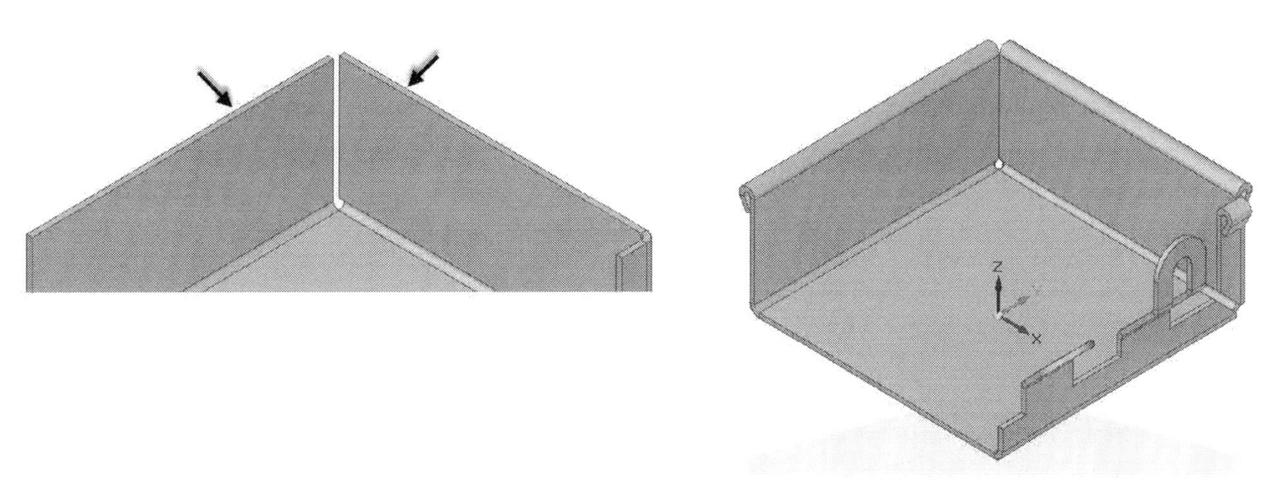

32. Rotate and orient the model, as shown below.

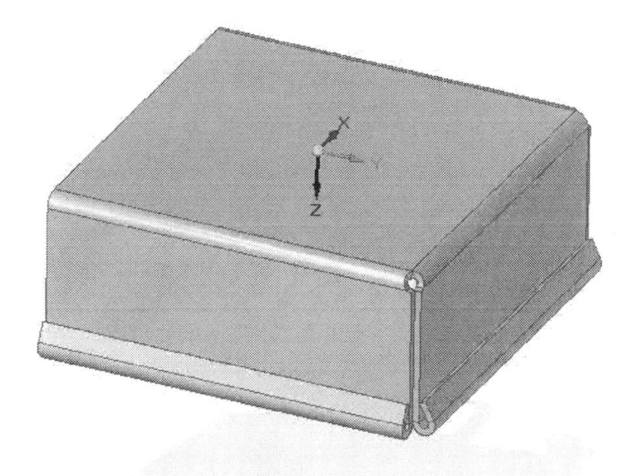

33. Activate the **Louver** command (click **Home > Sheet Metal > Dimple > Louver** on the ribbon) place the mouse pointer on the top face. Use the **N** key to change the orientation of the louver, as shown.

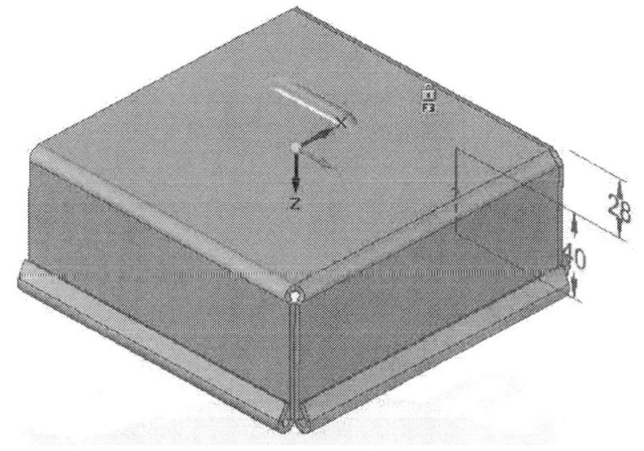

34. Press F3 on your keyboard to lock the plane.

35. On the command bar, click the **Louver Options** icon to open the **Louver Options** dialog. On this dialog, set the **Length**, **Depth** and **Height** values to 50, 10, and 5, respectively. Check the **Formed-end louver** option. Next, check the **Include rounding** option and set the Die radius to 1. Click **OK** to close the dialog.

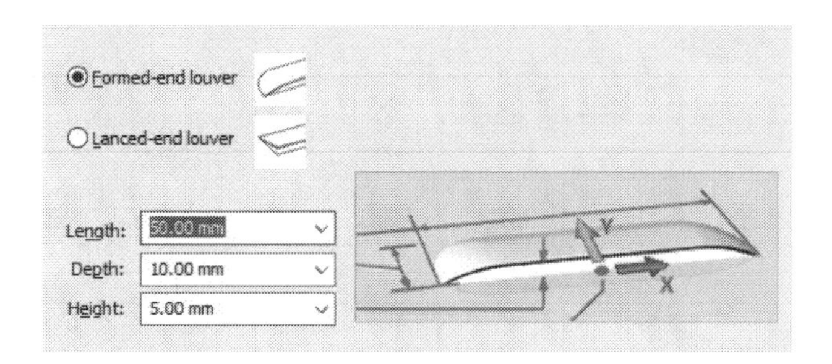

36. Place the mouse pointer on the left edge and press E twice. The location dimensions appear.
37. Type-in 76 and 120 in the dimension boxes. Press Enter to create the louver feature.

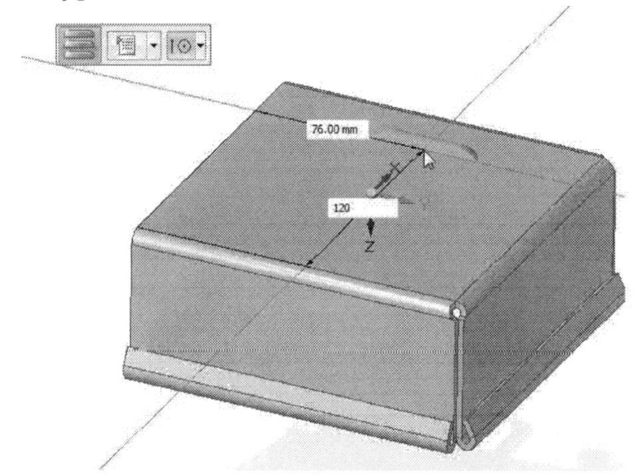

38. Select the louver feature and create a rectangular pattern.

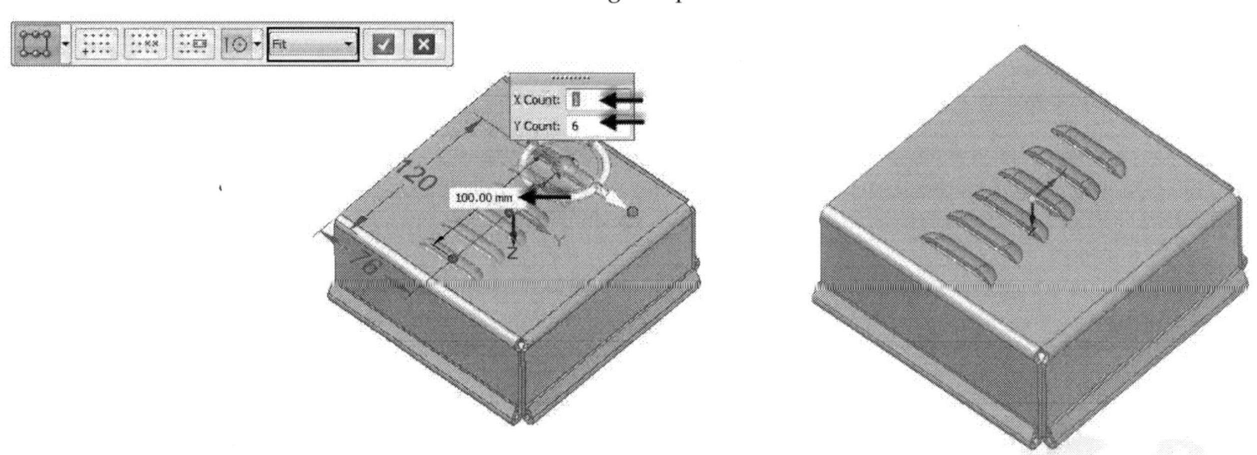

39. Change the view orientation of the sheet metal to Isometric.
40. On the ribbon, click **Tools > Model > Flatten** on the ribbon. The **Flat Pattern** command is activated.
41. Click on the top face of the tab feature.
42. Click on the front edge of the tab feature to define the x-axis of the flat pattern. The flat pattern is created.

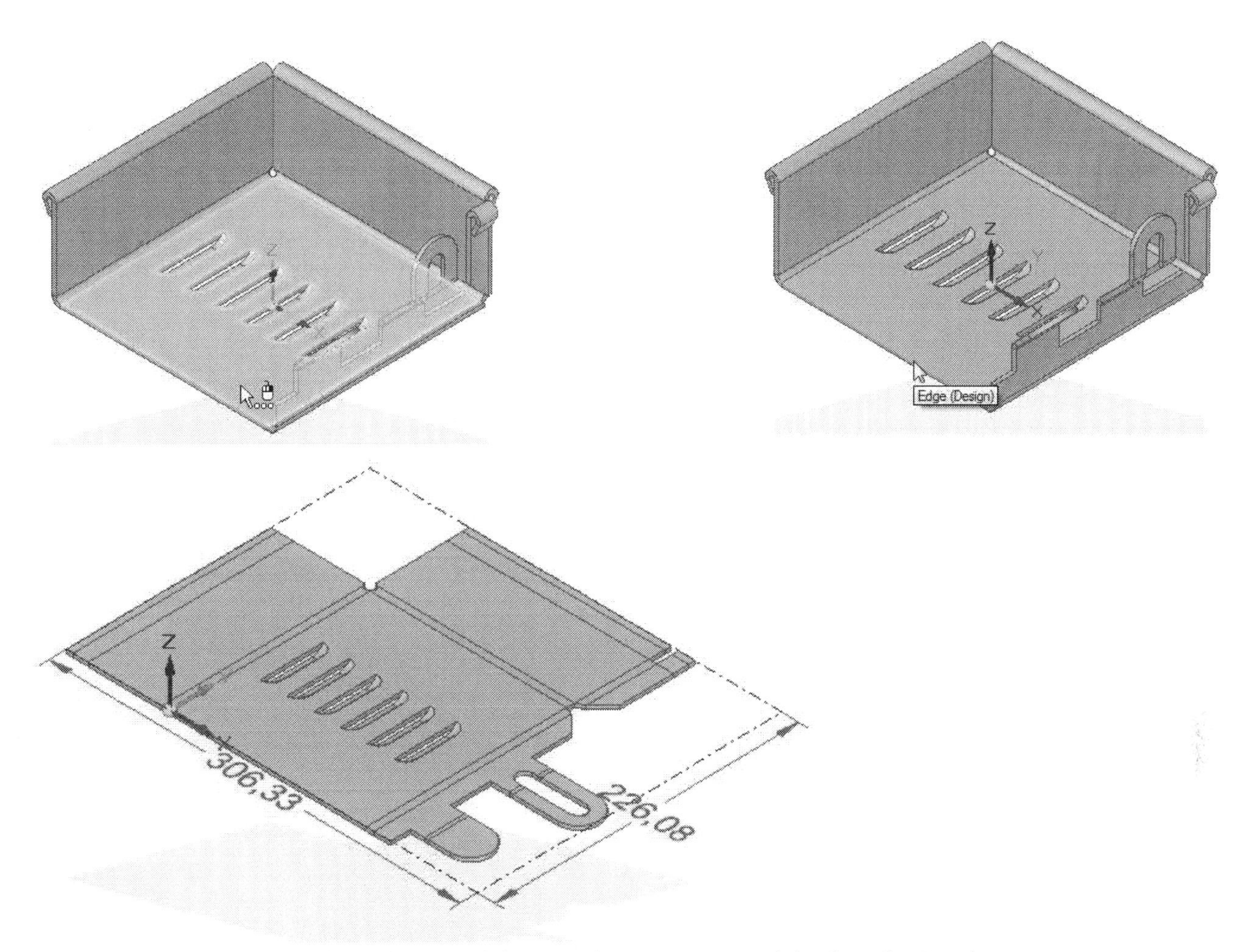

43. On the ribbon, click **Tools > Model > Synchronous** to switch back to the Synchronous environment.
44. Save and close the sheet metal part.

Questions

1. How do you insert a flat pattern into a drawing?
2. Describe parameters that can be specified on the **Material Table** dialog.
3. Define the term 'Neutral Factor'.
4. List any two parameters settings of a gage table that can be overridden when creating a feature.
5. What is the use of the **Cut** command?
6. Which command is used to apply rounds and chamfers to the corners of a sheet metal part?
7. List the types of hems that can be created in Solid Edge.
8. What is the use of the **Close 2-Corner** command?
9. What are the corner treatment options when closing a corner?
10. What is the difference between a dimple and drawn cutout?

Exercises

Exercise 1

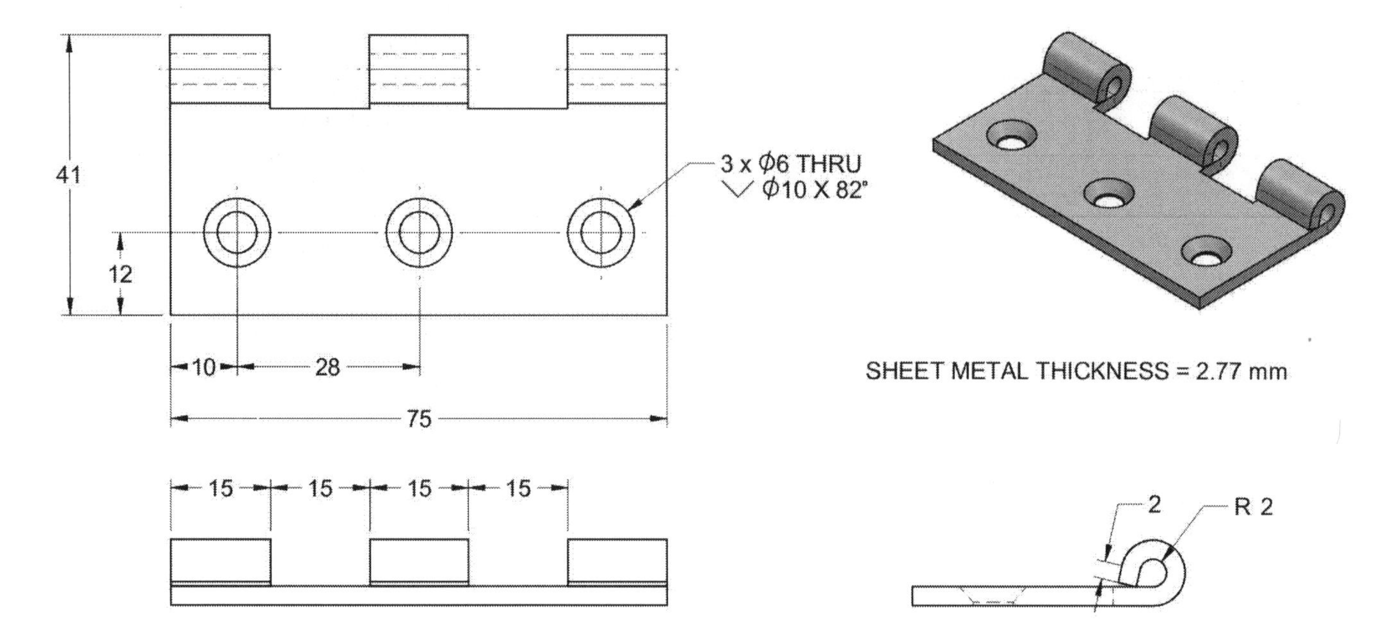

Exercise 2

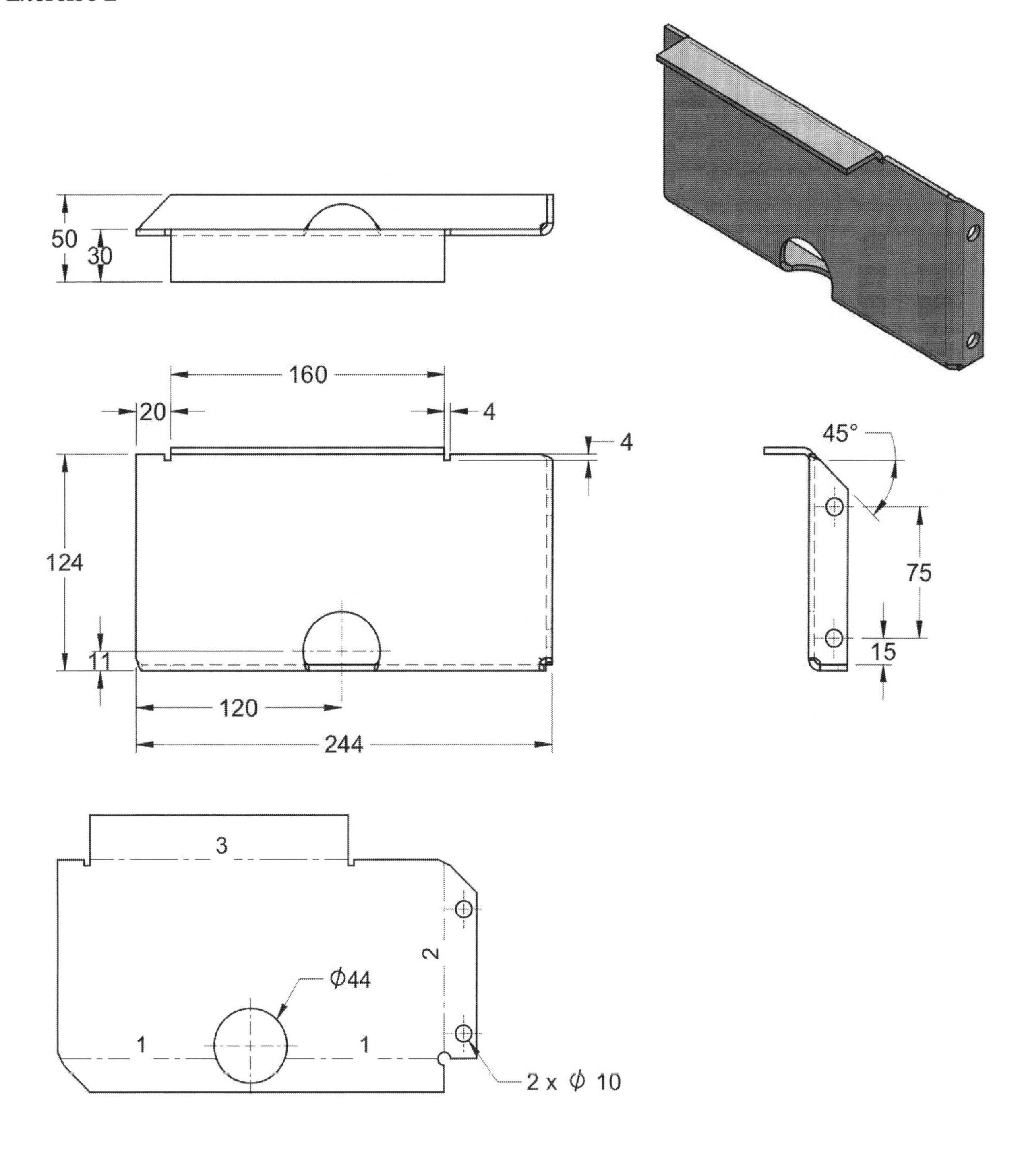

Sequence	Feature	Radius	Angle	Direction	Included Angle
1	Bend 1	3.58 mm	90.00 deg	Down	90.00 deg
2	Bend 2	3.58 mm	90.00 deg	Down	90.00 deg
3	Bend 3	3.58 mm	90.00 deg	Up	90.00 deg

Chapter 13: Surface Design

The topics covered in this chapter are:

- *Basic surfaces*
- *Curves*
- *Swept command*
- *BlueSurf*
- *Ruled*
- *Bounded*
- *Offset*
- *Copy*
- *Redefine*
- *Intersect*
- *Extend*
- *Replace Face*
- *Trim*
- *Extend*
- *Split*
- *Stitched*

Solid Edge Surfacing commands can be used to create complex geometries that are very difficult to create using standard extruded bosses, revolve bosses, and so on. They can also be used to edit and fix the broken imported parts. In this chapter, you learn the basics of surfacing commands that are mostly used. The surfacing commands are available in the **Surfacing** tab.

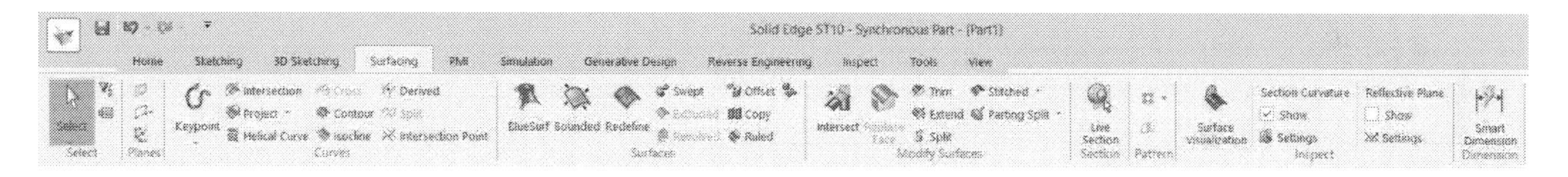

Solid Edge offers a rich set of surface design commands. A surface is an infinitely thin piece of geometry. For example, consider a cube shown in figure. It has six faces. Each of these face is a surface, an infinitely thin piece of geometry that acts as a boundary in 3D space. Surfaces can be simple or complex shapes.

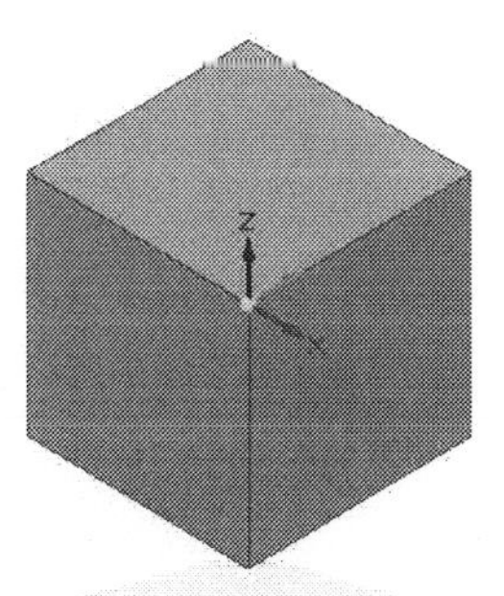

In solid modeling, when you have created solid features such as an Extruded feature or a Revolved feature, Solid

Edge creates a set of features (surfaces) that enclose a volume. The airtight enclosure is considered as a solid body. The advantage of using the surfacing commands is that you can design a model with more flexibility. You can create surfaces in Synchronous and Ordered environments. However, the Ordered environment offers history-based modeling which makes it easy to edit surfaces. This chapter explains the surface modeling commands by using them in the Ordered environment.

Extruded Surface

To create an extruded surface, first create an open or closed sketch and activate the **Extruded** command (on the ribbon, click **Surfacing > Surfaces > Extruded**). Select the sketch and right-click. Next, type-in a value in the **Distance** box available on the command bar, and press Enter. On the command bar, click the **Close Ends** button to create an extruded surface with closed ends. You can also use the **Treatment Step** button to apply draft or crown to the extruded surface. Click **Finish** to create the extruded surface.

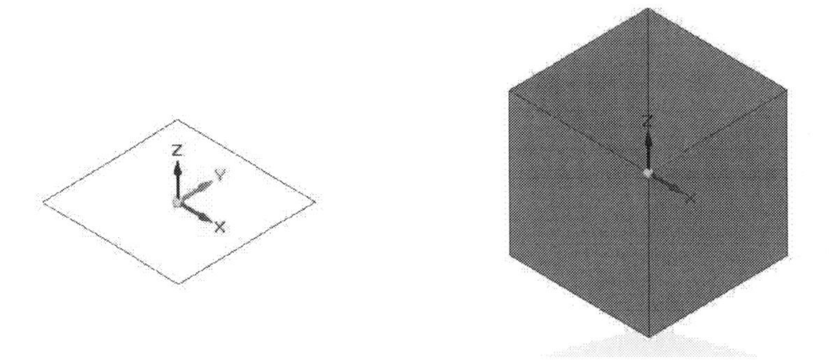

Revolved Surface

To create a revolved surface, first create an open or closed profile and the axis of revolution. Activate the **Revolved** command (on the ribbon, click **Surfacing > Surfaces > Revolved**). Select the sketch and right-click. Select the axis and type-in the angle of revolution in the **Angle** box or click the **Revolve 360** button. Click to define the side of the revolution, in case you have specified the angle value. Click **Finish** to complete the feature.

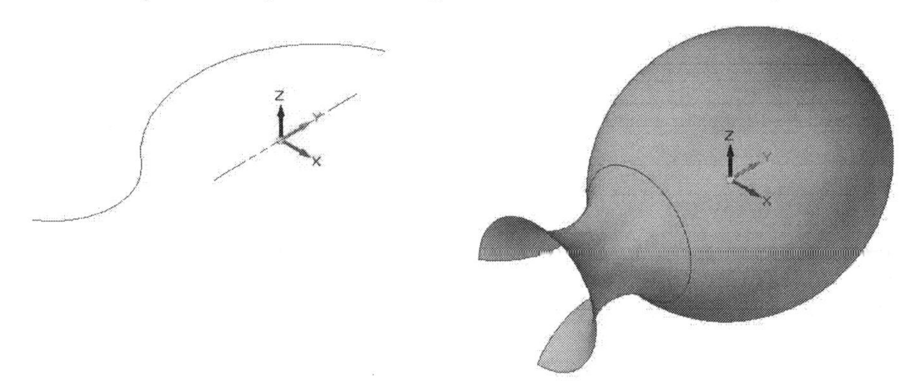

Even if you create an enclosed surface, Solid Edge will not recognize it as a solid body. You can examine this by activating the **Physical Properties** command (on the ribbon, click **Inspect > Physical Properties > Physical Properties**). On the **Physical Properties** dialog, click the **Change** button under the **Density** section. The **Material Table** dialog appears. On this dialog, select a material from the **Material** tree and click **Apply to Model**. Next, click **Update** on the **Physical Properties** dialog. You will notice that all the physical properties are displayed as zero.

This means that there exists no solid body. You will learn to convert a surface body into a solid
.later in this chapter.

Keypoint Curve

The **Keypoint Curve** command creates curves through selected keypoints. You can click in the graphics window to specify points or select existing points. Activate this command (on ribbon, click **Surfacing > Curves > Keypoint Curve**) and select keypoints from the graphics window. Click **Accept** and **Finish** on the command bar.

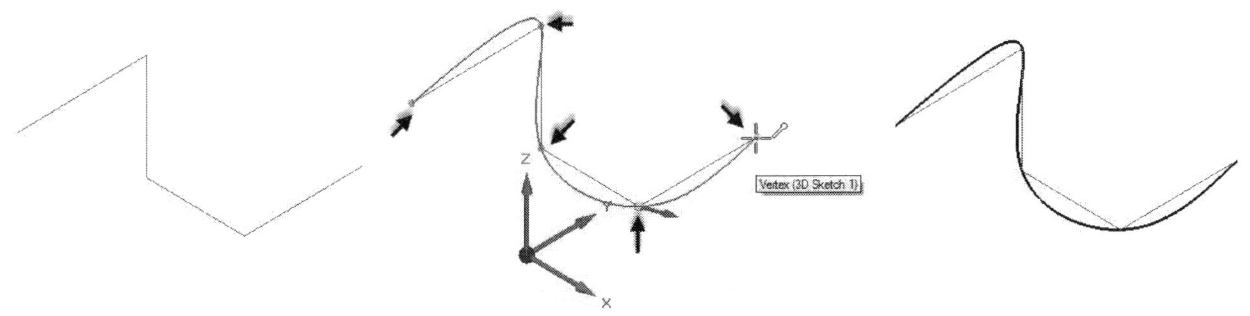

Curve by Table

The **Curve by Table** command creates curves by using the X, Y, Z points. These points can be defined using a spreadsheet. Create a spreadsheet by entering values in the A, B, and C columns and save it. The values in the A, B, C columns of the spreadsheet represent the X, Y, Z values. Activate the **Curve by Table** command (on ribbon, click **Surfacing > Curves > Keypoint** drop-down **> Curve by Table**). On the **Insert Object** dialog, select the **Create from file** option and click the **Browse** button. Go to the location of the spreadsheet and double-click on it. Click **OK** to close the dialog. You will notice that a curve appears.

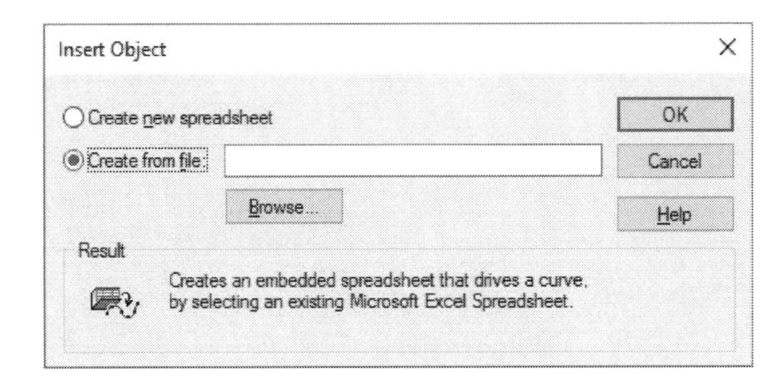

On the command bar, click the **Parameters Step** button. On the **Curve by Table Parameters** dialog, set the **Curve fit** type. You can select **Linear segments**, **Smoothing off**, and **Smoothing on**. The **Linear segments** option creates a linear curve. The **Smoothing off** option creates a curve directly passing through the points. The **Smoothing on** option creates a curve whose path is controlled by the **Tolerance** value.

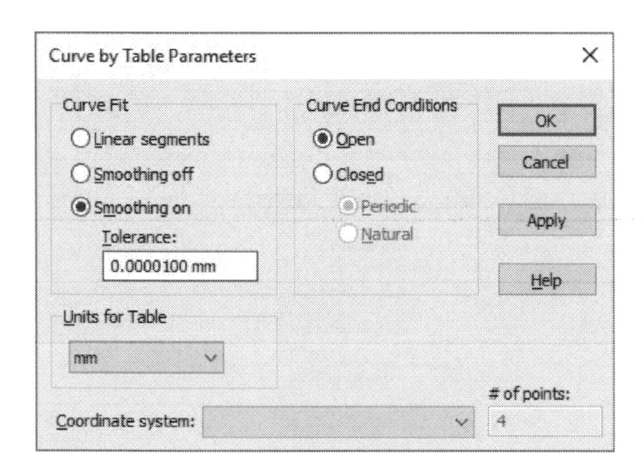

Next, set the **Curve End Conditions** options. You can define open or close curves. Specify the **Units for Table** and **Coordinate system**, and then click **OK**. Click **Finish** to complete the curve.

Helical Curve

This command allows you to create a helical curve.

Constant Pitch Helical Curve

A constant pitch helical curve has an equal distance between the turns. Activate this command (on ribbon, click **Surfacing > Curves > Helical Curve**); the **Axis Step** is activated and you need to define the axis of the helical curve. Select a keypoint, arbitrary point in the graphics window, line, circle, cylindrical face, cone face, or type-in values in the X, Y, and Z boxes on the command bar.

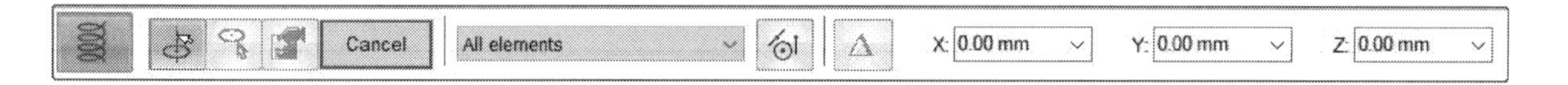

If you select a keypoint or an arbitrary point in the graphics window, then move the pointer in the graphics window. Click to specify the second point of the axis. After specifying the axis, the **Start Point Step** is activated and you need to specify the start point of the helical curve. Move the pointer in the direction perpendicular to the axis and click to specify the start point of the helical curve; the **Helical Curve Parameters** dialog pops up on the screen. On the dialog, select **Type > Constant Pitch**. Next, specify the helix creation method by selecting an option from the **Method** drop-down. The options in this drop-down are **Length and Turns**, **Length and Pitch**, and **Pitch and Turns**.

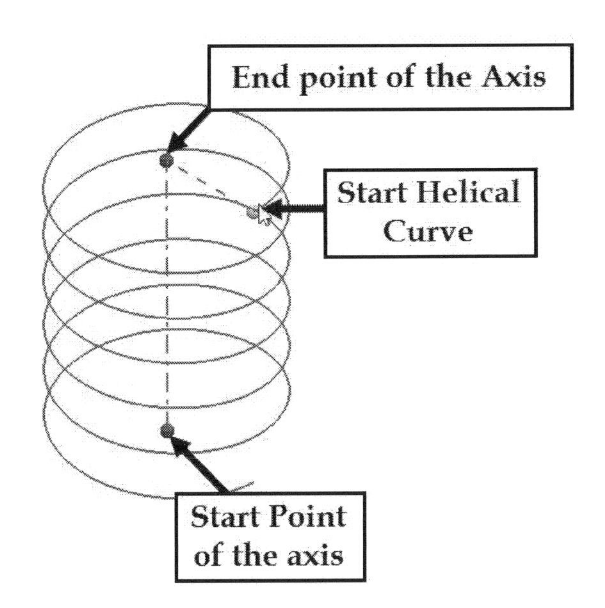

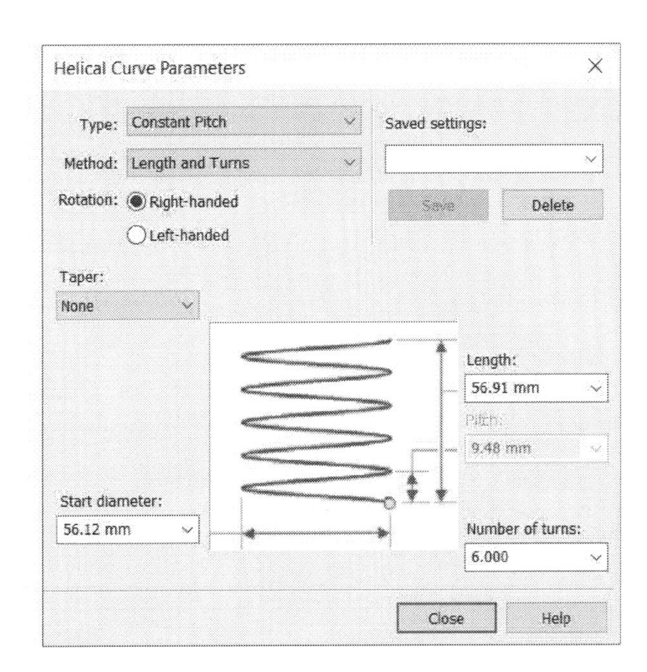

Length and Turns: In this method, you need to specify the total length of the helix and its number of turns in the **Length** and **Number of turns** boxes, respectively.

Length and Pitch: In this method, you need to specify the total length of the helix and the distance between the turns.

Pitch and Turns: In this method, you need to specify the distance between the turns and number of turns.

Next, specify the rotation direction of the helix. You can select the **Right-handed** or **Left-handed** direction of the rotation. Next, type-in a value in the **Start diameter** box. Use the arrow handle that appears on the helical curve to change its side. You can also apply taper to helix using the **Taper** drop-down. Next, click **Close** on the **Helical Curve Parameters** dialog. Next click **Finish** on the command bar to create the helical curve.

Variable Pitch Helical Curve

Activate the **Helical Curve** command (on ribbon, click **Surfacing > Curves > Helical Curve**); the **Axis Step** is activated and you need to define the axis of the helical curve. Select a line from the graphics window to define the axis. Move the pointer and click to define the start point of the helical curve. On the **Helical Curve Parameters** dialog, select **Type > Variable Pitch**. Next, select an option from the **Method** drop-down. For example, select the **Pitch and Turns** option. Next, specify the **End pitch**, **Pitch**, and **Number of turns**; the pitch value increases or decreases from the start point up to the end point of the helical curve. For example, if Pitch = 10, End pitch = 3, and Number of turns = 15, the pitch value varies from 10 to 3 (refer the **Helix** section of *Chapter 6: Sweep Features* for the rate of change in pitch value). Next, specify the **Start diameter** and the **Rotation** direction.

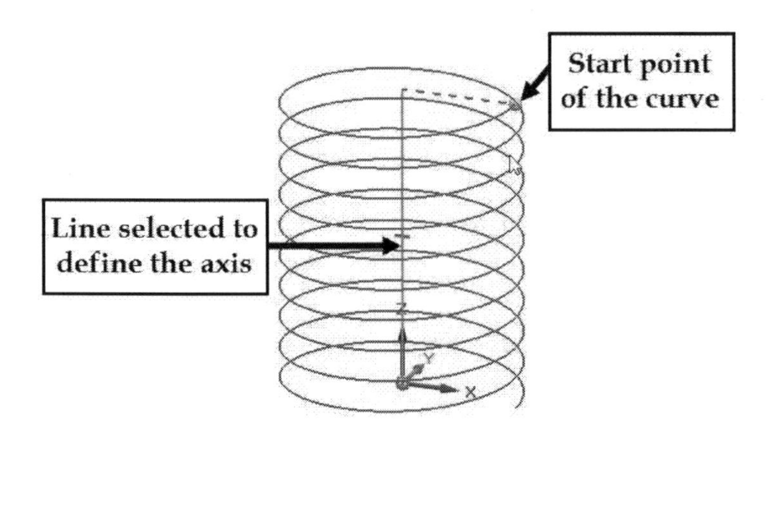

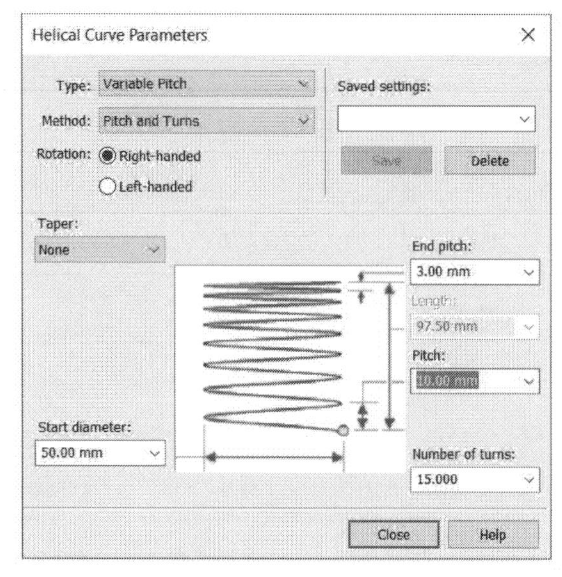

Click **Close** on the dialog, and then click **Finish** on the command bar.

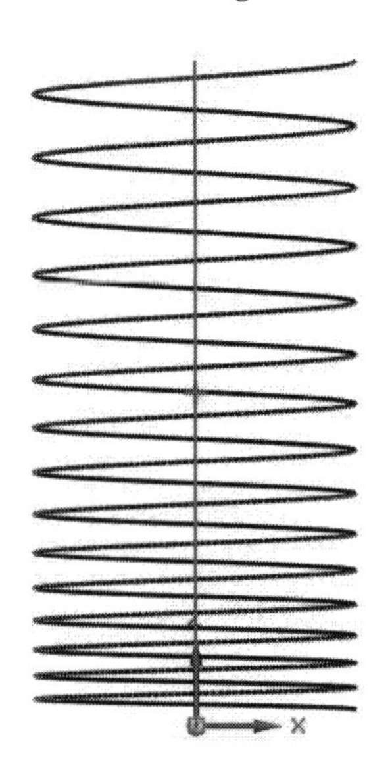

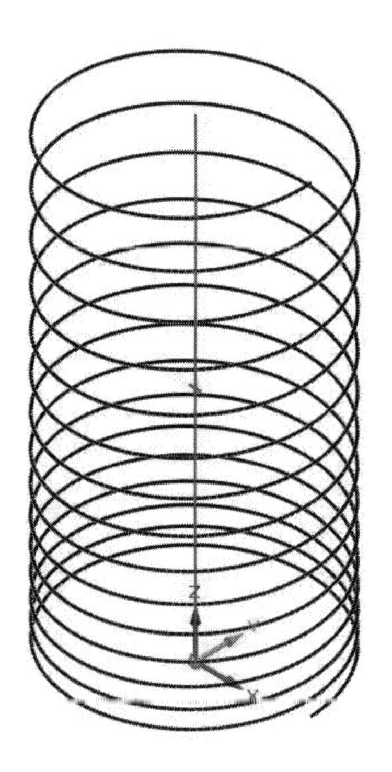

Compound Helical Curve

Activate the **Helical Curve** command (on ribbon, click **Surfacing > Curves > Helical Curve**); the **Axis Step** is activated and you need to define the axis of the helical curve. Select a circle from the graphics window; the center point of the circle acts as the axis point and a quadrant point on the circle is used to define the start point of the helical curve. On the **Helical Curve Parameters** dialog, select **Type > Compound**; a table appears at the bottom of the dialog. The parameters in the table are annotated in the figure displayed on the dialog. The table has four parameters and three point sets. You need to specify the parameters for the three point sets. However, you can add more point sets using the **Insert** button located at the bottom. Next, select an option from the **Method** drop-down. For example, select **Pitch and Turns** option and notice that **Length** fields of all the point sets are greyed out. Also, the **Turns** and **Diameter** fields of the first point are greyed out. You need to specify the **Pitch** values of the point sets. Also, specify the **Turns** and **Diameter** values for all the point sets except for the first point set. The following figure shows the values of the parameters in the table and output achieved.

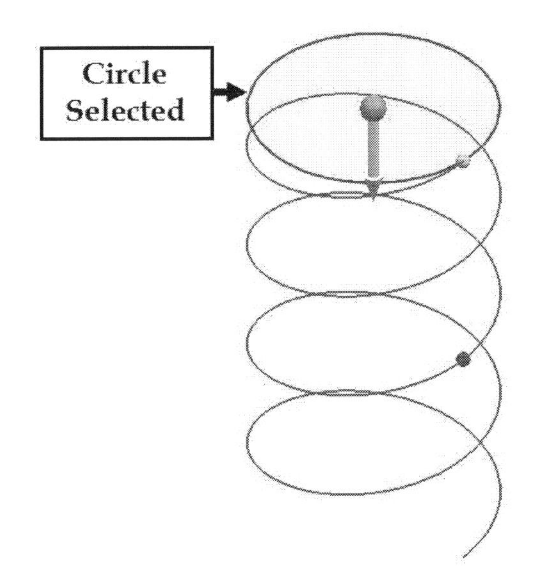

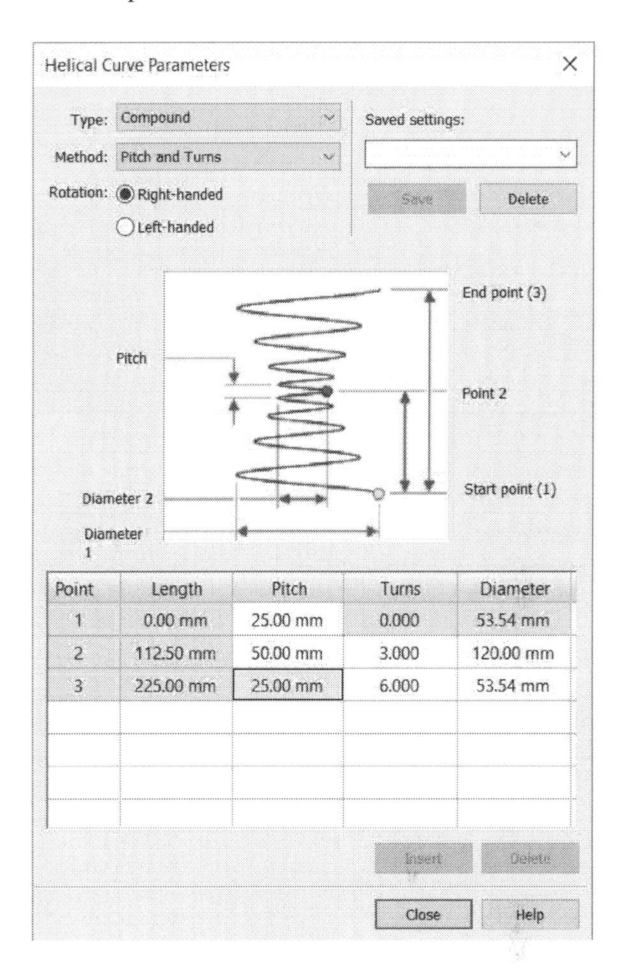

Point	Length	Pitch	Turns	Diameter
1	0.00 mm	25.00 mm	0.000	53.54 mm
2	112.50 mm	50.00 mm	3.000	120.00 mm
3	225.00 mm	25.00 mm	6.000	53.54 mm

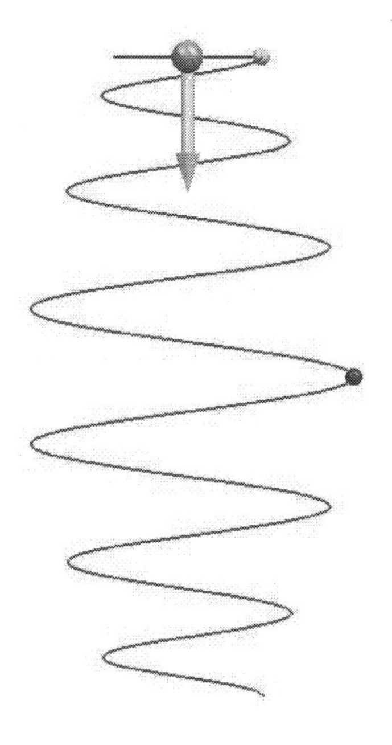

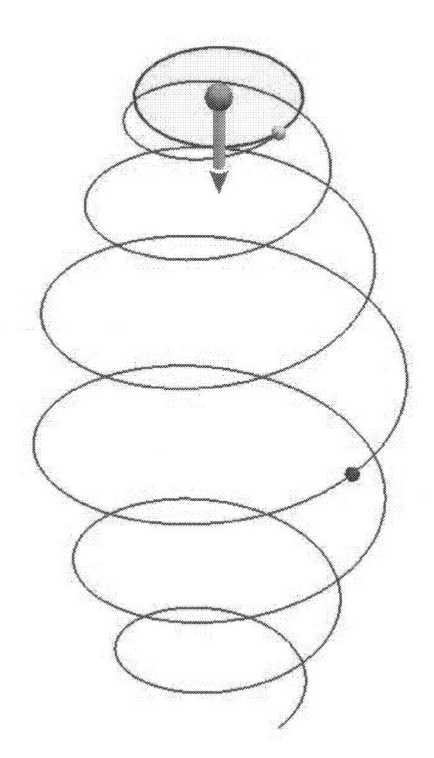

Click **Close** on the dialog, and then **Finish** on the command bar.

Spiral Curve

Activate the **Helical Curve** command (on ribbon, click **Surfacing > Curves > Helical Curve**); the **Axis Step** is activated and you need to define the axis of the spiral curve. Select a circle from the graphics window; the center point of the circle acts as the axis point and a quadrant point on the circle is used to define the start point of the spiral curve. On the **Helical Curve Parameters** dialog, select **Type > Spiral**. Next, specify the spiral curve creation method by selecting an option from the **Method** drop-down. The options in this drop-down are **End diameter and Turns, End diameter and Radial pitch**, and **Radial pitch and Turns**.

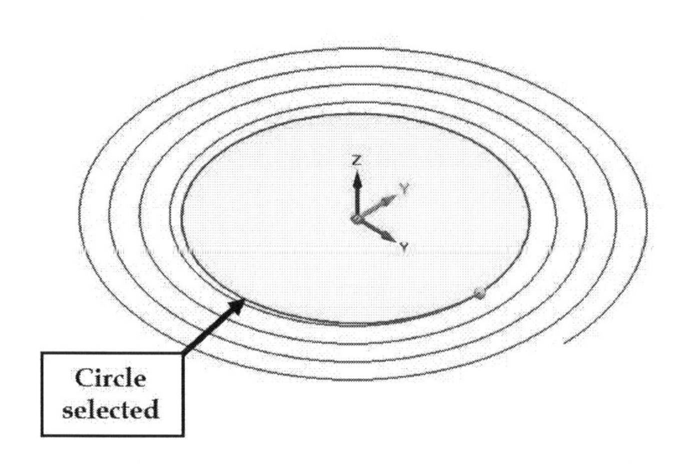

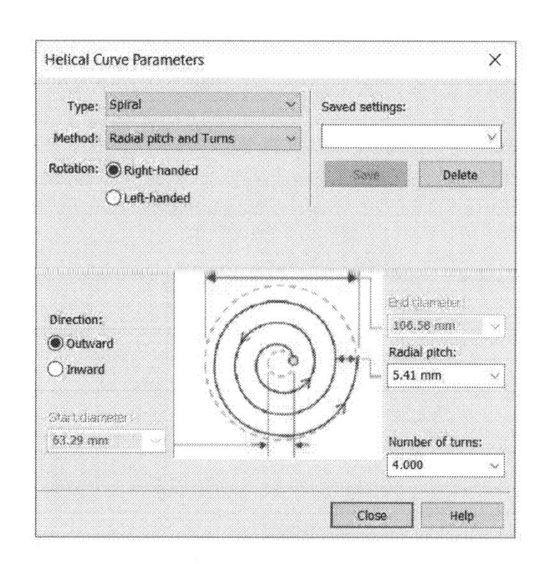

End diameter and Turns: In this method, you need to specify the outer or inner diameter of the spiral curve for the Outward or Inward spiral curve, respectively. Next, you need to specify the **Number of turns** value.

End diameter and Radial pitch: In this method, you need to specify the end diameter and the radial distance between the turns.

Radial pitch and Turns: In this method, you need to specify the radial distance between the turns and number of turns. Also, you need to specify the **Direction** of the spiral curve. It can be **Outward** or **Inward**.

Next, select the **Rotation** direction and specify the parameters (**End diameter**, **Radial pitch**, and **Number of turns**) depending on the option selected from the **Method** drop-down. Click **Close** on the dialog, and then click **Finish**.

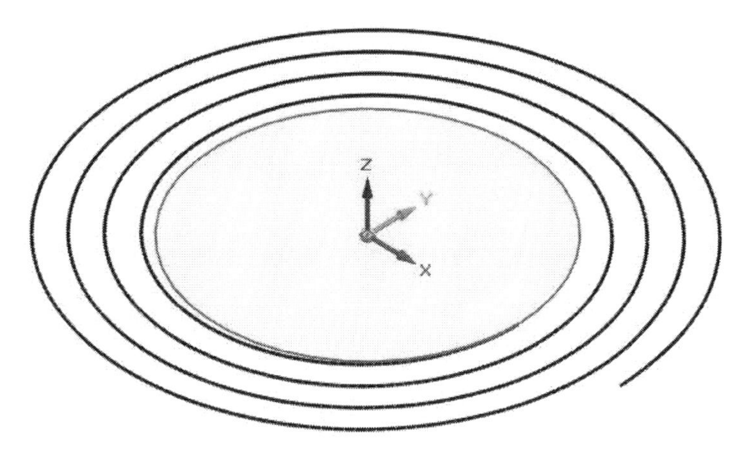

Intersection

The **Intersection** command creates a curve at the intersection of the surface and plane, or two surfaces, or solid and surface, or solid and plane. Activate the **Intersection** command (on the ribbon, click **Surfacing > Curves > Intersection**) and select the two intersecting surfaces. Click **Finish** to create the intersection curve.

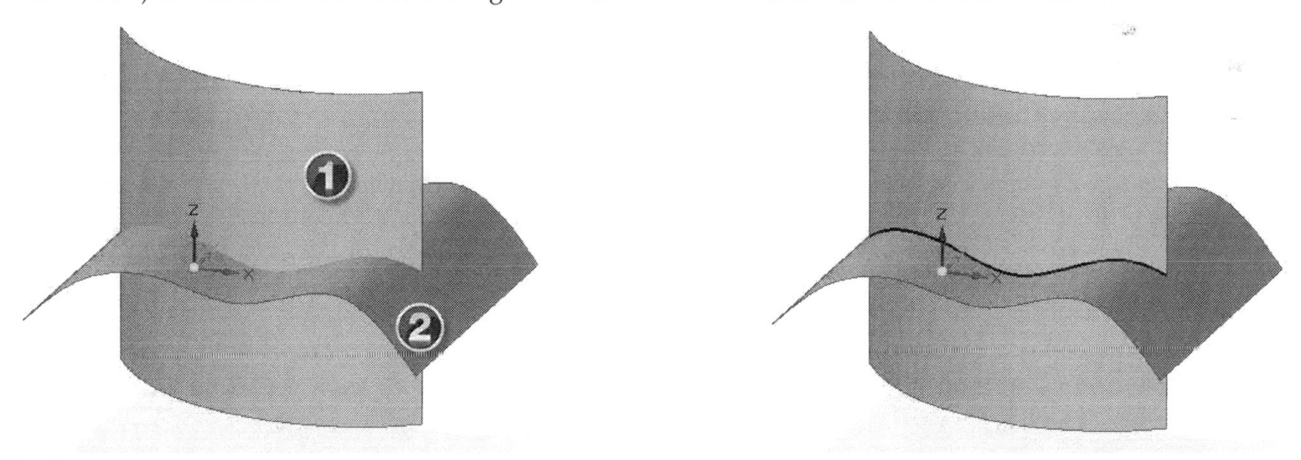

Project

The **Project** command takes a sketch or curve and maps it onto a surface. Click the **Project** button on the **Curves** panel and select the curve or sketch to project. Right-click and select the surface onto which the sketch/curve will be mapped. Right-click, and then define the side of the projection. If you have selected a curve to project, the **Projection Plane Step** button is activated and you need to select a plane to define the projection direction. The curve will be projected in the direction normal to the selected plane. Click **Finish** to complete the projection.

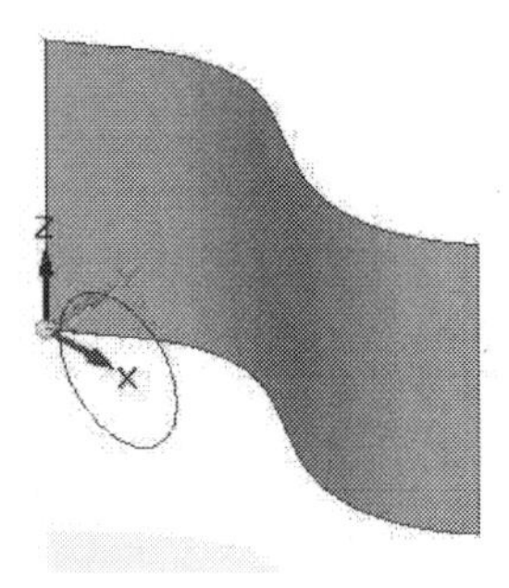

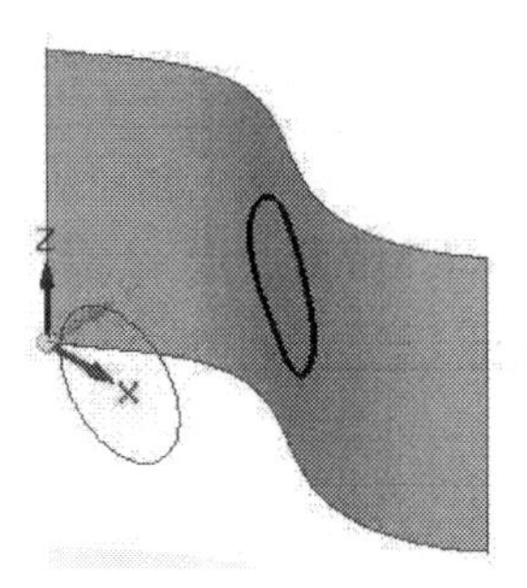

Cross

The **Cross** command is similar to the **Project** command except that it creates a curve by projecting one sketch/curve onto the another sketch/curve. Click the **Cross** button on the **Curves** panel and select the first sketch/curve. Click the **Accept** button and select the second curve/sketch. Click **Accept** and **Finish** to create the cross curve.

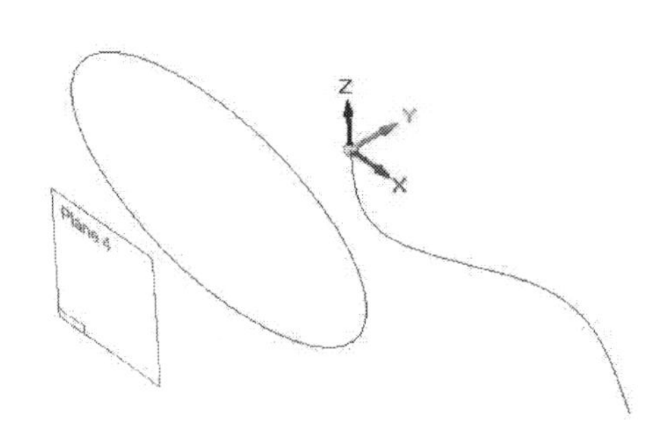

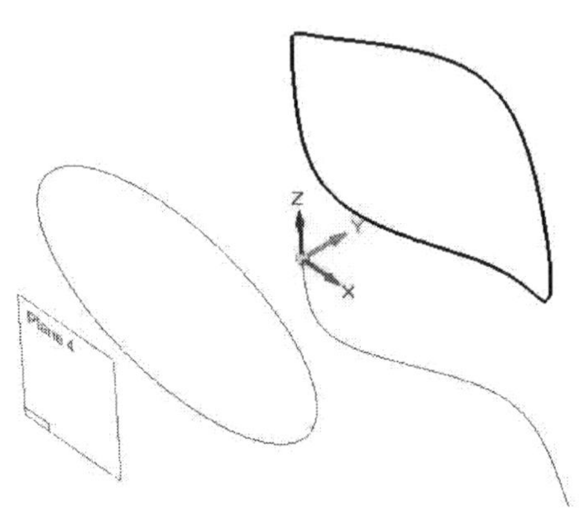

Wrap Sketch

The **Wrap Sketch** command wraps a sketch around the solid or surface body. First, create a sketch on the plane tangent to the surface onto which the sketch is to be wrapped. Next, click **Project > Wrap Sketch** on the **Curves** panel. Select the surface on to which you want to wrap the sketch, and then right-click. Select the sketch and right-click. Click **Finish** to wrap the sketch.

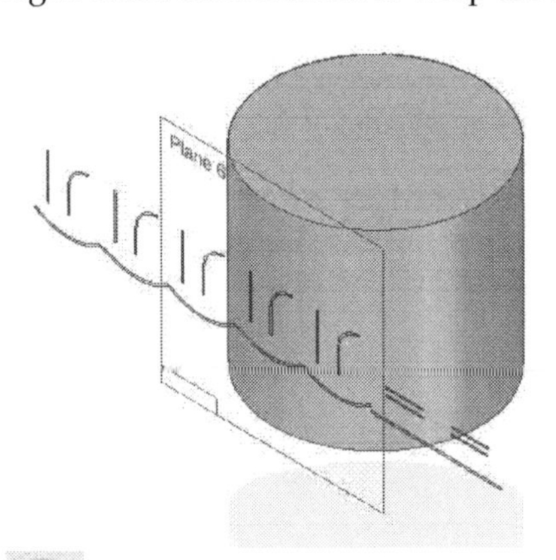

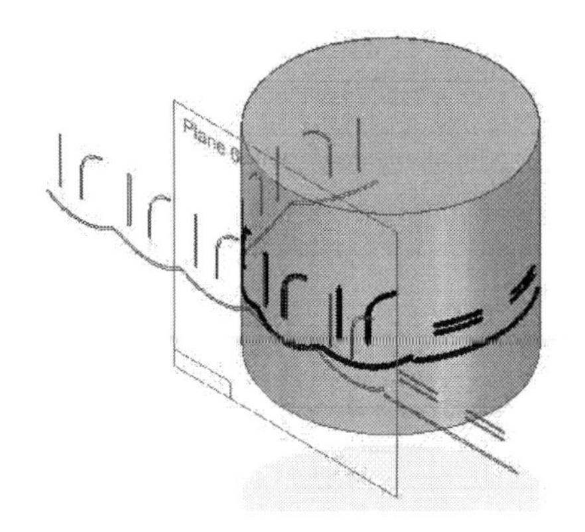

Contour

The **Contour** command creates curves on a surface. Activate this command (on the ribbon, **Surfacing > Curves > Contour**) select the surface. You can select a single or chain of surfaces. Click **Accept** after selecting the surface.

Start selecting points on the surface. On the command bar, click the **Close** button, if you want to close the curve. Click **Accept** and **Finish** to create the contour curve.

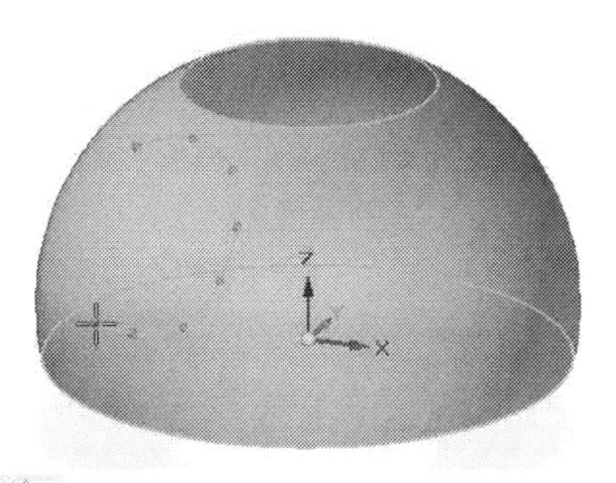

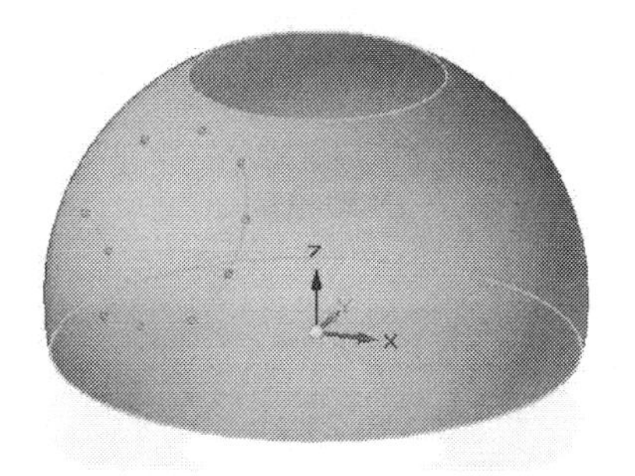

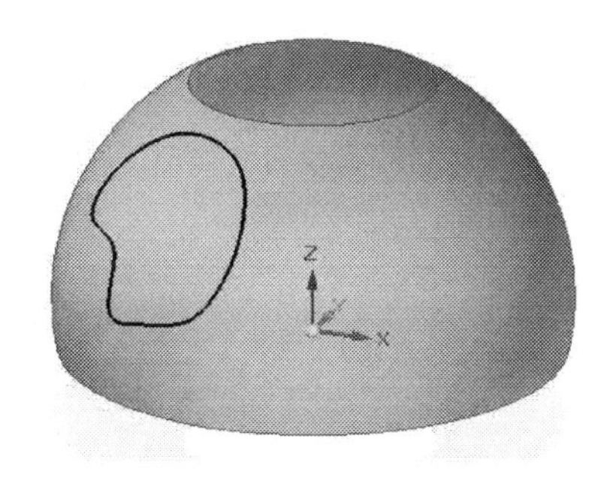

 Isocline

The **Isocline** command creates a curve on a surface by using a plane. You need to select a plane, and then specify an angle; a curve will be created at the point where the selected plane touches the surface when inclined at the specified angle. For example, click the **Isocline** button on the **Curves** panel and select the Front(XZ) plane. Select the solid or surface body. Type-in the inclination angle and click the arrow to define the side of the isoclines curve. Click **Accept** and **Finish** to create the curve.

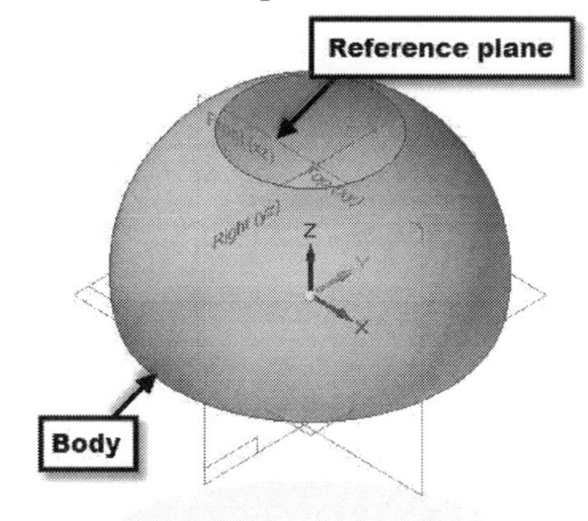

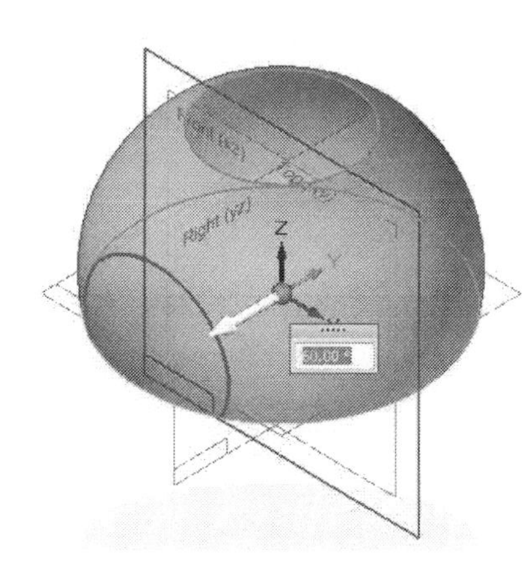

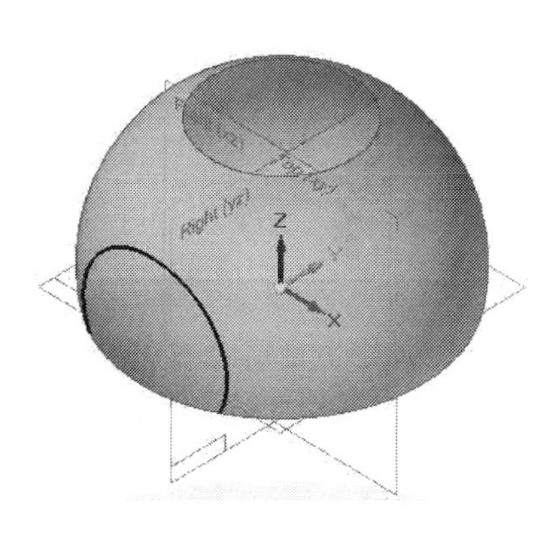

 Derived

The **Derived** command creates a curve from the selected edges of solid/surface geometry. Click the **Derived** button on the **Curves** panel and select the edges of the geometry. Click **Accept** and **Finish** to create the derived curve.

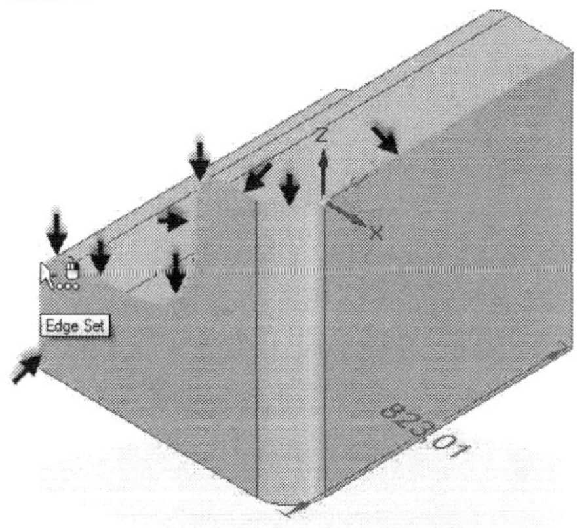

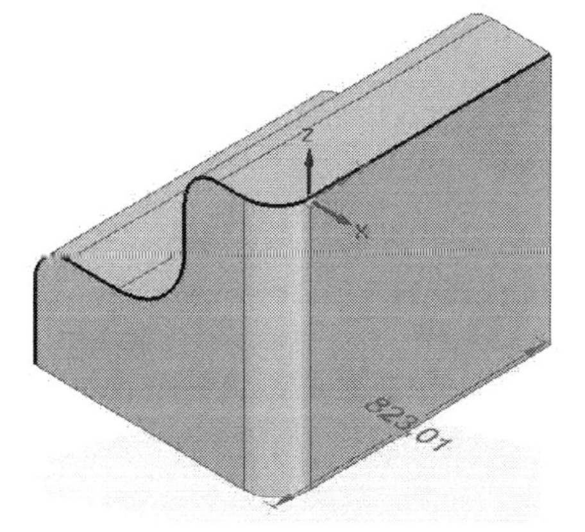

Split

The **Split** command splits a curve using an intersecting plane, curve, body, or point. Click the **Split** command on the **Curves** panel and select the curve. Right-click and select the intersecting elements. Click **Accept** and **Finish**.

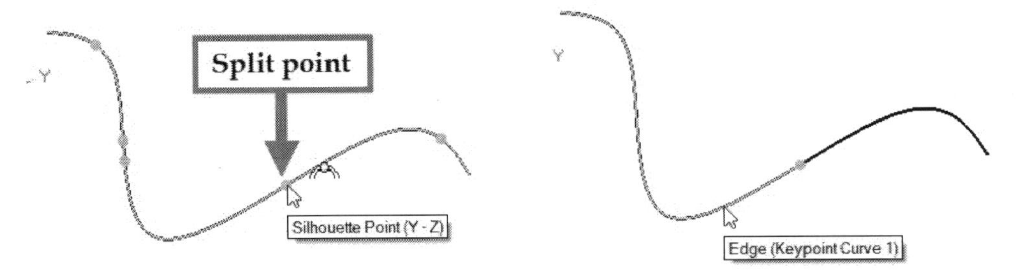

Intersection Point

The **Intersection Point** command creates points at the intersection of a curve/edge and another element. Click the **Intersection Point** button on the **Curves** panel. Select a curve/edge and right-click. Select a plane, axis or body. Click **Accept** and **Finish**.

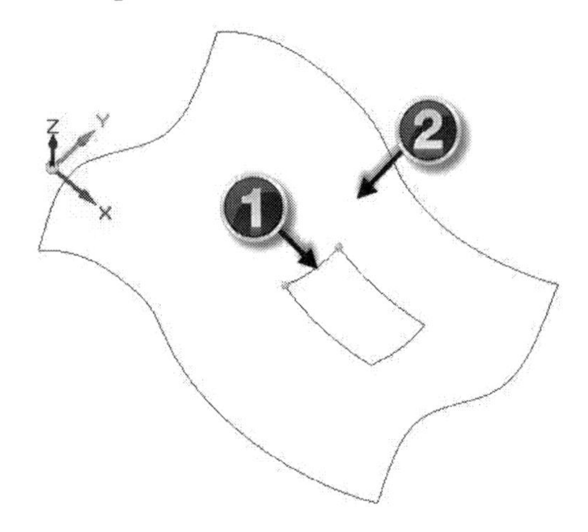

Swept Surfaces

The **Swept** command creates a surface by sweeping one or more cross-sections along guide curves. It also provides various options to control the shape along the guides. To create a swept surface, first create a sweep profile and a path. On the ribbon, click **Surfacing > Surfaces > Swept** . On the **Sweep Options** dialog, select the **Single path and cross section** option and click **OK**. Select the path and right-click. Select the cross-section and right-click. Click **Finish** on the command bar.

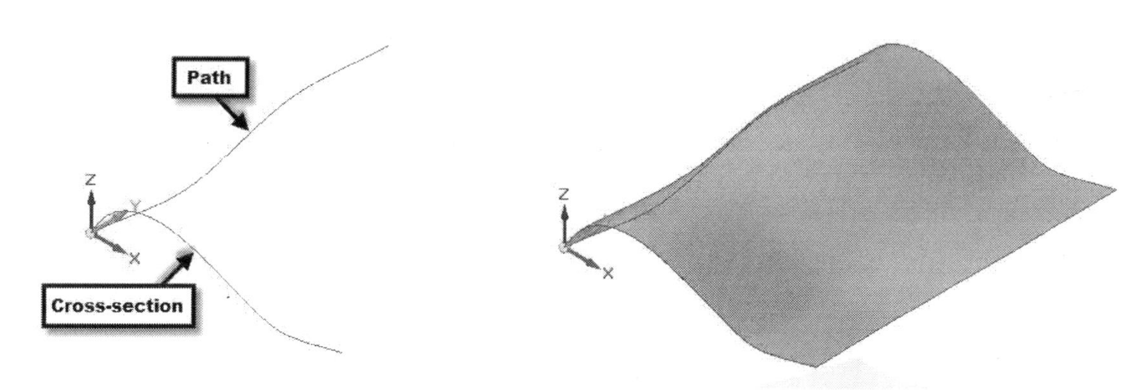

Various ways of creating swept surfaces are given next.

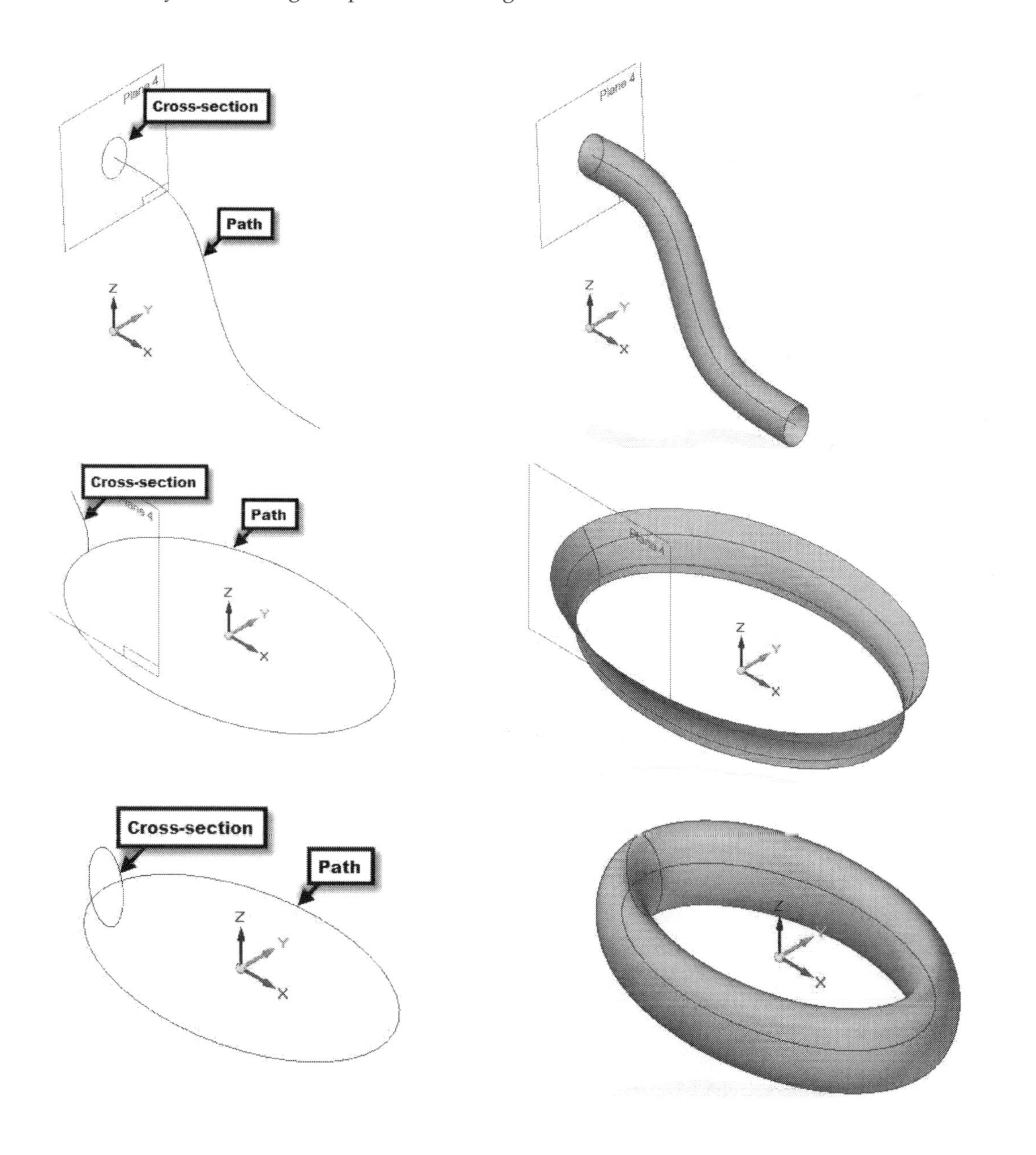

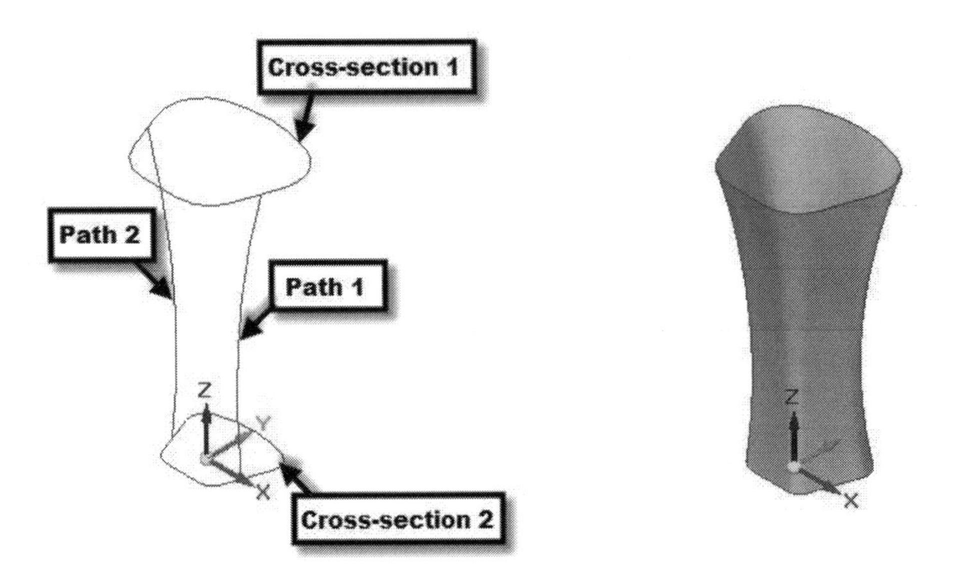

BlueSurf

This command creates a surface between two or more cross-sections. You can also add guide curves to specify the shape between two sections. Make sure that the guide curve is continuous without any sharp edges and touches the cross-sections as well.

BlueSurf between cross-sections

Activate the **BlueSurf** command (on the ribbon, click **Surfacing > Surfaces > BlueSurf**). Select the first cross-section and right-click. Likewise, select the second and third cross-sections.

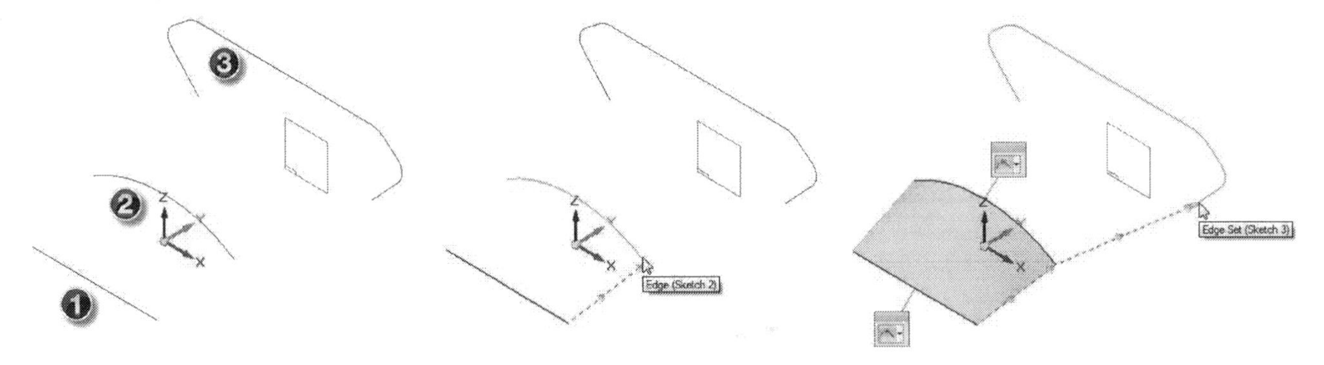

You can use the **Deselect All** icon to deselect all the selected cross sections.

Select any option from the **Tangency Control** handles attached to the cross-sections. Click **Next** and **Finish** on the command bar.

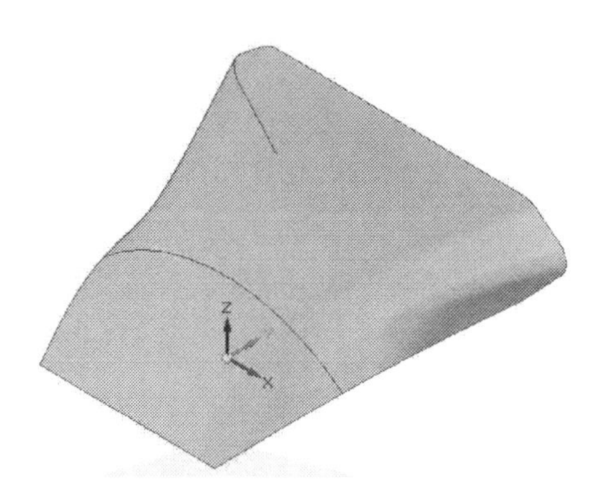

BlueSurf using Cross-sections and Guide Curves

Activate the **BlueSurf** command and select the cross-sections. Click **Accept** (green check) on the command bar after selecting each cross-section.

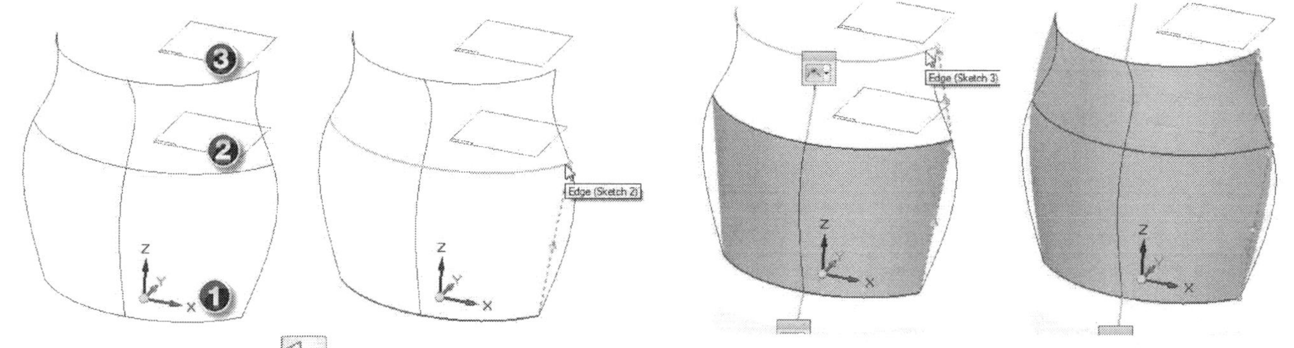

Click **Guide Curve Step** after selecting all the cross-sections. Now, select the guide curves one-by-one. Click **Accept** after each selection.

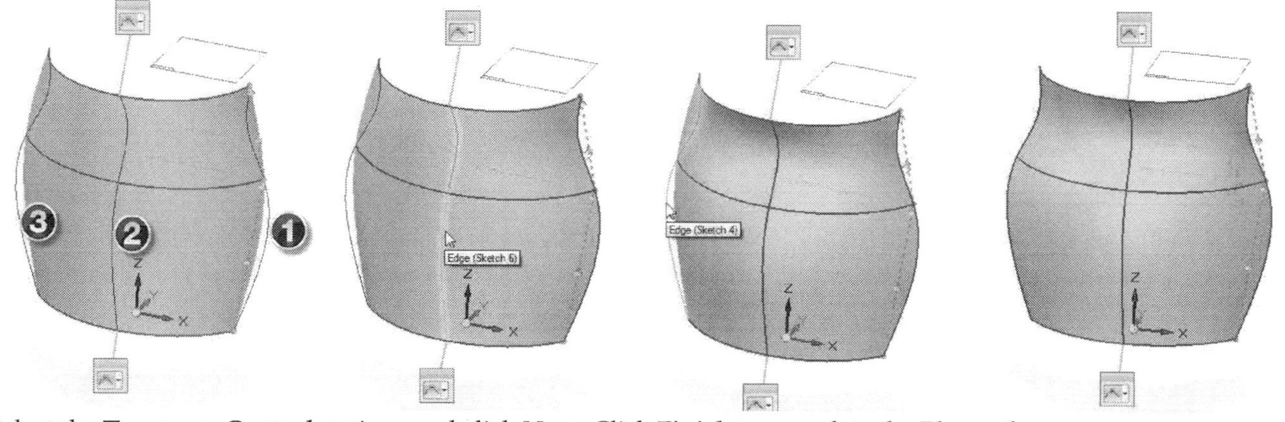

Select the **Tangency Control** options and click **Next**. Click **Finish** to complete the Bluesurf.

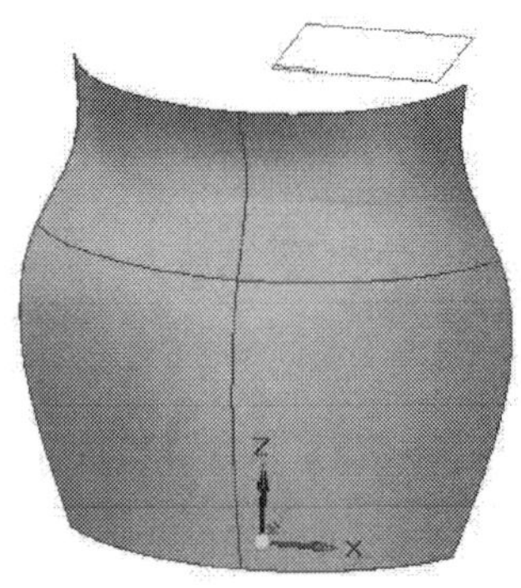

If you want to modify the shape of the Bluesurf by adding a new section, then click on the surface and select **Edit Definition**. Note that this option is available only when the surface is created in the **Ordered** mode. On the command bar, click the **Insert Sketch Step** button and define the location of the new cross-section plane. Click **Next** and **Finish** to complete the bluesurf. You will notice that a new sketch appears in the Pathfinder. Modify the shape of this sketch to modify the bluesurf.

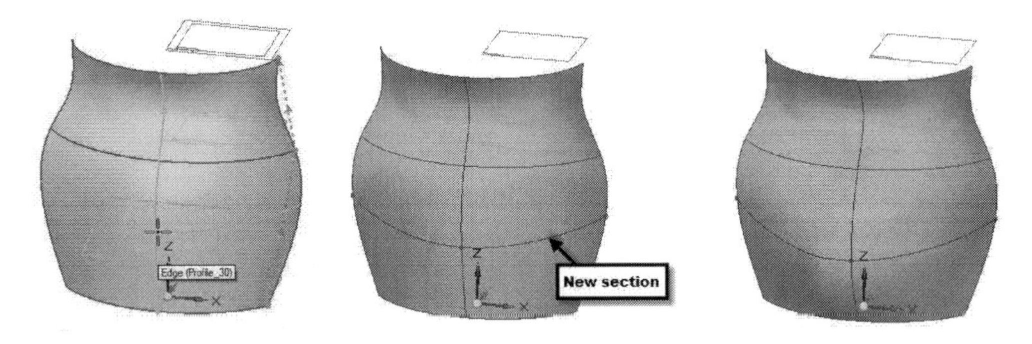

Bounded

The **Bounded** command can be used either to patch holes in models or to create complex surfaces. As a patching tool, the **Bounded** command is more robust than deleting holes. It provides more discrete control over the definition of the resultant patch. For example, consider the model shown in figure. You can see that a face is missing. In a case like this, the **Bounded** command can be used to fill the gap.

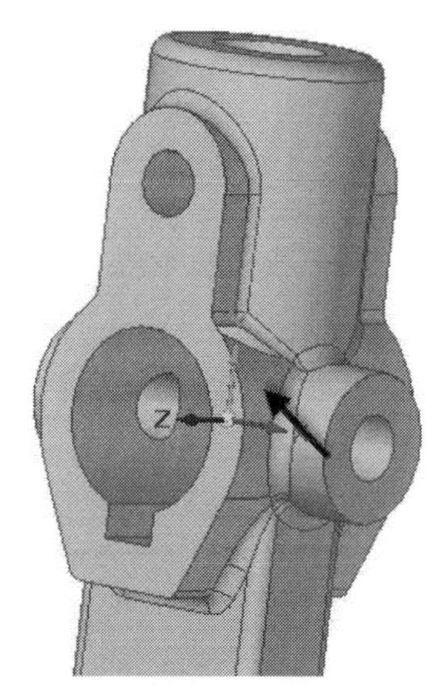

To create a bounded surface, click **Surfacing > Surfaces > Bounded**. Next, you need to select the patch boundaries. To select the patch boundaries, set the **Selection Type** to **Chain** and click on anyone of the open edges. Now, you need to set the tangent condition. You can use the **Tangency Control** handle attached to the selected boundary. The options in this handle are **Natural**, **Tangency Continuous**, and **Curvature Continuous**. Most of the gap edges should be tangent to the surrounding faces. For this example, bounded surface should be natural, as shown in Figure. Use the **Common Tangent Condition** button to apply the tangent condition to all edges.

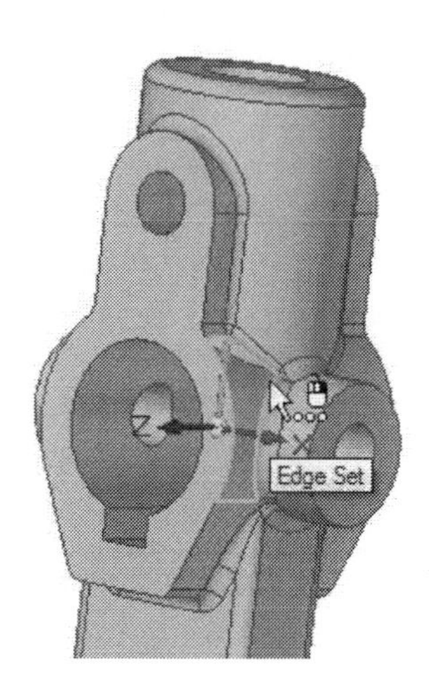

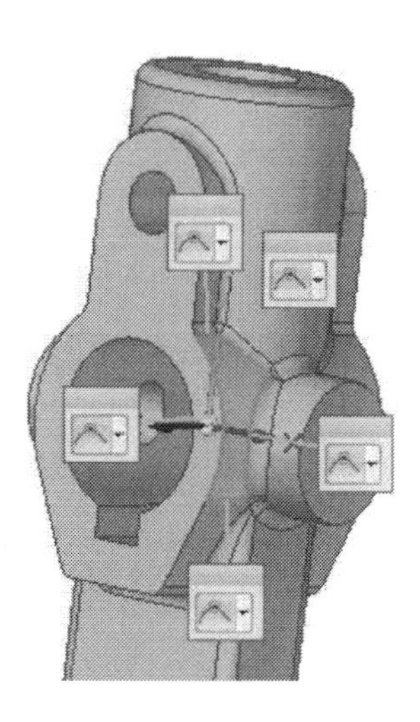

After specifying the required settings, click **Accept** and **Finish** to create the bounded surface, as shown below.

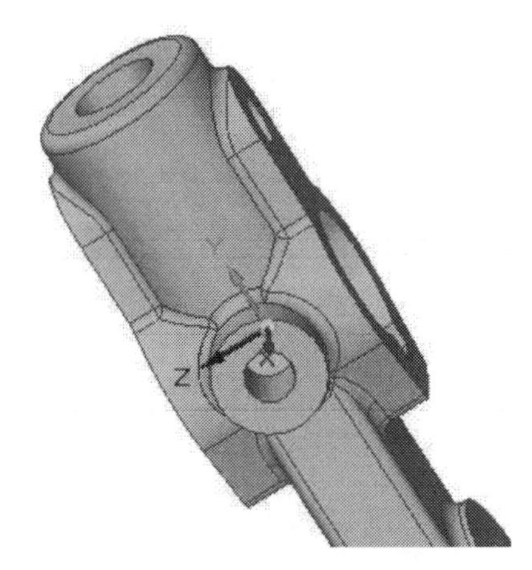

You can also use the **Bounded** command for creating a new surface. Activate this command and select the boundary. Right-click to accept the selection. Click the **Guide Curve Step** button and then select the guide curves. The preview of the bounded surface appears. After defining the required settings, click **Accept** and **Finish** to create the bounded surface

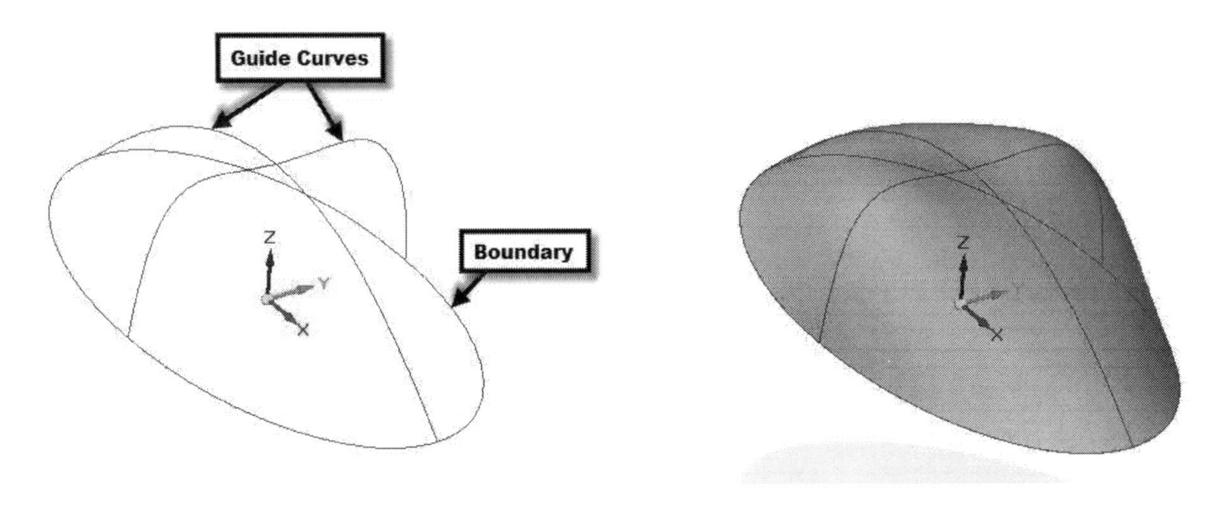

 Ruled Surfaces

The **Ruled** command creates surfaces attached to edges of existing surfaces. You can find the **Ruled Surface** command on the **Surfaces** panel. You can create five types of ruled surfaces using the options in the command bar. These five types of ruled surfaces are discussed next.

The first type is tangent ruled surface. To create a tangent ruled surface, select the **Tangent Continuous** option from the **Ruled Options** drop-down on the command bar. Select an edge from the model. You will notice that the preview of the ruled surface appears. The resultant surface will be tangent to the selected edge. In this case, the selected edge is associated with two reference surfaces (the vertical and the top surfaces). As a result, there will be two solutions available from the selected edge. Click the **Alternate Face/Side** button on the command bar to view the alternate solution. Enter a distance value in the **Distance** box. Click **Accept** to create the ruled surface.

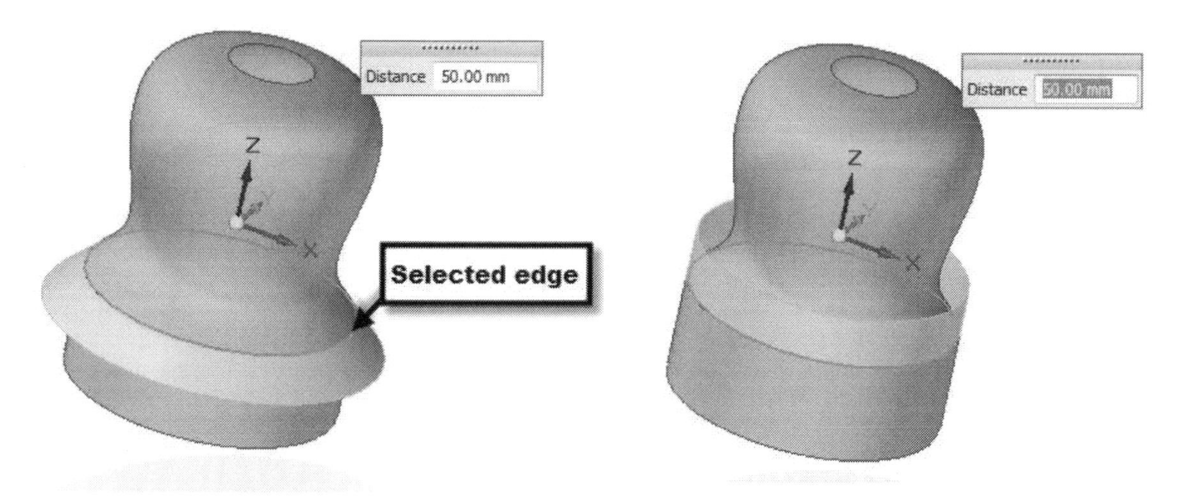

Select the **Normal to face** option from the **Ruled Options** drop-down to create a ruled surface normal to the supporting surface. Use the arrow to change the direction of the surface.

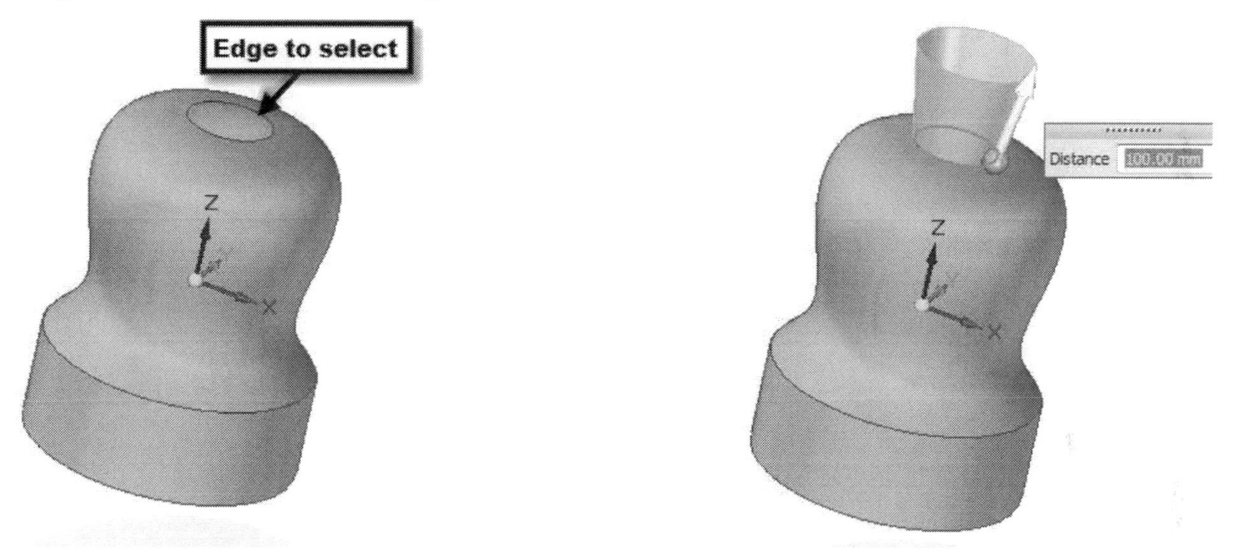

Use the **Natural** option to create a ruled surface without any constraining condition.

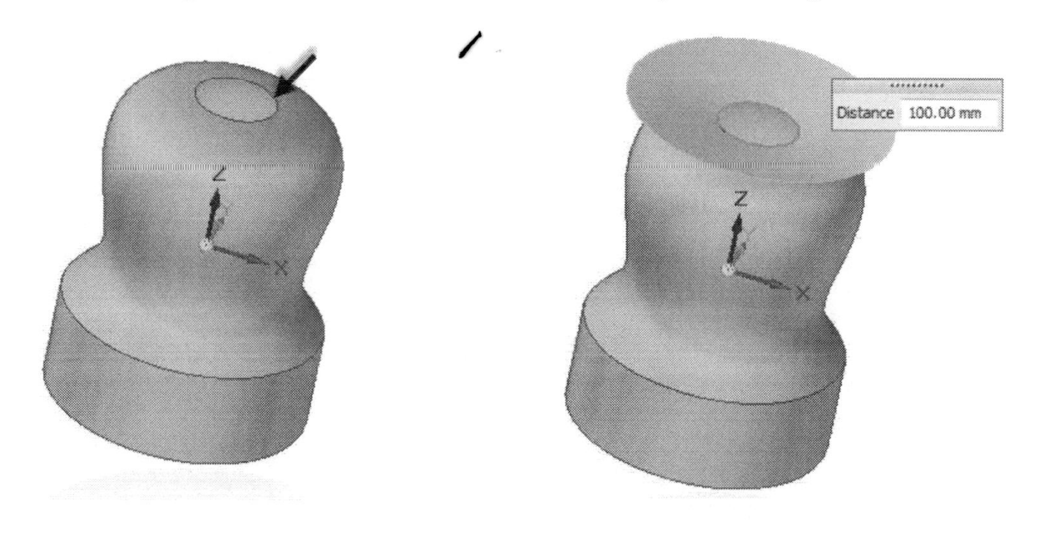

Use the **Along an axis** option to create a ruled surface by sweeping the selected edge along an axis. Select a sketch, edge, or curve to define the reference axis. Select the edge, and then define the **Distance** and **Angle** values.

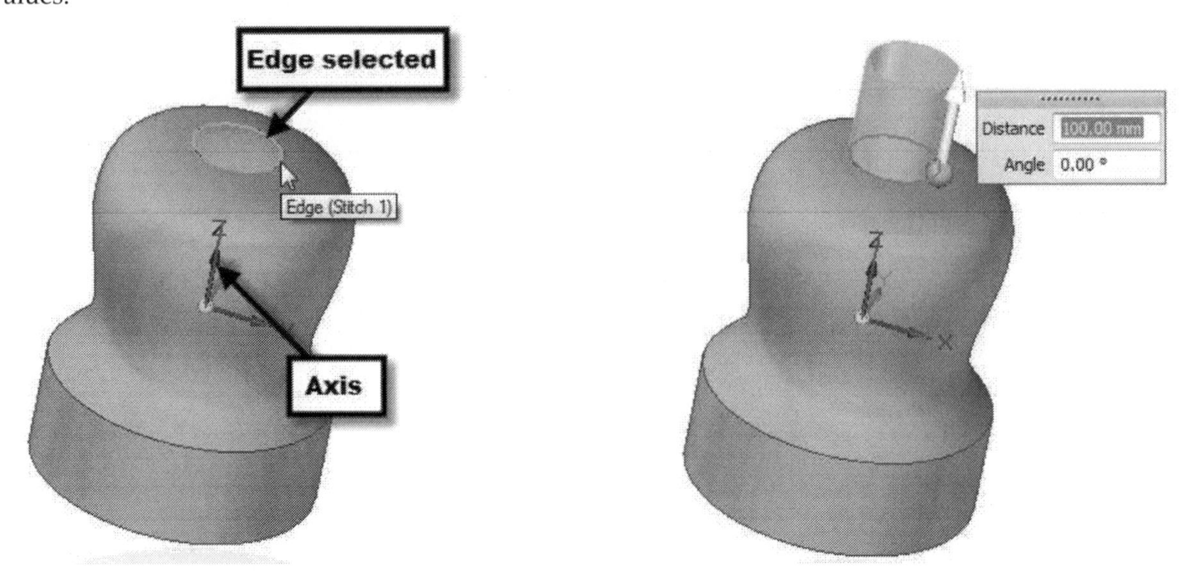

Select the **Tapered to plane** option from the **Ruled Options** drop-down to create a ruled surface at an angle to a plane. Select a planar face or plane to define the reference. Select an edge from the model.

Specify the distance and taper angle of the ruled surface in the **Distance** and **Angle** boxes. Click the arrow to change the direction of the ruled surface. Click **Accept** to create the ruled surface.

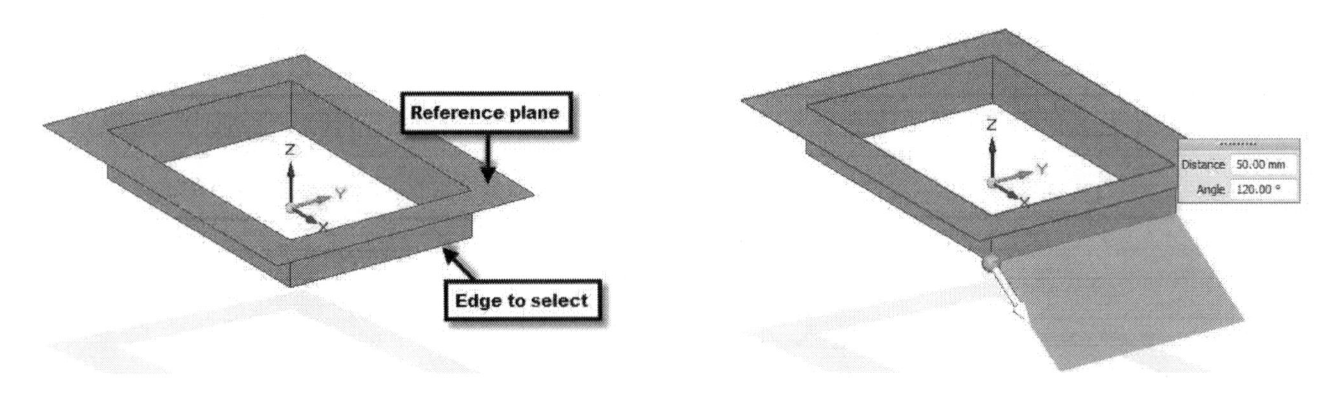

Offset

To create an offset surface, activate the **Offset** command (click **Surfacing > Surfaces > Offset** on the ribbon) and select the faces to offset. Right-click to accept the selection. Next, type-in a value in the **Distance** box and click to define the side of the offset surface. Click **Finish** to complete the offset surface.

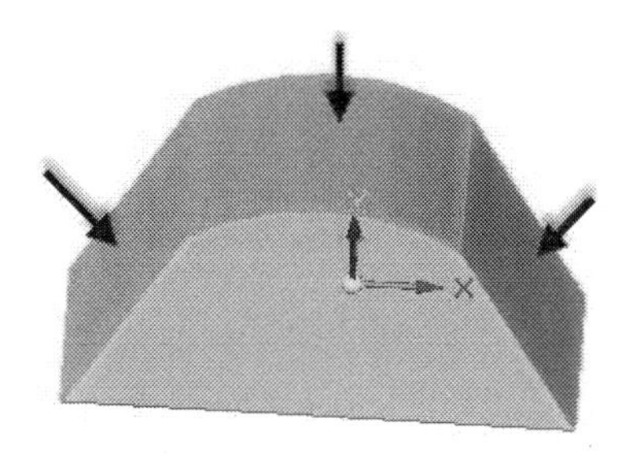
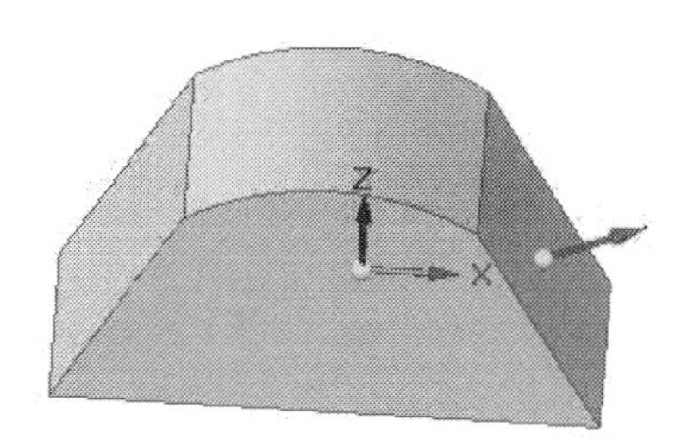
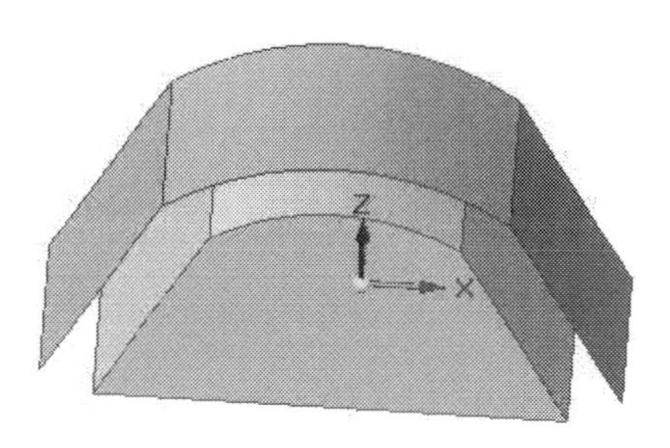

 Redefine

This command creates a surface by merging two or more closely connected surfaces. This command is very useful when a solid or surface body contains split faces, as shown in figure. Activate this command (on the ribbon, click **Surfacing > Surfaces > Redefine**) and select the closely connected faces. Right-click to accept the selection. On the command bar, click the **Options** button and check the **Replace faces on solid body** option in case of a solid body. Click **OK** to close the dialog. Click **Accept** and **Finish** to redefine the faces.

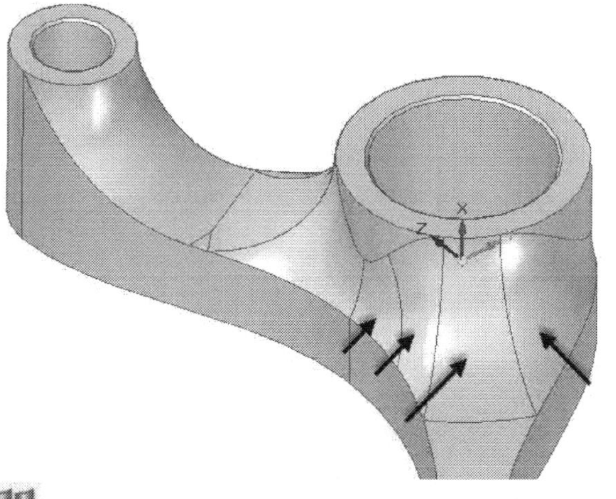
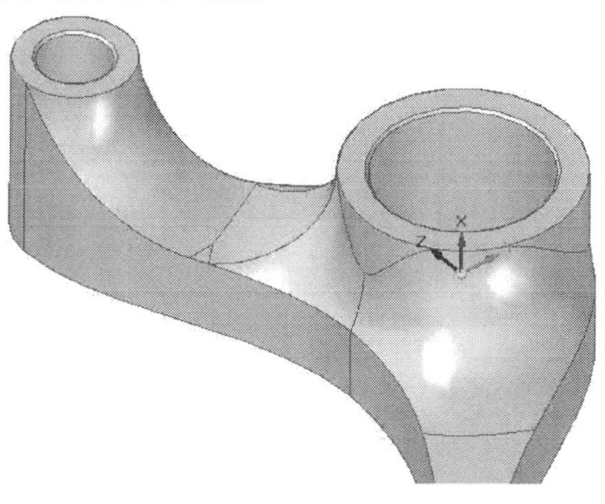

 Copy

This command creates a copy of existing surfaces. Activate this command (on the ribbon, click **Surfacing > Surfaces > Copy**) and select the surfaces to copy. Right-click to accept the selection, and then click **Finish**. Hide the original body to view the copied surface.

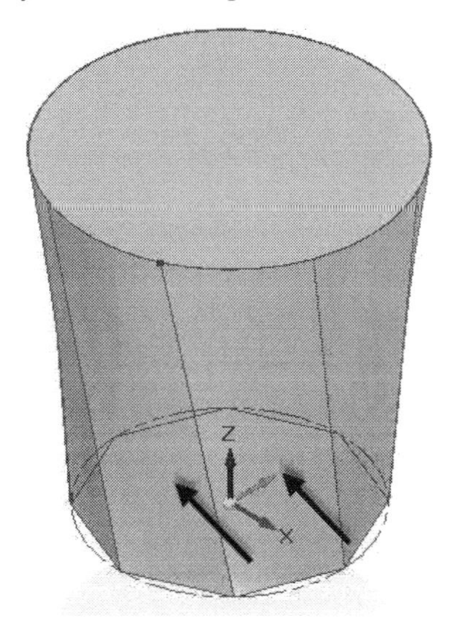
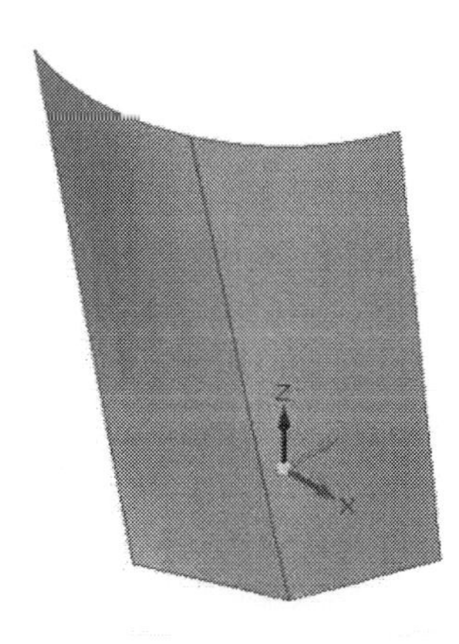

The **Remove Internal Boundaries** button will copy the surface by removing the internal boundaries.

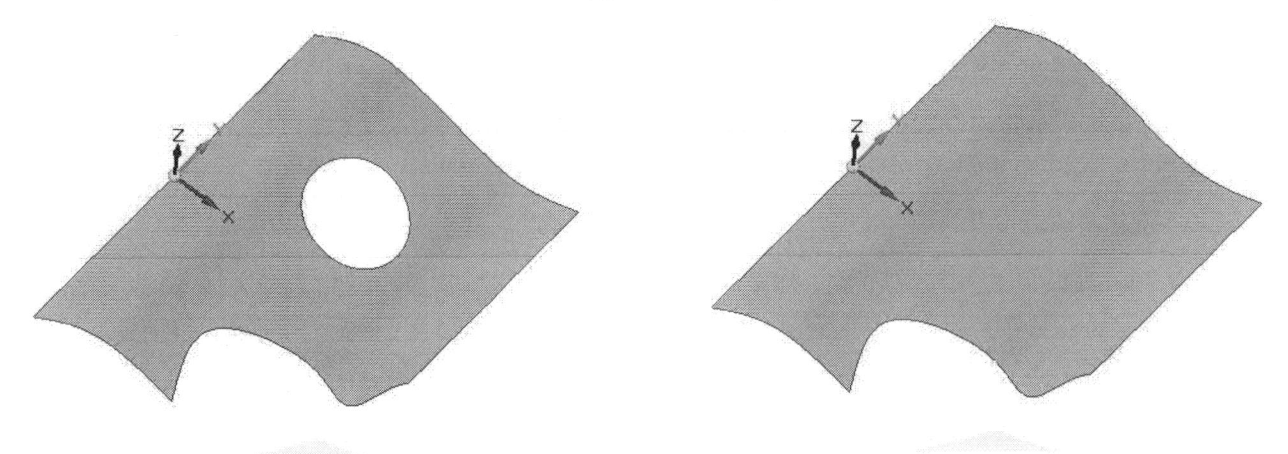

The **Remove External Boundaries** button will copy the surface by removing the external boundaries.

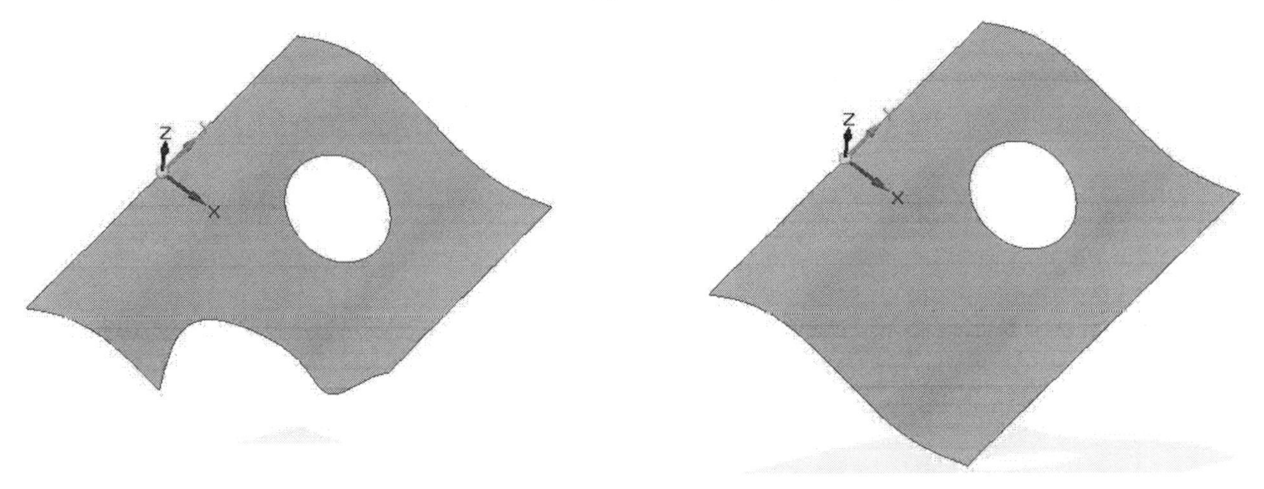

Creating Surface Blends

Surface blends have several uses. They can span across gaps between faces or can be useful in blending complex surfaces. For example, you can create a surface blend, which spans across a gap between two faces. To do this in the Ordered environment, activate the **Round** command (on the ribbon, click **Home > Solids > Round**) and click the **Round Options** button. On the **Round Options** dialog, select the **Surface blend** option and click **OK**. Select the first face and second face chain. Type-in a value in the **Radius** box and click **Accept**. Note that the radius value depends on the gap between the surfaces. Now, click to define the side of the blend. Make sure that the arrows point inwards.

Click the **Surface Blend Parameters** button on the command bar to open the **Surface Blend Parameters** dialog. The **Trim and stitch input faces** option on this dialog trims the selected faces up to the blend edges and stitches them. If you uncheck this option, the faces will not be trimmed. The **Trim output blend** option trims the blend to match the side edges of the selected faces.

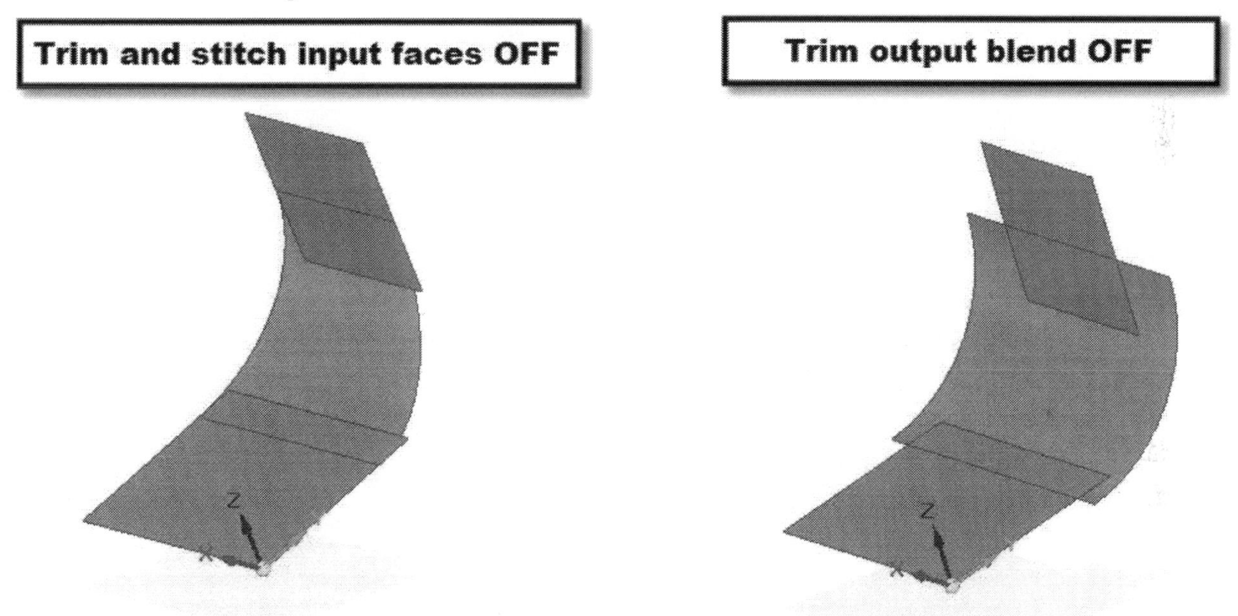

Click **Preview** and **Finish** to create the blend surface.

In the Synchronous environment, you can create surface blends using the **Blend** command (on the ribbon, click **Home > Solids > Round** drop-down > **Blend**).

Trim

This command trims a portion of a surface using a trimming tool. The trimming tool can be a surface, plane or a sketched entity. Activate this command (click **Surfacing > Modify Surfaces > Trim** on the ribbon) and select the target body. Right-click to accept the selection. On the command bar, set the **Selection type**, and then click on the trimming tool. Click **Accept** on the command bar. Select the region to remove and right-click. Click **Finish** to trim the surface.

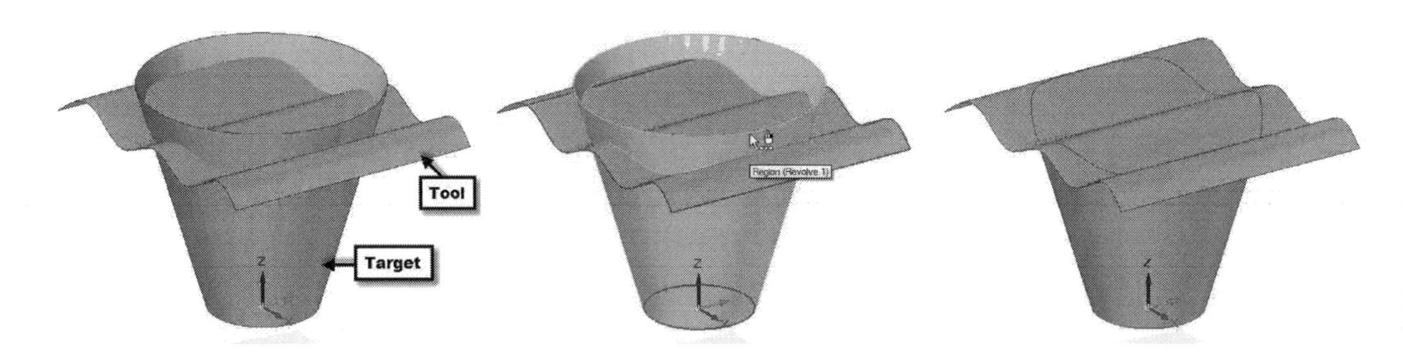

You can also trim a surface using a sketch. Activate the **Trim** command and select the target body. On the command bar, click **Accept**, and then click on the sketch. Right-click and select the region to remove. Click **Accept** and **Finish** to complete the trim operation.

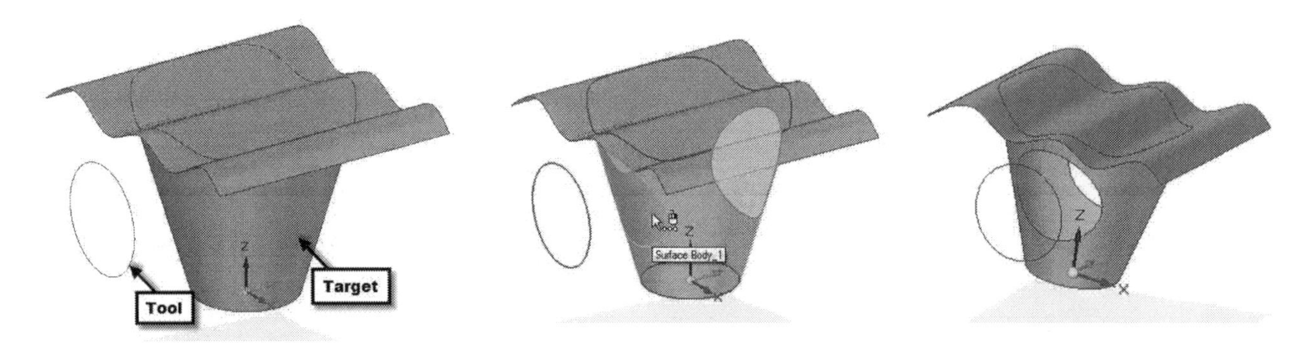

Extend

During the design process, you may sometimes need to extend a surface. You can extend a surface using the **Extend** command. Activate this command (On the ribbon, click **Surfacing > Modify Surfaces > Extend**) and click the surface to extend. Right-click to accept.

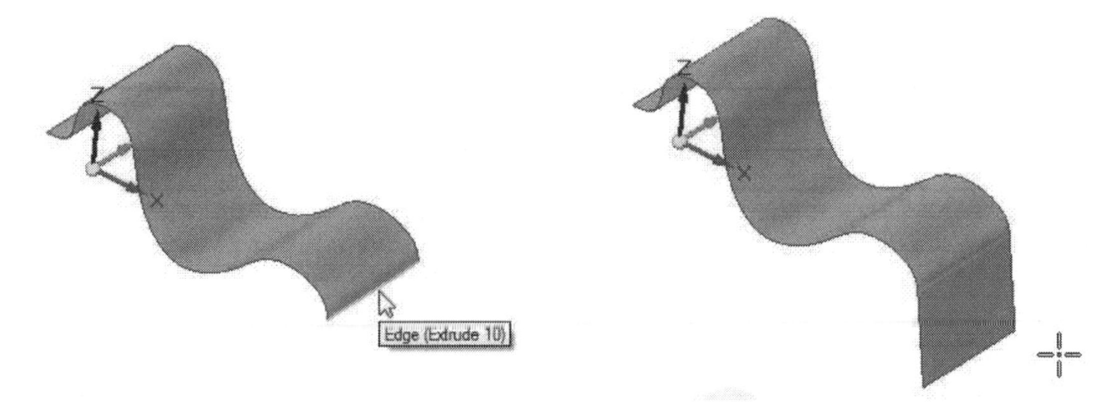

After selecting an edge, you can define the distance of the extension surface by using the **Finite Extent** and **Extend To** options. If you select the **Finite Extend** option, you can define the distance by entering a value in the **Distance** box. If you select the **Extend To** option, you can define the distance by selecting a boundary surface.

When the surface you have selected is not a planar one, you can decide the type of extension by using the **Extend Type** options. Use the **Curvature Continuous** option to extend the surface by maintaining the curvature of the original surface. If you select the **Linear** option, the extended surface will be created tangent to the original surface. The **Reflective** option extends the surface by reflecting the original surface. Click **Finish** after defining the distance of the extension.

Intersect

This command trims or extends a set of surfaces by the distance that you specify or up to another surface. Activate this command (on the ribbon, click **Surfacing > Modify Surfaces > Intersect**) and select two surfaces. If you want to extend a surface, then you need to select the surface to be extended and the boundary surface. Right-click and click on the edge to extend. Click **Accept** and **Finish** to extend the surface.

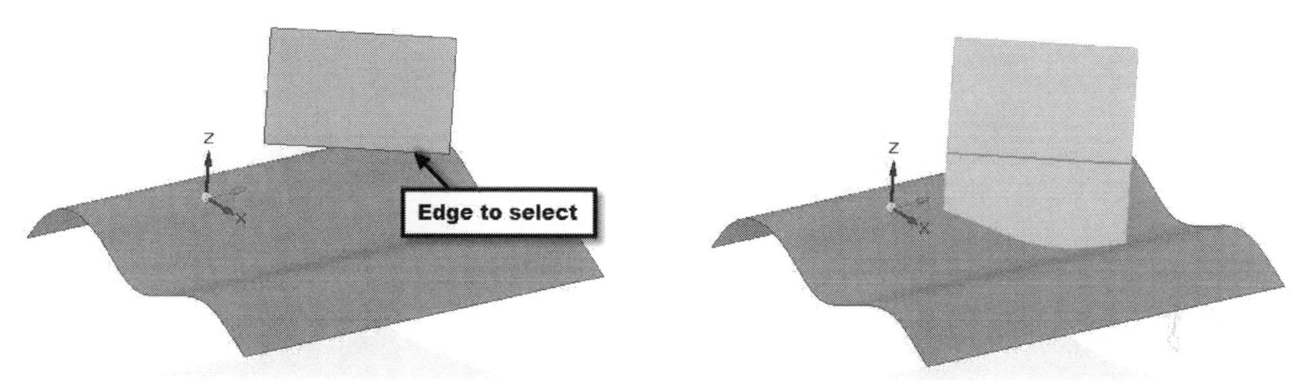

If you want to trim surfaces, activate the **Intersect** command and select two or more surfaces. Right-click and select the region to remove. On the command bar, click the **Stitch** button to stitch surfaces. Click **Accept** and **Finish** to trim the surface.

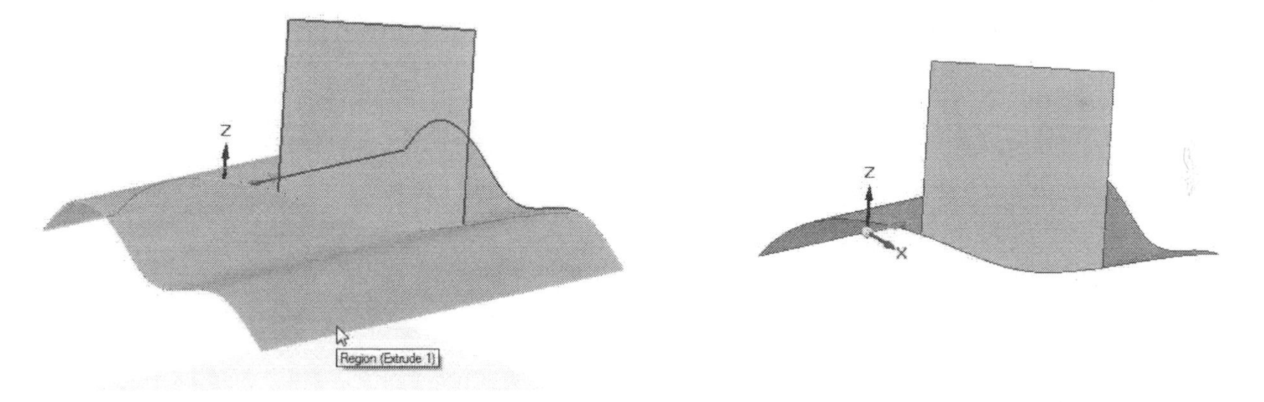

Create design Bodies

The **Intersect** command can be used to create design using two or more surfaces intersecting with each other and forming a closed volume. To do this, click activate the **Intersect** command and select **Intersect Options > Create design Bodies**. Select the intersecting surfaces and click Accept button on the command bar; the **Volume Regions** dialog appears displaying the regions detected from the intersecting surfaces. You can uncheck the regions that you do not want. Click the **Close** button on the **Volume Regions** dialog. Click the **Accept** button to create the design body. Next, hide the surface bodies to view the design body.

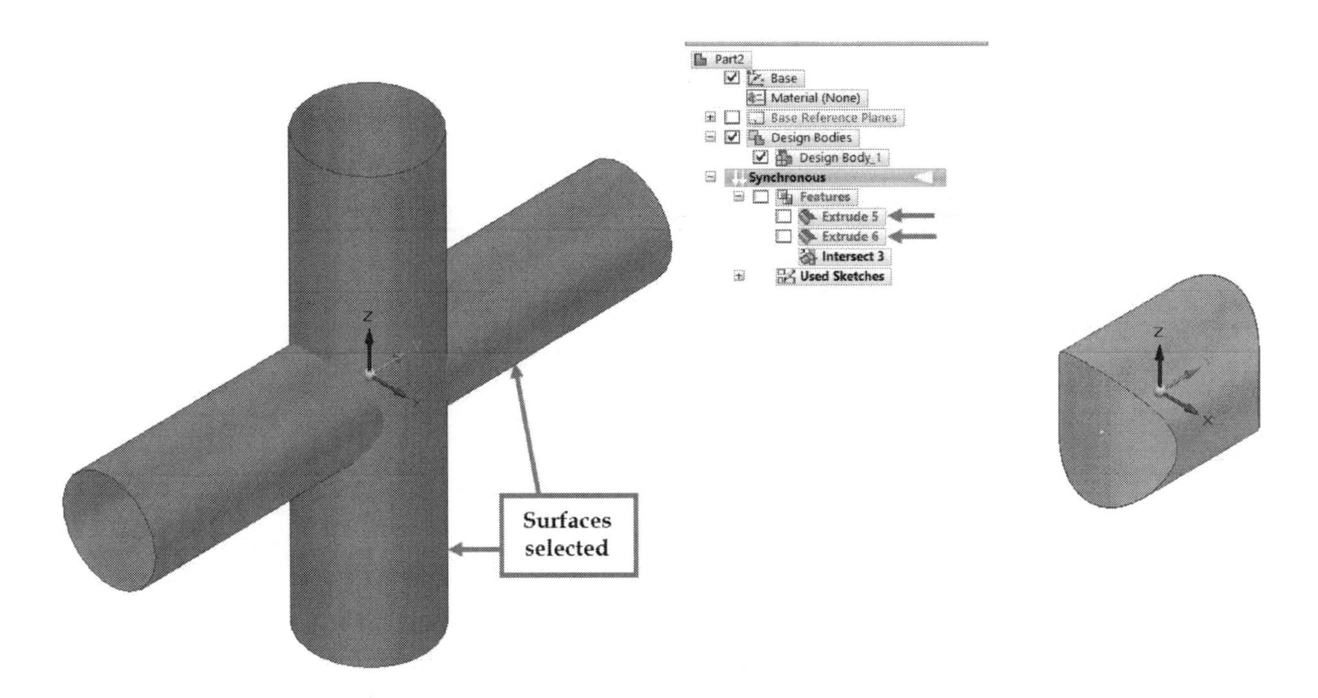

Auto-trim intersecting surfaces

The **Auto-trim** option available in the **Intersect Options** drop-down makes it easy to trim the unwanted portions of the intersecting surfaces. This option trims the unwanted portions such that a closed volume, closed loop, or continuous surface is formed out of the intersecting surfaces. For example, activate the **Intersect** command and select the three intersecting surfaces, as shown. Click the **Accept** button, and then select **Intersect Options > Auto-trim** on the command bar; notice that a closed volume of surfaces is created. Click the **Invert Selected Regions** icon, if you do not see the desired result. Next, click **Accept** on the command bar to complete the feature.

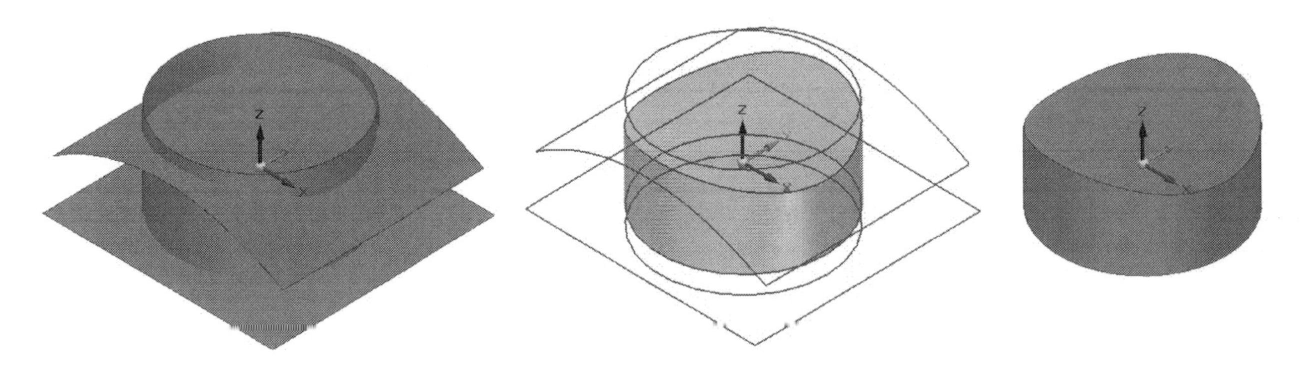

Stitched Surfaces

The surfaces created act as individual surfaces unless they are stitched together. The **Stitched** command lets you combine two or more surfaces to form a single surface. To stitch surfaces, activate this command (click **Surfacing > Modify Surfaces > Stitched** on the ribbon). On the **Stitched Surface Options** dialog, type-in a value in the **Stitched tolerance** box. The value you type in this box defines the tolerance gap. All the surfaces within the tolerance gap will be stitched. The **Heal stitched surfaces** option closes the gap between the stitched surfaces. Click **OK** and select the surfaces to stitch.

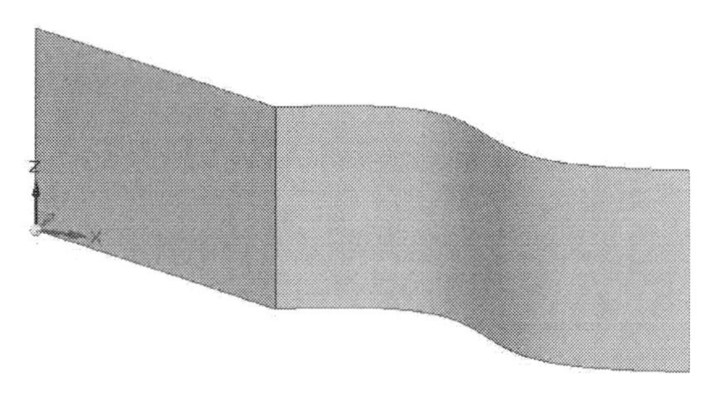

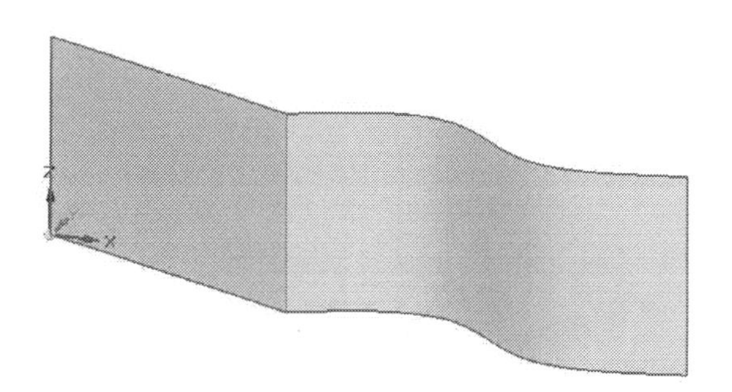

Click the **Accept** and **Finish** buttons to stitch the surfaces.

Thicken

Creating a solid from a surface can be accomplished by simply thickening a surface. To add thickness to a surface, activate the **Thicken** command (on the ribbon, click **Home > Solids > Add** drop-down **> Thicken**) and click on a face of the surface geometry. Enter the thickness value in the **Distance** box. Move the pointer inwards or outwards to define the side of material addition. Place pointer on the surface body and click to add material on both sides of the surface. Click **Finish** to thicken the surface.

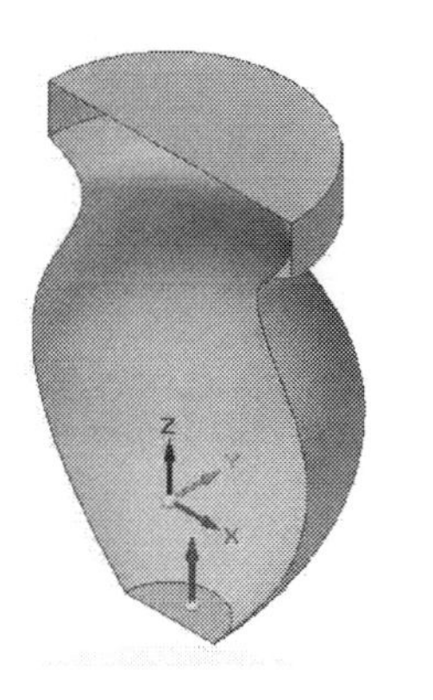

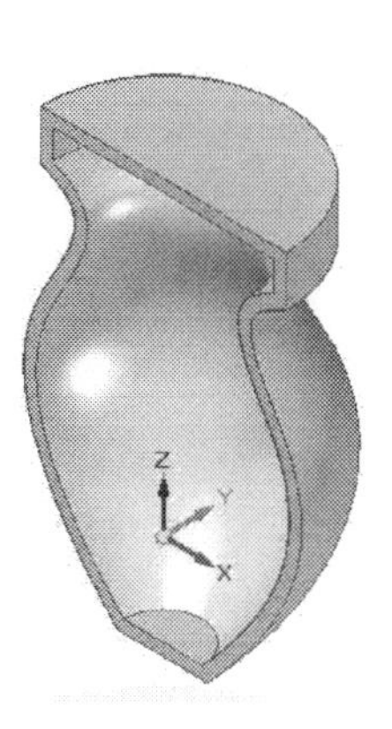

Replace Face

The **Replace Face** command replaces a face or group of faces with another face or group of faces. To replace a face, activate this command (on the ribbon, click **Surfacing > Modify Surfaces > Replace Face**). Select the faces to replace and click **Accept**. Select the replacement surface and click **Finish**.

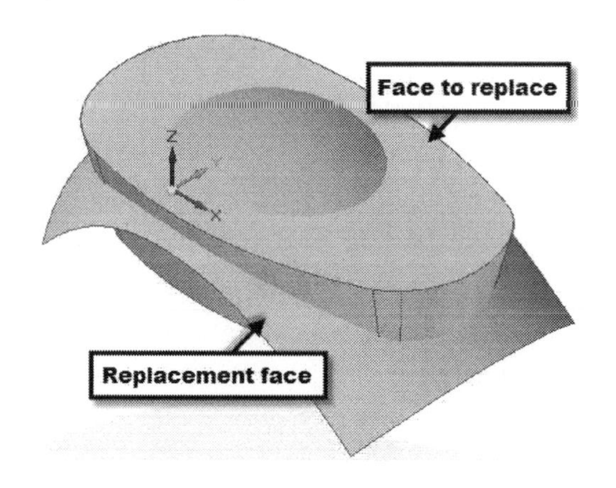

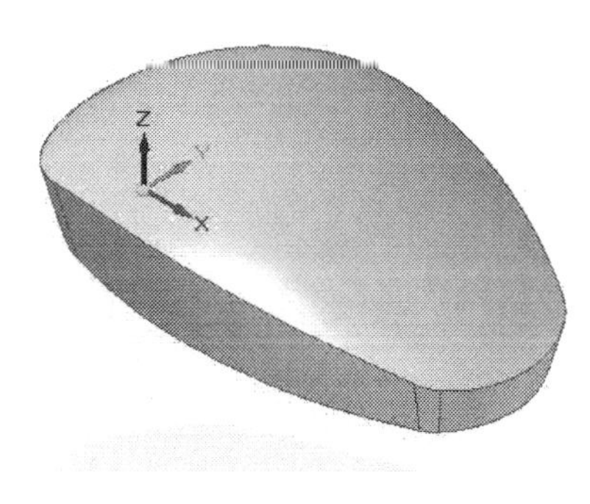

 Split

The **Split** command splits a face or a body using a plane, body, curve or sketch. Click the **Split** button on the **Modify Surfaces** panel and select a face or body. Right-click and select a splitting element. Click **Accept** and **Finish**.

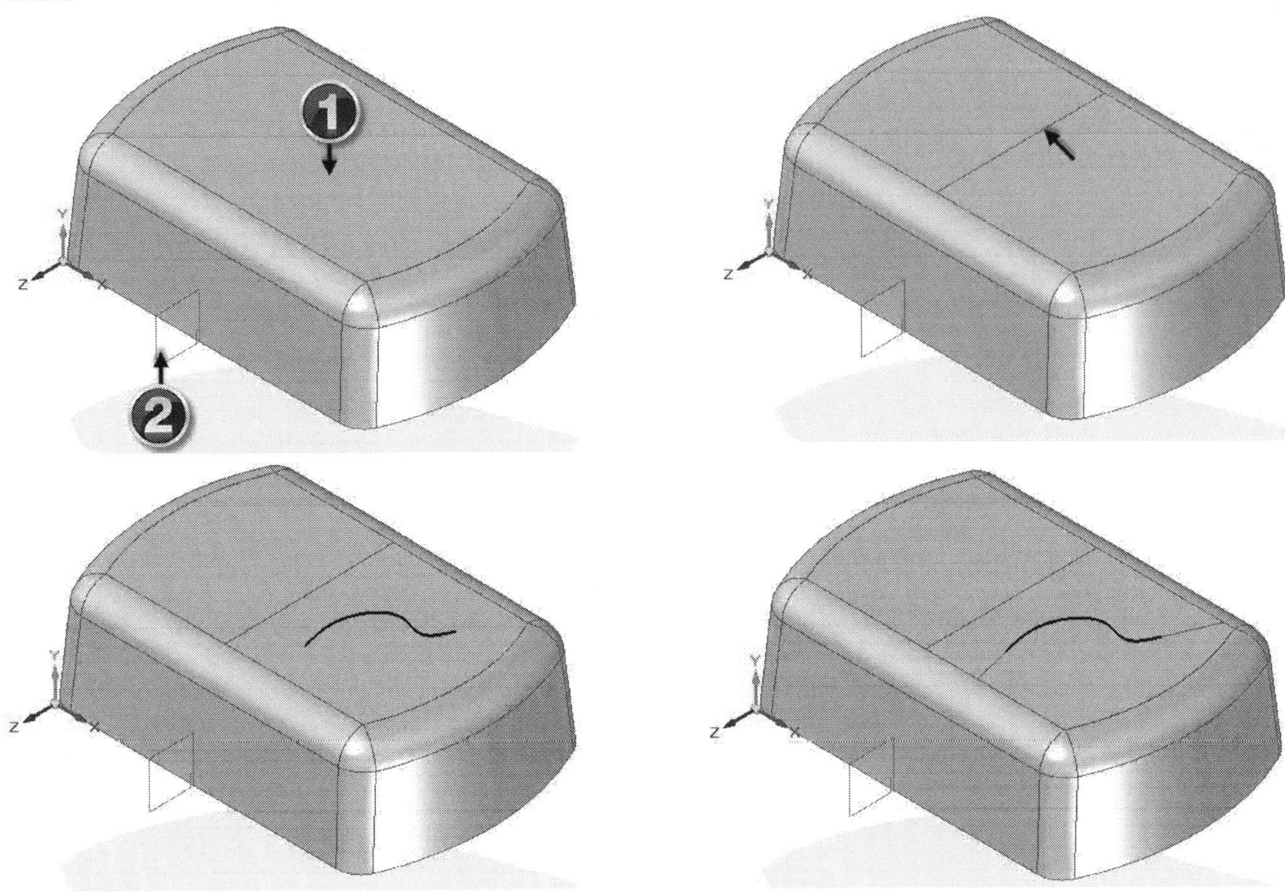

Example

In this example, you will construct the model shown below.

Drawing the Layout Curves

1. Start **Solid Edge 2019**.
2. Start a new part file using the **ISO Metric Part** template.
3. Right-click and select **Transition to Ordered** to switch to the ordered environment. In this tutorial, you will create the surface model in **Ordered** environment as you can edit the surfaces easily. You can also create this model in **Synchronous** environment.
4. Create a spreadsheet with the following values and save it as Curve1. You can also download this file from our website.

	A	B	C	D	E
1	-75	0	-20		
2	-65	0	18		
3	-67	0	32		
4	-80	0	125		
5	-66	0	160		
6	-45	0	182		
7	0	0	200		
8	60	0	182		
9	80	0	160		
10					
11					
12					
13					

5. On the ribbon, click **Surfacing > Curves > Keypoint** drop-down **> Curve by table** .
6. On the **Insert Object** dialog, select the **Create from file** option and click the **Browse** button. Go to the location of the **Curve1** spreadsheet, select it, and click **Open**. Click **OK** on the dialog.
7. On the Command bar, click the **Parameters Step** button and set the **Curve Fit** to **Smoothening on**. Set the **Curve End Conditions** to **Open**. Set the **Coordinate System** to **Base** and click **OK**.
8. Click **Finish** to create the curve, as shown below.

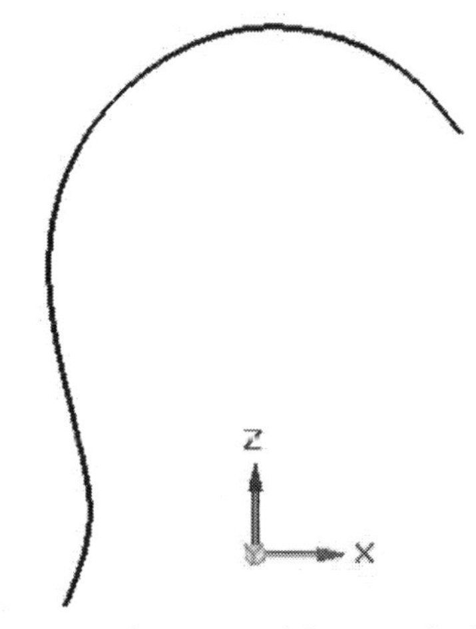

9. Create another spreadsheet with the following values and save it as Curve2.

	A	B	C	D
1	110	0	-5	
2	113	0	20	
3	110	0	45	
4	86	0	90	
5	60	0	155	
6	55	0	210	
7				
8				
9				
10				

10. On the ribbon, click **Surfacing > Curves > Keypoint** drop-down **> Curve by table** .
11. On the **Insert Object** dialog, click the **Browse** button and open the **Curve2** spreadsheet. Click **OK** on the dialog.
12. Click **Finish** to create the curve, as shown below.

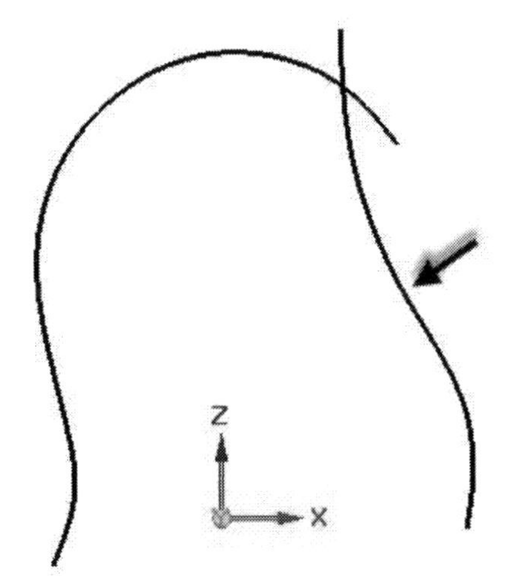

13. Create another spreadsheet with the following values and save it as Curve3.

	A	B	C	D
1	120	0	45	
2	110	0	65	
3	112	0	85	
4	117	0	120	
5	114	0	155	
6	95	0	185	
7	60	0	195	
8	35	0	175	
9				
10				
11				

14. Activate the **Curve by table** command and select the Curve3 spreadsheet. Click **Finish** to create the third curve.

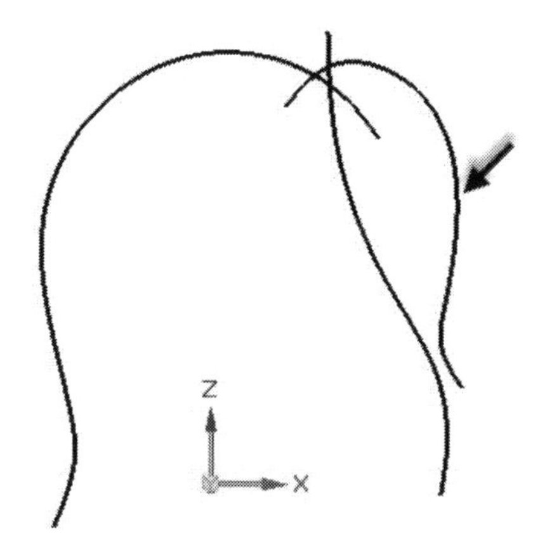

Creating the Front Surface

1. On the ribbon, click **Home > Sketch > Sketch** and select the XY plane
2. Create an arc and add dimensions to it. Click the **Close Sketch** button on the ribbon. Click **Finish** on the command bar.

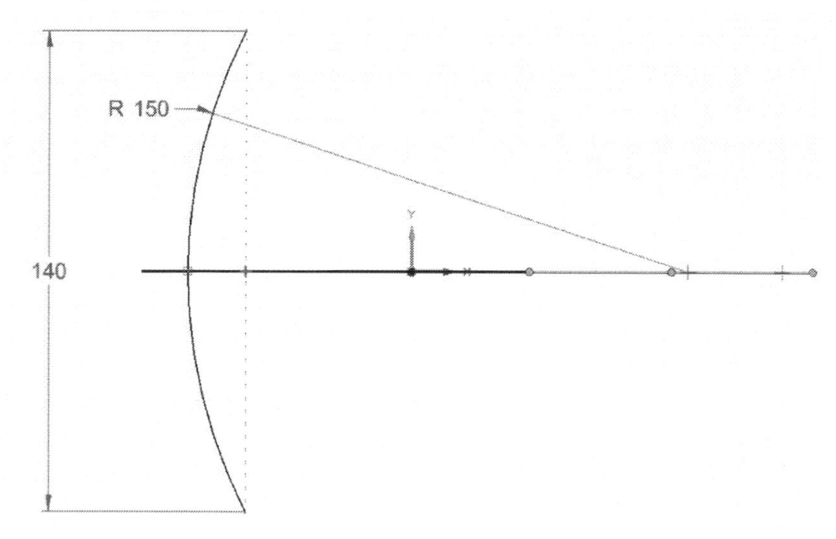

3. Create an arc on the YZ Plane and add dimensions to it. Finish the sketch.

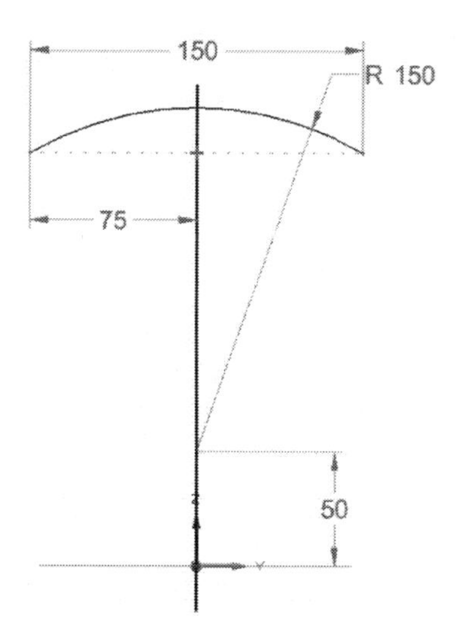

4. Create a plane normal to the first curve.

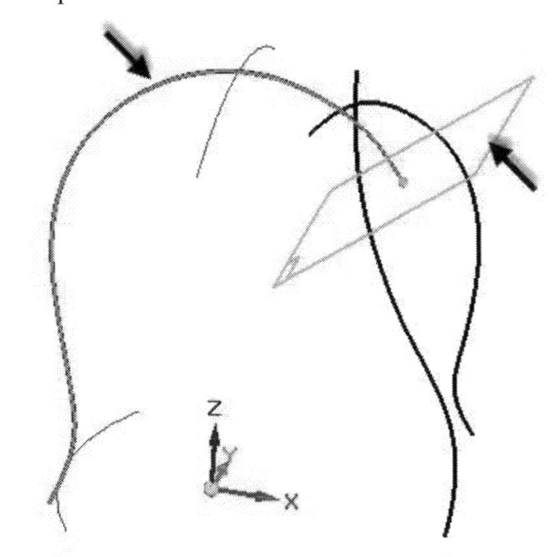

5. Create an arc on the plane normal to curve. Finish the sketch.

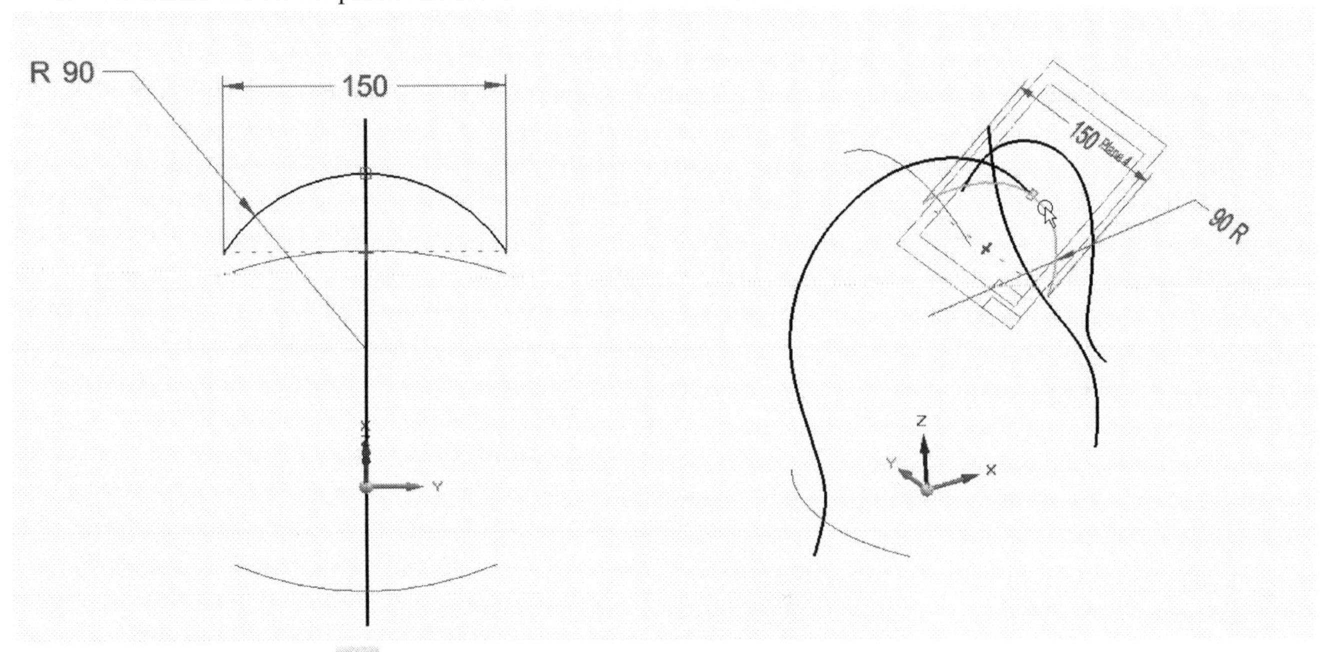

6. Activate the **Swept** command (on the ribbon, click **Surfacing > Surfaces > Swept**).
7. On the **Sweep Options** dialog, select the **Multiple paths and cross section** option. Select the **Along path** option from the **Face Merging** section and click **OK**.
8. Click on the first curve to define the path. Click the green check on the command bar to accept the selection.
9. On the command bar, click **Next** to activate the **Cross Section Step**. Select the arc located on the XY plane to define the first cross section. Make sure that you have selected the arc by clicking at the point, as shown in figure. Click the green check on the command bar.
10. Select the second arc by clicking at the point, as shown in figure. Click the green check to define the second cross section.
11. Likewise, select the third cross section. Click **Preview** to preview the swept surface.
12. Click **Finish** to complete the swept surface. Click **Cancel** to deactivate the command.

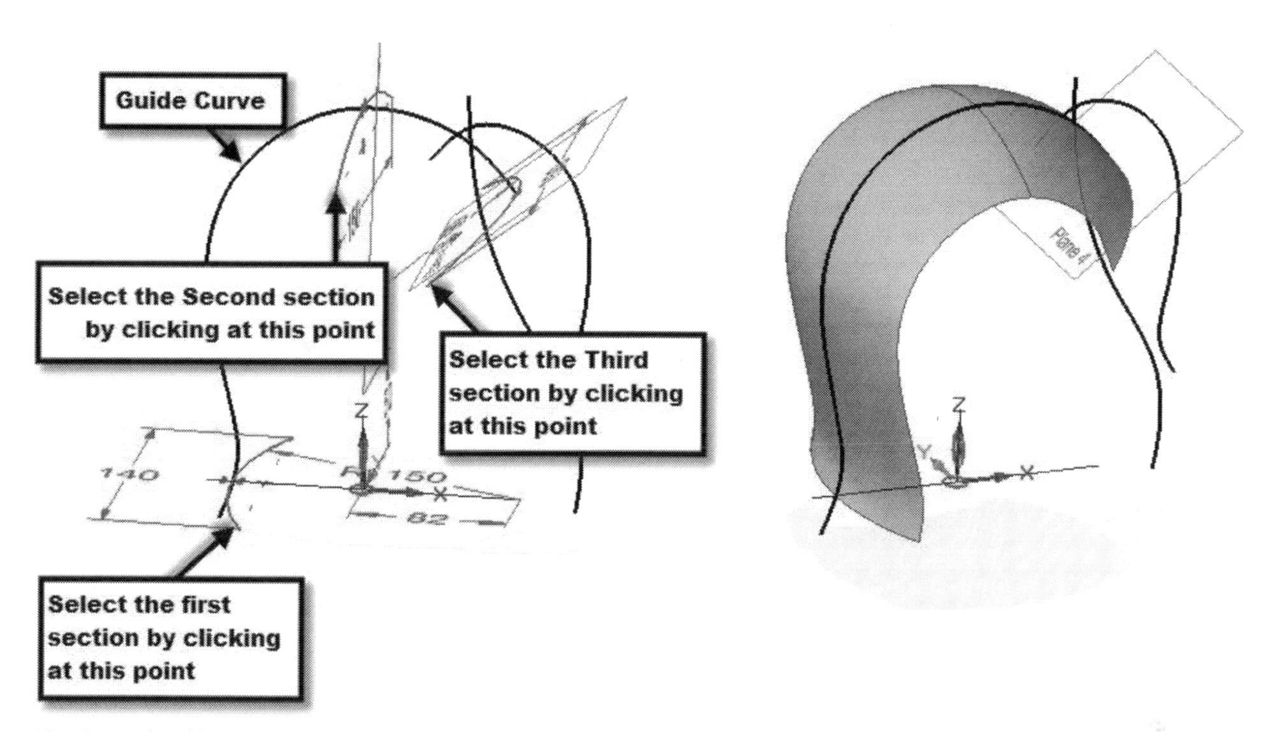

13. Save the file. As you are creating a complex geometry, it is advisable that you save the model after each operation.

Creating the Label surface

1. Create an arc on the XY plane. Finish the sketch.

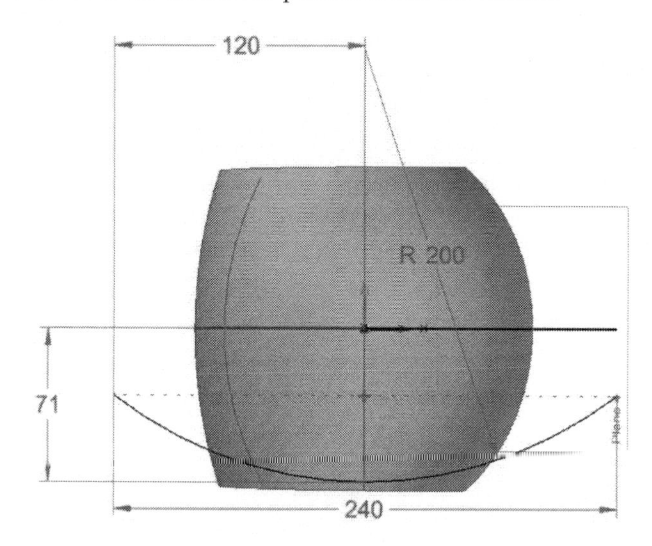

2. Activate the **Extruded** command (on the ribbon, click **Surfacing > Surfaces > Extruded**). On the command bar, click **Create-From Options** drop-down **> Select from Sketch**.
3. Select the sketch and click the green check on the command bar. Deactivate the **Symmetric Extent** button on the command bar. Type-in 220 in the **Distance** box on the command bar. Move the pointer upward and click to define the side of the extrusion. Click **Finish** to complete the extruded surface.
4. On the ribbon, click **Surfacing > Pattern > Mirror Copy Part**. Select the extruded surface and right-click.

5. Select the XZ plane and click **Finish** to mirror the extruded surface.

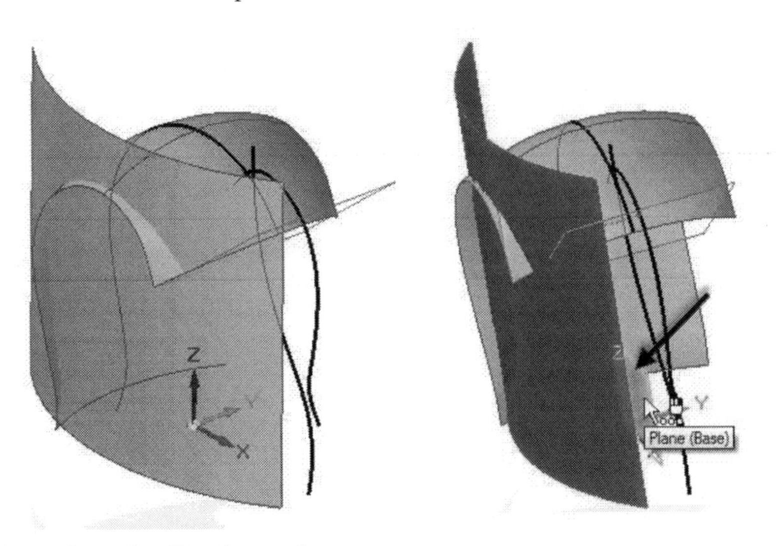

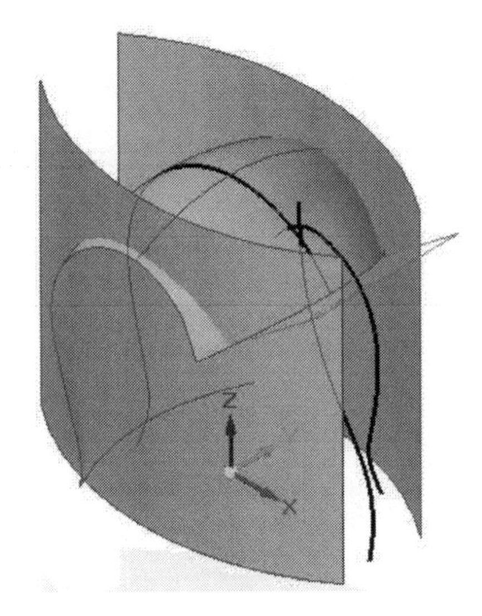

Creating the Back surface

1. Create an arc on the XY plane. Finish the sketch.

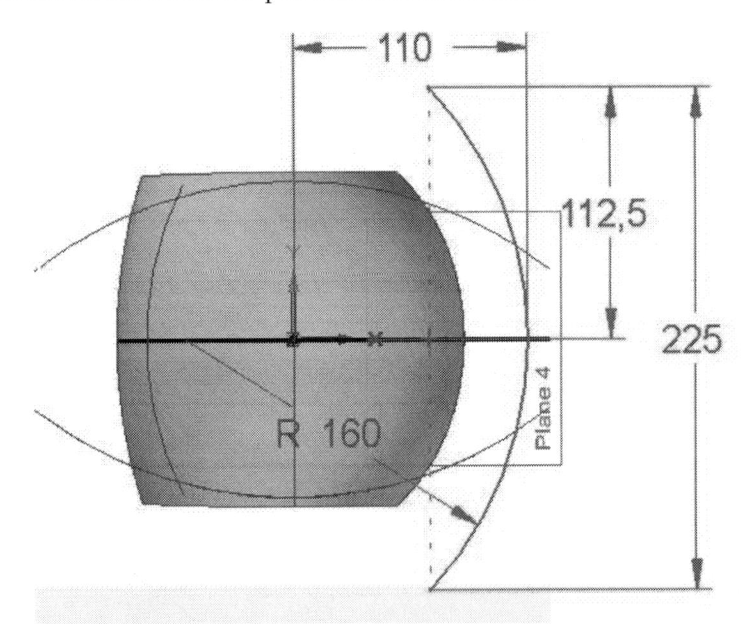

2. Activate the **Swept** command and select **Single path and cross section** option on the **Sweep Options** dialog. Click **OK**.
3. Select second curve to define the path of the swept surface. Click the green check to accept the selection
4. Select the arc and click the green check to define the cross section. Click **Finish** and **Cancel** to complete the swept surface.

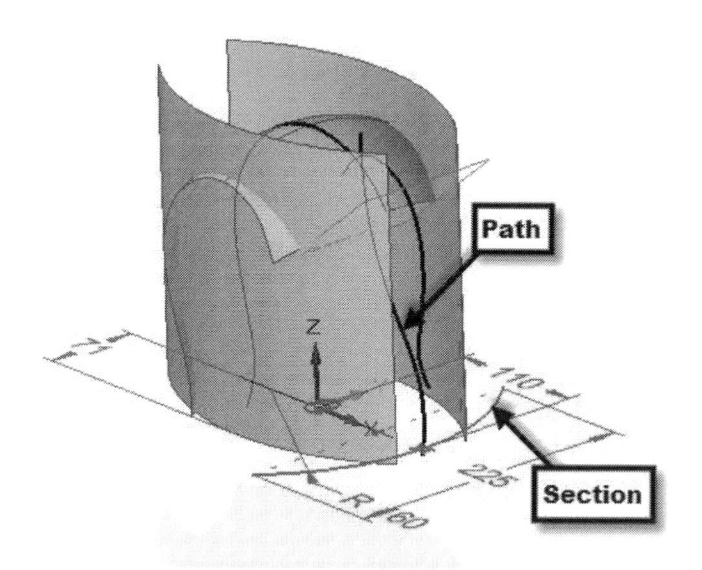

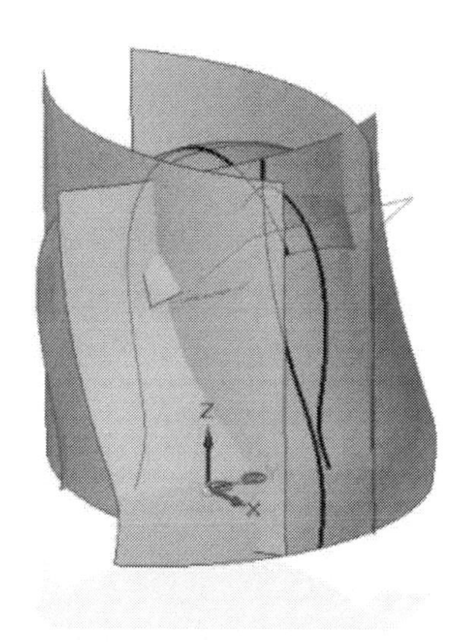

Trimming the Unwanted Portions

1. Activate the **Intersect** command (on the ribbon, click **Surfacing > Modify Surfaces > Intersect**).
2. Select the front swept surface and extruded surfaces. Click the green check.
3. Click on the portions to trim, as shown in figure.

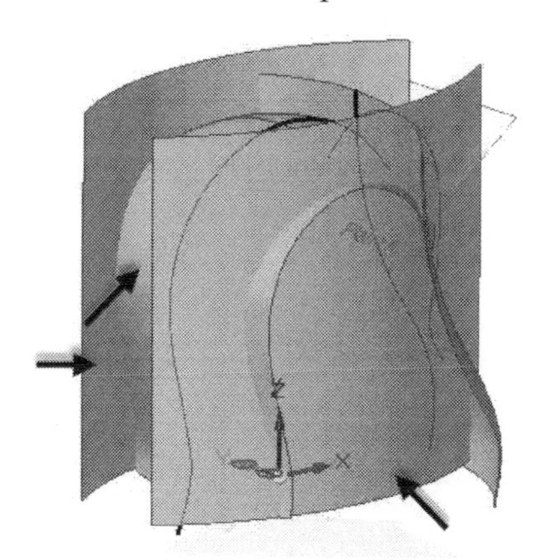

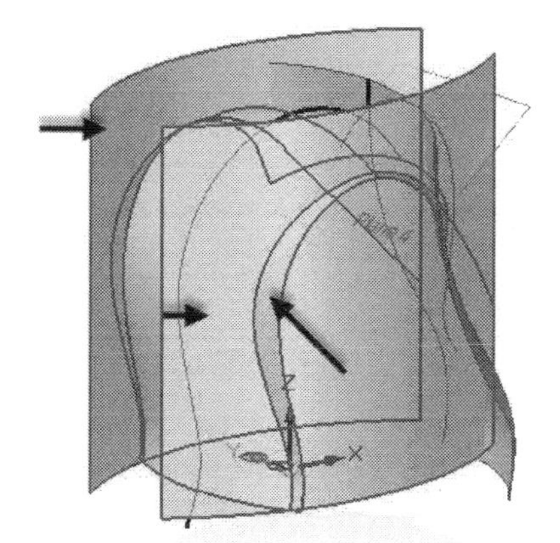

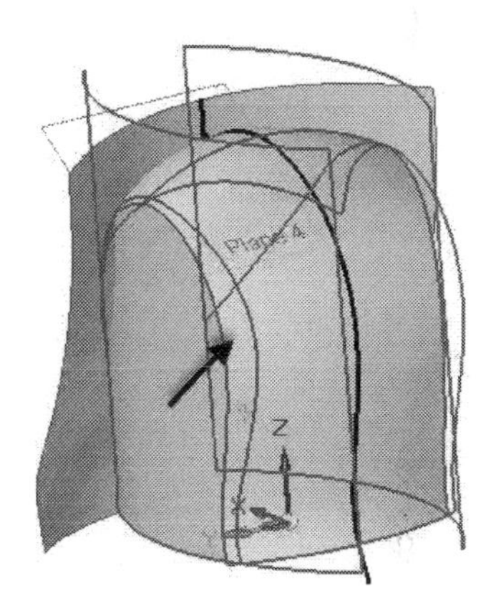

4. Select the **Stitch** button on the command bar. Click the green check to trim the selected portions. Click **Finish**.
5. With the **Intersect** command still active, select the back swept surface and the stitched surface. Click the green check.
6. Select the portions of the back swept surface and stitched surface, as shown below. Click the green check.

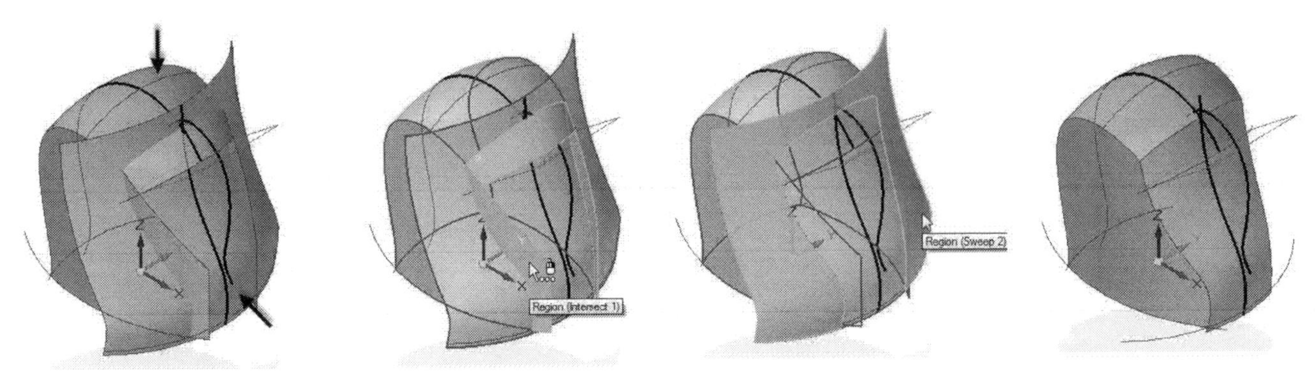

7. Click **Finish** and **Cancel**.
8. Activate the **Trim** command (on the ribbon, click **Surfacing > Modify Surfaces > Trim**) and click on the surface body. Click the green check.
9. Select the XY Plane, and then click the green check.
10. Select the portion of the target surface, as shown. Click the **Accept** button on the command bar.
11. Click **Finish** and **Cancel**.

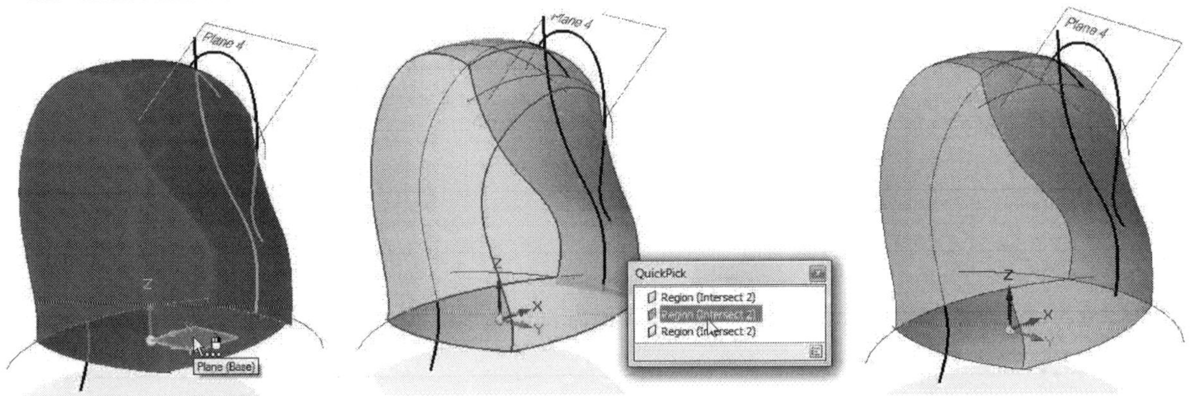

Creating the Handle Surface

1. Activate the **Normal to Curve** command and click on the lower end-point of the third curve. Left click to create the plane normal to the curve.

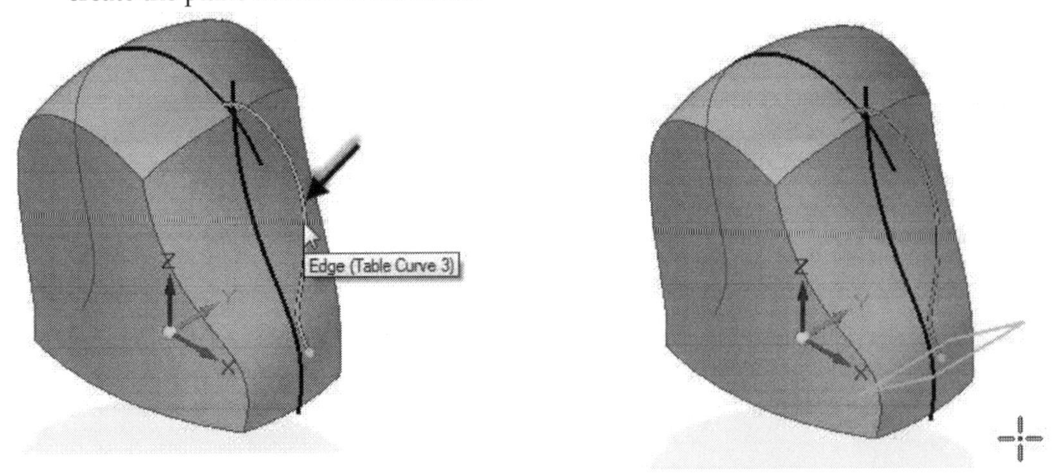

2. Start a sketch on the plane normal to the curve.
3. Create an ellipse on the sketch plane. Apply the **Horizontal** relation between the two quadrant points, as shown. Add dimensions and relations to the sketch.

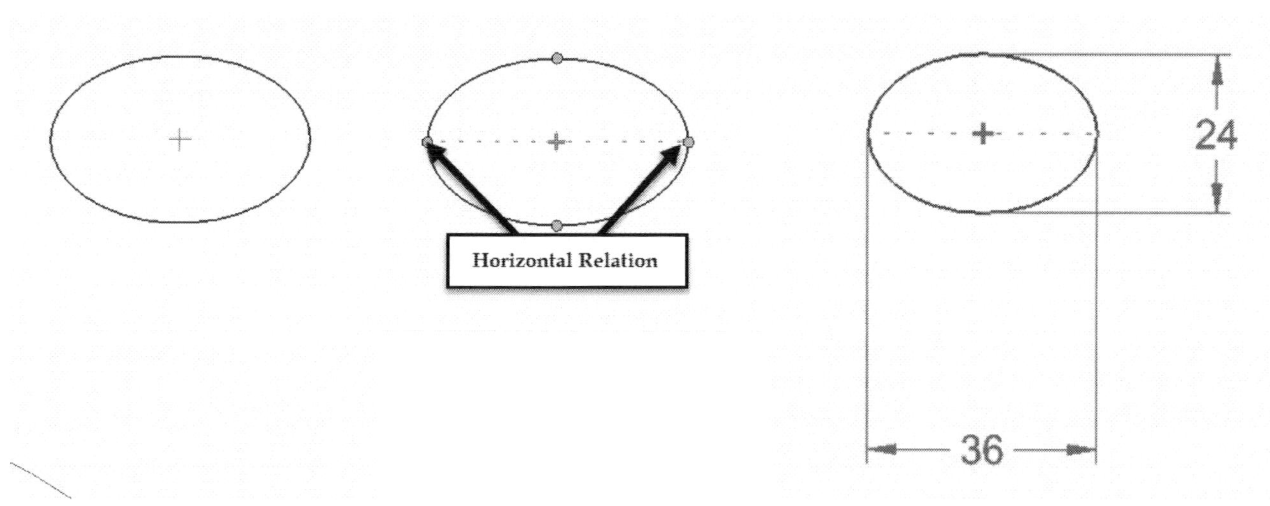

4. Make the upper quadrant point of the ellipse coincident with the end-point of the curve. Finish the sketch.

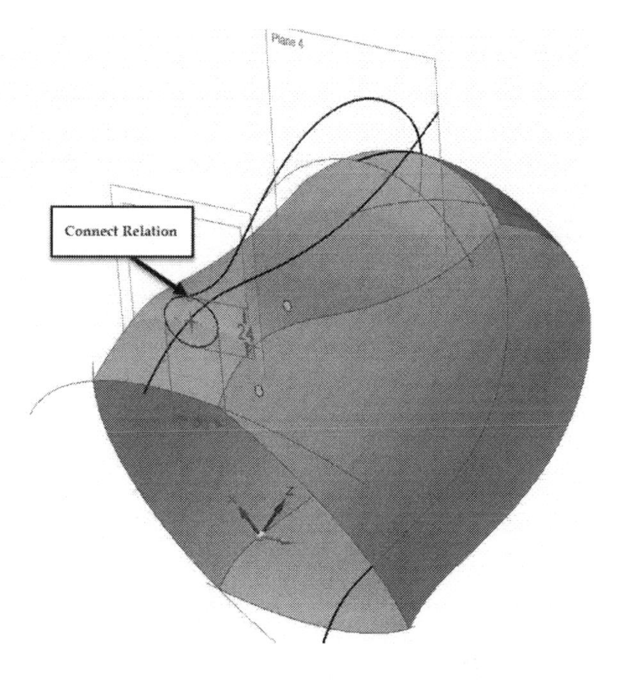

5. Activate the **Swept** command and create the handle surface (refer to the **Swept Surfaces** section).

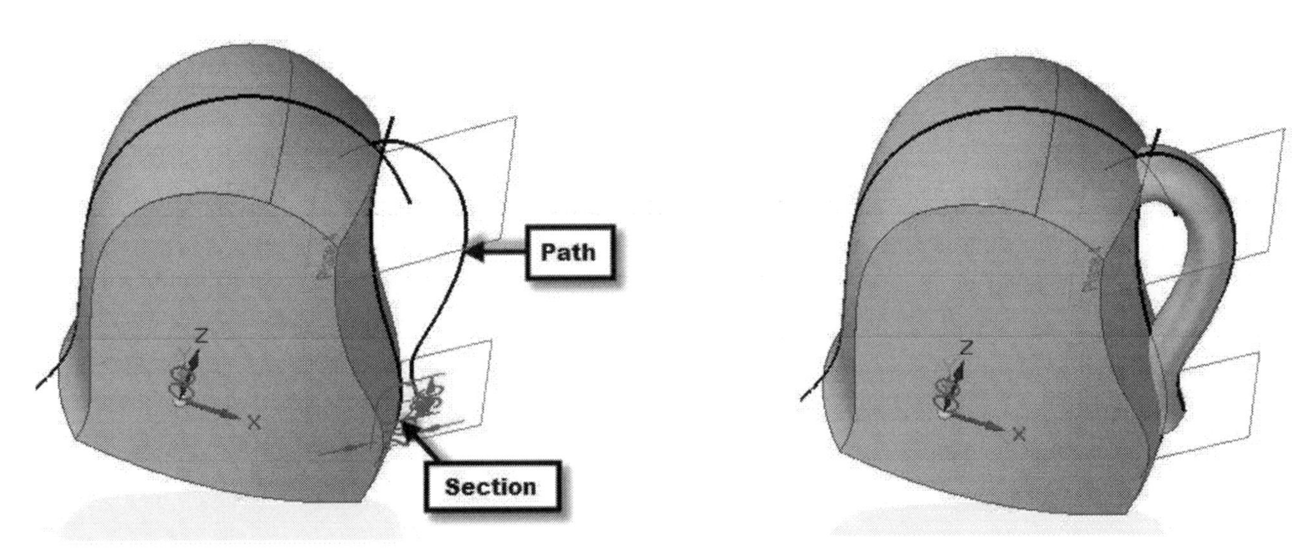

6. Activate the **Round** command and select the edge between the front and back surfaces. Type 25 in the Radius box and click the green check on the command bar. Click Preview, Finish, and Cancel and on the command bar.

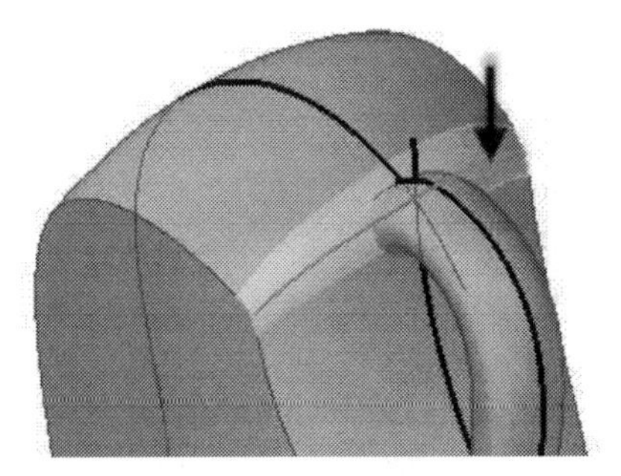

Trimming the Handle

1. On the ribbon, click **Surfacing > Planes > More Planes** drop-down **> Parallel**.
2. Select the YZ plane from the Base Coordinate System. Type-in 75 in the **Distance** box and press Enter. Move the pointer toward right and click.
3. Activate the **Trim** command and click on the Handle surface. Right-click to accept the selection.
4. Select the parallel plane and right-click to accept.
5. Select the region of the handle, as shown below. Click the green check on the command bar, and then click **Finish**.

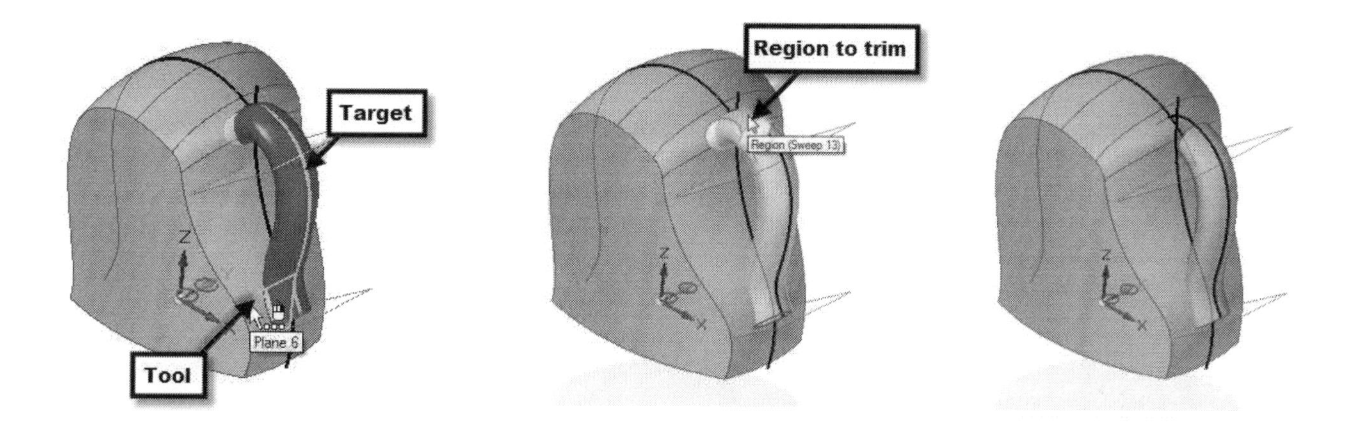

6. On the ribbon, **Surfacing > Planes > More planes > Normal to Curve** .
7. Select the path curve of the swept surface, and then its top endpoint.

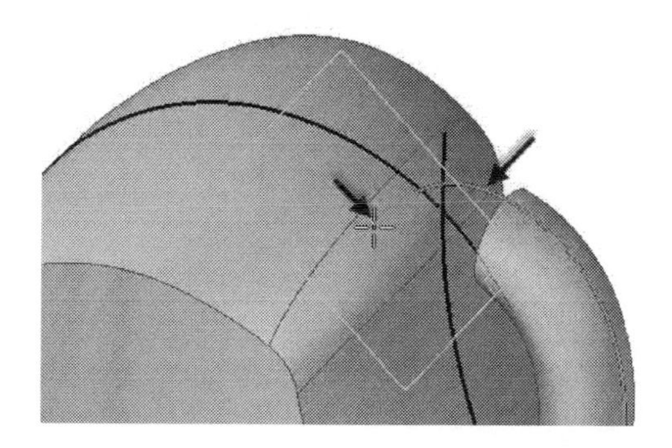

8. Start a sketch on the plane normal to the curve and draw an ellipse. Add dimensions to position the ellipse, and then finish the sketch.

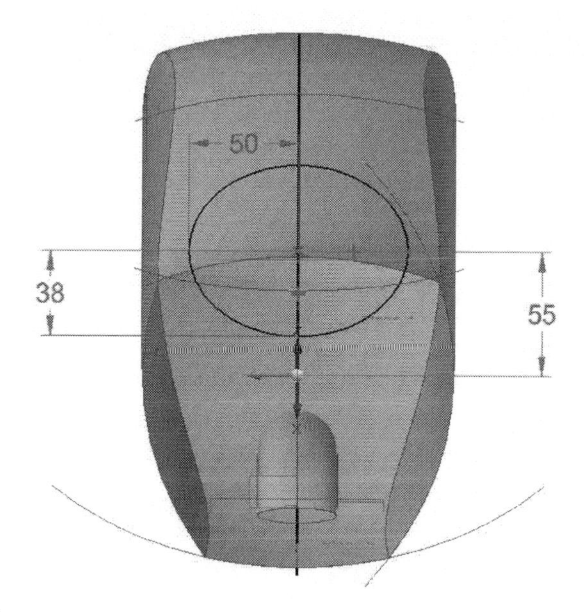

9. Activate the **Trim** command (on the ribbon, click **Surfacing > Modify Surfaces > Trim**) and click on the main body. Right-click to accept the selection.

10. Select the ellipse and right-click.
11. Select the surface region enclosed by the sketch. Right-click to accept the selection. Next, click **Finish** and **Cancel** on the command bar.

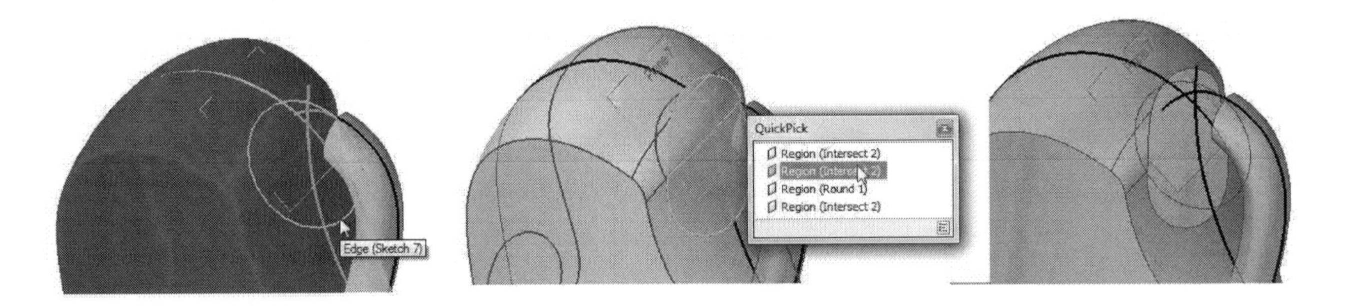

Blending the Top handle

1. Activate the **BlueSurf** command (on the ribbon, click **Surfacing > Surfaces > BlueSurf**) and click on the edges of the trimmed openings. Select the **Tangent Continuous** option from the drop-down attached to the handle edge. Select the **Natural** option from the drop-down attached to the main body edge.
2. Click **Next** to accept the selection. Click **Finish** on the command bar to create the BlueSurf surface. Click **Cancel** to deactivate the command.

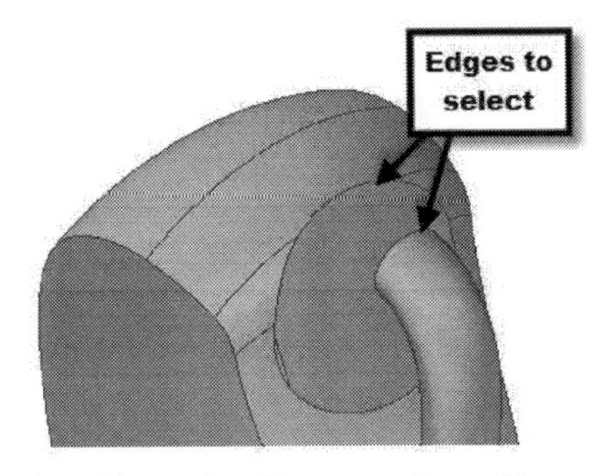

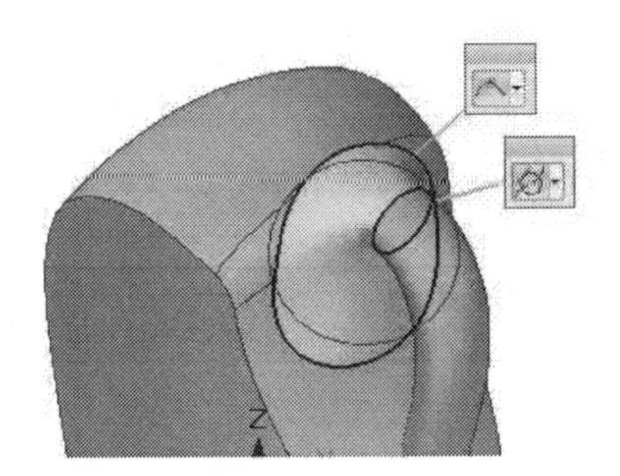

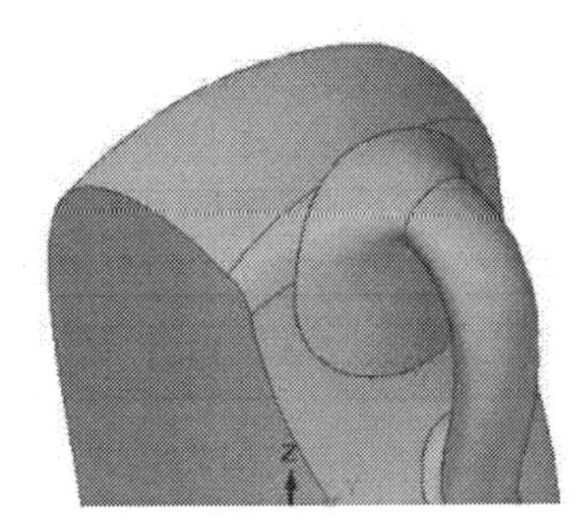

Blending the Bottom handle

1. On the ribbon, click **Surfacing > Planes > More Planes** drop-down **> Tangent** and select the handle surface.
2. On the Command bar, click **Keypoints** drop-down **> Silhouette**. Select the top silhouette point of the handle's lower edge. A plane tangent to the handle surface is created.

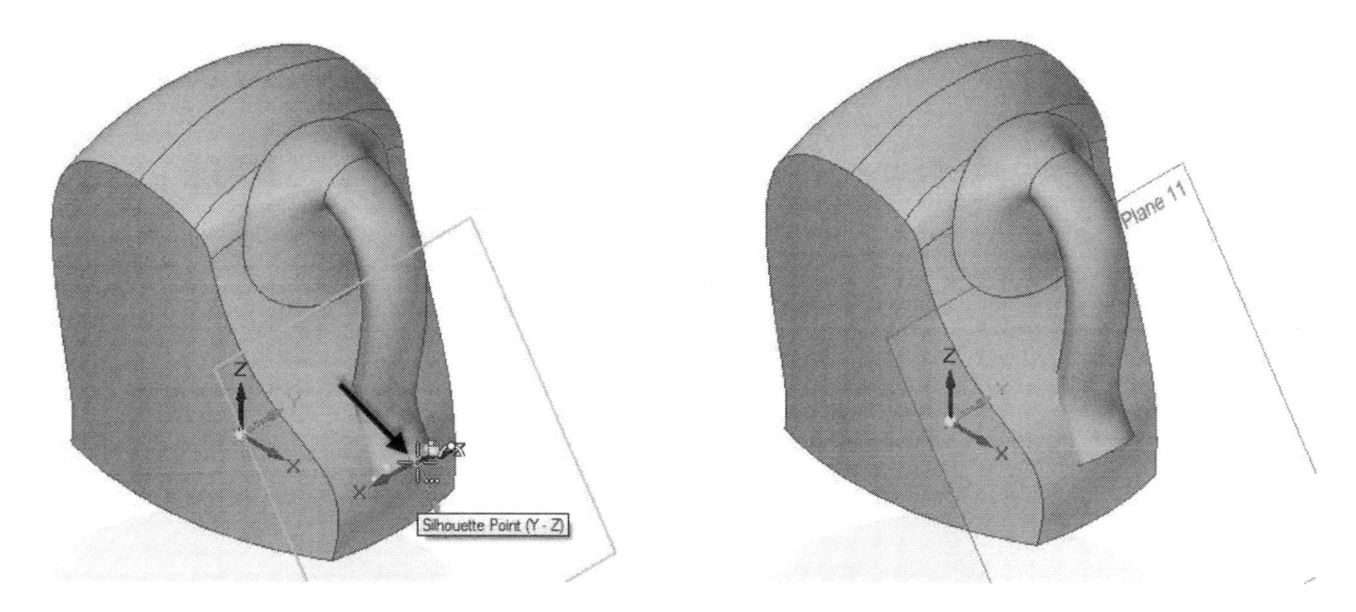

3. Create an ellipse on the new plane and trim it by half. Finish the sketch. Ensure that the sketch lies inside the handle surface.

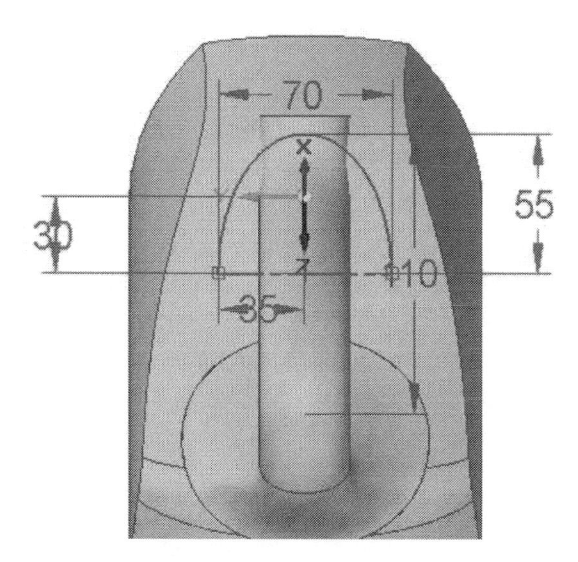

4. On the ribbon, click **Surfacing > Surfaces > Extruded**. Next, select **Create from Options > Select from Sketch** on the command bar.
5. Select the sketch and extrude it up to an arbitrary distance in both the directions.

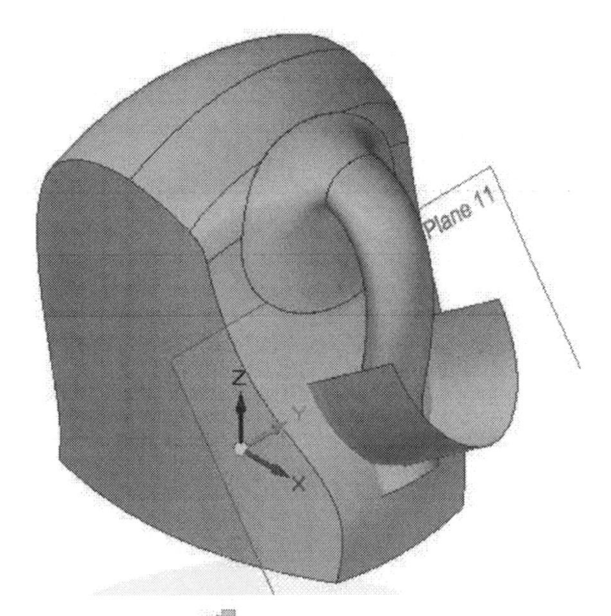

6. Activate the **Intersect** command.
7. Click on the handle and the extruded surfaces, and then right-click to accept.
8. Select the portions of the handle and extruded surfaces in the sequence shown below.
9. On the command bar, click the **Stitch** button, and then right-click. Click **Finish** to trim and stitch the surfaces.

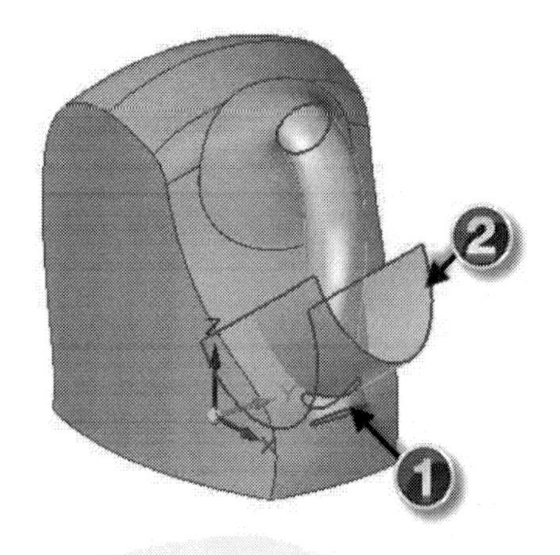

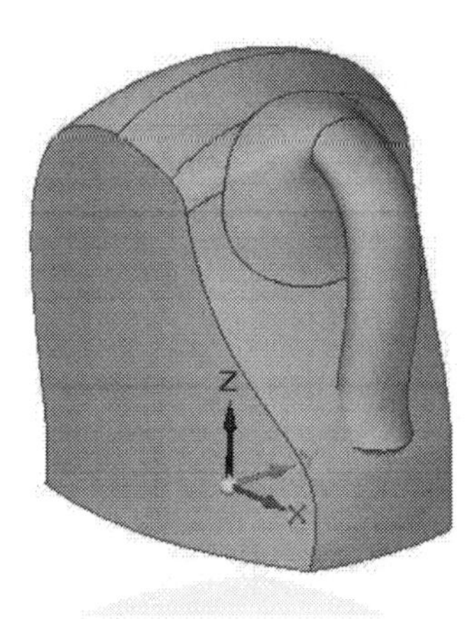

10. Select the handle and main body, and then right-click to accept the selection.
11. Rotate the model, select the intersecting portion, and stitched portion.
12. Make sure that the **Stitch** button is active. Click the **Accept** and **Finish** buttons to trim the surface.

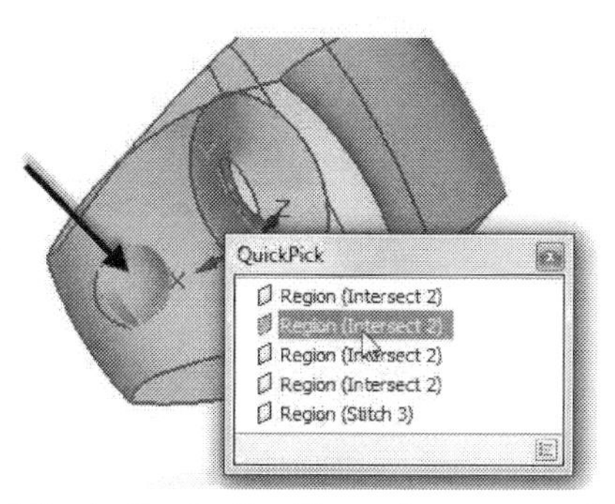

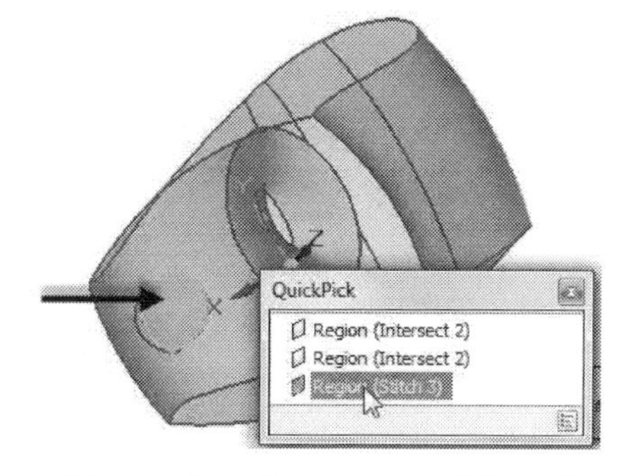

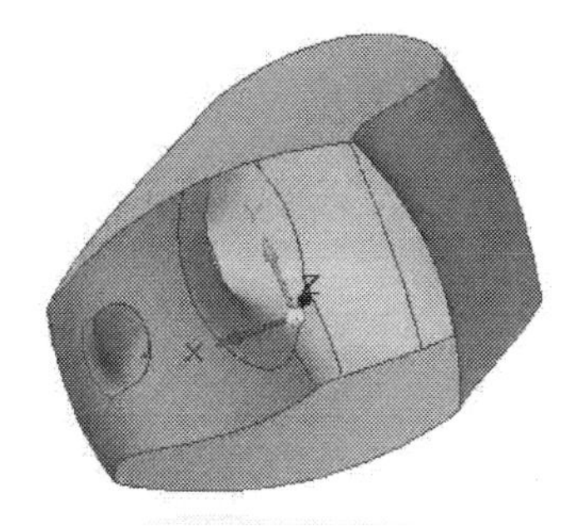

13. Activate the **Round** command (On the ribbon, click **Home > Solids > Round**) and round the edge of the handle. The round radius is 6 mm.
14. Round the intersection between the main surface and handle. The round radius is 5 mm.

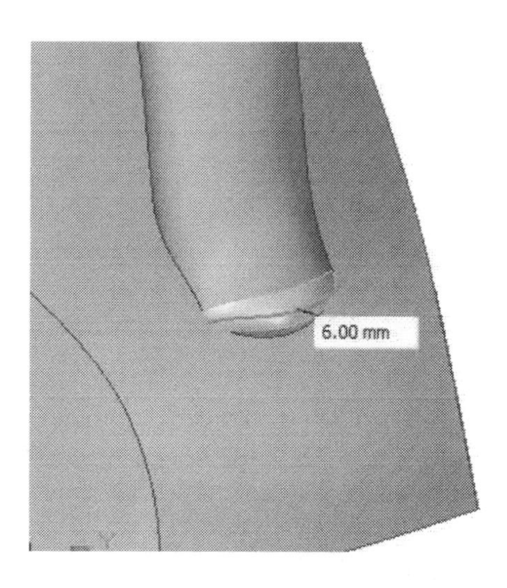

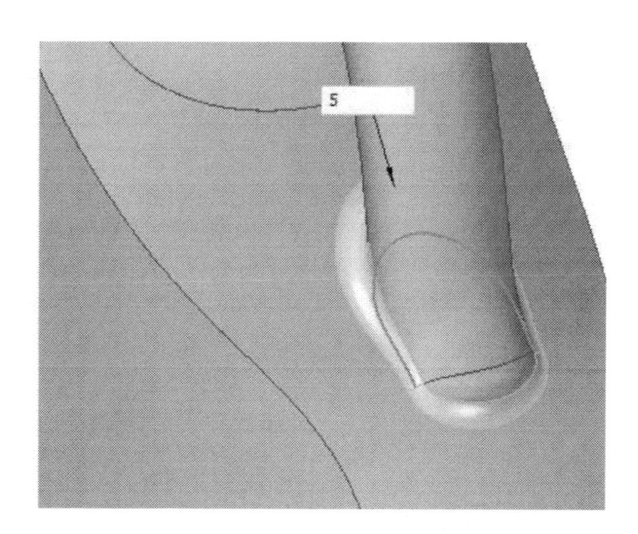

Creating the Neck and Spout

1. Draw a sketch on the XZ Plane for the revolved surface. Finish the sketch.

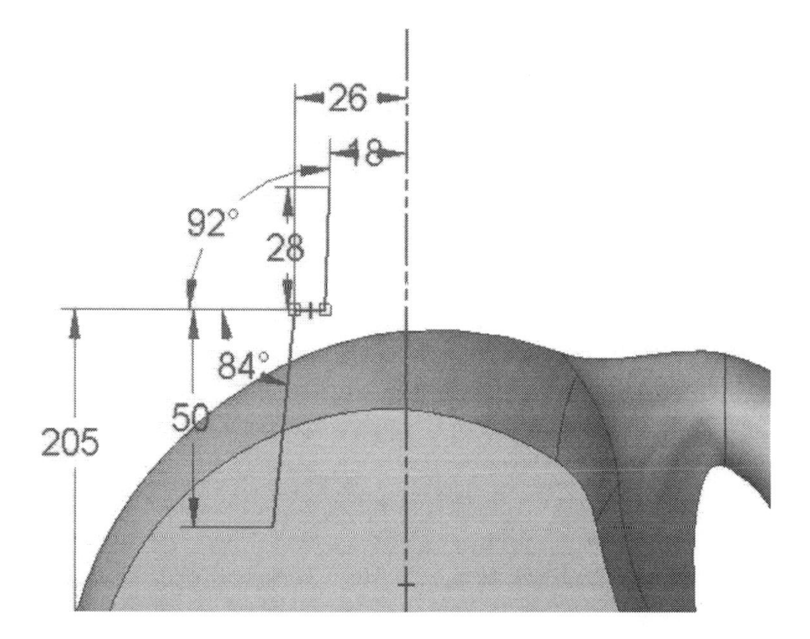

2. On the ribbon, click **Surfacing > Surfaces > Revolved**. On the command bar, set the **Create-From Options** type to **Select from Sketch**.
3. Select the sketch and right-click to accept.
4. Select the centerline and click the **Revolve 360** button on the command bar. Click **Finish** and **Cancel**.

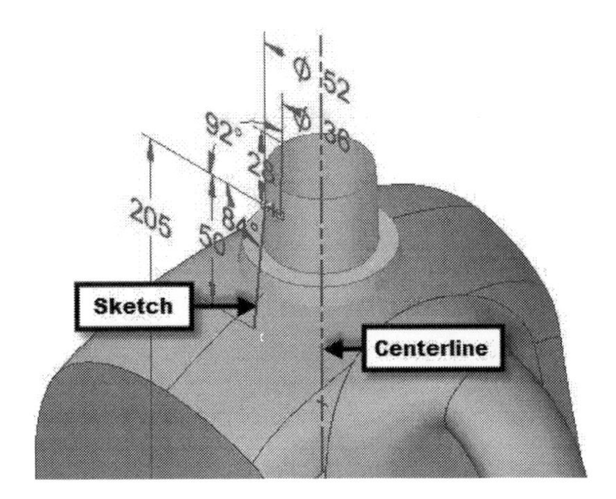

5. Activate the **Intersect** command, and trim the revolved and main surface. Stitch them together.

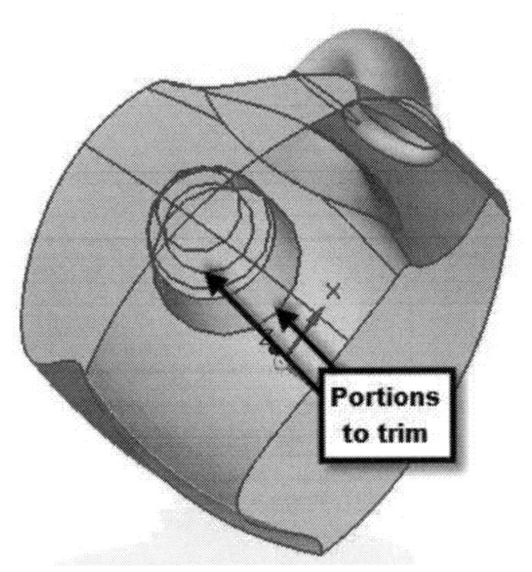

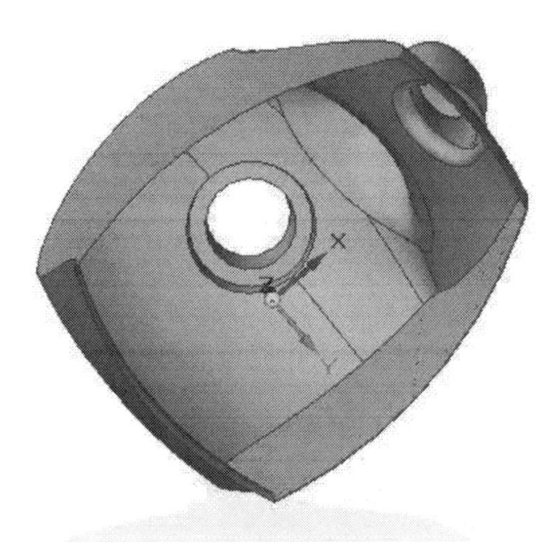

Rounding the Label Faces

1. Activate the **Round** command (on the ribbon, click **Home > Solids > Round**) and set the **Selection Type** to **Face**. Click on the front and back label faces.
2. Type-in **10** in the **Radius** box, and right-click. Click **Preview** and **Finish** to create the round.

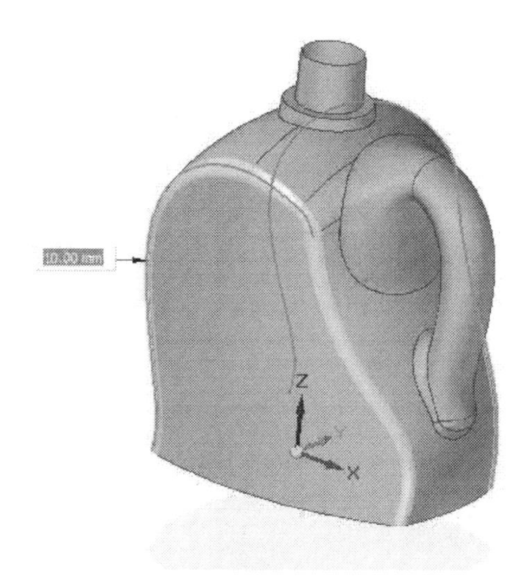

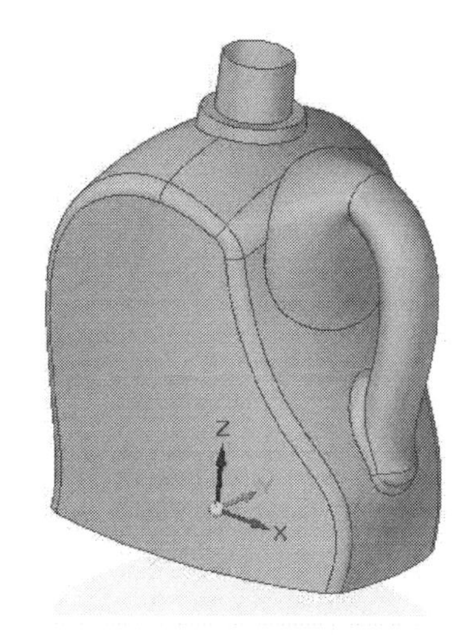

Creating the Bottom Face

1. Start a sketch on the XZ plane and draw a curve, as shown below. Click **Close Sketch** on the ribbon.

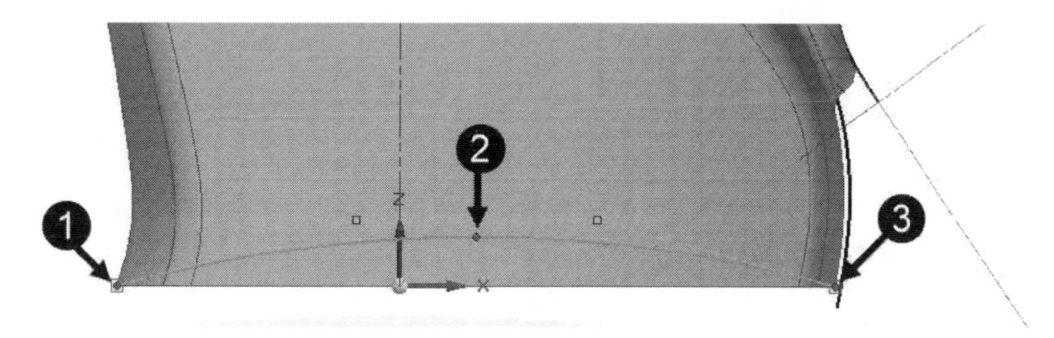

2. Activate the **Bounded** command (on the ribbon, click **Surfacing > Surfaces > Bounded**) and click on the edge-set at the bottom of the surface model. Right-click to accept the selection.
3. On the command bar, click the **Guide Curve Step** button and select the spline.
4. On the command bar, make sure that the **Common Tangent Condition** button is activated. Click the green check to accept the guide curve selection.
5. Click **Finish** and **Cancel**.

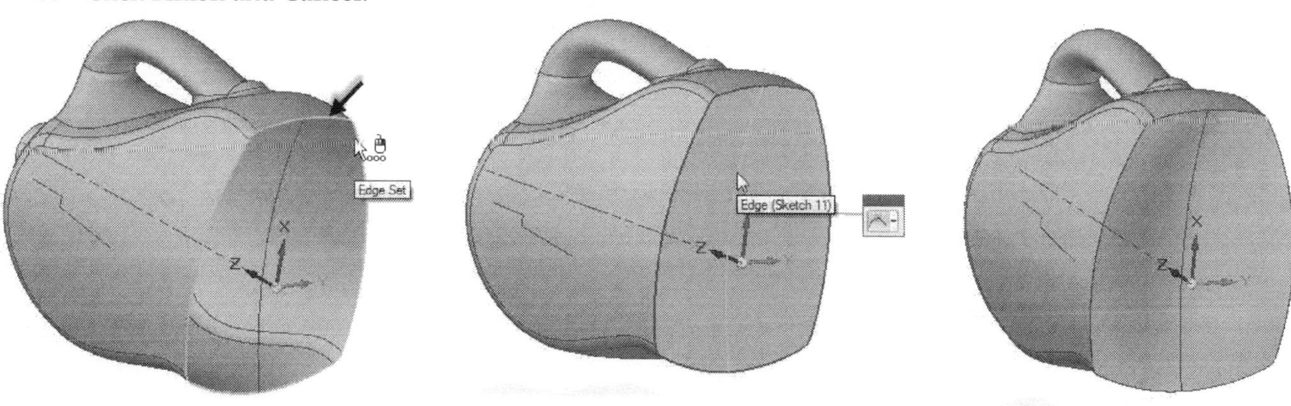

Rounding the Bottom Face

1. On the ribbon, click **Surfacing > Modify Surfaces > Stitched** . Click **OK** on the **Stitched Surface Options** dialog.
2. Select all the surfaces. Click the green check on the command bar.
3. Click **Finish** and **Cancel**.
4. Activate the **Round** command and select the **Selection Type** to **Loop**.
5. Select the edge set at the bottom and type-in 10 in the **Radius** box. Right-click and click **Preview**.
6. Click **Finish** and **Cancel**.

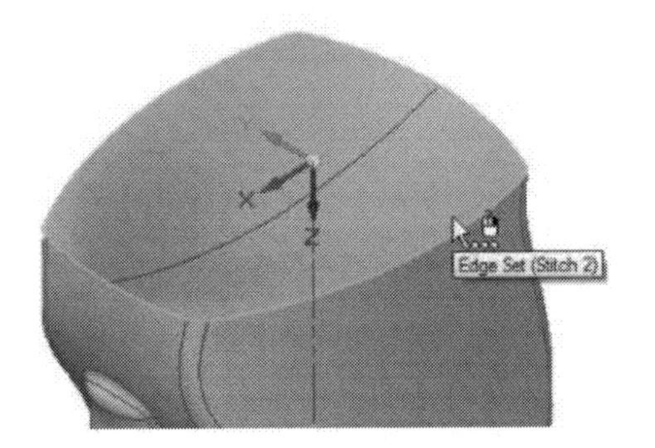

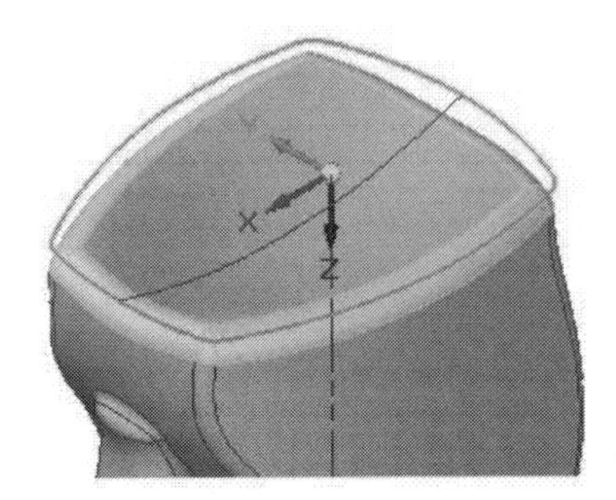

Blending the Bluesurface and Main body

1. Activate the **Round** command and click the **Round Options** button on the command bar. On the **Round Options** dialog, select the **Surface blend** option and click **OK**.
2. On the command bar, click the **Surface Blend Parameters** button and check the **Trim and stitch input faces** and **Trim output blend** options. Click **OK**.
3. Select the bluesurface and the main surface body connected to it.
4. Type-in 30 in the **Radius** box and click **Accept**.
5. Move the pointer such that the arrow points towards back. Click to define first side.
6. Move the pointer such that the arrow points upwards. Click to define the second side.
7. Click **Preview** and **Finish** to complete the blend surface.

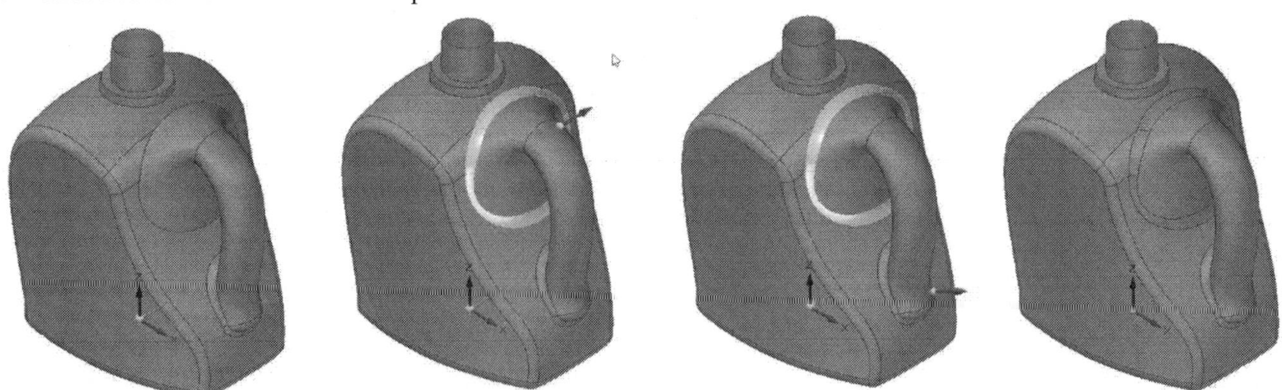

Adding thickness to the model

1. Activate the **Thicken** command (on the ribbon, click **Home > Solids > Add drop-down > Thicken**) and click on the surface body.
2. On the command bar, type-in 1.5 in the **Distance** box.
3. Move the pointer such that the arrow points outward, and then click to define the thickness side.
4. Click **Finish** to thicken the surface model.

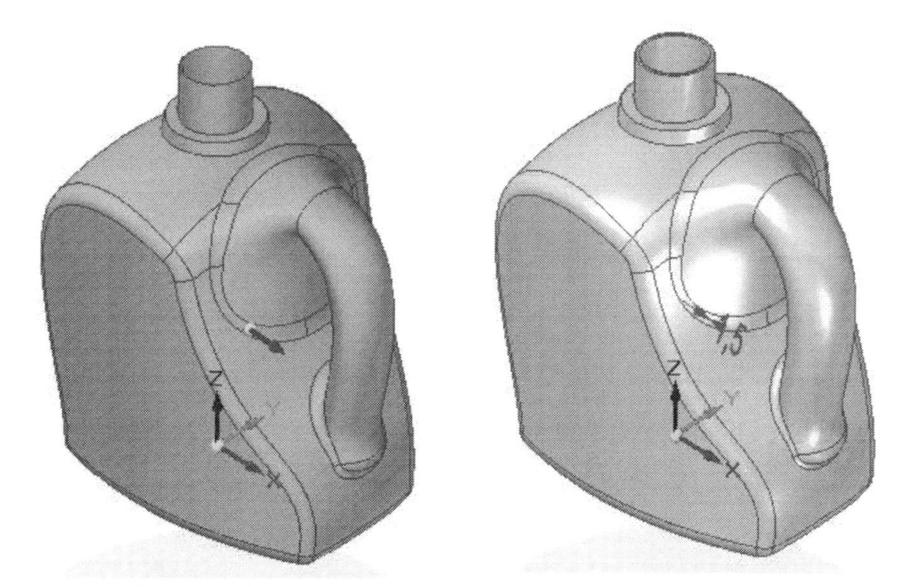

5. Activate the **Round** command, and then blend the sharp edges of the neck and spout. The blend radius is 1 mm.

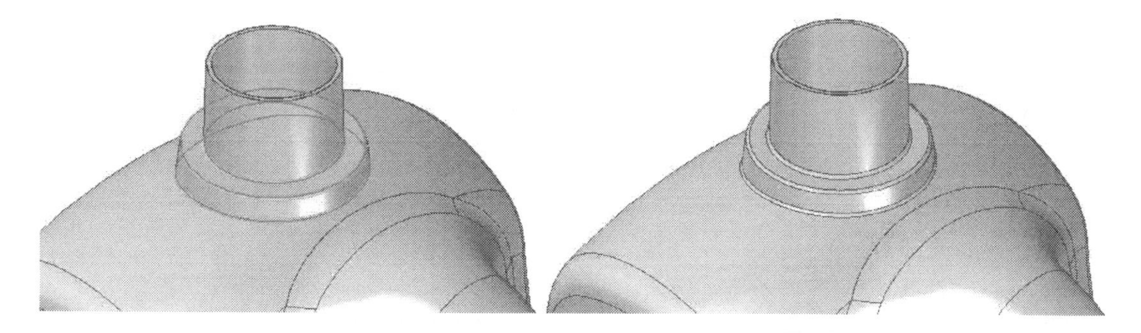

Creating threads

1. Create the cross-section and axis of the thread on the XZ Plane, as shown below. Click **Close Sketch**.

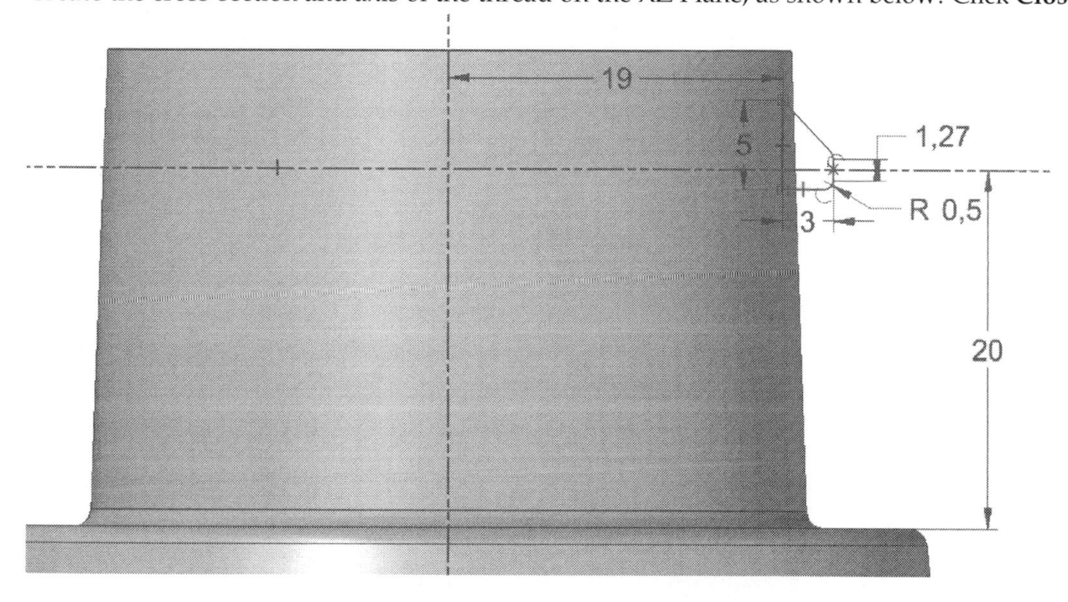

2. Activate the **Helix** command (on the ribbon, click **Home > Solids > Add drop-down > Helix**) and select **Select from sketch** option from the **Create-From Options** drop-down.
3. Select the cross-section and right-click.
4. Select the centerline to define the axis. Click on the top endpoint of the centerline to define the start point of the helical protrusion.
5. Set the **Helix method** to **Pitch & Turns**. Type-in 6 and 2 in the **Pitch** and **Turns** boxes, respectively. Click **Next** and **Preview** to view the helix.
6. Click **Finish** to complete the helical protrusion.

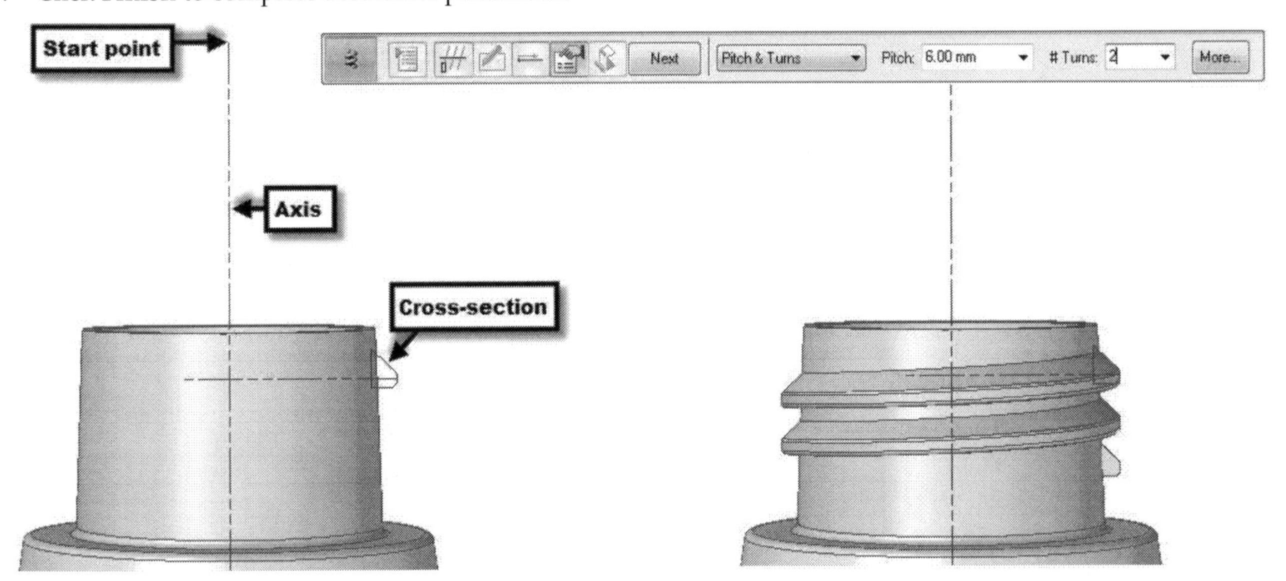

Embossing the label faces

1. On the ribbon, click **Surfacing > Planes > More Planes** drop-down **> Parallel**. Select the XZ plane, type 100 in the **Distance** box, and then press Enter. Next, move the pointer toward left and click.
2. Start a sketch on the newly created parallel plane and click **Home > Draw > Project to Sketch** on the ribbon.
3. On the **Project to Sketch Options** dialog, check the **Project with offset** option. Click **OK**.
4. Set the **Selection Type** to **Loop** and click on the inner edge of the round, as shown below. Click **Accept**.
5. Type-in **15** in the **Distance** box and click inside the loop. The edge loop will be projected. Next, press Esc to deactivate the **Project to Sketch** command.
6. Click on the offset constraints displayed on the sketch and press **Delete**.
7. Add 12 radius fillets to the corners of the sketch. Click **Close Sketch** on the ribbon.
8. Click **Finish** and **Cancel**.

9. On the ribbon, click **Home > Solids > Add Body** and click **OK** on the **Add body** dialog. A new body is created within the part file.
10. On the ribbon, click **Home > Solids > Extrude**. Select **Create-From Options** drop-down > **Select from Sketch** on the command bar. Select the sketch. Click **Accept**.
11. On the command bar, click the **From/To Extent** button and select the sketch plane to define the 'From' surface.
12. Rotate the model and select the inner face of the label face to define the 'To' surface.

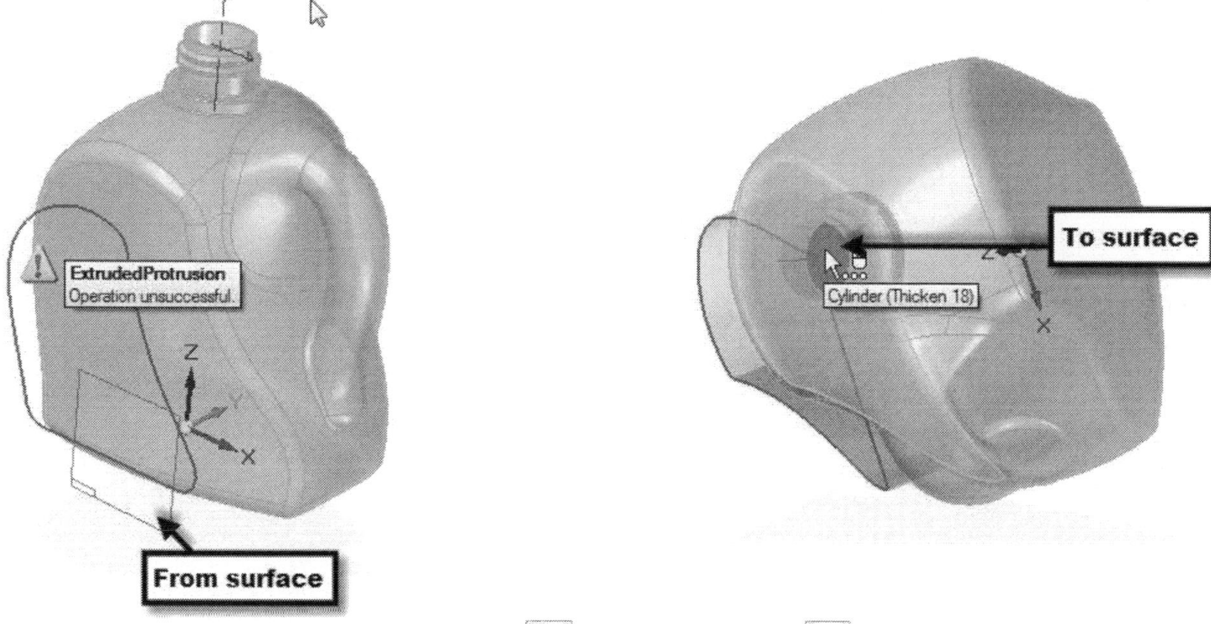

13. On the command bar, click **Treatment Step** and select the **Draft** button.
14. Type-in **5** in the **Angle 2** box. Use the **Flip 2** button to make sure that the arrow 2 points outwards.
15. Click **Preview** and **Finish** to create the *Extrude* feature.

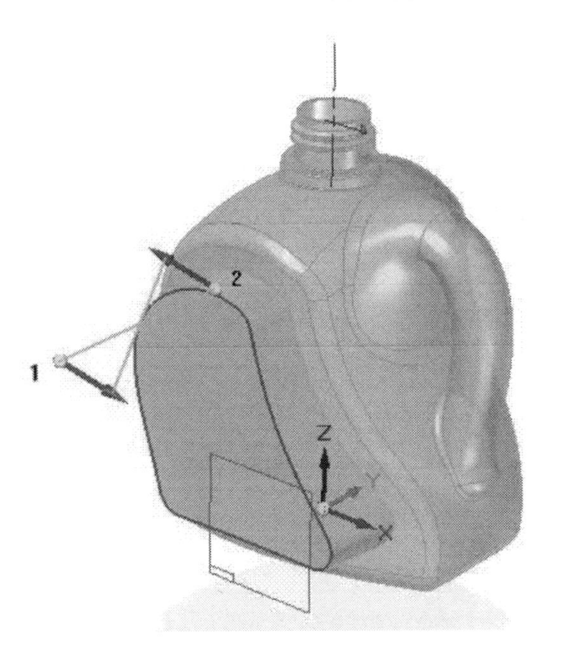

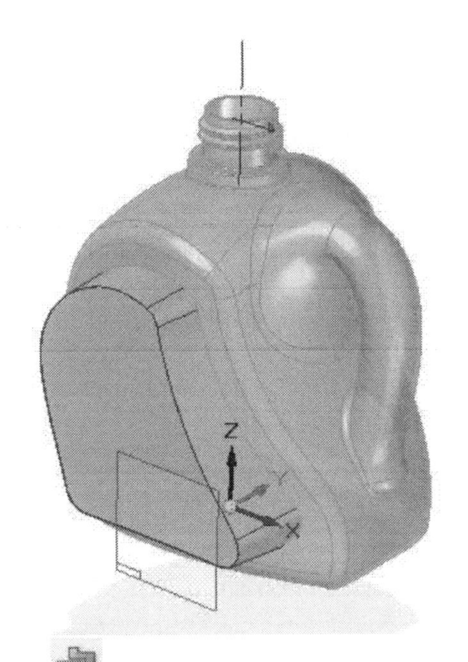

16. On the ribbon, click **Home > Solids > Thin Wall drop-down > Emboss** .
17. Select the target and tool bodies, as shown in figure. Click the **Direction** button on the command bar to reverse the direction of the emboss feature.
18. Type-in 0.5 and 1 in the **Clearance** and **Thickness** boxes, respectively.

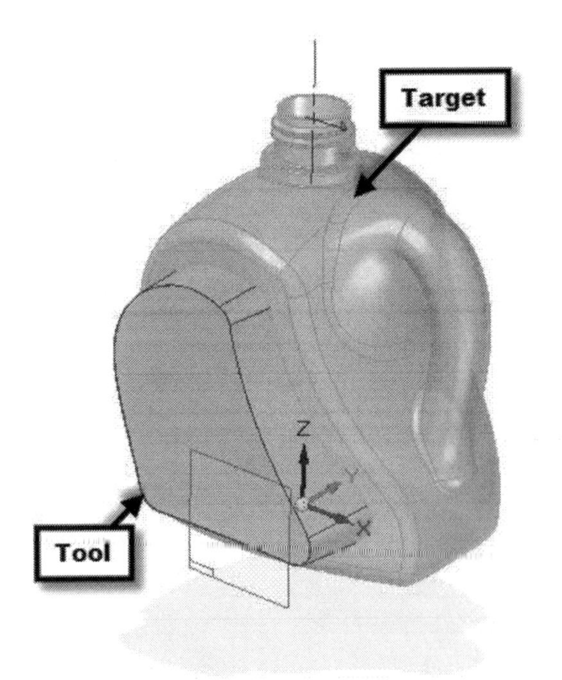

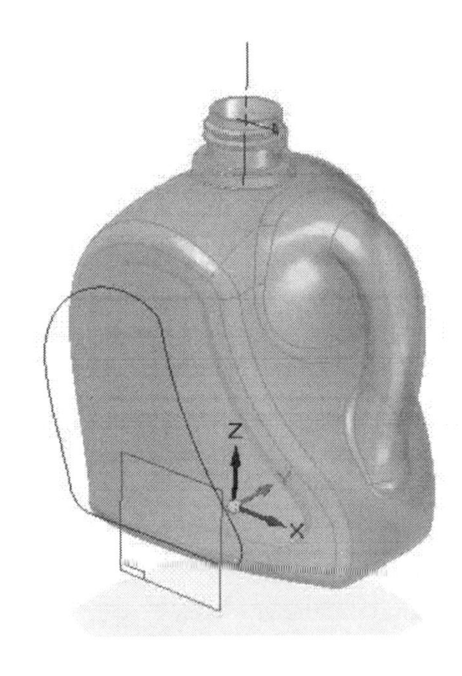

19. Click **Accept** to create the emboss feature. Press Esc to deactivate the **Emboss** command.
20. Under the PathFinder, uncheck the **Design Body_2** option to hide it.
21. Click the right-mouse button on **Design Body_1** and select **Activate Body**. This activates the main body.

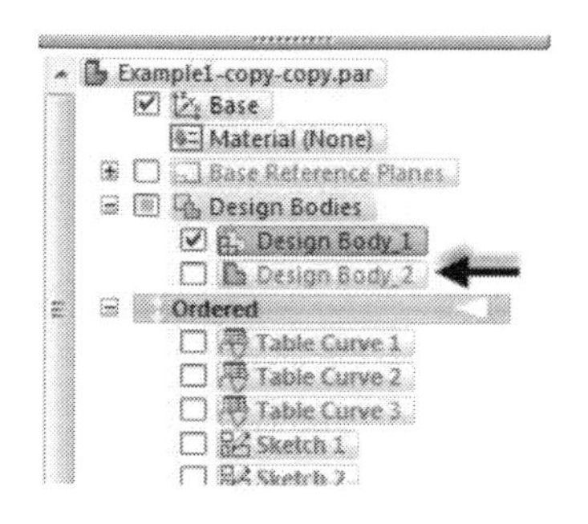

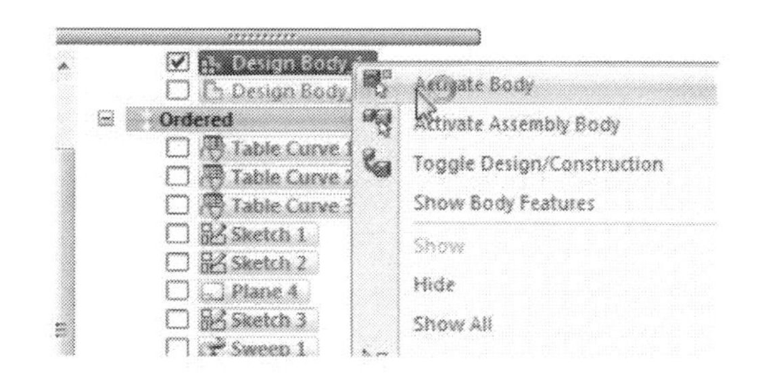

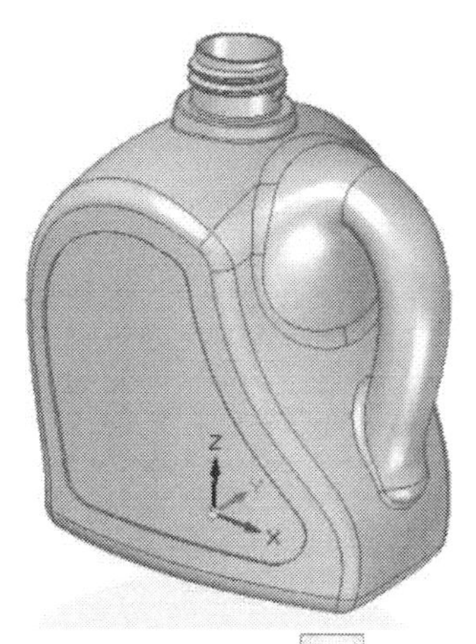

22. On the ribbon, click **Home > Pattern > Mirror drop-down > Mirror Copy Feature** and click the **Fast** button on the command bar.
23. Select the emboss feature and right-click.
24. Select the XZ plane from the Base Coordinate system and click **Finish**. The emboss feature is mirrored about the XZ plane.

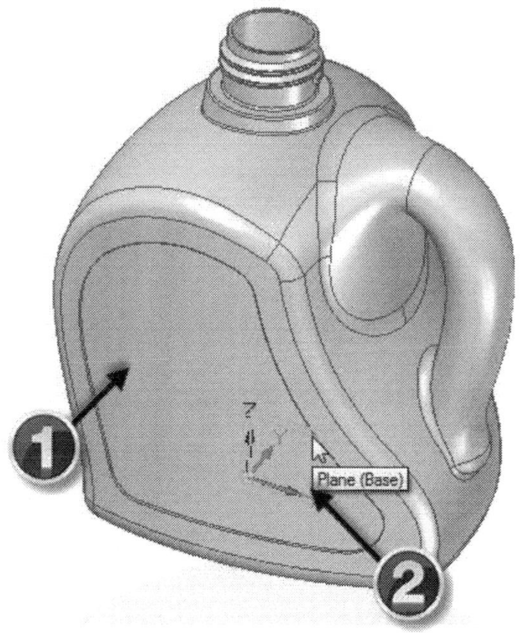

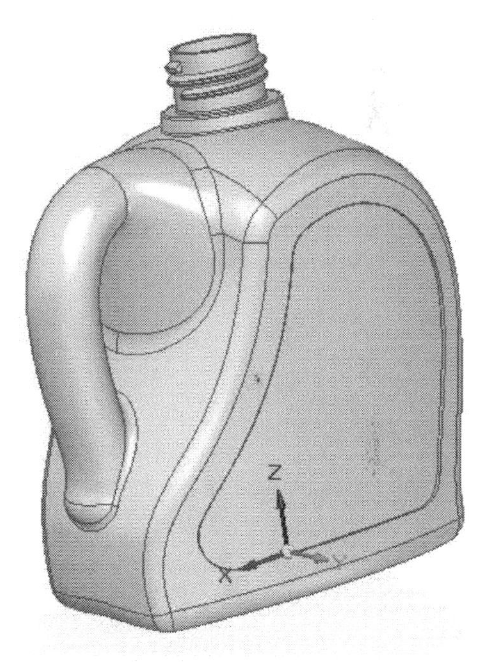

25. Activate the **Round** command and set the **Selection Type** to **Chain**.
26. Select the outer edges of the emboss features and type-in 3 in the **Radius** box. Click **Accept**.
27. Click **Preview** and **Finish** to complete the round feature.

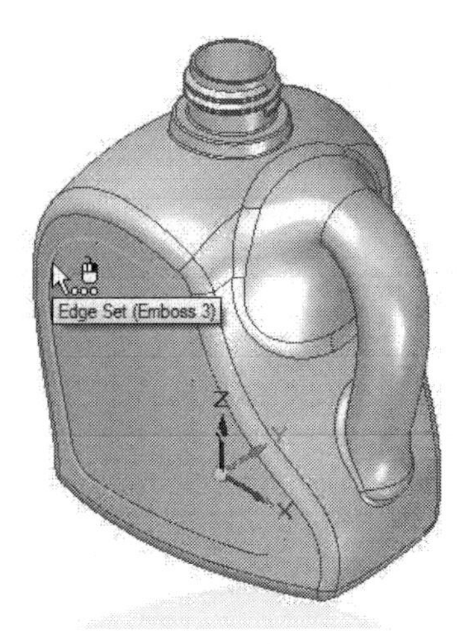

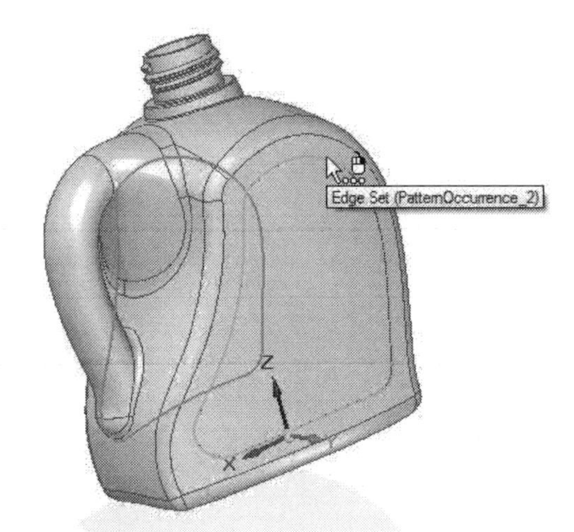

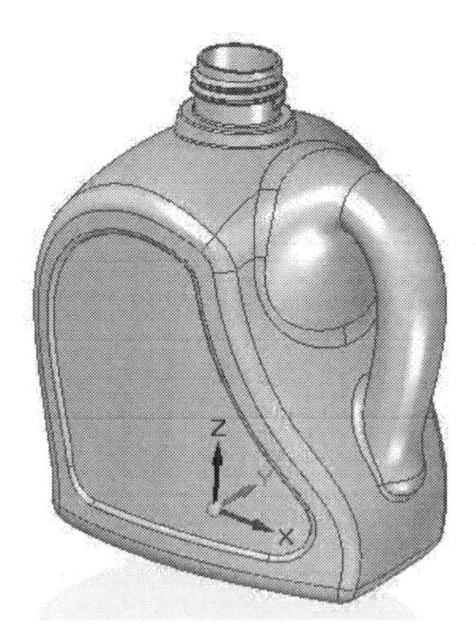

Measuring the Volume of the bottle

1. Activate the **Physical Properties** command (on the ribbon, click **Inspect > Physical Properties > Physical Properties**).
2. On the **Physical Properties** dialog, click the **Change** button under the **Density** section.
3. On the **Solid Edge Material Table** dialog, click **Material > Non-metals > Plastics > Polyethylene, low density**. Click **Apply to Model**. Click **Update** to display the physical properties of the bottle on the dialog.
4. View the **Volume** box on the **Physical Properties** dialog. Next, click **Close**.
5. Save and close the file.

Questions

1. What is the use of the **Stitched** command?
2. How many types of rounds can be created in Solid Edge?
3. Why do we use the **Boundary** command?
4. Which command can be used to bridge gap between two surfaces?
5. Name the command that can be used to perform a variety of operations.
6. How do you add thicknesses to a surface body?
7. Which command is used to extend surfaces from an edge?
8. Which command is used to offset faces?

Index

Made in the USA
Coppell, TX
26 January 2020